T0213919

Lecture Notes in Artificial Intelligence   9834

Subseries of Lecture Notes in Computer Science

More information about this series at http://www.springer.com/series/1244

Naoyuki Kubota · Kazuo Kiguchi
Honghai Liu · Takenori Obo (Eds.)

Intelligent Robotics and Applications

9th International Conference, ICIRA 2016
Tokyo, Japan, August 22–24, 2016
Proceedings, Part I

 Springer

Editors
Naoyuki Kubota
Tokyo Metropolitan University
Tokyo
Japan

Kazuo Kiguchi
Kyushu University
Fukuoka
Japan

Honghai Liu
University of Portsmouth
Portsmouth
UK

Takenori Obo
Tokyo Metropolitan University
Tokyo
Japan

ISSN 0302-9743   ISSN 1611-3349 (electronic)
Lecture Notes in Artificial Intelligence
ISBN 978-3-319-43505-3   ISBN 978-3-319-43506-0 (eBook)
DOI 10.1007/978-3-319-43506-0

Library of Congress Control Number: 2016946926

LNCS Sublibrary: SL7 – Artificial Intelligence

Printed on acid-free paper

This Springer imprint is published by Springer Nature
The registered company is Springer International Publishing AG Switzerland

Preface

The Organizing Committee of the 9th International Conference on Intelligent Robotics and Applications aimed to facilitate interactions among active participants in the field of intelligent robotics and mechatronics and their applications. Through this conference, the committee intended to enhance the sharing of individual experiences and expertise in intelligent robotics with particular emphasis on technical challenges associated with varied applications such as biomedical application, industrial automations, surveillance, and sustainable mobility.

The 9th International Conference on Intelligent Robotics and Applications was most successful in attracting 148 submissions addressing the state-of-the-art developments in robotics, automation, and mechatronics. Owing to the large number of valuable submissions, the committee was faced with the difficult challenge of selecting the most deserving papers for inclusion in these lecture notes and presentation at the conference. For this purpose, the committee undertook a rigorous review process. Despite the high quality of most of the submissions, a total of 114 papers were selected for publication in two volumes of Springer's *Lecture Notes in Artificial Intelligence* as subseries of *Lecture Notes in Computer Science*, with an acceptance rate of 77 %. The selected papers were presented at the 9th International Conference on Intelligent Robotics and Applications held during August 22–24, 2016, in Hachioji, Tokyo, Japan.

The selected articles were submitted by scientists from 22 different countries. The contribution of the Technical Program Committee and the reviewers is deeply appreciated. Most of all, we would like to express our sincere thanks to the authors for submitting their most recent work and to the Organizing Committee for their enormous efforts to turn this event into a smooth-running meeting. Special thanks go to the Tokyo Metropolitan University for their generosity and direct support. Our particular thanks are due to Alfred Hofmann and Anna Kramer of Springer for enthusiastically supporting the project.

We sincerely hope that these volumes will prove to be an important resource for the scientific community.

June 2016

Naoyuki Kubota
Kazuo Kiguchi
Honghai Liu
Takenori Obo

Preface

Organization

Advisory Committee

Jorge Angeles	MgGill University, Canada
Zhongqin Lin	Shanghai Jiao Tong University, China
Hegao Cai	Harbin Institute of Technology, China
Imre Rudas	Obuda University, Hungary
Tianyou Chai	Northeastern University, China
Shigeki Sugano	Waseda University, Japan
Jiansheng Dai	King's College London, UK
Guobiao Wang	National Natural Science Foundation of China, China
Toshio Fukuda	Meijo University, Japan
Kevin Warwick	University of Reading, UK
Fumio Harashima	Tokyo Metropolitan University, Japan
Bogdan M. Wilamowski	Auburn University, USA
Huosheng Hu	University of Essex, UK
Ming Xie	Nanyang Technological University, Singapore
Han Ding	Huazhong University of Science and Technology, China
Youlun Xiong	Huazhong University of Science and Technology, China
Oussama Khatib	Stanford University, USA
Huayong Yang	Zhejiang University, China

General Chair

Naoyuki Kubota	Tokyo Metropolitan University, Japan

General Co-chairs

Xiangyang Zhu	Shanghai Jiao Tong University, China
Jangmyung Lee	Pusan National University, Korea
Kok-Meng Lee	Georgia Institute of Technology, USA

Program Chair

Kazuo Kiguchi	Kyushu University, Japan

Program Co-chairs

Honghai Liu	University of Portsmouth, UK
Janos Botzheim	Tokyo Metropolitan University, Japan
Chun-Yi Su	Concordia University, Canada

Special Sessions Chair

Kazuyoshi Wada   Tokyo Metropolitan University, Japan

Special Sessions Co-chairs

Alexander Ferrein   University of Applied Sciences, Germany
Chu Kiong Loo   University of Malaya, Malaysia
Jason Gu   Dalhousie University, Canada

Award Committee Chair

Toru Yamaguchi   Tokyo Metropolitan University, Japan

Award Committee Co-chairs

Kentaro Kurashige   Muroran Institute of Technology, Japan
Kok Wai (Kevin) Wong   Murdoch University, Australia

Workshop Chair

Yasufumi Takama   Tokyo Metropolitan University, Japan

Workshop Co-chairs

Simon Egerton   Monash University, Malaysia
Lieu-Hen Chen   National Chi Nan University, Taiwan

Publication Chair

Takenori Obo   Tokyo Metropolitan University, Japan

Publication Co-chairs

Taro Nakamura   Chuo University, Japan
Chee Seng Chan   University of Malaya, Malaysia

Publicity Chair

Hiroyuki Masuta   Toyama Prefectural University, Japan

Publicity Co-chairs

Jiangtao Cao   Liaoning Shihua University, China
Simon X. Yang   University of Guelph, Canada
Mattias Wahde   Chalmers University of Technology, Sweden

Financial Chairs

Takahiro Takeda	Daiichi Institute of Technology, Japan
Zhaojie Ju	University of Portsmouth, UK

Track Chairs

Lundy Lewis	Southern New Hampshire University, USA
Narita Masahiko	Advanced Institute of Industrial Technology, Japan
Georgy Sofronov	Macquarie University, Australia
Xinjun Sheng	Shanghai Jiao Tong University, China
Kosuke Sekiyama	Nagoya University, Japan
Futoshi Kobayashi	Kobe University, Japan
Tetsuya Ogata	Waseda University, Japan
Ryas Chellali	Nanjing Tech University, China
Min Jiang	Xiamen University, China
Shinji Fukuda	Tokyo University of Agriculture and Technology, Japan
Boris Tudjarov	Technical University of Sofia, Bulgaria
Hongbin Ma	Beijing Institute of Technology, China
Sung-Bae Cho	Yonsei University, Korea
Nobuto Matsuhira	Shibaura Institute of Technology, Japan
Xiaohui Xiao	Wuhan University, China
Zhaojie Zu	University of Portsmouth, UK

Local Arrangements Chairs

Takenori Obo	Tokyo Metropolitan University, Japan
Takahiro Takeda	Daiichi Institute of Technology, Japan

Web Masters

Naoyuki Takesue	Tokyo Metropolitan University, Japan
Yihsin Ho	Tokyo Metropolitan University, Japan

Local Arrangements Committee

Shinji Fukuda	Tokyo University of Agriculture and Technology, Japan
Kazunori Hase	Tokyo Metropolitan University, Japan
Takuya Hashimoto	The University of Electro-Communications, Japan
Mime Hashimoto	Tokyo Metropolitan University, Japan
Narita Masahiko	Advanced Institute of Industrial Technology, Japan
Koji Kimita	Tokyo Metropolitan University, Japan
Taro Ichiko	Tokyo Metropolitan University, Japan
Hitoshi Kiya	Tokyo Metropolitan University, Japan

Lieu-Hen Chen   Tokyo Metropolitan University, Japan
Daigo Kosaka   Polytechnic University, Japan
Yu Sheng Chen   Tokyo Metropolitan University, Japan
Osamu Nitta   Tokyo Metropolitan University, Japan

Contents – Part I

Robot Vision and Sensing

Planning, Localization, and Mapping

Interactive Intelligence

Cognitive Robotics

Bio-inspired Robotics

Smart Material Based Systems

Mechatronics Systems for Nondestructive Testing

Contents – Part II

Intelligent Space

Sensing and Monitoring in Environment and Agricultural Sciences

Human Data Analysis

Robot Hand

Robot Control

Robust Backstepping Control for Spacecraft Rendezvous on Elliptical Orbits Using Transformed Variables

Yu Wang$^{(\boxtimes)}$, Haibo Ji, and Kun Li

Department of Automation, University of Science and Technology of China,
Hefei 230027, Anhui, People's Republic of China
{andyyu,zkdlk}@mail.ustc.edu.cn, jihb@ustc.edu.cn

Abstract. This paper is concerned with the problem of relative control for spacecraft rendezvous with the target spacecraft on an arbitrary elliptical orbit. A simplified dynamic model describing the relative motion between the chaser spacecraft and the target spacecraft is established via using transformed variables. Based on this simplified dynamic model, the relative motion is divided into in-plane motion and out-of-plane motion. A robust backstepping control scheme is designed to solve the rendezvous problem. Theoretical analysis and numerical simulation validate the effectiveness of the proposed method.

Keywords: Spacecraft rendezvous · Transformed variables · Simplified dynamic model · Robust backstepping control

1 Introduction

Spacecraft rendezvous technology has played a significant role for many space missions, such as the spacecraft maintenance, the rescue of astronaut, the replenishment for the platform of satellite and space stations [1,2]. Different kinds of effective control methods have been developed to figure out the rendezvous problem. For example, the problem of multi-objectives robust H_∞ control for a class of spacecraft rendezvous was studied in [3]; in [4], a parametric Lyapunov differential equation method was developed for the rendezvous maneuvers on elliptical orbit; adaptive control theory was investigated in [5]; and in [6], the neural network controller was put forward for rendezvous problem.

The current literature related to the relative motion dynamic model consider the target spacecraft on a circular orbit or an elliptical orbit and the chaser spacecraft on the target's neighbourhood. The elliptical orbit relative motion between the chaser and the target can be described by some nonlinear differential equations and for which the linearized equations are known as C-W equation

---

Y. Wang—This work was supported by the National High-tech R&D Program of China (863 Program 2014AA06A503) and the National Basic Research Program of China under Grant 973-10001.

© Springer International Publishing Switzerland 2016
N. Kubota et al. (Eds.): ICIRA 2016, Part I, LNAI 9834, pp. 3–13, 2016.
DOI: 10.1007/978-3-319-43506-0_1

[7] or T-H equation [8]. Both of the C-W equation and the T-H equation exist many uncertain model parameters. Successful rendezvous missions rely on accurate and reliable control for the relative motion. Hence, the accuracy of the dynamic model is the basic foundation for rendezvous. In [9], a simplification of the equation of the relative motion model is developed. This method utilizes the true anomaly θ of the target spacecraft as an independent variable instead of time t. In this new formulation the state variables $\mathbf{r} = [x, y, z]$, the relative distance between the chaser and the target and its time rate of change, $\dot{\mathbf{r}} = [\dot{x}, \dot{y}, \dot{z}]$ are replaced by variables $\tilde{\mathbf{r}} = [\tilde{x}, \tilde{y}, \tilde{z}]$ and $\tilde{\mathbf{r}}' = [\tilde{x}', \tilde{y}', \tilde{z}']$, respectively, which are called transformed variables. The transformed state vector $\tilde{\mathbf{r}} = [\tilde{x}, \tilde{y}, \tilde{z}]$ satisfies a linear differential equation.

In this paper, the relative control for spacecraft rendezvous with the target spacecraft on an arbitrary elliptical orbit is studied. Based on this simplified dynamic model, the main objective in this paper is to develop a control scheme to meet the spacecraft rendezvous requirements while minimizing the design complexity. There are three new features in this paper as follows.

- Compared with the linearized C-W equation or T-H equation with uncertain parameters such as $\omega, \dot{\omega}, \omega^2$ which are related to the varing orbit angular velocity, the uncertain papameter in the simplified dynamic model is only one (i.e., ρ). Finally, the control law proposed in this paper is independent of the uncertain parameter. Thus, the higher accuracy can be guaranteed in consideration of the system model uncertain.
- The relative motion dynamic model can be divided into two departments, the in-plane motion (namely, the $x - z$ subsystem) and the out-of-plane motion (namely, the y subsystem) which are independent. On this basis, the design of the controller is divided into two independent parts, which is significant for the application in actual engineering design.
- To reduce the polytopic uncertain, the robust H_∞ control theory is considered in [10–12]. However, these approaches are only available to the design of robust H_∞ controller for linear-invariant system. Based on the robust backstepping theory, the control law developed in this paper have solved the problem of the time-varying rendezvous system. Using Lyapunov stability method, a sufficient condition for the existence of a robust backstepping controller is acquired.

The remainder of this paper is organized as follows. Section 2 presents the problem, where the simplified relative motion dynamic model is established. The robust backstepping control approach is then proposed in Sect. 3 to solve the rendezvous problem. In Sect. 4, a numerical simulation is put forward to verify the effectiveness of the proposed method.

2   Problem Statement

2.1   Dynamic Model

In this section, the system dynamics are given. Figure 1 illustrates the schematic drawing of the spacecraft rendezvous configuration. Assume that the target

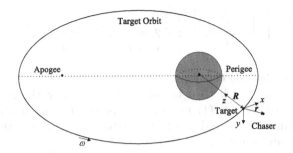

Fig. 1. Spacecraft rendezvous system and coordinates

spacecraft is on an eccentric orbit. Consider the target orbital coordinate system $C_0 = \{x - y - z\}$, where the origin of coordinatesis is attached to the center of the target spacecraft, and axis z orientates to the earth's core, axis x is points in the direction of motion of the target, perpendicular to the z axis, the y axis is perpendicular to the orbit plane and opposite the angular momentum vector. The relative motion dynamics of the rendezvous system are given by [9]

$$
\begin{bmatrix} \ddot{x} \\ \ddot{z} \\ \ddot{y} \end{bmatrix} = \begin{bmatrix} 2\omega\dot{z} + \dot{\omega}z + \omega^2 x - \frac{\mu x}{|\mathbf{R}+\mathbf{r}|} \\ \omega^2 z - 2\omega\dot{x} - \dot{\omega}x - \mu\left(\frac{z-R}{|\mathbf{R}+\mathbf{r}|^3} + \frac{1}{R^2}\right) \\ -\frac{\mu y}{|\mathbf{R}+\mathbf{r}|^3} \end{bmatrix} + \mathbf{u_f}
\tag{1}
$$

where $\mathbf{R}$ refers to the relative displacement vector of target spacecraft from the earth's core; $\mathbf{r} = [x\ y\ z]$ denotes the elements of the relative position between the chaser and the target; ω denotes the orbital angular velocity of the rotating coordinate system; μ signifies the gravity constant; $\|\mathbf{R}\| = R, \|\mathbf{r}\| = r, \mathbf{u_f} = [u_x\ u_z\ u_y]^T$ means the actual control acceleration input vectors applied to chaser.

Here h is represented as the orbital angular momentum of the target spacecraft. It's clear that $h = R^2\omega = constant$. Let $e \in [0, 1)$ be the eccentricity, θ be the true anomaly and $\rho \equiv 1 + ecos(\theta)$, then

$$
k = \frac{\mu}{h^{\frac{3}{2}}} = constant
\tag{2}
$$

Thus, there is

$$
\omega = \frac{h}{R^2} = k^2\rho^2
\tag{3}
$$

If the relative distance of the chaser with respect to the target is much smaller than the distance between the target and the earth's core, which means: $R \gg r$. When these relations are substituted into Eq. (1), the nonlinear system dynamics described by Eq. (1) will be linearized at the origin as follows [9]

$$\begin{bmatrix} \ddot{x} \\ \ddot{z} \\ \ddot{y} \end{bmatrix} = \begin{bmatrix} -k\omega^{3/2}x + 2\omega\dot{z} + \dot{\omega}z + \omega^2 x \\ 2k\omega^{3/2}z - 2\omega\dot{x} - \dot{\omega}x + \omega^2 z \\ -k\omega^{3/2}y \end{bmatrix} + \mathbf{u_f} \tag{4}$$

which is called as the T-H equations.

The strategy in [9] will then be adopted to simplify the relative motion dynamics described by Eqs. (1) and (4). According to this strategy, the true anomaly θ is utilized as an independent variable instead of the time variable t and adopt the transformation

$$\begin{bmatrix} \tilde{x}(\theta) \\ \tilde{z}(\theta) \\ \tilde{y}(\theta) \end{bmatrix} = \rho(\theta) \begin{bmatrix} x \\ z \\ y \end{bmatrix} \tag{5}$$

Since the true anomaly θ is monotonically increasing along with time t, then the independent variable is replaced by the true anomaly θ of the target from the time t. The derivative and second derivative of the relative position $\mathbf{r}$ are given by

$$\dot{r} = \omega r', \qquad \ddot{r} = \omega^2 r'' + \omega\omega' r' \tag{6}$$

Changing the independent variable, Eq. (4) can be converted with respect to the true anomaly θ as follows

$$\begin{cases} \omega^2 x'' + \omega\omega' = (\omega^2 - k\omega^{3/2})x + 2\omega^2 z' + \omega\omega' z + u_x \\ \omega^2 z'' + \omega\omega' z' = (\omega^2 + 2k\omega^{3/2})z - 2\omega^2 x' - \omega\omega' x + u_z \\ \omega y'' + \omega\omega' y' = -k\omega^{3/2}y + u_y \end{cases} \tag{7}$$

From the angular momentum, there are the following relations [9]

$$\begin{cases} \omega = (h/R^2) = (h/p^2)(1 + e\cos\theta)^2 = k^2\rho^2 \\ \omega' = 2k^2\rho\rho' = -2k^2 e\sin\theta\rho \end{cases} \tag{8}$$

Substituting (8) into (7), the relative motion dynamics can be simplified as

$$\begin{cases} \rho x'' - 2e\sin\theta x' - e\cos\theta x = 2\rho z' - 2e\sin\theta z + \dfrac{u_x}{k^4\rho^3} \\ \rho z'' - 2e\sin\theta z' - (3 + e\cos\theta)z = -2\rho x' + 2e\sin\theta x + \dfrac{u_z}{k^4\rho^3} \\ \rho y'' - 2e\sin\theta y' = -y + \dfrac{u_y}{k^4\rho^3} \end{cases} \tag{9}$$

According to Eq. (5), then the relative motion dynamics can be ultimately simplified as

$$\begin{cases} \tilde{x}'' = 2\tilde{z}' + u_x/k^4\rho^3 \\ \tilde{z}'' = 3\tilde{z}/\rho - 2\tilde{x}' + u_z/k^4\rho^3 \\ \tilde{y}'' = -\tilde{y} + u_y/k^4\rho^3 \end{cases} \tag{10}$$

Choose new state vectors $\mathbf{x}_1(\theta) = [\tilde{x}(\theta)\ \tilde{z}(\theta)\ \tilde{y}(\theta)]^T$, $\mathbf{x}_2(\theta) = [\tilde{x}'(\theta)\ \tilde{z}'(\theta)\ \tilde{y}'(\theta)]^T$ and control vector $\mathbf{u}(\theta) = [u_x\ u_z\ u_y]^T$. Then the rendezvous system dynamic model can be described as

$$\mathbf{x}_2'(\theta) = \mathbf{A}_1\mathbf{x}_1(\theta) + \mathbf{A}_2\mathbf{x}_2(\theta) + b(\theta)\mathbf{u}(\theta) \tag{11}$$

where $\mathbf{A}_1 = \begin{bmatrix} 0 & 0 & 0 \\ 0 & \frac{3}{\rho} & 0 \\ 0 & 0 & -1 \end{bmatrix}$, $\mathbf{A}_2 = \begin{bmatrix} 0 & 2 & 0 \\ -2 & 0 & 0 \\ 0 & 0 & 0 \end{bmatrix}$ and $b(\theta) = 1/k^4\rho^3$. Notice that the

rendezvous system is time variant as the parameter ρ is periodic.

It's clear that the in-plane motion (namely, the $x - z$ subsystem) and the out-of-plane motion (namely, the y subsystem) can be considered as two independent subsystems. As a result, we will design the controller for the two subsystems separately.

2.2 Control Objective

The entire spacecraft rendezvous procedure can be depicted as the transformation of state vectors $x(\theta)$ from the nonzero initial state $x(\theta_0)$ to the terminal state $x(\theta_f) = 0$, where $\theta_f = \theta(t_f)$, with t_f being the rendezvous time. The main control objective in this paper is to design a controller $\mathbf{u}(\theta)$, under which the chaser can reach the expectant position and approach the target gradually. That's to say, the state vectors $\mathbf{x}_1(\theta) = 0, \mathbf{x}_2(\theta) = 0$ can be reached finally.

3 Controller Design

3.1 Propose the Control Law

By defining the state vector $\mathbf{X} = [x_1, x_2, x_3, x_4]^T = [\tilde{x}(\theta), \tilde{z}(\theta), \tilde{x}'(\theta), \tilde{z}'(\theta)]^T$, the state-space equations of the in-plane motion subsystem model can be expressed as

$$\begin{cases} x_1' = x_3 \\ x_2' = x_4 \\ x_3' = 2x_4 + u_x/k^4\rho^3 \\ x_4' = \dfrac{3}{\rho}x_2 - 2x_3 + u_z/k^4\rho^3 \end{cases} \tag{12}$$

According to system (12), define a cluster of error variables as follows

$$\begin{cases} z_1 = x_1, & z_3 = \lambda_1 x_1 + x_3 \\ z_2 = x_2, & z_4 = \lambda_2 x_2 + x_4 \end{cases} \tag{13}$$

where parameters $\lambda_i > 0, i = 1, 2$ are positive parameters and are to be determined later.

Then, system (12) can be transformed into

$$
\begin{cases}
z_1' = -\lambda_1 z_1 + z_3 \\
z_2' = -\lambda_2 z_2 + z_4 \\
z_3' = -\lambda_1^2 z_1 - 2\lambda_2 z_2 + \lambda_1 z_3 + 2z_4 + u_x/k^4 \rho^3 \\
z_4' = 2\lambda_1 z_1 + \dfrac{3}{\rho} z_2 - \lambda_2^2 z_2 - 2z_3 + \lambda_2 z_4 + u_z/k^4 \rho^3
\end{cases}
\tag{14}
$$

By defining the state vector $\gamma = [x_5, x_6]^T = [\tilde{y}(\theta), \tilde{y}'(\theta)]^T$. According to system (11), the state-space of the out-of-plane subsystem model can be expressed as

$$
\begin{cases}
x_5' = x_6 \\
x_6' = -x_5 + u_y/k^4 \rho^3
\end{cases}
\tag{15}
$$

The robust backstepping controllers are proposed as follows

$$
\begin{cases}
\mathbf{u}(\theta) = \begin{bmatrix} \mathbf{u}_1(\theta) \\ u_2(\theta) \end{bmatrix}, \\
\mathbf{u}_1(\theta) = \begin{bmatrix} u_x \\ u_z \end{bmatrix} = -\begin{bmatrix} a_1 z_3 \\ a_2 z_4 \end{bmatrix}, \quad u_2(\theta) = u_y = -a_3 x_6
\end{cases}
\tag{16}
$$

where $a_i > 0, i = 1, 2, 3$. The parameters a_1, a_2, a_3 which need to be designed are given by

$$
\begin{cases}
a_1 = k^4 (1+e)^3 \left(\eta_3 + \dfrac{1}{2} + \dfrac{\lambda_1^4}{2} + \lambda_2^2 + \lambda_1 \right) \\
a_2 = k^4 (1+e)^3 \left(\eta_4 + \dfrac{1}{2} + \dfrac{\lambda_2^4}{2} + \lambda_1^2 + \dfrac{1}{(1-e)^2} + \lambda_2 \right) \\
a_3 > 0
\end{cases}
\tag{17}
$$

where parameters $\eta_3, \eta_4 > 0$.

3.2   The Stability Analysis

Theorem 1. *For the rendezvous system (12) and (15), respectively, for the in-plane motion and out-of-plane motion, taking the control law (16), the closed-loop system of system (12) and (15) are asymptotically stable in the presence of the uncertain time variant model parameters. The proof of the theorem is as follows.*

Proof. (i) **In-Plane Motion**
Consider the Lyapunov function

$$
V = \frac{1}{2} \sum_{i=1}^{4} z_i^2
\tag{18}
$$

the derivative of V along the trajectory of system (12) is as follows

$$V' = \sum_{i=1}^{4} z_i z_i'$$

$$= z_1(-\lambda_1 z_1 + z_3) + z_2(-\lambda_2 z_2 + z_4) + z_3\Big(-\lambda_1^2 z_1 - $$

$$2\lambda_2 z_2 + \lambda_1 z_3 + 2z_4 + \frac{u_x}{k^4 \rho^3}\Big) + z_4\Big(2\lambda_1 z_1 + $$

$$\frac{3}{\rho}z_2 - \lambda_2^2 z_2 - 2z_3 + \lambda_2 z_4 + \frac{u_z}{k^4 \rho^3}\Big)$$

$$= -\lambda_1 z_1^2 - \lambda_2 z_2^2 + z_1 z_3 - \lambda_1^2 z_1 z_3 + z_2 z_4 - \lambda_2^2 z_2 z_4$$

$$2\lambda_2 z_2 z_3 + \lambda_1 z_3^2 + 2\lambda_1 z_1 z_4 + \frac{3}{\rho}z_2 z_4 + \lambda_2 z_4^2 + $$

$$\frac{1}{k^4 \rho^3} z_3 u_x + \frac{1}{k^4 \rho^3} z_4 u_z \tag{19}$$

as $\rho = 1 + e\cos\theta$ and according to *Young's inequality*, substituting (16) into (19), then (19) can be further transformed as follows:

$$V' \le -\lambda_1 z_1^2 - \lambda_2 z_2^2 + \frac{1}{2}z_1^2 + \frac{1}{2}z_3^2 + \frac{1}{2}z_1^2 + \frac{1}{2}\lambda_1^4 z_3^2 + \frac{1}{2}z_2^2$$

$$+ \frac{1}{2}z_4^2 + \frac{1}{2}z_2^2 + \frac{1}{2}\lambda_2^4 z_4^2 + z_2^2 + \lambda_2^2 z_3^2 + \lambda_1 z_3^2 + z_1^2 + $$

$$\lambda_1^2 z_4^2 + \frac{9}{4}z_2^2 + \frac{1}{(1-e)^2}z_4^2 + \lambda_2 z_4^2 - \frac{a_1}{k^4(1+e)^3}z_3^2 - $$

$$\frac{a_2}{k^4(1+e)^3}z_4^2$$

$$= -(\lambda_1 - 2)z_1^2 - \Big(\lambda_2 - \frac{17}{4}\Big)z_2^2 - \Big(\frac{a_1}{k^4(1+e)^3} - $$

$$\frac{1}{2} - \frac{\lambda_1^4}{2} - \lambda_2^2 - \lambda_1\Big)z_3^2 - \Big(\frac{a_2}{k^4(1+e)^3} - \frac{1}{2} - \frac{\lambda_2^4}{2} - $$

$$\lambda_1^2 - \frac{1}{(1-e)^2} - \lambda_2\Big)z_4^2 \tag{20}$$

Let $\lambda_1 = \eta_1 + 2$, $\lambda_2 = \eta_2 + \frac{17}{4}$ and $\eta_1, \eta_2 > 0$ are positive parameters. Substituting (17) into (20), then there is

$$V' \le -\sum_{i=1}^{4} \eta_i z_i^2 \le -\eta \sum_{i=1}^{4} z_i^2 \tag{21}$$

where $\eta = min(\eta_1, \cdots, \eta_4) > 0$.

As V is a continuously differentiable, positive-definite and radially unbounded function, using LaSalle's Theorem [13], (21) signifies that $lim_{t\to\infty, -\pi \le \theta \le \pi} z_i = 0$, which guarantees the global asymptotic stability of the system (12) in the presence of the uncertain time variant model parameters.

(ii) Out-of-Plane Motion

According to system (15) and substituting (16), then the state-space of the out-of-plane subsystem model can be expressed as

$$
\begin{bmatrix} x_5 \\ x_6 \end{bmatrix}' = \begin{bmatrix} 0 & 1 \\ -1 & -\frac{a_3}{k^4 \rho^3} \end{bmatrix} \begin{bmatrix} x_5 \\ x_6 \end{bmatrix}
\tag{22}
$$

Let $A = \begin{bmatrix} 0 & 1 \\ -1 & -a \end{bmatrix}$, $a = \frac{a_3}{k^4 \rho^3}, b = \frac{a_3}{k^4(1+e)^3}, c = \frac{a_3}{k^4(1-e)^3}$, and it's clear that $0 < b < a < c$. Let $B = \begin{bmatrix} 0 & 1 \\ -1 & -b \end{bmatrix}$. Because B is a Hurwitz matrix, then there is a symmetric positive definite matrix P_1 so that

$$
P_1 B + B^T P_1 = -2I
\tag{23}
$$

Let $P = P_1 + \beta I, \beta > 0$ and is determined later and $d = (c - b) \parallel P_1 \parallel$. Choosing the Lyapunov function $V_1 = \gamma^T P \gamma$, then

$$
\begin{aligned}
V_1' &= \gamma^T (A^T P + PA)\gamma \\
&= \gamma^T (A^T P_1 + P_1 A)\gamma + \beta \gamma^T (A^T + A)\gamma \\
&= \gamma^T (B^T P_1 + P_1 B)\gamma + \gamma^T \left(\begin{bmatrix} 0 & 0 \\ 0 & -(a-b) \end{bmatrix} P_1 + \right. \\
&\quad \left. P_1 \begin{bmatrix} 0 & 0 \\ 0 & -(a-b) \end{bmatrix} \right) \gamma + 2\beta \gamma^T A \gamma \\
&\leq -2 \parallel \gamma \parallel^2 + 2d \parallel \gamma \parallel \cdot \mid x_6 \mid -a\beta x_6^2 \\
&\leq - \parallel \gamma \parallel^2 + d^2 \mid x_6 \mid^2 - b\beta x_6^2 \\
&= - \parallel \gamma \parallel^2
\end{aligned}
\tag{24}
$$

where $\beta = \frac{d^2}{b}$. $\qquad\qquad\square$

4 Simulation Results

In this section, numerical simulations are given to demonstrate the effectiveness of the proposed method to a specific elliptical orbit rendezvous mission. Assume that the target is in the geosynchronous transfer orbit, which is a provisional orbit to launch a spacecraft into the geosynchronous orbit [14]. The orbital elements are as follows: the semi-major axis $a = 24,616 \, \text{km}$, the eccentricity $e = 0.73074$ and the orbital period $T = 38,436 \, \text{s}$. In order to clarify, the parameters are listed in Table 1. According to the results obtained in Sect. 3, the total controller is designed as

$$
\mathbf{u}(\theta) = \begin{bmatrix} \mathbf{u}_1(\theta) \\ u_2(\theta) \end{bmatrix} = \begin{bmatrix} -a_1(\lambda_1 x_1 + x_3) \\ -a_2(\lambda_2 x_2 + x_4) \\ -a_3 x_6 \end{bmatrix}
\tag{25}
$$

Table 1. The orbital parameters of the target spacecraft

Parameters	Symbols	Values
Semi-major axis	a	$2.4616 \times 10^7 \, \mathrm{m}$
Eccentricity	e	0.73074
Angular momentum	h	$6.762 \times 10^{10} \, \mathrm{m}^2/\mathrm{s}$
Constant k	k	$2.267 \times 10^{-2}/\mathrm{s}^{1/2}$
Period	T	$38,436 \, \mathrm{s}$
Gravitational constant	μ	$3.98 \times 10^{14} \, \mathrm{m}^3/\mathrm{s}^2$

For simulation purposes, setting the control coefficients $\eta_1 = \cdots = \eta_4 = 1$. Suppose that at the initial time, the true anomaly is $\theta_0 = 0$. Besides, choosing the initial values of states in the target orbital coordinate system as $X(0) = [x_1(0), x_2(0), x_3(0), x_4(0), x_5(0), x_6(0)]^T = [800, 1000, 3, -5, -800, 3]^T$, representing that on the initial time of the rendezvous mission, the distances between the chaser spacecraft and the target spacecraft in all three directions(i.e., x, z and y axis) are respectively $800\,\mathrm{m}$, $1000\,\mathrm{m} - 800\,\mathrm{m}$ and the relative velocity in all three directions are respectively, $3\,\mathrm{m/rad}$, $-5\,\mathrm{m/rad}$, $3\,\mathrm{m/rad}$.

By simulation, the state trajectories (i.e., the relative position and the relative velocity between the chaser and the target) and the control input of the closed-loop rendezvous system (12) and (15) consisting the simplified dynamic model and the controller given by Eq. (16) are, respectively, shown in Figs. 2, 3 and 4. All these state variables and the control laws converge to zero finally. It indicates that the whole closed-loop rendezvous system (11) is asymptotically stable, which validates the proposed controller (16) can achieve a good performance. In fact, the rendezvous mission is implemented at approximately $\theta_f = 4.6\,(\mathrm{rad})$. The rendezvous time can be calculated as follows

$$t_f = \frac{\theta_f - \theta_0}{2\pi} T = 2.814 \times 10^4 \, \mathrm{s} \tag{26}$$

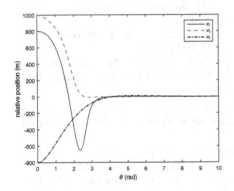

Fig. 2. Relative position

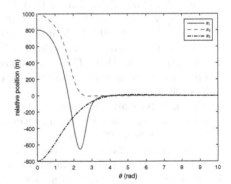

Fig. 3. Relative velocity

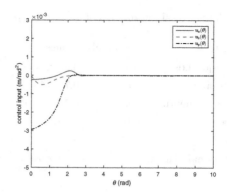

Fig. 4. The control input

5   Conclusions

In this paper, for the relative motion control of spacecraft rendezvous on an arbitrary elliptical orbit, a simplified dynamic model is investigated by using transformed variables. Based on this simplified model, a robust backstepping control law is proposed. Using Lyapunov theory, the proposed controller is validated to meet the control object. Simulation results are provided to validate the performance of the controller. Future work will focus on the extension of the presented simplified dynamic model to other spacecraft rendezvous problems such as fuel optimization or the application of other control theory for spacecraft rendezvous.

References

1. Zhang, D.W., Song, S.M., Pei, R.: Safe guidance for autonomous rendezvous and docking with a noncooperative target. In: Proceedings of the AIAA Guidance, Navigation, and Control Conference, pp. 1–19 (2010)
2. Wu, S.N., Zhou, W.Y., Tan, S.J.: Robust control for spacecraft rendezvous with a noncooperative target. Sci. World J. **2013**, 7 (2013)
3. Gao, H., Yang, X., Shi, P.: Multi-objective robust control of spacecraft rendezvous. IEEE Trans. Control Syst. Technol. **17**(4), 794–802 (2009)
4. Zhou, B., Lin, Z., Duan, G.R.: Lyapunov differential equation approach to elliptical orbital rendezvous with constrained controls. J. Guidance Control Dyn. **34**(2), 345–358 (2011)
5. Singla, P., Subbarao, K., Junkins, J.L.: Adaptive output feedback control for spacecraft rendezvous and docking under measurement uncertainty. J. Guidance Control Dyn. **29**(4), 892–902 (2006)
6. Youmans, E.A., Lutze, F.H.: Neural network control of space vehicle intercept and rendezvous maneuvers. J. Guidance Control Dyn. **21**(1), 116–121 (1998)
7. Clohessy, W.H., Wiltshire, R.S.: Terminal guidance system for satellite rendezvous. J. Aerosp. Sci. **27**(9), 653–658 (1960)

8. Carter, T.E.: State transition matrices for terminal rendezvous studies: brief survey and new example. J. Guidance Control Dyn. **21**(10), 148–155 (1998)
9. Yamanaka, K., Ankersen, F.: New state transition matrix for relative motion on an arbitrary elliptical orbit. J. Guidance Control Dyn. **25**(1), 60–66 (2002)
10. Green, M., Glover, K., Limebeer, D.J.N., Doyle, J.C.: A J-spectral factorization approach to $H\infty$ control. SIAM J. Control Optim. **28**, 1350–1371 (1990)
11. Khatibi, H., Karimi, A.: $H\infty$ controller design using an alternative to Youla parameterization. IEEE Trans. Autom. Control **55**(9), 2119–2123 (2010)
12. Choi, H.H., Chung, M.J.: Robust observer-based $H\infty$ controller design for linear uncertain time-delay systems. IEEE Trans. Autom. Control **33**, 1749–1752 (1997)
13. Khalil, H.K., Grizzle, J.W.: Nonlinear Systems. Prentice hall, Upper Saddle River (2002)
14. Shibata, M., Ichikawa, A.: Orbital rendezvous and flyaround based null controllability with vanishing energy. J. Guidance Control Dyn. **30**(4), 934–945 (2007)

Sampled Adaptive Control for Multi-joint Robotic Manipulator with Force Uncertainties

Hao Zhou, Hongbin Ma$^{(\boxtimes)}$, Haiyang Zhan, Yimeng Lei, and Mengyin Fu

School of Automation, Beijing Institute of Technology,
Beijing 100081, People's Republic of China
mathmhb@qq.com

Abstract. This paper addresses force estimation and trajectory tracking control for robotic manipulator in the presence of uncertain external load force at end effector. One-step Guess method using one step history data sampled from actual continuous-time plant at a constant sampling interval is developed to estimate the unknown fixed or time-varying force. A discrete-time adaptive controller based on estimation of load force is designed to track desired joint trajectory. System simulation of a 6 DOF manipulator is carried out with the help of robotic toolbox in MATLAB, which demonstrates performances of the proposed scheme dealing with both fixed and variable forces, compared with traditional control method.

Keywords: Robotic manipulator · Force estimation · Adaptive control · One-step guess

1   Introduction

Nowadays intelligent robots have come into our daily life as well as industrial process. The fact that robots do repeated work and deal with single kind of problem can not meet the increasing demand of robotic intelligence any more. On the one hand, in modern factory robotic arms are supposed to face uncertain tasks such as assembling or carrying some objects with unknown weight or even time-varying weight. On the other hand, smart robots need to possess excellent perception in unknown outside environments, for example, sense of external force or weight without force sensor in order to make optimal decisions. Therefore those motivate us to settle estimation of external force and position tracking control for robotic manipulator in the presence of uncertain load force at its end effector.

Position tracking of robotic manipulator always seems to be a fundamental and difficult task in robot control, especially in the presence of external disturbances and modal uncertainties. Various kinds of control methods have been used to address this problem, including proportional plus derivative (PD) control [8], iterative learning scheme [5], sliding PID control [12], repetitive and adaptive motion control [4].

This work is partially supported by National Natural Science Foundation (NSFC) under Grant 61473038.

N. Kubota et al. (Eds.): ICIRA 2016, Part I, LNAI 9834, pp. 14–25, 2016.
DOI: 10.1007/978-3-319-43506-0_2

To achieve high precise tracking of robot with load uncertainties, robust control that can reject load uncertainties have been studied extensively in literatures. A input-output robust control design, which could guarantee tracking performance in the presence of load variation as well as other disturbances was firstly introduced in [13]. A benchmark problem for robust feedback control for a flexible manipulator was presented in [11]. The robust control problem of robot manipulators could be translated into a optimal control [7] where load uncertainties were first reflected in the performance index and this approach was illustrated with two-joint SCARA type robot. Adaptive control has great advantages in coping with uncertainties. In [10], an adaptive control scheme was proposed for rigid link robots, where control signal computations were performed continuously and the control coefficient computations are performed in discrete time. An adaptive control system, requiring calculating only one parameter the tip load, was designed in [3]. Force estimation is important for adaptive control of robotic manipulators with unknown load at end effector, since usually it constitutes one part of control torque. Besides, high precise estimation of force can replace force sensor with high cost in application of intelligent robot. An approach, providing force estimation as well as full state estimation in the presence of robot inertial parameter variations and measurement noise, was proposed in [1]. Some intelligent control methods have also been adopted such as artificial neural networks (ANN) [14] and switched adaptive control [15,16].

Among the existing control methods, discrete-time adaptive control methods for robotic manipulators with unknown load force at end effector, are still seldom concerned. During the past decades, we have witnessed extensive application of digital computers in control system due to availability of cheap chip and the advantages of digital signals over continuous signals. The practical implementation of theoretic control methods will benefit much from directly taking true plant as sampled system and then designing control scheme in view of discrete-time control system, such as testing real-time performance easily. However, it is difficult to design a satisfactory discrete-time control scheme for robotic system. Force estimation only using history information of joint angles and velocities is also seldom studied because of the modal complexities of robotic system. Since multi-joint will increase modeling and computation difficulties, robotic manipulators used for simulation in many previous literatures only have one or two links while many a manipulator of six degrees of freedom (DOF) or more could be seen in practice, especially in intelligent robot field.

Our study object is robotic manipulator with unknown external load force at end effector, the only uncertainty considered in this paper for simplicities. Discrete-time adaptive control method based on One-step Guess (OSG) [9] was first introduced for load uncertainties [6], and this paper extend this method to a general case. Mathematically external force timed by Jacobian matrix is added to robotic dynamic equation, instead of direct addition as in some literatures. Force estimation is obtained through OSG, by which the discrete-time adaptive controller can cope with position tracking. The performance of this scheme is demonstrated with simulations for PUMA560, a kind of 6 DOF manipulator, further than the work in [6]. Robotic toolbox (RVC) [2] in MATLAB contributes to dynamic calculations and simulations. This scheme has three main advantages: free of force sensor, convenience for digital implementation, high-precision tracking.

The remaining part of this paper is organized as follows. First, Sect. 2 introduces the dynamics of robotic manipulator with external load force and the problem to be studied. Section 3 presents the detailed design of discrete-time force estimation and adaptive controller and briefly analyzes convergence characteristics of trajectory errors. Then Sect. 4 illustrates the simulation results of a 6 DOF manipulator with OSG-based adaptive controller and force estimation. Finally Sect. 5 briefly summarizes our work and also presents the future work.

2   Problem Formation

The dynamic model of a serial robotic manipulator in the presence of load force at its end effector can be represented by the following equation in matrix form

$$M(q)\ddot{q} + C(q,\dot{q}) + G(q) = u + J^T f \tag{1}$$

where n is the degree of freedom, $q \in R^n, \dot{q} \in R^n, \ddot{q} \in R^n$ are respectively the vector of generalized joint coordinates, velocities and accelerations, $M(q) \in R^{n \times n}$ is the joint-space inertial matrix, $C(q,\dot{q}) \in R^{n \times n}$ is the Coriolis and centripetal coupling matrix, $G(q) \in R^n$ is the gravity loading, and $u \in R^n$ is the vector of generalized actuator forces associated with the generalized coordinates q. The last term gives the joint forces due to external payload f applied at the end effector and J is the manipulator Jacobian matrix.

As is well known, the manipulator Jacobian transforms joint velocity to an end effector spatial velocity and the Jacobian transpose transforms a wrench applied at the end effector to torques experienced at the joints. It is noted that both of two transforms hold respectively in the same coordinate frame, either both in the world coordinate frame or both in the end effector coordinate frame. Generally speaking, the world coordinate frame is adopted and hence we have the following relationship

$$u_d = J^T(q)f \tag{2}$$

where the elements of u_d are joint torques for revolute joints. Generalized force f is denoted by $f = [f_x \ f_y \ f_z \ T_x \ T_y \ T_z]^T$, that is, f can represent an arbitrary external force or torque applied at end effector in all possible directions. For example, $f = [f_x \ f_y \ 0 \ 0 \ 0 \ 0]^T$ represents a horizontal force.

In practice, the external force f in the above Eq. (1) is not always known in advance, which results in bad performance of some traditional methods such as PD feedforward control, especially in the case of large value or time-varying case. The ultimate goal of control system is to achieve trajectory tracking, which needs control system to estimate f using observed history information such as sampled joint angle values and velocities. Before designing computation methods of force estimation and control torque, system discretization should be first done since digital control system is widely used. Then we estimate the load force at past time and assume that the force vary small in the next sampling time which corresponds with most actual cases since generally speaking sampling interval is very small. By taking force estimation at the last time as current time force, we can design the adaptive controller to finish trajectory tracking.

3   Design of Estimation and Controller

In this section, we design force estimation and tracking controller for the above problem, mainly consisting of the following four parts:

1. Discretization of manipulator dynamic equation;
2. Designing estimation algorithm of external force based on sampled history information consisting of joint angle values and velocities;
3. Designing control signal at the current time according to estimation of force of last time instant;
4. Analyzing convergence characteristics of trajectory error.

3.1   Discretization of Dynamic Equation

First let $\bar{q} = [q \quad \dot{q}]^T$, $q \in R^n$ and $M(q)$ is usually invertible, then Eq. (1) can be rewritten in the following state space form

$$\dot{\bar{q}} = A(q, \dot{q})\bar{q} + B(q)u + F(q)f - Q(q)G(q) \tag{3}$$

where

$$A(q, \dot{q}) = \begin{bmatrix} 0_{n \times n} & I_{n \times n} \\ 0_{n \times n} & -M^{-1}(q)C(q, \dot{q}) \end{bmatrix}$$

$$B(q) = Q(q) = \begin{bmatrix} 0_{n \times n} \\ M^{-1}(q) \end{bmatrix}$$

$$F(q) = \begin{bmatrix} 0_{n \times n} \\ M^{-1}(q)J^T(q) \end{bmatrix} \tag{4}$$

The sampling period is denoted as T and then at time $t = (k-1)T$ the joint angle value and velocity are respectively $x_k = q(t_k)$ and $\dot{x}_k = \dot{q}(t_k)$. Let the space state of discrete-time system be $\bar{x}_k = [x_k \quad \dot{x}_k]^T$, then from theory of discretization we can get

$$L_k = e^{A(q(t_k), \dot{q}(t_k))T}$$

$$H_k = \int_{(k-1)T}^{kT} e^{A(q, \dot{q})t} B(q) dt$$

$$R_k = \int_{(k-1)T}^{kT} e^{A(q, \dot{q})t} F(q) dt$$

$$S_k = \int_{(k-1)T}^{kT} e^{A(q, \dot{q})t} Q(q) dt \tag{5}$$

where L_k, H_k, R_k and S_k are counter-part matrices corresponding to the continuous time matrices $A(q, \dot{q})$, $B(q)$, $F(q)$ and $Q(q)$ in the Eq. (3). Hence the discrete-time space state equation for robotic manipulator is as follows

$$\bar{x}_{k+1} = L_k \bar{x}_k + H_k u_k + R_k f_k - S_k G(x_k) \tag{6}$$

Since only the sampled values $x_k = q(t_k)$ and $\dot{x}_k = \dot{q}(t_k)$ at the sampling time instant t_k can be obtained as well as history values, the exact values of L_k, H_k, R_k and S_k can not be calculated using Eq. (5). Instead, we can use the following formula:

$$\hat{L}_k = e^{A_k T}$$

$$\hat{H}_k = \int_{(k-1)T}^{kT} e^{A_k t} B(x_k) dt$$

$$\hat{R}_k = \int_{(k-1)T}^{kT} e^{A_k t} F(x_k) dt$$

$$\hat{S}_k = \int_{(k-1)T}^{kT} e^{A_k t} Q(x_k) dt \tag{7}$$

where $A_k = A(x_k, \dot{x}_k)$, which is determined at each sampling time. As a result, L_k, H_k, R_k and S_k can be calculated through Runge-Kutta method or other numerical integral methods. The estimated errors can be denoted as:

$$\tilde{H}_k = H_k - \hat{H}_k, \quad \tilde{L}_k = L_k - \hat{L}_k$$

$$\tilde{R}_k = R_k - \hat{R}_k, \quad \tilde{S}_k = S_k - \hat{S}_k \tag{8}$$

which will generate modal calculation errors but can be small enough if the sampling period is small enough.

3.2   Force Estimation

From Eq. (6), at time $t_{k-1} = (k-1)T$, we have

$$\bar{x}_{k-1} = L_{k-1} \bar{x}_{k-1} + H_{k-1} u_{k-1} + R_{k-1} f_{k-1} - S_{k-1} G(x_{k-1}) \tag{9}$$

Then

$$R_{k-1} f_{k-1} = \bar{x}_{k-1} - L_{k-1} \bar{x}_{k-1} - H_{k-1} u_{k-1} + S_{k-1} G(x_{k-1}) \tag{10}$$

which can be taken as the constraint equation of f_{k-1}. Then we denote the right hand side of Eq. (10) by $P(\bar{x}_k, \bar{x}_{k-1})$, that is,

$$P(\bar{x}_k, \bar{x}_{k-1}) = \bar{x}_{k-1} - L_{k-1} \bar{x}_{k-1} - H_{k-1} u_{k-1} + S_{k-1} G(x_{k-1}) \tag{11}$$

The constraint Eq. (10) is equivalent to

$$R_{k-1} f_{k-1} = P(\bar{x}_k, \bar{x}_{k-1}) \tag{12}$$

Generally R_{k-1} is not a square matrix and thus not invertible. Hence we could adopt the least-square method or regularized least-square method to solve the above Eq. (12). Besides, as previously mentioned, only estimation values $\hat{L}_{k-1}$, $\hat{H}_{k-1}$, $\hat{R}_{k-1}$ and $\hat{S}_{k-1}$ could be used at time t_k. Based on the two points, force estimation of f_{k-1} can be given as follows

$$\hat{f}_{k-1} = (\hat{R}_{k-1}^T \hat{R}_{k-1} + Q_f^T Q_f)^{-1} \hat{R}_{k-1}^T \hat{P}(\bar{x}_k, \bar{x}_{k-1}) \tag{13}$$

where

$$\hat{P}(\bar{x}_k, \bar{x}_{k-1}) = \bar{v}_k - \hat{L}_{k-1}\bar{v}_{k-1} - \hat{H}_{k-1}u_{k-1} + \hat{S}_{k-1}G(v_{k-1}) \tag{14}$$

and Q_f is a matrix for fine-tuning the estimation such that $Q_f \hat{f}_{k-1} = 0$, which can reflect a prior knowledge on the unknown force.

Since the change of external force during one sampling interval is assumed to be very small, hence estimated value $\hat{f}_{k-1}$ can serve as a priori estimation of f_k for designing control signal u_k, that is,

$$\check{f}_k = \hat{f}_{k-1} = (\hat{R}_{k-1}^T \hat{R}_{k-1} + Q_f^T Q_f)^{-1} \hat{R}_{k-1}^T \hat{P}(\bar{x}_k, \bar{x}_{k-1}) \tag{15}$$

although the unknown force f_k is unavailable at sampling time $t_k = kT$.

3.3   Adaptive Controller Design

The idea of adaptive controller (indirect approach) consists in replacing the unknown parameter by its estimation. After obtaining a prior estimation of f_k, we can design the controller for the Eq. (6) from which the following equation can be obtained

$$H_k u_k = \bar{x}_{k+1} - L_k \bar{x}_k - R_k f_k + S_k G(x_k) \tag{16}$$

Denote the desired reference trajectory of joint vector at sampling time t_{k+1} by $\bar{x}_{k+1}^*$. The actual joint vector is $\bar{x}_k$ at time t_k. The ideal control signal u_k can lead to the result that $\bar{x}_{k+1} = \bar{x}_{k+1}^*$. The right hand side of Eq. (16) can set as

$$V(\bar{x}_{k+1}, \bar{x}_k) = \bar{x}_{k+1} - L_k \bar{x}_k - R_k f_k + S_k G(x_k) \tag{17}$$

Likewise, only the estimation values at time t_k can be used then the following equation

$$\hat{V}(\bar{x}_{k+1}, \bar{x}_k^*) = \bar{x}_{k+1}^* - \hat{L}_k \bar{x}_k - \hat{R}_k f_k + \hat{S}_k G(x_k) \tag{18}$$

is estimation of $V(\bar{x}_{k+1}, \bar{x}_k)$.

In order to track the desired reference trajectory at time $t_{k+1} = (k+1)T$, in other words, $\bar{x}_{k+1} = \bar{x}_{k+1}^*$, the control signal at time $t_k = kT$ should be the following regularized least-square form

$$u_k = (\hat{H}_k^T \hat{H}_k + Q_u^T Q_u)^{-1} \hat{H}_k^T \hat{V}(\bar{x}_{k+1}, \bar{x}_k^*) \tag{19}$$

where Q_u is a matrix for fine-tuning the components of vector u_k, which might improve the performance of control. This controller is based on One-step Guess method which estimates unknown force using only the information of last time instant and hence results in fast adaption.

3.4   Tracking Characteristics

The Eq. (10) is rewritten as

$$R_{k-1}f_{k-1} = \bar{x}_{k-1} - L_{k-1}\bar{x}_{k-1} - H_{k-1}u_{k-1} + S_{k-1}G(x_{k-1}) \tag{20}$$

which can obtain the ideal estimation of force. However, we use the following equation to estimate f

$$\hat{R}_{k-1}\hat{f}_{k-1} = \bar{x}_{k-1} - \hat{L}_{k-1}\bar{x}_{k-1} - \hat{H}_{k-1}u_{k-1} + \hat{S}_{k-1}G(x_{k-1}) \tag{21}$$

From the above Eqs. (20) and (21), we get

$$R_{k-1}f_{k-1} - \hat{R}_{k-1}\hat{f}_{k-1} = -\tilde{L}_{k-1}\bar{x}_{k-1} - \tilde{H}_{k-1}u_{k-1} + \tilde{S}_{k-1}G(x_{k-1}) \tag{22}$$

The actual space state equation and desired equation are respectively

$$\bar{x}_{k+1} = L_k\bar{x}_k + H_ku_k + R_kf_k - S_kG(x_k) \tag{23}$$

and

$$\bar{x}_{k+1}^* = \hat{L}_k\bar{x}_k + \hat{H}_ku_k + \hat{R}_k\hat{f}_{k-1} - \hat{S}_kG(x_k) \tag{24}$$

By subtracting Eq. (24) from Eq. (23), we obtain

$$\bar{x}_{k+1} - \bar{x}_{k+1}^* = \tilde{L}_k\bar{x}_k + \tilde{H}_ku_k + R_kf_k - \hat{R}_k\hat{f}_{k-1} - \tilde{S}_kG(x_k) \tag{25}$$

where $\tilde{H}_k = H_k - \hat{H}_k, \tilde{L}_k = L_k - \hat{L}_k, \tilde{R}_k = R_k - \hat{R}_k, \tilde{S}_k = S_k - \hat{S}_k$.
For time instant $t = (k-1)T$, we obtain

$$\bar{x}_k - \bar{x}_k^* = \tilde{L}_{k-1}\bar{x}_{k-1} + \tilde{H}_{k-1}u_{k-1} + $$
$$R_{k-1}f_{k-1} - \hat{R}_{k-1}\hat{f}_{k-2} - \tilde{S}_{k-1}G(x_{k-1}) \tag{26}$$

By substituting Eq. (22) into Eq. (26), we get

$$\bar{x}_k - \bar{x}_k^* = \hat{R}_{k-1}\hat{f}_{k-1} - R_{k-1}f_{k-1} + R_{k-1}f_{k-1} - \hat{R}_{k-1}\hat{f}_{k-2}$$
$$= \hat{R}_{k-1}(\hat{f}_{k-1} - \hat{f}_{k-2}) \tag{27}$$

From the above Eq. (27), we can conclude that the position trajectory error $\|\bar{x}_k - \bar{x}_k^*\|$ will converge to zero if $\|\hat{f}_{k-1} - \hat{f}_{k-2}\| \to 0$, which can be easily achieved by estimation Eq. (13).

4   Simulation Examples

This section validates the above proposed controller with dynamic simulation, carried out in MATLAB with the help of RVC, a toolbox dealing with robotics and machine vision. As a comparison, the simulation results using PD feedforward controller are also illustrated in this section. In this paper, the plant is PUMA560, a well-known 6 DOF industrial robotic manipulator with unknown load force at the end effector, which results in uncertainties in this robot control system. This manipulator depicted in Fig. 1 has six revolute joints, that is, $n = 6$.

If we are more interested in estimation of external force, then we might use some *a prior* knowledge to set Q_f so that we can get more precise estimation. The whole simulation system consists of a continuous-time robotic plant and a discrete-time controller, either PD feedforward one or OSG-based adaptive controller one. The OSG-based adaptive controller is given by the Eq. (19), while the generic PD feedforward controller is given by

$$U_{ff} = M^*(q^*)\ddot{q}^* + C(q^*, q^*)\dot{q}^* + G(q^*) + \{K_v(\dot{q}^* - \dot{q}) + K_p(q^* - q)\} \qquad (28)$$

where q^* and $\dot{q}^*$ are respectively desired joint angle and velocity, and K_v and K_p are velocity and position gain (or damping) matrices respectively. Before adding external force at the end effector, the control gain $K_v = 100 * I_{6\times6}$ and $K_p = I_{6\times6}$ of PD feedforward controller have been well adjusted in order that original control parameters can guarantee a satisfactory result of position trajectory. In this way, we can compare the two kinds of controller in dealing with unknown load force. The sampling time interval is $T = 0.02\,\mathrm{s}$ and the default unit of f is Newton (N).

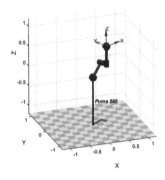

Fig. 1. PUMA560 in MATLAB using RVC

4.1   Fixed Case

The load force applied at the end effector is unknown and fixed, for example, a constant external force $f = [50\ 100\ 0\ 0\ 0\ 0]^T$ N that is a fixed horizon force. We obtain the response curves of PD feedforward control in Fig. 2(a) and OSG-based adaptive controller in Fig. 2(b). The trajectories of joint 4 to joint 6 are not presented in Fig. 2 since they are tracked well both in these two controllers. From these two figures, we can see that the tracking errors in OSG-based adaptive controller converge to zero while the errors of the first joint to the third joint in PD forward control scheme are far from zero.

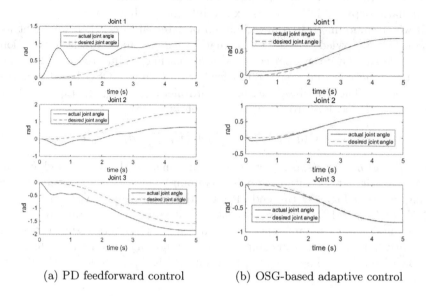

(a) PD feedforward control   (b) OSG-based adaptive control

Fig. 2. Position trajectory results in fixed case

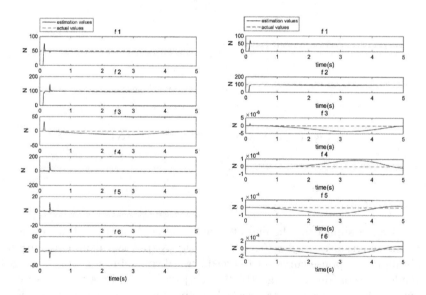

Fig. 3. Force estimation in fixed case

Also the estimation curve of load force is depicted in Fig. 3(a), where each subfigure, from f_1 to f_6, represents one component of force f. We set $Q_f = 0_{6 \times 6}$ of Eq. (13) in this simulation. The estimation result of f_3 during tracking is as not good as the other components of load force, which might be a result of singularity of $\hat{R}_{k-1}$ in Eq. (12). However, this estimation deviation has little impact on tracking precision of all joints.

A *prior* knowledges of direction of load force can be used and in the case of $f = [50\ 100\ 0\ 0\ 0\ 0]^T$, we set $Q_f = \mathrm{diag}\{0, 0, 10, 10, 10, 10\}$. Then estimation curves of load force are illustrated in Fig. 3(b), better than Fig. 3(a).

4.2   Time-Varying Case

In practice, external load force might be time-varying and unknown, for example, $f = [50,\ 60+10\sin(2\pi t),\ 0,\ 0,\ 0,\ 0]^T$. The tracking trajectories are illustrated in Fig. 4, from which we conclude that OSG-based adaptive control can also deal with time-varying load force well despite of the presence of small estimation deviations of force. The trajectories of joint 4 to joint 6 are not presented in Fig. 4 due to the same reason. The estimation curve of load force is depicted in Fig. 5(a), where only the estimations of f_1 to f_3 are depicted here. Likewise, *a prior* knowledge of the direction of force can be used and the estimation curves of force are depicted in Fig. 5(b) reflecting better estimation with *a prior* knowledge.

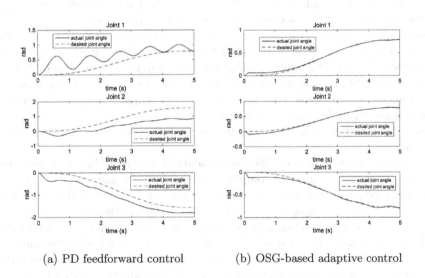

(a) PD feedforward control   (b) OSG-based adaptive control

Fig. 4. Position trajectory in time-varying case

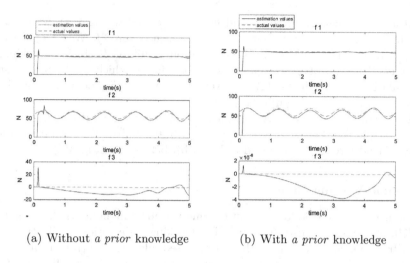

(a) Without *a prior* knowledge   (b) With *a prior* knowledge

Fig. 5. Force estimation in time-varing case

5   Conclusion

In this paper, we have presented a novel scheme of discrete-time force estimation and tracking control based on one-step-guess for robotic manipulator with
unknown load force applied at end effector. The history information of joint angle
values and velocities sampled from the true arm system is used to estimate the
unknown fixed or time-varying force, and a discrete-time adaptive controller
based on force estimation is designed to achieve position tracking. Dynamic
simulations for a 6 DOF robot manipulator are carried out in MATLAB with
RVC toolbox. Simulation results have demonstrated that this control approach
could obtain a remarkable tracking performance compared with tradition control scheme. In addition, the estimation method of unknown force also has a
considerably high precision, which could be used for intelligent robotic sense of
external force.

While the performance of OSG-based adaptive controller has been validated
through computer simulation, complete theoretical proof and experimental verification of physical system are also required which will be the goal of the future
work. Meanwhile, the proposed scheme mainly considers the uncertainties from
external load force. However, a true robotic system also suffers from other disturbances such as friction disturbance, which could result in bad performance. A
further development of the control scheme is to study how OSG-based adaptive
controller deal with friction disturbance as well as unknown external force.

References

1. Chan, L.P., Naghdy, F., Stirling, D.: Extended active observer for force estimation and disturbance rejection of robotic manipulators. Robot. Auton. Syst. **61**(12), 1277–1287 (2013)
2. Corke, P.: Robotics, Vision and Control. Springer, Heidelberg (2011)
3. Feliu, J.J., Feliu, V., Cerrada, C.: Load adaptive control of single-link flexible arms based on a new modeling technique. IEEE Trans. Robot. Autom. **15**(5), 793–804 (1999)
4. Kaneko, K., Horowitz, R.: Repetitive and adaptive control of robot manipulators with velocity estimation. IEEE Trans. Robot. Autom. **13**(2), 204–217 (1997)
5. Kuc, T.Y., Nam, K.H., Lee, J.S.: An iterative learning control of robot manipulators. IEEE Trans. Robot. Autom. **7**(6), 835–842 (1991)
6. Li, J.P., Ma, H.B., Yang, C.G., Fu, M.Y.: Discrete-time adaptive control of robot manipulator with payload uncertainties. In: IEEE International Conference on Cyber-Technology in Automation, Control and Intelligent Systems, Shenyang, pp. 8–12, June 2015
7. Lin, F., Brandt, R.D.: An optimal control approach to robust control of robot manipulators. IEEE Trans. Robot. Autom. **14**(1), 69–77 (1998)
8. Lozano, R., Valera, A., Albertos, P., Albertos, P., Nakayama, T.: PD control of robot manipulators with joint flexibility, actuators dynamics and friction. Automatica **35**(10), 1697–1700 (1999)
9. Ma, H.B., Rong, L.H., Wang, M.L., Fu, M.Y.: Adaptive tracking with one-step-guess estimator and its variants. In: Proceedings of the 2011 30th Chinese Control Conference (CCC 2011), Yantai, pp. 2521–2526, July 2011
10. Middleton, R.H.: Adaptive control for robot manipulators using discrete time identification. IEEE Trans. Autom. Control **35**(5), 633–637 (1990)
11. Moberg, S., Ohr, J., Gunnarsson, S.: A benchmark problem for robust feedback control of a flexible manipulator. IEEE Trans. Control Syst. Technol. **17**(6), 1398–1405 (2009)
12. Parra-Vega, V., Hirzinger, G., Liu, Y.H.: Dynamic sliding pid control for tracking of robot manipulators: theory and experiments. IEEE Trans. Robot. Autom. **19**(6), 967–976 (2003)
13. Qu, Z.H.: Input-output robust tracking control design for flexible joint robots. IEEE Trans. Autom. Control **40**(1), 78–83 (1995)
14. Teixeira, R.A., Braga, A.D., De Menezes, B.R.: Control of a robotic manipulator using artificial neural networks with on-line adaptation. Neural Process. Lett. **12**(1), 19–31 (2000)
15. Wang, X., Niu, R., Chen, C., Zhao, J.: Switched adaptive control for a classof robot manipulators. Trans. Inst. Measur. Control **36**(3), 347–353 (2014)
16. Wang, X., Zhao, J.: Switched adaptive tracking control of robot manipulators with friction and changing loads. Int. J. Syst. Sci. **16**(6), 955–965 (2015)

Rapid Developing the Simulation and Control Systems for a Multifunctional Autonomous Agricultural Robot with ROS

Zhenyu Wang$^{(\boxtimes)}$, Liang Gong, Qianli Chen, Yanming Li, Chengliang Liu, and Yixiang Huang

Institute of Mechatronics and Logistic Equipment,
School of Mechanical Engineering, Shanghai Jiao Tong University,
Shanghai 200240, People's Republic of China
{silent_180,gongliang_mi,chenqianli,ymli,
chlliu,huang.yixiang}@sjtu.edu.cn

Abstract. Building customized control system for specific robot is generally acknowledged as the fundamental section of developing auto robots. To simplify the programming process and increase the reuse of codes, this research develops a general method of developing customized robot simulation and control system software with robot operating system (ROS). First, a 3D visualization model is created in URDF (unified robot description format), and is viewed in Rviz to achieve motion planning with Movelt! software package. Second, the machine vision provided by camera driver package in ROS enables the use of tools for image process, 3D point cloud analysis to reconstruct the environment to achieve accurate target location. Third, the communication protocols provided by ROS like serial, Modbus support the communication system development. To examine the method, we designed a tomato harvesting dual-arm robot, and conducted farming experiment with it. This work demonstrates the advantages of ROS when applied in robot control system development, and offers a plain method of building such system with ROS.

Keywords: Robot Operating System (ROS) · Dual-arm multifunctional robot · Rapid system development · Autonomous agricultural robot · Rviz

1 Introduction

With the rapid development of modern agriculture of high efficiency, the vital position and the function of automation control technology has been widely acknowledged. Faced with the pressure of agricultural products output and market competition, agriculture tend to develop with higher efficiency and accuracy, combined with automated mechanical equipment [1]. In agricultural aspect, production lines equipped with machines have been widely applied in planting crops of large scale, for instance wheat, cotton, etc. At the same time, it's clear that there are many process of planting crops requiring to be operated with higher accuracy and flexibility, due to its complex environment and changeable conditions, like growing tomatoes. Within automated mechanical equipment, agricultural robots stand out for its high controllability and

© Springer International Publishing Switzerland 2016
N. Kubota et al. (Eds.): ICIRA 2016, Part I, LNAI 9834, pp. 26–39, 2016.
DOI: 10.1007/978-3-319-43506-0_3

flexible kinematic characteristics, which meet the requirements above perfectly. Taking these into consideration, improving accuracy and controllability of agricultural robots with excellent control systems is gaining more and more significance in modern automation researches [2].

Robot Operating System (ROS) is a collection of software frameworks for robot software development, providing operating system-like functionality on a heterogeneous computer cluster. ROS provides standard operating system services such as hardware abstraction, low-level device control, implementation of commonly used functionality, message-passing between processes, and package management. Running sets of ROS-based processes are represented in a graph architecture where processing takes place in nodes that may receive, post and multiplex sensor, control, state, planning, actuator and other messages [3].

Designed to increase the reuse of codes, ROS is completely open-sourced and compatible with multi programming languages. Over 2000 existing program packages are available in ROS freely.

The main characteristics of ROS include: Open sourced, multilingual support, library integration, plentiful tools kit, and point to point communication. The computation graph level is shown in Fig. 1.

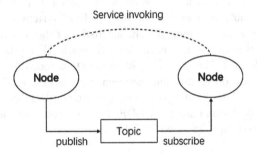

Fig. 1. The computation graph level of ROS

The computation graph reflects the connection way when the processes cooperate to process the statistics (point to point). Concepts concerned include node, service, topic, and messages, etc.

Nodes: A node is an executable that uses ROS to communicate with other nodes.

Messages: ROS data type used when subscribing or publishing to a topic.

Topics: Nodes can publish messages to a topic as well as subscribe to a topic to receive messages.

Master: Name service for ROS (i.e. helps nodes find each other)

Service: The method of communication between nodes, which allows nodes to send the request and answer.

Such point-to-point communication method is playing an important role in multi processes and multi hosts. When the multi hosts connect with different kinds of networks, there might be the risk of data transporting jam in central data server. While for point to point communication, there is no central data server, so it can avoid such problem to ensure the stability of multi processes and hosts.

This paper is organized as follows. Section 2 highlights the key ROS modules for developing an intelligent robot simulation and control systems. Section 3 describes the ROS deployment on a newly developed multifunctional agricultural robot. And conclusion is given in Sect. 4.

2 Customizing Robotics Modules in ROS

In general, there are four essential aspects for developing specific simulation and control systems for an intelligent robot, i.e. the operating system architectural model, the motion planning module, the machine vision module and built-in communication module.

2.1 Building Description Model Based on URDF

Robot visualization models enable the users to be informed of the current situation of the robot, decreasing the workload and the error rate. In ROS, the visualization is achieved with Rviz, and the general format of the robot description model is in URDF (unified robot description format). Based on XML, URDF is a language designed to describe the robot simulation model universally in ROS system, including the shape, size and colour, kinematic and dynamic characteristics of the model [4].

The basic way of building the visualization model in URDF is writing and compiling the URDF file. In URDF grammar, robot structure should be divided into links and joints. The connection relationship between parts is described by <parent> and <child>. To precisely describe the parts with parameters, ROS provides XACRO to allow users to use calculation macro in URDF. The common commands used to define the connection relationships are listed in Table 1.

Table 1. Common commands used in URDF

Commands	Grammar
Name the robot	robot name = "***"
Define the part	link name = "link***"/
Define the connection node	joint name = "joint***" type = "***"
Define the connection relationship	parent link = "link****"/ child link = "link****"/

When the structure of robots is relatively complex, the complexity of writing URDF file increases greatly. To ensure the accuracy of description, the often-used method is using 3D modelling software like Solidworks, Unigraphics NX, and transform the model into URDF by applying the plug-in like "solidworks to urdf". The suggested format of the 3D model is *.gae, while using stl format file is also acceptable in ROS as it contains the main grid statistics.

After writing the URDF file, the check_urdf tool can be used to check the grammar. If there exist no mistakes, this URDF file can be viewed in Rviz to visualize the robot model now.

When the URDF file needs to be applied in robots, users need to publish the robot conditions to tf, robot_state_publisher serves as the basis. The relevant parameters include the urdf xml robot description, and the joints information source with sensor_msgs/JointState format. The process of compiling URDF will generate the *. launch file, as the executive program in ROS.

In short, to realize the visualization, users need to write the urdf description file, and a node to publish and transform the information supported by robot_state_publisher.

2.2 Motion Planning Visualization Based on Rviz and Moveit

Rviz is the built-in visualization tool in ROS, providing the 3D simulation environment. With properly-set URDF file, using Rviz to visualize the robot model will be of no difficulty. When it comes to the kinematic analysis, the precise kinematic definitions needing can be acquired from the URDF file, and the relevant operation is supported by Moveit. The visualization environment provided by Rviz is shown in Fig. 2.

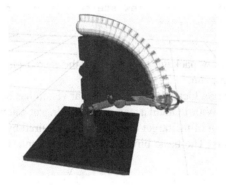

Fig. 2. The visualization environment provided by Rviz

Moveit is a universally-used integrated tool kit in ROS, as the core of motion control system, in charge of the calculation process of the positive and athwart kinematics of the robot's kinematic model. While the motion planning algorithm is based on the third-party library, OMPL (open motion plan library) [5].

To apply Moveit! in robot control system, the most convenient way is using the application assistant, which allows users to finish relevant configuration in steps. The main steps include importing the URDF file, setting the collision matrix, and adding the links and joints to the concerned motion planning group [6].

2.3   Implement Machine Vision with ROS

Machine vision is the technology and methods used to provide imaging-based automatic inspection and analysis for such applications as automatic inspection, process control, and robot guidance in industry. ROS integrates the driver kit OpenNI of Kinect, and can use OpenCV library to operate various image processes.

The common package used in ROS for image operating include camera_calibration, which serves as a calibration tool for the camera. The main algorithm it uses is the calibration method of chess board put forward by Zhengyou Zhang, which is also the general calibration method applied in OpenCV. The package image_view allows users to check the camera photos and give corresponding advice for the robot [7].

Here is the frequently-used usage:

```
rosrun camera_calibration cameracalibrator.py --size 8x6 --sq
uare 0.108 image:=/my_camera/image camera:=/my_camera
```

Use an 8 × 6 chessboard with 108 mm squares to calibrate the camera.

```
rosrun image_view stereo_view stereo:=<stereo namespace> imag
e:=<image topic identifier>
```

Apply the image_view package to look over the stereo photo acquired from the camera for further correction.

To achieve better target location effect, binocular vision has been the mainstream in machine vision aspect. With binocular cameras shooting at known positions, the relative position information of the target object can be acquired by analyzing the images of the binocular cameras. The basic binocular vision principle is shown in Fig. 3.

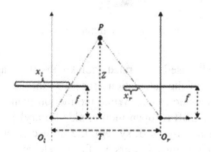

Fig. 3. Binocular vision principle

Suppose that we use two cameras to estimate the position of the target P. The two cameras have parallel optical axis, and the distance between the two cameras is T. The red thick lines stand for the image planes. Both focus of the cameras is f. According to the similar triangle principle, Z can be induced by:

$$\frac{T - (x_l - x_r)}{Z - f} = \frac{T}{Z} \Rightarrow Z = \frac{fT}{x_l - x_r}$$

ROS provides the relevant camera drivers packages, so users can select the corresponding branch to adapt need.

2.4 Devices Communication Principle

The communication within computers using ROS is generally based on TCP/IP protocol, which provides convenience in system building as the communication through internet nodes has been fully developed. With the master process running on the control computer, each node is able to interact with the other node in form of messages [8].

The mechanical arm motion statistics acquired from motion planning is transported with JointTrajectory message, which is a kind of track statistics in PVT (position, velocity, time) format. It stands for the position, the instantaneous velocity at the position, and the time taken to reach this position of each mechanical arm concerned. Through Modbus TCP Protocol, the PVT motion statistics is transported to multi-axis controller. Each axis finishes the planned motion according to the statistics after interpolation operation.

In telecommunications, RS-232 is a standard for serial communication transmission of data. It formally defines the signals connecting between a DTE (data terminal equipment) such as a computer terminal, and a DCE (data circuit-terminating equipment or data communication equipment), such as a modem. The RS-232 standard is commonly used in computer serial ports. In ROS, the package ROS-serial is a protocol for wrapping standard ROS serialized messages and multiplexing multiple topics and services over a character device such as a serial port or network socket. Classified by the different clients, ROS-serial provides various library aimed at Arduino, windows, Linux, etc. Usually, serial package is used to realize the communication between RS-232 serial and the device running Windows or Linux [9].

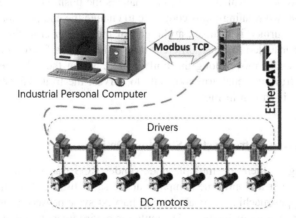

Fig. 4. The Modbus TCP communication base on EtherCAT bus

The Modbus package provides a wrapper from the Modbus TCP communication to standardized ROS messages. Programs of users use API library to run a Modbus server, in which there are the holding register and coil register. The upper machine acts as a client of Modbus, realizing communication and information interaction through reading the value of the register in Modbus server. The corresponding value leads the user's programs to realize motion control. The Modbus TCP communication is shown in Fig. 4.

3 Deployment Instance: Tomato Harvesting Dual-Arm Robot BUGABOO

It is a challenging task to develop an autonomous agricultural robot to fulfil multiple purposes due to the fact that the unstructured environment leads to difficulties for machine vision to identify targets and for intelligent manipulation to avoid obstacles. A humanoid agricultural robot, BUGABOO, is designed at Shanghai Jiao Tong University to perform various tasks such as plant disease monitoring, pesticide spraying and fruit harvesting. In this section the tomato harvesting task of the agricultural robot is selected as a symbolic case to show that the ROS facilitates a rapid development of simulation and control systems for BUGABOO.

3.1 The Mechanical System for BUGABOO

To finish the autonomous tomato harvesting task with accuracy and efficiency, this designed structure is as follows,

BUGABOO has two 3 DOF upper limbs mimicking the human arms, and a rotating platform serving as the waist. The serial arms with same structure are installed on the waist platform symmetrically. The 3 degrees of freedom include: the DOF of the lifting joint vertically, the DOF of the rotating joint of bigger arm, the DOF of the rotating joint of smaller arm. The waist-shape platform can rotate around the axis perpendicular with the ground to change the overall direction of robot. The single arm structure is like that of SCARA robot, in which the lift joint changes the position of the end in vertical direction, and the two rotating joints cooperate to change the end position in horizontal direction. Flange surface is installed at the end of the smaller arm; thus different end actuators can be installed to finish different tasks. The base is fixed on the automated trail car, which could move along the trail in field. When the robot recognizes and locates the ripe fruit, the dual-arm cooperate to harvest the fruit. The structure of the robot main body is shown in Fig. 5.

3.2 The General Control System Structure Design

As shown in Fig. 6, there are multiple nodes under ROS framework running on the monitoring computer and airborne computer at the same time. All the nodes together constitute the upper machine software part of harvest system. According to the function, the structure mainly include the 3D simulation environment based on ROS built-in

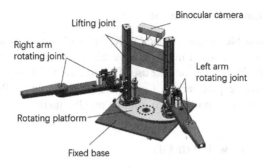

Fig. 5. Structure of the robot main body

visualization tool RViz, mechanical arm motion planning function library set based on MoveIt!, machine vision processing based on OpenCV open-source library, task level state machine programming based on SMACH library, interactive control interface based on the wxPython (encapsulated in Python sizers cross-platform GUI library), etc. Besides, there are also some function modules concerned with the bottom hardware, such as camera driver, and lower machine communication serial port and Modbus TCP procedures, etc. These function modules continuously produce messages when running, and at the same time also have demand for other information or services. ROS framework provides a good message-swapping and service invocation mechanism.

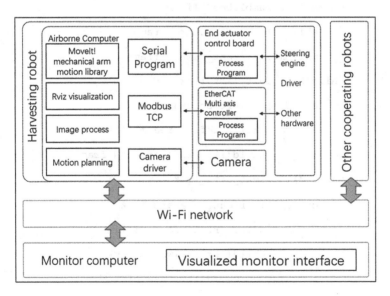

Fig. 6. Whole software system structure

3.3 Software Running Environment

Running ROS is available in Linux, Mac OS X, Android and Microsoft Windows, but users generally choose Ubuntu Linux, because this is the official recommendation to support the best operating system, and the use of Ubuntu is completely free of charge. The existing ROS versions are quite various. Considering the running stability, we choose the first version with 5 years support, indigo running on Ubuntu14.04 as the ROS version.

3.4 Modelling URDF with *.stl File

To ensure the accuracy of the model, we choose to build the 3D model with professional modelling software, Solidworks to get the required *.stl file, and import it into the URDF file as meshes.

With the URDF file finished, we use the tool in Rviz check_urdf to check if there is any grammar mistake. The corresponding output is shown as below:

```
robot name is: lr1_robot
---------- Successfully Parsed XML ------
root Link: support_link has 1 child(ren)
    child(1):  waist_link
        child(1):  bumblebee2_link
            child(1):  left_camera_frame
            child(2):  right_camera_frame
        child(2):  left_lift_link
            child(1):  left_upperarm_link
                child(1):  left_forearm_link
                    child(1):  left_arm_end_frame
        child(3):  right_lift_link
            child(1):  right_upperarm_link
                child(1):  right_forearm_link
                    child(1):  sawwrist_link
                        child(1):  saw_link
                        child(2):  right_arm_end_frame
```

To view the model more intuitively, we can use the tool urdf_to_graphiz to show the tree structure of the robot model, as Fig. 7 shows.

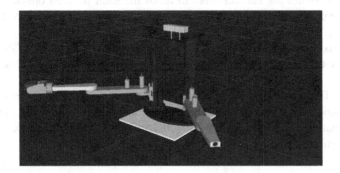

Fig. 7. The visualization model of the robot

3.5   Visualization and Motion Planning

With Rviz and MoveIt!, we can easily get the visualization model. Given the start point and end point, OMPL library support to finish the motion planning, "lr1" being the name of the robot. The simulation interface is shown in Fig. 8.

```
roslaunch lr1_move demo.launch

roslaunch lr1_description display.launch

roslaunch lr1_robot_moveit_config move_group.launch
```

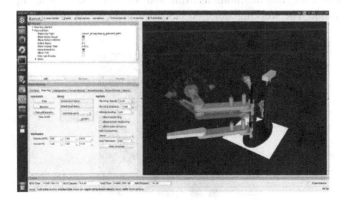

Fig. 8. MoveIt! interface in RViz and the arm trajectory simulation

3.6   Machine Vision

In fruits harvest mission, the relative position and posture of fruit relative to the robot are needed to control the end actuators to reach the ideal position operating harvest. Binocular stereo vision is similar to human visual system, and enjoys good accuracy and efficiency. Here we choose the binocular camera Bumblebee2.

There exists the matched software package for Bumblebee2 in ROS, thus the configuration of the camera is much easier. Project uses Bumblebee2 camera on Ubuntu Linux, uses libdc1394 library to control camera and capture the images, and then use the Triclops library to correct image and complete the image matching and depth calculation. Program runs as a ROS node, and publish the image, depth point cloud information as ROS topic, for the subscription of other ROS node. The basic logic is shown in Fig. 9.

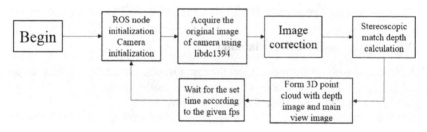

Fig. 9. The general procedure of stereo vision

3.7   Field Experimentation

To examine the working effect of the harvest dual-arm robot, we conducted a series of experiments in farming base. We took the robot to operate the independent harvesting process, realized the full autonomous fruit and vegetable harvesting operations by agricultural robot. The working situation is shown in Figs. 10 and 11.

Fig. 10. The Operating scenario of BUGABOO

Fig. 11. The operating details driven by ROS

Field experimentations demonstrate that the tomato harvesting dual-arm robot BUGABOO works with excellent stability as the data communication is based on TCP/IP protocol. Decent location accuracy is ensured by the binocular stereo vision. This experiment proved the advantages of ROS when applied in autonomous robot control system, which provides good reference for further improvement of the control system (Table 2).

Table 2. General index of BUGABOO

Parameters	Values
Weight	40 kg
Total power	2.2 KW
Work breadth	350 mm × 350 mm × 580 mm
Total degree of freedom	12
Average harvesting efficiency	45 s/each
Method of target location	Binocular stereo vision

3.8   Compare with MFC-Based Control System

At the beginning of program, we tried to build the control system on Windows platform, using Visual Studio development tool to write the C++ program and user interface.

As shown in Fig. 12, the interface can be divided into four part, including the photo view part, the 3-D simulation part, and two parameter-control parts. In the developing process, we found that there were many disadvantages compared with ROS:

- The program frame with single project is hard to develop when several programmers cooperate. As a single executive file is formed with all the sub-programs, debugging process takes a lot of time. While in ROS frame, the function module can be divided completely in different projects (nodes), which increases the developing efficiency.

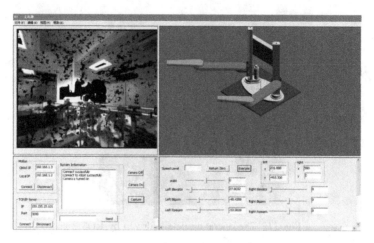

Fig. 12. Early control interface in Windows

- As the algorithm of the robot researching is changing constantly, the codes from Internet require to be edited greatly to adapt to the existing program frame, which is not worthy for selecting algorithms at the beginning step. While ROS has open-sourced internet community, in which there are many excellent codes and algorithm, and many algorithm libraries provide support to ROS.
- Multi-process communication is relatively messy in Windows platform. ROS provides the practical communication standard by each node, which could avoid the conflict due to dependencies and synchronous conditions.

4 Conclusion

In summary, this paper describes a general method of building robot control system software applying ROS, mainly including modelling with URDF, visualization with Rviz, motion planning with MoveIt!, and vision with OpenCV library, communication with serial and Modbus.

To illustrate the method more concretely, the paper takes the tomato harvesting dual-arm robot running ROS as example. Experiments prove that the robot control system built with ROS enjoys convenience of use and good stability.

In the research process, various kinds of problems arise. Further work is needed to improve the performance of the control system developed by ROS. Such as developing a performance evaluation system based on the data from field experiments to measure the reposition precision of the robot. And try to apply the control system developed by ROS in complex mechanical systems to make use of the advantages in complex communication of ROS module program.

With robot technology developing, the application of ROS in robot control system is expected to be wider and more efficient.

Acknowledgements. This research is founded by MOST of China under Grant No. 2014 BAD08B01 and No. 2015BAF13B02, and partially supported by the National High Technology Research and Development Program of China under Grant No. 2013AA102307.

References

1. Bac, C.W., Henten, E.J., Hemming, J., et al.: Harvesting robots for high-value crops. In: State of the Art Review and Challenges Ahead, pp. 888–911 (2014)
2. Wang, Y.H., Lee, K., Cui, S.X., Risch, E.: Research on agricultural robot and applications. In: Southern Plains Agricultural Research Center, College Station (2014)
3. Quigley, M., Conley, K., Gerkey, B., et al.: ROS: an open-source robot operating system. In: ICRA Workshop on Open Source Software (2009)
4. Cao, Z.W., Ping, X.L., Chen, S.L., Jiang, Y.: Research on method of developing robot model based on ROS (2015)
5. Yousuf, A., Lehman, W., Mustafa: Introducing kinematics with robot operating system (ROS). In: ASEE Annual Conference and Exposition (2015)
6. Chitta, S., Sucan, I., Cousins, S.: Moveit![ROS topics]. IEEE Robot. Autom. Mag. **19**(1), 18–19 (2012)
7. Bradski, G., Kaehler, A.: Learning OpenCV: Computer vision with the OpenCV library. O'Reilly Media Inc., Sebastopol (2008)
8. Hoske, M.T.: ROS Industrial aims to open, unify advanced robotic programming. Control Eng. **60**(2), 20 (2013)
9. https://en.wikipedia.org/wiki/RS-232

Design of 3D Printer-Like Data Interface for a Robotic Removable Machining

Fusaomi Nagata[1(✉)], Shingo Yoshimoto[1], Kazuo Kiguchi[2], Keigo Watanabe[3], and Maki K. Habib[4]

[1] Tokyo University of Science, Yamaguchi, 1-1-1 Daigaku-Dori, Sanyo-Onoda 756-0884, Japan
nagata@rs.tusy.ac.jp
[2] Kyushu University, 744 Motooka, Nishi-ku, Fukuoka 819-0395, Japan
[3] Okayama University, 3-1-1 Tsushima-naka, Kita-ku, Okayama 700-8530, Japan
[4] American University in Cairo, AUC Avenue, P.O. Box 74, New Cairo 11835, Egypt

Abstract. In this paper, a 3D printer-like data interface is proposed for the machining robot. The 3D data interface enables to control the machining robot directly using STL data without conducting any CAM process. This is done by developing a robotic preprocessor that helps to remove the need for the conventional CAM process by converting directly the STL data into CL data. The STL originally means Stereolithography which is a file format proposed by 3D Systems, and recently is supported by many CAD/CAM softwares. The STL is widely used for rapid prototyping with a 3D printer which is a typical additive manufacturing system. The STL deals with a triangular representation for a curved surface geometry. The developed interface allows to control the machining robot through a zigzag path generated according to the information included in STL data. The effectiveness and potential of the developed approach are demonstrated through actual experimental machining results.

Keywords: Machining robot · CAD/CAM · STL data · 3D printer-like data interface · CL data · Robotic preprocessor

1 Introduction

In manufacturing industries, there exist two representative systems for prototyping. One is the conventional removal manufacturing systems, such as NC milling or NC lathe machines which can precisely perform metalworking. Figure 1(a) shows a typical conventional machining process. The second is the additive manufacturing systems, such as optical shaping apparatus or 3D printer which enables to quickly transform a design concept into a real model. As for removal machining, Lee introduced a machining automation using an industrial robot [1]. The robot had double parallel mechanism and consequently performed a large work space as well as a high stiffness to reduce deformation and vibration. Schreck et al. launched Hard Material Small-Batch Industrial Machining Robot (HEPHESTOS) project, in which the objective was focused on developing

© Springer International Publishing Switzerland 2016
N. Kubota et al. (Eds.): ICIRA 2016, Part I, LNAI 9834, pp. 40–50, 2016.
DOI: 10.1007/978-3-319-43506-0_4

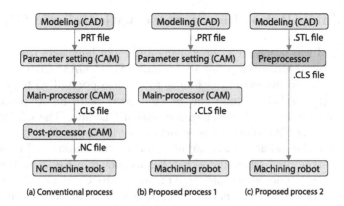

Fig. 1. Comparing the proposed two processes with the conventional process.

robotic manufacturing methods in order to give rise to a cost-efficient solution in hard materials machining [2].

As one of examples of post-processor in CAM system, CL data written in ISO format produced by main-processor in CAM system could be transformed into G-codes files (Numerical Control files) and an industrial 5-axis machine with a nutating table could be actually controlled using the NC files [3]. However, it is not easy but complicated for the CAM system to generate each robot language according to different industrial robot makers. The authors have developed an industrial machining robotic system for foamed polystyrene materials [4]. In the machining robot, the developed robotic CAM system called the direct servo system provided a simple interface for NC data and CL data, without using any robot languages between operators and the machining robot [5]. The direct servo system simplified the machining process without a post-processor as shown in Fig. 1(b). However, a CAM process to generate CL data after the design process has to be further passed through in order to machine the designed model using the robot. Although the authors searched related papers, e.g., [6,7] to make the process easier, any suitable system was not seen.

In this paper, a robotic preprocessor is proposed for the machining robot to directly convert STL data into CL data as the proposed process 2 shown in Fig. 1(c), and this helps to remove the need for having the conventional CAM process. The STL originally means Stereolithography which is a file format proposed by 3D Systems and recently is supported by many CAD and CAD/CAM softwares. It is also known as Standard Triangulated Language in Japan. The STL is widely used for rapid prototyping with a 3D printer which is a typical additive manufacturing system [8]. The STL deals with a triangular representation of a 3D surface geometry [9,10]. The robotic preprocessor allows the machining robot to be controlled along continuous triangular polygon mesh included in STL data or along a zigzag path generated by analyzing triangle patches in the data. The effectiveness and potential of this unique machining system are demonstrated through actual machining experiments.

2   Machining Robot Incorporated with Robotic CAM

A robotic CAM was proposed without using any robot languages to enhance the affinity between an industrial robot RV1A and a CAD/CAM Creo [5]. Figure 2 shows the developed industrial machining robot incorporated with the robotic CAM. The tip of a ball endmill can be controlled as to follow position and orientation components in cutter location within CL data. The CL data were generated by using CAD/CAM Creo provided by PTC Inc. Our current interest is to enable the industrial machining robot to run through STL data that consists of unstructured triangulated patches as shown in Fig. 3. In this paper, a preprocessor is proposed to convert STL data into CL data and integrated with the developed industrial machining robotic system to execute an assigned machining job using CL data. Accordingly, the system can implement its task and control its sequence of machining actions based on STL data.

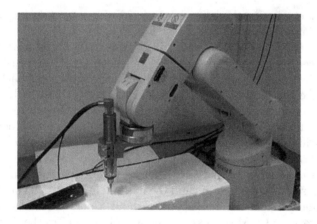

Fig. 2. Machining robot RV1A for foamed polystyrene materials.

3   Preprocessor Based on Triangle Patches in STL Data

The proposed preprocessor for the industrial machining robotic system consists of three processes, (a) essential process, (b) smart process, and (c) advance process. As for (c) advance process, the details are described in the next section.

3.1   Essential Process

The STL is a file format proposed by 3D Systems and recently is supported by many CAD and CAD/CAM softwares. STL is known as Standard Tessellation Language using triangular patches $\boldsymbol{P}_i = [\boldsymbol{n}_i^T \; \boldsymbol{v}_{1i}^T \; \boldsymbol{v}_{2i}^T \; \boldsymbol{v}_{3i}^T]^T$ as shown in Fig. 3, in which $\boldsymbol{n}_i = [n_{xi} \; n_{yi} \; n_{zi}]^T$ is normal vector, $\boldsymbol{v}_{1i} = [x_{1i} \; y_{1i} \; z_{1i}]^T$,

$v_{2i} = [x_{2i}\ y_{2i}\ z_{2i}]^T$ and $v_{3i} = [x_{3i}\ y_{3i}\ z_{3i}]^T$ are position vectors of vertexes forming a triangular patch, respectively. i denotes the number of the triangular patch in STL file. The content of a triangulated patch in a binary file format is composed of 80 characters header, number of triangulated patches, normal vector to triangle face, position vectors of three vertexes describing each triangulated patch, and attribute byte count, which is written as

```
char[80]    //Header
uint32      //Number of triangles i in a STL file
float[3]    //Normal vector n(1)
float[3]    //Vertex vector v₁(1)
float[3]    //Vertex vector v₂(1)
float[3]    //Vertex vector v₃(1)
uint16      //Attribute byte count

  ⋮              ⋮

float[3]    //Normal vector n(i)
float[3]    //Vertex vector v₁(i)
float[3]    //Vertex vector v₂(i)
float[3]    //Vertex vector v₃(i)
uint16      //Attribute byte count

  ⋮              ⋮
```

As can be seen from Fig. 3, three "GOTO" statements with the same normal vector can be generated as CL data from one STL triangulated patch as below,

GOTO/Vertex vector $v_1^T(i)$, Normal vector $n^T(i)$
GOTO/Vertex vector $v_2^T(i)$, Normal vector $n^T(i)$
GOTO/Vertex vector $v_3^T(i)$, Normal vector $n^T(i)$

Figure 4 shows an example of plotted output of all converted data (STL data → CL data → NC data), in which the original STL data was generated by CAD/CAM Creo. The NC data is used as unique desired trajectory based on continuous triangular polygon mesh. As can be seen, if the conversion is

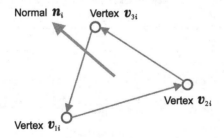

Unstructured triangulated patch

Fig. 3. STL file format proposed by 3D systems.

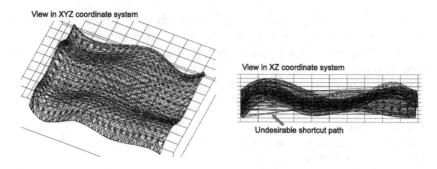

Fig. 4. Output of NC data directly converted from STL data.

conducted without any step and consideration, undesirable shortcut paths frequently appear. However, such problems will be avoided when the STL data are used for additive layered machining with a 3D printer. But, when the NC data directly obtained from the STL data is applied to the removal machining with a ball endmill, serious mis-removal cutting tends to occur according to the undesirable shortcut paths.

3.2   Smart Process

To overcome the problem of mis-removal cutting, a smart process is developed within the robotic machining preprocessor. Figure 5 shows the Window dialogue designed for the robotic machining preprocessor and post-processor. There are two parameters for the smart process in the robotic machining preprocessor that should be defined as shown in Fig. 6. One is the height h [mm] of tool escape in z-direction. The other is the distance $d_i = \|\boldsymbol{v}_{1i} - \boldsymbol{v}_{3(i-1)}\|$ [mm] from a vertex $\boldsymbol{v}_{3(i-1)}$ to the one $\boldsymbol{v}_{1i}$ in the next triangle patch. If d_i is longer than

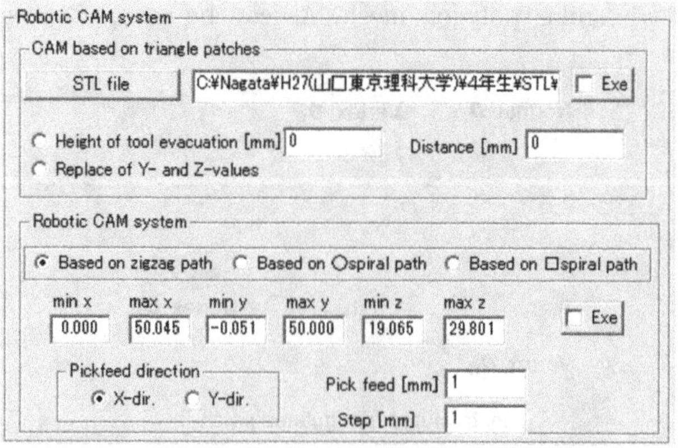

Fig. 5. Dialogue designed for robotic machining preprocessor to process STL data.

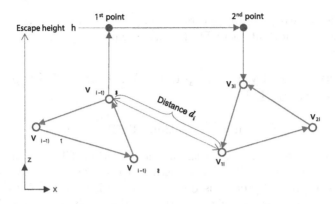

Fig. 6. Two points with the height h are added to avoid mis-removal cuttings.

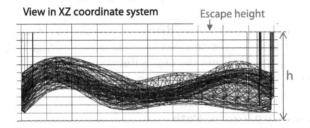

Fig. 7. An example of paths corrected by the smart process.

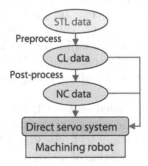

Fig. 8. Robotic CAM system available for NC data, CL data and STL data.

the reference value d_r of distance set in the dialogue, then the preprocessor inserts two positions for escaping. The first position is for the escape just upper direction. Also, the second position is for the movement from the current patch to the next one. Figure 7 shows an example of paths improved by the smart process, in which h and d_r were empirically set to 20 mm and 10 mm, respectively, by checking the sizes of the model and the triangulated patches in the STL data.

With this process, the proposed machining robot can run based on not only NC data and CL data but also STL data as shown in Fig. 8.

4   Preprocessor by Intelligently Analyzing Triangle Patches in STL Data

4.1   Automatic Dimension Extraction of STL Data

In the two processes explained in the previous section, CL data are basically generated along original STL data consisting of many triangle patches. In this subsection, (c) advance process is described, in which the CL data for robot control are generated along a zigzag path by intelligently analyzing STL data. The significant advantage of this approach is to eliminate the need to use any commercial CAM, and accordingly, 3D printer-like data interface can be smartly realized.

First of all, dimensions of the STL data are extracted by retrieving all patches in a STL file and they are set to two constants $v_{min} = [x_{min}\ y_{min}\ z_{min}]^T$ and $v_{max} = [x_{max}\ y_{max}\ z_{max}]^T$. From the next subsection, a base zigzag path viewed in xy-plane is designed considering the extracted dimensions.

4.2   Design of Base Zigzag Path

This subsection explains how the base zigzag path is designed. Two effective machining parameters, i.e., pick feed and step, are respectively set to p_f and s_p by referring the dimensions. s_p is a constant pitch viewed in xy-plane between two adjacent points $c_j = [c_{xj}\ c_{yj}\ 0]^T$ and $c_{j+1} = [c_{x(j+1)}\ c_{y(j+1)}\ 0]^T$ on a zigzag path. The scallop height is caused by the machining with a ball endmill. Although the height depends on the pick feed and the ball radius of an endmill, they are experimentally determined while avoiding the interference between the endmill and the designed STL model.

One pass point $c(j)$ is appended into a CL file as a "GOTO" statement. $j(1 \leq j \leq m)$ is the number of the pass point generated from STL data. m is the total number of triangle patches in a STL file. Figure 9 illustrates a base zigzag path drawn within STL data consisting of multiple triangulated patches. As can be seen, c_{xj} and c_{yj} are located just on the zigzag path, so that remained height c_{zj} has only to be determined by analyzing triangle patches in STL data.

4.3   Generation of CL Data Along Base Zigzag Path

Figure 10 shows an example of generation of pass points $c_j = [c_{xj}\ c_{yj}\ c_{zj}]^T$ in a patch (2). The number within the patch depends on the length of the step in Fig. 9. The dashed line in the upper figure shows one section of the base zigzag paths $c_j = [c_{xj}\ c_{yj}\ 0]^T$ viewed in xy-plane and the chained line in the lower figure draws the generated path $c_j = [c_{xj}\ c_{yj}\ c_{zj}]^T$ along the triangulated patch viewed in yz-plane. The pass points for constructing CL data are generated along

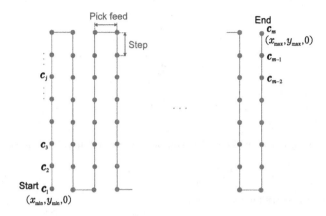

Fig. 9. Base zigzag path $c_j = [c_{xj} \; c_{yj} \; 0]^T$ along STL data consisting of multiple triangulated patches.

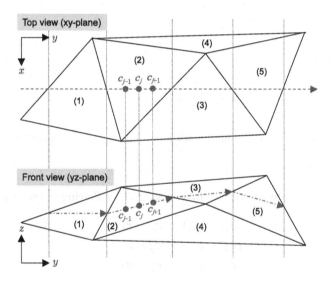

Fig. 10. Generation of pass points. Upper and lower figures show the top and front views of five adjacent triangulated patches, respectively.

the chained line. The pass points in other patches such as (1), (3), (4), (5) can be similarly obtained.

To realize the preprocessor based on STL data without any CAM process, the z-component c_{zj} must be calculated just on a triangulated patch. In order to calculate c_{zj}, first of all, a triangulated patch, in which the point c_j viewed in xy-plane is included, is searched in the target STL data. Figure 11 shows the scene where the pass point c_j is located within a patch. Note that c_{j-1} and c_{j+1} may be also within the triangle patch as shown in Fig. 10. Whether c_j is located

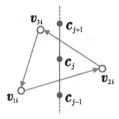

Fig. 11. Pass point c_j is located within a triangle patch viewed in xy-plane.

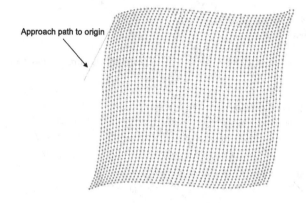

Approach path to origin

Fig. 12. Regular and precise zigzag path $c_j = [c_{xj}\ c_{yj}\ c_{zj}]^T$ generated by the pre-processor.

in the patch or not can be known by checking the following outer products.

$$(v_{2i} - v_{1i}) \times (c_j - v_{1i}) \tag{1}$$
$$(v_{3i} - v_{2i}) \times (c_j - v_{2i}) \tag{2}$$
$$(v_{1i} - v_{3i}) \times (c_j - v_{3i}) \tag{3}$$

If c_j is located within the patch, then the above three equations have the same sign. After finding the first triangle satisfying this condition, the equation of the plane including the triangle is determined by the perpendicular condition to the normal vector n_i, which leads to

$$n_{xi}(x - x_{1i}) + n_{yi}(y - y_{1i}) + n_{zi}(z - z_{1i}) = 0 \tag{4}$$

By respectively substituting c_{xj} and c_{yj} into x and y in Eq. (4), if $n_{zi} \neq 0$ then z-directional component c_{zj} can be calculated by

$$c_{zj} = z_{1i} - \frac{1}{n_{zi}}(n_{xi}(c_{xj} - x_{1i}) + n_{yi}(c_{yj} - y_{1i})) \tag{5}$$

where c_{xj} and c_{yj} are extracted from the base path shown in Fig. 9.

Consequently, by repeating the above calculations, all pass points $c_j = [c_{xj} \; c_{yj} \; c_{zj}]^T$ ($1 \leq j \leq m$) can be obtained. Figure 12 shows the regular and precise zigzag path (CL data) generated from a STL file by the preprocessor.

4.4   Removal Machining Experiment

In the earlier subsections, the preprocessor that can generate a regular and precise zigzag tool path without conducting any CAM process has been proposed. In this subsection, a machining experiment is conducted using the tool path generated by the preprocessor. Figure 13 shows the successful machining scene using the CL data made along a zigzag path (left side) and the resultant surface (right side). The feasibility and effectiveness were confirmed from the actual experiment using the machining robot.

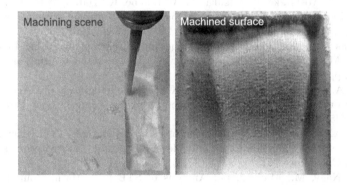

Fig. 13. Machining scene using CL data generated by preprocessor and its resultant surface.

5   Conclusions

The STL file format was designed for fabbers in 1989. Fabbers means specialists who can perform 3D rapid prototyping from digital data, e.g., using a 3D printer. The 3D printer is recognized as a typical additive manufacturing system. In this paper, a robotic preprocessor has been proposed for the machining robot to convert STL data into CL data forming a zigzag path. The STL means Stereolithography which is a file format proposed by 3D Systems and recently is supported by many CAD/CAM softwares. The robotic preprocessor has allowed the machining robot to be controlled along continuous triangular polygon mesh included in STL data or along a zigzag path obtained by analyzing triangle patches in the data. The effectiveness and promise of this unique machining system are demonstrated through actual machining experiments. The noteworthy point is that the machining robot has a promising data interface without CAM process like a 3D printer.

Acknowledgments. This work was supported by JSPS KAKENHI Grant Number 25420232.

References

1. Lee, M.K.: Design of a high stiffness machining robot arm using double parallel mechanisms. In: Proceedings of 1995 IEEE International Conference on Robotics and Automation, vol. 1, pp. 234–240 (1995)
2. Schreck, G., Surdilovic, D., Krueger, J.: HEPHESTOS: hard material small-batch industrial machining robot. In: Proceedings of 41st International Symposium on Robotics (ISR/Robotik 2014), pp. 1–6 (2014)
3. My, C.A.: Integration of CAM systems into multi-axes computerized numerical control machines. In: Proceedings of 2010 Second International Conference on Knowledge and Systems Engineering (KSE), pp. 119–124 (2010)
4. Nagata, F., Otsuka, A., Watanabe, K., Habib, M.K.: Fuzzy feed rate controller for a machining robot. In: Proceedings of the 2014 IEEE International Conference on Mechatronics and Automation (IEEE ICMA 2014), pp. 198–203 (2014)
5. Nagata, F., Yoshitake, S., Otsuka, A., Watanabe, K., Habib, M.K.: Development of CAM system based on industrial robotic servo controller without using robot language. Rob. Comput.-Integr. Manuf. **29**(2), 454–462 (2013)
6. Al-Ahmari, A., Moiduddin, K.: CAD issues in additive manufacturing. In: Comprehensive Materials Processing. Advances in Additive Manufacturing and Tooling, vol. 10, pp. 375–399 (2014)
7. Matta, A.K., Ranga Raju, D., Suman, K.N.S.: The integration of CAD/CAM and rapid prototyping in product development: a review. Mater. Today Proc. **2**(4/5), 3438–3445 (2015)
8. Brown, A.C., Beer, D.D.: Development of a stereolithography (STL) slicing and G-code generation algorithm for an entry level 3-D printer. In: Proceedings of IEEE African Conference 2013, pp. 1–5 (2013)
9. Szilvasi-Nagy, M., Matyasi, G.: Analysis of STL files. Math. Comput. Model. **38**(7/9), 945–960 (2003)
10. Iancu, C., Iancu, D., Stancioiu, A.: From CAD model to 3D print via STL file format. Fiability Durability **1**(5), 73–81 (2010)

Combined Model-Free Decoupling Control and Double Resonant Control in Parallel Nanopositioning Stages for Fast and Precise Raster Scanning

Jie Ling, Zhao Feng, Min Ming, and Xiaohui Xiao[✉]

School of Power and Mechanical Engineering, Wuhan University, Wuhan, 430072, China
{jamesling,fengzhaozhao7,mingmin_whu,xhxiao}@whu.edu.cn

Abstract. A design of double resonant control combined with a model-free decoupling filter (MFDF) is presented in this paper. The design is demonstrated using the proposed MFDF to decouple a parallel multi-input multi-output (MIMO) system into several single-input single-output systems and applying a double resonant controller for vibration damping and cross coupling reduction in nanopositioners. Raster scan results of simulations based on an identified MIMO transfer function of a nanopositioning stage over an area of 4 μm × 0.4 μm with small RMS errors are demonstrated. Comparisons with using the double resonant controller alone show the effectiveness of the proposed controller.

Keywords: Decoupling control · Resonant control · Vibration damping · Cross coupling reduction · Nanopositioner

1 Introduction

Since its invention, the atomic force microscope (AFM) has emerged as the workhorse tool for studying, interrogating, and manipulating objects and matter at the nanoscale [1–4]. In AFMs, parallel piezo-actuated flexure-based nanopositioning stages are commonly used for positioning optics and many other micro and nanoscale systems [5–7]. However, piezo-actuated stages themselves suffer from the inherent drawbacks produced by the inherent creep and hysteresis nonlinearities [8, 9]. On the other hand, the raster scan trajectory is conventionally used in AFMs, The triangular signal excites the mechanical resonance modes of the PTS. This limits the positioning accuracy of PTSs for high speed surface imaging [10]. On top of that, the signal applied to the X-axis will corrugate the traced trajectory in the X-Y plane due to the presence of the cross-coupling effect in high frequency raster scanning [11, 12].

Various control approaches have been proposed to improve the performance of the AFM at high scanning rates. In high-speed and short distance AFM scanning, the creep and hysteresis nonlinearities are not the prime concern. Therefore, the following review focus on vibration damping and cross-coupling effect reduction. These works can be divided into feedforward and feedback categories [1, 5, 7, 10]. Feedforward control techniques [13, 14] are popular because of their noise efficiency, packageability, and low cost [15]. Model inversion methods are applied for vibration compensation in [16, 17].

© Springer International Publishing Switzerland 2016
N. Kubota et al. (Eds.): ICIRA 2016, Part I, LNAI 9834, pp. 51–62, 2016.
DOI: 10.1007/978-3-319-43506-0_5

However, the dynamics of nanopositioners change with the sample weight, ageing, and temperature [18]. Iterative learning control (ILC) [19–21] technique provides good tracking performance, but the performance of ILC is sample dependent [18].

There are also feedback control methods taking vibration damping and cross-coupling reduction into consideration. Some fix-structure damping controllers have been proposed, such as positive position feedback (PPF) [22], polynomial-based pole placement [23], positive velocity and position feedback (PVPF) controller [24], resonant control (RC) [25], and integral resonant control (IRC) [26]. Such controllers are effective to handle with vibration damping in single input single output (SISO) systems. For parallel MIMO systems, cross coupling effect at high scanning speed cannot be ignored. A high bandwidth MIMO H∞ controller is designed, regarding the cross coupling effect as external disturbance to improve tracking performance [11]. But the order of H∞ controllers depend on the order of systems. This will increase the complexity in the design process for high-order systems. The implementation of high-order controllers requires advanced DSP systems [15]. The MIMO damping controllers using reference model matching approach [18, 27] and mixed negative-imaginary approach [28] have been proposed to damp the first resonant mode as well as minimize cross-coupling effect simultaneously. However, the reference model matching approach relies on the optimal searching process. The initial values of parameters in order matrices are not easy to determine.

The motivation of this paper is to eliminate vibration and cross coupling effect in parallel nanopositioners. For vibration elimination, the double resonant controllers (IRC and RC) in [27] are adopted as their low orders and simple structure. Being different form the reference model matching approach for MIMO systems in [27], the effect of cross coupling is reduced through model-free decoupling filters (MFDF) applied to each axis. The advantages of this combination are the simplicity of controller design and the ease of implementation. Our contribution lies in improving the double resonant controllers for coupled parallel nanopositioners by introducing the MFDFs into the control loop of each axis.

The reminder of this paper is organized as follows. The system description of a parallel nanopositioner is given in Sect. 2. Section 3 discusses the model-free decoupling filter and double resonant controllers design. The simulation results with the analysis are presented in Sect. 4. Finally, the paper is concluded in Sect. 5.

2 System Description

A 2-DOF parallel piezo-actuated nanopositioning stage was used as the controlled objective shown in Fig. 1. Each of the x- and y-axes is actuated by a PZT with a stroke of 100 μm. The displacement of each axis is detected by a capacitive sensor with the close loop resolution of 10 nm. The normalized transfer function of the MIMO system from the identification process is obtained using sinusoidal sweep response method as shown in Eq. (1).

$$
\begin{cases}
G_{xx} = \dfrac{146.6s^5 + 7.9 \times 10^5 s^4 + 9.8 \times 10^8 s^3 + 2.1 \times 10^{12} s^2 + 7.3 \times 10^{14} s + 9.4 \times 10^{17}}{s^6 + 1009s^5 + 3.8 \times 10^6 s^4 + 1.8 \times 10^9 s^3 + 3.5 \times 10^{12} s^2 + 7.1 \times 10^{14} s + 9.4 \times 10^{17}} \\[2mm]
G_{xy} = \dfrac{104.1s^5 - 3.6 \times 10^4 s^4 + 8.9 \times 10^7 s^3 - 1.7 \times 10^{11} s^2 + 8.2 \times 10^{13} s - 1.6 \times 10^4}{s^6 + 1009s^5 + 3.8 \times 10^6 s^4 + 1.8 \times 10^9 s^3 + 3.5 \times 10^{12} s^2 + 7.1 \times 10^{14} s + 9.4 \times 10^{17}} \\[2mm]
G_{yx} = \dfrac{104.1s^5 - 3.6 \times 10^4 s^4 + 8.9 \times 10^7 s^3 - 1.7 \times 10^{11} s^2 + 8.2 \times 10^{13} s - 1.6 \times 10^4}{s^6 + 1009s^5 + 3.8 \times 10^6 s^4 + 1.8 \times 10^9 s^3 + 3.5 \times 10^{12} s^2 + 7.1 \times 10^{14} s + 9.4 \times 10^{17}} \\[2mm]
G_{yy} = \dfrac{146.6s^5 + 7.9 \times 10^5 s^4 + 9.8 \times 10^8 s^3 + 2.1 \times 10^{12} s^2 + 7.3 \times 10^{14} s + 9.4 \times 10^{17}}{s^6 + 1009s^5 + 3.8 \times 10^6 s^4 + 1.8 \times 10^9 s^3 + 3.5 \times 10^{12} s^2 + 7.1 \times 10^{14} s + 9.4 \times 10^{17}}
\end{cases} \quad (1)
$$

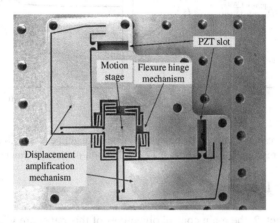

Fig. 1. A 2-DOF parallel nanopositioning stage

Vibration Problem. The frequency response of the system was obtained as displayed in Fig. 2. It can be seen that the first order of mechanical resonance mode occurs at the frequency of 123 Hz. For AFMs, a raster scan trajectory is the most widely used as scan trajectory because of the simplicity of the image reconstruction [18]. However, with increasing scan speeds, the high-frequency components of the trajectory reference signals will excite the mechanical resonant modes of the nanopositioner and introduce unwanted residual vibrations and tracking errors [7]. Therefore, vibration damping is important for high speed raster scan in AFMs.

Cross Coupling Effect. As is known, the cross coupling in parallel structures between the x- and y-axes can be more difficult to deal with compared to serial mechanisms [5]. Therefore, like the existing works in [12, 29, 30], the stage in Fig. 1 was designed to be decoupled with cross coupling as low as possible. As depicted in Fig. 2, the cross coupling in non-diagonal plots are achieved as −65 dB to −20 dB at low frequency (from 1 Hz to 70 Hz). However, the magnitude tends to be positive with the increase of scan speed, which results in strong cross coupling effect on imaging. This limits the positioning accuracy of the stage.

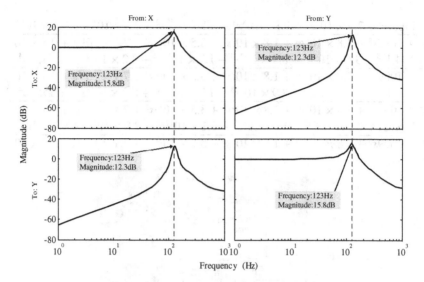

Fig. 2. Frequency response of the MIMO system. The resonant peak is 15.8 dB at 123 Hz for diagonal frequency responses and 12.3 dB at 123 Hz for non-diagonal frequency responses.

3 Controller Design

As analyzed in Sect. 2, the main control objectives of this paper are vibration damping and cross coupling reduction. In order to achieve these goals, a combination of double resonant controller and decoupling finite impulse response (FIR) filter is designed in this section.

3.1 Design of Double Resonant Controller

The double resonant control is firstly proposed in [18] and discussed deeply in [15, 27]. This method contains an IRC controller to damp the first resonant mode and a RC controller to broaden the close loop bandwidth.

Integral Resonant Controller. IRC is a feedback control technique suitable for damping highly resonant structures [26]. It is a combination of integral controller and a feed through term with its simplified structure given in Fig. 3. The integral controller is wrapped around the controlled objective to achieve damped close loop system. The feed through term d is selected to achieve a zero-pole interlacing property instead of pole-zero interlacing property for the system [27]. To build an IRC, the controlled system needs to be reduced into a second-order system [26] with the dynamics displayed as

$$G(s) = \frac{\Gamma}{s^2 + 2\xi_p \omega_p s + \omega_p^2} \qquad (2)$$

where $\Gamma > 0$ is the low frequency gain, ω_p denotes the natural frequency, and ξ_p is the damping coefficient.

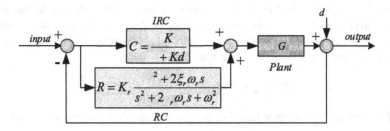

Fig. 3. Double resonant controller. K_r, ξ_r and ω_r are the static gain, damping constant and resonant frequency of R, respectively [18].

The feed through term d can be achieved through Eq. (3) and the integral gain K can be obtained by Eq. (4).

$$d = -\frac{4\Gamma}{3\omega_p^2} \tag{3}$$

$$K = \frac{\sqrt{2\omega_p}}{d} \tag{4}$$

Resonant Controller. The introducing of integral component of IRC leads to a decreased bandwidth of the closed-loop system. Therefore, a high-pass resonant controller R needs to be added into the control system given in Fig. 3. The combined closed-loop transfer function can be expressed as

$$T_{cl}(s) = \frac{G(s)(C(s) + R(s))}{1 + G(s)(C(s) + R(s))}. \tag{5}$$

In this paper, the MIMO system in (1) is regarded as two SISO systems. A double resonant controller described in Fig. 3 is applied to each SISO loop for vibration damping. The cross coupling effect is reduced by a decoupling controller discussed in Sect. 3.2. There are five parameters in the double resonant controller with the initial values of $C(s)$ chosen according to Eq. (3). With respect to the chosen of parameter values in the high-pass resonant controller $R(s)$, the resonant frequency $\omega_r = 2\pi\omega_d$ can be estimated according to the ideal close-loop transfer function set as

$$T(s) = \frac{1}{\frac{1}{2\pi\omega}s + 1} \tag{6}$$

where ω is the ideal close-loop bandwidth in Hz. Here, ω was chosen as 120 Hz based in the frequency response shown as Fig. 2.

Some trials need to be conducted to determine the static gain and the damping constant, making sure that the close-loop (5) is stable.

Optimization Process. Following the initial selection of controllers is the parameter optimization process. The optimization was carried out by using the simulated annealing algorithm from the MATLAB optimization toolbox [18]. The *fminsearch* command was adopted for optimization. The objective function is

$$\|E(s)\|_\infty = \|T(s) - T_{cl}(s)\|_\infty \tag{7}$$

where $E(s)$ denotes for the error function, $\|\cdot\|_\infty$ represents the infinity norm of a transfer function.

The optimized results are $K = -314$, $d = -2$, $K_r = 0.1$, $\xi_r = 0.6$, $\omega_r = 753.6$ for each axis control loop as the x- and y-axes are designed symmetry. Taking x-axis as the example, the closed-loop and open-loop bode plots are shown in Fig. 4. The Fig. 4a describes the step response. Figure 4b is the bode diagram, which tells that the designed closed-loop bandwidth is close to the first resonant peak as shown in Fig. 2.

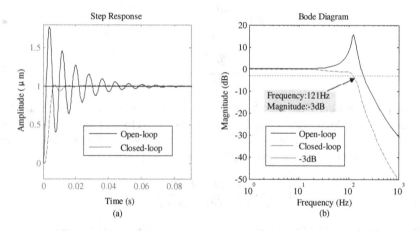

Fig. 4. Designed results of double resonant controller for x-axis.

3.2   Design of Model-Free Decoupling Controller

In general, the model-based decouplers are infinite impulse response (IIR) model with the need for accurate identified model and model structure. For instance, the model-based decoupler for x-axis can be derived as

$$D_x(s) = \frac{G_{xy}(s)}{G_{xx}(s)} \tag{8}$$

However, the non-minimum phase zeros can be an obstacle to solve the decouplers via IIR model. Another method is to use a finite impulse response (FIR) filter with the advantage of no model structure to be chosen, i.e. modeling-free approach. In this paper, the FIR Model-free Decoupling Controller (MFDF) design can be treated as (1) nonparametric frequency-domain system estimation and (2) IDFT transformation. It should be mentioned that Step (1) is conducted based on [31], and Step (2) is our contribution to introduce the method into decoupling controller design.

Nonparametric Frequency-Domain System Estimation. The synthesis of MFDF is based on empirical transfer-function estimate (ETFE) [31] of the plant. A two-run method for the plant is adopted. Here, we use the pseudo-random binary signal (PRBS) that is deterministic and spectrally white as the input to excite system. The x-axis input is PRBS and y-axis has no input for the first run. The second run is reverse. Then, two sets of data can be obtained as describe in Fig. 5.

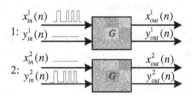

Fig. 5. The two-run block diagram for ETFE.

In the frequency samples $k \in [0, M-1]$, the ETFT plants from x- to x-axis and y- to x-axis (see Fig. 2) are denoted $\hat{G}_{xx}(k)$, $\hat{G}_{xy}(k)$ respectively,

$$\hat{G}_{xx}(k) = \frac{x_{out}^1(k)}{x_{in}^1(k)} \tag{9}$$

$$\hat{G}_{yx}(k) = \frac{x_{out}^2(k)}{y_{in}^2(k)} \tag{10}$$

where $x_{out}^1(k)$, $y_{in}^1(k)$, $x_{out}^2(k)$, $y_{in}^2(k)$ are the discrete Fourier transforms (DFT) with the superscript denoting experiment number and the subscript denoting data flow. Hereto, the ETFT of decoupler for x-axis can be expressed as

$$\hat{D}_x(k) = \frac{x_{out}^2(k)}{x_{out}^1(k)}, \tag{11}$$

$$\begin{cases} x_{out}^1(k) = \sum_{n=0}^{M-1} x_{out}^1(n)\, e^{-j2\pi kn/M} \\ x_{out}^2(k) = \sum_{n=0}^{M-1} x_{out}^2(n)\, e^{-j2\pi kn/M} \end{cases} \tag{12}$$

and n stands for the time samples.

IDFT Transformation. To obtain the FIR decoupler, the inverse discrete Fourier transform (IDFT) is implemented for the unit impulse response $d_x(n)$

$$d_x(n) = \frac{1}{M} \sum_{k=0}^{M-1} \hat{D}_x(k)\, e^{j2\pi kn/M},\tag{13}$$

and the FIR filter then expressed in the z-domain as

$$D_{xfir}(z^{-1}) = \sum_{n=0}^{M-1} d_x(n)z^{-n}.\tag{14}$$

In this paper, the decoupler design was implemented through the MATLAB. The *etfe* command was used for frequency-domain system estimation, and the *impulseest* command was used for IDFT transformation. The designed result of MFDF is shown in Fig. 6 comparing with derived results using model information through Eq. (8). It can be observed that the designed MFDF is anastomotic to the model-based decoupler, especially in high-frequency domain.

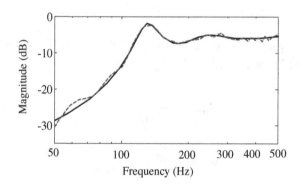

Fig. 6. Comparisons of model-based decoupler and the designed MFDF for x-axis. The solid dark line (–) is model-based decoupler and the blue dash line (–) is the MFDF. (Color figure online)

3.3   Overall Design Procedure

Hereto, we propose the following design procedure.

1. Use the two-run method in Fig. 5 to collect two set of the experiment data, and design MFDF for x- and y-axes.
2. Design the double resonant controller for each axis independently. Select the initial values for the IRC and RC according to Eqs. (3), (4) and (6). Perform the optimization process to achieve the final controller.

Compared with design methods in [18, 27], we decreased the number of parameters in the MIMO double resonant controller from 13 to 5 for 2-DOF motion systems through adding the MFDF to reduce cross coupling effect instead of the order matrices (8 parameters). Finally, the control scheme can be depicted as Fig. 7.

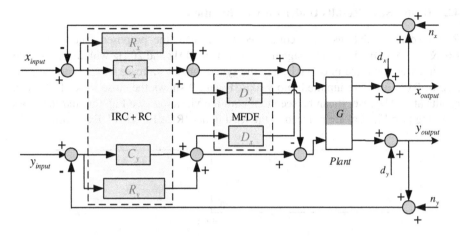

Fig. 7. Combined MFDF and double resonant controller scheme.

4 Evaluation

4.1 Cross Coupling Reduction by Adding MFDF

Before raster scan simulations, the cross coupling reduction was analyzed through one-channel input. Results are shown in Fig. 8. It can be seen that the slightest cross coupling effect is achieved by the combination of IRC&RC&MFDF.

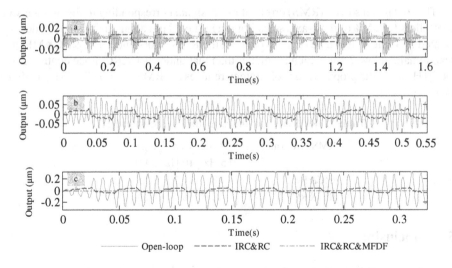

Fig. 8. Cross coupling outputs of y-axis by inputting triangular signals to the x-axis at 5 Hz (Fig. 9a), 15 Hz (Fig. 9b) and 25 Hz (Fig. 9c) with a distance of 4 μm under open-loop, IRC&RC and IRC&RC&MFDF, respectively.

4.2   Raster Scan Results Under the Combination

Raster scan simulations were conducted to evaluate the proposed combination of IRC&RC&MFDF. Triangular signals under 5 Hz, 10 Hz and 20 Hz with the amplitude of 4 µm were inputted into x-axis, and the synchronized staircase waves with the stair step of 0.05 µm were inputted into y-axis. Figure 9 shows the close look of the scan results at 20 Hz case. It can be seen that both the vibration (see Fig. 9b) and the cross coupling (see Fig. 9c) are reduced to the least under IRC&RC&MFDF control.

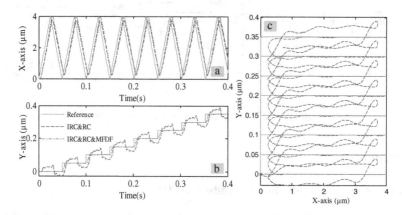

Fig. 9.   Raster scan results. (a) X-axis tracking versus time. (b) Y-axis tracking versus time. (c) XY-plane tracking

The root mean square (RMS) errors of raster signal corresponding to the two sets of controllers for 75 % of the x-axis scanning range (i.e., 3 µm along the x-axis) are documented in Table 1. It can be observed that all the RMS errors under IRC&RC&MFDF scan remain below 7 nm, which is smaller than that under IRC&RC alone control. For the 20 Hz scan, the proposed control strategy reduces the RMS error by 78 % (from 30.5 to 6.8 nm).

Table 1.   RMS errors of the raster tracking performance.

RMS error (nm)	Raster scan signal		
	5 Hz	10 Hz	20 Hz
IRC&RC	9.8	17.2	30.5
IRC&RC&MFDF	6.2	6.4	6.8

5   Conclusions

The main goal of this paper was to damp vibration and reduce cross coupling simultaneously for high-speed raster scan in nanopositioners. This was done by: (1) applying an IRC to damp the first mechanical resonant of the structure and a RC to broaden the closed-loop system bandwidth and (2) adding a MFDF designed through nonparametric frequency-domain system estimation and inversed DFT transformation, which was our contribution in this work. Comparisons between double resonant controller alone with IRC&RC&MFDF control were made through simulations to evaluate the proposed method. Results proved that the proposed IRC&RC&MFDF achieved a 78 % RMS error improvement under 20 Hz raster scan from IRC&RC alone control, i.e., the better tracking performance for high-speed rater scanning.

The ongoing work involves the consideration of external disturbance as well as noise, and the implementation of experiments.

Acknowledgment. This research was sponsored by National Natural Science Foundation of China (NSFC, Grant No. 51375349).

References

1. Devasia, S., Eleftheriou, E., Moheimani, S.O.R.: A survey of control issues in nanopositioning. IEEE Trans. Control Syst. Technol. **15**(5), 802–823 (2007)
2. Ando, T.: High-speed atomic force microscopy coming of age. Nanotechnology **23**(6), 062001 (2012)
3. Pantazi, A., Sebastian, A., Antonakopoulos, T.A., et al.: Probe-based ultrahigh-density storage technology. IBM J. Res. Dev. **52**(4.5), 493–511 (2008)
4. Paul, P.C., Knoll, A.W., Holzner, F., et al.: Rapid turnaround scanning probe nanolithography. Nanotechnology **22**(27), 275306 (2011)
5. Yong, Y.K., Moheimani, S.O.R., Kenton, B.J., et al.: Invited review article: high-speed flexure-guided nanopositioning: Mechanical design and control issues. Rev. Sci. Instrum. **83**(12), 121101 (2012)
6. Yong, Y.K., Aphale, S.S., Moheimani, S.O.R.: Design, identification, and control of a flexure-based XY stage for fast nanoscale positioning. IEEE Trans. Nanotechnol. **8**(1), 46–54 (2009)
7. Tuma, T., Sebastian, A., Lygeros, J., et al.: The four pillars of nanopositioning for scanning probe microscopy: the position sensor, the scanning device, the feedback controller, and the reference trajectory. Control Syst. **33**(6), 68–85 (2013)
8. Gu, G.Y., Zhu, L.M., Su, C.Y., et al.: Modeling and control of piezo-actuated nanopositioning stages: a survey. IEEE Trans. Autom. Sci. Eng. **13**(1), 313–332 (2016)
9. Janocha, H., Kuhnen, K.: Real-time compensation of hysteresis and creep in piezoelectric actuators. Sens. Actuators A Phys. **79**(2), 83–89 (2000)
10. Clayton, G.M., Tien, S., Leang, K.K., et al.: A review of feedforward control approaches in nanopositioning for high-speed SPM. J. Dyn. Syst. Measur. Control **131**(6), 061101 (2009)
11. Yong, Y.K., Liu, K., Moheimani, S.O.R.: Reducing cross-coupling in a compliant XY nanopositioner for fast and accurate raster scanning. IEEE Trans. Control Syst. Technol. **18**(5), 1172–1179 (2010)
12. Li, Y., Xu, Q.: Development and assessment of a novel decoupled XY parallel micropositioning platform. IEEE/ASME Trans. Mechatron. **15**(1), 125–135 (2010)
13. Croft, D., Devasia, S.: Vibration compensation for high speed scanning tunneling microscopy. Rev. Sci. Instrum. **70**(12), 4600–4605 (1999)
14. Schitter, G., Stemmer, A.: Identification and open-loop tracking control of a piezoelectric tube scanner for high-speed scanning-probe microscopy. IEEE Trans. Control Syst. Technol. **12**(3), 449–454 (2004)

15. Das, S.K., Pota, H.R., Petersen, I.R.: Damping controller design for nanopositioners: a mixed passivity, negative-imaginary, and small-gain approach. IEEE/ASME Trans. Mechatron. **20**(1), 416–426 (2015)
16. Croft, D., Shed, G., Devasia, S.: Creep, hysteresis, and vibration compensation for piezoactuators: atomic force microscopy application. J. Dyn. Syst. Measur. Control **123**(1), 35–43 (2001)
17. Leang, K.K., Devasia, S.: Feedback-linearized inverse feedforward for creep, hysteresis, and vibration compensation in AFM piezoactuators. IEEE Trans. Control Syst. Technol. **15**(5), 927–935 (2007)
18. Das, S.K., Pota, H.R., Petersen, I.R.: A MIMO double resonant controller design for nanopositioners. IEEE Trans. Nanotechnol. **14**(2), 224–237 (2015)
19. Ter Braake, J.: Iterative Learning Control for High-Speed Atomic Force Microscopy. TU Delft, Delft University of Technology (2009)
20. Barton, K.L., Hoelzle, D.J., Alleyne, A.G., et al.: Cross-coupled iterative learning control of systems with dissimilar dynamics: design and implementation. Int. J. Control **84**(7), 1223–1233 (2011)
21. Ling, J., Feng, Z., Xiao, X.: A position domain cross-coupled iteration learning control for contour tracking in multi-axis precision motion control systems. In: Liu, H., Kubota, N., Zhu, X., Dillmann, R. (eds.) ICIRA 2015. LNCS, vol. 9244, pp. 667–679. Springer, Heidelberg (2015)
22. Mahmood, I.A., Moheimani, S.O.R.: Making a commercial atomic force microscope more accurate and faster using positive position feedback control. Rev. Sci. Instrum. **80**(6), 063705 (2009)
23. Aphale, S.S., Bhikkaji, B., Moheimani, S.O.R.: Minimizing scanning errors in piezoelectric stack-actuated nanopositioning platforms. IEEE Trans. Nanotechnol. **7**(1), 79–90 (2008)
24. Bhikkaji, B., Ratnam, M., Fleming, A.J., et al.: High-performance control of piezoelectric tube scanners. IEEE Trans. Control Syst. Technol. **15**(5), 853–866 (2007)
25. Pota, H.R., Moheimani, S.O.R., Smith, M.: Resonant controllers for smart structures. Smart Mater. Struct. **11**(1), 1–8 (2002)
26. Bhikkaji, B., Moheimani, S.O.R.: Integral resonant control of a piezoelectric tube actuator for fast nanoscale positioning. IEEE/ASME Trans. Mechatron. **13**(5), 530–537 (2008)
27. Das, S.K., Pota, H.R., Petersen, I.R.: Multivariable negative-imaginary controller design for damping and cross coupling reduction of nanopositioners: a reference model matching approach. IEEE/ASME Trans. Mechatron. **20**(6), 3123–3134 (2015)
28. Das, S.K., Pota, H.R., Petersen, I.R.: Resonant controller design for a piezoelectric tube scanner: a mixed negative-imaginary and small-gain approach. IEEE Trans. Control Syst. Technol. **22**(5), 1899–1906 (2014)
29. Li, Y., Xu, Q.: Modeling and performance evaluation of a flexure-based XY parallel micromanipulator. Mech. Mach. Theor. **44**(12), 2127–2152 (2009)
30. Aphale, S.S., Devasia, S., Moheimani, S.O.R.: High-bandwidth control of a piezoelectric nanopositioning stage in the presence of plant uncertainties. Nanotechnology **19**(12), 125503 (2008)
31. Ljung, L.: System identification: theory for the user. PTR Prentice Hall Information and System Sciences Series (1999)

Towards the Development of Fractional-Order Flight Controllers for the Quadrotor

Wei Dong[1,2], Jie Chen[1], Jiteng Yang[3], Xinjun Sheng[1(✉)], and Xiangyang Zhu[1]

[1] State Key Laboratory of Mechanical System and Vibration, School of Mechanical Engineering, Shanghai Jiaotong University, Shanghai 200240, China
xjsheng@sjtu.edu.cn
[2] State Key Laboratory of Fluid Power and Mechatronic Systems, Zhejiang University, Zhejiang 310058, China
[3] Department of Precision Instrument, Tsinghua University, Beijing 100084, China

Abstract. The criterion for the development and associated parameter tunning of a class of fractional-order proportional-integral-derivative controllers, regarding the attitude stabilization as the inner control loop, is proposed for the quadrotor in this work. To facilitate this development, the dynamic model of the quadrotor is firstly formulated, and the transfer function of the inner loop is presented based on the real-time flights conducted in previous researches. With the obtained transfer function model, a class of fractional order controllers, including fractional order proportional-derivative controllers and proportional-integral controllers are developed accordingly. For each controller, the parameter tunning methods are addressed in details. To verify the effectiveness of this development, numeric simulations are conducted at last, and the results clearly verify the superiority of the fractional order controllers over conventional proportional-integral-derivative controllers in real-time flight of the quadrotor.

Keywords: Quadrotor · Fractional order controller · Paramter tunning · Flight control

1 Introduction

The agilities and versatilities of the quadrotor attract lots of researchers in recent years [1–4]. In this progress, to enhance the performance of the quadrotor, the flight control, trajectory generation and simultaneous localization and mapping are extensively studied [5–9]. In particular, the flight control is the basic but indispensable element for the quadrotor to fulfill their specific missions in real world.

Numbers of well developed flight controllers, such as linear quadratic (LQ) controller [5], sliding-mode controller [10], linear matrix inequalities (LMI) based controller [11] and disturbance observer based controller [12], were proposed during the last decade. Those controllers have effectively improve the performance of the quadrotor in real-time flights. Unfortunately, the dominant flight controller

© Springer International Publishing Switzerland 2016
N. Kubota et al. (Eds.): ICIRA 2016, Part I, LNAI 9834, pp. 63–74, 2016.
DOI: 10.1007/978-3-319-43506-0_6

of the quadrotor is still the classical proportional-integral-derivative (PID) controller [13,14]. This is because with its three-term functionality covering treatment to both transient and steady-state responses, the PID control provides the a simple and efficient solution for real world applications. However, the pure PID technique shows limited capabilities in disturbance rejection [12], which is the main reason that researchers insistently pursue alternative control strategies. Considering those facts, numbers of studies have tried to directly improve the performance of the flight control based on the PID technique. According to those studies, two methods demonstrate promising capabilities. The first one introduces the tracking differentiator, extended state observer, and utilizes the nonlinear proportional-derivative control to improve the performance of the flight control [15]. The second one directly introduces the fractional calculus into the proportional-integral-derivative technique [16].

Comparatively speaking, the second one provides a more explicit solution, which is similar to its traditional counterpart, i.e., the PID controller. In addition, the fractional calculus, with integrals and derivatives of real order instead of integer order, can be properly further utilized in modeling, which is a significantly more comprehensive description for the specific objects, e.g., the quadrotor in this work. This is because objects, such as the quadrotor controlled by this work, might be of fractional order. Therefore, the results may improve the effectiveness of the simulation compared to the traditional methods. In such a case, it will be also logically more suitable to utilize the fractional order controllers (FOCs) to control those objects [16].

FOCs have showed promising capabilities in many applications that suffer from the classical problems of overshoot and resonance, as well as time diffuse applications such as thermal dissipation and chemical mixing [16,17]. The FOCs could better handle the tracking process with a fractional order calculator, as it provides a powerful instrument for the description of memory and hereditary effects in various substance [16]. Therefore, better robustness and stabilities could be achieved with the FOCs.

In view of the state-of-the-art, this work is motivated to develop a class of FOCs and the associated parameter tunning methods for the quadrotor to enhance its robustness. To facilitate this development, the dynamics model of the quadrotor is firstly formulated, and the transfer function of the attitude is presented based on the previous researches. With the transfer function model, a class of FOCs, including PD^μ, FO (PD), PI^λ, and FO (PI) controllers are developed. For each controller, the parameter tunning methods are addressed in details. To verify the effectiveness of this development, extensive numeric simulations are conducted at last.

The reminder of this paper is organized as follows. First, the Quadrotor dynamics is introduced, and a transfer function is properly adopted to describe the attitude control loop in Sect. 2. Then the stabilized attitude is treated as the pseudo control input of the position control loop, and the design criterion for the FOCs is presented in Sect. 3. In Sect. 4, numeric simulations are provided to verify the effectiveness of the developed FOCs. At last, Sect. 5 concludes this work.

2 Quadrotor Dynamics

To facilitate the following development, the dynamic model of the quadrotor is firstly presented in this section. With this model, the pseudo control variables for the translational flight control are determined, and a first order time-delay transfer function is adopted according to real-time experiments in previous researches.

2.1 Rigid Body Dynamics

The free body diagram and coordinate frames of the quadrotor are shown in Fig. 1. Based on this illustration, four control inputs can be defined as $U_1 = F_1 + F_2 + F_3 + F_4$, $U_2 = (F_2 - F_4)L$, $U_3 = (F_3 - F_1)L$, $U_4 = M_1 - M_2 + M_3 - M_4$. where L is the length from the rotor to the center of the mass of the quadrotor, and F_i and M_i are the thrust and torque generated by rotor i ($i \in \{1, 2, 3, 4\}$).

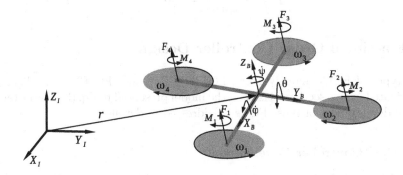

Fig. 1. Free body diagram

In the near hovering state ($\phi \approx 0$, $\theta \approx 0$), the dynamical model of the quadrotor with respect to the inertial coordinates can be then expressed as [7]

$$\ddot{x} = \frac{U_1}{m}(\theta \cos \psi + \phi \sin \psi), \quad \ddot{y} = \frac{U_1}{m}(\theta \sin \psi - \phi \cos \psi),$$
$$\ddot{z} = \frac{1}{m}U_1 - g, \quad \ddot{\phi} = \frac{U_2}{I_{xx}}, \quad \ddot{\theta} = \frac{U_3}{I_{yy}}, \quad \ddot{\psi} = \frac{U_4}{I_{zz}}. \tag{1}$$

where ϕ, θ, and ψ are roll, pitch and yaw, respectively; x, y, and z are the position of the quadrotor in the inertial coordinates; m, I_{xx}, I_{yy}, and I_{zz} are the mass and moments of inertia of the quadrotor, respectively; and g is the gravity constant.

In this way, z, ϕ, θ, and ψ are linearly related to U_i ($i \in \{1, 2, 3, 4\}$). The roll and pitch angle can be then taken as the pseudo control inputs to stabilize x and y. The desired attitude angles can be then explicitly calculated with given translational accelerations as follows

$$\ddot{\eta}^* \triangleq \begin{bmatrix} \theta^* \\ \phi^* \end{bmatrix} = (\frac{U_1}{m}G)^{-1} \begin{bmatrix} \ddot{x}^* \\ \ddot{y}^* \end{bmatrix} = \frac{m}{U_1}G \begin{bmatrix} \ddot{x}^* \\ \ddot{y}^* \end{bmatrix} \tag{2}$$

where θ^*, ϕ^*, $\ddot{x}^*$, and $\ddot{y}^*$ denote the desired values for θ, ϕ, $\ddot{x}$, and $\ddot{y}$ respectively.

2.2 Dynamics of the Pseudo Control Variables

The pseudo control variables $\eta = [\theta, \phi]^T$ utilized in Eq. (2), are commonly stabilized by a inner-loop controller, such as the proportional-derivative (PD) controller [12]. In such a case, real-time experimental identification approaches can be adopted to determine the transfer function from η^* to η, which could be presented in the form of $P_a(s) \approx \frac{1}{Ts+1}e^{-\tau s}$ [18].

In the near-hovering state, $\frac{U_1}{m}G$ can be treated as a constant, therefore integrating (2) and substituting it into $P_a(s)$, the transfer function taking the attitude command as input and the speed as the output is $P(s) = \frac{K}{s(Ts+1)}e^{-Ls}$.

The parameters in this transfer function could be identified from series of random flights, which has been addressed in details in [18]. In this work, the following parametric model is adopted

$$P(s) = \frac{1.4}{s(0.05s + 1)}e^{-0.15s} \tag{3}$$

3 Fractional Order Controller Design

Four types of FOCs, namely PD^μ, FO(PD), PI^λ, and FO(PI) controllers are investigated in this section. Based on the model presented in Eq. (3), the criterion for the development of those FOCs is addressed in details.

3.1 PD^μ Controller Development

The PD^μ controller is commonly designed in the following form [17]

$$C(s) = K_p(1 + K_d s^\mu) \tag{4}$$

The PD^μ FOC described by Eq. (4) can be rewritten as

$$C(j\omega) = K_p[(1 + K_d\omega^\mu \cos\frac{\mu\pi}{2}) + jK_d\omega^\mu \sin\frac{\mu\pi}{2}] \tag{5}$$

considering the fact $(j\omega)^\mu = \omega^\mu(\cos\frac{\mu\pi}{2} + i\sin\frac{\mu\pi}{2})$ [19].

The phase and gain of Eq. (5) are

$$\arg[C(j\omega)] = \tan^{-1}\frac{\sin\frac{(1-\mu)\pi}{2} + K_d\omega^\mu}{\cos\frac{(1-\mu)\pi}{2}} - \frac{(1-\mu)\pi}{2} \tag{6}$$

$$|C(j\omega)| = K_p\sqrt{(1 + K_d\omega^u \cos\frac{\mu\pi}{2})^2 + (K_d\omega^\mu \sin\frac{\mu\pi}{2})^2} \tag{7}$$

Similarly, the phase and gain of the original system, i.e., Eq. (3), are

$$\arg[P(j\omega)] = -\tan^{-1}(\omega T) - \frac{\pi}{2} - \omega L, \quad |P(j\omega)| = \frac{K}{\omega\sqrt{1 + (\omega T)^2}} \tag{8}$$

The phase and gain of the open-loop $G(s) = C(s)P(s)$ are

$$\arg[G(j\omega)] = \tan^{-1}\frac{\sin\frac{(1-\mu)\pi}{2} + K_d\omega^\mu}{\cos\frac{(1-\mu)\pi}{2}} + \frac{\mu\pi}{2} - \tan^{-1}(\omega T) - \pi - \omega L \quad (9)$$

$$|G(j\omega)| = \frac{K_p K}{\omega}\sqrt{\frac{(1 + K_d\omega^u \cos\frac{\mu\pi}{2})^2 + (K_d\omega^\mu \sin\frac{\mu\pi}{2})^2}{1 + (\omega T)^2}} \quad (10)$$

Similar to [20,21], three specifications are interested by this work in the design of the FOC PD^μ controller. These specifications are proposed as follows:

(i) Proper phase margin ϕ_m should be achieved at $\omega = \omega_c$, i.e., $\arg[G(j\omega)]_{\omega=\omega_c} = -\pi + \phi_m$.

(ii) To guarantee the robustness to the variation in the gain of the plant, the following specification is imposed $\frac{d(\arg[G(j\omega)])}{d\omega}|_{\omega=\omega_c} = 0$

(iii) The gain at crossover frequency should be $|G(j\omega_c)|_{dB} = 0$

According to specification (i), the relationship between K_d and μ can be estimated as [17]

$$K_d = \frac{1}{\omega_c^\mu}\tan[\phi_m + \tan^{-1}(\omega_c T) - \frac{\mu\pi}{2} + \omega_c L]\cos\frac{(1-\mu)\pi}{2} - \frac{1}{\omega_c^\mu}\sin\frac{(1-\mu)\pi}{2} \quad (11)$$

To meet the specification (ii) about the robustness to gain variation, one can obtain

$$A\omega_c^{2\mu}K_d^2 + BK_d + A = 0 \quad (12)$$

which is equivalent to

$$K_d = \frac{-B \pm \sqrt{B^2 - 4A^2\omega_c^{2\mu}}}{2A\omega_c^{2\mu}} \quad (13)$$

where $B = 2A\omega_c^\mu \sin\frac{(1-\mu)\pi}{2} - \mu\omega_c^{\mu-1}\cos\frac{(1-\mu)\pi}{2}$. In such a case, one can solve K_d and μ simultaneously utilizing Eqs. (11) and (13).

In view of specification (iii), the equation about K_p can be obtained as

$$|G(j\omega_c)| = \frac{K_p K\sqrt{(1 + K_d\omega_c^\mu \cos\frac{\mu\pi}{2})^2 + (K_d\omega_c^\mu \sin\frac{\mu\pi}{2})^2}}{\omega_c\sqrt{1 + (\omega_c T)^2}} = 1. \quad (14)$$

In this way, the control gain K_p could be explicitly solved as

$$K_p = \frac{\omega_c\sqrt{1 + (\omega_c T)^2}}{K\sqrt{(1 + K_d\omega_c^\mu \cos\frac{\mu\pi}{2})^2 + (K_d\omega_c^\mu \sin\frac{\mu\pi}{2})^2}} \quad (15)$$

3.2   FO (PD) Controller

The FO (PD) controller is designed in the form of [17]

$$C_2(s) = K_{p2}(1 + K_{d2}s)^\mu \tag{16}$$

which can be rewritten as

$$C_2(j\omega) = K_{p2}(1 + K_{d2}(j\omega))^\mu \tag{17}$$

In this way, the phase and gain of Eq. (17) are

$$\arg[C_2(j\omega)] = \mu \tan^{-1}(\omega K_{d2}), \ |C_2(j\omega)| = K_{p2}(1 + (K_{d2}\omega)^2)^{\frac{\mu}{2}} \tag{18}$$

The open-loop transfer function $G_2(s)$ is obtained as $G_2(s) = C_2(s)P(s)$. In this way, the phase and gain of $G_2(s)$ are

$$\arg[G_2(j\omega)] = \mu \tan^{-1}(\omega K_{d2}) - \tan^{-1}(\omega T) - \frac{\pi}{2} - \omega L \tag{19}$$

$$|G_2(j\omega)| = \frac{K_{p2}K(1 + (K_{d2}\omega)^2)^{\frac{\mu}{2}}}{\omega\sqrt{1 + (\omega T)^2}} \tag{20}$$

The parameter tunning for the FO (PD) controller, as well as the following controllers, is the same with the PD^μ controller. The meet the specification (i), the relationship between K_d and μ can be expressed as

$$K_{d2} = \frac{1}{\omega_c}\tan(\frac{1}{\mu}(\phi_m - \frac{\pi}{2} + \tan^{-1}(T\omega_c) + \omega_c L)) \tag{21}$$

To meet the specification (ii), the relationship between K_{d2} and μ can be expressed as

$$\omega_c^2 A K_{d2}^2 - \mu K_{d2} + A = 0 \Longrightarrow K_{d2} = \frac{\mu \pm \sqrt{\mu^2 - 4(A\omega_c)^2}}{2(A\omega_c)^2} \tag{22}$$

To meet the specification (iii), K_{p2} can be obtained as

$$K_{p2} = \frac{\omega_c\sqrt{(T\omega_c)^2 + 1}}{K(1 + (k_{d2}\omega_c)^2)^{\frac{\mu}{2}}} \tag{23}$$

3.3 PI^λ Controller

The PI^λ controller is designed in the form as follows [21]

$$C_3(s) = K_{p3}(1 + \frac{K_i}{s^\lambda}) \tag{24}$$

The phase and gain of Eq. (24) is

$$\arg[C_3(j\omega)] = -\tan^{-1}[\frac{K_i\omega^{-\lambda}\sin(\frac{\lambda\pi}{2})}{1 + K_i\omega^{-\lambda}\cos(\frac{\lambda\pi}{2})}] \tag{25}$$

$$|C_3(j\omega)| = K_p\sqrt{(1 + K_i\omega^{-\lambda}\cos(\frac{\lambda\pi}{2}))^2 + (K_i\omega^{-\lambda}\sin(\frac{\lambda\pi}{2}))^2} \qquad (26)$$

The phase and gain of the open-loop transfer function are

$$\arg[G_3(j\omega)] = -\tan^{-1}[\frac{K_i\omega^{-\lambda}\sin(\frac{\lambda\pi}{2})}{1 + K_i\omega^{-\lambda}\cos(\frac{\lambda\pi}{2})}] - \tan^{-1}(\omega T) - \frac{\pi}{2} - \omega L \qquad (27)$$

$$|G_3(j\omega)| = \frac{KK_{p3}\sqrt{(1 + K_i\omega^{-\lambda}\cos(\frac{\lambda\pi}{2}))^2 + (K_i\omega^{-\lambda}\sin(\frac{\lambda\pi}{2}))^2}}{\omega\sqrt{\omega^2T^2 + 1}} \qquad (28)$$

To satisfy the specification (i), one can obtain

$$\frac{K_i\omega_c^{-\lambda}\sin(\frac{\lambda\pi}{2})}{1 + K_i\omega_c^{-\lambda}\cos(\frac{\lambda\pi}{2})} = \tan(\tan^{-1}(T\omega_c) + L\omega_c + \phi_m - \frac{\pi}{2}) \qquad (29)$$

Then the relationship between K_i and λ can be established as

$$K_i = \frac{C}{\omega_c^{-\lambda}\sin(\frac{\lambda\pi}{2}) - C\omega_c^{-\lambda}\cos(\frac{\lambda\pi}{2})} \qquad (30)$$

where $C = \tan(\tan^{-1}(T\omega_c) + L\omega_c + \phi_m - \frac{\pi}{2})$.
To satisfy the specification (ii), one can obtain

$$\frac{K_i\lambda\omega_c^{\lambda-1}\sin(\frac{\lambda\pi}{2})}{\omega_c^{2\lambda} + 2K_i\omega_c^{\lambda}\cos(\frac{\lambda\pi}{2}) + K_i^2} = A \implies K_i = \frac{-F \pm \sqrt{F^2 - 4A^2\omega_c^{-2\lambda}}}{2A} \qquad (31)$$

where $F = 2A\omega_c^{-\lambda}\cos(\lambda\pi/2) - \lambda\omega_c^{-\lambda-1}\sin(\lambda\pi/2)$.
To satisfy the specification (iii), one can obtain

$$K_{p3} = \frac{\omega_c\sqrt{\omega_c^2T^2 + 1}}{K\sqrt{(1 + K_i\omega_c^{-\lambda}\cos(\frac{\lambda\pi}{2}))^2 + (K_i\omega_c^{-\lambda}\sin(\frac{\lambda\pi}{2}))^2}} \qquad (32)$$

3.4 FO (PI) Controller

The FO (PI) controller is designed in the form as follows [21]

$$C_4(s) = (K_{p4} + \frac{K_i}{s})^\lambda \qquad (33)$$

The phase and gain of this controller is

$$\arg[C_4(j\omega)] = -\lambda\tan^{-1}(\frac{K_i}{K_{p4}\omega}), \quad |C_4(j\omega)| = (K_{p4}^2 + \frac{K_i^2}{w^2})^{\frac{\lambda}{2}} \qquad (34)$$

The phase and gain of the open-loop $G_4(s)$ is

$$\arg[G_4(j\omega)] = -\lambda\tan^{-1}(\frac{K_i}{K_{p4}\omega}) - \tan^{-1}(\omega T) - \frac{\pi}{2} - L\omega \qquad (35)$$

$$|G_4(j\omega)| = \frac{K(K_{p4}^2 + \frac{K_i^2}{\omega_c^2})^{\frac{\lambda}{2}}}{\omega\sqrt{1+(\omega T)^2}} \qquad (36)$$

To satisfy the specification (i), one can obtain

$$\frac{K_i}{K_{p4}} = D = \omega_c \tan(-(\phi_m - \frac{\pi}{2} + \tan^{-1}(\omega_c T) + L\omega_c)/\lambda) \qquad (37)$$

To satisfy the specification (ii), one can obtain

$$\frac{\lambda K_i K_{p4}}{(K_{p4}\omega_c)^2 + K_i^2} = A \qquad (38)$$

To satisfy the specification (iii), one can obtain

$$K_{p4}^2 + \frac{K_i^2}{\omega_c^2} = E = (\frac{\omega_c}{K}\sqrt{1+(\omega_c T)^2})^{\frac{2}{\lambda}} \qquad (39)$$

From Eqs. (37), (38) and (39), one can obtain

$$\lambda = A\frac{\omega_c^2 + D^2}{D}, \; K_{p4} = \sqrt{\omega_c^2 E \omega_c^2 + D^2}, \; K_i = K_{p4}D \qquad (40)$$

4 Simulations

To verify the effectiveness of the developed FOCs for the quadrotor, extensive numeric simulations are conducted in this section.

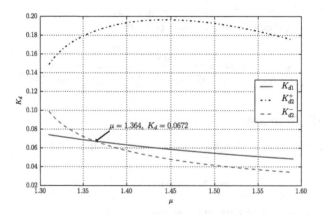

Fig. 2. The plot of K_d vs. μ.

To demonstrate the merits of the FOCs over their conventional counterparts, this work first investigates whether the PID controller could satisfy the specifications (i) to (iii). In view of Eqs. (9) and specification (ii), for a classic PD controller, one can obtain

$$\frac{\mathrm{d}(\arg[G(j\omega)])}{\mathrm{d}\omega}\Big|_{\omega=\omega_c} = \frac{K_d}{1+(K_d\omega_c)^2} - \frac{T}{1+(T\omega_c)^2} - L = 0 \qquad (41)$$

The solution is $K_d = \frac{1\pm\sqrt{1-4\omega_c^2 A^2}}{2\omega_c^2 A}$, where $A = \frac{T}{1+(\omega_c T)^2} + L$. In such a case, the phase of $G(j\omega)$ is obtained as

$$\arg[G(j\omega_c)] = \tan^{-1}(K_d\omega_c) - \tan^{-1}(\omega_c T) - \tfrac{\pi}{2} - \omega_c L \qquad (42)$$

This means $\arg[G(j\omega_c)]$ is a constant with determined K_d. As a result, specifications (i) and (ii) cannot be satisfied simultaneously for traditional PD controller.

In contrast, the parameters of the FOCs can be analytically solved with the aforementioned specifications. In this work, the PD^μ controller is adopted to demonstrate this feature, which is obviously the same with the other three kinds of FOCs. As formulated in Sect. 3, the PD^μ controller can be properly designed utilizing Eqs. (11), (13) and (15). By assigning $\phi_m = 70°$, $\omega_c = 5$, the parameters K_d and μ can be determined based on the graphic illustration. As shown in Fig. 2, the K_d and μ are explicitly determined as the intersection point, then K_p is evaluated by using Eq. (15). In this way, the control gains are determined as $K_d = 0.0672$, $\mu = 1.364$, and $K_p = 4.3$.

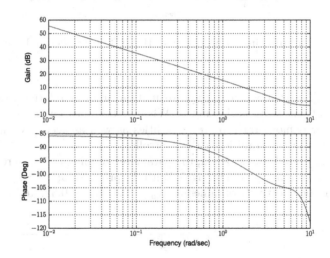

Fig. 3. The frequency response of open-loop plant with the PD^μ controller.

The bode plot of the corresponding controller is illustrated in Fig. 3. It can be seen that both the phase margin ϕ_m and the gain crossover frequency criterion ω_c (specification ii) are properly satisfied.

With the aforementioned parameters, the performance of the PD^μ controller is compared to the classic PD controller. The parameters of the conventional PID controller are selected by utilizing the ITAE criterion and the system Simulation techniques [22]. As the PD^μ is actually a finite dimensional linear filter due to the fractional order differentiator [20]. A band-limit implementation is important in practice, and a finite dimensional approximation method, namely Oustaloup Recursive Algorithm, is utilized in this work [20]. The comparative simulation results are illustrated in Fig. 4, where n is the order of the transfer function used in the approximation [20].

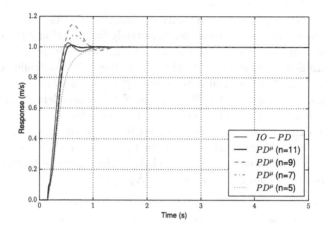

Fig. 4. The comparative simulation results of the PD^μ controller and the $IO - PD$ controller.

It can be seen that when the approximation order is relatively small, the performance of the FOC varies much. When n becomes larger, say $n = 11$, PD^μ controller demonstrates better stabilities as well as accuracy, thus shows its superiority over the traditional PD controller.

As an additional demonstration, the parameter tunning and step response of the PI^λ controller is illustrated in Fig. 5. With the proposed approach, the desired parameter can be effectively obtained as $K_p = 2.4$, $K_i = 0.32$, and $\mu = 0.202$.

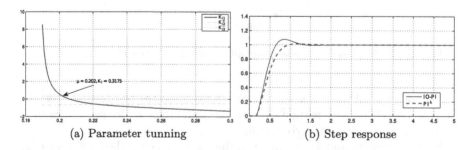

(a) Parameter tunning   (b) Step response

Fig. 5. The comparative simulation results of the PI^λ controller and the $IO - PD$ controller.

With the tunned parameters, the response of the quadrotor compared to the PI controller is illustrated in Fig. 5(b). It can be seen that the PI^λ controller demonstrate faster convergence compared to the conventional PI controller.

5   Conclusion

This work has developed a class of FOCs and the associated parameter tunning methods for quadrotors, regarding the attitude as the pseudo control input. To facilitate this development, the transfer function of the attitude is first presented based on previous researches. With the obtained transfer function model, a class of FOCs, including PD^μ, FO(PD), PI^λ, and FO (PI) controllers are developed accordingly. For each controller, the parameter tunning methods are addressed in details. To verify the effectiveness of this development, comparative numeric simulations are carried out. The results show that with proper implementation, the FOCs demonstrate better robustness and stabilities over their conventional counterpart.

In future, the fractional calculus would be adopted to more accurately describe the quadrotor model, and the proposed controller is considered to implement into real-time flights to improve the performance of quadrotor in their specific missions.

Acknowledgments. This work was funded by the special development fund of Shanghai Zhangjiang Hi-Tech Industrial Development Zone (No. 201411-PD-JQ-B108-009) and Open Foundation of the State Key Laboratory of Fluid Power Transmission and Control (No. GZKF-201510).

References

1. Burri, M., Oleynikova, H., Achtelik, M., Siegwart, R.: Real-time visual-inertial mapping, re-localization and planning onboard MAVs in unknown environments. In: Proceedings of the IEEE/RSJ International Conference on Intelligent Robots and Systems, Hamburg, pp. 1872–1878, September 2015
2. Dong, W., Gu, G.Y., Ding, Y., Xiangyang, Z., Ding, H.: Ball juggling with an under-actuated flying robot. In: Proceedings of the IEEE/RSJ International Conference on Intelligent Robots and Systems, Hamburg, pp. 68–73, September 2015
3. Hehn, M., D'Andrea, R.: A flying inverted pendulum. In: Proceedings of IEEE International Conference on Robotics and Automation, pp. 763–770 (2011)
4. Kumar, V., Michael, N.: Opportunities and challenges with autonomous micro aerial vehicles. Int. J. Robot. Res. **31**(11), 1279–1291 (2012)
5. Bouabdallah, S., Noth, A., Siegwart, R.: PID vs LQ control techniques applied to an indoor micro quadrotor. IN: Proceedings of the IEEE/RSJ International Conference on Intelligent Robots and Systems, vol. 3, pp. 2451–2456 (2004)
6. Droeschel, D., Nieuwenhuisen, M., Beul, M., Holz, D., Stückler, J., Behnke, S.: Multilayered mapping and navigation for autonomous micro aerial vehicles. J. Field Robot. (2015). doi:10.1002/rob.21603. Article first published online: 5 June 2015

7. Dydek, Z.T., Annaswamy, A.M., Lavretsky, E.: Adaptive control of quadrotor UAVs: a design trade study with flight evaluations. IEEE Trans. Autom. Sci. Eng. **21**, 1400–1406 (2013)
8. Mellinger, D., Kumar, V.: Minimum snap trajectory generation and control for quadrotors. In: Proceedings of IEEE International Conference on Robotics and Automation, pp. 2520–2525 (2011)
9. Richter, C., Bry, A., Roy, N.: Polynomial trajectory planning for aggressive quadrotor flight in dense indoor environments. In: Proceedings of the International Symposium on Robotics Research, Singapore, 1–16 December 2013
10. Bouabdallah, S.: Design and control of quadrotors with application to autonomous flying. Ph.D. thesis (2007)
11. Ryan, T., Kim, H.: Lmi-based gain synthesis for simple robust quadrotor control. IEEE Trans. Autom. Sci. Eng. **10**(4), 1173–1178 (2013)
12. Dong, W., Gu, G.Y., Zhu, X., Ding, H.: High-performance trajectory tracking control of a quadrotor with disturbance observer. Sens. Actuators A Phys. **211**, 67–77 (2014)
13. Lim, H., Park, J., Lee, D., Kim, H.: Build your own quadrotor: open-source projects on unmanned aerial vehicles. IEEE Robot. Autom. Mag. **19**(3), 33–45 (2012)
14. Michael, N., Mellinger, D., Lindsey, Q., Kumar, V.: The grasp multiple micro-UAV testbed. IEEE Robot. Autom. Mag. **17**(3), 56–65 (2010)
15. Peng, C., Tian, Y., Bai, Y., Gong, X., Zhao, C., Gao, Q., Xu, D.: ADRC trajectory tracking control based on PSO algorithm for a quad-rotor. In: Proceedings of the IEEE Conference on Industrial Electronics and Applications, pp. 800–805 (2013)
16. Podlubny, I.: Fractional-order systems and pi/sup/spl lambda//d/sup/spl mu//-controllers. IEEE Trans. Autom. Control **44**(1), 208–214 (1999)
17. Luo, Y., Chen, Y.: Fractional order [proportional derivative] controller for a class of fractional order systems. Automatica **45**(10), 2446–2450 (2009)
18. Dong, W., Gu, G., Zhu, X., Ding, H.: Modeling and control of a quadrotor UAV with aerodynamic concepts. In: ICIUS 2013: International Conference on Intelligent Unmanned Systems (2013)
19. Palka, B.: An Introduction to Complex Function Theory. Undergraduate Texts in Mathematics. Springer, London (2012)
20. Li, H., Luo, Y., Chen, Y.Q.: A fractional order proportional and derivative (FOPD) motion controller: tuning rule and experiments. IEEE Trans. Control Syst. Technol. **18**(2), 516–520 (2010)
21. Malek, H., Luo, Y., Chen, Y.: Identification and tuning fractional order proportional integral controllers for time delayed systems with a fractional pole. Mechatronics **23**(7), 746–754 (2013)
22. Chen, Y., et al.: System Simulation Techniques with MATLAB and Simulink. Wiley, London (2013)

Design and Implementation of Data Communication Module for a Multi-motor Drive and Control Integrated System Based on DSP

Qijie Yang, Chao Liu, Jianhua Wu[✉], Xinjun Sheng, and Zhenhua Xiong

State Key Laboratory of Mechanical System and Vibration,
School of Mechanical Engineering, Shanghai Jiao Tong University,
Shanghai 200240, China
{yangqijie,aalon,wujh,xjsheng,mexiong}@sjtu.edu.cn

Abstract. This paper focuses on the design and implementation of data communication module for a multi-motor drive and control integrated system based on DSP. A kind of overall design for data communication is firstly presented due to the integration of DSP and W5300. Secondly, the real-time data communication under RTX is introduced. In this paper, UDP protocol is adopted and data sending is realized by using socket programming. After that, the complete communication function is given and the data communication between DSP and W5300 is implemented. The data communication method between DSP and FPGA is realized through a dual port RAM afterwards. Finally, Ethernet communication test is conducted based on the self-designed multi-motor drive and control integrated system to verify the reliability and real-time performance of this data communication module.

Keywords: Data communication · Multi-motor drive and control integrated system · RTX · TMS320F28335 · TCP/IP protocol stack

1 Introduction

Recently with the development of computer technology and the urgent need of openness, extensibility for robot controller, motion controller based on PC has been a mainstream trend, because PC machine has the advantage of versatility, low cost and good communication function. At the same time there are a number of graphics data interfaces in the Windows operating system and it owns a broad customer base. Therefore it is a good choice to use a Windows PC to meet the needs of open architecture controller. Hong et al. [4] presented a PC-based open robot control system. Zhao et al. [13] did a research on open-CNC system based on PC.

On the other hand, with the development of microelectronics and large scale integrated circuit technology, further integration of systematic drive part, even the combination of control part and drive part, has made it possible to realize

© Springer International Publishing Switzerland 2016
N. Kubota et al. (Eds.): ICIRA 2016, Part I, LNAI 9834, pp. 75–86, 2016.
DOI: 10.1007/978-3-319-43506-0_7

complex and compact high-performance controller for modern industrial application. Klas et al. [7] proposed an integrated architecture for industrial robot programming and control. Zeng et al. [12] developed a kind of FPGA modular for a multi-motor drive and control integrated system.

The block diagram of multi-motor drive and control integrated system is shown in Fig. 1. A multi-motor drive and control integrated system can be used to control and drive robots with multi motors. Technology of architecture that motion control and drive part are closely integrated together can avoid many shortcomings in traditional multi-motor control structures, such as: (1) Too many wires make it difficult for system debugging; (2) The data transmission based on analog signals behaves badly on resisting disturbance; (3) Various modules can restrict each other sometimes, etc. This drive and control integrated system can save much space and avoid hardware repetition. As a result, the total cost can be greatly reduced.

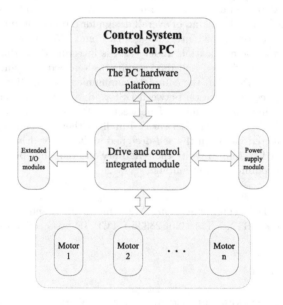

Fig. 1. A multi-motor drive and control integrated system

The hardware design scheme of multi-motor drive and control integrated system can not only support Ethernet communication but also control and drive multiple permanent magnet synchronous motors simultaneously. The communication part of the scheme is based on a chip with hardware TCP/IP protocol. DSP+FPGA architecture is used in servo control section and intelligent power module IPM is the core part of drive section.

This paper focuses on the design and implementation of data communication module for multi-motor drive and control integrated system. Floating-point DSP TMS320F28335 (hereinafter referred to as F28335) of TI company is chosen as the master control chip and hardware protocol stack chip W5300 is used to implement TCP/IP communication with PC.

This article is organized as follows. In Sect. 2, the module function and overall design is given. Section 3 describes the realization of communication under RTX in detail. In Sect. 4 communication of joint driving control layer is implemented. The communication method between DSP and FPGA is given in Sect. 5. In Sect. 6 Ethernet communication tests are conducted to test the proposed module. Finally, a conclusion is given in Sect. 7.

2   Communication Module Description

Data communication module is an important part of multi-module drive and control integrated system with the main function to realize data communication between PC and chips of hardware system. Data communication includes TCP/IP communication with PC and communication with information processing chip FPGA on the control panel. Furthermore, all of the communication is a two-way street. Figure 2 shows the function of data communication module.

Integration structure of data communication module for multi-motor drive and control integrated system is shown in Fig. 3. This module is mainly composed

Fig. 2. The function of data communication module

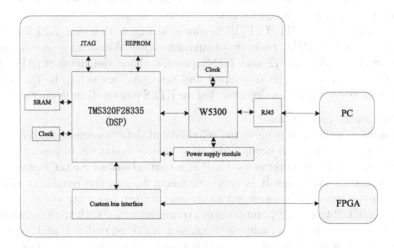

Fig. 3. Integration structure of data communication module

of F28335 and its affiliated circuit, TCP/IP communication module and interface circuit with master computer. As the master chip F28335 controls W5300's TCP/IP communication with PC [9], and data exchange with FPGA is implemented using dual port RAM memory shared between F28335 and FPGA. Outside SRAM is used for data cache. System state parameters and position loop tuning parameters are saved in the EEPROM. When the system is running, status monitoring parameters can be sent to PC, and the position loop tuning parameters can be programmed to control position loop which is calculated in DSP F28335.

3   The Realization of Real-Time Communication Under RTX

Standard Windows-driven PCs and RTX real-time software are key components of a standard, pre-integrated RTOS platform for building the many complex hard real-time applications that also require the sophisticated user interface Windows. RTX is unique in that it supports the creation of a single, integrated system that executes a Windows-based HMI and the real-time system in parallel with SMP architecture on a single PC. Therefore, the real-time communication between PC and the control panel can be realized.

(1) RT-TCP/IP protocol stack
RTX provides complete RT-TCP/IP protocol stack to support IPv4 and IPv6. Network card is operated under the environment of RTSS and RTSS process has access to the network card through the Winsock API [1]. Special Network card needs to be installed to achieve Ethernet communication under RTSS, and kept fully independent from Network card under Windows. A kind of mechanism is provided by RTX to switch PCI interface card from Windows equipment to RTX equipment, and real-time network card drivers for some mainstream Network cards on the market are also provided.

The architecture of RT-TCP/IP is shown in Fig. 4. Relevant API functions provided by RT-TCP/IP make it compatible with Winsock2.0 on the great degree. Because the Win32 and RTSS process share the network API, RT-TCP/IP application can be developed and debugged according to the Win32 process to achieve rapid development before RTSS compilation is operated.

(2) RTSS socket programming
TCP communication can achieve the reliability of data transmission, but in the process of execution it will occupy too many system resources to meet the real-time motion control requirements. UDP is a connectionless packet communication protocol [8]. Although it is very convenient to use, the reliability of data transmission cannot be guaranteed easily and packet loss may occur [5]. But when using UDP to realize point-to-point communication with the transmission speed below a certain value, reliable transmission can be realized, and there will be no data loss and confusion [11]. In this paper, UDP protocol is adopted and data sending is realized by using socket programming.

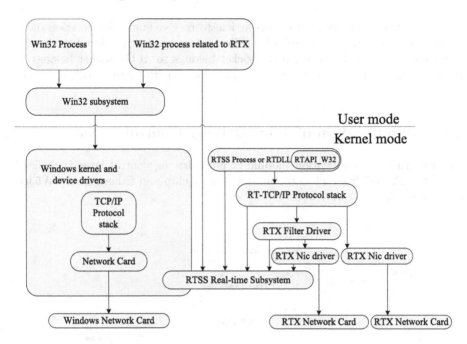

Fig. 4. RT-TCP/IP architecture

RTX supports ioctlsocket function to convert a socket to a non-blocking mode, and supports the following I/O models: blocking model, select model, WSAEventSelect model and overlapping model.

In the software implementation of this paper, because the real-time trajectory planning generation and off-line trajectory generation are distributed in two threads, a blocking socket can be defined in each thread which makes function call simple and the real-time performance guaranteed. In data receiving threads, select model is chosen to realize the data reception because state data sent from joint driving control layer need to be received. Select function is used by select model to judge whether data can be read on a socket, or whether to write data. When it is used, firstly FD-SET structure is set through related macro (such as FD-CLR FD-SET, FD-ISSET, etc.), and then select function is called to monitor the socket. The select function prototype is as follows:

int select(

int nfds, // Generally ignored, maintain compatible with the Berkeley socket application

fd-set *readfds, // Check readability of the socket

fd-set *writefds, // Check writability of the socket

fd-set *exceptfds, // other state instead of Reading and writing state

const struct timeval *timeout // the longest waiting time when waiting for I/O operation to complete

);

The socket is mainly used in this article to receive status information data, so the socket is added to the read-fds collection and called in select function for further judging which collection the socket belongs to. If the socket belongs to the read-fds, it suggests that the socket receiving function can be called to read data.

4   Communication of Joint Driving Control Layer

The Ethernet communication module of joint driving control layer is realized with DSP F28335 from TI as the master control chip and Ethernet chip W5300 as a slave chip.

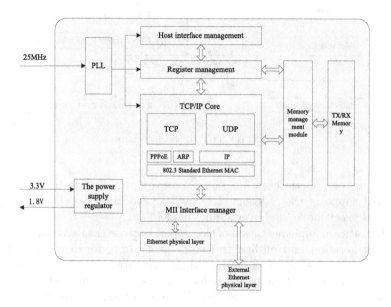

Fig. 5. Internal architecture of W5300

As a kind of Ethernet chip of WIZnet, W5300 supports hardware TCP/IP protocol stack and enables 8 sockets to work at the same time. The internal architecture of W5300 is shown in Fig. 5. According to the use of W5300, TX/RX storage space size (128 KB) of each socket can be allocated flexibly, and third-party PHY interface is supported as well [2]. Bus interface is adopted between master control chip DSP F28335 and W5300. It supports directly or indirectly addresses access patterns and 8/16 bits of data bus [3]. The total transfer rate is as high as 50 mbps. The hardware connection schematic between DSP F28335 and W5300 is shown in Fig. 6.

Polling and interrupts are two kinds of implementation methods for DSP to perform Ethernet communication through W5300. Polling means that DSP

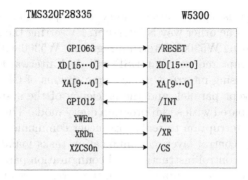

Fig. 6. Hardware connection schematic between TMS320F28335 and W5300

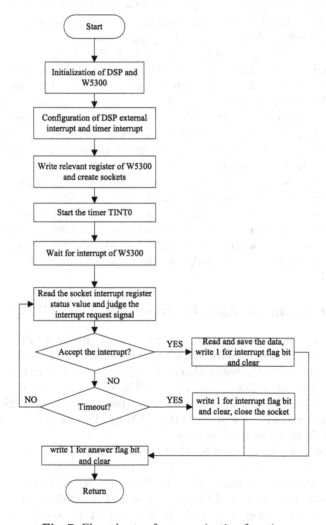

Fig. 7. Flow charts of communication function

inquires W5300's status register at a certain cycle to judge and deal with related function called [10]. The other way is interrupt. By setting the DSP's interrupts enabled and configuring W5300's interrupt registers, W5300 will put forward the corresponding interrupt request that DSP will deal afterwards when Ethernet incidents occur. By using interrupt mode, some actions of CPU and peripheral equipment can be kept parallel, and the efficiency of the system can be obviously increased compared with serial program query mode. Therefore this article adopts the way of interruption to realize Ethernet communication.

For joint driving control layer, communication tasks mainly include: (1) to receive real-time PC control instruction and configuration parameters; (2) regularly send state data to host computer. From the perspective of system function, the real-time requirements of the former is higher, because the joint driving control layer needs to finish real-time closed-loop control according to the control instruction. Therefore, the data receiving interrupt is prior to data sending interrupt when reflected on the interrupt priority level.

In this paper, CPU timer 0 (TINT0) of DSP F28335 is adopted to realize regularly sending state data and its CPU priority is 5, priority of PIE set is 7. Data reception is realized using external interrupt (XINT1) through GPIO, and its CPU interrupt priority is 5, PIE set priority is 4. Through writing the interrupt selection register (GPIOXINT1SEL), interrupt configuration register (XINT1CR), and setting the interrupt entrance function address, Ethernet communication can be realized. The realization process of main function and receiving interrupt function is shown in Fig. 7.

In addition, reading and writing access sequence and time interval can be configured through the XTIMINGx, XINTCNF2 and XBANK register when making access to W5300 through external interface of DSP F28335. By default, the above parameters are maximum and greater than timing requirements of W5300 being read and writing properly. Reasonable configuration can reduce the cycle of DSP's access to W5300, thus the Ethernet communication rate can be optimized.

5   Data Communication Between DSP and FPGA

The communication between data communication module and FPGA is realized by a dual port RAM which can be shared with both DSP and FPGA. A dual port RAM can be applied to the realization of multiprocessor interface technology [6]. In this paper, the data collected and processed by FPGA is written to this RAM, and then the DSP will read the data. After the data processing by DSP and PC, the speed command will be written to the RAM by DSP and FPGA can get this command to do speed loop calculation.

Custom bus interface is used for the interface between DSP and FPGA with 8 bit data line, 8 bit address line, 1 kHz synchronous clock, read enable, write enable, chip select and ground wire. The dual port RAM is connected to the external extension interface (XINTF) zone0 of DSP F28335 through two pieces of dual voltage conversion chip (16t245). It is important to note that the output enable pin (/OE) of 16t245 which controls the synchronous clock must be grounded to enable the output of synchronous clock signal. If the output enable pin is connected to the chip select pin like the other interface signals, it will lead to that the synchronous clock signal cannot be recognized by DSP F28335.

6   Ethernet Communication Test

High requirements for real-time and reliability of Ethernet communications are needed in this project, because standard Ethernet UDP protocol is adopted to complete communications between PC software and the integration module of control and driving, and the trajectory planning and interpolation of robot are completed by PC control software at the same time. Real-time and relia-bility performances of data sending from PC to joint driving control layer are tested respectively under the following experimental conditions. The hardware test platform of a multi-motor drive and control integrated system is shown in Fig. 8. A 4-channel driving board is used to drive 4 motors to test the reliability and real-time performance of this module.

Fig. 8. Hardware test platform of a multi-motor drive and control integrated system

In the PC, RTSS applications are developed in Visual Studio 2010 and UDP packets are sent to the joint driving control layer using a blocking socket. A 48 byte single packet is sent 5000 times with a sending cycle of 2 ms, and the above experiment is repeated for 50 times.

In the testing program of DSP+W5300, the interrupt can be received and responded in real time, and state data with a packet size of 64 byte is sent from DSP to PC software every 10 ms. GPIO pin level is written in the entrance and exit of receiving interrupt function to judge the response state of interrupt. In addition, the correctness of the data received needs to be checked in the

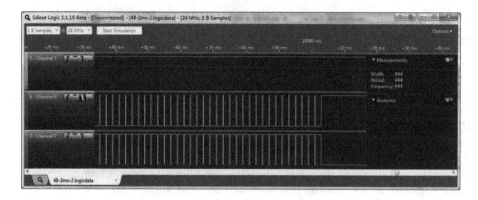

Fig. 9. Single sampling result by logical analyzer

program, and another GPIO pin level needs to be lowered if there is an error. In this article the logic analyzer is used to observe the change of the pin level.

(1) Reliability test
In this experiment the reliability of UDP transmission is determined according to the packet loss of UDP packets under a certain transmission speed. A single sampling result of Logic analyzer is shown in Fig. 9. Channel 6 is the level signal to interrupt pin, and Channel 7 is the level signal to in and out flag pin of interrupt function, and Channel 5 is level signal to packet error flag pin. Results can be exported to the disk file, and finally the response time can be analyzed through MATLAB. Experimental data analysis results show that there is no lost package and data received is the same with data that PC has sent.

(2) Real-time performance test
For real-time Ethernet communication, real-time performance of W5300's interrupt request signal and response latency of DSP to W5300's interrupt request

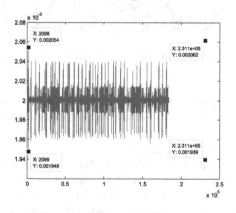

Fig. 10. Interrupt request signal measurement

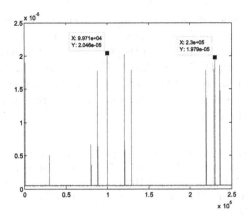

Fig. 11. Delay time of DSP's responses to W5300 interrupt requests

signal are measured respectively, and thus the real-time performance can be judged based on the measurement results.

In this experiment, the interrupt request signal can be directly obtained by W5300's interrupt signal pin, and real-time performance of W5300's interrupt request signal can be analyzed based on this. The lag time between interrupt request level signal and IO signal of interrupt entrance function is measured to evaluate DSP F28335's response time to interrupt.

The analysis is carried on after measurement data is imported to MATLAB. The measurement results of interrupt request signal's real-time performance is shown in Fig. 10, and delay analysis of DSP to interrupt request is shown in Fig. 11. It can be seen that the biggest interrupt request time error is 62 us in the experiment and the maximum response delay of DSP to accept interrupt is 20 us. Therefore the requirements of real-time control can be met based on the results.

(3) Measurement error analysis
The sampling frequency of logic analyzer is 24 MHZ, so the resolution is 41.67 ns. The time delay error analyzed above can reflect the data reception and processing performance of DSP F28335 and W5300, though it contains time error during the sending and transmission process of PC as well.

7   Conclusions

In this paper, the data communication module of a multi-motor drive and control integrated system is successfully designed and implemented, which is essential to conduct data transmission between PC and the control panel. Firstly, a kind of integration structure of data communication module composed mainly of DSP TMS320F28335 and W5300 is given. And then, the realization of communication under RTX is introduced in detail and RTSS socket programming is adopted to receive status information data in this paper. The communication between DSP and W5300 of joint driving control layer is implemented with the communication function which will be run once the system is initialized. After that, the communication between DSP F28335 and FPGA is given in detail. At last, the Ethernet communication test of this self-designed data communication module is carried out, which shows that the reliability and real-time performance of data communication can be guaranteed.

Acknowledgements. This research was supported in part by National Natural Science Foundation of China under Grant 51575355, China Postdoctoral Science Foundation under Grant 2015M80325, National High-tech Research and Development Project of China under Grant 2015BAF01B02 and National Key Basic Research Program of China under Grant 2013CB035804.

References

1. Adams, K., Huang, S.: RTX: A real-time operating system environment for CNC machine tool control. In: IFAC SYMPOSIA SERIES, pp. 55–55. PERGAMON PRESS (1993)
2. Fei, S., Xuzhe, F.: Design of the fast ethernet interface based on hardware protocol stack chip. Ind. Instrum. Autom. **4**, 016 (2012)
3. Haoran, Q., Jianping, Z.: Design of ethernet interface based on W5300. Electron. Meas. Technol. **7**, 035 (2012)
4. Hong, K.S., Choi, K.H., Kim, J.G., Lee, S.: A PC-based open robot control system: PC-ORC. Robot. Compu. Integr. Manuf. **17**(4), 355–365 (2001)
5. Jiao, Y.T., Ren, Y.F., Li, N.N., Wang, X.S.: Reliable ethernet interface design and implementation. Control Instrum. Chem. Ind. **1**, 022 (2012)
6. Li, S.K., Wang, X.S.: Design of ethernet interface based on FPGA and W5300. In: Key Engineering Materials, vol. 503, pp. 402–405. Trans Tech Publ (2012)
7. Nilsson, K., Johansson, R.: Integrated architecture for industrial robot programming and control. Robot. Auton. Syst. **29**(4), 205–226 (1999)
8. Shunhu, H.: Design method of real-time simulation system based on RTX platform. Comput. Appl. Softw. **4**, 055 (2009)
9. Wenjuan, Z., Zhongyu, W., Weihu, Z., Yawei, W.: Design of laser tracker data communication and processing module based on DSP. Microcomput. Appl. **17**, 020 (2011)
10. Wu, H., Yan, S.G., Xue, S.X.: Design and implementation of ethernet data transfer system based on W5300. Electron. Des. Eng. **9**, 032 (2012)
11. Wu, Y., Xiong, Z.H., Ding, H.: Data acquisition and control systembased on RTX and MFC packaging platform. Syst. Eng. Electron. **9**, 029 (2004)
12. Zeng, X., Liu, C., Sheng, X., Xiong, Z., Zhu, X.: Development and implementation of modular FPGA for a multi-motor drive and control integrated system. In: Liu, H., Kubota, N., Zhu, X., Dillmann, R. (eds.) ICIRA 2015. LNCS, vol. 9244, pp. 221–231. Springer, Heidelberg (2015)
13. Zhao, C., Qin, X., Tang, H.: Research on open-CNC system based on PC. Mech. Sci. Technol. **24**(9), 1108–1113 (2005)

A Novel Continuous Single-Spindle Doffing Robot with a Spatial Cam and Multiple Grippers

Sicheng Yang[1], Yue Lin[1], Wenzeng Zhang[1(✉)], and Liguo Cao[2]

[1] Department of Mechanical Engineering, Tsinghua University, Beijing 100084, China
wenzeng@tsinghua.edu.cn
[2] Huaxiang Textile Machinery, Qingdao 266041, China

Abstract. Doffing machines for ring spinning machines include single-spindle doffers and group doffers. Group doffers are efficient but complex and expensive to build. Present single-spindle doffers are easy to build and low in cost, but it contacts and hurts yarns in cops and even damages spindles after a long ride. This paper developed a novel continuous single-spindle doffing robot with a spatial cam and multiple grippers for ring spinning machines, called the SCMG robot. The SCMG robot mainly consists of a doffing device, a moving and locating device. The doffing device is composed of 14 grippers moving along a fixed 3D trajectory of a spatial cam. The moving and locating device adopts a positioning wheel meshing with the bottom of spindles. The mechanism principle, structure and force analysis of the SCMG robot are introduced in detail. Experimental results show that the SCMG robot can pull the cops out of the spindle vertically one by one. The grippers grasp the top of the cops without contacting the yarn. The SCMG robot is highly efficient and reliable.

Keywords: Doffing robot · Continuous single-spindle doffer · Spatial cam · Gripper · Ring spinning machine

1 Introduction

Ring spinning is a useful and important technology. During the ring spinning process, the operator needs to doff the cops of ring spinning machines. Currently, the doffing operation in over 60 % of the spinning corporations is accomplished by manual [1], shown in Fig. 1a. Automation of doffing operation is significant to the spinning industry with the rising of labor cost. Generally, doffing machines for ring spinning machines include single-spindle (or wagon) doffers and group doffers.

Group doffers, shown in Fig. 1b, have a high degree of automation and high production efficiency. However, the large cost limits their development. Group doffers have to be fixed to the frames of ring spinning machines, respectively. One group doffer can only serve one ring spinning machine. It is inevitable to modify the original spinning system largely in order to install group doffers on conventional ring spinning machines.

The common features of wagon doffers include low manufacture cost, simple structure, low energy levels, high efficiency and more. Different from group doffers, wagon doffers do not need to be fixed to the frames of ring spinning machines and can move

© Springer International Publishing Switzerland 2016
N. Kubota et al. (Eds.): ICIRA 2016, Part I, LNAI 9834, pp. 87–98, 2016.
DOI: 10.1007/978-3-319-43506-0_8

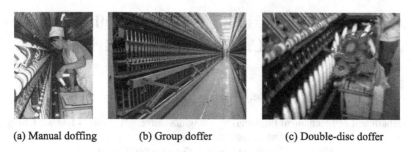

| (a) Manual doffing | (b) Group doffer | (c) Double-disc doffer |

Fig. 1. Doffing types of ring spinning machines

flexibly. Wagon doffers move along the rails of ring spinning machines only during the doffing process, so one wagon doffer can serve several ring spinning machines. A wide range of wagon doffers are developed since 1950s [2–6]. Double-disc doffer is a typical representative of them [7], as shown in Fig. 1c. Double-disc doffer doffs single cops every time and can achieve continuous doffing. The doffing efficiency of double-disc doffer is very high. But in the doffing process, the double discs will contact the yarn and damage spindles. Recently, some novel wagon doffers are developed [8, 9]. Most of them adopt the structure of the orthogonal coordinate robot. During the doffing process, these wagon doffers doff several cops (about 10 cops) at the same time and can pull the cops out vertically without contact with the yarn.

Group doffers are efficient but complex and expensive to build. Present single-spindle doffers are easy to build and low in cost, but it contacts and hurts yarns in cops and even damages spindles after a long ride.

This paper developed a novel continuous single-spindle doffing robot with a spatial cam and multiple grippers for ring spinning machines, called the SCMG robot. The SCMG robot mainly consists of a doffing device, a moving and locating device. The doffing device is composed of 14 grippers moving along a fixed 3D trajectory of a spatial cam. The moving and locating device adopts a positioning wheel meshing with the bottom of spindles. The SCMG robot combines the advantages of the double-disc wagon doffer and the group doffer. On one hand, the SC robot works with single-spindle and continuous doffing, which is helpful to promote doffing efficiency. On the other hand, the SCMG robot can pull cops out of spindles vertically. During the doffing process, the gripper of the SCMG robot grasps the top of cops without touching yarns.

2   Principle and Structure of the SCMG Robot

2.1   Mechanism Principle of the SCMG Robot

Generally, the doffing operation includes two steps: doffing cops and donning bobbins, as shown in Fig. 2a. There have been some effective measures to donning bobbins in the spinning industry. This paper concentrates on the methods of doffing cops.

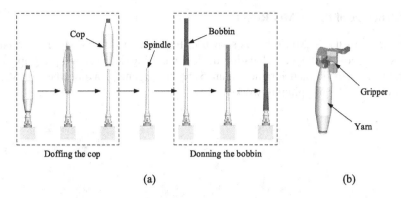

Fig. 2. The doffing operation

Some requirements have to be satisfied during the process of doffing cops:

(a) The relative motion trend in the horizontal direction between the cop and the spindle is not allowed for protecting spindles before the cop leaves the spindle totally;
(b) Grippers cannot touch the yarn, so as not to damage the yarn.

The grippers grasp the top of cops in order to avoid damaging the yarn, because there are no yarns in the top of cops, as shown in Fig. 2b.

The mechanism principle of the SCMG robot is shown in Fig. 3. The grippers are fixed to the sliding blocks. On one hand, the sliding block is sleeved on the vertical rail and can move up or down along the vertical rail. On the other hand, the sliding block is sleeved on the space curve guide rail and has to move along the track of the space curve guide rail. The vertical guide rails are fixed to horizontal conveyors through fixed hinges. The distance (or arc length) between two adjacent vertical guide rails is equal. Therefore, the vertical guide rails can move along with horizontal conveyors. The sliding blocks follow the horizontal movement of vertical guide rails. Meanwhile, the sliding blocks move up or down due to the constraint of the space curve guide rail. Therefore, the grippers achieve the three-dimensional motion.

The horizontal component of the velocity of grippers is equal to the velocity v_1 of horizontal conveyors in the linear segment. When the velocity v_1 of horizontal conveyors is equal to the velocity v_0 of the frame of the SCMG robot, the gripper and the spindle can remain relative static state during the doffing process.

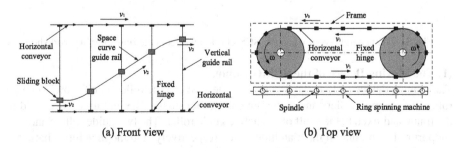

(a) Front view   (b) Top view

Fig. 3. The mechanism principle of the SCMG robot

2.2 Structure of the SCMG Robot

The structure of the SCMG robot is shown in Fig. 4 shows the SCMG robot mainly consists of a moving device, a locating device, a doffing device, multiple grippers, a bobbin insertion device and a frame. Figure 5 shows the general layout when the SCMG robot works in the ring spinning workshop.

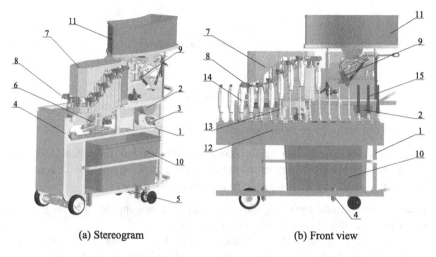

(a) Stereogram   (b) Front view

Fig. 4. The structure of the SCMG robot. 1-frame; 2-first motor; 3-active guide roller; 4-following guide roller; 5-support roller; 6-locating device; 7-doffing device; 8-gripper; 9-bobbin insertion device; 10-cop basket; 11-bobbin basket; 12-ring spinning machine; 13-spindle; 14-cop; 15-bobbin.

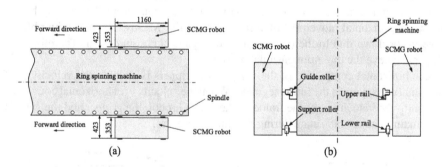

Fig. 5. The general layout when the SCMG robot works in the ring spinning workshop

(1) The Moving Device of the SCMG Robot.

The moving device include the first motor, one active guide roller, one following guide roller, two support rollers and matching connecting shafts. The first motor is placed in the frame and fixed to the shaft of the active guide roller. The two guide rollers and two support rollers mount on the matching shafts respectively, and the matching shafts are set on the frame. The active and following guide rollers contact with the upper rail of the ring spinning machine. The two support rollers contact with the lower rail of the ring spinning machine.

(2) The Locating Device of the SCMG Robot.

The locating device can insure the positional accuracy of the SCMG robot, so that the grippers can grasp cops in a proper position. The locating device is mainly composed of the toothed disc, the locating input shaft, the locating output shaft, the locating transmission mechanism, the first motor and two friction wheels, as shown in Fig. 6. The toothed disc is fixed to the locating input shaft and can mesh with the bottom of spindles of ring spinning machines. The locating input shaft connects with the input terminal of the locating transmission mechanism. The locating transmission mechanism is placed in the bottom of the doffing device and the output terminal of the locating transmission mechanism connects with the locating output shaft. The first friction wheel is fixed to the locating output shaft. The second friction wheel is arranged on the locating transmission mechanism and contacts with the first friction wheel. The output shaft of the second motor is fixed to the second friction wheel.

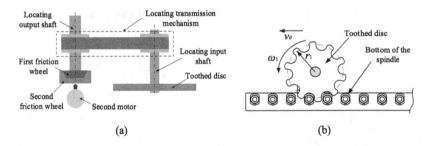

Fig. 6. The mechanism principle of the locating device of the SCMG robot

The function of the second motor is to provide the locating output shaft with extra power, so that the load of the toothed disc can be reduced. The power transmission between the second motor and the locating output shaft is accomplished with the two friction wheels. The transmission ratio between the toothed disc and the locating output shaft is constant, while the transmission ratio between the two friction wheels is not. Hence, it is not necessary to keep the two friction wheels at the same rotating speed. What need to do is to ensure that the rotating speed of the second friction wheel is faster than that of the first friction wheel, so the second motor can provide the locating output shaft with extra power rather than extra resistance.

(3) The Doffing Device of the SCMG Robot.

The doffing device (shown in Fig. 7) mainly consists of the first driving shaft, the second driving shaft, the chains, the first chain wheels, the second chain wheels, the spatial cam, the vertical guide rails, the sliding blocks, the track rollers and the base. The base is mounted on the frame. The spatial cam is fixed to the base. The first and the second cams are fixed to the spatial cam. The first driving shaft is placed in the base and fixed to the locating output shaft. The second driving shaft is placed in the base. The first chain wheels are fixed to the first driving shaft. The second chain wheels are movably sleeved on the second driving shaft. The chain is set between the first chain wheel and the second chain wheel. The vertical guide rails are fixed to the chain. The distance (or arc length) between two adjacent vertical guide rails is equal to the distance between two adjacent spindles. The sliding blocks are movably sleeved on the vertical guide rails respectively and can slide along the vertical direction. The gripper and the track roller are fixed to the sliding block respectively. The movable claw is placed in the gripper. The track rollers contact with the spatial cam.

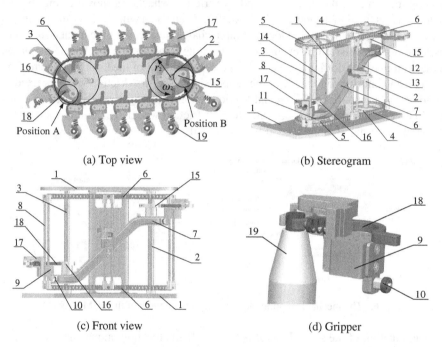

(a) Top view   (b) Stereogram

(c) Front view   (d) Gripper

Fig. 7. The doffing device of the SCMG robot. 1-base; 2-first driving shaft; 3-second driving shaft; 4-first chain wheel; 5-second chain wheel; 6-chain; 7-spatial cam; 8-vertical guide rail; 9-sliding block; 10-track roller; 11-lower horizontal segment of the spatial cam; 12-upper horizontal segment of the spatial cam; 13-upslope segment of the spatial cam; 14-downhill segment of the spatial cam; 15-first cam; 16-second cam; 17-gripper; 18-movable claw; 19-cop.

When the doffing device works, the grippers will move along the track of the spatial cam. Generally, the path of the spatial cam can be divided into four parts, i.e., the lower horizontal segment, the upper horizontal segment, the upslope segment and the downhill segment, as shown in Fig. 7b and c. Moreover, there are circular arc transition between two adjacent segments.

3   Doffing Process of the SCMG Robot

This section introduces the doffing process of the SCMG robot in detail.

The first motor starts and drives the active guide roller to rotate. So the frame moves forward on the upper rail of the ring spinning machine. The toothed disc is pulled by spindles and starts to rotate. The toothed disc drives the first driving shaft and the first chain wheels to rotate via the locating transmission mechanism. Then the vertical guide rails move along with the rotational motion of the chain. The sliding blocks follow the horizontal movement of the vertical guide rails, on the other hand, the sliding blocks move up or down due to the constraint of the spatial cam. Therefore, the grippers achieve the three-dimensional motion.

As shown in Fig. 8, v_0 is the velocity of the frame, v_1 is the velocity of grippers. In order to ensure that grippers can pull the cops out of the spindle vertically, the horizontal component of the velocity of grippers should be equal to the velocity of the frame identically.

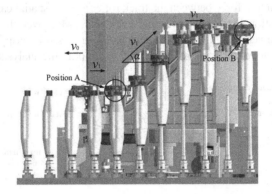

Fig. 8. The doffing process of the SCMG robot

Analyzing Figs. 6, 7 and 8, one obtains:

$$\omega_1 r_1 = v_0 \tag{1}$$

$$\omega_2 r_2 = v_0 \tag{2}$$

Combining the Eqs. (1–2), one gets:

$$i = \frac{\omega_1}{\omega_2} = \frac{r_2}{r_1} \tag{3}$$

There, i is the transmission ratio of the locating transmission mechanism, r_1 is the pitch diameter of the toothed disc, r_2 is the pitch diameter of the first driving wheel.

When the transmission ratio of the locating transmission mechanism satisfies the Eq. (3), grippers and spindles can remain relative static state during the doffing process.

As shown in Figs. 7 and 8, when a gripper reaches position A, the movable claw of the gripper is pushed by the second cam. Then the gripper opens and grasps the top of a cop. After that, the gripper moves into the upslope segment of the spatial cam.

The gripper rises up with the cop together. When the gripper moves into the upper horizontal segment of the spatial cam, the cop leaves the spindle totally. The gripper keeps moving forward. The movable claw is pushed by the first cam in position B, the gripper opens again and releases the cop. The cop basket collects the cop. The gripper continues to move along the path of the spatial cam until it reaches the position A again, and then the gripper begins the next doffing cycle. During the whole doffing process, the gripper neither contacts the yarn nor damages spindles.

4 Force Analysis of the Doffing Process

The doffing device is a critical subsystem of the SCMG robot, which has a great effect on the operation efficiency and the doffing result of the SCMG robot. Therefore, it is essential to analyze the force imposed on the doffing device. The force analysis of the doffing device is shown in Fig. 9. When the track roller is at the horizontal segment of the spatial cam, the rolling friction force between the track roller and the spatial cam is too small to influence the operational efficiency and can be neglected. Moreover, the friction between the gripper and the vertical guide rail is neglected. The mass of a gripper is assumed to converge at the center of the track roller in order to simplify the analysis model.

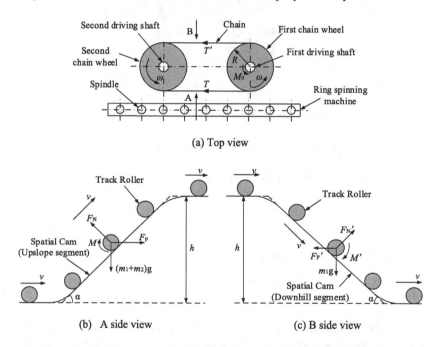

(a) Top view

(b) A side view   (c) B side view

Fig. 9. Force analysis of the doffing device of the SCMG robot

Let: m_1- the total mass of a gripper, a sliding block and a track roller, kg;
  m_2 - the mass of a cop, kg;
  F_N - the support force to track roller imposed by spatial cam, N;

R - the radius of the track roller, mm;

M - the rolling friction torque to track roller imposed by the spatial cam, Nm;

F_P - the tractive force to the gripper imposed by the chain, N;

v - the velocity of the gripper, m · s^{-1};

T - the tensile force of the chain, N;

M_0 - the actuated moment of the first driving shaft, Nm;

R - the radius of the first chain wheel, mm

l_0 - the distance between two adjacent spindles, mm;

h - the vertical height of the upslope (or downhill) segment of the spatial cam, mm;

μ_1 - the coefficient of rolling friction between the spatial cam and track rollers, mm;

α - the angle between the upslope (or downhill) segment of the spatial cam and the horizontal plane, (°).

According to the force analysis, the rolling friction torque imposing on the track rollers is as follows:

$$\begin{pmatrix} M \\ M' \end{pmatrix} = \begin{pmatrix} \mu_1 F_N \\ \mu_1 F'_N \end{pmatrix} \tag{4}$$

As shown in Fig. 7b, according to the force balance of the track roller in the upslope segment of the spatial cam, one obtains:

$$\begin{pmatrix} F_P & F_N & 1 \end{pmatrix} = \begin{pmatrix} \dfrac{M}{r}\sin\alpha & \dfrac{M}{r}\cos\alpha & 1 \end{pmatrix} \begin{pmatrix} r/\mu_1 & -1/\cos\alpha & 0 \\ -1 & 0 & 0 \\ 0 & (m_1+m_2)g/\cos\alpha & 1 \end{pmatrix} \tag{5}$$

Likewise, according to the force balance of the track roller in the downhill segment of the spatial cam, as shown in in Fig. 7c, one obtains:

$$\begin{pmatrix} F'_P & F'_N & 1 \end{pmatrix} = \begin{pmatrix} \dfrac{M'}{r}\cos\alpha & \dfrac{M'}{\mu_1}\sin\alpha & 1 \end{pmatrix} \begin{pmatrix} -1 & 0 & 0 \\ 1 & \dfrac{-\mu_1}{r\cos\alpha} & 0 \\ 0 & \dfrac{m_1 g}{\cos\alpha} & 1 \end{pmatrix} \tag{6}$$

Combining Eqs. (4–6), the following relationship is arrived at:

$$\begin{pmatrix} F'_P & F_P & 1 \end{pmatrix} = \begin{pmatrix} \dfrac{m_1 g}{r+\mu_1\tan\alpha} & \dfrac{(m_1+m_2)g}{r-\mu_1\tan\alpha} & 1 \end{pmatrix} \begin{pmatrix} r\tan\alpha-\mu_1 & 0 & 0 \\ 0 & r\tan\alpha+\mu_1 & 0 \\ 0 & 0 & 1 \end{pmatrix} \tag{7}$$

In order to calculate the actuated moment M_0 of the first driving shaft, it is necessary to figure out the maximum number of the track rollers locating on the upslope and downhill segment of the spatial cam. When setting the maximum number as n, one gets:

$$n = \left\lceil \frac{h}{l_0 \tan \alpha} \right\rceil, ([x] = \min\{k \in Z | x \le k\}) \tag{8}$$

The cop has to rise 180 mm at least when leaving the spindle totally, i.e., the vertical height of the upslope (or downhill) segment of the spatial cam has to be greater than or equal to 180 mm. Here assume $h = 200$ mm and $l_0 = 70$ mm. Considering the compactness of the doffing device, one assumes $2 \le n \le 6$, so the value range of the angle α is:

$$25.5° \le \alpha \le 70.7° \tag{9}$$

$$T = nF_P \tag{10}$$

$$T' = nF_P' \tag{11}$$

$$(M_0 \ 1) = (R \ R)\begin{pmatrix} T & 0 \\ -T' & 1/R \end{pmatrix} \tag{12}$$

Combining Eqs. (7–12), the following relationship is arrived at:

$$M_0 = gR\left\lceil \frac{h}{l_0 \tan \alpha} \right\rceil [\frac{r \tan \alpha + \mu_1}{r - \mu_1 \tan \alpha}(m_1 + m_2) - \frac{r \tan \alpha - \mu_1}{r + \mu_1 \tan \alpha}m_1] \tag{13}$$

Considering the compactness of the spatial cam, the value range of the radius of the track roller has to be limited in [4 mm, 12 mm], i.e., $r \in$ [4 mm, 12 mm].

In order to figure out the relationship among the actuated moment M_0, the angle α and the radius r of the track roller, setting some variables' value in Eq. (13) is necessary. Assume $m_1 = 1$ kg, $m_2 = 0.5$ kg, $R = 66$ mm, $\mu_1 = 0.05$ mm, $g = 9.8$ m · s⁻². So the relationship among the actuated moment M_0, the angle α and the radius r of the track roller can be described as Fig. 10.

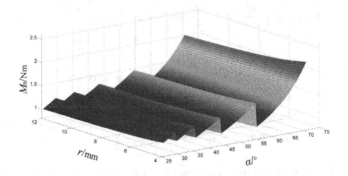

Fig. 10. The relationship among the moment M_0, the angle α and the radius r.

Figure 10 shows that the relationship between the actuated moment M_0 and the angle α is similar to the piecewise function. When $M_0 = M_{0,\min}$, the angle α can be several possible value within the limited range, as follows:

$$\alpha = \{25.5°, 29.9°, 35.9°, 43.9°, 55.1°\} \tag{14}$$

The designers can choose one of the values according to the demand of the configuration. In addition, there is a negative correlation between the radius r of the track roller and the actuated moment M_0 of the first driving shaft. Therefore, the designers should choose the track roller with the larger radius in the admitted range.

5   Doffing Experiments of the SCMG Robot

Figure 11 shows the doffing applications (experiments) of the SCMG robot. The SCMG robot has been applied to the ring spinning factory for the purpose of evaluating the performances. The key performances of manual doffing, the SCMG robot and double-disc doffer are shown in Table 1.

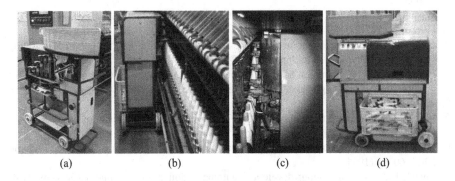

(a)   (b)   (c)   (d)

Fig. 11.   The doffing applications of the SCMG robot

Table 1.   The key performances of several typical doffing robot [1, 10]

Indicators	Manual doffing	Double-disc doffer	SCMG robot
Doffing rate	100 %	96 %	99 %
Doffing time (240 spindles)	about 150 s	about 50 s	60 s
Damage yarns or not?	No	Yes	No
Damage spindles or not?	No	Yes	No

The Table 1 shows that the doffing efficiency of the SCMG robot is close to that of double-disc doffer and much higher than that of manual doffing. Different from the double-disc doffer, the SCMG robot does not damage spindles and yarns in the doffing process.

6   Conclusion

This paper developed a novel continuous single-spindle doffing robot with a spatial cam and multiple grippers for ring spinning machines, called the SCMG robot. The SCMG robot mainly consists of a doffing device, a moving and locating device. The doffing device is composed of 14 grippers moving along a fixed 3D trajectory of a spatial cam. The moving and locating device adopts a positioning wheel meshing with the bottom of spindles. Experimental results show that the SCMG robot can pull the cops out of the spindle vertically one by one. The grippers grasp the top of the cops without touching yarns. The SCMG robot is highly efficient and reliable.

Acknowledgement. This Research was supported by National Natural Science Foundation of China (No. 51575302).

References

1. Gao, X.: Function study of a new auto-doffing spinning frame. China Textile Leader (01) (2011)
2. Watanabe, E.: Toyoda continuous auto-doffer. J. Textile Mach. Soc. Japan **11**, 213–216 (1965)
3. Inoo, O.: Pony doffer: an automatic doffing apparatus for single-spindle, continuous doffing system. J. Textile Mach. Soc. Japan **11**, 223–236 (1965)
4. Hori, J., Kinari, T., Shintaku, S.: Development of an auto-doffer for covering machine part 1: apparatus for drawing spandex into hollow spindles. J. Textile Mach. Soc. Japan **44**(4), 82–86 (1998)
5. Han, W.: On the understanding of smart doffing trolley. Sci. Technol. Innovation **08**, 42–43 (2014)
6. Cao, M.: Introduction of the abroad novel automatic doffer. Shanghai Text. Sci. Technol. **53**(2), 60–61 (1991)
7. Zheng, K.: Discussion about developing automatic doffer for the spinning system. Text. Mach. (04) (2010)
8. Wang, H., Fei, H.: Design and analysis of two-way clutch for spinning automatic doffers. Shandong Text. Sci. Technol. **28**(2), 13–15 (2000)
9. Ruan, Y., Suo, S., Jia, J., et al.: The mechanical analysis of the doffing process and the mechanical design of the robot doffer in the ring spinning process. Text. Accessories **40**(4), 1–5 (2013)
10. Li, S., Zhang, Y.: S12 intelligent doffer produces considerable economic benefit. China Textile News, 2015-10-26, 007 (2015)

Robot Mechanism

Design of Wireframe Expansion and Contraction Mechanism and Its Application to Robot

Yuki Takei and Naoyuki Takesue[✉]

Graduate School of System Design, Tokyo Metropolitan University,
6-6 Asahigaoka, Hino-shi, Tokyo, Japan
ntakesue@tmu.ac.jp
http://www.tmu.ac.jp/

Abstract. In this paper, a novel component that enables expansion for wide workspace and contraction for portability is developed. A prototype of potable robot using the expansion and contraction component is fabricated. The robot structure is made of wireframe based on Mandala (Flexi-Sphere), a geometric toy from ancient India. Experiments are carried out and the transformation and the flexible deformation of the robot by wire driving is confirmed. A potential of the robot is shown.

Keywords: Portable robot · Expansion and contraction · Flexible · Wireframe · Transformation

1 Introduction

Recently, many kinds of robots become closer to people such as vacuum cleaning robots, communication robots, drones and so on. Robots that have the portability and the wide workspace as well as the safety are required.

Since an inflatable robot arm using air pressure [1,2] is light-weight and flexible, it is a promising robot close to humans. However, a compressor is needed and air leakage is concerned.

Some robots that use the characteristic of geometric shape are proposed. Most of the geometric robots use the characteristic of changing the shapes. In [3], the robot exterior using Origami theory is changed. A thread-actuated origami robot made with paper was developed in [4]. Moreover, manipulation and locomotion were demonstrated. In [5], the tensegrity structure is used to move by rolling. The designing method that makes it easy to fold up a solid figure is developed [6]. Many kinds of geometric robots are presented which allow flexible locomotion such as an above tensegrity robot [5] and snake-like robots [7,8].

We focused on Mandala's structural transformation. Mandala (Flexi-Sphere) is a geometric toy from ancient India. The original Mandala has 3 degrees of freedom and can transform to several shapes such as sphere, cylinder, disk, and gourd-shaped as shown in Fig. 1. The 3 dof consists of 1 dof at the middle layer for expansion and contraction and 2 dof at the both ends for open and close. In our previous research [9], we increase the dof from 3 to 9 in order to change the shape more variously.

© Springer International Publishing Switzerland 2016
N. Kubota et al. (Eds.): ICIRA 2016, Part I, LNAI 9834, pp. 101–110, 2016.
DOI: 10.1007/978-3-319-43506-0_9

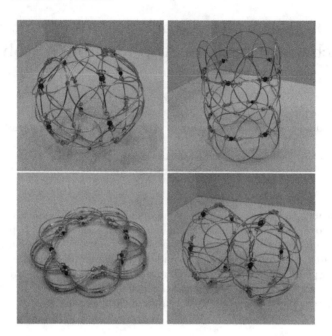

Fig. 1. Mandala – a wireframe toy from ancient India

In this paper, a novel wireframe component that enables expansion for wide workspace and contraction for portability is developed, which is inspired from the original Mandala. A prototype of potable robot using the expansion and contraction component is fabricated. Since the robot structure is made of wireframe, the robot can be light-weight and flexible. Experiments are carried out and the transformation and the flexible deformation of the robot by wire driving is confirmed. A potential of the robot is shown.

2   Parameter Analysis of Mandala

To find the configuration and size appropriate to the specifications for the expansion and contraction component, we parameterize Mandala geometrically. Figure 2 shows the elements of Mandala. We name the center circle of Mandala, which is a perfect circle, "CC" and the semi-circle around CC, which is oval, "SC", respectively. The geometric parameters of Mandala are shown in Fig. 3. The definitions are listed in Table 1.

The appropriate values of parameters are derived to achieve large ratio of expansion and contraction as follows. First, according to the characteristics of Mandala, we assume the following conditions.

$$0 < \alpha < \pi \tag{1}$$

$$0 < \beta < \alpha/2 \tag{2}$$

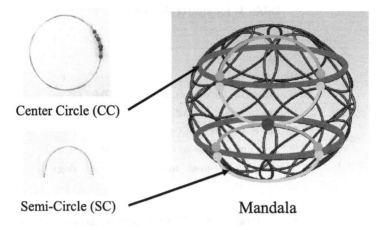

Fig. 2. Elements of Mandala – CC and SC

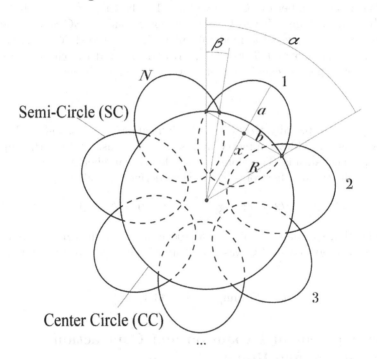

Fig. 3. Parameters for Mandala

In addition, to set the position of joints between CC and SC to be equal interval on CC, and the following equations are assumed.

$$\beta = \alpha/3 \tag{3}$$

$$N(\alpha - \beta) = 2\pi \tag{4}$$

Table 1. Parameter definitions

Symbol	Parameter	Value	Unit
N	Number of SC	9	
R	Radius of CC	50.0	mm
a	Radius of major axis of SC	70.0	mm
b	Radius of minor axis of SC	25.0	mm
x	Distance of centers of CC and SC	43.3	mm
α	SC's occupied angle in CC	60.0	deg
β	SC's duplicative angle in CC	20.0	deg

Next, we consider the number of SC N. Ordinary Mandala has 7 to 9 SCs in a CC. We made some prototypes of Mandala that had the different number of SC. When the number of SC in a CC was 11, the shape of Mandala became relatively hard to change. It was because the interference of SC made it difficult to rotate around CC. For the moderate motion, we decided $N < 12$. Because of the conditions of α and β, Eqs. (1), (3) and (4), and the characteristics of Mandala prototype, the candidates of N were chosen as below:

$$N = 5, 7, 9, 11$$

In case of $N = 9$, especially, the SCs can be divided into 3 parts, and the Mandala becomes symmetry. The symmetry makes it easy to use and set the supports and driving mechanisms. Therefore, we decided the number $N = 9$.

As a result, the numbers of α and β were derived as follows:

$$\alpha = \pi/3 = 60\,[\text{deg}], \qquad \beta = \pi/9 = 20\,[\text{deg}]$$

We decided that the diameter of CC was 100 [mm] (the radius $R = 50$ [mm]) and the radius of major axis of SC was $a = 70$ [mm]. Then, b and x were derived as follows:

$$b = 25\,[\text{mm}] \qquad x = 43.3\,[\text{mm}]$$

3   Development of Expansion and Contraction Mechanism and Robot

3.1   Expansion and Contraction Component

A prototype of expansion and contraction component was fabricated based on the parameters listed in Table 1. The component was made of stainless steel wire whose diameter was $\phi 2.0$. Figure 4(a) shows the top view of the component. Coil springs were used to keep the interval of joints between CC and SC, as shown in Fig. 4(b).

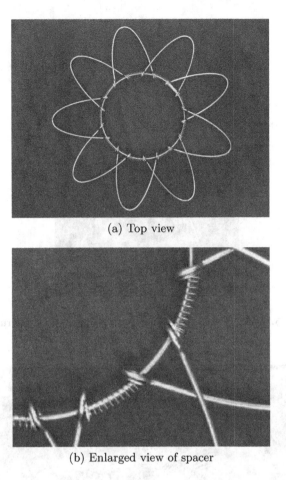

(a) Top view

(b) Enlarged view of spacer

Fig. 4. Appearance of component with coil springs (spacers)

Torsion springs were placed at the connecting joints between SCs, which is illustrated as a blue point in Fig. 2, to avoid the singular point, as shown in Fig. 5. When the prototype robot is cylindrical shape, connecting joints bend slightly out because of the torsion springs.

3.2 Expansion and Contraction Robot

Finally, a robot using four components was developed as shown in Fig. 6. The base was made by using 3D printer. The specifications obtained from the experiment using the actual prototype robot are listed in Table 2. The ratio of expansion and contraction is nine in the direction of z axis. Its representative shapes are (1) disk, (2) cylinder and (3) gourd-shaped, and the pictures are shown in Fig. 7.

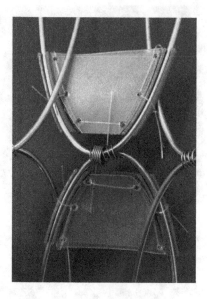

Fig. 5. Appearance of joint between SCs with torsion spring

Fig. 6. Prototype robot (top view of Fig. 7(1))

Table 2. Specifications obtained from the experiment using the actual prototype robot

	Disk	Gourd-shaped	Ratio
Dimension of x and y axis [mm]	$\phi 230$	$\phi 110$	0.48
Size of z axis [mm]	60	540	9.00
Volume [mm³]	$2.49 10^6$	$5.04 10^6$	2.02

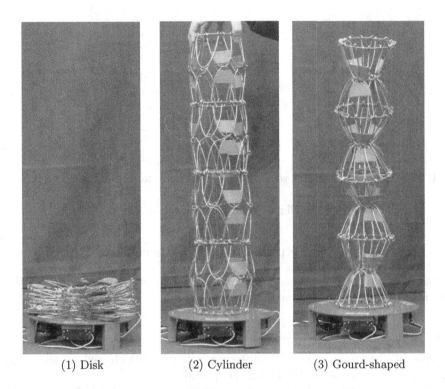

(1) Disk   (2) Cylinder   (3) Gourd-shaped

Fig. 7. Representative shapes of prototype robot (side view)

The prototype robot is motorized to control the motion. Four RC servo motors altered to allow unlimited rotation are employed. The motors rotate reels to pull nylon strings whose diameters are $\phi 0.369$. In other words, the robot is wire driven. The wire arrangements are shown in Fig. 8.

One motor is driven for changing the robot shape to the gourd-shaped. Other three motors are driven to control the tip position of the robot. The experiments are carried out in the following sections.

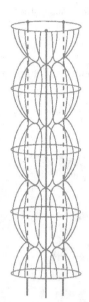

(1) Wire drive for gourd-shaped   (2) Wire drive to control the tip position

Fig. 8. Wire arrangement

4   Experiments of Expansion and Contraction Robot

In this section, some experiments using the prototype robot described in the previous section are carried out.

4.1   Expansion from State of Contraction

In case of the contraction, the prototype robot is disk-shaped as shown in Fig. 7(1), and it is easy to be carried, i.e. portable.

Once the robot is carried to the target place, the robot should be expanded. In this study, the support by human is required to expand the robot as shown in Fig. 7(2). While the human user keeps the robot to be the expansion state, one motor reels the wires up to change the shape from cylinder to gourd-shaped. Finally, the robot transforms to gourd-shaped as shown in Fig. 7(3), and the robot can be self-supported.

4.2   Contraction from State of Expansion

In contrast to the previous subsection, to change the shape from gourd-shaped to cylinder, the motor unfastens the wires. The support by human is needed in the same way. As described in the previous section, torsion springs were employed at the joints between SCs to make it easy to change to contraction. The mechanism, however, didn't work well this time.

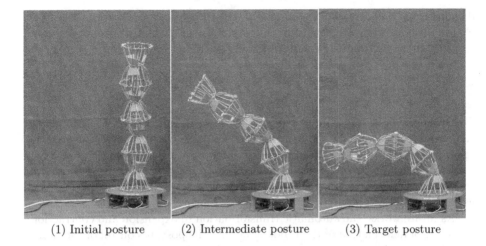

(1) Initial posture   (2) Intermediate posture   (3) Target posture

Fig. 9. Motion of prototype robot

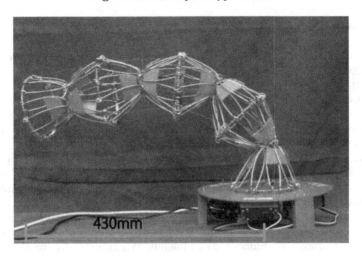

430mm

Fig. 10. Reach of prototype robot

4.3   Deformation Motion in State of Expansion

As mentioned above, when the robot is gourd-shaped, it can be self-supported. Moreover, it is deformable because of mechanical play of joints between CC and SCs. The characteristics can be utilized to control the posture of the prototype robot with three wires. The prototype robot has no sensor such as rotary encoder and tension sensor. The tensions of three wires were manually operated with motors as a preliminary experiment. Figure 9 shows an example of changing postures. Figure 9(1) represents the initial posture of gourd-shape. Figure 9(2) demonstrates the intermediate posture to the target posture shown in Fig. 9(3). Figure 10 shows the reach length at the target posture. The tip of robot is moved to 430 [mm] away from the center of setting position.

5   Conclusions

To develop a light and compact portable robot, we focused on the Mandala's structural transformation and deformation. Mandala was parameterized and the appropriate values were found to achieve large ratio of expansion and contraction. We developed the expansion and contraction component and a prototype robot. Expansion and contraction experiments and deformation motion were carried out, and a potential of the robot was shown.

As future works, the automatic transformation to expansion and contraction, modeling of deformation of the robot structure, the posture control of the robot, and the evaluation of performance are needed.

References

1. Sanan, S., Moidel, J.B., Atkeson, C.G.: Robots with inflatable links. In: Proceedings of the 2009 IEEE/RSJ International Conference on Intelligent Robots and Systems, pp. 4331–4336 (2009)
2. Kim, H.-J., Tanaka, Y., Kawamura, A., Kawamura, S., Nishioka, Y.: Development of an inflatable robot arm systems controlled by a joystick. In: Proceedings of the 24th IEEE International Symposium on Robot and Human Interactive Communication, RO-MAN 2015, pp. 664–669 (2015)
3. Sato, K., Aoki, T.: Development of flexible mobile robot with closed type crawler belt–Design of crawler belt by theory of "Origami"–. In: Proceedings of the 2014 JSME Conference on Robotics and Mechatronics, 2A2-D07 (2014). (in Japanese)
4. Hoff, E.V., Jeong, D., Lee, K.: OrigamiBot-I: a thread-actuated origami robot for manipulation and locomotion. In: Proceedings of the 2014 IEEE/RSJ International Conference on Intelligent Robots and Systems, IROS 2014, pp. 1421–1426 (2014)
5. Hirai, S., Koizumi, Y., Shibata, M., Minghui, W., Bin, L.: Active shaping of a tensegrity robot via pre-pressure. In: Proceedings of the 2013 IEEE International Conference on Advanced Intelligent Mechatronics, AIM 2013, pp. 19–25 (2013)
6. Kase, Y., Mitani, J., Kanamori, Y., Fukui, Y.: Flat-foldable axisymmetric structures with open edges. In: The 6th International Meeting on Origami in Science, Mathematics and Education, 6OSME (2014)
7. Primerano, R., Wolfe, S.: New rolling and crawling gaits for snake-like robots. In: Proceedings of the 2014 IEEE/RSJ International Conference on Intelligent Robots and Systems, IROS 2014, pp. 281–286 (2014)
8. Iwamoto, N., Yamamoto, M.: Jumping motion control planning for 4-wheeled robot with a tail. In: Proceedings of the 2015 IEEE/SICE International Symposium on System Integration, SII 2015, pp. 871–876 (2015)
9. Aka, T., Takesue, N.: Development of flexi-sphere robot. In: Proceedings of the 15th SICE System Integration Division Annual Conference, 2H1-2 (2014). (in Japanese)

Practical Robot Edutainment Activities Program for Junior High School Students

Noriko Takase[1(✉)], János Botzheim[1,2], Naoyuki Kubota[1], Naoyuki Takesue[1], and Takuya Hashimoto[3,4]

[1] Graduate School of System Design, Tokyo Metropolitan University,
6-6 Asahigaoka, Hino, Tokyo 191-0065, Japan
takase-noriko2@ed.tmu.ac.jp, {botzheim,kubota,ntakesue}@tmu.ac.jp
[2] Department of Automation, Széchenyi István University,
1 Egyetem tér, Győr 9026, Hungary
[3] Department of Mechanical Engineering and Intelligent Systems,
Graduate School of Informatics and Engineering, The University
of Electro-Communications, 1-5-1 Chofugaoka, Chofu, Tokyo 182-8585, Japan
tak@rs.tus.ac.jp
[4] Faculty of Engineering, Tokyo University of Science, 6-3-1 Niijuku,
Katsushika, Tokyo 125-8585, Japan

Abstract. In this paper, we describe the approach of the research activities in order to take advantage of the creativity and thinking abilities in practical research of the robot for junior high school female students. The students mainly understand the main idea and the definition of robots and they created them based on information and advice provided by our university and teachers. As a result, the students created the robots by their unique imagination.

Keywords: Robot edutainment · Education

1 Introduction

Recently, the decline in the students' positive attitude towards science has become a problem in Japan. We can realize, that the high difficulty of the science study in junior high schools is one of the main reasons of the decline [6], and the curriculum towards the examination study also makes science harder to enjoy and understand for high school students [4]. According to previous research, we can understand, that mostly the female students' interest and motivation in science has decreased compared to the male students [5]. The Japanese government puts great effort in order to solve this issue by promoting the science subjects to female students.

Yamawaki Junior and Senior High School aims to encourage female students to become active leaders and be a productive member in the society. As one element for achieving this aim, they started the Science Island Project (SI Project) in 2011. In SI Project, they put more effort on experimental trials to visually

© Springer International Publishing Switzerland 2016
N. Kubota et al. (Eds.): ICIRA 2016, Part I, LNAI 9834, pp. 111–121, 2016.
DOI: 10.1007/978-3-319-43506-0_10

show the results of experiments and make it more interesting. The SI Project also plans and implements a variety of scientific efforts. They have some scientific facilities such as various laboratories, outdoor experimental fields, and so on. Every May, they participate in a school trip to Iriomote Island, during the trip they collaborate with a research institute which is located in Iriomote Island. They research and observe the ecology of organisms and plants in the island. Yamawaki Junior and Senior High School collaborates with some universities in the research fields of robotics, biology, computer science, and so on. Tokyo Metropolitan University has been supporting robot research and design, and has been carrying out robot experimental workshops at summer.

Recently, Japan Science and Technology Agency (JST) carried out Science Partnership Program (SPP) [3]. SPP aimed to support observation, experiment, exercise, and problem solving learning activities regarding science, technology, and mathematics. We also received support from the SPP until 2014 and we had been carrying out robot experimental workshops during summer [7,8]. However, in recent years, it did not lead to further study and in order to support the interest of the students in science education, it only provided science activities only for short period of time.

Thus, JST started to implement continuous research activity program "Promotion of Pre-University Research Activities in Science" [2]. This program aims to increase the students' learning motivation and ability to use the future's scientific and technological resources and also to improve the research experiences of teachers through cooperation with different universities and research institutes.

Yamawaki Junior and Senior High School received support for this program from fiscal year 2015 and they cooperated with Tokyo Metropolitan University and with The University of Electro-Communications and they started practical science research activities program that targeted junior high school third grade students. In robotics, biology, computer science, we have been conducting practical science research activities program (scientific research challenge program) for one year that have been carried out by junior high school third grade students and teachers.

In this paper, we describe the research activities approach to take advantage of the creativity and thinking abilities in practical research of the robot that have been carried out by the students and the teachers. The structure of the paper is as follows. In Sect. 2 we introduce and propose an effective approach of the robot research practical creation activities for junior high school students. Section 3 presents the robots that were created by junior high school students. Section 4 shows the process and result of the research program in the first year. Conclusions are drawn in Sect. 5.

2   Approach of Robot Research Practical Creation Activities for Junior High School Students

In this program, junior high school students participated in the research activities. The goal of the program is that the junior high school students acquire skills to identify challenges, to solve problems and to be able to go into further details based on the results that were obtained. The junior high school students decided the structure and the contents of the research activities by themselves and they carried out the activities, and researched the scientific methods.

The practical research creation activities of the robot were carried out by junior high school students at the third grade stage, so they had some limitations. We realized, that we had to teach them how to create and design a robot properly.

We tried to take advantage from the creativity and thinking skills of the students. First, we asked the students about their idea of the robot they want to build, and about their goals they would like to achieve by using the robot. The students used white boards and bill board paper, they collected their ideas, and they planned the robot based on their imagination.

The universities provided the necessary parts for the students and teachers in order to create the robot, based on the student's idea. We supported the preparation of the robot by programming and circuit design. We used the budget paid by JST to purchase parts, equipment, and materials.

First, in the creation activities, at the programming and circuit design, the students and teachers created the robot, based on the reference book and the manual they were reading. We gave advice and recommendations to students and to teachers when they had problems with the robot. However instead of solving their problem, we encouraged the students to find the solution by themselves.

3   Manufactured Robots

One goal of this program is to create a robot in order to support assisting the ecology research in Iriomote Island. Therefore, we planned to create an underwater and a soil investigating robot.

In the robot creation we used Arduino [1], which is a micro computer device being able to handle digital and analogue input and output data, and it is possible to control the device by connecting the breadboard circuit and a variety of input and output devices. Arduino can be controlled by programming in C language. Arduino is easy to be used even for beginners including junior high school students.

3.1   Underwater Robot

The purpose of the underwater robot is to examine the ecology and water quality in the sea and in the river. The functions and elements which are mounted in an underwater robot are as follows: the movement mechanism in the water, the remote control device of the robot, a camera which can take pictures under the water and send the image to the PC, and a container for storing the device and preventing flood.

The movement mechanism contains the following ideas: wheels, obstacle avoidance, left and right turning. We can control the movement mechanism by two motors. The remote control applies an infra-red remote control. The control board of the movement mechanism and the infra-red remote control uses Arduino Uno which connected to a breadboard circuit that incorporates the motor control device and the infra-red receiver. We can switch ON and OFF each of the motor screws by infra-red remote control. In order to remotely send the picture to PC from far away locations such as underwater, we apply a compact wireless camera with LEDs that can be connected to PC with Wi-Fi connection.

In order to store all of the above equipment, we used a mason jar as a container for storing the device and preventing flood. We placed the devices into a mason jar, and sealed it strictly, and we also bonded the screws on the side of the bottle using a sucker in order avoid floating. The complete underwater robot is illustrated in Fig. 1.

Fig. 1. Underwater robot

3.2   Soil Investigating Robot

The purpose of the soil investigator robot is to explore the ecology and microorganisms and status in the soil. In this time, we assume that we dig a hole to certain depth in advance and the robot can explore the composition of the soil. The functions and elements which are mounted into the soil robot are as follows: an elevating device to vertically move the robot in the soil, a container for storing the device, sensors to measure the temperature and humidity in the soil, wireless devices for data reception, and a camera which can take pictures of the soil and send the image to the PC.

LEGO Mindstorms were used as the motivating power of the elevating device. The line that was connected to the device at the head was hooked on the groove of the LEGO Mindstorms' rotary motor. Elevating by the pulley formula was used to elevate the device by the button operation of the control box of LEGO Mindstorms. Moreover, in order not to leave to a certain distance from the soil and in order to facilitate the withdrawal of the device, we created a guide rail using a curtain rail and wood.

A capsule type container was made from plastic is applied into the robot in order to match the width of the guide rail to the container for storing the device. We used Arduino nano as the control device of temperature and humidity sensors and a wireless device in order to downsize the container.

In order to remotely send the picture to PC from far away location like deep down in the soil, we apply a compact wireless camera with LEDs that can be connected to PC with Wi-Fi connection.

The complete soil robot is illustrated in Fig. 2.

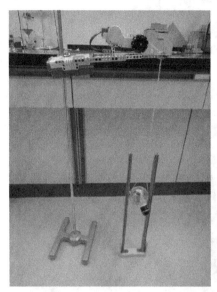

Fig. 2. Soil robot

4   Process and Results of the Robot Research Practical Activities

The first year of this program was conducted from April of 2015 to March of 2016. Eight junior high school female individuals participated, and they were divided into 3 groups. In April, we started with the orientation and the introduction of the program. Based on the students' plan, they collected ideas about the robots they wanted to create. As the target of this program, in order to carry out scientific research activities cooperating with the biological field, which is another research field in this project, we decided to create the robot to be used to support the assisting in the ecology research in Iriomote Island. As a result of this project, the groups created underwater robot and soil investigating robot.

First, the students gathered their ideas about the robots and they wrote them on white board and on bill board papers (Fig. 3). Based on the ideas, junior high school's teachers, university cooperators, and the students proposed

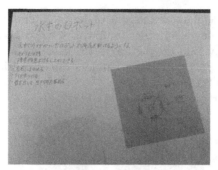

(a) Underwater Robot Idea

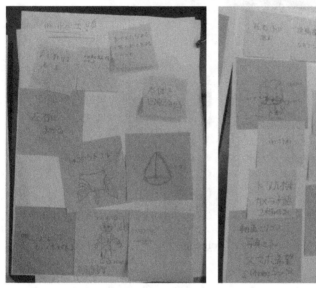

(b) Soil Robot Idea 1

(c) Soil Robot Idea 2

Fig. 3. Robot ideas

and prepared the devices and equipment for implementing the robots within the budgets.

Full-fledged robot research practical creation activities was started in June (Fig. 4). As basic lecture, the students learned programming with a sample program of Arduino.

4.1 Process of Underwater Robot Creation

The underwater robot group successfully created the robot using the available equipments and mounted devices and following the instruction of web pages and reference books (Fig. 5). The students used the following equipments to achieve

Fig. 4. Scene of the robot research practical activities

their goal: they used a waterproof spray to the container for waterproof, they changed and connected short codes in order not to be disconnected from the breadboard circuit, and the device was stored in the mason jar.

The problems were malfunction of wireless camera and malfunction of infra-red receiver and how to store the devices in the mason jar.

In wireless camera, there were the following problems: difficulty of wireless connection with PC, and the duration of the camera's battery. It is recommended that the camera has high-quality wireless connection and long-life battery, but there are issues with the limited budget and compatibility. In infra-red receiver, the following problems occurred: incorrect reactions related to electricity and circuit because the circuit of the receiver was shared with the circuit of the motor driver, and it affected the motor and the circuit including the magnet. These problems were resolved by fixing the program code so the circuit could receive the remote control and could stop the response of the infra-red and moving the motors after a few seconds by making the infra-red circuit be different from the motor circuit. By putting the devices into the mason jar, the following problems occurred: malfunction of camera, direction and arrangement of the circuit and device, disconnection of codes of the circuit. We could not repair the malfunction of the devices because the mason jar was sealed like an envelope in order to prevent the flooding of the devices. We tried out the following method to find the solution: we used short codes, we fixed the codes and the devices, we set infra-red receiver in the position where the receiver can always see the infra-red.

4.2 Process of Soil Robot Creation

The soil investigator robot group successfully created the program and mounted equipments and devices by using manuals, websites and reference books. The students devised the following things to design the soil robot: they proposed the guide rail from themselves in order to facilitate the guide of the movement of the device, they made the foundation of the guide rail using woods, they did experiments outside pro-actively to improve and test the devices in a real natural environment (Fig. 6).

Fig. 5. Creation of underwater robot

During the experiment, the problem was that none of the devices were put in small capsule type container and the width of the guide rail did not fit the container too. When we put all of devices (camera, micro computer, circuit) into the container, the following problems occurred: the container was cramped and the view of the camera was blocked by the container. To resolve this problem, we attached the camera outside of the container. The width of the guide rail was re-adjusted many times to fit the capsule.

Fig. 6. Soil robot experiment

4.3   Presentation of the Result

In this program, it is recommended that the participants present the results of their research activities at the school, inside and outside. This year, we arranged the opportunity to present the result of their research activity in a cultural festival in October and they had to submit a report about their scientific research in March (Fig. 7) in Yamawaki Junior and Senior High School. The students presented the aim, the background and the purpose of their activity, their robots' features. They had to demonstrate their robots, it was also compulsory to show a video about the research, they had to share their impressions, reflections and

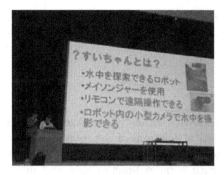

Fig. 7. Reporting of scientific research challenge program

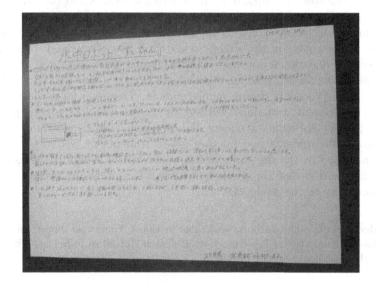

Fig. 8. Succession of the underwater robot

they also had to say a few words about their future work. This was the first time for the students using a PC for their presentation. Even though, the majority of the presentations was proceeded smoothly including other fields group of the research activities program there were also not very good presentations. In the presentation, they presented the following things: they gained the skill to perform and to design, they realized the importance of team work, the perseverance to solve the given problems, and most importantly, they felt the importance and the difficulties of their experiment based on failure and success in the activities. Which leads us to the conclusion, that the students got the interest and the curiosity to take part in more scientific research. At the end of this program, the junior high school third grade students who participated in the program, wrote their robots' features and future work's details of the robot on the bill board paper for the juniors who will participate in the next time in the program in order to support their work (Fig. 8).

5   Conclusion

In this paper, we described the research activities we tested, to take advantage of the creativity and thinking skills in practical research of the robot that have been carried out by the students and the teachers. As the result of the activities, we were able to create robots equipped with a radio functionality and movement functionality by unique ideas coming from junior high school third grade students with the cooperation of universities. Among the activities, it was also realized that the students worked on the research activities and the students are interested in scientific research depending on the topic.

In future works, we are going to improve the robots. In particular, we will do actual use of the robots in the ecology research in Iriomote Island. In addition, as an effective educational approach, it was possible to improve the knowledge and interest in the robotic field among students. The students can work with the learned scientific contents during regular classes. They can form their opinion and ideas in cooperative manner working in groups. We applied this program for junior high school female students. We can apply similar program for junior high school male students and another grade students as well. The students will be interested in the designing appearances of the robots. If we get more detailed data and evaluation of the influence of the program, we need to take the questionnaire to the students.

Acknowledgments. The authors would like to thank Yamawaki Junior and Senior High School, "Promotion of Pre-University Research Activities in Science" by Japan Science and Technology Agency (JST).

References

1. Arduino official homepage. http://arduino.cc/
2. Promotion of pre-university research activities in science. http://www.jst.go.jp/cpse/jissen/
3. Science partnership program. http://www.jst.go.jp/cpse/spp/
4. Gen, E., Shigeo, H., Hiroya, Y.: Development of an educational fish-like 1-DOF gliding locomotion robot with passive wheels: an educational tool to bridge a classroom lecture and a hands-on experience for highschool students. Robot. Soc. Jpn. **31**, 124–132 (2013)
5. Inada, Y.: A practical study on the improvement of girls' feelings and attitudes toward science learning: the case of "electric current" in lower secondary science. J. Res. Sci. Educ. **54**, 149–159 (2013)
6. Naganuma, S.: A study of research trends in "Decline in students' positive attitude toward science": focusing on its current conditions and causes. Jpn. Soc. Sci. Educ. **39**, 114–123 (2015)

7. Narita, T., Tajima, K., Takase, N., Zhou, X., Hata, S., Yamada, K., Yorita, A., Kubota, N.: Reconfigurable locomotion robots for project-based learning based on edutainment. In: Watanabe, T., Watada, J., Takahashi, N., Howlett, R.J., Jain, L.C. (eds.) Intelligent Interactive Multimedia: Systems & Services. SIST, vol. 14, pp. 375–384. Springer, Heidelberg (2011)
8. Takesue, N.: Lessons to make objective robots using arduino microcontrollers. In: Proceedings of the 2014 JSME Conference on Robotics and Mechatronics, Toyama, Japan, 25–29 May 2014

Modeling and Analysis on Position and Gesture of End-Effector of Cleaning Robot Based on Monorail Bogie for Solar Panels

Chengwei Shen[1], Lubin Hang[1(✉)], Jun Wang[1], Wei Qin[1],
Yabo Huangfu[1], Xiaobo Huang[1], and Yan Wang[2]

[1] Shanghai University of Engineering Science, Shanghai 201620, China
hanglb@126.com
[2] Shanghai Jiaotong University, Shanghai 200240, China

Abstract. The dust particles on solar panel surface have a serious influence on the consistency and efficiency of photovoltaic power station, a new cleaning robot based on monorail bogie technology using for automatic cleaning of solar panel is presented in this paper. Position and gesture of the end-effector are critical to the quality and efficiency of work. According to the mechanical structure and motion mechanism of the bogie, five hypotheses which simplify the robot-rails system as a double masses-spring-damper model are proposed. The governing motion equations of the robot during travel process are established, and then the corresponding position and gesture of end-effector within motion range are determined by analyzing dynamics responses of bogie under the input of end-effector's motion. The simulation model under this defined function is built and curves of position and gesture are plotted based on Simulink. A prototype is fabricated and tested by Leica laser tracker, which shows that the position and gesture of the end-effector are related to its working position.

Keywords: Cleaning robot for solar panel · Monorail bogie technology · End-effector · Robot-rails system · Dynamics responses · Position and gesture

1 Introduction

Environment conflict appears particularly prominent along with mineral resource's gradual depletion. As a renewable sources of energy, solar power has been attached more and more importance. Solar industry, which has become one of tendencies in the field of new energy industry, is developing fast in many countries [1]. Solar panel is the core component of the solar power generation system and its photoelectricity conversion efficiency will affect the performance of system directly. At present, research on photoelectricity conversion efficiency of solar panel has concentrated on power generation technology [2] and application of materials [3].

Yet dust particles, in the air, on solar panel also reduce photoelectricity conversion efficiency [4], especially in the large system of PV power station. Under these circumstances, it is necessary to clean solar panel while vibration of panels and generation of water spots should be avoided during cleaning process. Automated cleaning system

© Springer International Publishing Switzerland 2016
N. Kubota et al. (Eds.): ICIRA 2016, Part I, LNAI 9834, pp. 122–133, 2016.
DOI: 10.1007/978-3-319-43506-0_11

of the large photovoltaic array has not yet universal by far, so mechanical design and research on clean of solar panel has great value in application.

For cleaning of large panels and special equipment, robot "Sky wash" [5], carried by vehicle chassis, cleans different types of plane via a multi-joint manipulator, which is used by Lufthansa Flight. Cleanbot-III, designed by City University of Hong Kong, can crawl over the panel and cross the barrier via a chassis composed of vacuum-absorb machines [6]. Fraunhofer Institute for Production Technology composed cleaning technology with rail engineering technology [7] and developed auto-cleaning system SFR I, this machine can do cleaning work while moving guide track, as shown in Fig. 1. In China, Guan [8] designed a solar panel cleaning machine and optimized the structure of manipulator aiming at cleaning problems of solar panel. Ecoppia, an Israeli company that sets out to robotic solar cleaning solution, developed the robot "ECOPPIA E4" in 2014 [9]. This robot combines a powerful, soft microfiber cleaning system with controlled airflow, as is shown in Fig. 2. E4 leverages Eco-Hybrid technology to recover energy when the robot descends the solar panel, and later reuse this energy to optimize performance.

Fig. 1. Auto-cleaning system SFR I

Fig. 2. ECOPPIA E4

In this paper, a robot using flexible bracket of wheels for cleaning solar panel is designed, which can move along the rails of panel support and clean the panel one by one. Quality and efficiency of work can be guaranteed and not affected by potholes on the ground. However, this flexible bracket will affect the position and gesture of robot, under these circumstances, the robot-rails system is simplified as a planar double masses-spring-damper model based on five relative hypothesis and the motion equations of robot during walking status are established. The dynamics responses of the bogie, under the input of end-effector's motion, are analyzed. Then the corresponding position and gesture of end-effector within motion range are determined. A prototype is fabricated, then tested by Leica laser tracker, some conclusions are drawn.

2 Mechanism of Cleaning Robot

The robot using rail-tracked mechanism can guarantee reliability even under poor working conditions. Monorail vehicle has the character of adapting to complex land structure and it becomes the most popular transportation in mountain cities, as shown in

Fig. 3. The monorail bogie, different from wheeled locomotion mechanism in structure [10], moves on the track beam via a pair of running wheels with steering wheels and stabilizing wheels located on the bottom of vehicle body moving on the two sideways of track, as shown in Fig. 4. This ensures the safe and stable operation of vehicle and no risk of running out of the track.

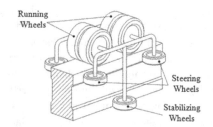

Fig. 3. Monorail transportation   **Fig. 4.** Structure of monorail bogie

The cleaning robot for running on the solar panel support, which can be seen in Fig. 5, consists of groups of wheels, bogie

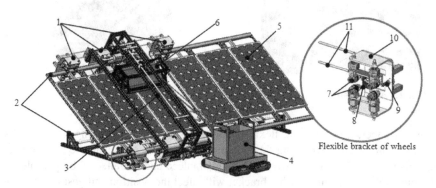

1-Group of wheels 2-Panel support 3-Bogie 4-Water recycle system 5-Solar panel
6-End-effector 7-Clamping wheels 8-Running wheel 9-Springs 10-Bracket 11-Rods

Fig. 5. Assembly of cleaning robot (Color figure online)

and end-effector.

The end-effector is the cleaning device of robot. The movable water tank can supply the end-effector with cleaning liquid and a trio of solar panels is fixed on the panel support. Running wheel and clamping wheels are fitted on the bracket using flexible structure. The close-up which is marked with red circle is shown in Fig. 5. Under work condition, the two-sides clamping wheels contact reliably with metal rails of panel support to guarantee that robot can move directly with running wheels. When the interval of two panel supports is detected, double groups of wheels in front can be stuck out by the extensive rods to stride over the interval. The end-effector, driven by screw which is located on the bogie, reciprocates on the linear guide of bogie according to working requirements and cleans the surface of solar panel.

To realize the optimum performance of the cleaning robot, the horizontal of bogie need to keep parallel to the surface of solar panel approximately. The solar panel to be cleaned is large and travel range of end-effector is long, so the size of load carried by each wheel of bogie will vary with movement of end-effector while tires' contact with metal rail is guaranteed by the spring between each wheel and bogie, which has inevitable consequences for the position and gesture of bogie, and these of end-effector will also be affected. Therefore, these positions and gestures should be analyzed.

3 Kinematic Model of Cleaning Robot

In the field of robot control, parallel manipulators analysis and vehicle dynamics, mechanical system is simplified as a multi-body connected with spring and damper [11] to analyze its characteristic based on spring-damper unit [12]. Position and gesture of end-effector are closely related to its working condition and structural parameters of robot. The robot-rails system is simplified as a mass-spring-damper model and equations of motion for cleaning robot are established to analyze the motion of bogie, then the corresponding position and gesture of end-effector are determined.

3.1 Model Simplification

The cleaning robot for solar panel is a complex mechanical system, five hypotheses are proposed to simplify model as following.

(1) Bogie, screw and linear guide are described as a mass, and masses of wheels are negligible.
(2) End-effector is simplified as a mass, moves on the bogie, and frictional resistance is negligible.
(3) Each wheel contacts with rails all the time, and no slippage of tire happens.
(4) Yaw angle of bogie is small.

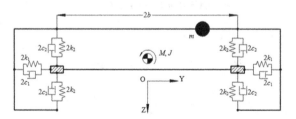

Fig. 6. Robot-rails system

(5) Horizontal and vertical movements of bogie are decoupled [13].

The angle, between panel of most PV power station and ground, is small so the panels can be considered parallel to the ground. During working process of the cleaning robot, the bogie travels at a smooth and low speed. Bogie and rails of panel supports remain relatively constant and static without considering that bogie strides over the interval of panel supports. Ignoring the tire dynamics performance [14], the wheel contacts with rails by the spring, this can be described as a point contact. Then the connection between the point and bogie is represented by a spring-damper unit [15], which is a linear conceptual model. Based on the five hypotheses above, the robot-rails system is simplified as a planar double masses-spring-damper model, as shown in Fig. 6.

In this model, the bogie has three freedoms of translation along Y, translation along Z and rotation about X and the end-effector has only one freedom of translation Y on the linear guide of bogie. M is mass of bogie, J is rotary inertia of bogie about its center of mass, m is mass of end-effector, $2b$ is length of linear guide, k_1 is equivalent stiffness of running wheel, c_1 is equivalent damping of running wheel, k_2 is equivalent stiffness of clamp wheel, c_2 is equivalent damping of running wheel. Furthermore, motion of end-effector on the bogie is the main factor affecting position and gesture of bogie. The corresponding position and gesture of end-effector can be determined by calculating dynamics responses of bogie under the input of end-effector's motion based on the equations of motion for robot.

3.2   Equations of Motion for Robot

Relative displacement of end-effector to bogie can be expressed as

$$y_m - y = S(t) \tag{1}$$

Where $S(t)$ can be described as the motion excitation of the robot-rails system.

According to Newton's laws of motion, the differential equations of motion for robot can be determined as follows.

Horizontal Motion of Robot

As shown in Fig. 7, the static equilibrium position is chosen to be the initial position. By letting y be the displacement of M, y_m be the displacement of m from the static equilibrium, then

$$M\ddot{y} + m\ddot{y}_m = f_{1L} + f_{1R} \tag{2}$$

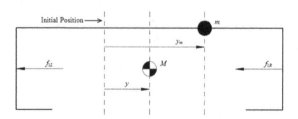

Fig. 7. Robot model under horizontal force

Where f_{1L} and f_{1R} are the horizontal equivalent spring-damper force carried by bogie, and

$$f_{1L} = f_{1R} = -2c_1\dot{y} - 2k_1 y \tag{3}$$

Substituting Eq. (1) into Eq. (2), then Eq. (2) can be written as

$$(M+m)\ddot{y} + 4c_1\dot{y} + 4k_1 y = -m\ddot{S}(t) \tag{4}$$

According to Duhamel's principle [16], the response of the system for the zero initial condition is

$$y(t) = \frac{1}{(M+m)\omega_{d1}} \int_0^t \left(-m\ddot{S}(t)\right)e^{-\zeta_1\omega_{n1}(t-\tau)} \sin \omega_{d1}(t-\tau)d\tau \tag{5}$$

Where $\omega_{n1} = \sqrt{\frac{4k_1}{M+m}}$, $\zeta_1 = \frac{4c_1}{2(M+m)\omega_{n1}}$ and $\omega_{d1} = \sqrt{1-\zeta_1^2}\omega_{n1}$.

Vertical Motion of Robot

As shown in Fig. 8, the position that the robot statically contacts the rails of panel support is chosen to be the initial position. By letting z be the displacement and φ be the rotation of bogie from the initial position. $\sin\varphi \approx \varphi$ and $\cos\varphi \approx 1$ due to the small yaw angle of bogie, then

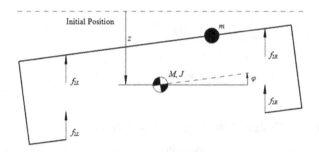

Fig. 8. Robot model under vertical force

Equation of vertical motion for robot can be written as

$$(M+m)\ddot{z} = 2f_{2L} + 2f_{2R} + (M+m)g \tag{6}$$

Where f_{2L} and f_{2R} are the vertical equivalent spring-damper force carried by bogie, and

$$\begin{cases} f_{2L} = -2k_2(z+b\phi) - 2c_2\left(\dot{z}+b\dot{\phi}\right) \\ f_{2R} = -2k_2(z-b\phi) - 2c_2\left(\dot{z}-b\dot{\phi}\right) \end{cases} \tag{7}$$

Then Eq. (6) can be written as

$$(M+m)\ddot{z}+8c_2\dot{z}+8k_2z = (M+m)g \tag{8}$$

The general solution is given by the equation

$$z(t) = A_1e^{s_1t} + A_2e^{s_2t} + \frac{(M+m)g}{8k_2} \tag{9}$$

Where A_1 and A_2 are the coefficients to be determined, and

$$s_{1,2} = -\frac{8c_2}{2(M+m)} \pm \sqrt{\left(\frac{8c_2}{2(M+m)}\right)^2 - \frac{8k_2}{M+m}}.$$

Equation of rotation for robot can be written as

$$J\ddot{\phi} = 2f_{2L}b - 2f_{2R}b - mg(y_m - y) \tag{10}$$

Substituting Eqs. (1) and (7) into Eq. (11), then

$$J\ddot{\phi} + 8c_2b\dot{\phi} + 8k_2b^2\phi = -mgS(t) \tag{11}$$

The response can be written as

$$\phi(t) = \frac{1}{J\omega_{d2}} \int_0^t (-mS(t))e^{-\zeta_2\omega_{n2}(t-\tau)} \sin\omega_{d2}(t-\tau)d\tau \tag{12}$$

Where $\omega_{n2} = \sqrt{\frac{8k_2b^2}{J}}$, $\zeta_2 = \frac{8c_2b}{2J\omega_{n2}}$ and $\omega_{d2} = \sqrt{1-\zeta_2^2}\omega_{n2}$.

3.3   Position and Gesture Modeling for End-Effector

Position and gesture of bogie will affect these of end-effector. Based on the proposed model above, the position and gesture of end-effector can be expressed as

$$\begin{bmatrix} y_m \\ z_m \\ \phi_m \end{bmatrix} = \begin{bmatrix} 1 & & \\ & 1 & -S(t) \\ & & 1 \end{bmatrix} \begin{bmatrix} y \\ z \\ \phi \end{bmatrix} + \begin{bmatrix} S(t) \\ 0 \\ 0 \end{bmatrix} \tag{13}$$

For the end-effector, horizontal displacement y_m, vertical displacement z_m and yaw angle ϕ_m are the corresponding position and gesture of it within motion range, which can be determined in Eq. (13) by substituting the initial conditions and prosperities.

4 Simulation

4.1 Input of End-Effector's Motion

Relative displacement of end-effector to bogie is the main factor of translation and motion of bogie. The end-effector reciprocates on the linear guide located on the bogie and process of one motion trip is divided into three stages: accelerated motion, uniform motion and decelerated motion, which can be described via a piecewise function. So the mathematic function of one motion trip can be expressed as

$$S(t) = \begin{cases} -b + \frac{1}{2}at^2 & t_1 \leq t < t_2 \\ v(t-6) & t_2 \leq t < t_3 \\ b - \frac{1}{2}a(t-12)^2 & t_3 \leq t \leq t_4 \end{cases} \tag{14}$$

Where a is relative acceleration of end-effector to bogie and $a = 0.1$ m/s^2, v is relative velocity of end-effector to bogie and $v = 0.1$ m/s, b is one-second of length of linear guide and $b = 0.55$ m. t_1, t_2, t_3 and t_4 are chosen to be 0 s, 1 s, 11 s and 12 s. The curve of function is plotted as shown in Fig. 9.

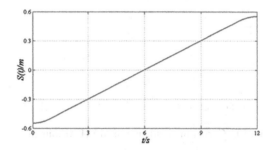

Fig. 9. Input of end-effector's motion

4.2 Simulink Model

Simulink in MATLAB is usually applied in the dynamics modeling and simulation. Based on Eqs. (4), (8) and (11), the simulation model is established in the Simulink. The simulation parameters: $M = 75$ kg, $J = 100$ kg·m^2, $m = 25$ kg, $k_1 = 1200$ N/m, $c_1 = 200$ Ns/m, $k_2 = 3000$ N/m, $c_2 = 300$ Ns/m. The duration of simulation is 12 s, position and gesture of bogie and end-effector, are determined with ODE 45 algorithm and curves of simulation are plotted, as shown in Fig. 10.

Under the defined working function of end-effector, the simulation results show that small change of Y displacement of bogie takes place when end-effector accelerates and decelerates. The change of Z displacement of bogie is unrelated to the motion of end-effector. Yaw angle of end-effector varies with rotation of bogie and the maximum of angel appears when end-effector locates on both ends of the linear guide of bogie. The corresponding position and gesture of end-effector is related to its working position.

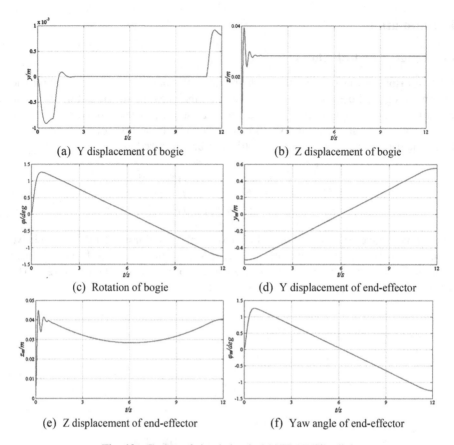

Fig. 10. Curves of simulation in MATLAB/Simulink

5   Experiment

A prototype of cleaning robot for solar panel is built based on the structural parameters, as is shown in Fig. 11(a), and details of the group of wheels can be seen in Fig. 11(b). After debugging PLC control system based on the working function of end-effector above, the prototype of cleaning robot can stride over the interval between panel supports without interference and end-effector washes up the surface of solar panels automatically according to the design requirements.

To test operation parameters of the prototype of cleaning robot and compare with the theory value to make further analysis, the Leica laser tracker is used to measure the yaw angle of end-effector, as is shown in Fig. 12. In addition, the reflector, as is show in Fig. 13, is the "probe" of the tracker. In the course of the experiment, the reflector is put on three different points of bogie, respectively. The coordinates of them are measured followed. Then the plane of bogie, when the end-effector moves on the linear guide of it, can be determined by the following equation:

$$
\begin{vmatrix} y_2 - y_1 & y_3 - y_1 \\ z_2 - z_1 & z_3 - z_1 \end{vmatrix} \cdot (x - x_1) + \begin{vmatrix} z_2 - z_1 & z_3 - z_1 \\ x_2 - x_1 & x_3 - x_1 \end{vmatrix} \cdot (y - y_1)
$$
$$
+ \begin{vmatrix} x_2 - x_1 & x_3 - x_1 \\ y_2 - y_1 & y_3 - y_1 \end{vmatrix} \cdot (z - z_1) = 0 \tag{15}
$$

Where (x_1, y_1, z_1), (x_2, y_2, z_2) and (x_3, y_3, z_3) are the coordinates of these three different points.

(a) Robot in working condition   (b) Close-up of the group of wheels

Fig. 11. Prototype of cleaning robot

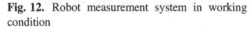

Fig. 12. Robot measurement system in working condition

Fig. 13. Close-up of the group of wheels

The three planes of bogie when end-effector is on the beginning position, the middle position and the end position of linear guide are measured. The measurement data is shown in Table 1. The rotation angle of bogie can be calculated as follows

$$
\theta = \arccos\left(\frac{|A_i A_j + B_i B_j + C_i C_j|}{\sqrt{A_i^2 + B_i^2 + C_i^2} \cdot \sqrt{A_j^2 + B_j^2 + C_j^2}} \right), 0° \leq \theta \leq 180° \tag{16}
$$

Where (A_i, B_i, C_i) and (A_j, B_j, C_j) are the normal vectors of the plane π_i and π_j, which are determined by Eq. (16). Moreover, this angle is also the yaw angle of end-effector according to analysis in Part 3.

Table 1. Experimental data

Position	Point	x/mm	y/mm	z/mm
Beginning	1st	2629.49	1652.31	−565.69
	2nd	2414.81	1698.09	−563.50
	3rd	2161.88	480.73	−660.74
Middle	1st	2629.96	1653.76	−569.12
	2nd	2415.01	1699.58	−566.90
	3rd	2164.62	484.25	−658.84
End	1st	2630.78	1654.59	−571.56
	2nd	2417.63	1699.91	−569.42
	3rd	2165.17	486.25	−657.59

After substituting these data above into Eqs. (15) and (16), the yaw angles of end-effector on the beginning position and the end position are 0.17° and 0.24°, respectively, which are smaller than the results obtained in Part 4. The reason for this situation is that the springs which have bigger stiffness coefficients are assembled to prevent the bogie from turning out of metal rails of panel support. The yaw angle of end-effector becomes smaller, consequently. Through experiments with a series of springs, the stiffness and damping coefficients of springs affect the position and gesture of end-effector. Intuitively, the conclusion that gesture of end-effector changes as its working position changes can be obtained.

6 Conclusions

(1) As the solution to automatic cleaning of solar panel, a cleaning robot based on monorail bogie is described and its operational principle is presented.
(2) Five hypotheses are proposed based on the mechanical structure and working condition of bogie, then the robot-rails system is simplified as a double masses-spring-damper model and equations of motion for cleaning robot are established.
(3) The general solutions of differential equations of motion for robot are given, position and gesture of end-effector are determined.
(4) The simulation model under the defined working function of end-effector is built using Simulink. The results show that motion of end-effector has a minimal effect on the position of bogie and the position and gesture of end-effector are related to its working position.
(5) A prototype is fabricated, and then tested using the Leica laser tracker. As a consequence, the results of experiment have proved that the design of this cleaning robot and the analysis model are correct and practical.

Acknowledgements. The authors would like to acknowledge the financial support of the Natural Science Foundation of China under Grant 51475050, Shanghai Science and Technology Committee under Grant 12510501100, the Natural Science Foundation of Shanghai City under Grant 14ZR1422700, and the Technological Innovation Project of Shanghai University of Engineering Science under Grant E1-0903-15-01012.

References

1. Chen, F., Wang, L.: On distribution and determinants of PV solar energy industry in China. Resour. Sci. **34**(2), 287–294 (2012)
2. Mastromauro, R.A., Liserre, M., Dell'Aquila, A.: Control issues in single-stage photovoltaic systems: MPPT, current and voltage control. IEEE Trans. Ind. Inform. **8**(2), 241–254 (2012)
3. Lu, L., Xu, T., Chen, W., et al.: The role of N-doped multiwall carbon nanotubes in achieving highly efficient polymer bulk heterojunction solar cells. Nano Lett. **13**, 2365–2369 (2013)
4. Guangshuang, M., Dedong, G., Shan, W., et al.: Mechanics modeling of dust particle on solar panel surface in desert environment. Trans. Chin. Soc. Agric. Eng. **30**(16), 221–229 (2014)
5. Schraft, R.D., Wanne, M.C.: The aircraft cleaning robot "SKYWASH". Ind. Robot **20**(6), 21–24 (1993)
6. Zhu, J., Sun, D., Tso, S.K.: Development of a tracked climbing robot. J. Intell. Robot. Syst. **35**(4), 427–443 (2002)
7. Bräuning, U., Orlowski, T., Hornemann, M.: Automated cleaning of windows on standard facades. Autom. Constr. **9**(5–6), 489–501 (2000)
8. Shixue, G.: Design and Optimization of the Manipulator Parts of Cleaning Machine of Solar Panel. Lanzhou University of Technology (2014)
9. The homepage of Ecoppia Empowering Solar. http://www.ecoppia.com/ecoppia-e4
10. Liqun, P.E.N.G., Dawen, L.I.N., Xinglei, W.U., et al.: Experimental design and research of straddle-type monorail vehicle bogie traction mechanism. Railw. Locomotive Car **34**(2), 70–73 (2014)
11. Ivanchenko, I.: Substructure method in high-speed monorail dynamic problems. Mech. Solids **43**(6), 925–938 (2008)
12. Bruni, S., Goodall, R., Mei, T.X., et al.: Control and monitoring for railway vehicle dynamics. Veh. Syst. Dyn. **45**(7), 743–779 (2007)
13. Wanming, Z.: Vehicle-Track Coupling Dynamics, 3rd edn. Science Press, Beijing (2007)
14. Alkan, V., Karamihas, S.M., Anlas, G.: Experimental analysis of tire-enveloping characteristics at low speed. Veh. Syst. Dyn. **47**(5), 575–587 (2009)
15. Jun, Ma., Kun, Y.: Modeling simulation and analysis of radial spring tire model. Agric. Equip. Veh. Eng. **52**(12), 33–37 (2014)
16. Meirovitch, L.: Fundamentals of Vibrations. The McGraw-Hill Companies Inc., New York (2001)

PASA Hand: A Novel Parallel and Self-Adaptive Underactuated Hand with Gear-Link Mechanisms

Dayao Liang, Jiuya Song, Wenzeng Zhang[✉], Zhenguo Sun, and Qiang Chen

Department of Mechanical Engineering, Tsinghua University, Beijing 100084, China
wenzeng@tsinghua.edu.cn

Abstract. This paper proposes a novel underactuated robotic finger, called the PASA finger, which can perform parallel and self-adaptive (PASA) hybrid grasping modes. A PASA hand is developed with three PASA fingers and 8 degrees of freedom (DOFs). Each finger in the PASA hand has two joints, mainly consists of an actuator, an accelerative gear system, a spring, a parallel four-link mechanism and a mechanical limit. Two extra actuators in the palm of the PASA hand independently control the base rotation of two fingers. The PASA hand executes multiple grasping modes based on the dimensions, shapes and positions of objects: (1) a parallel pinching (PA) grasp for precision grasp like industrial grippers; (2) a self-adaptive (SA) enveloping grasp for power grasp like traditional self-adaptive hands; (3) a parallel and self-adaptive (PASA) grasping mode for hybrid grasp like human hands. The switch through different grasping modes is natural without any sensors and control. Kinematics and statics show the distribution of contact forces and the switch condition of PA, SA, and PASA grasping modes. Experimental results show the high stability of the grasps and the versatility of the PASA hand. The PASA hand has a wide range of applications.

Keywords: Robot hand · Underactuated finger · Parallel and self-adaptive grasp

1 Introduction

A human hand is the most flexible and stable gripper until now. It can both grasp small and flat objects by using its distal phalanges, and grasp large objects by enveloping all the phalanges. With many degrees of freedom (DOFs), a human hand can realize different grasp modes.

To imitate human hand, dexterous hands were designed and developed, which have many joints and driven by many motors. The Utah/MIT Dexterous Hand [1], Stanford/JPL Hand [2], DLR Hand [3] and Gifu hand II [4] are outstanding examples. Soft robot hands [5] are also highly researched.

However, due to the limitation of mechanical dimension and grasping force need, it is really difficult to develop a robot hand which has as many as grasping modes as human hand and large grasping force. Furthermore, the development of robot hands is also limited by control scheme and sensor system.

To deal with these difficulties, underactuated hands were developed. In the research area of robotic hands, the underactuated mechanism is defined as: the number of

© Springer International Publishing Switzerland 2016
N. Kubota et al. (Eds.): ICIRA 2016, Part I, LNAI 9834, pp. 134–146, 2016.
DOI: 10.1007/978-3-319-43506-0_12

actuators is less than the number of degrees of freedom (DOFs). Underactuated hands have been researched since 1970s. Many kinds of underactuated hands have been developed to match the different needs in industrial applications, such as SARAH hand [6], tendon mechanisms hand [7], multi-finger underactuated hand [8], passive adaptive underactuated hand [9], underactuated prosthetic hand [10], and coupled self-adaptive hand [11]. Some methods are also used to study underactuated hands, see [12–14].

Traditional underactuated hands successfully solve the difficulties of controlling; but they only have few grasping modes. Laval University developed an underactuated hand [15] which can both realize pinching grasp and enveloping grasp. This hybrid grasping mode is proved to be a great success in multiple grasping tasks. [15] presented another underactuated grasper with multiple grasping mode.

Hybrid grasping mode combines the flexibility of dexterous hands and the stability and simplicity of traditional underactuated hands. Therefore, it is a popular research direction nowadays. This paper introduces a parallel and self-adaptive underactuated finger with a novel gear-link mechanism. Depending on the dimension, position and shape of objects, before the contact force is applied on the proximal phalanx, it executes parallel motion. After the force is applied, it envelopes the object. If the force is applied on the distal phalanx, the parallel mechanism remains unchanged and executes a pinching grasp. Different from the existing hands with the hybrid grasping mode, the PASA finger uses one actuator to control two joints, uses underactuated mechanism to switch different grasping modes automatically.

The second part of this paper introduces the concept of the parallel and self-adaptive (PASA) underactuated hand; the third part analyzes the structure, grasping-force distribution and the condition of pinching grasp and enveloping grasp of the PASA hand; the forth part shows the experimental results of different grasping processes.

2   Concept of Parallel and Self-Adaptive Underactuated Hand

Self-adaptive underactuated hands are suitable for grasping many kinds of objects. But most of them are difficult to grasp objects which are thin or small, as they can only process an enveloping grasp. When the objects are thin or small, the point contacts between the objects and phalanges are not stable.

Pinching is a very common way to grasp objects for humans. It can hold objects tightly such as paper and coin. As pinching and enveloping have their own advantages, it will improve its versatility if a robotic hand has both these two grasping modes.

As Fig. 1 shows, the parallel and self-adaptive underactuated grasping mode is a hybrid of the traditional self-adaptive enveloping grasp and the pinching grasp. Depending on the dimension, position and the shape of objects, the finger with the parallel and self-adaptive underactuated grasping mode executes an enveloping grasp or a parallel pinching grasp. The SARAH hand [16] designed by Laval University used this hybrid grasping mode, which proves its high value for industrial and civil applications.

To realize this hybrid grasping mode, the turn point from pinching grasp to enveloping grasp is the key. Before the lower part of the finger touches the object, the distal phalange and the base should form a parallel mechanism. As long as a contact force

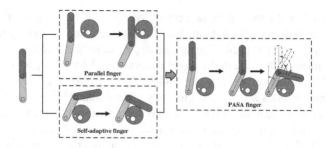

Fig. 1. Concept of parallel and self-adaptive underactuated finger.

applies on the lower part, the grasping mode transition is triggered, and a part of the parallel mechanism is fixed to the base. Hence, the parallel mechanism should be destroyed to execute an enveloping grasp. The next part of this paper will introduce the structure of a novel parallel and self-adaptive underactuated hand with gear-link mechanisms.

3   Architecture and Analysis of the PASA Hand

3.1   The Self-adaptive Rotation and Translation Systems of the PASA Finger

The translation system mainly consists of a parallel wheel, a link, a proximal phalanx, a distal phalanx, a gear set, a driving gear and a driven gear, which is fixed to the distal phalanx, as Fig. 2a shows. The driving force is transferred from the driving gear to the driven gear by the gear set. The connecting lines among the rotary axis of parallel wheel, proximal phalanx, link and distal phalanx form a parallelogram. If the parallel wheel does not rotate, the link keeps the distal phalanx always parallel to the last moment of the distal phalanx action.

The underactuated self-adaptive rotation system mainly includes a parallel wheel, a limiting block, a sliding block, a steel wire, a spring and an actuator. As it is shown in Fig. 2b, the sliding block is fixed to the parallel wheel and the limiting block is fixed to

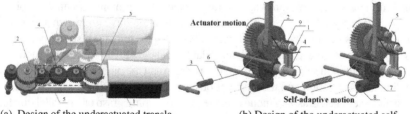

(a) Design of the underactuated translation system. 1-distal phalanx; 2- driving
gear; 3-driven gear; 4-gear set; 5-link.

(b) Design of the underactuated self-
adaptive rotation system. 1-proximal shaft;
2- parallel wheel; 3-spring; 4-limiting
block; 5-sliding block; 6-steel wire; 7-
worm wheel; 8-worm; 9- torsional spring.

Fig. 2. Architecture of the PASA Hand (Color figure online)

the base. One end of the steel wire is connected to the spring, the other end is connected to the parallel wheel. The driving force is transmitted to the driving gear through a worm, a worm wheel and other transmission mechanism. In the initial state, the sliding block abuts the limiting block by the force of spring. Once the proximal phalanx touches an object, the ensemble of gear set is stopped, but each gear continues rotating. The parallel wheel rotates in the direction of the purple arrow in Fig. 2b because of the translation system. As a result, a self-adaptive motion is produced while the actuator motion is taking place.

Based on the above analysis, take the ensemble of gear set as a reference system, the driven has two rotary directions: the green arrow direction in Fig. 2a (self-adaptive enveloping grasp) and the red arrow direction in Fig. 2a (parallel pinching grasp).

Figure 3 illustrates the function of the torsional spring. The torsional spring is added to enhance the grasping forces, which can also eliminate the effects of gear clearances.

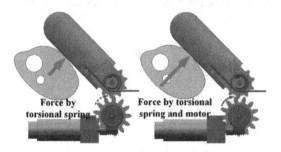

Fig. 3. Function of the torsional spring.

3.2 Integrated Design of the PASA Hand

The structure of the PASA finger is shown in Fig. 4. Each PASA finger mainly consists of a parallel wheel, a link, two phalanges, two shafts, a driving gear, a driven gear, a gear set, a sliding block, a limiting block, a spring and a torsional spring.

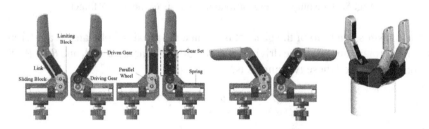

Fig. 4. The motion and structure of the PASA finger and the PASA Hand.

In the initial-state, the sliding block contacts with the limiting block. The spring keeps the sliding block contact with the limiting block. The link keeps the distal phalanx always parallel to the last moment of the distal phalanx action since the sliding block is immobile.

As Fig. 4 shows, the proximal phalanx rotates forward since the driving gear rotates forward. The distal phalanx is parallel to the last moment of the distal phalanx because of the spring force, which is pinching process. Once the proximal phalanx touches an object, the driving gear continually rotates and drives the distal phalanx rotating forward. As a result, the parallel wheel rotates and the spring is strained, which is enveloping process.

3.3 Grasping-Force Distribution Analysis

This part focus on the grasping forces distribution of the PASA finger. For the reason of simplification, the gravity force of the finger and objects and the friction between phalanges are neglected, and the contact forces are applied on points.

As Fig. 5 shows, β is the rotary angle of the driving gear, θ_1 and θ_2 are the rotary angles of the proximal phalanx and the distal phalanx. The rotary angle of the parallel wheel is same as the distal phalanx because of the transmission link.

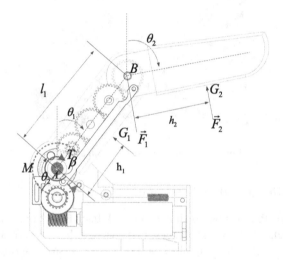

Fig. 5. Grasping-forces distribution analysis of the PASA finger.

The transmission ratio of the gear set is a. Take the proximal phalanx as a reference system, the rotary angles of the driving gear and driven gear are $\beta - \theta_1$ and $\theta_2 - \theta_1$. The relationship between these two angles is:

$$a(\beta - \theta_1) = \theta_2 - \theta_1 \tag{1}$$

Or

$$\theta_2 + (a - 1)\theta_1 = a\beta \tag{2}$$

When the finger executes a pinching grasp, $\theta_2 = 0$, and θ_1 can be described as:

$$\theta_1 = a\beta/(a - 1) \tag{3}$$

To realize a self-adaptive grasp, the rotary angle of proximal phalanx must be positive. One achieves $a > 1$.

As it is shown in Fig. 5, M and T are the torques produced by the spring and driving gear, $\vec{F}_1$ and $\vec{F}_2$ are the contact force vectors by the objects.

$$\vec{F}_1 = (-F_1 \cos \theta_1, F_1 \sin \theta_1) \tag{4}$$

$$\vec{F}_2 = (-F_2 \cos \theta_2, F_2 \sin \theta_2) \tag{5}$$

G_1, G_2 are the contact points. h_1 and h_2 are the distances between the contact points and the rotary axis. The vectors of the contact points $\vec{G}_1$, $\vec{G}_2$ can be described as

$$\vec{G}_1 = (h_1 \sin\theta_1, \ h_1 \cos\theta_1) \tag{6}$$

$$\vec{G}_2 = (l_1 \sin\theta_1 + h_2 \sin\theta_2, l_1 \cos\theta_1 + h_2 \cos\theta_2) \tag{7}$$

Since the rotary angle of the parallel wheel is also θ_2, the spring torque M is in direct proportion with θ_2. We use $k\theta_2 + M_0$ to represent M.

According to Lagrange's equation, one obtains

$$-\frac{\delta L}{\delta \vec{q}} = \vec{0} \tag{8}$$

Where

$$L = -V = -[-T, \frac{1}{2}k\theta_2 + M_0]\begin{bmatrix} \beta \\ \theta_2 \end{bmatrix} + [\vec{F}_1, \vec{F}_2]\begin{bmatrix} \vec{G}_1^t \\ \vec{G}_2^t \end{bmatrix} \tag{9}$$

Combine Eqs. (8) and (9), one achieves

$$[-T, k\theta_2 + M_0]\begin{bmatrix} \delta\beta \\ \delta\theta_2 \end{bmatrix} = [\vec{F}_1, \vec{F}_2]\begin{bmatrix} \delta\vec{G}_1^t \\ \delta\vec{G}_2^t \end{bmatrix} \tag{10}$$

Where

$$\delta\vec{G}_1 = (h_1 \cos \theta_1, \ -h_1 \sin \theta_1)\delta\theta_1 \tag{11}$$

$$\delta\vec{G}_2 = (l_1 \cos \theta_1 \delta\theta_1 + h_2 \cos \theta_2 \delta\theta_2, -l_1 \sin \theta_1 \delta\theta_1 - h_2 \sin \theta_2 \delta\theta_2) \tag{12}$$

$$[-T, \ \tau\theta_2 + M_0]\begin{pmatrix} \delta\beta \\ \delta\theta_2 \end{pmatrix} = [-T, \ k\theta_2 + M_0]\begin{bmatrix} \frac{a-1}{a} & \frac{1}{a} \\ 0 & 1 \end{bmatrix}\begin{bmatrix} \delta\theta_1 \\ \delta\theta_2 \end{bmatrix} \tag{13}$$

$$[\vec{F}_1, \vec{F}_2]\begin{pmatrix} \delta\vec{G}_1' \\ \delta\vec{G}_2' \end{pmatrix} = [F_1, F_2]\begin{bmatrix} -h_1 & 0 \\ -l_1\cos(\theta_2 - \theta_1) & -h_2 \end{bmatrix}\begin{bmatrix} \delta\theta_1 \\ \delta\theta_2 \end{bmatrix} \tag{14}$$

Here we introduce two Jacobian matrix

$$\mathbf{J}_1 = \begin{bmatrix} \dfrac{a-1}{a} & \dfrac{1}{a} \\ 0 & 1 \end{bmatrix} \tag{15}$$

$$\mathbf{J}_2 = \begin{bmatrix} -h_1 & 0 \\ -l_1\cos(\theta_2 - \theta_1) & -h_2 \end{bmatrix} \tag{16}$$

Finally, one obtains the relationship between spring force, driving force and contact forces:

$$[F_1, F_2] = [-T, k\theta_2 + M_0]\mathbf{J}_1\mathbf{J}_2^{-1} \tag{17}$$

Where

$$\mathbf{J} = \mathbf{J}_1\mathbf{J}_2^{-1} = \begin{bmatrix} -\dfrac{1}{h_1}(1 - \dfrac{1}{a} - \dfrac{l_1\cos(\theta_2 - \theta_1)}{ah_2}) & -\dfrac{1}{ah_2} \\ \dfrac{l_1\cos(\theta_2 - \theta_1)}{h_1h_2} & -\dfrac{1}{h_2} \end{bmatrix} \tag{18}$$

From Eq. (18), one can study the relation between the contact forces and the distances of the contact points. As Fig. 6 shows, obviously, with the distances augment, the corresponding contact forces decrease.

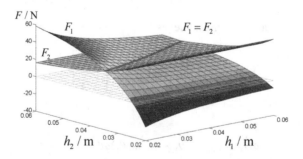

Fig. 6. Contact-force distribution by h_1, h_2, where $\theta_1 = 30^0, \theta_2 = 60^0, T = 3\mathrm{N} \cdot \mathrm{m}, a = 3, l_1 = 0.06\mathrm{m}, M_0 = 0.006\mathrm{N} \cdot \mathrm{m}, k = 0.0344\mathrm{N} \cdot \mathrm{m}$.

The surfaces F_1 and F_2 intersect in a curve. It is almost a straight line, which means the contact forces are well-distributed.

If h_2 is too small, F_1 will become negative, which is an interesting phenomenon. Figure 7 shows this "retraction" phenomenon.

Fig. 7. "Retraction" phenomenon of a finger.

The forces distribution is also related to the rotary angles of two phalanges. If h_1, h_2, T, M_0, k, a, l_1 are fixed, F_2 has nothing to do with θ_1, and it only depends on θ_2. However, F_1 depends on both two rotary angles. As Fig. 8 shows, compared to F_1, F_2 is almost a constant.

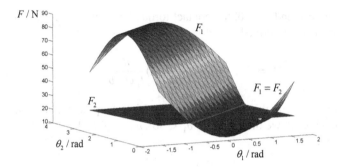

Fig. 8. Contact-force distribution by θ_1, θ_2, where $h_1 = 0.04\text{m}, h_2 = 0.04\text{m}, T = 3\text{N} \cdot \text{m}, a = 3, l_1 = 0.06\text{m}, M_0 = 0.006\text{N} \cdot \text{m}, k = 0.0344\text{N} \cdot \text{m}.$

In general, if θ_1 is small and θ_2 is large, F_1 is larger than F_2. Contrarily, if θ_1 is large and θ_2 is small, F_1 is smaller than F_2. (But not absolutely)

3.4   The Switch Condition of the Pinching and Enveloping Grasp

In order to find out the conditions where the parallel structure performs pinching motion, and the conditions where the parallel structure is switched to the enveloping motion, this part analyzes the static equilibrium when $F_1 = 0$ and $\theta_2 = 0$. As Fig. 9 shows, here $k\theta_2 + M_0$ is replaced by $-M'$, which is the contact torque between the limiting block and the sliding block.

Similar to the previous part, one obtains the expression of the contact forces:

$$F_1 = T\frac{1}{h_1}(1 - \frac{1}{a} - \frac{l_1 \cos(\theta_2 - \theta_1)}{ah_2}) - M'\frac{l_1 \cos(\theta_2 - \theta_1)}{h_1 h_2} \tag{19}$$

$$F_2 = T\frac{1}{ah_2} + M'\frac{1}{h_2} \tag{20}$$

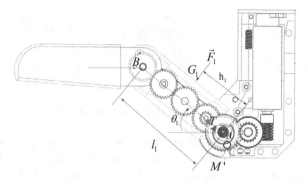

Fig. 9. Condition of pinching and enveloping grasp of the PASA finger.

Because $F_1 = 0$ and $\theta_2 = 0$, the contact torque between the limiting block and the sliding block can be described as

$$M' = T(1 - \frac{1}{a} - \frac{l_1 \cos \theta_1}{ah_2})\frac{h_2}{l_1 \cos \theta_1} \tag{21}$$

The condition where the parallel structure performs a pinching motion is $M' > 0$. As a robotic hand, we limit $-90^0 < \theta_1 < 90^0, a > 1, T > 0$.
One achieves

$$h_2 > \frac{l_1 \cos \theta_1}{a - 1} \tag{22}$$

If h_2 satisfies Eq. (22), the PASA finger executes pinching grasp mode. Otherwise, the PASA finger executes enveloping grasp mode. This condition is determined by a, l_1 and θ_1, where a and l_1 is the system parameters and can be optimized. θ_1 changes once the size, shape and position of objects change.

$h_2 > l_1/(a - 1)$ can be defined as a pinching zone and $h_2 < -l_1/(a - 1)$ can be defines as an enveloping zone. In the interval $-l_1/(a - 1) \leq h_2 \leq l_1/(a - 1)$, the grasping mode changes according to the rotary angle of the proximal phalanx θ_1.

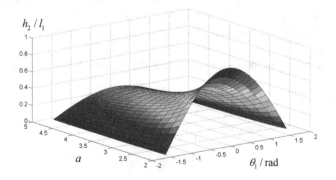

Fig. 10. Relation between h_2/l_1 and a, θ_1.

As a multi-grasping-mode finger, the pinching zone should not be too small, otherwise it will lose its practical significance. So parameter a should be much bigger than 1. For example, when $a = 3.5$, the pinching zone is $h_2 > 0.4l_1$.

Figure 10 shows the relation between h_1/l_1 and a, θ_1.

4   Experiments of the PASA Hand

To evaluate the performances of the PASA hand, this part conducts sevral experiments on the forces distribution, the influence of parameters and the versatility of the PASA hand.

As it is shown in Fig. 11, the prototype of PASA hand has three fingers, two of them can rotate around the arm. The actuators are ESCAP 16G214EMR19. The rotate speed of the driving gear is 10 r/min.

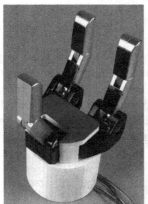

Fig. 11. The PASA hand.

4.1   Force Distribution

As a self-adaptive hand, the PASA hand envelopes objects when the force is applied on the proximal phalanx before the distal phalanx. If the contact force of the proximal phalanx is large but the distal phalanx doesn't contact the objects, the object will probably deform and the grasp is weak. So it is important to evaluate the contact force of the proximal phalanx when the distal phalanx begins to rotate. We call this force the "triggering force".

As Fig. 12a shows, the "triggering force" decreases when h_1 augments, which means larger objects are easier to envelope. It is also influenced by the rotary angle of the proximal phalanx.

Figure 12b shows the variation of the contact forces in an enveloping grasp process. After the distal phalanx rotates, the contact force of the proximal phalanx augments as the rotary angle of the distal phalanx becomes larger. When the distal phalanx comes into contact with the object, the contact force of the distal phalanx arguments instantly, and becomes larger than the contact force of the first phalanx.

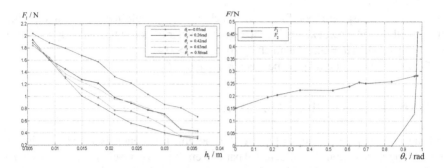

(a) Contact forces on the proximal phalanx
when the distal phalanx begins to rotate.

(b) Contact forces after the distal phalanx
rotates.

Fig. 12. Experiments of force distribution

4.2 Objects Grasping

To evaluate the versatility of the PASA hand, we use the PASA hand to grasp different objects. Depending on the objects, it executes pinching grasp or enveloping grasp automatically. When the object is located on the upper side of the finger, the PASA hand probably pinches it. When the object is large and located on the lower side of the finger, the PASA hand executes enveloping grasp. For some special objects, it can also execute pinching and enveloping grasps at the same time, as it is shown in the third picture of Fig. 13a.

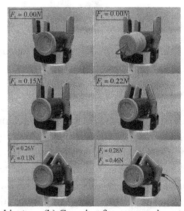

(a) Grasping experiments for different objects.   (b) Grasping forces experiments.

Fig. 13. Experiments of different objects with the PASA hand.

Figure 13b shows three enveloping processes. The results show that the contact forces of the distal phalanx are larger than the contact force of the proximal phalanx when the enveloping processes are finished. A larger distal contact force will make the contact between the object and the base more stable.

5   Conclusions

This paper proposes a novel underactuated robotic finger (the PASA finger), which can perform parallel and self-adaptive (PASA) hybrid grasping modes. A PASA hand is developed with three PASA fingers and 8 degrees of freedom (DOFs). The PASA hand executes multiple grasping modes depending on the dimensions, shapes and positions of objects: (1) a parallel pinching (PA) grasp for precision grasp; (2) a self-adaptive (SA) enveloping grasp for power grasp; (3) a parallel and self-adaptive (PASA) grasping mode for hybrid grasp. Kinematics and statics show the distribution of contact forces and the switch condition of PA, SA, and PASA grasping modes. Experimental results show the high stability of the grasps and the versatility of the PASA hand. The PASA hand has a wide range of applications.

Acknowledgement. This Research was supported by National Natural Science Foundation of China (No. 51575302).

References

1. Jacobsen, S.C., Iversen, E.K., Knutti, D.F., et al.: Design of the Utah/MIT dextrous hand. In: Proceedings of the IEEE International Conference on Robotics and Automation, pp. 1520–1532 (1986)
2. Loucks, C.S.: Modeling and control of the stanford/JPL hand. In: 1987 International Conference on Robotics and Automation, pp. 573–578 (1987)
3. Butterfass, J., Grebenstein, M., Liu, H., et al.: DLR-hand II: next generation of a dextrous robot hand. In: IEEE International Conference on Robotics and Automation (ICRA), vol. 1, pp. 109–114 (2001)
4. Kawasaki, H., Komatsu, T., Uchiyama, K., et al.: Dexterous anthropomorphic robot hand with distributed tactile sensor: Gifu hand II. In: IEEE International Conference on Systems, Man, and Cybernetics (SMC), pp. 782–787 (1999)
5. Paek, J., Cho, I., Kim, J.: Microrobotic tentacles with spiral bending capability based on shape-engineered elastomeric microtubes, 01 August 2015. http://www.nature.com/srep/2015/150611/srep10768/abs/srep10768.html#supplementary-informationAB
6. Thierry, L., Gosselin, C.M.: Simulation and design of underactuated mechanical hands. Mech. Mach. Theor. 33(1–2), 39–57 (1998)
7. Che, D., Zhang, W.: GCUA humanoid robotic hand with tendon mechanisms and its upper limb. Int. J. Soc. Robot. 3(1), 395–404 (2011)
8. Zhang, W., Tian, L., Liu, K.: Study on multi-finger under-actuated mechanism for TH-2 robotic hand. In: IASTED International Conference on Robotics and Applications, pp. 420–424 (2007)
9. Zhang, W., Chen, Q., Sun, Z., et al.: Under-actuated passive adaptive grasp humanoid robot hand with control of grasping force. In: IEEE International Conference on Robotics and Automation (ICRA), pp. 696–701 (2003)
10. Massa, B., Roccella, S., Carrozza, M.C., et al.: Design and development of an underactuated prosthetic hand. In: IEEE International Conference on Robotics and Automation (ICRA), pp. 3374–3379 (2002)
11. Li, G., Liu, H., Zhang, W.: Development of multi-fingered robotic hand with coupled and directly self-adaptive grasp. Int. J. Humanoid Rob. 9(4), 1–18 (2012)

12. Birglen, L., Gosselin, C.M.: Kinetostatic analysis of underactuated fingers. IEEE Trans. Robot. Autom. **20**(1), 211–221 (2004)
13. Birglen, L., Gosselin, C.M.: On the force capability of underactuated fingers. In: IEEE International Conference on Robotics and Automation (ICRA), pp. 1139–1145 (2003)
14. Birglen, L., Gosselin, C.M.: Geometric design of three-phalanx underactuated fingers. J. Mech. Des. **128**(1), 356–364 (2006)
15. Townsend, W.: The BarrettHand grasper – programmably flexible part handling and assembly. Ind. Robot Int. J. **27**(3), 181–188 (2000)
16. Demers, L.A., Lefrancois, S., Jobin, J.: Gripper having a two degree of freedom underactuated mechanical finger for encompassing and pinch grasping. US Patent US8973958 (2015)

Topological Structure Synthesis of 3T1R Parallel Mechanism Based on POC Equations

Tingli Yang[1], Anxin Liu[2(✉)], Huiping Shen[1], and Lubin Hang[3]

[1] Changzhou University, Changzhou, China
{yangtl, shp65}@126.com
[2] PLA University of Science and Technology, Nanjing, China
liuanxinn@163.com
[3] Shanghai University of Engineering Science, Shanghai, China
hanglb@126.com

Abstract. A systematic method for topological structure synthesis of PM based on POC equations is introduced. The complete synthesis process includes the following several steps: (a) synthesize all candidate SOC and HSOC branches based on the POC equation for serial mechanisms, (b) determine branch combination schemes and geometrical conditions for branches to be assembled on two platforms based on POC equation for PMs, (c) check the obtained PMs for design requirements and obtain all usable PMs. Topological structure synthesis of 3T1R PM which can be used in SCARA robot is discussed in detail to illustrate to procedure of this method. 18 types of 3T1R PMs containing no prismatic pair are obtained.

Keywords: Parallel mechanism (PM) · Topological structure synthesis · Position and orientation characteristic (POC) · Single open chain (SOC) · Hybrid single open chain (HSOC)

1 Introduction

The SCARA parallel robot can achieve 1 rotation and 3 translations and has been used widely in industries such as sorting, packaging and assembling. The SCARA serial robot first developed by Hiroshi [1] has 4 dofs. It has a cantilever design. So it features low rigidity, low effective load and high arm inertia. It is very hard for its end effector to move at high speed.

The SCARA parallel robot FlexPicker developed by ABB Corporation [2] has one output platform and one sub-platform (see Fig. 10). It is obtained by adding a RUPU branch to the original Delta parallel robot [3]. The SCARA parallel robot Par4 developed by Pierrot [4–6] has one output platform and 2 sub-platforms (see Fig. 9a). Relative motion between the two sub-platforms is converted to rotation of the output platform. This robot features high flexibility, high speed and high rigidity. Based on Par4 robot, Adept Corporation developed the SCARA parallel robot Quattro, which boasts the highest operation speed [7–9]. Some innovative work has also been done by Huang Tian on this type of SCARA parallel robot [10–13]. The SCARA parallel robot X4 has one output platform and 4 parallel-connected branches (see Fig. 7a). It features simple structure and can realize a rotation within the range of ±90° [14].

© Springer International Publishing Switzerland 2016
N. Kubota et al. (Eds.): ICIRA 2016, Part I, LNAI 9834, pp. 147–161, 2016.
DOI: 10.1007/978-3-319-43506-0_13

Obviously, the most important part of the above SCARA parallel robots is the 3T1R PM involved. In this paper, topological structure synthesis of 3T1R PM is discussed systematically using the method based on POC equations [15–23] in order to provide the readers with more comprehensive understanding about the topological structure of 3T1R PM and to obtain some more new 3T1R PMs with potential practical use.

The topological structure synthesis method based on POC equations introduced in this paper is the further development of our early work on topological structure synthesis of PMs and now becomes a systematic and easy-to-follow method for topological structure synthesis of PMs due to some new achievements of this paper, such as: (a) general method for topological structure synthesis of simple branches based on POC equation for serial mechanisms; (b) general method for topological structure synthesis of complex branches based on replacement of sub-SOCs in a SOC branch by topologically equivalent sub-PMs; (c) general method for determination of geometrical conditions for assembling branches between the fixed platform and the moving platform; and (d) general criteria for selection of driving pairs.

2 Theoretical Bases

2.1 POC Equation for Serial Mechanism

The POC equation for serial mechanism [15, 16, 21, 23] is

$$M_S = \bigcup_{i=1}^{m} M_{J_i} = \bigcup_{j=1}^{k} M_{sub-SOC_j} \tag{1}$$

where, M_S—POC set of the end link relative to the frame link, m—number of kinematic pairs, M_{J_i}—POC set of the ith kinematic pair (Table 1), $M_{sub-SOC_j}$—POC set of the jth sub-SOC (12 sub-SOCs containing R and P pairs only are listed in Table 2).

Table 1. POC set of kinematic pair

P pair	R pair	H pair
$\begin{bmatrix} t^1(\|P) \\ r^0 \end{bmatrix}$	$\begin{bmatrix} t^1(\perp(R,\rho)) \\ r^1(\| R) \end{bmatrix}$	$\begin{bmatrix} t^1(\| H) \cup t^1(\perp(H,\rho)) \\ r^1(\| H) \end{bmatrix}$
$Dim\{M_J\} = 1$		

Note: "‖" means "parallel to", "⊥" means "perpendicular to".

"Union" operation rules for Eq. (1) include 8 linear operation rules and 2 nonlinear criteria [21, 23].

2.2 POC Equation for PM

The POC equation for PM [15, 16, 21, 23] is

Table 2. POC set of sub-SOC

Sub-SOCs	$SOC\{-R\parallel R-\}$ $SOC\{-R\perp P-\}$	$SOC\{-R\parallel R\parallel R-\}$ $SOC\{-R\parallel R\perp P-\}$ $SOC\{-P\perp R\perp P-\}$	$SOC\{\Diamond(P,P,\cdots,P)\}$	$SOC\{-R\vert P-\}$	$SOC\{-RR\overparen{RR}-\}$	$SOC\{-\overparen{RRR}-\}$
POC set (M_s)	No.1	No. 2	No. 3	No. 4	No. 5	No. 6
	$\begin{bmatrix} t^2(\perp R) \\ r^3(\parallel R) \end{bmatrix}$	$\begin{bmatrix} t^2(\perp R) \\ r^3(\parallel R) \end{bmatrix}$	$\begin{bmatrix} t^2 \\ r^0 \end{bmatrix}$	$\begin{bmatrix} t^1(\parallel P)\cup t^1(\perp(R,\rho)) \\ r^1(\parallel R) \end{bmatrix}$	$\begin{bmatrix} t^2(\perp\rho) \\ r^2 \end{bmatrix}$	$\begin{bmatrix} t^2(\perp\rho) \\ r^3 \end{bmatrix}$
	$\mathrm{Dim}\{M_s\}=2$	$\mathrm{Dim}\{M_s\}=3$	$\mathrm{Dim}\{M_s\}=2$	$\mathrm{Dim}\{M_s\}=2$		$\mathrm{Dim}\{M_s\}=3$
	No. 1*	No. 2*	No. 3*	No. 4*	No. 5*	No. 6*
	$\begin{bmatrix} t^1(\perp(R,\rho)) \\ r^1(\parallel R) \end{bmatrix}$	$\begin{bmatrix} t^2(\perp R) \\ r^1(\parallel R) \end{bmatrix}$	$\begin{bmatrix} t^2 \\ r^0 \end{bmatrix}$	$\begin{bmatrix} t^1(\parallel P) \\ r^1(\parallel R) \end{bmatrix}$	$\begin{bmatrix} t^0 \\ r^2 \end{bmatrix}$	$\begin{bmatrix} t^0 \\ r^3 \end{bmatrix}$
	$\mathrm{Dim}\{M_s\}=2$	$\mathrm{Dim}\{M_s\}=3$	$\mathrm{Dim}\{M_s\}=2$	$\mathrm{Dim}\{M_s\}=2$		$\mathrm{Dim}\{M_s\}=3$

Note: The base point o' of end link of the sub-SOC lies on the pair axis for No.1*–No.6*.

$$M_{Pa} = \cap_{j=1}^{(v+1)} M_{b_j} \tag{2}$$

where, M_{Pa}—POC set of the moving platform, M_{b_j}—POC set of the end link in the jth branch, v—number of independent loop.

"Intersection" operation rules for Eq. (2) include 12 linear operation rules and 2 nonlinear criteria [21, 23].

According to Eq. (2), there is

$$M_{b_j} \supseteq M_{Pa} \tag{3}$$

2.3   DOF Formula

(1) The DOF formula [22] is

$$F = \sum_{i=1}^{m} f_i - \sum_{j=1}^{v} \xi_{L_j} \tag{4a}$$

$$\xi_{L_j} = \dim\cdot\left\{ \left(\cap_{j=1}^{j} M_{b_j}\right) \bigcup M_{b_{(j+1)}} \right\} \tag{4b}$$

where, F—DOF of mechanism, f_i—DOF of the ith kinematic pair, ξ_{L_j}—number of independent displacement equations for the jth independent loop, M_{b_j}—POC set of the end link in the jth branch.

(2) Criteria for driving pair selection [22]

For the mechanism with DOF = F, select and lock (make the two links connected by a kinematic pair into one integral link) F kinematic pairs. Only if DOF of the obtained new mechanism is 0, the selected F kinematic pairs can all be used as driving pairs simultaneously.

3   Method for Topological Structure Synthesis of Branch

3.1   Method for Topological Structure Synthesis of Simple Branch

The branch containing no loop is called simple branch. It is also referred to as SOC branch. Main steps of SOC branch topological structure synthesis include:

Step 1. Determine POC set and DOF of SOC branch

(1) Determine POC set of SOC branch (M_S) according to Eq. (3).
(2) Determine DOF of SOC branch according to dimension of M_S.

Step 2. Determine kinematic pair combination scheme of SOC branch
According to Eq. (1), the number of kinematic pairs of SOC branch containing only R and P pairs shall meet the following requirements:

(1) DOF of SOC branch

$$F = m_R + m_P \tag{5}$$

where, F – DOF of branch, m_R – number of R pair, m_P – number of P pair.
(2) Range of R pair number

$$m_R \geq \dim\{M_S(r)\} \tag{6}$$

where, $\dim\{M_S(r)\}$ – number of independent rotation elements in M_S.
(3) Range of P pair number

$$m_P \leq \dim\{M_S(t)\} \tag{7}$$

where, $\dim\{M_S(t)\}$ – number of independent translation elements in M_S.

Step 3. Determine sub-SOCs contained in the SOC branch
With M_S and the kinematic pair combination scheme being known, determine the sub-SOCs contained in SOC branch according to Eq. (1), Tables 1 and 2. Connect these sub-SOCs in tandem and obtain the desired SOC branch.

Step 4. Check POC set of the obtained SOC branch
For each SOC branch obtained in step 3, check whether its POC set complies with the design requirement according to Eq. (1).

Example 4.1. Topological structure synthesis of SOC branch with 4 DOFs (3T1R)

Step 1. Basic functions of the SOC branch

(1) There is $M_S = \begin{bmatrix} t^3 \\ r^1 \end{bmatrix}$. Select an arbitrary point on the end link as base point o'.

(2) $F = \dim\{M_S\} = 4$.

Step 2. Determine kinematic pair combination scheme of SOC branch

Since $\dim\{M_S(r)\} = 1$ and $\dim\{M_S(t)\} = 3$, there must be $m_R \geq 1$ and $m_P \leq 3$ according to Eqs. (5, 6 and 7). So, there are 3 pair combination schemes according to Eq. (1) and Table 2: 3R1P, 2R2P and 1R3P.

Step 3. Determine sub-SOCs contained in the SOC branch

Case 1: 3R1P

According to Eq. (1) and Table 2, this branch contains $sub - SOC\{-R \parallel R \parallel R-\}$ and a P pair. The SOC branch shall be $SOC\{-R \parallel R \parallel R - P-\}$.

Case 2: 2R2P

According to Eq. (1) and Table 2, this branch contains $sub - SOC\{-R \parallel R - P-\}$ and a P pair. The SOC branch shall be $SOC\{-R \parallel R - P - P-\}$.

Case 3: 1R3P

According to Eq. (1) and Table 2, this branch contains $sub - SOC\{-R - P - P-\}$ and a P pair. The SOC branch shall be $SOC\{-R - P - P - P-\}$.

Step 4. Check POC set of the obtained SOC branch

3.2 Method for Topological Structure Synthesis of Hybrid Branch

The branch containing one or more loop(s) is called hybrid branch. It is also referred to as HSOC branch.

If two kinematic chains have same POC set, they are considered as topologically equivalent. Similarly, if a sub-PM (PM) and a sub-SOC have same POC set, the sub-PM and the sub-SOC are topologically equivalent. If a SOC branch and a HSOC have the same POC set, the HSOC is topologically equivalent to the SOC branch.

Two-branch sub-PM, i.e. single-loop mechanism, is usually used in HSOC branch. These sub-PMs, together with their topologically equivalent sub-SOCs and there POC sets are listed in Table 3.

Table 3. Two-branch sub-PMs and their topologically equivalent sub-SOCs

No.	1	2	3	4	5
Sub-PMs					
POC set	$\begin{bmatrix} t^1(\parallel(ad)) \\ r^0 \end{bmatrix}$	$\begin{bmatrix} t^1(\parallel bc) \\ r^1(\parallel bc) \end{bmatrix}$	$\begin{bmatrix} t^1(\parallel(ad))\cup t^1(\perp(ad)) \\ r^1(\parallel(ad)) \end{bmatrix}$	$\begin{bmatrix} t^2(\perp R_{11}) \\ r^0 \end{bmatrix}$	$\begin{bmatrix} t^1 \\ r^0 \end{bmatrix}$
Equivalent sub-SOC					

Generally, a HSOC branch can be obtained by replacing a sub-SOC in a SOC branch with a topologically equivalent sub-PM. According to Eq. (1), this HSOC branch and the original SOC branch have the same POC set.

For example, the SOC branch $\{-R \parallel R \parallel C-\}$ in Fig. 1a can be re-expressed as $\{-R \parallel R \parallel R \parallel P-\}$. POC set of its sub-SOC $\{-R \parallel R \parallel P-\}$ is $\begin{bmatrix} t^1(\parallel P) \cup t^1(\perp R) \\ r^1(\parallel R) \end{bmatrix}$.

So this sub-SOC is topologically equivalent to No. 3 sub-PM in Table 3.

Replace sub-SOC $\{-R \parallel R \parallel P-\}$ with this No. 3 sub-PM and obtain the HSOC branch in Fig. 1b. The branch in Fig. 1a and the branch in Fig. 1b have the same POC set.

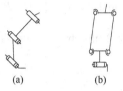

(a)   (b)

Fig. 1. Two topologically equivalent (3T1R) branches

4 Geometrical Conditions for Assembling Branches

Since POC set of each branch contains more elements than (or at least the same number of elements as) POC set of the PM, these branches shall be so assembled between the two platforms that some elements of their POC sets are eliminated to obtain the desired POC set of the PM. The basic equations used to determine geometrical conditions for assembling branches will be discussed in the following two different cases.

Case 1. After the first k branches are assembled, a sub-PM will be obtained. If POC set of this sub-PM still contains any element that is not included in POC set of the PM, the $(k + 1)^{th}$ branch shall be so assembled between the two platforms that this unwanted element shall be eliminated during intersection operation between POC sets.

A. If a rotational element is to be eliminated, the following three intersection operation equations shall be used.

$$\left[r^1(\parallel R_i) \right]_{bi} \cap \left[r^1(\parallel R_j) \right]_{bj} = \left[r^0 \right]_{Pa}, \quad if\ R_i \nparallel R_j \tag{8a}$$

$$\left[r^1(\parallel R_i) \right]_{bi} \cap \left[r^2(\parallel \Diamond(R_{j1}, R_{j2})) \right]_{bj} = \left[r^0 \right]_{Pa}, \quad if\ R_i \nparallel (\Diamond(R_{j1}, R_{j2})) \tag{8b}$$

$$\left[r^2(\parallel \Diamond(R_1, R_2)) \right]_{bi} \cap \left[r^2(\parallel \Diamond(R_{j1}, R_{j2})) \right]_{bj}$$
$$= \left[r^1(\parallel (\Diamond(R_{i1}, R_{i2}) \cap (\Diamond(R_{j1}, R_{j2}))) \right]_{Pa}, if\ (\Diamond(R_{i1}, R_{i2})) \nparallel (\Diamond(R_{j1}, R_{j2})) \tag{8c}$$

B. If a translational element is to be eliminated, the following three intersection operation equations shall be used.

$$\left[t^1(\|\ P_i^*)\right]_{bi} \cap \left[t^1(\|\ P_j^*)\right]_{bj} = \left[t^0\right]_{Pa}, \textit{if } P_i^* \nparallel P_j^* \tag{9a}$$

$$\left[t^1(\|\ P_i^*)\right]_{bi} \cap \left[t^1(\|\ P_j^*)\right]_{bj} = \left[t^0\right]_{Pa}, \textit{if } P_i^* \nparallel P_j^* \tag{9b}$$

$$\left[t^2(\|\ \Diamond(P_1^*, P_2^*))\right]_{bi} \cap \left[t^2(\|\ \Diamond(P_{j1}^*, P_{j2}^*))\right]_{bj}$$
$$= \left[t^1(\|\ (\Diamond(P_{i1}^*, P_{i2}^*) \cap (\Diamond(P_{j1}^*, P_{j2}^*)))\right]_{Pa}, \textit{if } (\Diamond(P_{i1}^*, P_{i2}^*)) \nparallel (\Diamond(P_{j1}^*, P_{j2}^*)) \tag{9c}$$

Note: P* – translation of P pair or derivative translation of R pair or H pair.

Case 2. After the first k branches are assembled, a sub-PM will be obtained. If POC set of this sub-PM is identical to POC set of the PM, the $(k + 1)^{th}$ branch shall be so assembled between the two platforms that no element shall be eliminated during intersection operation between POC sets.

C. The following intersection operation equations shall not eliminate any rotational element.

$$\left[r^1(\|\ R_i)\right]_{bi} \cap \left[r^1(\|\ R_j)\right]_{bj} = \left[r^1(\|\ R_i)\right]_{Pa}, \textit{if } R_i \|\ R_j \tag{10a}$$

$$\left[r^1(\|\ R_i)\right]_{bi} \cap \left[r^2\Diamond((R_{j1}, R_{j2}))\right] = \left[r^1(\|\ R_i)\right]_{Pa}, \textit{if } R_i \|\ (\Diamond(R_{j1}, R_{j2})) \tag{10b}$$

$$\left[r^2(\|\ \Diamond(R_1, R_2))\right]_{bi} \cap \left[r^2(\|\ \Diamond(R_{j1}, R_{j2}))\right]$$
$$= \left[r^2(\|\ \Diamond R_{i1}, R_{i2})\right]_{Pa}, \textit{if } (\Diamond(R_{i1}, R_{i2})) \|\ (\Diamond(R_{j1}, R_{j2})) \tag{10c}$$

$$\left[r^1(\|\ R_i)\right]_{bi} \cap \left[r^3\right]_{bj} = \left[r^1(\|\ R_i)\right]_{Pa} \tag{10d}$$

$$\left[r^2(\|\ \Diamond(R_{i1}, R_{i2}))\right]_{bi} \cap \left[r^3\right]_{bj} = \left[r^2(\|\ \Diamond(R_{i1}, R_{i2}))\right]_{Pa} \tag{10e}$$

$$\left[r^3\right]_{bi} \cap \left[r^3\right]_{bj} = \left[r^3\right]_{Pa} \tag{10f}$$

D. The following intersection operation equations shall no eliminate any translational element.

$$\left[t^1(\|\ P_i^*)\right]_{bi} \cap \left[t^1(\|\ P_j^*)\right]_{bj} = \left[t^1(\|\ P_i^*)\right]_{Pa}, \textit{if } P_i^* \|\ P_j^* \tag{11a}$$

$$\left[t^1(\|\ P_i^*)\right]_{bi} \cap \left[t^2(\|\ \Diamond(P_{j1}^*, P_{j2}^*))\right] = \left[t^1(\|\ P_i^*)\right]_{Pa}, \textit{if } P_i^* \|\ (\Diamond(P_{j1}^*, P_{j2}^*)) \tag{11b}$$

$$\left[t^2(\|\ \Diamond(P_1^*, P_2^*))\right]_{bi} \cap \left[t^2(\|\ \Diamond(P_{j1}^*, P_{j2}^*))\right]$$
$$= \left[t^2(\|\ \Diamond P_{i1}^*, P_{i2}^*)\right]_{Pa}, \textit{if}\ (\Diamond(P_{i1}^*, P_{i2}^*))\ \|\ (\Diamond(P_{j1}^*, P_{j2}^*)) \tag{11c}$$

$$\left[t^1(\|\ P_i^*)\right]_{bi} \cap \left[t^3\right]_{bj} = \left[t^1(\|\ P_i^*)\right]_{Pa} \tag{11d}$$

$$\left[t^2(\|\ \Diamond(P_{i1}^*, P_{i2}^*))\right]_{bi} \cap \left[t^3\right]_{bj} = \left[t^2(\|\ \Diamond(P_{i1}^*, P_{i2}^*))\right]_{Pa} \tag{11e}$$

$$\left[t^3\right]_{bi} \cap \left[t^3\right]_{bj} = \left[t^3\right]_{Pa} \tag{11f}$$

By the way, if POC set of the sub-PM obtained after the first k branches are assembled is already identical to POC set of the PM, the $(k+1)^{th}$ branch can be replaced by any simple branch whose POC set contains at least all elements of POC set of the PM, e.g. the $SOC\{-S-S-R-\}$ branch.

5 Topological Structure Synthesis of 3T1R PM

Step 1 Design requirements. $M_{Pa} = \begin{bmatrix} t^3 \\ r^1 \end{bmatrix}$; DOF = 4; No P pair or only driving P pair shall be contained; Each branch contains only one driving pair and all driving pairs must be allocated on the same platform.

Step 2 Determine POC set of SOC branch. According to Eq. (3), POC set of the SOC branch shall be $M_{b_j} = \begin{bmatrix} t^3 \\ r^1 \end{bmatrix}, \begin{bmatrix} t^3 \\ r^2 \end{bmatrix}$ or $\begin{bmatrix} t^3 \\ r^3 \end{bmatrix}$

Step 3 Topological structure synthesis of SOC branch.

(1) Topological structure synthesis of SOC branch with POC set $\begin{bmatrix} t^3 \\ r^1 \end{bmatrix}$

As discussed in Example 4.1, three candidate SOC branches are obtained (see Table 4).

(2) Topological structure synthesis of SOC branch with POC set $\begin{bmatrix} t^3 \\ r^2 \end{bmatrix}$

Four candidate SOC branches are obtained based on a synthesis process similar to that described in Example 4.1 (see Table 4).

(3) Topological structure synthesis of (3T3R) SOC branch with POC set $\begin{bmatrix} t^3 \\ r^3 \end{bmatrix}$

As we all know, the simplest such SOC branch is $SOC\{-S-S-R-\}$.

Table 4. Feasible SOC branches and corresponding HSOC branches

M_{b_j}	SOC branch	HSOC branch
$\begin{bmatrix} t^3 \\ r^1 \end{bmatrix}$	(1) $SOC\{-R \parallel R \parallel R - P-\}$	(1) $HSOC\{-R \parallel R \parallel R - P^{(4R)}-\}$
	(2) $SOC\{-R \parallel R - P - P-\}$	(2) $HSOC\{-R \parallel R^{(4S)} - P^{(4S)} - P^{(4S)}-\}$
		(3) $HSOC\{-R(\parallel \Diamond(P^{(4R)}, P^{(4R)}) \parallel R-\}$
		(4) $HSOC\{-R \parallel R - (\Diamond(P^{(5R1C)}, P^{(5R1C)})-\}$
	(3) $SOC\{-R - P - P - P-\}$	(5) $HSOC\{-R - 2\{R \parallel R^{(4S)} - P^{(4S)} - P^{(4S)}-\}-\}$
		(6) $HSOC\{-R - Delta\,PM-\}$
$\begin{bmatrix} t^3 \\ r^2(\parallel \Diamond(R, R')) \end{bmatrix}$	(4) $SOC\{-R \parallel R \parallel R - R \parallel R-\}$	
	(5) $SOC\{-R \parallel R \parallel R - P - R-\}$	(7) $HSOC\{-R \parallel R \parallel R - P^{(4R)} - R-\}$
	(6) $SOC\{-R \parallel R - P - R \parallel R-\}$	(8) $HSOC\{-R \perp R^{(2R2S)} - P^{(2R2S)} - R \parallel R-\}$
	(7) $SOC\{-R - P - P - R \parallel R-\}$	(9) $HSOC\{-R - P^{(4S)} - P^{(4S)} - R^{(4S)} \parallel R-\}$
		(10) $HSOC\{-\Diamond(P^{(5R1C)}, P^{(5R1C)}) \parallel R \parallel R \perp R-\}$
$\begin{bmatrix} t^3 \\ r^3 \end{bmatrix}$	(8) $SOC\{-S - S - R-\}$	

Step 4 Topological structure synthesis of HSOC branch. Sub-SOCs containing P pairs can be replaced by the following sub-PMs.

(1) $sub - SOC\{-P-\} = sub - PM : \{-P^{(4R)}-\}$.
(2) $sub - SOC\{-R - P-\} = sub - PM : \{-R^{(2R2S)} - P^{(2R2S)}-\}$.
(3) $sub - SOC\{-R - P - P-\} = sub - PM : \{-R^{(4S)} - P^{(4S)} - P^{(4S)}-\}$.
(4) $sub - SOC\{-P - P - P-\} = sub - PM : \{-2SOC\{-R \parallel R^{(4S)} - P^{(4S)} - P^{(4S)}-\}-\}$.
(5) $sub - SOC\{-P - P-\} = sub - PM : \{-\Diamond(P^{(5R1C)}, P^{(5R1C)})-\}$.
(6) $sub - SOC\{-P - P - P-\} = sub - PM : \{-Delta\,PM-\}$.

After topologically equivalent replacement, 6 types of 3T1R branches and 5 types of 3T2R branches which contain no P pairs are obtained (see Table 4), as shown respectively in Figs. 2 and 3.

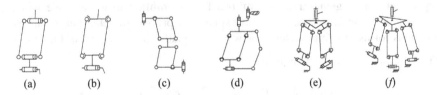

(a)   (b)   (c)   (d)   (e)   (f)

Fig. 2. Six types of 3T1R branches without P pairs

Step 5 Branch combination scheme. Based on the 2 SOC branches (No. 4 and No. 8 in Table 4) and 10 HSOC branches (refer to Table 4), different branch combination schemes which can generate the desired POC set and DOF of the PM can be obtained. During branch combination, we shall ensure that the branch shall have simple structure and certain symmetry. Each branch shall have only one driving pair and all driving pairs shall be allocated on the same platform. Table 5 shows some of these branch combination schemes.

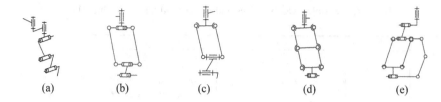

(a)	(b)	(c)	(d)	(e)

Fig. 3. Five types of 3T2R branches without P pairs

Table 5. Branch combination scheme of 3T1R PM

Branch type		Branch combination scheme
Table 4-SOC-(4)	I	$4 - SOC\{-R \parallel R \parallel R - R \parallel R-\}$
	2	$2 - SOC\{-R \parallel R \parallel R - R \parallel R-\} \oplus 2 - SOC\{-S - S - R\}$
Table 4-HSOC-(7)	3	$4 - HSOC\{-R \parallel R \parallel R - P^{(4R)} - R-\}$
	4	$2 - HSOC\{-R \parallel R \parallel R - P^{(4R)} - R-\} \oplus 2 - SOC\{-S - S - R-\}$
Table 4-HSOC-(8)	5	$4 - HSOC\{-R \bot R^{(2R2S)} \parallel P^{(2R2S)} - R \parallel R-\}$
	6	$2 - HSOC\{-R \bot R^{(2R2S)} \parallel P^{(2R2S)} - R \parallel R-\} \oplus 2 - SOC\{-S - S - R-\}$
Table 4-HSOC-(9)	7	$4 - HSOC\{-R - P^{(4S)} - P^{(4S)} - R^{(4S)} \parallel R-\}$
	8	$2 - HSOC\{-R - P^{(4S)} - P^{(4S)} - R^{(4S)} \parallel R-\} \oplus 2 - SOC\{-S - S - R-\}$
Table 4-HSOC-(1)	9	$4 - HSOC\{-R \parallel R \parallel R - P^{(4S)}-\}$
	10	$2 - HSOC\{-R \parallel R \parallel R - P^{(4S)}-\} \oplus 2 - SOC\{-S - S - R-\}$
Table 4-HSOC-(5)	11	$2 - HSOC\{-R \parallel R^{(4S)} - P^{(4S)} - P^{(4S)}-\}-\}$
	12	$1 - HSOC\{-R - 2\{R \parallel R^{(4S)} - P^{(4S)} - P^{(4S)}-\}-\} \oplus 2 - SOC\{-S - S - R-\}$
Table 4-HSOC-(6)	13	$1 - HSOC\{-R - Delta\,PM-\} \oplus 1 - SOC\{-S - S - R-\}$
Table 4-HSOC-(3)	14	$2 - HSOC\{-R(\parallel \Diamond(P^{(4R)}, P^{(4R)}) \parallel R-\} \oplus 2 - SOC\{-S - S - R-\}$
Table 4-HSOC-(4)	15	$1 - HSOC\{-R \parallel R - (\Diamond(P^{(5R1C)}, P^{(5R1C)})-\} \oplus 2 - SOC\{-S - S - R-\}$
Table 4-HSOC-1O)	16	$2 - HSOC\{-\Diamond(P^{(5R1C)}, P^{(5R1C)}) \parallel R \parallel R\bot R-\}$

Step 6 Determine geometrical conditions for assembling branches. We will use scheme 1 in Table 5 as an example to explain the process to determine geometrical conditions for assembling branches between two platforms, as shown in Fig. 4a.

(1) Select 4 identical SOC branches

$$SOC\{-R_{j1} \parallel R_{j2} \parallel R_{j3} - R_{j4} \parallel R_{j5}-\}, j = 1, 2, 3, 4$$

(2) Select an arbitrary point on moving platform as the base point o'.
(3) Determine POC set of SOC branch

$$M_{bj} = \begin{bmatrix} t^3 \\ r^2(\parallel \Diamond(R_{j3}, R_{j4}) \end{bmatrix}, j = 1, 2, 3, 4,$$

(4) Establish POC equation for PM

Substitute the desired POC set of the PM and POC set of each branch into Eq. (2) and obtain

$$\begin{bmatrix} t^3 \\ r^1 \end{bmatrix} \Leftarrow \begin{bmatrix} t^3 \\ r^2(\| \diamond(R_{13}, R_{14}) \end{bmatrix} \cap \begin{bmatrix} t^3 \\ r^2(\| \diamond(R_{23}, R_{24}) \end{bmatrix} \cap \begin{bmatrix} t^3 \\ r^2(\| \diamond(R_{33}, R_{34}) \end{bmatrix} \cap \begin{bmatrix} t^3 \\ r^2(\| \diamond(R_{43}, R_{44}) \end{bmatrix}$$

Where, "$\Leftarrow$" means the POC set on the left side of the equation is to be obtained by intersection operation of all the POC sets on the right side of the equation

(5) Determine geometrical conditions for assembling the first two branches.

In order for the moving platform to obtain the desired POC set, intersection of POC sets of the first two branches must eliminate one rotational element. According to Eq. (8c), when $R_{13} \nparallel R_{23}$ and $R_{15} \parallel R_{25}$, the moving platform has only one rotation ($\| R_{15}$) and 3 translations. So the above POC equation can be rewritten as

$$\begin{bmatrix} t^3 \\ r^1 \end{bmatrix} \Leftarrow \begin{bmatrix} t^3 \\ r^1(\| R_{15}) \end{bmatrix} \cap \begin{bmatrix} t^3 \\ r^2(\| \diamond(R_{33}, R_{34}) \end{bmatrix} \cap \begin{bmatrix} t^3 \\ r^2(\| \diamond(R_{43}, R_{44}) \end{bmatrix}$$

Where, the first POC set on right side of the equation is POC set of the sub-PM form obtained after the first two branches are assembled.

(6) Determine geometrical conditions for assembling the other two branches

Since POC set of the sub-PM formed by the first two branches is already identical to the desired POC set of the PM, assembly of the other two branches shall not change POC set of the PM. According to Eq. (10b), geometrical conditions for assembling these two branches are: $R_{15} \parallel R_{25} \parallel R_{35} \parallel R_{45}$, $R_{11} \parallel R_{31}$ and $R_{21} \parallel R_{41}$.

(7) Determine allocation scheme of the 4 branches on platforms

According to geometrical conditions for assembling branches, there may be several different allocation schemes of the 4 branches on platforms. One feasible allocation scheme is:

$$SOC\{-R_{j1} \parallel R_{j2} \parallel \overset{\frown}{R_{j3} \perp R_{j4}} \parallel R_{j5}-\}, R_{15} \parallel R_{25} \parallel R_{35} \parallel R_{45}, R_{11} \parallel R_{31}, R_{21} \\ \parallel R_{41}, R_{21} \parallel R_{41},$$

(Axes of R_{11}, R_{21}, R_{31} and R_{41} lie in the same plane and form a square)

The PM assembled based on this allocation scheme is shown in Fig. 4a. The PMs shown in Fig. 4b and c also satisfy the geometrical assembling conditions.

(8) Discussion

Since POC set of the sub-PM formed by the first two branches and the two platforms is already identical to the desired POC set of the PM, the other two branches can be replaced by two simpler $SOC\{-S - S - R-\}$ branches, see Fig. 4d.

For each branch combination scheme in Table 5, corresponding PM(s) can be obtained similarly, as shown in Figs. 4, 5, 6, 7, 8, 9, 10, 11 and 12.

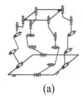

(a)   (b)   (c)   (d)

Fig. 4. Four PMs with No.6 SOC in Table 4

Fig. 5. Two PMs with No.7 HSOC in Table 4

Fig. 6. Two PMs with No.8 HSOC in Table 4

Fig. 7. Two PMs with No.9 HSOC in Table 4

Fig. 8. Two PMs with No.1 HSOC in Table 4

Fig. 9. Two PMs with No.5 HSOC in Table 4

Fig. 10. PM with No.6 HSOC in Table 4

Fig. 11. PM with No.3 HSOC in Table 4

Fig. 12. Two PMs with No.4 HSOC in Table 4

Step 7 Check DOF of the PM. Check the above obtained 18 PMs for their DOFs. Take the PM in Fig. 4a as example, Eqs. (4a, b) are used to calculate its DOF.

(1) Check POC set of branch: POC set of each branch is

$$M_{bj} = \begin{bmatrix} t^3 \\ r^2(\parallel \Diamond(R_{j3}, R_{j4}) \end{bmatrix}, j = 1, 2, 3, 4.$$

(2) Calculate DOF of the PM: Substitute M_{bj} into Eq. (4b) and obtain $\xi_{L1} = 6$, $\xi_{L2} = 5$, $\xi_{L3} = 5$. Then substitute ξ_{L_j} into Eq. (4a) and obtain DOF of the PM

$$F = \sum_{i=1}^{m} f_i - \sum_{j=1}^{v} \xi_{L_j} = 20 - (6 + 5 + 5) = 4$$

All the other 17 PMs can be checked in the same way.

Step 8 Select driving pairs. According to criteria for driving pair selection (refer to Sect. 2.3), check whether the 4 R pairs on (Fig. 5) the fixed platform (Fig. 4a) can all be selected as driving pairs.

(1) Suppose the 4 R pairs $(R_{11}, R_{21}, R_{31}, R_{41})$ to be driving pairs and lock them, a new PM can be obtained. Topological structure of each branch is $SOC\{-R_{j2} \parallel R_{j3} - R_{j4} \parallel R_{j5}-\}, j = 1, 2, 3, 4.$
(2) According to Eq. (1), POC set of each branch is

$$M_{bj} = \begin{bmatrix} t^3 \\ r^2(\parallel \Diamond(R_{j3}, R_{j4}) \end{bmatrix}, j = 1, 2, 3, 4.$$

(3) Determine DOF of the new PM
     Substitute M_{bj} into Eq. (4b) and obtain $\xi_{L1} = 6$, $\xi_{L2} = 5$, $\xi_{L3} = 5$. Then there is

$$F^* = \sum_{i=1}^{m} f_i - \sum_{j=1}^{v} \xi_{L_j} = 16 - (6 + 5 + 5) = 0$$

(4) Since DOF of the obtained new PM is $F^* = 0$, the 4 R pairs (Fig. 6) on the fixed platform can be selected as (Fig. 8) driving pairs simultaneously according to Sect. 2.3.

All the other 18 PMs can be checked in the same (Fig. 11) way and find that the 4 R pairs on the fixed (Fig. 12) platform of each PM can be selected as driving pairs simultaneously.

6 Conclusions

(1) 18 types of 3T1R PMs with no P pairs are obtained using the topological structure synthesis of PMs based on POC equations, as shown in Figs. 4, 5, 6, 7, 8, 9, 10, 11, 12. Among these 18 types of PMs, 3 types of 3T1R PMs shown in Figs. 7a, 9a and 10 respectively have been used in SCARA robot design. The other 15 types of 3T1R PMs also feature simple structure and may have the possibility to be used in new SCARA robot design.

(2) The topological structure synthesis method based on POC equations discussed in this paper is the further development of our early work on topological structure synthesis of PMs. New achievements of this paper include: (a) general method for topological structure synthesis of simple branches based on POC equation for serial mechanisms; (b) general method for topological structure synthesis of complex branches based on replacement a sub-SOC in a SOC branch with a topologically equivalent sub-PM; (c) general method for determination of geometrical conditions for assembling branches between platforms (for the same geometrical condition, there may be several different branch assembling schemes); (d) general criteria for selection of driving pairs.

(3) The three basic equations (Eqs. (1, 2 and 4a, b)) unveil the internal relations among topological structure, POC set and DOF. They are the theoretical basis for mechanism structure analysis and synthesis.

Acknowledgment. The authors acknowledge the financial support of the national science foundation of China (NSFC) with grant numbers of 51375062, 51365036, 51475050 and 51405039.

References

1. Mitsubishi SCARA robots-RH Series [EB/OL], 16 May 2015. http://www.meau.com/eprise/main/sites/public/Products/Robots/Robot_Videos/default
2. ABB. IRB 360 FlexPicker [EB/OL], 16 May 2015. http://new.abb.com/products/robotics/industrial-robots/irb-360
3. Clavel, R.: Device for displacing and positioning an element in space. Switzerland, WO1987003528, 18 June 1987
4. Pierrot, F., Reynaud, C., Fournier, A.: Delta—a simple and efficient parallel robot. Robotica 8(2), 105–109 (1990)
5. Fanuc. M-1iA series robots [EB/OL], 16 May 2015. http://robot.fanucamerica.com/products/robots/product.aspx
6. Pierrot, F., Nabat, V., Company, O., et al.: Optimal design of a 4-DOF parallel manipulator: from academia to industry. IEEE Trans. Rob. 25(2), 213–224 (2009)
7. Pierrot, F.: High-speed parallel robot with four degrees of freedom: France, WO2006087399A1, 24 Aug 2006
8. An articulated traveling plate [EB/OL], 16 May 2015. http://www.adept.com/products/robots/parallel/quattro-s650h/downloads

9. Adept Quattro s650H [EB/OL], 16 May 2015. http://www.adept.com/products/robots/parallel/quattro-s650h/general
10. Huang, T., Zhao, X.,Wang, P.,et al.: Rod-wheel combined PM with three translations and a rotation: China, CN102152306A. Accessed 17 Aug 2011
11. Huang, T., Wang, M., Liu, S., et al.: A PM with three translations and one rotation: China, CN201907121U, 27 July 2011
12. Huang, T., Mei, J., Wang, H., et al.: A high speed PM with three translations and one rotation: China, CN101863024A, 20 Oct 2010
13. Yuhang, L., Jiangping, M., Songtao, L., et al.: Dynamic dimensional synthesis of a 4-DOF high-speed parallel manipulator (in Chinese). Chin. J. Mech. Eng. 50(19), 32–40 (2014)
14. Xinjun, L., Fugui, X., Jinsong, W.: Current opportunities in the field of mechanisms in China (in Chinese). Chin. J. Mech. Eng. 51(13), 2–12 (2015)
15. Yang, T.-L., Jin, Q., Liu, A.-X., et al.: Structural synthesis and classification of the 3-DOF translational parallel robot mechanisms based on the units of single-opened-chain (in Chinese). Chin. J. Mech. Eng. 38(8), 31–36 (2002)
16. Jin, Q., Yang, T.-L.: Theory for topology synthesis of parallel manipulators and its application to three-dimension-translation parallel manipulators. ASME J. Mech. Des. 126, 625–639 (2004)
17. Jin, Q., Yang, T.-L.: Synthesis and analysis of a group of 3-degree-of-freedom partially decoupled parallel manipulators. ASME J. Mech. Des. 126, 301–306 (2004)
18. Yang, T.-L.: Theory and Application of Robot Mechanism Topology (in Chinese). China machine press, Beijing (2004)
19. Shen, H.-P., Yang, T.-L., et al.: Synthesis and analysis of kinematic structures of 6-DOF parallel robotic mechanisms. Mech. Mach. Theory 40, 1164–1180 (2005)
20. Shen, H.-P., Yang, T.-L., et al.: Structure and displacement analysis of a novel three-translation PM. Mech. Mach. Theory 40, 1181–1194 (2005)
21. Yang, T.-L., Liu, A.-X., Luo, Y.-F., et al.: Position and orientation characteristic equation for topological design of robot mechanisms. ASME J. Mech. Des. 131, 021001-1-17 (2009)
22. Yang, T.-L., Sun, D.-J.: A general DOF formula for PMs and multi-loop spatial mechanisms. ASME J. Mech. Robot. 4(1), 011001-1-17 (2012)
23. Yang, T.-L., Liu, A.-X., Shen, H.-P., et al.: On the correctness and strictness of the POC equation for topological structure design of robot mechanisms. ASME J. Mech. Robot. 5(2), 021009-1-18 (2013)

Structural Analysis of Parallel Mechanisms Using Conformal Geometric Algebra

Lubin Hang[1(✉)], Chengwei Shen[1], and Tingli Yang[2]

[1] Shanghai University of Engineering Science, Shanghai 201620, China
hanglb@126.com
[2] Changzhou University, Changzhou 213016, China

Abstract. The operable description of parallel mechanisms is the key to auto-matic derivation of structural analysis and synthesis. Conformal geometric algebra is introduced to describe robot mechanisms in this paper. A group of basis bivectors $\{e_{23}, e_{31}, e_{12}, e_{1\infty}, e_{2\infty}, e_{3\infty}\}$ is established to express position and orientation characteristics of the joint axis. The union of each joint's char-acteristics of serial mechanisms is defined via outer product operation and the intersection of each limb's characteristics of parallel mechanisms is defined by shuffle product operation, respectively. In this work, a new algebraic symbol algorithm for end-effectors' characteristics is proposed and proved validity via two case studies of 3-RRR and 4-URU PMs which is suitable for computer-aided derivation of mechanisms position and orientation characteristics.

Keywords: Conformal geometric algebra · Parallel mechanism · Position and orientation characteristic · Symbolic computation · Automatic derivation

1 Introduction

The digital design for type of parallel mechanisms (PMs) has been a critical part of developing software for mechanical innovation design. The method based on Position and Orientation Characteristics sets (POC sets) proposes the concept of dimensional constraint types to describe geometric relations of joint axes, which can represents mechanism topology structure by concise symbols and it is applied to type synthesis of various PMs [1–4]. For the automatic derivation based on this method, operable rep-resentation of mechanism structure is the key to the problem.

Gao [5] has developed the integrated software for computer aided type synthesis of PMs in the a framework of G_F sets. Li [6, 7] adopts the twists in the form of Geometric Algebra to describe the branches of PMs and proposes a mobility analysis approach for limited-DOF PMs based on the outer product operation. Husty et al. [8] introduce an algebraic approach via Study's kinematic mapping of the Euclidean group to reveal global kinematic behavior properties of 3-RPS parallel manipulator. Based on this approach, Nurahmi [9] describes a 4-RUU parallel manipulator by a set of constraint equations and computes the primary decomposition which can deal with the charac-terization of the operation modes.

As a new mathematical tool of geometric description and operation, Conformal Geometric Algebra (CGA) provides the unified indication independent of coordinate

© Springer International Publishing Switzerland 2016
N. Kubota et al. (Eds.): ICIRA 2016, Part I, LNAI 9834, pp. 162–174, 2016.
DOI: 10.1007/978-3-319-43506-0_14

for geometric entities and realizes direct operation of geometries via its invariant system [10]. Its efficient algorithm of elimination and simplification contributes to the more concise and compact expressions, which has been widely applied in robotics [11–14]. Still, CGA is rarely used for robotic structure analysis.

In this paper, conformal geometric algebra is introduced to structural analysis of parallel mechanisms for automatic derivation of position and orientation characteristics. The group of basis bivectors $\{e_{23}, e_{31}, e_{12}, e_{1\infty}, e_{2\infty}, e_{3\infty}\}$ in unified form is derived in 5-dimensional conformal space to describe revolute and prismatic joint axes. The union and intersection of mechanism analysis are defined as mathematical operation in CGA. Finally, the motion output characteristics of 3-RRR and 4-URU PMs are obtained via this proposed method to shows its applicability.

2 Basics of Conformal Geometric Algebra

2.1 5-Dimensional Conformal Space

Conformal geometric algebra, a unified and compact homogeneous algebraic framework for computational geometry, which enables to realize extremely complicated symbolic geometric computations and advantageous application in geometric modeling and computing. As the extension of geometric algebra [15], the elementary operator of conformal geometric algebra is geometric product, which can be expressed as

$$ab = a \cdot b + a \wedge b \tag{1}$$

Where $a \cdot b$ means the inner product of vectors a and b, $a \wedge b$ means the outer product of them. This operator can be applied to n-dimensional space.

As for the 3-dimensional Euclidean space $\mathbf{R}^3$, geometric algebra has a group of orthogonal basis vectors $\{e_1, e_2, e_3\}$. The 5-dimensional conformal space are established via introducing two additional orthogonal basis vectors $\{e_+, e_-\}$ with positive and negative signature in Minkowski space to $\mathbf{R}^3$, which is called as $\mathbf{G}^{4,\,1}$. These two additional vectors have the following properties

$$e_+^2 = 1, \ e_-^2 = 1, \ e_+ \cdot e_- = 0 \tag{2}$$

Another group of basis vectors $\{e_0, e_\infty\}$ with geometric meaning [16] can be defined as

$$e_0 = \frac{1}{2}(e_- - e_+), \ e_\infty = e_- + e_+ \tag{3}$$

Where e_0 can represents the 3-dimensional origin and e_∞ can represents the infinity.

According to the above definition, $\{e_1, e_2, e_3, e_0, e_\infty\}$ is a group of basis vectors which forms the 5D conformal space and these five basis vectors have the following relation

$$\begin{cases} e_1^2 = e_2^2 = e_3^2 = 1 \\ e_0^2 = e_\infty^2 = 0 \\ e_0 \cdot e_\infty = -1 \end{cases} \quad (4)$$

The exclusive Euclidean space can be defined by the exclusive infinity e_∞, which means the representation of Euclidean geometry can be independent of coordinate [17], and it is different from the traditional model. Based on this, the representation of geometric entities and rigid body motion are more intuitive and explicit, which should be advantageous to describe the links and joints of mechanisms.

2.2   Basis Bivectors for Line

There are two representations of geometric entities: one is based on inner product null space and the other is on outer product null space, as shown in Table 1 [11]. The representation based on inner product null space means a geometric entity is generated by the intersection of relevant geometric entities. As for the representation based on outer product null space, the geometric entity can be defined by the points lie on itself.

Table 1. Representation of geometries

Geometry entities	Inner product null space	Outer product null space
Point	$P = x + 0.5 \cdot x^2 e_\infty + e_0$	None
Sphere	$S = P - 0.5 \cdot r^2 e_\infty$	$S^* = P_1 \wedge P_2 \wedge P_3 \wedge P_4$
Plane	$\pi = n + d e_\infty$	$\pi^* = P_1 \wedge P_2 \wedge P_3 \wedge e_\infty$
Circle	$C = S_1 \wedge S_2$	$C^* = P_1 \wedge P_2 \wedge P_3$
Line	$L = \pi_1 \wedge \pi_2$	$L^* = P_1 \wedge P_2 \wedge e_\infty$
Point pair	$Pp = S_1 \wedge S_2 \wedge S_3$	$Pp^* = P_1 \wedge P_2$

Here, the superscription "*" denotes the dualization operator, which can be obtained as

$$A^* = AI^{-1} = -AI = -A(e_0 \wedge e_1 \wedge e_2 \wedge e_3 \wedge e_\infty) \quad (5)$$

According to the standard representation of a plane in Table 1, any plane in 5D space can be described as

$$\pi_i = a_i e_1 + b_i e_2 + c_i e_3 + d_i e_\infty \quad (6)$$

Where (a_i, b_i, c_i) refers to the 3D normal vector of the plane π_i and d_i refers to the distance from the plane π_i to the origin.

Then the line of intersection of these two planes, as shown in Fig. 1, can be expressed based on the inner product null space.

$$L = (a_1e_1 + b_1e_2 + c_1e_3 + d_1e_\infty) \wedge (a_2e_1 + b_2e_2 + c_2e_3 + d_2e_\infty) \qquad (7)$$

Furthermore, a line in 5D space can be define by

$$L = \alpha e_{23} + \beta e_{31} + \gamma e_{31} + p e_{1\infty} + q e_{2\infty} + r e_{3\infty} \qquad (8)$$

Where

$$e_{23} = e_2 \wedge e_3, \; e_{31} = e_3 \wedge e_1, \; e_{32} = e_3 \wedge e_2, \; e_{1\infty} = e_1 \wedge e_\infty, \; e_{2\infty} = e_2 \wedge e_\infty, \; e_{3\infty} = e_3 \wedge e_\infty$$
$$e_{23}^2 = e_{31}^2 = e_{12}^2 = e_{1\infty}^2 = e_{2\infty}^2 = e_{3\infty}^2 = 1$$

The e_{ij} refers to the 2-blade obtained by the outer product of e_i and e_j, which denotes the 2-dimensional space that is directional but formless, as shown in Fig. 2. In Eq. (8), vector (α, β, γ) means the Euclidean direction of the line and vector (p, q, r) is the movement vector. Based on the above derivation, any line in a 5D conformal space can be identified by using these six 2-blades $\{e_{23}, e_{12}, e_{31}, e_{1\infty}, e_{2\infty}, e_{3\infty}\}$ as basis bivectors with bivectors $\{e_{23}, e_{12}, e_{31}\}$ for the orientation and $\{e_{1\infty}, e_{2\infty}, e_{3\infty}\}$ for the position, which can also be extended to structural analysis of PMs.

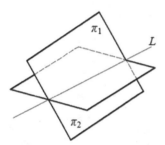

Fig. 1. Line of intersection of planes

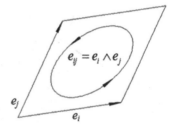

Fig. 2. Outer product of e_i and e_j

3   Representation of POC

In this section, the mathematical description of relative translation and rotation of two movable links is given based on the above group of basis bivectors representing the position and the orientation of a joint. Corresponding to mechanism topological structure, the operation of a POC equation through outer product and shuffle product is defined hence the representation of POC sets with CGA is established.

3.1   POC Matrix of Kinematic Pairs

General kinematic pairs include revolute joint (R), prismatic joint (P), universal joint (U), cylinder joint (C), spherical joint (S) and so on. Here we introduce the POC sets of

R pair and P pair since any kinematic pair in mechanisms can be treated as the combination of certain numbers of revolute or prismatic joints.

(1) POC matrix of revolute joint

The revolute joint is illustrated in Fig. 3. Point O on the moving link is chosen as origin e_0 and the moving coordinate system is attached to the moving link is established. From the line definition using conformal geometric algebra, the axis l_R of a revolute joint can be described as

$$l_R = \alpha e_{23} + \beta e_{31} + \gamma e_{31} + p e_{1\infty} + q e_{2\infty} + r e_{3\infty} \qquad (9)$$

Where $\alpha e_{23} + \beta e_{12} + \gamma e_{31}$ means the moving link has one rotation around the axis of R pair and $p e_{1\infty} + q e_{2\infty} + r e_{3\infty}$ means the moving link also has a derivative translation perpendicular to the axis of R pair and radius vector ρ, i.e., $(p, q, r) = (\alpha, \beta, \gamma) \times \rho$.

In addition, the POC matrix of R pair can be defined as

$$M_R|_{CGA} = \begin{bmatrix} \alpha e_{23} + \beta e_{31} + \gamma e_{31} \\ \{p e_{1\infty} + q e_{2\infty} + r e_{3\infty}\} \end{bmatrix} \text{ or } M_R|_{CGA} = \begin{bmatrix} \{\alpha e_{23} + \beta e_{31} + \gamma e_{31}\} \\ p e_{1\infty} + q e_{2\infty} + r e_{3\infty} \end{bmatrix} \qquad (10)$$

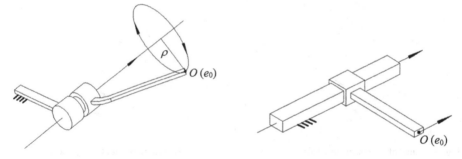

Fig. 3. Revolute joint   **Fig. 4.** Prismatic joint

Where { } refers to dependent motion.

(2) POC matrix of prismatic joint

The prismatic joint is illustrated in Fig. 4. Point O on the moving link is chosen as origin e_0 and the moving coordinate system is attached to the moving link is established. The moving link can only translate without rotation, so its axis can be described by above three position bivectors as

$$l_P = l e_{1\infty} + m e_{2\infty} + n e_{3\infty} \qquad (11)$$

Where (l, m, n) refers to direction of P pair in Euclidean space and $l^2 + m^2 + n^2 = 1$. Furthermore, the POC matrix of P pair can be defined as

$$M_\text{P}|_\text{CGA} = \begin{bmatrix} 0 \\ le_{1\infty} + me_{2\infty} + ne_{3\infty} \end{bmatrix} \tag{12}$$

Based on the above definition and analysis, the polynomials composed of position and orientation bivectors of a joint axis can describe relative motion between two links connected by a kinematic pair, and this can extend to the relationship of any two links in a mechanism. As for indicating motion characteristics of link of kinematic pair by using the group of basis bivectors $\{e_{23}, e_{12}, e_{31}, e_{1\infty}, e_{2\infty}, e_{3\infty}\}$, $\{e_{23}, e_{12}, e_{31}\}$ represents basis bivectors of rotation and $\{e_{1\infty}, e_{2\infty}, e_{3\infty}\}$ represents basis bivectors of translation, respectively.

3.2 POC Equation of Serial Mechanism

The motion output of the end-effector can be regarded as composition of motion characteristics of every joint in a serial mechanism, which means the POC set of the end-effector of serial mechanism is the union of all of the joints' POC sets. In conformal geometric algebra, the union of any two vectors A and B can be performed via outer product, so the union operation of joints' POC sets is defined as

$$M_S|_\text{CGA} = \overset{m}{\underset{i=1}{\wedge}} M_{Ji}|_\text{CGA} \tag{13}$$

Where $M_S|_\text{CGA}$ is the POC set of end-effector in the form of CGA, $M_{Ji}|_\text{CGA}$ is the POC set of the ith joint in the form of CGA, and Eq. (13) is the POC equation of a serial mechanism. The result of outer product operation is called Blade, so $M_S|_\text{CGA}$ can be considered as the blade of POC in a serial mechanism, or open chain's blade of POC.

3.3 POC Equation of Parallel Mechanism

A typical parallel mechanism has a base and a moving platform connected in parallel by several limbs. The movement of the moving platform is constrained by individual limbs. So the motion characteristics of a moving platform are the intersection of all limbs' characteristics and each limb can be considered as an open chain in series. Corresponding to the operation in conformal geometric algebra, the intersection of any two vectors A and B is defined as shuffle product in Ref. [18], which can be obtained as

$$A \vee B = (A^* \wedge B^*)J \tag{14}$$

In Eq. (14) the dualization operation is similar to the form of Eq. (5), but the difference is that J refers to the pseudoscalar of the maximum subspace spanned by A and B, instead of the pseudoscalar $e_0 \wedge e_1 \wedge e_2 \wedge e_3 \wedge e_\infty$ of the whole space. A^* and B^* can be expressed as

$$A^* = AJ^{-1} = -AJ, \quad B^* = BJ^{-1} = -BJ \tag{15}$$

Then the intersection of all the limbs' blades of POC is defined as

$$M_{Pa}|_{CGA} = \bigvee_{j=1}^{n} M_{Sj}|_{CGA} \qquad (16)$$

Where $M_{Pa}|_{CGA}$ is the POC set of moving platform in the form of CGA, $M_{Sj}|_{CGA}$ is the POC set of the jth limb in the form of CGA respectively and Eq. (16) is the POC equation of parallel mechanism. Similarly, $M_{Pa}|_{CGA}$ is called as moving platform's blade of POC. Corresponding to mechanism theory, the so-called pseudoscalar J operated in Eq. (16) should be the maximal dimensional blade spanned by all limbs, composed of basis bivectors.

3.4   Structrual Analysis Procedure

Based on the above conformal geometric algebra description of kinematic pairs, serial and parallel mechanisms, the derivation procedure of position and orientation characteristics of robot mechanism can be summarizes as follows:

(1) According to mechanism topological structure, choose the origin e0 and establish the moving coordinate system.
(2) Based on the outer product, obtain the position and orientation characteristics of serial kinematic chains by the union operation.
(3) Establish the POC equation of parallel mechanism based on the blade of position and orientation characteristic of each limb, which can be expressed as the shuffle product operation.
(4) Choose the maximal dimensional blade spanned by all limbs and obtain the position and orientation characteristics of the end-effector.

The above procedure can also be seen in Fig. 5.

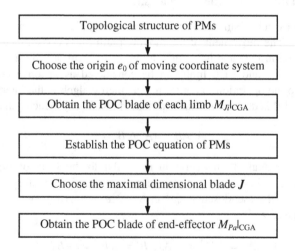

Fig. 5. Derivation procedure of POC of parallel mechanism structural analysis

4 Examples

To further explain and verify the method demonstrated above, the end-effectors of 3-RRR and 4-URU parallel mechanisms are described using CGA and the characteristics of their end-effectors are analyzed excluding singularities in this section.

4.1 3-RRR Parallel Mechanism

Figure 6 shows a type of 3-RRR parallel mechanism, which is composed of a base and a moving platform connected by three RRR chains, and the three axes of the joints in every limbs are orthogonal and all axes of these nine joints intersect at a common point.

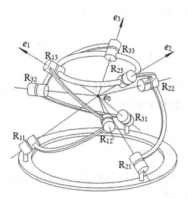

Fig. 6. 3-RRR parallel mechanism

(1) Establish the moving coordinate system

A moving coordinate system is established so that the origin e_0 is selected at the common point and the axes e_1, e_2 and e_3 coincide with R_{13}, R_{23} and R_{33} pair respectively, as shown in Fig. 6.

(2) Obtain the POC blade of each limb

According to CGA description of kinematic pair R and serial mechanism, the POC blades of ith limb of this PM can be obtained by

$$M_{Si}|_{CGA} = M_{R_{i1}}|_{CGA} \wedge M_{R_{i2}}|_{CGA} \wedge M_{R_{i3}}|_{CGA} \tag{17}$$

According to the expression of revolute joint axis using CGA, and then Eq. (17) can be expressed as

$$\begin{cases} M_{S1}|_{CGA} = e_{31} \wedge e_{12} \wedge e_{23} \\ M_{S2}|_{CGA} = e_{12} \wedge e_{23} \wedge e_{31} \\ M_{S3}|_{CGA} = e_{23} \wedge e_{31} \wedge e_{12} \end{cases} \tag{18}$$

(3) Establish the POC equation of parallel mechanism

The POC blade of moving platform is denoted by the intersection of all the limbs, based on the shuffle product operation in CGA, this can be expressed as

$$M_{Pa}|_{\text{CGA}} = M_{S1}|_{\text{CGA}} \vee M_{S2}|_{\text{CGA}} \vee M_{S3}|_{\text{CGA}} \qquad (19)$$

(4) Obtain POC blade of the moving platform

In Eq. (18), the POC blades of each limb are equal, so the result of Eq. (19) can be obtained without the selection of the maximal dimensional blade, which can be expressed as

$$M_{Pa}|_{\text{CGA}} = e_{23} \wedge e_{31} \wedge e_{12} \qquad (20)$$

The POC matrix of the moving platform is

$$M_{Pa}|_{\text{CGA}} = \begin{bmatrix} e_{23} \wedge e_{31} \wedge e_{12} \\ 0 \end{bmatrix} \qquad (21)$$

Which means the moving platform of 3-RRR PM can realize three rotational DOFs, or full rotational capability and no translation.

4.2   4-URU Parallel Mechanism

The 4-URU PM, as is shown in Fig. 7, is composed of a base and a moving platform connected by four URU chains.

The U joint can be regard as the composition of two orthogonal R joints, then the three middle R joint in each limb are parallel. These four R joints located on the base are parallel to each other and perpendicular to plane of base, and so do those four R

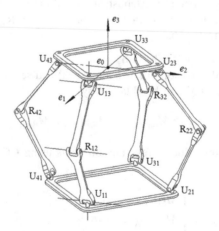

Fig. 7. 4-URU parallel mechanism

joints of moving platform. The shapes of the base and the manipulator are all rhombus and they are parallel to each other in the assembly condition.

(1) Establish the moving coordinate system

The intersection point of the connections of centralities of the diagonal U joints located on the moving platform is selected as the origin e_0 and a moving coordinate system is established so that the axis e_1 coincides with the connection of centralities of U_{13} and U_{33} and the axis e_2 coincides with the connection of centralities of U_{23} and U_{43}. The axis e_3 is perpendicular to the moving platform.

(2) Obtain the POC blade of each limb

For the first limb $U_{11}R_{12}U_{13}$, the POC blades of these three kinematic pairs can be expressed as

$$\begin{cases} M_{U_{11}}|_{CGA} = (e_{12} + q_{11}e_{2\infty}) \wedge (e_{31} + p_{11}e_{1\infty} + r_{11}e_{3\infty}) \\ M_{R_{12}}|_{CGA} = e_{31} + p_{12}e_{1\infty} + r_{12}e_{3\infty} \\ M_{U_{13}}|_{CGA} = (e_{31} + r_{13}e_{3\infty}) \wedge (e_{12} + q_{13}e_{2\infty}) \end{cases} \tag{22}$$

Where $p_{11}, q_{11}, r_{11}, p_{12}, r_{12}, q_{13}$ and r_{13} are scalar coefficients related to the positions of joint axes. The POC blade of $U_{11}R_{12}U_{13}$ is the outer product of each joint, which can be obtained as

$$M_{S1}|_{CGA} = e_{31} \wedge e_{12} \wedge e_{1\infty} \wedge e_{2\infty} \wedge e_{3\infty} \tag{23}$$

According to the symmetry of this mechanism, the POC blade of the third limb $U_{31}R_{32}U_{33}$ is same as the first limb

$$M_{S3}|_{CGA} = M_{S1}|_{CGA} \tag{24}$$

For the second limb $U_{21}R_{22}U_{23}$, the POC blades of these three kinematic pairs can be expressed as

$$\begin{cases} M_{U_{21}}|_{CGA} = (e_{12} + p_{21}e_{1\infty}) \wedge (e_{23} + q_{21}e_{2\infty} + r_{21}e_{3\infty}) \\ M_{R_{22}}|_{CGA} = e_{23} + q_{22}e_{2\infty} + r_{22}e_{3\infty} \\ M_{U_{23}}|_{CGA} = (e_{23} + r_{23}e_{3\infty}) \wedge (e_{12} + p_{23}e_{1\infty}) \end{cases} \tag{25}$$

Where $p_{21}, q_{21}, r_{21}, q_{22}, r_{22}, p_{23}$ and r_{23} are scalar coefficients related to the positions of joint axes. The POC blade of $U_{21}R_{22}U_{23}$ is the outer product of each joint, which can be obtained as

$$M_{S2}|_{CGA} = e_{23} \wedge e_{12} \wedge e_{1\infty} \wedge e_{2\infty} \wedge e_{3\infty} \tag{26}$$

According to the symmetry of this mechanism, the POC blade of the forth limb $U_{41}R_{42}U_{43}$ is same as the second limb

$$M_{S4}|_{CGA} = M_{S2}|_{CGA} \tag{27}$$

(3) Establish the POC equation of parallel mechanism

The POC blade of the moving platform is shuffle product of POC blades of every limb, which can be expressed as

$$M_{Pa}|_{CGA} = M_{S1}|_{CGA} \vee M_{S2}|_{CGA} \vee M_{S3}|_{CGA} \vee M_{S4}|_{CGA} \tag{28}$$

Due to Eqs. (24) and (27), the follow relation can be obtained

$$\begin{aligned} M_{S1}|_{CGA} \vee M_{S3}|_{CGA} &= e_{31} \wedge e_{12} \wedge e_{1\infty} \wedge e_{2\infty} \wedge e_{3\infty} \\ M_{S2}|_{CGA} \vee M_{S4}|_{CGA} &= e_{23} \wedge e_{12} \wedge e_{1\infty} \wedge e_{2\infty} \wedge e_{3\infty} \end{aligned} \tag{29}$$

Then Eq. (28) can be written as

$$M_{Pa}|_{CGA} = M_{S1}|_{CGA} \vee M_{S2}|_{CGA} \tag{30}$$

(4) Obtain POC blade of the moving platform

The maximal dimensional blade J spanned by $M_{S1}|_{CGA}$ and $M_{S2}|_{CGA}$ can be chosen as

$$J = e_{23} \wedge e_{31} \wedge e_{12} \wedge e_{1\infty} \wedge e_{2\infty} \wedge e_{3\infty} \tag{31}$$

According to Eq. (15), the dualization of $M_{S1}|_{CGA}$ and $M_{S2}|_{CGA}$ can be obtained as

$$\begin{cases} M_{S1}|_{CGA}{}^* = e_{23} \\ M_{S2}|_{CGA}{}^* = e_{31} \end{cases} \tag{32}$$

The POC blade of the moving platform can be expressed as

$$M_{Pa}|_{CGA} = (e_{23} \wedge e_{31})(e_{23} \wedge e_{31} \wedge e_{12} \wedge e_{1\infty} \wedge e_{2\infty} \wedge e_{3\infty}) \tag{33}$$

Calculating Eq. (33), then

$$M_{Pa}|_{CGA} = e_{12} \wedge e_{1\infty} \wedge e_{2\infty} \wedge e_{3\infty} \tag{34}$$

The POC matrix of the moving platform is

$$M_{Pa}|_{CGA} = \begin{bmatrix} e_{1\infty} \wedge e_{2\infty} \wedge e_{3\infty} \\ e_{12} \end{bmatrix} \tag{35}$$

The result indicates that the moving platform of 4-URU parallel mechanism has one rotation DOF which is perpendicular to the base and three translation DOFs.

5 Conclusions

In this paper, conformal geometric algebra is introduced to structural analysis of PMs and some conclusions can be drawn as follows:

1. The basis vectors in 5-dimensional conformal space are formed into six independent bivectors which can describe position and orientation of joint axis and also corresponding to the motion characteristics of moving link.
2. The end link's characteristics of serial mechanism are defined as the union of each joint, operated by outer product. The moving platform's characteristics of parallel mechanism are defined as the intersection of each limb, operated by shuffle product. The motion output characteristic is expressed as the blades connected by "∧"s and "+"s.
3. The motion characteristics of 3-RRR and 4-URU PMs are analyzed by this proposed method which is different from the traditional algorithms and shows intuition and concise.
4. Due to the unity of this group of basis bivectors, they can be further applied to rigid body motion and kinematics analysis.

Acknowledgement. The authors would like to acknowledge the financial support of the Natural Science Foundation of China under Grant 51475050 and also thank the reviewers for their suggestions and comments, which have helped to improve the quality of this paper.

References

1. Jin, Q., Yang, T.L.: Theory for topology synthesis of parallel manipulators and its application to three-dimension-translation parallel manipulators. ASME J. Mech. Des. **126**(4), 625–639 (2004)
2. Yang, T.L., Liu, A.X., Jin, Q., Luo, Y.F., Shen, H.P., Hang, L.B.: Position and orientation characteristic equation for topological design of robot mechanisms. ASME J. Mech. Des. **131**(2), 021001 (2009)
3. Yang, T.L., Liu, A.X., Shen, H.P., Luo, Y.F., Hang, L.B., Shi, Z.X.: On the correctness and strictness of the position and orientation characteristic equation for topological structure design of robot mechanisms. ASME J. Mech. Robot. 5(2), 021009 (2013)
4. Yang, T.L., Liu, A.X., Luo, Y.F.: Theory and Application of Robot Mechanism Topology. Science Press, Beijing (2012)
5. Meng, X., Gao, F.: A framework for computer-aided type synthesis of parallel robotic mechanisms. Proc. Inst. Mech. Eng. Part C J. Mech. Eng. Sci. **228**(18), 3496–3504 (2014)
6. Chai, X., Li, Q.: Mobility analysis of two limited-DOF parallel mechanisms using geometric algebra. In: Zhang, X., Liu, H., Chen, Z., Wang, N. (eds.) ICIRA 2014, Part I. LNCS, vol. 8917, pp. 13–22. Springer, Heidelberg (2014)
7. Li, Q., Chai, X.: Mobility analysis of limited-DOF parallel mechanisms in the framework of geometric algebra. ASME J. Mech. Robot. 8(4), 041005 (2015)
8. Schadlbauer, J., Walter, D.R., Husty, M.L.: The 3-RPS parallel manipulator from an algebraic viewpoint. Mech. Mach. Theor. **75**, 161–176 (2014)

9. Nurahmi, L., Caro, S., Wenger, P., Schadlbauer, J., Husty, M.: Reconfiguration analysis of a 4-RUU parallel manipulator. Mech. Mach. Theor. **96**, 269–289 (2016)
10. Hestenes, D., Li, H., Rockwood, A.: New algebraic tools for classical geometry. In: Sommer, G. (ed.) Geometric Computing with Clifford Algebras, pp. 3–26. Springer, Heidelberg (2001)
11. Zamora, J., Bayro-Corrochano, E.: Inverse kinematics, fixation and grasping using conformal geometric algebra. In: Proceedings of 2004 IEEE International Conference on Intelligent Robots and Systems, vol. 4, pp. 3841–3846, October 2004
12. Bayro-Corrochano, E.: Robot perception and action using conformal geometric algebra. In: Handbook of Geometric Computing, pp. 405–458. Springer, Heidelberg (2005)
13. Wei, Y., Jian, S., He, S., Wang, Z.: General approach for inverse kinematics of nR robots. Mech. Mach. Theor. **75**, 97–106 (2014)
14. Kim, J.S., Jeong, J.H., Park, J.H.: Inverse kinematics and geometric singularity analysis of a 3-SPS/S redundant motion mechanism using conformal geometric algebra. Mech. Mach. Theor. **90**, 23–36 (2015)
15. Li, H., Hestenes, D., Rockwood, A.: Generalized homogeneous coordinates for computational geometry. In: Sommer, G. (ed.) Geometric Computing with Clifford Algebras, pp. 27–59. Springer, Heidelberg (2001)
16. Hildenbrand, D.: Geometric computing in computer graphics and robotics using conformal geometric algebra (Doctoral dissertation, Hildenbrand, Dietmar) (2006)
17. Hongbo, L.: Conformal geometric algebra - a new framework for computational geometry. J. Comput. Aided Des. Comput. Graph. **17**(11), 2383–2393 (2005)
18. Selig, J.M.: Geometric fundamentals of robotics. Springer Science & Business Media, New York (2005)

Kinematics Analysis of a New Type 3T1R PM

Wei Qin[1], Lu-Bin Hang[1(✉)], An-Xin Liu[2], Hui-Ping Shen[3],
Yan Wang[4], and Ting-Li Yang[3]

[1] Shanghai University of Engineering Science, Shanghai, China
hanglb@126.com
[2] PLA University of Science and Technology, Nanjing, China
[3] Changzhou University, Changzhou, China
[4] Shanghai Jiao Tong University, Shanghai, China

Abstract. A new type of four degree of freedom (DOF) symmetric parallel mechanism (PM) with Schönflies motion is proposed based on the topological structure synthesis. The type synthesis process of the PM is carried out by using the theory of POC set. The new mechanism possesses the properties of folding characteristic, extensibility and large coverage of workspace in XY plane. The forward kinematics is solved by Sylvester's dialytic elimination in Maple. This research has certain theoretical significance for the synthesis and analysis of the other PM.

Keywords: PM · 3T1R · Kinematics analysis · Sylvester's dialytic elimination

1 Introduction

The PM has been widely concerned and studied since it is proposed, for the reasons that it has the characteristics of simple structure, high rigidity, high bearing capacity, high positioning accuracy and easy to control [1], and it is widely used in the field of assembly line, parallel machine tool, flight simulators and so on. The research of the PM is mainly focus on the process sectors of assembly line like high speed capture, precise positioning assembly and material handling. Therefore, some important efforts have been carried out in recent years on new type of 3T1R-PM. In 1999, a new kind of 3T1R PM suit for the industrial handling is proposed by L Rolland [2]. A new 3T1R PM is proposed by Jin Qiong [3] based on the theory of Single Open Chain. Another novel PM with Schönflies motion has been proposed high-speed pick-and-place manipulation in industrial lines in 2015 by Xie Fugui [4].

In this paper, a branch is added to the original Delta PM based on the theory of POC set to construct a novel 3T1R PM. The 3T1R PM proposed is different from the Delta PM which is invented by Clavel [5] in 1988, and can realize the Schönflies motion. The new PM possesses spatial folding characteristics and planar extensibility, and also it takes a small motion space in z direction.

The paper is organized as follows. The type synthesis of 3T1R PM is given in Sect. 2. The kinematic analysis is carried out in Sect. 3. The example analysis of the new mechanism's forward kinematics is carried out in Sect. 4, and 4 real solutions of the example are obtained. Finally, conclusions are drawn in Sect. 5.

© Springer International Publishing Switzerland 2016
N. Kubota et al. (Eds.): ICIRA 2016, Part I, LNAI 9834, pp. 175–185, 2016.
DOI: 10.1007/978-3-319-43506-0_15

2   Type Synthesis of PM

2.1   Design Requirements and the Expected Output Matrix of PM

The expected motion, which the mechanism should realize, is a 4 DOF motion of three-dimension translation and one-dimension rotation, the POC set of the PM should be derived as $M_{P_a} = \begin{bmatrix} t^3 \\ r^1 \end{bmatrix}$. The methodology of position and orientation characteristic sets and its relevant definition in this paper can be referred to Ref. [6] (Fig. 1).

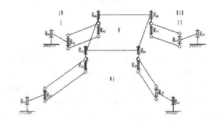

Fig. 1. The mechanism sketch of 3T1R PM

2.2   The SOC Branches' Structure Type of the PM

In order to achieve a PM with symmetric and simple structure, there is only one driving pair in each branch and all the branches set to be the same. The POC set of each SOC branch [7] is $M_i = \begin{bmatrix} t^3 \\ r^1(\parallel R) \end{bmatrix}_i, (i = 1, 2, 3, 4)$, according to the POC set theory [8], each branch consists of two revolute joints R_{i1}, R_{i4}, and a parallel quadrilateral connecting rod, which is equivalent as $HSOC\{-R_{i1}\|R(-P^{(4R)})\|R_{i4}-\}$, then the branch combination scheme of 3T1R PM is denoted by $4\text{-}HSOC\{-R^{i1}\|R(-P^{(4R)})\|R_{i4}-\}$.

2.3   PM Synthesis with the Output Motion Characteristic Matrix

1) **The POC sets of each branch end.**

   The POC sets of SOC_i (i = 1,2,3,4) branch is $M_i = \begin{bmatrix} t^3 \\ r^1(\parallel R) \end{bmatrix}_1$

2) **The POC sets of the PM.**

   Substitute the POC matrix of each branch into formula $M_{P_a} = \cap_{i=1}^{\nu+1} M_{b_i}$ [9], then the POC sets of the PM is obtained.

$$M_{P_a} = \begin{bmatrix} t^3 \\ r^1(\parallel R) \end{bmatrix} \cap \begin{bmatrix} t^3 \\ r^1(\parallel R) \end{bmatrix} \cap \begin{bmatrix} t^3 \\ r^1(\parallel R) \end{bmatrix} \cap \begin{bmatrix} t^3 \\ r^1(\parallel R) \end{bmatrix} \Rightarrow \begin{bmatrix} t^3 \\ r^1 \end{bmatrix}$$

Where the symbol $\Rightarrow$ represents the POC set on the left side of the equation is to be obtained by intersection operation of all the POC sets on the right side of the equation.

2.4 The DOF Analysis of the PM

The POC set of the moving platform is constrained by the DOF (degrees of freedom) of the mechanism, so to analyze whether the DOF of the PM can meet the design demand (DOF = 4) is necessary.

(1) **The number of independent displacement equation ξ_{Li}.**
According to the formula [10] of the number of the independent loop ξ_{Li}, The first independent loop ξ_{L1}, which consists of branch I and II:

$$\xi_{L_i} = \dim_\bullet \left\{ \sum_{i=1}^{2} M_{b_i} \right\} = \dim_\bullet \left\{ \left[\begin{matrix} t^3 \\ r^1(\| R) \end{matrix} \right] \cup \left[\begin{matrix} t^3 \\ r^1(\| R) \end{matrix} \right] \right\} = \dim_\bullet \left\{ \left[\begin{matrix} t^3 \\ r^1 \end{matrix} \right] \right\} = 4$$

In the same way, $\xi_{L2} = 4$, $\xi_{L3} = 4$,

(2) **The calculation of the PM's DOF.**
After the parallelogram mechanism in the PM has been equivalently transformed, there are 12 revolute joints and 4 translation pairs, the total number of the kinematic pairs is 16. There are three basic loops in the mechanism, The first one is the spatial loop $-R_{11}\|R_{12}(-P^{(4R)})\|R_{13}R_{21}\|R_{22}(-P^{(4R)})\|R_{23}-$, the second and third branches can be denoted by $-R_{31}\|R_{32}(-P^{(4R)})\|R_{33}-$, and $-R_{41}\|R_{42}(-P^{(4R)})\|R_{43}-$ respectively, so $\xi_{L1} = 4$, $\xi_{L2} = 4$ and $\xi_{L3} = 4$. By using the formula of DOF [11], we can obtain:

$$F = \sum_{i=1}^{m} f_i - \sum_{j=1}^{v} \xi_{L_j} = 16 - 3 * 4 = 4 \tag{1}$$

Where m—the number of kinematic pairs
f_i—the DOF of the ith kinematic pair
v—the number of the independent loops
ξ_{Li}—the number of the independent loop of the jth basic loop

According to the DOF analysis of the mechanism above, the degree of freedom of the 3T1R PM can meet the design requirement (Fig. 2).

2.5 Coupling Degree of the PM's BKC

The coupling degree [12] indicates complexity of the kinematic and dynamic problem of multi-loop BKC, and it can be regarded as one of the indexes for selecting optimal

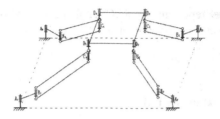

Fig. 2. The mechanism sketch of the equivalent 3T1R PM

structure type of topological design of mechanism, so the analysis of the PM's coupling degree is given based on the formula [6]:

$$
\begin{cases}
\Delta_j = m_j - I_j - \xi_{L_j} \\
\kappa = \frac{1}{2}\min_{\bullet}\left\{\left(\sum_{j=1}^{v}|\Delta_j|\right)\right\}
\end{cases}
\tag{2}
$$

Where m_i—the number of the jth SOC's kinematic pair

f_i—the DOF of the ith kinematic pair

I_j—the number of the jth SOC's driving pair

ξ_{Li}—the number of the independent loop of the jth basic loop

So, we can obtain that $\Delta_1 = 2{>}0$, $\Delta_2 = -1 < 0$, $\Delta_3 = -1 < 0$. It means the first SOC can add the DOF of the PM by 2, while the second and third SOC can add a constraint to the PM respectively, and reduce the DOF of the PM by 1.

Substitute Δ_i of each SOC into Eq. (2) then we can obtain that $\kappa = 2$, so the kinematics and dynamics of the PM need to analyzed based on multi-loops.

2.6 Analysis of Folding Characteristic and Extensibility

In order to ensure that the PM has good spatial retractable characteristics and can be folded when it doesn't work, the analysis of folding characteristic of the PM should be carried out. And to make sure that the mechanism has a large coverage of working space, the analysis of the mechanism's extensibility is necessary.

Since the axes of the four revolute joints R_i in the four branches are vertical to the base, and the 3T1R PM has the special geometric characteristics in its dimension, the PM has a spatial folding characteristic and planar extensibility, as is shown in Figs. 3 and 4. The structural sketch of the 3T1R PM in a folding state is shown in Fig. 3. The structural sketch of the 3T1R PM in an extension state is shown in Fig. 4.

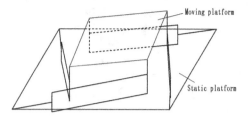

Fig. 3. The structural sketch of the 3T1R PM in a folding state

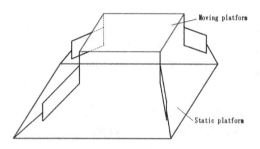

Fig. 4. The structural sketch of the 3T1R PM in an extension state

3 Kinematic Analysis of 3T1R PM

3.1 The Establishment of Kinematic Equation

The static and moving coordinate systems are established respectively. The static coordinate system O-XYZ is attached to the base with the X-axis passing through the origin O of the static coordinate system and the center of the revolute joint R_{11}, and the Y-axis passing through the origin O of the static coordinate system and the center of the revolute joint R_{21}, The moving coordinate system o'-xyz is attached to the moving platform with the x-axis passing through the origin o' of the moving coordinate system and the center of the revolute joint R_{14}, and the y-axis passing through the origin O of the static coordinate system and the center of the revolute joint R_{24}. The direction of Z-axis and z-axis can be determined by the right hand criterion. Suppose the rotation angle of the four revolute joints R_{11}, R_{21}, R_{31} and R_{41} in base can be denoted by $\theta_i (i = 1,2,3,4)$. As is shown in Fig. 6, the circumradius of the base and moving platform are denoted by R and r respectively, the center of revolute joints R_{i1}, R_{i2}, R_{i3} and R_{i4} are denoted by A_i, B_i, C_i and D_i, and the length of each linkage in each branch are denoted by a, b, c (Fig. 5).

The vector expression of each point in the moving coordinate system can be transformed into the static coordinate system according to the transformation matrix of the static and moving coordinates.

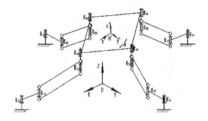

Fig. 5. The mechanism sketch of the 3T1R PM with the coordinate system

$$\overrightarrow{OB_i} = \overrightarrow{OA_i} + \overrightarrow{A_iB_i} \qquad\qquad \overrightarrow{OC_i} = \overrightarrow{OO'} + \overrightarrow{O'D_i} + \overrightarrow{D_iC_i}$$

$$\overrightarrow{OA_i} = R\begin{bmatrix} \cos(\frac{i-1}{2}\pi) \\ \sin(\frac{i-1}{2}\pi) \\ 0 \end{bmatrix}, \quad = \begin{bmatrix} R\cos(\frac{i-1}{2}\pi) - a\cos(\frac{i-1}{2}\pi - \theta_i) \\ R\sin(\frac{i-1}{2}\pi) - a\sin(\frac{i-1}{2}\pi - \theta_i) \\ 0 \end{bmatrix}, \quad = \begin{bmatrix} X_0' + r\cos(\alpha + \frac{i-1}{2}\pi) \\ Y_0' + r\sin(\alpha + \frac{i-1}{2}\pi) \\ Z_0' - c \end{bmatrix}.$$

Let $\varphi_i = \frac{i-1}{2}\pi, (i = 1, 2, 3, 4)$, so that the subsequent derivation process can be simplified, and the vector expressions can be expressed as:

$$\overrightarrow{OA_i} = R\begin{bmatrix} \cos\varphi_i \\ \sin\varphi_i \\ 0 \end{bmatrix}, \quad \overrightarrow{OB_i} = \begin{bmatrix} R\cos\varphi_i - a\cos(\varphi_i - \theta_i) \\ R\sin\varphi_i - a\sin(\varphi_i - \theta_i) \\ 0 \end{bmatrix}, \quad OC_i = \begin{bmatrix} X_0' + r\cos(\alpha + \varphi_i) \\ Y_0' + r\sin(\alpha + \varphi_i) \\ Z_0' - c \end{bmatrix}$$

The vector loop relationship of each branch can be established based on the geometry constraint relationship:

$$\overrightarrow{B_iC_i} = \overrightarrow{OC_i} - \overrightarrow{OB_i} \tag{3}$$

Since $|\overrightarrow{B_iC_i}| = b$, then the kinematics equation can be expressed as:

$$(X_0' + r\cos(\alpha + \varphi_i) - R\cos\varphi_i + a\cos(\varphi_i - \theta_i))^2 + (Y_0' + r\sin(\alpha + \varphi_i)$$
$$- R\sin\varphi_i + a\sin(\varphi_i - \theta_i))^2 + (Z_0' - c)^2 = b^2 \tag{4}$$

3.2   Inverse Position Problem

The inverse position problem is to obtain the four revolute angle $\theta_i (i = 1,2,3,4)$ of driving pairs when the position and orientation parameters X_0', Y_0', Z_0', α are known [13].

Let $D_i = X_0' + r\cos(\alpha + \varphi_i) - R\cos\varphi_i$, $E_i = Y_0' + r\sin(\alpha + \varphi_i) - R\sin\varphi_i$, $F = Z_0' - c$, then Eq. (4) can be expressed as:

$$(D_i + a\cos(\varphi_i - \theta_i))^2 + (E_i + a\sin(\varphi_i - \theta_i))^2 + F^2 = b^2 \tag{5}$$

Expand Eq. (5), then we can obtain:

$$D_i^2 + E_i^2 + F^2 + a^2 - b^2 + 2aD_i\cos(\varphi_i - \theta_i) + 2aE_i\sin(\varphi_i - \theta_i) = 0$$

$$\theta_i = \varphi_i - 2\arctan\frac{2aE_i \pm \sqrt{4a^2(E_i^2 + D_i^2) - (D_i^2 + E_i^2 + F^2 + a^2 - b^2)^2}}{2aD_i + 2aE_i}$$

Where $D_i = X_0' + r\cos(\alpha + \varphi_i) - R\cos\varphi_i$

 $E_i = Y_0' + r\sin(\alpha + \varphi_i) - R\sin\varphi_i$

 $F = Z_0' - c$

 $\varphi_i = \frac{i-1}{2}\pi, (i = 1, 2, 3, 4),$

3.3 Forward Position Problem

The forward position problem is to obtain the position and orientation parameters X_0', Y_0', Z_0', α, when the four revolute angle $\theta_i(i = 1,2,3,4)$ of the driving pairs are known.

According to the projection relationship of the mechanism in the coordinate plane o'-xy, the position equation of the four DOF PM's moving platform in any time can be obtained.

Let $p_i = -R\cos\varphi_i + a\cos(\varphi_i - \theta_i)$, $q_i = -R\sin\varphi_i + a\sin(\varphi_i - \theta_i)$, then Eq. (4) can be expressed as:

$$\left(X_0' + r\cos(\alpha + \varphi_i) + p_i\right)^2 + \left(Y_0' + r\sin(\alpha + \varphi_i) + q_i\right)^2 + \left(Z_0' - c\right)^2 = b^2 \tag{6}$$

Then the four kinematics equations can be expressed as:

$$(X_{0'} + r\cos(\alpha + \varphi_1) + p_1)^2 + (Y_{0'} + r\sin(\alpha + \varphi_1) + q_1)^2 + (Z_{0'} - c)^2 = b^2 \tag{7}$$

$$(X_{0'} + r\cos(\alpha + \varphi_2) + p_i)^2 + (Y_{0'} + r\sin(\alpha + \varphi_2) + q_2)^2 + (Z_{0'} - c)^2 = b^2 \tag{8}$$

$$(X_{0'} + r\cos(\alpha + \varphi_3) + p_3)^2 + (Y_{0'} + r\sin(\alpha + \varphi_3) + q_3)^2 + (Z_{0'} - c)^2 = b^2 \tag{9}$$

$$(X_{0'} + r\cos(\alpha + \varphi_4) + p_4)^2 + (Y_{0'} + r\sin(\alpha + \varphi_4) + q_4)^2 + (Z_{0'} - c)^2 = b^2 \tag{10}$$

According to Eqs. (7)–(10), three independent linear equations are obtained by eliminating $(Z_0'-c)^2$ and b^2 [14]:

$$\begin{cases} A_1 \bullet X_{O'} + B_1 \bullet Y_{O'} + C_1 = 0 \\ A_2 \bullet X_{O'} + B_2 \bullet Y_{O'} + C_2 = 0 \\ A_3 \bullet X_{O'} + B_3 \bullet Y_{O'} + C_3 = 0 \end{cases} \tag{11}$$

Where $A_1 = -2p_4 - 2r_2 \sin \alpha + 2r_2 \cos \alpha + 2p_1$

$A_2 = 2p_2 - 2p_4 - 4r_2 \sin \alpha$

$A_3 = -2p_4 - 2r_2 \sin \alpha - 2r_2 \cos \alpha + 2p_3$

$B_1 = -2q_4 + 2r_2 \sin \alpha + 2r_2 \cos \alpha + 2q_1$

$B_2 = 4r_2 \cos \alpha + 2q_2 - 2q_4$

$B_3 = -2q_4 - 2r_2 \sin \alpha + 2r_2 \cos \alpha + 2q_3$

$C_1 = 2r_2q_4 \cos \alpha + 2r_2p_1 \cos \alpha - 2r_2p_4 \sin \alpha + 2r_2q_1 \sin \alpha + p_1^2 + q_1^2 - p_4^2 - q_4^2$

$C_2 = -2r_2p_4 \sin \alpha - 2r_2p_2 \sin \alpha + 2r_2q_4 \cos \alpha + 2r_2q_2 \cos \alpha + p_2^2 + q_2^2 - p_4^2 - q_4^2$

$C_3 = 2r_2q_4 \cos \alpha - 2r_2p_3 \cos \alpha - 2r_2p_4 \sin \alpha - 2r_2q_3 \sin \alpha + p_3^2 + q_3^2 - p_4^2 - q_4^2$

By using Sylvester's dialytic elimination [15], a univariate nonlinear equation in α can be obtained from

$$\begin{bmatrix} A_1 & B_1 & C_1 \\ A_2 & B_2 & C_2 \\ A_3 & B_3 & C_3 \end{bmatrix} \begin{bmatrix} X_{O'} \\ Y_{O'} \\ 1 \end{bmatrix} = \begin{bmatrix} 0 \\ 0 \\ 0 \end{bmatrix}$$

$$\Rightarrow \begin{vmatrix} A_1 & B_1 & C_1 \\ A_2 & B_2 & C_2 \\ A_3 & B_3 & C_3 \end{vmatrix} = 0 \tag{12}$$

Expand Eq. (12) by using the software Maple, then use the half-angle transformation $t = \tan \frac{\alpha}{2}$, $\cos \alpha = \frac{1-t^2}{1+t^2}$, $\sin \alpha = \frac{2t}{1+t^2}$, and multiply the equation by $(1 + t^2)^2$, a fourth-degree polynomial equation in s can be obtained:

$$H_4 t^4 + H_3 t^3 + H_2 t^2 + H_1 t + H_0 = 0 \tag{13}$$

According to Ref. [16], Eq. (13) has a closed-form solution, after the revolute angle α of moving platform is solved, the output parameters of the moving platform can be solved then.

4   Examples Analysis of the Forward Kinematics

Example: as is shown in Fig. 6, the PM's structure parameters are set as: $R = 150$ mm, $r = 35$ mm, $a_i = 50$ mm, $b_i = 102.98$ mm, $c_i = 77.5$ m. The input revolute angle of each drive pair is set as: $\theta_1 = -58.16°$, $\theta_2 = 110.50°$, $\theta_3 = -10.71°$, $\theta_4 = -123.89°$ respectively.

Following the analysis in above sections, 4 real solutions of the example is obtained and listed in Table 1. The 4 configurations corresponding to the real solutions are plotted in Fig. 7.

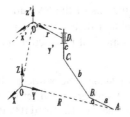

Fig. 6. The kinematic diagram of the branch

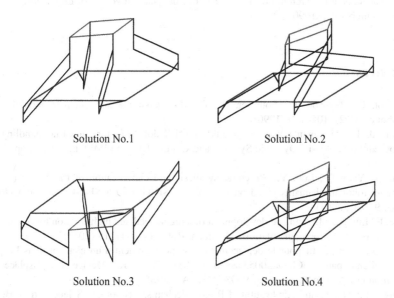

Solution No.1   Solution No.2

Solution No.3   Solution No.4

Fig. 7. Real configurations of the PM

Table 1. The real solutions of the forward problem

Number	X_0/mm	Y_0/mm	Z_0/mm	$\alpha/°$
1	19.99999	5.99999	135.00000	14.99°
2	20.34549	7.75201	40.57131	103.27°
3	19.99999	5.99999	−35.00000	14.99°
4	20.34549	7.75201	59.42869	103.27°

5 Conclusion

1. A new type of 3T1R PM is proposed based on the POC set, it can realize the Schönflies motion and can meet the design demand.
2. The new 3T1R PM proposed in the paper possesses the properties of folding characteristic, extensibility and large coverage of workspace in XY plane.

3. By solving the kinematic equations of the PM, an example with 4 real solution of the forward kinematics are obtained. It provides a certain theoretical basis for the application of the PM in the practical production and processing.

Acknowledgement. The authors would like to acknowledge the financial support of the national science foundation of China (NSFC) with grant number of 51475050, Shanghai Science and Technology Committee with grant number of 12510501100, the Natural Science Foundation of Shanghai City with grant number of 14ZR1422700, and the Technological In-novation Project of Shanghai University of Engineering Science (SUES) with grant number of E109031501012, and the Graduate student research innovation project of Shanghai University of Engineering Science with grant number of E109031501016.

References

1. Pierrot, F., Reynaud, C., Fournier, A.: DELTA: a simple and efficient parallel robot. Robotica **8**(02), 105–109 (1990)
2. Rolland, L.: The manta and the kanuk: novel 4-dof PMs for industrial handling. In: Proceedings of ASME Dynamic Systems and Control Division IMECE, vol. 99, pp. 14–19 (1999)
3. Jin, Q., Yang, T., Luo, Y.: Structural synthesis and classification of the 4DOF (3T-1R) parallel robot mechanisms based on the units of SINGLE-OPENED-Chain. China Mech. Eng. **9**, 020 (2001)
4. Xie, F., Liu, X.J.: Design and development of a high-speed and high-rotation robot with four identical arms and a single platform. J. Mech. Rob. **7**(4), 041015 (2015)
5. Clavel, R.: Dispositif pour le deplacement et le positionnement d'un element dans l'espace, Switzerland patent, Ch1985005348856 (1985): Clave R.: Device for displace and positioning an element in space, WO8703528A1 (1987)
6. Tingli, Y.: Theory and Application of Robot Mechanism Topology. Science Press, Beijing (2012)
7. Yang, T.-L.: Theory and Application of Robot Mechanism Topology (in Chinese). China machine press, Beijing (2004)
8. Lubin, H., Yan, W., Tingli, Y.: Analysis of a new type 3 translations- 1 rotation decoupled parallel manipulator. Chin. Mech. Eng. **15**(12), 1035–1037 (2004)
9. Yang, T.L., Liu, A.X., Luo, Y.F., et al.: Position and orientation characteristic equation for topological design of robot mechanisms. ASME J. Mech. Des. **131**(2), 021001-1–021001-17 (2009)
10. Yang, T.L., Liu, A.X., Shen, H.P., et al.: On the correctness and strictness of the POC equation for topological structure design of robot mechanisms. ASME J. Mech. Rob. **5**(2), 021009-1–021009-18 (2013)
11. Yang, T.L., Sun, D.J.: A general DOF formula for PMs and multi-loop spatial mechanisms. ASME J. Mech. Rob. **4**(1), 011001-1–011001-17 (2012)
12. Yang, T.L., Liu, A.X., Luo, Y.F., et al.: Ordered structure and the coupling degree of planar mechanism based on single-open chain and its application. In: ASME 2008 International Design Engineering Technical Conferences and Computers and Information in Engineering Conference. American Society of Mechanical Engineers, pp. 1391–1400 (2008)
13. Lubin, H., Yan, W., Tingli, Y.: Analysis of a new type 3 translations-1 rotation decoupled parallel manipulator. Chin. Mech. Eng. **26**(4), 1035–1041 (2004)

14. Salgado, O., Altuzarra, O., Petuya, V., et al.: Synthesis and design of a novel 3T1R fully-parallel manipulator. J. Mech. Des. **130**(4), 042305 (2008)
15. Hu, Q., Cheng, D.: The polynomial solution to the Sylvester matrix equation. Appl. Math. Lett. **19**(9), 859–864 (2006)
16. Prasolov, V.V.: Polynomials. Springer, Heidelberg (2009)

Power Efficiency-Based Stiffness Optimization
of a Compliant Actuator for Underactuated Bipedal Robot

Qiang Zhang, Xiaohui Xiao[(✉)], and Zhao Guo

School of Power and Mechanical Engineering, Wuhan University, Wuhan 430072, China
{zhangqiang007,xhxiao,guozhao}@whu.edu.cn

Abstract. Introducing compliant actuation to robotic joints can obtain better disturbance rejection performance and higher power efficiency than conventional stiff actuated systems. In this paper, inspired by human joints, a novel compliant actuator applied to underactuated bipedal robot is proposed. After modeling the stiffness of the compliant actuator, this paper gives the configuration of the bipedal robot actuated by compliant actuators. Compared with the elastic structure of MABEL, the compliant element of our robot is simplified. Based on the dynamics of the compliant actuator-driven bipedal robot, a feedback linearization controller is presented to implement position control of the compliant actuator for power efficiency analysis and stiffness optimization. Co-simulations of MATLAB and ADAMS are performed under the defined control trajectory by altering actuator stiffness. The simulation results indicate that, compared with the actuator maintaining very high stiffness like a rigid actuator, the power efficiency of the compliant actuator is improved, and the stiffness optimized to 375 N•m/rad can reach the highest power efficiency.

Keywords: Compliant actuator · Underactuated bipedal robot · Feedback linearization · Trajectory tracking · Power efficiency

1 Introduction

In the growing field of humanoid robotics, the design and implementation of actuators have been an essential technique for robots motion. The actuators applied to traditional bipedal walking robots are preferred to be as stiff as possible to make precise position movements or trajectory tracking control easier (faster systems with high bandwidth) [1]. But in recent years, the researchers have realized the importance of compliant actuators due to their ability to minimize shocks, improve power efficiency and reject external disturbances. When compared with classical robotic actuators, the motion tracking control and power efficiency optimization of compliant actuator are the main barriers for its widespread application.

Some bipedal walking robots actuated by compliant actuator adapt intrinsically to interaction constraints, store and release elastic potential energy into/from compliant actuation. [2–4] can generate motions with power peaks which could not be obtained using stiff structures and actuators. The beneficial influence of the intrinsic compliance on specific aspects of bipedal robot have been also highlighted and studied [5, 6] by

© Springer International Publishing Switzerland 2016
N. Kubota et al. (Eds.): ICIRA 2016, Part I, LNAI 9834, pp. 186–197, 2016.
DOI: 10.1007/978-3-319-43506-0_16

using actuators with preset or variable level of compliance. By adapting the compliance of the MACCEPA (the mechanically adjustable compliance and controllable equilibrium position actuator), a controllable and energy efficient walking motion is obtained in the testbed Veronica [7]. A systematic method was proposed to optimally tune the joint elasticity of multi-dof compliant SEA (series elastic actuation) robots based on resonance analysis and energy storage maximization criteria [8]. Throughout these researches, there are two pre-conditions: (1) the bipedal walking robots are mostly based on active control for every joint; (2) the robots with round feet or flat feet can accomplish static and dynamic walking successfully. However, the drawbacks of bipedal robot based on active control joints consist of large energy consumption, low energy efficiency and unnatural gait.

Because of the high maneuverability and the low cost of resource, underactuated bipedal walking robot has been studied prevalently [9, 10]. The underactuated bipedal walking does not rely on large feet and slow movement for achieving stability of a walking gait. Some famous underactuated bipedal walking robots, such as RABBIT [11], MABEL [12] and AMBER [13], were published recently. But only a few underactuated bipedal robots have taken the compliant actuator or compliant joint into consideration. MABEL, for example, achieves stability, efficiency, and speed through a combination of the novel design of its drivetrain and the analytical method being developed to control it [14], but the structure of its compliant element is complex, which leads to its low energy efficiency.

To address this issue, in this paper, a novel compliant actuator is designed to apply to an underactuated bipedal robot, and the stiffness of the knee joint actuator is optimized to acquire highest power efficiency during robot walking on a level ground. In Sect. 2, the modeling and description of the novel compliant actuator and the underactuated bipedal robot is presented, respectively. Section 3 presents the compliant actuator position tracking based on static feedback linearization method. In Sect. 4, the trajectory tracking validity of proposed control approach and the actuator stiffness optimization are executed through different co-simulations of MATLAB and ADAMS, respectively. Finally, Sect. 5 presents conclusions and future work.

2 Modeling

2.1 Description of the Compliant Actuator

The mechanical implementation of the compliant actuator based on the concept of SEA is shown in Fig. 1. Particular attention has been paid to satisfying the dimensional and structure matching the underactuated bipedal robot. Though the stiffness of one certain actuator is fixed, it could become variable through replacing linear springs with different stiffness. In case of power efficiency, requirements can still be guaranteed if the actuator stiffness is set to an appropriate value. In order to minimize dimensions while achieving high levels of rotary stiffness, a four-spoke input component, an outer sleeve output element and eight linear springs have been designed and fabricated. The compliant actuator is fixed between the robot calf and thigh through inner sleeve and fixture, respectively. Inside, the inner and outer sleeves are assembled by two deep groove ball bearings.

Additionally, the four-spoke element fixed to the axis of the planetary reducer is coupled to the outer sleeve by means of the eight springs which are arranged as shown in Fig. 1.

Fig. 1. The CAD prototype of the compliant actuator

Based on the mechanical description of the compliant actuator, the stiffness model of the four-spoke arrangement is also presented. As is shown in Fig. 2, the eight linear springs inserted in the actuator experience a pre-contraction equal to half of the maximum acceptable deflection. Here the angle deflection θ_s is designed changing from $-10°$ to $10°$. Deflection that larger than the maximum of allowable value are not permitted by means of mechanical pin based locks.

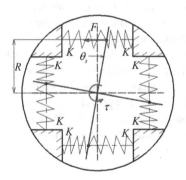

Fig. 2. Compression of springs as a result of the module deflection

Figure 2 illustrates that because of the spring contraction along their main axis, the deflection of the actuator generates torques. Considering one of the antagonist linear spring pairs in Fig. 1, when the four-spoke component is deflected from the equilibrium position with angle θ_s, the forces generated by each spring along the main axis are given by:

$$F_1 = K(x_p + x(\theta_s)), F_2 = K(x_p - x(\theta_s)) \tag{1}$$

where x_p is the pre-contraction and $x(\theta_s) = R\sin(\theta_s)$ is the contraction of the two springs along the main axis. Due to the range of angle deflection θ_s, the contraction of spring can be simplified as $x(\theta_s) = R\theta_s$. K is the spring stiffness and R the length of the spoke arm. The combined force applied to each spoke is expressed as:

$$F = F_1 - F_2 = 2KR\theta_s \tag{2}$$

The corresponding torque generated on each spoke because of the axial forces of one pair springs is equal to:

$$T = FR = 2KR^2\theta_s \tag{3}$$

Thus, the combined torque on the compliant actuator considering the axial forces of four pairs of springs is:

$$T_{total} = 4T = 8KR^2\theta_s \tag{4}$$

By directly differentiating the torque equation, the rotary stiffness of the four-spoke component which is due to the axial deflection of the springs can be expressed as:

$$K_A = \frac{\partial T_{total}}{\partial \theta_s} = 8KR^2 \tag{5}$$

From above equations, a linear relation can be figured out between the torque generated from the compliant actuator and the angle deflection.

2.2 Underactuated Bipedal Robot Configuration

A simplified model of a planar bipedal robot is given in Fig. 3(a), which is composed of a hip (q_3), two knees (q_2, q_4), two thighs (l_2, l_3), two calves (l_1, l_4), but no ankle. Thus, it has six DOFs (degrees of freedom). And the three joints (q_2, q_3, q_4) are controllable actively and driven by compliant actuator independently introduced before. Correspondingly, the CAD prototype established in this research is shown in Fig. 3(b). The stand leg is assumed to act as a passive pivot in the sagittal plane, so the leg end is modelled as a point contact with a DOF and no actuation. The underactuated DOF at foot ("foot" indicates the point on the bottom of calf) acts as the classical DOF at bipedal robot ankle.

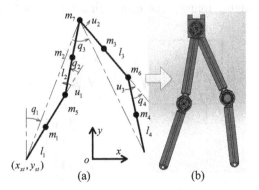

Fig. 3. Underactuated bipedal robot model (a. Simplified rod robot model; b. CAD prototype)

The structure description and the definition of the generalized coordinates $q_r = [q_1, q_2, q_3, q_4, x_{st}, y_{st}]'$ are indicated in Fig. 3, where x_{st} and y_{st} are the Cartesian coordinates of the stance foot [13]. In particular, the masses are distributed, and the position angles of joints are computed counter-clockwise except q_1. It is noted that the position angle for the stance leg is computed from the indicated vertical line to the virtual line connecting the hip and the stance foot, and the position angle for hip is also similar.

2.3 Dynamic Model of Compliant Actuator-Based Robot

According to the compliant actuator mentioned above, a simplified schematic model shown in Fig. 4 is proposed to explore its dynamic characteristics. The device is characterized by a driving motor serially connected to the driven link through deformable transmissions. In Fig. 4, because of the rotary compliance, the deflection θ_s of the transmission can be represented by the link angle q and the motor angle θ ($\theta_s = q - \theta$). The potential energy $U_e(\theta_s)$ is associated to the deflection θ_s, and the flexibility torque of the transmission can be expressed as:

$$\tau_e(\theta_s) = \frac{\partial U_e(\theta_s)}{\partial \theta_s} \tag{6}$$

In general, considering a bipedal robot consists of an open kinematic chain with N compliant actuators, $q \in R^N$ and $\theta \in R^N$ represent the N-dimensional vectors of link and motor position. Moreover, $\theta_s = q - \theta$ donates the vector of transmissions deformation. Ignoring the viscous effects at the motor and link side, the dynamic model of N-dimension takes the form:

$$M(q)\ddot{q} + H(q, \dot{q}) = \tau_e(\theta_s) \tag{7}$$

$$B\ddot{\theta} + \tau_e(\theta_s) = \tau \tag{8}$$

where $M(q) > 0$ is the robot links inertia matrix, the constant diagonal matrix $B > 0$ contains the motor inertia, $H(q, \dot{q})$ is the vector of Coriolis and gravitational terms, $\tau_e(\theta_s)$ is the vector of flexibility torques that couple link Eq. (7) and the motor Eq. (8).

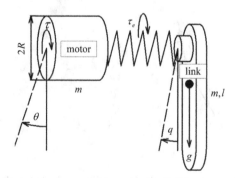

Fig. 4. Schematic model of the compliant actuator

In general, an analytic formulation of $\tau_e(\theta_s)$ can be obtained from the expression of the elastic potential energy $U_e(\theta_s) = \theta_s^T K \theta_s / 2$, the energy leads to linear elasticity torque vector $\tau_e(\theta_s)$ and constant device stiffness (diagonal) matrix σ:

$$\tau_e(\theta_s) = \frac{\partial U_e(\theta_s)}{\partial \theta_s} = K\theta_s, \quad \sigma = \frac{\partial \tau_e(\theta_s)}{\partial \theta_s} = K \tag{9}$$

3 Feedback Linearization Position Control of Compliant Actuator

The analysis on the feedback linearization control of robots with actuator of constant elasticity carried out in previous work [15] is considered as a starting to form a general approach to the solution of the feedback linearization problems for robots with variable joint stiffness.

In the following, the algorithm takes the static feedback linearization of the system dynamics into account to achieve simultaneous joint link position control. And the destination of feedback linearization is to cancel external nonlinear behaviors of the compliant actuator using the input. External means that only the behavior of some outputs can be linearized, generally as many as the inputs. Obviously all other outputs and the internal part of the system will remain nonlinear. In the compliant actuator present above, there is only one input, the motor torque τ in Fig. 4, thus only one output can be feedback linearized. We suppose that there is no coupling between the stiffness of the actuators, in other words, that the stiffness of the i-th actuator is influenced only by the internal linear springs.

To realize robot joint trajectory tracking, the first problem to be solved is to find the desired actuator position trajectory. Based on the walking dynamic model of underactuated bipedal robot built in Sect. 2, the off-line desired angular trajectory of every robot joint is planned in [16]. Here the link angle position vector q has been chosen as desired output for each compliant actuator:

$$y = q \tag{10}$$

since in classical application the control of the link angle position is requested.

With reference to the actuator model Eqs. (7) and (8), the output differentiation is expressed as:

$$
\begin{aligned}
y &= q \\
\dot{y} &= \dot{q} \\
\ddot{y} &= \ddot{q} = \frac{1}{M(q)}\left(\tau_e(\theta_s) - H(q)\right) \\
y^{[3]} &= \frac{1}{M(q)}\left(K(\dot{\theta} - \dot{q}) - \dot{H}(q)\right) \\
y^{[4]} &= \frac{1}{M(q)}\left(\frac{K}{B}(\tau - M(q)\ddot{q} - H(q)) - K\ddot{q} - \ddot{H}(q)\right) \\
&= \frac{K}{M(q)B}\tau + b(x)
\end{aligned}
\tag{11}
$$

where b is a function of the state $x = (\theta, q, \dot{\theta}, \dot{q})$. The number of derivation is equal to the dimension of the state of the system, thus no zero dynamics are present. The system can be trivially linearized by inverting as:

$$\tau = \frac{M(q)B}{K}(v_1 - b(x)) \tag{12}$$

where v_1 is the input of the new closed loop system $v_1 = q^{[4]}$.

The following is the introduction of control strategy proposed in this paper. The static feedback linearization defined in the previous section allows to control the actuator positions by means of an independent linear controller. More in general, the feedback linearization control design should be completed by specifying linear control laws v_1 in the compliant actuator that stabilize the system to the desired task, expressed in term of link and stiffness behavior. In order to asymptotically reproduce a desired trajectory $q_d(t)$ for the link angle position, we set:

$$v_1 = q_d^{[4]} + \sum_{i=0}^{3} K_{q,i}(q_d^{[i]} - q^{[i]}) \tag{13}$$

with diagonal gain matrices $K_{q,i}$, $i = 0, \dots, 3$ such that:

$$s^4 + k_{q,3j}s^3 + k_{q,2j}s^2 + k_{q,1j}s + k_{q,0j} = 0 \tag{14}$$

with $j = 1, \dots, 3$ are Hurwitz polynomials, where $k_{q,ij}$ is the j-th term of the main diagonal of the gain matrix $K_{q,i}$, while $q_d^{[i]}$, $i = 0, \dots, 4$ are the vector of the desired joint link position and their time derivative up to the 4-th order. The gains $k_{q,ij}$ are chosen so that the Hurwitz polynomial has all roots in the left-hand side of the complex plane. The actual values of the control gains in Eq. (14) can be chosen, e.g., by pole placement techniques, yielding exponential convergence of the trajectory tracking errors to zero. The high-order derivatives of q that appear in Eq. (13) and its first derivative, can be directly evaluated as functions of the state of the system by means of Eq. (11). Thus, there is no need to differentiate with respect to time the measured state variables.

4   Simulation of the Planar Robot Knee Joint

Studies show that the leg of bipedal robot consume more energy in stance phase than in swing phase [17]. Thus, the support leg knee joint in underactuated bipedal robot has been taken into account. The co-simulation is performed by combining the CAD proto-type robot model (shown in Fig. 3) established in ADAMS and the position control strategy built in MATLAB. Being the most essential element of the bipedal robot, the compliant actuator has a variable parameter, which is the stiffness of linear spring (eight all same). The stiffness is consistent and selective for analyzing the trajectory tracking performance and the actuator power efficiency, respectively.

4.1 Validity of Control Approach on Trajectory Tracking

In the simulation scheme, the control strategy has been chosen in Eq. (13) and the matrix $K_{q,i}$, $i = 0, \ldots, 3$ is obtained from the solution of the CARE (Continuous Algebraic Riccati Equation) with a diagonal state weight matrix. The parameters of compliant actuator with fixed stiffness and bipedal robot are reported in Table 1.

Table 1. Parameters of actuator and robot

Actuator parameters		Robot parameters	
K (N/m)	10^4	g (m/s^2)	9.8
M (kg·m^2)	9.375×10^{-6}	$l_1 \sim l_4$ (m)	0.30
B (kg·m^2)	0.033	$m_1 \sim m_6$ (kg)	0.85
R (m)	0.025	m_7 (kg)	1.7

During SSP, the support leg of bipedal robot is assumed as an inverted pendulum, so is the thigh based on the knee joint rotation center. Considering that the relative angular position between calf and thigh q_2 is equal to link angular position q of compliant knee actuator, q will be the controlled parameter to realize tracking desired output trajectory. Figure 5 illustrates the position trajectory tracking of the compliant actuator link with and without feedback linearization control strategy.

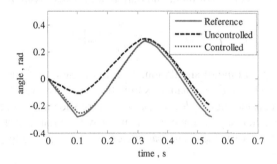

Fig. 5. Comparison of actuator link position trajectory tracking with and without controller

As shown in Fig. 5, compared with the condition without controller (the dashed line), the feedback linearization control strategy applied to the compliant knee actuator significantly improve the trajectory tracking performance (the dotted line). Looking into the time less than 0.3 s, the effect with controller is much more rapid and accurate than that without controller. Furthermore, the position trajectory tracking errors with and without controller are calculated, respectively, as shown in Fig. 6. Here, the error's variety of the controlled tracking is less than the uncontrolled one. And their error's RMS are 0.0297 and 0.0995, respectively.

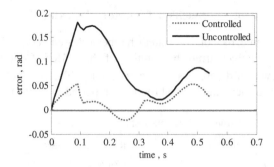

Fig. 6. Comparison of position trajectory tracking error with and without controller

4.2   Power Efficiency Analysis and Actuator Stiffness Optimization

As shown in modeling section, q represents the link angular position of the actuator in Eq. (7), and θ corresponds to the motor drive position in Eq. (8). In addition, the input torque τ equals the transformed drive torque.

For the power analysis of the knee compliant actuator, the power required to perform the motion of input $P_{in} = \tau \cdot \dot{\theta}$ and output $P_{out} = \tau_e \cdot \dot{q}_l$ are investigated based on Eqs. (7), (8) and (9). Here, the elastic element is supposed to have a negligible damping constant. The motion energies of input and output are further given by:

$$E_{in} = \int_{tm} |P_{in}| dt, \ E_{out} = \int_{tm} |P_{out}| dt \tag{15}$$

where tm is the elapsed time of simulation. The investigations of the CAD prototype model are performed by inverse dynamics simulations, based on feedback linearization control approach, considering the desired out trajectory shown in Fig. 5. Additionally, to compare power efficiency of compliant actuator with variable stiffness, the spring stiffness K is set for five values. The parameters K and K_A are reported in Table 2 according to Eq. (5).

Table 2.   Parameters of spring and actuator stiffness

Parameters	Values				
Spring stiffness K (N/m)	10^2	10^3	10^4	10^5	10^6
Actuator stiffness K_A (N·m/rad)	0.375	3.75	37.5	375	3750

For each simulation condition with a specific actuator stiffness, they have the consistent desired output trajectory (in Fig. 5). Based on the feedback linearization control approach, five groups drive torque τ and five groups drive position θ corresponds to τ are calculated. The drive torque curve under each condition is shown in Fig. 7. Although five simulation conditions contain different actuator stiffness, there is no apparent difference among five groups drive torque.

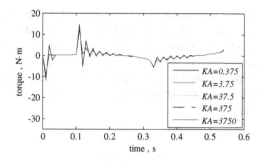

Fig. 7. Drive torque of the compliant actuator with different stiffness

On the contrary, actuator stiffness has significant influence on drive position θ, which is presented in Fig. 8. The drive position variation trends under each simulation condition are similar to the desired output trajectory shown in Fig. 5 with phase lead. Additional, as the actuator stiffness increases, the drive position curve gets close to desired output, however, it's interesting that the peak value of the drive position decreases first and then increases approaching desired output peak value.

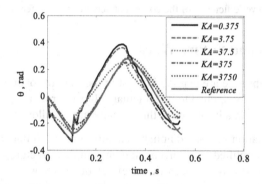

Fig. 8. Drive position θ of the compliant actuator with different stiffness

According to the drive torque, drive position, elasticity torque and actual output position above, the motion energy of input and output under each simulation condition is calculated based on Eq. (15). Consequently, the power efficiency is obtained to execute the optimization by modifying the optimal actuator stiffness. The co-simulation results are demonstrated in Table 3.

Table 3. Motion energy and power efficiency

Parameters	Values				
Actuator stiffness K_A (N·m/rad)	0.375	3.75	37.5	375	3750
Energy of input (J)	7.25	7.13	6.69	6.84	7.01
Energy of output (J)	0.19	0.59	4.11	5.12	4.97
Power efficiency (%)	2.62	8.27	61.43	74.85	70.90

The results indicate that the input energy of drive torque varies within a small range, but the output energy of deformable transmission has significant difference among these simulation conditions. As the actuator stiffness increases, the power efficiency increases first and then decreases, which is illustrated in Fig. 9. Compared with the stiff actuator, the compliant actuator power efficiency is improved in a certain stiffness region. Here, the stiffness can be optimized to the optimal value 375 N•m/rad for acquiring the highest actuator power efficiency.

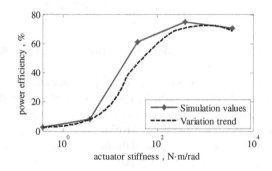

Fig. 9. Power efficiency of the compliant actuator with different stiffness

5   Conclusions and Future Work

Actuators with compliance are gaining interest in the field of robotics. In this paper, a novel compliant actuator applied to underactuated bipedal robot is proposed. Through co-simulations, the main advantages of this actuator are as follows:

(1) The implementation of this compliant actuator applied to underactuated bipedal robot is more simplified and convenient than complex structure of MABEL.

(2) The actuator stiffness has significant effect on the power efficiency. As the stiffness increases, the power efficiency increases first and then decreases. Compared with the actuator maintaining very high stiffness like a rigid actuator, the compliant actuator stiffness is optimized to 375 N•m/rad for acquiring highest power efficiency.

Currently, the actuator proposed here is passive-compliant, and it is very complex to change the stiffness through artificially replacing linear spring with different stiffness. In the future work, the practical compliant actuator with active-adjusted stiffness applied to underactuated bipedal robot will be investigated, and a self-adaptive control strategy will be executed to realize stiffness and position control simultaneously.

Acknowledgment.   This research is sponsored by National Natural Science Foundation of China (NSFC, Grant No. 51175383).

References

1. Ham, R., Sugar, T.G., Vanderborght, B., et al.: Compliant actuator designs. IEEE Robot. Autom. Mag. **16**(3), 81–94 (2009)
2. Ishikawa, M., Komi, P.V., Lepola, G.V., et al.: Muscle tendon interaction and elastic energy usage in human walking. J. Appl. Physiol. **99**(2), 603–608 (2005)
3. Laffranchi, M., Tsagarakis, N. G., Cannella, F., et al.: Antagonistic and series elastic actuators: a comparative analysis on the energy consumption. In: IEEE International Conference on Intelligent Robots and Systems, St. Louis, USA, pp. 5678–5684 (2009)
4. Roberts, T.J., Marsh, R.L., Weyand, P.G., et al.: Muscular force in running turkeys: the economy of minimizing work. Science **275**(5303), 1113–1115 (1997)
5. Hurst, J.W.: The electric cable differential leg: a novel design approach for walking and running. Int. J. Humanoid Rob. **8**(2), 301–321 (2011)
6. Li, Z., Tsagarakis, N.G., Caldwell, D.G.: A passivity based admittance control for stabilizing the compliant humanoid COMAN. In: IEEE-RAS International Conference on Humanoid Robots, Osaka, Japan, pp. 44–49 (2012)
7. Ham, R.V., Vanderborght, B., Damme, M.V., et al.: MACCEPA, the mechanically adjustable compliance and controllable equilibrium position actuator: design and implementation in a biped robot. Robot. Auton. Syst. **55**(10), 761–768 (2007)
8. Tsagarakis, N.G., Morfey, S., Cerda, G.M., et al.: Compliant humanoid coman: optimal joint stiffness tuning for modal frequency control. In: IEEE International Conference on Robotics and Automation, Karlsruhe, Germany, pp. 673–678 (2013)
9. Grizzle, J.W., Chevallereau, C., Sinnet, R.W., et al.: Models, feedback control, and open problems of 3D bipedal robotic walking. Automatica **50**(8), 1955–1988 (2014)
10. Collins, S., Ruina, A., Tedrake, R., et al.: Efficient bipedal robots based on passive-dynamic walkers. Science **307**(5712), 1082–1085 (2005)
11. Chevallereau, C., Gabriel, A., Aoustin, Y., et al.: Rabbit: a testbed for advanced control theory. IEEE Control Syst. Mag. **23**(5), 57–79 (2003)
12. Grizzle, J., Hurst, J., Morris, B., et al.: MABEL, a new robotic bipedal walker and runner. In: American Control Conference, St. Louis, USA, pp. 2030–2036 (2009)
13. Yadukumar, S.N., Pasupuleti, M., Ames, A.D.: Human-inspired underactuated bipedal robotic walking with AMBER on flat-ground, upslope and uneven terrain. In: IEEE/RSJ International Conference on Intelligent Robots and System, Vilamoura, Portugal, pp. 2478–2483 (2012)
14. Sreenath, K., Park, H.W., Poulakakis, I., et al.: A compliant hybrid zero dynamics controller for stable, efficient and fast bipedal walking on MABEL. Int. J. Robot. Res. **30**(9), 1170–1193 (2011)
15. Isidori, A.: Nonlinear Control Systems. Springer Science and Business Media, London (2013)
16. Luca, A.D., Farina, R., Lucibello, P.: On the control of robots with visco-elastic joints. In: IEEE International Conference on Robotics and Automation, Barcelona, Spain, pp. 4297–4302 (2005)
17. Wang, Y., Ding, J., Xiao, X.: Periodic stability for 2-D biped dynamic walking on compliant ground. In: International Conference on Intelligent Robotics and Applications, Portsmouth, UK, pp. 369–380 (2015)

Analysis of the Stiffness of Modular Reconfigurable Parallel Robot with Four Configurations

Qisheng Zhang, Ruiqin Li$^{(\boxtimes)}$, Qing Li, and Jingjing Liang

School of Mechanical and Power Engineering, North University of China, Taiyuan 030051, China
{1020709642,695558745,944719179}@qq.com, liruiqin@nuc.edu.cn

Abstract. This paper studies the static stiffness of a kind of Modular Reconfigurable Parallel robot (MRP robot for short). The MRP robot can be reconstituted to four different configurations. The 3D entity models of the MRP robot of all configurations are established by UG software, according to the modular modeling method and certain simplified rules. The stiffness model of the MRP robot is established. The factors affecting stiffness of the MRP robot are obtained. The static stiffness and stress distribution of the MRP robot are obtained with different forces in the initial position of various configurations by using ANSYS. The static stiffness in z direction (perpendicular to the static base) of each configuration is larger than the static stiffness in x and y directions (in the static base). This shows that the main stiffness is located in z direction. While the stiffness in x, y directions are close to each other. The main stiffness of four kinds of configurations is different. The main stiffness of 6-SPS configuration is significantly greater than that of other three kinds of configurations. The weaker links of the MRP robot are related to the position of the hinges and the connecting position of the moving platform and the screw. The overall stiffness of the MRP robot can be obviously improved by increasing the stiffness of the module which has great influence on the stiffness. The results provide a theoretical basis for the design of the MRP robot.

Keywords: Modular reconfigurable robot (MRP) · Parallel robot · Configuration · Stiffness

1 Introduction

Modular Reconfigurable Parallel robot (MRP robot for short) consists of a series of modules such as joint, connecting rod, the end-effector with different size and functional characteristics. The MRP robot can be changed to different configurations through simple and quick assembly and disassembly among modules in the way of building blocks [1]. The MRP robot has the flexibility characteristic. It can better meet the demands of configuration variation. However, stiffness, precision and ratio of load to self-weight of the MRP robot are not satisfactory which are limited by its own structure. Therefore, it is an important research topic to make the MRP robot have better reconfigurable ability, strong processing capacity and good operating performance.

The static stiffness is one of the important performance indices of parallel robot. It is beneficial to improve the efficiency of the robot, machining accuracy and surface

© Springer International Publishing Switzerland 2016
N. Kubota et al. (Eds.): ICIRA 2016, Part I, LNAI 9834, pp. 198–209, 2016.
DOI: 10.1007/978-3-319-43506-0_17

quality by improving the static stiffness of robot. Many studies on the stiffness of the parallel robot have been done. Many methods are adopted including finite element analysis method, analytical model of static stiffness and static stiffness performance analysis, etc. [2]. The stress and deformation at the end of parallel robot is not simple linear superposition of the deformation produced by limbs and frame, but a coupling nonlinear function with many limbs. Therefore, it is much more complex to analyze the static stiffness of parallel robot. Some simplification and optimization of the model have been done before modeling when analytic method and performance analysis used to analyze static stiffness. However, these simplification methods can't accurately solve the stiffness of the robot and the error is large. In view of the complexity of geometric shape and boundary conditions of mechanical structure, finite element analysis method is usually used to modeling and calculating the stiffness of the robot [3].

In recent years, many scholars have studied the stiffness of parallel manipulators with various configurations by using finite element method. Wang Nan et al. [4] have made static and dynamic characteristics analysis of a 3-DOF 3-SPS/S type parallel machine tool, and gained the static stiffness performance by using finite element analysis software. Yan Binkuan et al. [5] have made static analysis of a 3-SPS/S type parallel machine tool, got the stiffness characteristics in each direction and characteristics of the first 6 order natural frequency. Li Xingshan et al. [6] have set up 3D model of 2TPT-PTT hybrid parallel machine tool, and built finite element model of this type parallel machine tool in workbench, studied the static stiffness with different forces. The results indicate that the static stiffness in z direction is larger than it in x and y directions and the hooke joint and parallel mechanism are the important factors affecting the stiffness of the whole mechanism. Chen Guangwei et al. [7] have established finite element model of whole static stiffness for a new type gantry plane parallel mechanism of parallel machine tool, and got distribution of stiffness of moving platform under the generalized workspace. Li et al. [8] established an improved 3-PRC model of flexible parallel manipulator mechanism, and analyzed stiffness and static mechanics of the model by using ANSYS software, gained the stiffness change trend of related to structural parameters. All above researches on the parallel mechanism are based on the finite element software, while all elastic of elastic parts is ignored such as frame, hinge, ball screw etc.

In view of the fact that ignoring elastic of transmission system in the process of finite element analysis such as hinge, screw and bearings etc. lead to larger errors, this paper will establish the stiffness model for an existing MRP robot experiment platform. The strategy to simplify 3D model of MRP robot is proposed. In Ansys Workbench platform, the overall static stiffness of the MRP robot with four configurations at initial position will be analyzed.

2 Structure and Parameters of the MRP Robot

2.1 Structure Characteristics of the MRP Robot

The MRP robot can change into four different configurations, through changing connection modes of moving platform, linkage, ball screw pairs, hooke hinge into different

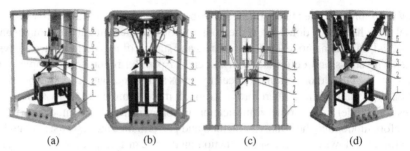

(a)   (b)   (c)   (d)

1-Body support; 2-Moving platform; 3-Hooke hinge; 4-Linkage; 5-Moving module;
6-Vertical pillar

Fig. 1. Four kinds of different configurations of the MRP robot

ones, as shown in Fig. 1(a) ~ (d), which are called 6-PSS slider type, 6-PSS scissors type, 3-Delta slider type, 6-SPS telescopic type, respectively.

As shown in Fig. 1(a), 6-PSS slider type parallel robot consists of body support, six kinematic chains and driver module. Six pairs of driver module are fixed into three group vertical pillars. Each pillar contains 2 sets of driver modules. Each driver module connects to a set of linkage, and at linage end connects to the moving platform.

As shown in Fig. 1(b), 6-PSS scissors type parallel robot configuration consists of six sets of driver modules which are fixed on the top base plate. Each driver module connects to a set of linkage, and working platform connects to the end of linkage.

As shown in Fig. 1(c), 3-Delta slider type parallel robot consists of body support and three kinematic chains and driver module. Under this kind of configuration, three sets of driver modules are fixed in vertical pillar, each driver module connects two sets of linkage, which forms parallelogram mechanism, and the end of linkage connects to moving platform.

As shown in Fig. 1(d), when change into 6-SPS telescopic type parallel robot configuration, one end of six set of driver modules are fixed on the top base plate, the other side connect to linkage with parts of ball screw socket within the linkage, and keep the ball screw and linkage in same axis, which make the length of the linkage variable. The end of linkage connected to the moving platform. Driver modules are driven by servo motors, and it makes lead screw nut pair move, make the wire mother slide along the axis of the ball screw through cooperative movement of the linkage, drive moving platform working and achieve the desired trajectory.

2.2   Structural Parameters of the MRP Robot

The outside framework of the MRP robot is 1316 mm, the static platform constitutes of hexagon with 1156 mm circumscribed circle diameter. The diameter of the moving platform is 316 mm. The body weight is 240 kg. The work scope of prismatic pairs is limited to {-100 mm, 100 mm}. The length of the linkage is 321 mm.

3 Theoretical Basis of Stiffness Analysis

The main factor affecting the stiffness of the MRP robot is the stiffness of transmission system, when the influence of gravity of linkage and stiffness of hinge are ignored. Supposing the force applying on the center of the moving platform is $F_p = [\, F_x \; F_y \; F_z \; m \,]$, m is torque, driving force is $F_d = [\, F_1 \; F_2 \; F_3 \,]$, the following equation can be gotten.

$$F_d = K_t \Delta l \tag{1}$$

Where, K_t is the transmission stiffness of the parallel robot. It can be expressed as

$$K_t = diag(K_{ii}), \; i = 1, 2, 3$$

Where, K_{ii} is the transmission stiffness of ith limb. Δl is displacement deformation along driving direction caused by transmission stiffness, corresponding with which terminal deformation is $\Delta O_p = [\, \Delta x \; \Delta y \; \Delta z \,]^T$.

Under the condition of static equilibrium, in order to make virtual displacement of drive as $\delta l = [\, \delta l_1 \; \delta l_2 \; \delta l_3 \,]^T$, corresponding with δl make terminal deformation as $\delta O_p = [\, \delta x \; \delta y \; \delta z \,]^T$, and get:

$$\Delta l = J \Delta O_p \tag{2}$$

$$\delta l = J \delta O_p \tag{3}$$

By using the principle of virtual work, yield:

$$F^T \delta O_p = F_d^T \delta l \tag{4}$$

Substituting Eqs. (1), (2) and (3) into Eq. (4), then

$$F = J^T K_t J \cdot \Delta O_p \tag{5}$$

Let $K = J^T K_t J$, it is called the static stiffness matrix of the MRP robot. Equation (5) is the static stiffness model of the parallel robot. From Eq. (5), the stiffness matrix of the end-effector of the MRP robot consists of the stiffness of each link and the Jacobian matrix J. The Jacobian matrix changes over the position and orientation of the robot. Thus the stiffness matrix also depends on the position and orientation of the robot.

For a given position and orientation, the deformation size of the moving platform is related to the direction of the force. Static stiffness of the parallel robot can be calculated by the force F which is applied at the center of moving platform, and the displacement of the point is applied by force.

$$K = F / \Delta O_p \tag{6}$$

The static stiffness of the parallel robot in all directions can be calculated by using this method.

4   Stiffness Analysis of the MRP Robot

4.1   Modeling and Simplification Strategy of the MRP Robot

The MRP robot can change into four different configurations. Each configuration is made up of many parts. Considering the relationships among various configurations, all configurations are constituted of the same parts, components and modules, which can be gained by different combination. The method of modular modeling is used to build the model of the robot in order to make full use of the parametric modeling advantages of UG software and the modular characteristics of the mechanism. All of the components are treated as individual modules, which can be independent as a unit such as the moving platform, linkage, kinematic pair, robot body, etc.

Various configurations are built through connecting these modules to the moving platform and static platform with the aid of hooke hinges. Equivalent replacing, simplifying and modifying the model in UG software follow the following principles in order to facilitate subsequent finite element analysis.

(1) Merging all of the parts which contact to each other without relative movement. Deleting the parts which do not force or a little force, such as servo motor, screw, nut, bearing cover, etc. The influence of these parts for the whole structure can be ignored in the process of stress analysis.
(2) Deleting all the features of holes and chamfers in the parts. These features are not affect the results of stiffness analysis. However, it is very significant to occupy computer resources when dividing grid.
(3) Simplifying rules for elastic components and Hooke hinges. Under a certain position and orientation, Hooke hinge is simplified to a 2-DOF joint. The contact surface of screw and screw nut of prismatic pair are coupled which can't be ignored due to the existence of gap.

4.2   Pre-process for Finite Element Analysis

4.2.1   Defining Material Properties of the Robot Parts

The materials of driving parts and Hooke hinges are structural steel, modulus of elasticity $E = 2.09 \times 10^{11}$Pa, Poisson's ratio $\gamma = 0.269$ and density $\rho = 7890$kg/m^3. The rest parts of the parallel robot are 45 steel, modulus of elasticity $E = 2.06 \times 10^{11}$Pa, Poisson's ratio $\gamma = 0.3$ and density $\rho = 7890$kg/m^3.

4.2.2   Meshing

The meshing quality of model directly affects the precision of the calculation results and computing time. Therefore it is a key factor in finite element analysis.

The sizes of robot body and static platform are bigger compared to other parts, which has less influence on overall stiffness. Selecting 50 mm as element size. The main forced parts select 2 mm as element size such as screw and hinge. The contact areas of the parts need further refined. In general, with the increase of the number of grid, the precision of the calculation results will be improved, but the computing time will be increased.

4.2.3   Loading and Solving

According to the actual working situation of the MRP robot, robot body of each configuration is fixed to the ground. The stiffness of the MRP robot system under different configurations, different positions and orientations is different from Eq. (5). The paper studies the stiffness of the four configurations in the initial position and orientation.

Down milling type of milling cutter is usually adopted when the MRP robot of each configuration is located in the initial position and orientation. This is equivalent to a planar milling machine movement. The excitation force applied on the robot comes from milling cutter. The numerical of the milling force can be seen as amplitude of sine excitation. The calculating equation can be obtained from reference [9]. The main milling force for end milling plane can be expressed as follows.

$$F_m = 6e10^4 \times \frac{P_m}{V_m} \tag{7}$$

Where, F_m is the main milling force, i.e., the component of milling force along the main movement direction of milling cutter, N.

P_m is the milling power, KW.

V_m is the milling velocity, m/min.

In the process of milling of the four configurations, the milling power of 3-Delta type parallel robot is $P_m = 0.3KW$ and the milling velocity is $V_m = 6$ m/min. Thus $F_m = 3000N$ can be obtained by using Eq. (7). The milling power of other configurations is $P_m = 0.6KW$ and the milling velocity is $V_m = 6$m/min. Thus $F_m = 6000N$ can be obtained by using Eq. (7).

$$F_x = 0.35F_m$$
$$F_y = 0.525F_m \tag{8}$$
$$F_z = 0.9F_m$$

The component forces F_x, F_y and F_z of the milling force F_m along x, y and z direction of each configuration can be obtained, respectively by substituting F_m from Eq. (7) into Eq. (8). The milling force F_m along x, y, z direction are applied to the center of the moving platform. The component forces of 3-Delta configuration are $F_x = 1050N$, $F_y = 1575N$ and $F_z = 2700N$. The component forces of other configurations are $F_x = 2100N$, $F_y = 3150N$, $F_z = 5400N$.

4.3 Results Analysis

Four kinds of different configurations are imported into ANSYS Workbench platform, respectively. The total contour of displacement and stress are gained of each configurations of the MRP robot at the initial position and orientation, which is gotten by applying the forces along *x, y, z* axis direction, as shown in Figs. 2, 3, 4 and 5.

4.3.1 Results of Four Kinds of Configurations

Figure 2 shows contours of displacement and stress at center of the moving platform when the force $F_x = 2100$ N, $F_y = 3150$ N, $F_z = 5400$ N is applied at the center of the moving platform, respectively for 6-PSS slider type configuration. Figure 2(a) and (b) shows the stress and displacement contours, respectively when the force $F_x = 2100$ N is applied at the center of the moving platform along *x* axis direction. Figure 2(c) and (d) shows the stress and displacement contours, respectively when the force $F_y = 3150$ N is applied at the center of the moving platform along *y* axis direction. Figure 2(e) and (f) shows the stress and displacement contours, respectively when the force $F_z = 5400$ N is applied at the center of the moving platform along *z* axis direction.

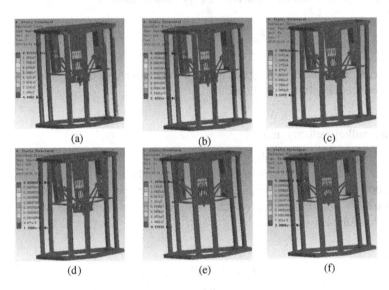

(a)   (b)   (c)

(d)   (e)   (f)

Fig. 2. Displacement and stress of 6-PSS slider configuration

Figure 3 shows contours of displacement and stress at mid-point of moving platform on the condition that the force $F_x = 2100$ N, $F_y = 3150$ N, $F_z = 5400$ N is applied at center of moving platform respectively under 6-PSS scissors type configuration. Figure 3(a) and (b) shows the stress and displacement contours, respectively on the condition that the force $F_x = 2100$ N is applied in the midpoint of working platform along *x* axis direction. Figure 3(c) and (d) shows the stress and displacement contours, respectively on the condition that the force $F_y = 3150$ N is applied in the midpoint of working platform along *y* axis direction. Figure 3(e) and (f) shows the stress and displacement contours respectively on the condition that the force $F_z = 5400$ N is applied in the midpoint of working platform along z axis direction.

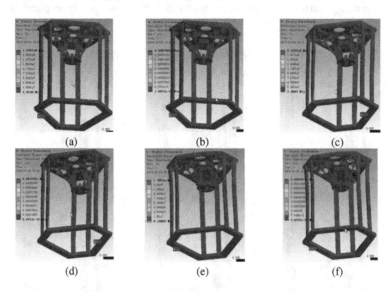

(a)   (b)   (c)

(d)   (e)   (f)

Fig. 3. Displacement and stress of 6-PSS scissors configuration.

Figure 4 shows contours of displacement and stress at mid-point of moving platform on the condition that the force $F_x = 1050$ N, $F_y = 1575$ N, $F_z = 2700$ N is applied at center of moving platform, respectively under 3-Delta type parallel robot configuration.

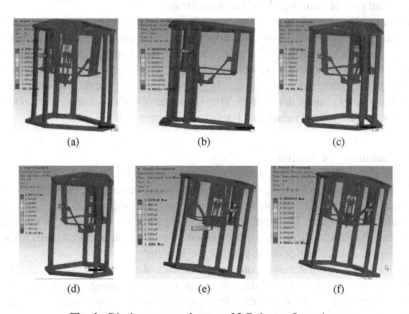

(a)   (b)   (c)

(d)   (e)   (f)

Fig. 4. Displacement and stress of 3-Delta configuration

Figure 5 shows contours of displacement and stress at mid-point of moving platform on the condition that the force $F_x = 2100$ N, $F_y = 3150$ N, $F_z = 5400$ N is applied at center of moving platform, respectively under 6-SPS telescopic type parallel robot configuration.

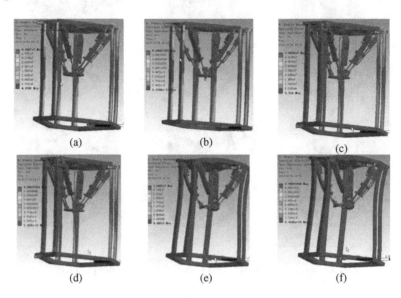

Fig. 5. Displacement and stress of 6-SPS telescopic configuration

4.3.2 Stiffness of Four Kinds of Configurations
Under four different configurations in Fig. 1, the whole deformation (denoted as Δ) and stress values, the deformation values of midpoint of moving platform (denoted as $U_x/U_y/U_z$ in x, y, z direction, respectively), the maximum stress (σ_{max}) and the displacement values of the maximum deformation are obtained along x, y, z directions at the center of the moving platform (denoted as δ_{xmax}, δ_{ymax}, δ_{zmax}, respectively). Then the parts and location with weak stiffness can be determined. Thus the value of static stiffness can be obtained by substituting the gained values into Eq. (6), as shown in Table 1.

4.3.3 Summary of Analysis
From calculating and analyzing, the following results can be obtained.

(1) For the same configuration at initial position and orientation, the stiffness along three axes of the static coordinate system is different. The stiffness in x and y direction are similar to each other. However, the stiffness in z direction is an order of magnitude larger than that in x, y directions. The stiffness in z direction is called main stiffness, which are the characteristics of the parallel mechanism.

(2) For four kinds of different configurations, the orders of magnitude in each direction are the same, but the values of the stiffness are different. Among all the main stiffness of all the four configurations, 6-SPS telescopic type has the largest stiffness, while 3-Delta type has the smallest stiffness. The difference between 6-PSS scissors type and

Table 1. Deformation and stiffness of the MRP robot under the different forces

Configuration	Force /N	Δ_{max} /e-5 m	σ_{max} /e7 pa	δ_{xmax} /e-5 m	δ_{ymax} /e-5 m	δ_{zmax} /e-5 m	$U_x/U_y/U_z$ /e-6 m	Stiffness /e5 N/m
6-PSS Slider	$F_x = 2100$	178	9.97	178	13.1	15.0	1780/−2.5/12.3	10.8
	$F_y = 3150$	268	17.1	18.0	268	23.4	−4.37/2680/17.8	11.8
	$F_z = 5400$	43.5	13.3	25.1	26.8	43.5	1.99/1.16/435	124
6-PSS Scissors	$F_x = 2100$	191	15.7	191	3.01	36.2	1890/−1.95/−8.54	11.1
	$F_y = 3150$	286	21.5	3.99	286	50.7	1.78/2830/43.9	11.1
	$F_z = 5400$	52.0	14.0	7.64	12.7	51.4	−23.1/71.5/511	106
3-Delta Slider	$F_x = 1050$	107	8.26	107	9.75	12.2	1070/−1.57/5.84	9.81
	$F_y = 1575$	160	13.4	160	20.2	61.0	−2.36/1600/8.65	9.84
	$F_z = 2700$	31.6	11.0	18.6	21.5	31.6	0.725/0.587/316	854
6-SPS Telescopic	$F_x = 2100$	137	4.26	137	3.95	8.27	1290/−4.04/2.11	1.63
	$F_y = 3150$	206	5.86	2.94	206	13.8	−1.03/1950/3.06	1.62
	$F_z = 5400$	9.28	3.09	4.02	5.06	9.27	1.94/1.7/92.8	582

6-PSS slider type has little difference. 6-SPS telescopic type parallel robot has a wider application range. For 3-Delta type configuration, each kinematic pair need to drive two links, which is made up of ball screw and have a relative smaller stiffness. It is consistent with the actual situation. For 6-SPS telescopic type configuration, a steel plate is added at the bottom of the mobile pairs, which caused increasing of the overall stiffness. While under others configurations, ball screw with larger deformation is just fixed at the bottom of the mobile pairs. Therefore, it is helpful to enhance the overall stiffness of the robot system by improving the stiffness of modules which have large influence on stiffness.

(3) The location with the maximum deformation under four configurations is obtained. For 6-SPS telescopic type configuration, the largest deformation locate at the areas, where is hooke joint that installed on fixed platform connecting to telescopic rod mobile pairs. And the stiffness there is relatively smaller, it is weaker links of this kind of configuration. Thus the hinge stiffness will affect the stiffness of the configuration. By analyzing the location with the maximum deformation of other configurations, the maximal displacement occurs at the position that screw of motion pair and sliding platform connecting areas, where is also weak link of the robot. The deformation of linkage and frame is small can be neglected under all four configurations. Under each configuration, the smallest deformation is at upper parts, while it is large at the lower parts. It is increasing gradually from top to down. The deformation trend is in accordance with the actual working situation. The analysis results are reasonable.

(4) The maximum stress is 215 MPa but still far less than the allowable stress of the material. The machine will not damage in actual processing conditions under each configuration. Therefore, the design is reasonable. The maximum stress occurs in the position with smaller stiffness. Thus, it is necessary to strengthen some weak parts.

5   Conclusions

The 3D entity models of the MRP robot under four different configurations are built by using UG software according to certain rules. The finite element analysis is carried out in ANSYS Workbench software. The deformation data, distribution, change regular of stress and strain of the MRP robot under four different configurations are gained when the robot is applied by external forces in x, y and z direction. The conclusions can be drawn as follows.

(1) The stiffness along the main stiffness direction is one order of magnitude larger than the other directions under various configurations. There is only very small difference between the other two directions.
(2) For the MRP robot, the stiffness in all directions is varied from each other under different configurations. The stiffness of the MRP robot can be improved by improving the stiffness of modules.
(3) The weaker link of the MRP robot is different from each other for varied configurations. All of which is relevant to the hinge, screw and the position connecting to sliding platform. Therefore, the stiffness of hinge and the screw prismatic pairs will affect the overall stiffness of the MRP robot. The weakness of each configuration at initial position is found out, which is in accordance with the actual situation.
(4) Under 6-SPS telescopic configuration, a steel plate is added to the bottom of prismatic pairs which consist of telescopic rods, and the main stiffness of this configuration is greatly improved. Therefore, the overall stiffness of the MRP robot can be significantly enhanced by improving the stiffness of modules which has large influence on stiffness.

The results provide a theoretical basis to design, improve and optimize the robot structure in the future.

Acknowledgements. The authors gratefully acknowledge the financial and facility support provided by the National Natural Science Foundation of China (Grant No. 51275486).

References

1. Jiang, Y., Wang, H., Pan, X., et al.: Autonomous online identification of configurations for modular reconfigurable robot. Chin. J. Mech. Eng. **47**(15), 17–24 (2011). (in Chinese)
2. Ai, Q., Huang, W., Zhang, H., et al.: Review of stiffness and statics analysis of parallel robot. Adv. Mech. **42**(5), 583–592 (2012). (in Chinese)
3. Wei, Y., Wang, Z.: Finite element analysis on the stiffness of the structure of parallel machine tool. Mach. Electron. **10**(4), 16–19 (2004). (in Chinese)
4. Wang, N., Zhao, C., Gao, P., et al.: Parallel manipulator 3-SPS/S static and dynamic stiffness performance study. Mach. Des. Manuf. **8**, 213–215 (2013). (in Chinese)
5. Yan, B., Zhang, N.: 3-SPS-S 3-DOF parallel mechanism static and dynamic characteristics analysis. Mech. Res. Appl. **26**(3), 35–36 (2013). (in Chinese)
6. Li, X., Cai, G.: Static stiffness analysis of hybrid parallel machine tools based on workbench. Manuf. Technol. Mach. Tool (4), 60–62 (2011). (in Chinese)

7. Chen, G., Wang, J.: The finite element analysis and optimization of static stiffness of a new parallel kinematic machine. Mach. Des. Manuf. (12), 4–6 (2006). (in Chinese)
8. Li, Y.M., Xu, Q.S.: Stiffness and statics analysis of a compact 3-PRC parallel micromanipulator for micro/nano scale manipulation. In: IEEE International Conference on Robotics and Biomimetics, Sanya, China, pp. 59–64 (2007)
9. Shanghai association of metal cutting technology. Metal cutting manual. Shanghai science and technology Press, Shanghai (2004). (in Chinese)

Kinematic Analysis and Simulation of a Ball-Roller Pair for the Active-Caster Robotic Drive with a Ball Transmission

Masayoshi Wada$^{(\boxtimes)}$ and Kosuke Kato

Tokyo University of Agriculture and Technology, 2-24-16, Koganei, Tokyo 184-8588, Japan
mwada@cc.tuat.ac.jp

Abstract. In this paper, kinematics analysis and simulation of the ball-roller pair is presented. The authors grope has been developed a special kind of transmission with a dual-ball configuration for transmit drive powers to wheel shaft and steering shaft of an active-caster mechanism (ACROBAT). In the transmission design, two balls are required where one ball is used for combining the two motor powers to rotate the ball in 2D way while another ball is for distributing the combined power to the wheel and steering. The transmission design enables an omnidirectional robot with three active casters to be controlled its 3D motion by three motors with no redundancy.

To reduce the number of friction drive between balls and rollers in the first design, we are now planning to build a new type of ACROBAT with a single ball transmission. For this purpose, we analyze a kinematics of a ball-roller pair mechanism and verify the kinematics and motions of proposed mechanism by simulations.

From the results, it is confirmed that proposed transmission is applicable for ACROBAT mechanism for realizing the non-redundant omnidirectional motions.

Keywords: Active-caster · Holonomic and omnidirectional mobile robot · Ball transmission

1 Introduction

The holonomic and omnidirectional mobile robots contribute for simplifying robot control architectures with path planning and 3DOF (position and orientation) independent feedback for the trajectory tracking control in 2D plane since non-holonomic constraints do not have to be taken into account. This feature gives many advantages not only automated control systems but also human operators. Therefore this class of omnidirectional mobile system is popular for soccer robots [1] and is also applied to wheelchairs [2] and a personal mobility [3], etc.

For creating the holonomic motions of the omnidirectional robot, omnidirectional wheel mechanisms such as Universal-wheel or Mechanum wheel are widely used in laboratory or conditioned indoor environments. However, these mechanisms do not show enough performances in the real-world environments such as dusty or wet outdoor

© Springer International Publishing Switzerland 2016
N. Kubota et al. (Eds.): ICIRA 2016, Part I, LNAI 9834, pp. 210–221, 2016.
DOI: 10.1007/978-3-319-43506-0_18

grounds with irregularities, steps, gaps, and so on. To overcome these problems, the authors group has proposed active-caster drive mechanisms [4]. The feature of the active-caster is that normal tires, rubber or pneumatic, can be used for transmit the traction force on the ground as normal wheel systems. Therefore very smooth traveling with no vibration is realized. Also the payload capability is very large since contact point is not a single point.

To drive the active-caster mechanism, at least two motors are required to control its wheel rotation and the steering rotation which control law is given by Eq. (1) including a function of wheel orientation ϕ which is detected by an absolute angle sensor. Therefore if one build an omnidirectional robot with three active-casters, it is required to install three absolute angle sensors and six motors. These requirements could be a cause of complexity of the system architecture and controller for dealing with the redundantly actuated mechanism (Fig. 1).

$$\begin{bmatrix} \omega_w \\ \omega_s \end{bmatrix} = \begin{bmatrix} \dfrac{\cos\theta}{r} & 0 \\ 0 & \dfrac{\sin\theta}{s} \end{bmatrix} V \tag{1}$$

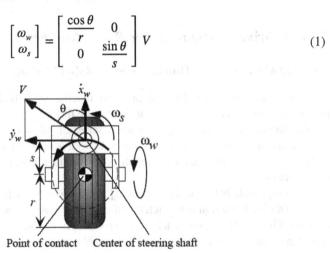

Point of contact   Center of steering shaft

Fig. 1. Active-caster omnidirectional control [4]

Where, θ: wheel orientation
    r: wheel radius
    s: caster offset
    ω_w: rotation of wheel axis
    ω_s: rotation of steering axis

To avoid the disadvantages of the original active-caster wheel, we have introduced a ball transmission to distribute traction power mechanically to a wheel axis and a steering axis in an appropriate ratio. By using the ball transmission, absolute angle sensors are not required for the wheel control. Further, one of the features of the ball transmission is that the multiple balls in line can be driven by common actuator. Based on this idea, an omnidirectional robot with three-ACROBATs in which three pairs of rollers are driven by three common motors could be one of the possible configurations, whose schematic is shown in Fig. 2. However the first transmission design with dual-ball configuration involves 5points of friction drive. To reduce the point of friction drive between balls and rollers in the first design, we are now planning to build a new type of ACROBAT with a single ball transmission. For the purpose, we study existing ball drive mechanisms and propose a new type of ball transmission.

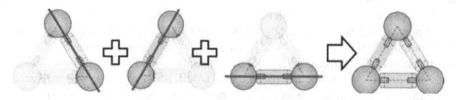

Fig. 2. Concept of 3-wheel robot (three pairs of a common drive unit)

2 Ball Drive Mechanisms for Robot Motion

2.1 Ball Mechanism for Omnidirectional Robot Motions

Here, we mention about the existing ball mechanism which were developed for providing omnidirectional motions of the mobile robots.

The ball track mechanism [5], rotates a series of balls arranged in line by two roll bars and a chained carrier which are driven by independent motors. Each ball is driven in 2DOF which determines the rolling direction on the plane. However, some balls slip when robot rotates.

Reconfigurable ball-wheel platform [6], equips four ball wheel mechanisms each of which 1DOF ball rotation is restricted while another 1DOF is actively controlled by the actuator. Final 1DOF is completely free to rotate. The mechanism design allows the ball wheel mechanism to replace the omni-wheels, such as Universal-wheel, with no unwanted vibration due to contact switching of free rollers.

Ball balancer [7] includes a single big ball driven by three omni-wheels. The robot with three omniwheels is located on the top of the big ball and balanced its posture by controlling the three omniwheels.

All of those ball wheel mechanisms transmit the traction force on the ground by the ball(s). Therefore the ball might raise dust on the road through the ball surface which results in the slip or jerking. Furthermore a ball makes single contact to the ground then the traction force and payload capability are limited. On our design, balls are used for transmit the power for propelling of the robot, but normal wheel provides traction force to the ground.

2.2 The ACROBAT Ball Transmission

First, we will briefly explain about our original ball transmission mechanism, which is implemented in ACROBAT [8]. This dual-ball transmission is composed by two parts, A and B. The part A includes a ball A and two actuators for drive the ball A via two rollers contacting to the ball A on its large circle. As the rollers rotate about the horizontal axes, the ball A rotates about a horizontal axis where the rotation about the vertical axis is restricted by rollers. The part B includes ball B whose traction force is distributed to another pair of rollers, c and d. One of the roller is connected to a wheel axis the other is connected to a steering axis for driving these axes. The ball A and B make contacts to transmit traction power from part A to B as shown in Fig. 3.

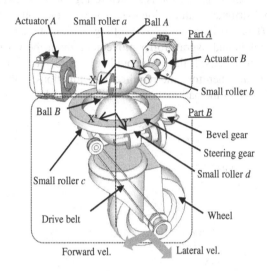

Fig. 3. The first design of dual-ball transmission of ACROBAT [8]

Figure 4 shows a top view of the lower layer of the dual ball transmission mechanism which includes ball B, roller c and d, and a wheel. The two small rollers are arranged in right angle to contact to the ball at the horizontal large circle. When the ball rotate about the y_a-axis as is in the Fig. 4 which generates velocity V at the top of the ball, two rollers are rolling on the ball surface whose trajectories are shown by red and blue lines. Therefore, two roller rotates in the different ball surface velocities at the roller contact points, which are represented in Eq. (1)

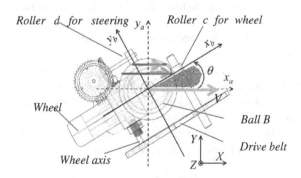

Fig. 4. Velocity distridution by a ball and two rollers [8] (Color figure online)

Thus, the requirement for the active-caster control is to drive the wheel shaft velocity and the steering shaft velocity by the ratio of cosine and sine.

3 Kinematic Analysis of Ball-Roller Pair

Concept of the single ball transmission is shown in Fig. 5. A fundamental idea of the ball-roller pair is to distribute a ball motion to two axes by one roller which contacts on the ball surface NOT on the large circle. The rolling motion of the roller may be transmitted to the wheel axis and overall roller movement about the vertical axis may be transmitted to the steering axis. If this idea could be possible, a roller can be located at any position on the ball, therefore the required two functions: combining the motor motions and distribution of the ball motions, can be realized by using just one ball.

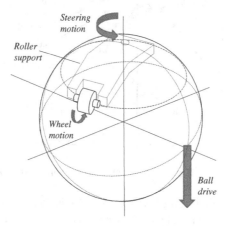

Fig. 5. Concept view of the single ball transmission

Here let us consider about a ball-roller pair. Figure 6 shows a ball and a roller which makes a contact with a single point (illustrated as A in the figure). The coordinate system is prepared which origin is at the center of the ball where XY plane is parallel to the ground. Now, a roller is contacting to the ball surface whose contacting point is located at (x,y,z) at the moment. Note here that a roller shaft is parallel to the XY plane at all times and height of the shaft maintained to be constant. No slip allowed on the sideways direction of the roller.

As the ball rotates, the roller cannot stay in a constant position except the some specific locations. For instance, at a roller position as shown in Fig. 6, the roller is rotated by the ball motion in its rolling direction while it changes own location around the z-axis by the ball motion in lateral direction of the roller.

Suppose that the ball rotates about the X-axis with an angler velocity ω. The surface velocity on the ball V in the figure which is represented by R and ω as,

$$V = R\omega \tag{2}$$

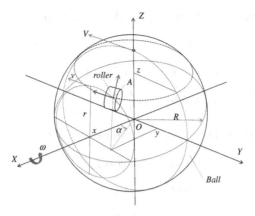

Fig. 6. The configuration of ball-roller pair

Now, we think a small circle perpendicular to X-axis which includes the contact point A. The contact point A is determined by parameters, R, α, β, γ, and x, r, z shown in Fig. 7.

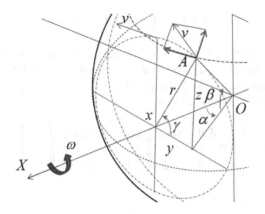

Fig. 7. Angles and parameters for contact point A

Now the surface velocity on the small circle v is represented by

$$v = r\omega \tag{3}$$

From Eqs. (2), (3) we get the relationships of V and v as,

$$v = \frac{r}{R}V \tag{4}$$

Note that the magnitude of the velocity v is identical every point on the small circle while the direction of the velocity varies. Therefore, at the contact point A, The surface velocity is also v while the direction of the velocity is along the tangential direction of the small circle.

The velocity components at contact point A along x, y and z axes can be derived as,

$$v_{Ax} = 0$$
$$v_{Ay} = -v \sin \gamma = v \frac{R}{r} \sin \beta$$
$$v_{Az} = v \cos \gamma = v \frac{x}{r} \tan \alpha$$

(5)

Where note that the following relationships for get Eq. (5).

$$r \sin \gamma = -R \sin \beta = z$$

(6)

$$r \cos \gamma = x \tan \alpha$$

(7)

$$\beta = \sin^{-1} \left(\frac{z}{R} \right)$$

(8)

$$\gamma = \sin^{-1} \left(\frac{z}{r} \right)$$

(9)

$$\alpha = \tan^{-1} \left(\frac{r}{x} \cos \gamma \right)$$

(10)

From these three vector components in x,y and z direction, we can derive three vector components along tangential direction on the ball surface at point contact A. For the purpose, vector components Eq. (5) are coordinate-transformed by rotating α about the X axis then rotate β about the Z axis as,

$$\begin{bmatrix} v_{AR} \\ v_{A\alpha} \\ v_{A\beta} \end{bmatrix} = \begin{bmatrix} \cos \beta & 0 & -\sin \beta \\ 0 & 1 & 0 \\ \sin \beta & 0 & \cos \beta \end{bmatrix} \begin{bmatrix} \cos \alpha & -\sin \alpha & 0 \\ \sin \alpha & \cos \alpha & 0 \\ 0 & 0 & 1 \end{bmatrix} \begin{bmatrix} v_{Ax} \\ v_{Ay} \\ v_{Az} \end{bmatrix}$$

(11)

Where,

v_{AR}: velocity component perpendicular to the ball surface
$v_{A\alpha}$: velocity component in roller rolling direction
$v_{A\beta}$: velocity component in roller sideways direction

Then we get the three vector components all of these are tangential direction on the ball surface as follows.

$$v_{AR} = v(\cos \beta \sin \alpha \sin \gamma - \sin \beta \cos \gamma) = 0$$
$$v_{A\alpha} = -v \cos \alpha \sin \gamma = -\frac{R}{r} v \sin \beta \cos \alpha$$
$$v_{A\beta} = v(\sin \beta \sin \alpha \sin \gamma + \cos \beta \cos \gamma) = \frac{R}{r} v \sin \alpha$$

(12)

Note here the following relationships for get Eq. (12).

$$\ell = R \cos \beta \tag{13}$$

$$x = \ell \cos \alpha \tag{14}$$

$$\frac{x \cos \beta}{R \cos \alpha} = \frac{\ell^2}{R^2} \tag{15}$$

Here that R and β are constants determined by the mechanical design while r and α are variables continuously vary depending on a roller location. However, from Eqs. (2), (4), we get,

$$\frac{R}{r} v = R\omega = V \tag{16}$$

By Eq. (16), we derive the simple vector relationships as,

$$\begin{aligned} v_{AR} &= 0 \\ v_{A\alpha} &= -V \sin \beta \cos \alpha \\ v_{A\beta} &= V \sin \alpha \end{aligned} \tag{17}$$

For the active-caster control, it is required that wheel axis and steering axis of the caster mechanism have to be driven to satisfy the specific relationships such a way that wheel velocity and the steering velocity maintain the ratio of sine to cosine as shown in Eq. (1). To realize this requirement, those axes are driven by independent motors in the original active caster mechanism or distribute ball velocity to those axis by a dual-ball transmission in the first design of ACROBAT.

In Eq. (17), we can find that the ratio of $v_{A\alpha}$ and $v_{A\beta}$ satisfies the required relationships of cosine and sine which is a function of roller orientation angle a.

4   Simulation

To verify the effectiveness of the kinematic analysis of ball-roller pair, we tested the proposed configuration by a simulation using a motion analyze function in Solid Works 3D. We do not have to implement the kinematic model presented in the previous section, but just we create 3D CAD model of the mechanism in this case a ball and a roller, and give some conditions to the model (contact condition, contact load) and constant motion to a ball (constant rotation about X-axis whose velocity on the surface is V). Solid Works automatically creates an internal dynamic model of each mechanical element and calculates mechanical motion. Figure 8 shows a schematic view of the 3D CAD model. Note that the model is shown in upside-down to show the roller motion clearly. The caster wheel on the top of the figure might contact to the ground and provide traction force. Major parameters for the simulation are listed in Table 1.

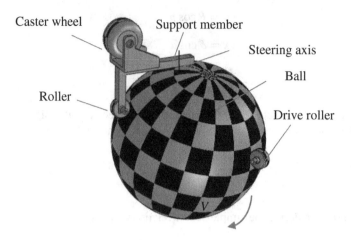

Fig. 8. 3D Simulation model (illustrated upside-down)

Table 1. Parameters in the ball-roller simulation

Parameters	Value	
	Value	Unit
R	Ball diameter	0.2 m
V	Ball surface velocity	1.0 m/s
	Roller diameter	0.05 m
z	Height of contact point	0.05 m
β	Angle of contact point	30°

4.1 Ball Drive Model

In the simulation, a drive roller is modeled as shown in Fig. 8. To drive a ball by a roller, the roller has to contact on the some point on the horizontal large circle since sideways slips are not allowed on the contact point between the ball and the roller. By the non-slip condition, ball rotation about the vertical axis is restricted by the drive roller and it is supposed that the rotation axis of ball is always parallel to the horizontal plane. To change direction of the ball rotation, another roller should be installed on the same large circle in the actual transmission design. However, only one roller is modeled to drive a ball in a constant angular velocity in this simulation.

4.2 Roller Model for Power Distribution

The other roller is located at the midpoint of the ball surface between the top and the large circle where the drive roller contacts. This 2nd roller is for distributing the power to wheel and steering axis for actuating the caster wheel to create omnidirectional motion. This roller is supported by the member that allows the roller to rotate about the vertical axis (this axis is a steering axis of the caster wheel) which intersects with the center of the ball. Therefore the roller can maintain its contact to the ball with its contact height to be constant.

The support member is free to rotate about the steering axis but the angle of the support member is actively actuated by a traction force acting in sideways of the roller from the ball surface. This motion is used for steering motion of the active-caster.

The rolling motion of the roller is also created by the ball rotation. The motion is transmitted to caster wheel rotation by gear or belt drive with some ratio (this gear or belt is not shown in the simulation model). This motion is used for wheel motion of the active-caster.

4.3 Simulation Results

Figure 9 shows one of the simulation results which show a surface velocity components at contact point of the distributed roller in rolling direction and sideway direction respectively. Small dots are the results by Solid Works while dotted lines are theoretical curve calculated by the derived kinematics Eq. (17). The results by Solid Works include noise like performances in some area, but it is due to dynamic friction model in the simulator. However, the simulation results are well agreed with the theoretical values derived by kinematics analysis.

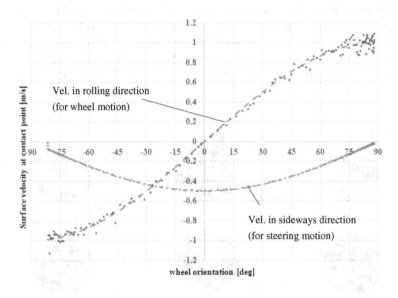

Fig. 9. Simulation results (surface vel. against orientation α)

Figure 10 shows profiles of surface velocity components against time. It is seen that wheel velocity is negative at the initial of the simulation with rotating the steering gradually. At about 0.5 s, wheel velocity goes to positive and this moment, the steering velocity shows maximum value and decrease gradually after the peak. This peak point shows a "flip motion" of a caster which is often shown in a passive caster. After the flip motion, the wheel velocity converges to 1[m/s] and the steering velocity to 0[m/s], which means the caster wheel and the roller maintain staying at constant position on the ball surface.

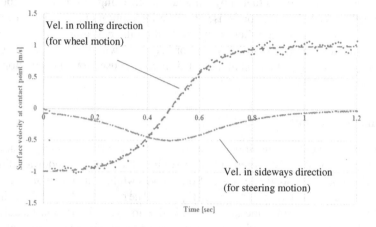

Fig. 10. Simulation results (velocity profiles against time)

For better understanding this motion by images, snapshots of the 3D images of the simulation are shown in Fig. 11. In the figure, images from 0 s to 2.5 s are shown. At Fig. 11(b), flip motion of the caster is seen where distribution roller makes right angle position to the drive roller. After this moment, The distribution roller and the caster wheel moves towards the drive roller, then converge and stay at the top of the drive roller.

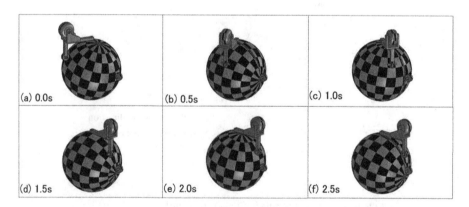

Fig. 11. 3D animations of the simulation (drive velt between a roller and a wheel is not shown)

5 Conclusion

In this paper we presented the kinematic analysis and simulation of a ball-roller pair for new transmission design of the active-caster mechanism, ACROBAT. To reduce the number of friction drive point on the dual-ball transmission of the first design of ACROBAT, single ball transmission is proposed. In the new design, power combining and power distribution is realized by a single ball. To confirm that the required motion is possible, the kinematics of a ball-roller pair was analyzed and the performances were tested by 3D simulation using Solid Works. By the simulation results, it was confirmed that an expected power distribution was realized by the single ball transmission design together with the correctness of the kinematic analysis. Based on this study we will design and build the prototype ACROBAT wheel mechanism with a single ball transmission.

References

1. Ishida, S., Shimpuku, N., Ishi, K., Miyamoto, H.: Holonomic omnidirectional vehicle with ball wheel drive mechanism and application to RoboCup soccer middle size league. Jpn Soc. Fuzzy Theory Intell. Inform. **26**(3), 669–677 (2014). (in Japanese)
2. Terashima, K., Kitagawa, H., Miyoshi, T., Urbano, J.: Frequency shape control of omni-directional wheelchair to increase user's comfort. In: Proceedings of IEEE 2004 International Conference on Robotics and Automation, pp. 3119–3124 (2004)
3. Honda U3-X. http://www.honda.co.jp/robotics/u3x/
4. Wada, M., Mori, S.: Holonomic and omnidirectional vehicle with conventional tires. In: Proceedings of the 1996 IEEE International Conference on Robotics and Automation, pp. 3671–3676 (1996)
5. West, M., Asada, H.: Design of a holonomic omnidirectional vehicle. In: Proceedings of the 1992 IEEE International Conference on Robotics and Automation, pp. 97–103, May 1992
6. Wada, M., Asada, H.H.: Design and control of a variable footprint mechanism for holonomic and omnidirectional vehicles and its application to wheelchairs. IEEE Trans. Robot. Autom. **15**(6), 978–989 (1999)
7. Kumagai, M., Ochiai, T.: Development of a robot balanced on a ball – application of passive motion to transport. In: Proceedings of the 2009 IEEE International Conference on Robotics and Automation, pp. 4106–4111 (2009)
8. Wada, M., Inoue, Y., Hirama, T.: A new active-caster drive system with a dual-ball transmission for omnidirectional mobile robots. In: Proceedings of the 2012 IEEE International Conference on Intelligent Robots and Systems, pp. 2525–2532 (2012)

Robot Vision and Sensing

Fast Hierarchical Template Matching Strategy for Real-Time Pose Estimation of Texture-Less Objects

Chaoqiang Ye, Kai Li, Lei Jia, Chungang Zhuang, and Zhenhua Xiong[✉]

State Key Laboratory of Mechanical System and Vibration, School of Mechanical Engineering,
Shanghai Jiao Tong University, Shanghai 200240, China
{649182890,leica8244,jerryjia,cgzhuang,mexiong}@sjtu.edu.cn

Abstract. This paper proposes a fast template matching strategy for real-time pose estimation of texture-less objects in a single camera image. The key novelty is the hierarchical searching strategy through a template pyramid. Firstly, a model database whose templates are stored in a hierarchical pyramid need to be generated offline. The online hierarchical searching through the template pyramid is computed to collect all candidate templates which pass the similarity measure criterion. All of the template candidates and their child neighbor templates are tracked down to the next lower level of pyramid. This hierarchical tracking process is repeated until all template candidates have been tracked down to the lowest level of pyramid. The experimental result shows that the runtime of template matching procedure in the state-of-the-art LINE2D approach [1] after applying our fast hierarchical searching strategy is 44 times faster than the original LINE2D searching method while the database contains 15120 templates.

Keywords: Template matching · Template searching · Real-time · Pose estimation · Hierarchical searching · LINE2D

1 Introduction

Object recognition and pose estimation in three-dimensional space plays a significant role in robotic applications that range from home-service robots to intelligent manipulation and assembly. In most of industrial applications, pose of objects ought to be estimated precisely in advance so that they can be operated by a robotic end effector. Typically, industrial objects are texture-less and usually have a highly reflective surfaces, together with changing conditions in lightning, scale, rotation or occlusion, which significantly increase the difficulty in object recognition and pose estimation.

In recent years, many researchers around the world have paid their attention to the recognition of textured objects for which discriminative appearance features can be extracted. For 2D image, SIFT [2], SURF [3], ORB [4], FREAK, [5] and HOG [6] are the most popular features which is invariant to changes in scale, rotation and illumination. The position and orientation matrix of the objects in the scene can be computed by the pairs of feature points between the object model and the scene. However, when applied to texture-less objects, discriminative appearance features detectors typically

© Springer International Publishing Switzerland 2016
N. Kubota et al. (Eds.): ICIRA 2016, Part I, LNAI 9834, pp. 225–236, 2016.
DOI: 10.1007/978-3-319-43506-0_19

fail to extract enough feature points to establish the object model and find the corresponding feature points in the scene.

Template matching is a very popular approach related with recognition and pose estimation of texture-less objects. In this approach, object images in various poses are captured in advance to establish the object template database. For online recognition, we can sweep sliding windows of several discrete sizes over the entire scene image with a small pixel step and search for a match against all stored object template images in the database. Different from the discriminative appearance, these template images do not have scale, rotation invariance. As a result, template matching approach usually needs a large amount of template images to fulfill the whole pose space, which causes poor real-time performance of this approach.

Hinterstoisser et al. [1] have proposed an efficient template matching based method named LINE2D/LINE3D/LINE-MOD to recognize the pose of texture-less objects. The LINE2D uses the image gradients cue only, while LINE3D uses the surface normals cue in depth image only, and LINE-MOD uses both. This method follows the template matching procedure to get initial estimated pose corresponding to the template with highest matching score. Then this initial estimated pose is used as the starting iterative pose for subsequent pose refinement with the Iterative Closest Point (ICP) algorithm [7]. With data structures optimized for fast memory access and a highly linearized and parallelized implementation using special SSE hardware instructions, the method can achieve very fast matching for every single template to the scene. As a result, this method is capable of real-time matching with thousands of templates to get the initial estimated pose. The total time is the sum of time consumed both in template matching procedure and ICP algorithm. The less difference between initial estimated pose and the real pose, the less computational time by the ICP algorithm to get convergence. However, the precision of the initial estimated pose depends on the distribution density of the sample template. The real-time template matching performance is expected to degrade noticeably for a large object database, since its time complexity is linear with the number of loaded templates in the database. In [8], Hinterstoisser et al. have pointed out that the LINE-method needs about 3000 templates per object to fulfill the entire recognition space with the scale step of 10 cm and the rotation step of 15°. This template distribution is sparse and limits the precision of the initial estimated pose.

In this paper, we propose a fast hierarchical search strategy based on the basic template matching procedure in LINE2D to reconcile the conflict between the precision of initial estimated pose and the number of templates. With this strategy, we can find the template that is most consistent to the ground truth in a hierarchical way rather than go through all templates in the dataset one by one, which benefits us for significantly shortening the runtime required in template matching and helps us to estimate object pose more precisely in a larger template database without the loss of real-time performance. This fast hierarchical search strategy can also easily extend to LINE3D and LINEMOD to speed up the template matching and promote the precision of estimated pose.

The outline of this paper is as follows: Sect. 2 gives a review of the background knowledge and notations related with LINE2D method. In Sect. 3, fast hierarchical template matching strategy is introduced in detail. We have applied the proposed strategy to real experiments to recognize a common industrial workpiece and estimate its pose. The results are showed in Sect. 4. Finally, Sect. 5 draws some conclusive remarks.

2   Background and Notations

2.1   Robust Similarity Measure

A template T can be defined as a pair $T = (O, P)$, where O is the reference image of an object to detect and P is a feature list that specifies the detail information of every single feature and its location relative to reference image O. Steger [9] proposed a similarity measure which is robust to small translation and deformations to compare a template with a region at location c in a scene image I:

$$\varepsilon_{steger}(I, T, c) = \sum_{r \in P} |\cos(ori(O, r) - ori(I, c + r))| \tag{1}$$

where $ori(O, r)$ and $ori(I, c + r)$ are the gradient orientations at location r in image O and $c + r$ in image I respectively. Based on this idea, Hinterstoisser et al. [1] proposed a more efficient similarity measure by using the maximum over a small neighborhood:

$$\varepsilon(I, T, c) = \sum_{r \in P} (\max_{t \in R(c+r)} |\cos(ori(O, r) - ori(I, t))|) \tag{2}$$

where $R(c + r)$ defines the neighborhood of size T centered on location $c + r$ in the scene image. By finding this maximum, the local neighborhood of each gradient in the template and scene image gets better aligned.

The similarity score S of the reference image O at the location c in image I is defined as follows:

$$S(I, T, c) = \frac{\varepsilon(I, T, c)}{size(P)} \tag{3}$$

where $size(P)$ is the number of features in feature list P. By finding the similarity score S which is exceeding a given threshold, the matching position of a template in an image is found.

2.2   Computing the Gradient Orientations

Just as mentioned in [1], LINE2D method computes the gradients on three channels of every single pixel in the input image separately and only take pixels whose maximum gradient norm exceeds a given threshold into consideration. These pixels can be regarded as edge pixels and represented as follows:

$$C(x): \underset{C \in \{R, G, B\}}{\arg \max} \left\| \frac{\partial C}{\partial x} \right\| > t \tag{4}$$

where R, G, B are the RGB channels of the corresponding colorful input image and t is the related threshold.

However, the number of pixel points that pass the criterion expressed in Eq. (4) is usually much larger than that of pixel points required to represent edges. Consequently, the thick edges in the input image are detected, which increases the robustness of algorithm to small deformations and scale difference to a certain extent on one hand. On the other hand, it also magnifies the disturbance aroused by the cluttered surroundings and causes some false positive results. In order to detect edge pixel and get its corresponding gradient orientation precisely, we utilize CANNY [10] operator to detect edge pixels in advance. Then we compute the gradient orientation map $Ig(x)$ at edge pixel detected by CANNY operator which satisfies Eq. (4):

$$Ig(x) = ori(C(x)) \qquad (5)$$

We only consider the orientation rather than the direction of gradient in edge pixels to quantize the orientation map and divide the orientation space into 8 equal subspaces as shown in Fig. 1.

(a) orientations quantizing   (b) original image   (c) LINE2D edge image   (d) our edge image

Fig. 1. (**a**): Quantizing the gradient orientations. Every colorful sector represents one orientation. The black arrow direction belongs to the third orientation. (**b**): A scene image which contains a common black industrial part and a calibration pattern. (**c**): The edge pixels and their corresponding orientation detected by LINE2D. Different color in edge pixels represents different orientations. Obviously, the detected edge is very thick. (**d**): The edge pixels and their corresponding orientation detected with our method. The precise thin edges are detected successfully. (Color figure online)

2.3   Spreading the Orientations

In order to avoid applying the max operator in Eq. (2), LINE2D adopts a new binary representation of the gradients around each edge pixel location to construct orientation map. We use a binary string with 8 bits to represent the gradient orientation and each individual bit of this string corresponds to one quantized orientation. Then we can spread each gradient orientation within a neighborhood around every major gradient orientation in the input scene image to obtain a new orientation map. The spreading operation is to spread the central orientation to every pixel location in the neighborhood and can be computed quickly by using the logical OR operation. With this spreading operation, the template matching method increases its robustness to scale changes and rotations. The neighborhood range depends on the magnitude T of the size. This magnitude T is useful for our online fast hierarchical searching. See more details in Sect. 3. Figure 2 shows the spreading process in detail when $T = 3$.

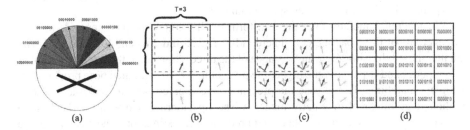

(a)   (b)   (c)   (d)

Fig. 2. Spreading process with $T = 3$. (a) The binary representation of every orientation. (b) The major orientations detected in the input scene image before spreading. (c) The orientations in the input scene image after spreading. The spreading operation is occurred in the neighborhood with size $T = 3$ of every major orientation. (d) The corresponding binary representation of the orientations after the spreading operations. The spreading operations can be computed quickly with logical *OR* operations by using this binary representation.

3   Fast Hierarchical Template Matching Strategy

3.1   Hierarchical Template Database Generation

The hierarchical template database is automatically generated from CAD model of the object to be detected. Every single template image is created by placing a virtual camera around the CAD model at a given viewpoint and projecting the model into the image plane of the virtual camera. The intrinsic parameters of the virtual camera are consistent with the real camera so that the virtual template projecting process can precisely simulate the real imaging process. The viewpoints location is defined under a spherical coordinates system by giving the spherical parameters α (latitude), β (longitude) and r (radius). To simplify the representation of pose and create a more vivid demonstration of the

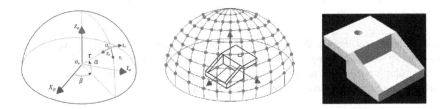

Fig. 3. Left: Spherical coordinates system of viewpoints location. The pose of objects in object coordinate system (X_o, Y_o, Z_o) with respect to the camera coordinate system (X_c, Y_c, Z_c) is defined by sphere radius (r), latitude (α) and longitude (β). **Middle:** Example of viewpoints distribution at a given sphere radius r: yellow vertices represent the virtual camera centers used to generate templates. **Right:** A template image generated by our method with $r = 430$ mm, $\alpha = 36°, \beta = 130°$. This CAD model is the prototype of the industrial workpiece used in our experiment (See more details in Sect. 4). (Color figure online)

hierarchical template database generating process, we only take above three parameters related with poses into consideration and just set the in-plane rotation angle to zero. Figure 3 shows the definition of the spherical coordinates system and an example of viewpoints distribution at a given sphere radius r.

We generate templates by using a uniform subdivision of the pose space defined by the intervals of three parameters: sphere radius (r), latitude (α) and longitude (β). The smaller the subdivision angle and distance of the three parameters, the larger the number of templates and eventually the higher the precision of final estimated pose produced by template matching. To increase the precision of estimated pose as much as possible without degrading real-time performance, we set up a hierarchical template database to reduce the total number of templates required for each estimation. The basic ideas are as follows.

Three parameters related to the pose of template can be regarded as the three dimensions in the 3D pose space. Each pose defined by its sphere radius (r), latitude (α) and longitude (β) can be regarded as a single point in the 3D pose space. For each point, its adjacent points have similar parameters. As a result, they will have similar matching scores in the template matching. Based on this fact, we select the cubic area in the 3D pose space every 27 pose points on one level template pyramid and just keep the center pose point of the cubic area to generate one pose point on the next level template pyramid. This step achieves a substantial reduction in the number of templates from one level template pyramid to the next level. In our implementation, we only need to construct 3 level template pyramids to reduce the number of templates from 15120 to 126. Figure 4 illustrates the generation process of hierarchical template database.

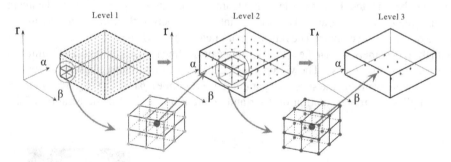

Fig. 4. Hierarchical template database generation process from level 1 to level 3. Every template pose is drawn as a point in α-β-r 3D pose space. We only keep the center point of every 27 pose points in cubic area to generate one pose point in next pyramid. The number of template points is significantly reduced after several level pyramid generation.

3.2   Online Fast Hierarchical Search

For the template pyramid generated as shown in Fig. 4, the template density of different level is different. The higher the pyramid level is, the sparser the template is. Just as Sect. 2.3 mentioned, the operation of spreading gradient orientations increases robustness to scale changes and rotations. In order to increase the robustness of the hierarchical search and to ensure the convergence of the algorithm, we propose different scene orientation images for corresponding level of template pyramid by spreading the gradient orientations with different neighborhood size T. The matching of templates on each template pyramid level is computed at its corresponding scene orientation image. We generate the orientation image with large neighborhood size T of spreading operation for templates on high pyramid level and small neighborhood size T for templates on low pyramid level. Figure 5 shows the orientation images with different neighborhood size T for corresponding level pyramid of template database.

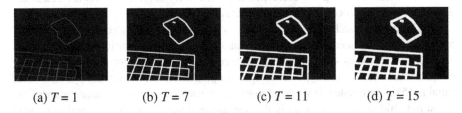

(a) $T = 1$ (b) $T = 7$ (c) $T = 11$ (d) $T = 15$

Fig. 5. Spreading the orientations with different neighborhood size T for corresponding level pyramid of template database. **(a)** Original orientation image without spreading the orientations. **(b)** Orientation image with spreading neighborhood size $T = 7$ for level 1 pyramid. **(c)** Orientation image with spreading neighborhood size $T = 11$ for level 2 pyramid. **(d)** Orientation image with spreading neighborhood size $T = 15$ for level 3 pyramid.

The online hierarchical searching is computed from the highest pyramid level to the lowest pyramid level and starts at the highest level. All templates on the highest pyramid are matched by applying the similarity measure (2) between the template and the corresponding orientation image. We set several similarity score threshold Si for each pyramid level to pick the candidate templates. The subscript i corresponds to index i of

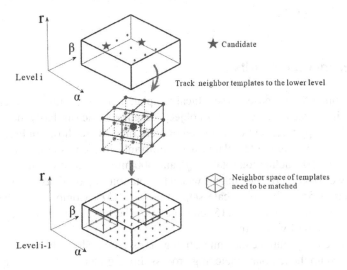

Fig. 6. Illustration of the fast hierarchical search strategy.

pyramid level. All templates with similarity score that exceeds its corresponding similarity score threshold Si are stored in a list of candidates. We track all of the candidate templates on the next lower level of pyramid and measure the similarity between all neighbor templates of every candidate and the corresponding orientation image. We collect all the templates whose similarity score exceeds its corresponding Similarity score threshold Si to build a new candidate list and track these candidates on the next lower level pyramid again. This process is repeated again and again until all template candidates have been tracked down to the lowest pyramid level or no templates with similarity score exceeding the corresponding Similarity score threshold are collected. Figure 6 draws the process of this hierarchical search strategy.

This kind of hierarchical searching is very fast because each tracking down to next lower pyramid only needs to match a small amount of template within the neighborhood of candidates to achieve a much higher recognition accuracy of the corresponding pyramid level. That is to say, in one of our implementation, the lowest pyramid level contains 15120 templates. If we want to get the estimated pose with the accuracy of the lowest pyramid level, we can apply this hierarchical searching strategy and only need to match a few hundred templates rather than match all of the 15120 templates one by one. Obviously, the runtime of this hierarchical searching strategy is related with the number of candidates on every pyramid level. Similarity score threshold of every pyramid level is the critical parameter which determines the searching runtime and should be chosen carefully. Smaller threshold score means that more candidates are needed to be collected on every pyramid level and the corresponding searching runtime would become longer. A good set of thresholds is that collects false candidate templates as little as possible and preserve all right candidate templates at the same time. Note that this fast hierarchical search strategy is essentially independent of the number of parameters related with pose. If we take the in-plane rotation angle into consideration, we can generate the hierarchical template database in 4D pose space whose dimensions are represented by four parameters related with pose and apply the online fast hierarchical search strategy similarly.

4   Experimental Results

We have applied the proposed hierarchical template matching strategy to real experiments to estimate the pose of 4 types of objects based on basic similarity measurements of LINE2D. The pictures and CAD model of these objects are shown in Fig. 7.

Firstly, we generate the template image uniformly in the pose range of 0–90° tilt rotation (α), 0–360° inclination rotation (β) and 500 mm–680 mm scaling (r). To cover this range, we collect 15120 templates with a tilt rotation step of 3° ($\Delta\alpha$), an inclination rotation step of 5° ($\Delta\beta$) and a scale step of 30 mm (Δr). Compared to the original LINE2D which only contains 3115 templates [8], the templates in our database is a denser distribution of viewpoints.

Our hierarchical template matching strategy is essentially by reducing the searching space to speed up the template matching process. For a given number of templates, the

(a) industrial workpiece   (b) plastic part

(c) cup   (d) pen container

Fig. 7. 4 types of objects used in the experiments.

more layers in pyramid, the smaller the searching space. To estimate the proposed strategy performance with different size of searching space and find the suitable number of layers in pyramid, we constructed five different template pyramids with different layers to perform template matching. There are 15120 templates in the lowest layer of these five pyramid. The number of templates in the highest layer is different. There are 4 templates on the highest layer for 5 layer-height pyramid, 27 templates for 4 layer-height pyramid, 126 templates for 3 layer-height pyramid, 1620 templates for 2 layer-height pyramid and 15120 templates for 1 layer-height pyramid. Note that the template matching in 1 layer-height pyramid is actually the original strategy applied in LINE2D.

For color gradient features extraction towards each template image, we keep 63 features uniformly located on the contour of the object silhouette by adopting the gradient feature extraction strategy introduced in [8]. Furthermore, we make the most of SSE hardware instructions to compute the response maps and linearize the parallelized memory for matching each single template to scene image which is proposed in [1]. For the online hierarchical searching, we compute five scene orientation images with spreading neighborhood size $T = 7, 11, 15, 15, 15$ for corresponding layer 1 to 5 of pyramid, respectively. The Similarity score threshold Si is set to 80 %, 75 %, 60 %, 50 % and 50 % for corresponding layers to pick the candidates need to be tracked in next layer of pyramid. Considering there are only one object in every picture, we sort the candidate list and only track top 4 candidates with high similarity score on the next lower layer. This step helps set the number of templates that need to be matched for each downward-searching and stabilize the runtime in every frame to be compared with original LINE2D method.

For every object, we randomly capture 20 pictures with different poses and match the templates in these pictures through above five types of template pyramid with different height. Then, we record the average matching score and average runtime of five types of searching pyramid. Figure 8 shows the results.

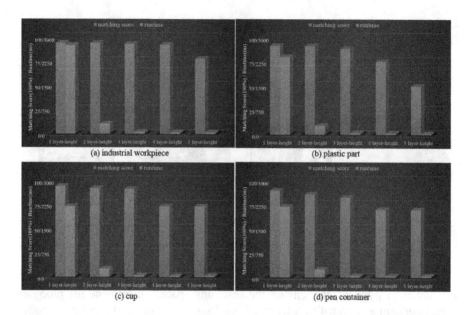

Fig. 8. Average matching score and runtime of five types of pyramid with different layers.

The pyramid with only 1 layer-height represents the original searching strategy in LINE2D by matching 15120 templates one by one, which expectedly achieve the highest matching score and longest runtime. The pyramid with 5 layer-height indicates the smallest searching space. Though the runtime of this pyramid is shortest, the matching score is also the lowest. A good trade-off between speed and robustness is 3 layer-height pyramid. Let's just take the experiments of 3 layer-height pyramid into discussion. In LINE2D searching method, we need to match 15120 templates to the scene image one by one. However, we only need to match $126 + 4 \times 27 + 4 \times 27 = 342$ templates out of 15120 templates with our hierarchical searching strategy. As a result, the average runtime of template matching procedure in LINE2D after applying our fast hierarchical searching strategy is about 44 times faster than the original LINE2D searching method while the database contains 15120 templates.

We also conduct three types of pyramid to perform another experiment to estimate the relationships between the distribution density of templates and the accuracy of estimated poses. We set the number of templates on the lowest layer of pyramid as 126, 1620, 15120 respectively and keep the number of templates on the top layer to 126. That's to say, we construct a pyramid with 1 layer-height for experiment with 126 templates, a pyramid with 2 layer-height for 1620 templates and a pyramid with 3 layer-height for 15120 templates. The object that need to be recognized in this experiment is the industrial workpiece (see Fig. 7(a)). The templates matching results by applying our fast hierarchical searching strategy in different template density databases are shown in Fig. 9.

(a) Original input image   (b) 126 templates   (c) 1620 templates   (d) 15120 templates

Fig. 9. Templates matching results for three kinds of databases with different template density. (a) The original input image for templates matching. (b) Templates matching result with 126 templates. The template with highest similarity score is presented in the bottom-left corner. (c) Templates matching result with 1620 templates. (d) Templates matching result with 15120 templates.

Figure 9 demonstrates that we can obtain the pose estimation results with higher accuracy by matching templates in a larger template database. Therefore, we can estimate object pose more precisely in a larger template database without the loss of real-time performance by applying our hierarchical template matching strategy.

Up to now, we didn't introduce any complicated clutter backgrounds to the experiments. Just as the explanation and discussions mentioned in [1], LINE2D will produce many false positive in cluttered environment. What we are concerned about is that how to speed up the templates matching procedure in a large database. If we are expected to recognize the pose of object in a cluttered environment, this fast hierarchical search strategy can also easily extend to a more robust algorithm named LINEMOD [1] to speed up the template matching and promote the precision of estimated pose.

The experiments were performed on standard computer with an Intel Processor Core(TM) with 3.2 GHz and 8 GB of RAM. The implementation of the original LINE2D, we use the publicly available source code in OpenCV 2.4.11.

5   Conclusions

We have proposed a fast hierarchical template matching strategy for real-time pose estimation of texture-less objects. This strategy is intended for speeding up the template matching based approaches that are applied in pose estimation of texture-less objects. We have introduced the hierarchical template database generation process and online fast hierarchical searching strategy through the template pyramid in detail. Then, we applied this strategy into the basic template matching procedure of a state-of-the-art pose estimation approach named LINE2D. In the experiment, we extracted 15120 templates to construct the templates database. The experiment result shows that the runtime of template matching procedure in LINE2D after applying our fast hierarchical searching strategy is 44 times faster than the original LINE2D searching method. In conclusion, this fast hierarchical template matching strategy can reconcile the conflict between the precision of estimated pose and the number of templates. We can construct a much denser template database which contains more templates covered in pose range to obtain more precise results without a typical loss in runtime performance by applying this fast hierarchical template matching strategy.

Acknowledgments. This research was supported in part by National Key Basic Research Program of China under Grant 2013CB035804 and National Natural Science Foundation of China under Grant U1201244.

References

1. Hinterstoisser, S., Cagniart, C., Ilic, S., Sturm, P., Navab, N., Fua, P., Lepetit, V.: Gradient response maps for real-time detection of textureless objects. IEEE Trans. Pattern Anal. Mach. Intell. **34**(5), 876–888 (2012)
2. Lowe, D.G.: Distinctive image features from scale-invariant keypoints. Int. J. Comput. Vis. **60**(2), 91–110 (2004)
3. Bay, H., Tuytelaars, T., Van Gool, L.: Surf: speeded up robust features. In: Leonardis, A., Bischof, H., Pinz, A. (eds.) ECCV 2006, Part I. LNCS, vol. 3951, pp. 404–417. Springer, Heidelberg (2006)
4. Rublee, E., Rabaud, V., Konolige, K., Bradski, G.: ORB: an efficient alternative to SIFT or SURF. In: 2011 IEEE International Conference on Computer Vision (ICCV), pp. 2564–2571. IEEE, November 2011
5. Alahi, A., Ortiz, R., Vandergheynst, P.: Freak: fast retina keypoint. In: 2012 IEEE Conference on Computer Vision and Pattern Recognition (CVPR), pp. 510–517. IEEE, June 2012
6. Dalal, N., Triggs, B.: Histograms of oriented gradients for human detection. In: IEEE Computer Society Conference on Computer Vision and Pattern Recognition, CVPR 2005, vol. 1, pp. 886–893. IEEE, June 2005
7. Fitzgibbon, A.W.: Robust registration of 2D and 3D point sets. Image Vis. Comput. **21**(13), 1145–1153 (2003)
8. Hinterstoisser, S., Lepetit, V., Ilic, S., Holzer, S., Bradski, G., Konolige, K., Navab, N.: Model based training, detection and pose estimation of texture-less 3D objects in heavily cluttered scenes. In: Lee, K.M., Matsushita, Y., Rehg, J.M., Hu, Z. (eds.) ACCV 2012, Part I. LNCS, vol. 7724, pp. 548–562. Springer, Heidelberg (2013)
10. Steger, C.: Occlusion, clutter, and illumination invariant object recognition. Int. Arch. Photogrammetry Remote Sens. Spat. Inf. Sci. **34**(3/A), 345–350 (2002)
11. Canny, J.: A computational approach to edge detection. IEEE Trans. Pattern Anal. Mach. Intell. **8**(6), 679–698 (1986)

Latent Force Models for Human Action Recognition

Zhi Chao Li, Ryad Chellali[✉], and Yi Yang[✉]

CEECS- Nanjing Robotics Institute, Nanjing Tech University,
Nanjing, People's Republic of China
{rchellali,yyang}@njtech.edu.cn

Abstract. Human action recognition is a key process for robots when targeting natural and effective interactions with humans. Such systems need solving the challenging task of designing robust algorithms handling intra and inter-personal variability: for a given action, people do never reproduce the same movements, preventing from having stable and reliable models for recognition. In our work, we use the latent force model (LFM [2]) to introduce mechanistic criteria in explaining the time series describing human actions in terms actual forces. According to LFM's, the human body can be seen as a dynamic system driven by latent forces. In addition, the hidden structure of these forces can be captured through Gaussian processes (GP) modeling. Accordingly, regression processes are able to give suitable models for both classification and prediction. We applied this formalism to daily life actions recognition and tested it successfully on a collection of real activities. The obtained results show the effectiveness of the approach. We discuss also our future developments in addressing intention recognition, which can be seen as the early detection facet of human activities recognition.

Keywords: Skeleton model · Latent force model · Gaussian processes regression · Feature modeling

1 Introduction

Human actions recognition is getting more attention the last years. This task is of interest in many fields such as robotics, computer vision, human-computer interaction, and natural language processing, etc. targeting applications in homecare, personal and manufacturing robotics, behavior analysis and many other domains. In our case, we are interested in human-robots interactions (HRI), where robots need to understand the actual contexts to better serve humans. Human actions are an important part of these contexts and actions recognition capability is a key feature for friendly and accepted personal robotics.

A lot of works have been done in HAR using videos [9]. The developed techniques use image sequences (2D information evolving in time) and analyze the spatiotemporal changes of human bodies appearance to infer human actions or activities. Recently, RGB-D cameras were introduced. This allows using more reliable data as they encode time series describing human postures and skeletons with avoiding classical issues such

© Springer International Publishing Switzerland 2016
N. Kubota et al. (Eds.): ICIRA 2016, Part I, LNAI 9834, pp. 237–246, 2016.
DOI: 10.1007/978-3-319-43506-0_20

as occlusions, view point changes, lightning, etc. In its generality, the human actions recognition problem can be seen as a general pattern recognition problem: basically, one needs to match the observation, a multivariate time series to a previously seen pattern and assign a label to it, i.e. an action.

Most of existing skeleton based techniques model explicitly the temporal dynamics of skeleton joints. The dynamics are expressed as local models such as ARMA in [10] or sequential state transition models (graphical models) such as HMM or DBN [11, 12].

In our work, we consider the human body as a dynamic system and we describe the dynamics through latent force models (LFM [2]) for which, the human body is seen as a dynamic system driven by latent forces. The LFM has been used mainly for prediction purposes. Here, we adapt it to handle recognition tasks. The LFM two main advantages: (1) as other dynamic systems based techniques, it allows understanding the skeleton times series as HA, (2) it introduces a mechanistic flavor, which can be advantageously used to enhance both the robustness and the interpretability of the sequences. Indeed, observed body movements are generated by latent forces (muscles) and any mechanistic model could be mapped to these forces even indirectly (the general case of LFM).

In the following, we explicit the LFM model and the way we derive it from the raw skeleton data. In Sect. 4, we show how the obtained LFM model is used for classification purposes. Mainly we will demonstrate how to combine the forces and the sensitivity parameters (the weights of the forces) in order the feed a simple linear SVM to perform the classification task. Finally, we present our experimental protocol including the used datasets (the MAD dataset from CMU and the daily action dataset we collected) as well as the evaluation methods. We finish by presenting the obtained results and discuss our future works.

2 Related Work

Action recognition research is very active since a decade. Pushed mainly by social networks industry, many impressive researches were achieved based on the analysis of 2D videos (see [9] for a good review). More recently and with the RGB-D cameras, the HAR issue changed slightly. Indeed, in RGB-D videos, accurate depth and skeleton [1] information are available. This simplifies the HRA by removing occlusions and viewpoint related ambiguities (perspective distortion, lack of Euclidian metrics) while providing absolute measurements. In addition, one can use two different cues: depth maps, which can be processed as normal RGB maps and skeletons joints positions, which may reduce the effects of single inputs.

In RGB-D/skeleton data, the geometry is exact. Taking this advantage, some authors proposed encoding HA according to some Euclidian groups formalisms: In [4], the Euclidian rotation-translation group SE(3) is mapped on the Lie group to have compact joints trajectories. The mapped trajectories are then warped with a DTW and a linear SVM are sufficient to perform HRA. More classical is the work in [8]. Authors used classical tools for time series to classify human actions with LDA classifiers.

Grammatical models have been used in [3]: "Bags of words" were constructed from local descriptors to generatively derive *actionlets*, which are considered as the atomic components of actions. Probabilistic graphical models were in fact the most used in HAR. They allow capturing the dynamics of the body with handling the difficult issues of inter and intra-personal discrepancies. In [4], deep belief networks and HMM are combined together to perform simultaneous segmentation and recognition. Piyathilaka et al. [6] developed Gaussian mixture-based HMM for activity recognition [6]. Jaeyong Sung used a hierarchical maximum entropy Markov model detect human actions [7]. In [5], authors present a method of learning latent structures in human actions. Their graphical model relies not only on the body dynamics but also on objects with which humans interact.

Linear dynamic systems (LDS) have been also used. In [11], the HA time series are modeled as an ARMA process and then projected on Grassman manifold, which allows clustering actions.

More recently, hierarchical recurrent neural networks were used to avoid engineering low level coding (i.e. features design) [12]. In their work, authors grouped body parts into subgroups, facilitating the learning process as well as taking advantage of the actual cross-correlation of the subparts when performing actions.

Different from previous works, we take inspiration from the findings of Laurence et al. [2] concerning probabilistic dynamic systems. We extend the latent force model (LFM) to use it for action recognition purposes.

The LFM describes human movements as a dynamic system for which the model can be derived through Gaussian processes regression. This model makes the assumption that latent forces generate body movements, i.e., the forces excite the body to generate the observed data. In the initial formulation, the LFM was used to predict the future body postures/movements. In our work, we use the same formalism with the inclusion of the full set of parameters (the forces and their relative weights) to perform a discriminative recognition using a simple linear SVM. We demonstrate that this formalism is efficient in HAR but not only. Indeed, our approach captures HA per se but also can give mechanistic hints about the movements, which can be of interest in explaining some aspects related to motor activity analysis, as it will be discussed in the conclusion.

3   Action Pattern Representation

This part focuses on the presentation of action and features used for recognizing. In the original work of Alvarez et al., the purpose of using LFM was forecasting. That is to say, given a time series at time t, the LFM is used to predict to future states of system. We modified the formalism in order to use it as a classification framework.

3.1   Action Definition

An action is a physical activity, where ones body perform a set of movements to achieve a predetermined goal such moving an object, changing posture, etc. Formally and considering skeleton based description, an action could be described as a time series of an N-vector, starting at time t_{start} and finishing at time t_{end}:

$$Y(t) = [y_1(t)\, y_2(t) \dots y_N(t)]$$
$$t \in [t_{start}, t_{end}]$$

Where the y_i are the R joints in the skeleton issued from the RGB-D sensor (Fig. 1).

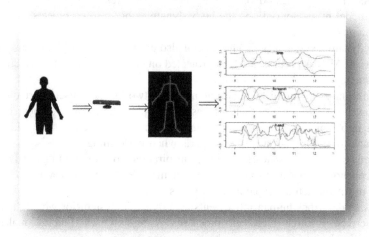

Fig. 1. Body postures time series

3.2   Latent Force Model

In Alvarez et al. introduced latent force models. Basically, the proposed method performs dimension reduction in time series.

$$Y = F.W^T + E$$

Under the assumption of normally distributed noise, the reduced representation or the latent structure F can be seen as a Gaussian process (GP):

$$p(F_r(t)|t; \theta) \sim \prod_{r=1}^{R} N\big(f_r(t)|0, K_r(t, t')\big)$$

Where the posterior F can be derived from the covariance matrix K.

This formulation is then extended to describe the time series as series issued by a dynamic system equivalent to a second order system: a mass spring-damper system excited by latent forces.

$$FS = \ddot{Y}M + \dot{Y} + YD + \sum$$

Where F is the forces matrix, M and C are the mass diagonal mass and damping matrices, D is the original system matrix. These hybrid models consider the human body in movement as a dynamic system driven by some latent forces, non-exactly related to the actual forces (i.e. muscles activity) but allowing involving interesting mechanistic principles to model body parts movements. This model has some similarities with the dynamic movements primitives, which is used in robotics to encode robots trajectories for learning actions by imitation.

Rewriting the differential Eq. (1), we have:

$$\frac{d^2 y_i(t)}{dt^2} + C_i \frac{dy_i(t)}{dt} + D_i y_i(t) = \sum_{r=1}^{R} S_{d,r} f_r(t) \tag{1}$$

Where every observed time series $y_i(t)$ *is related to* the R driving $f_r(t)$ latent forces and the $N*R$ constants $S_{n,r}$ sensitivities. The Knowledge of the latent forces, the sensitivities and the constants C and D, it possible to derive the dynamics of the system and to predict its outputs $y_i(t)$. This has been done in [4].

Assuming the latent forces as R independent GPs, it is shown that recovering new values $y_i^*(t)$ is possible through a Gaussian process regression. Indeed, the output covariance can be expressed as linear functions of the latent forces and the covariance of a general stochastic process. From an original sequence $y_i = \{(y_k, t_k) k = 1, \dots, n\}$, it is possible to predict new values y^* in the sequence. It is shown that $[y, y^*]^T$ satisfies the following distribution:

$$\begin{bmatrix} y \\ y^* \end{bmatrix} \sim N \left(0, \begin{bmatrix} \mathbf{K}_{yy} & \mathbf{K}_{yy*} \\ \mathbf{K}_{y*y} & \mathbf{K}_{y*y*} \end{bmatrix} \right)$$

Where:

$$y^* \sim N(\mu^*, \sigma^*),$$
$$\mu^* = \mathbf{K}_y^T \cdot \mathbf{K}_{yy}^T y$$
$$\sigma^* = - \mathbf{K}_{y*}^T \mathbf{K}_{yy}^T \mathbf{K}_{y*} + \mathbf{K}_{y*y*}$$

With:

$$K_{y_i y_j}(t, t') = \frac{\sum_{1}^{R} S_{ri}.S_{rj}\sqrt{\pi L_r^2}}{8\omega_i.\omega_j} K_{y_i y_j}^r(t, t'), \omega_p = \sqrt{4.D_p - C_p^2}$$

4 Action Recognition

In our case, the aim is to use the model for classification purposes. In other words, we use the vector $\theta = [C_i \quad D_i \quad S_{ir} \quad f_r \quad L_r]_{i=1,N;r=1,R}$ to encode actions and use it in a discriminative procedure to perform action recognition.

4.1 Hyper-parameters Learning

In order to derive the model's hyper-parameters, one needs to maximize the following logarithm marginal likelihood function:

$$\log(p(Y|F,\theta)) = -\frac{1}{2}Y^T K_Y^{-1} Y - \frac{1}{2}\log|K_y| - \frac{n}{2}\log 2\pi \qquad (2)$$

$K_y = K_{f_r} + \sigma_n^2 I$ is the covariance matrix for noisy inputs Y. K_{f_r} is the latent forces covariance without noise, while $(-0.5.Y^T.K_y^{-1}Y)$ expresses the predictions errors, $(-0.5.\log|K_Y|)$ is a penalty depending only on the covariance function and the inputs. $-\frac{n}{2}\log(2.\pi)$ is a normalization constant.

To optimize (2), we use a gradient descent over the vector

$$\theta = [C_i \quad D_i \quad S_{ir} \quad f_r]_{i=1,N;r=1,R} \qquad (3)$$

4.2 Implementation

The pseudo-code of our implementation is the following:

Step 1:
Given the sequence $Y = [y_1(t) \quad y_2(t) \quad \dots \quad y_N(t)]$
Step 2:
Initialize $\theta = [C_i \quad D_i \quad S_{ir} \quad f_r \quad L_r]_{i=1,N;r=1,R}$
Step 3:
For i=1,N
    For j=2,N

$$K_{y_i y_j}(t,t') = \frac{1}{8\omega_i.\omega_j} \sum_{r}^{R} S_{ri} S_{rj} \sqrt{\pi L_r^2} \, K_{y_i y_j}^r(t,t')$$

    end
end

Step 4:

With $\theta = \begin{bmatrix} C_i & D_i & S_{ir} & L_r \end{bmatrix}_{i=1,N;r=1,R}$

$$J^{new} = -\frac{1}{2}Y^T K_{Y'}^{-1} Y - \frac{1}{2}\log|K_{Y'}| - \frac{n}{2}\log(2.\pi)$$

$K_{y'} = K_y + \sigma^2.I$, σ is a noise

$\quad if \left| J^{new} - J^{old} \right| < \varepsilon$

$\quad else$

$$\theta^{new} = \theta^{old} - \alpha.\frac{\partial J(\theta)}{\partial \theta}, \text{goto Step 3}$$

Step 5:

For i=1,R

$\quad$ For j=1,N

$$K_{f_ry_j} = \frac{L_r}{q.4.\omega_j}\left[\gamma_r\left(\bar{\gamma},t,t'\right) - \gamma_j\left(\bar{\gamma},t,t'\right)\right]$$

$\quad\quad$ end

$\quad$ end

Step 6:

$F = K_{fy}K_{yy}.Y$

For more readability, we omitted detailing the γ functions (can be found in [2]).

In our case, we considered 13 joints ($N = 13$) for the skeleton and we used two latent forces ($R = 2$). Accordingly, the hyper-parameters vector components are the following:

$$\begin{cases} F = f_{1,2}(t) \\ C = C_{1:13} \\ D = D_{1:13} \\ S = S_{1,2:1,3} \end{cases}$$

4.3 Features Vector and Action Classification

For action recognition, we started considering only the latent forces. That is to say the time series $[f_{1,2}(t)]$. Unfortunately, this information was not sufficient to discriminate among actions. Mainly, we found that similar movements but performed by different body parts, generates similar forces, e.g., raising a hand and raising a leg. On the contrary, the constants of the system, namely, the sensitivities, the damping and the friction parameters were clearly different. This corresponds to the initial intuition that the sensitivities modulate/guide the energy towards some body parts, while the damping/friction represents the sharpness of the observed movement. Accordingly, we constructed our features vector as the concatenation of the forces time series together with the system constants.

$$X = [\, f_{1,2}(t) \quad S_{1,:1,3} \quad S_{1,:1,3} \quad D_{1,13} \quad C_{1,13} \quad]$$

This vector has been used to feed a Support Vector Machine (SVM) to perform the classification.

5   Results

Given human activity sequences, action recognition problem consists in solving two sub-problems: (1) segmenting the sequence into actions, (2) recognizing the segmented action. In a previous work [13, 15], the segmentation was addressed and is not considered here and we only focus on segmented sequences.

We evaluate our proposed Latent force-based features on two different datasets. One is the MAD dataset from CMU [14]. The second dataset includes daily actions we collected with a Kinect sensor.

To check the preliminary feasibility, we considered 8 different actions from both datasets. 20 people perform every single action in the MAD dataset twice. We extract latent force features using 13 original joints time series. The homemade dataset is very similar to the MAD. We focused on daily actions, such as drinking, wearing glass and stirring etc. Every action has been performed 20 times from 4 different people. Here as well, every sample has 13 features sequences also joint angles.

Table. 1. MAD dataset results: av precision 89.93 %

	Crouching	Jump and side	Left arm	Right Arm Pointing to the	Right Leg Kick to the	Cross Arms in the	Basketball	Both Arms Pointing to
Crouching	1.00							
Jump and side kick		0.89				0.11		
Left arm wave			1.00					
Right Arm Pointing to the				0.97		0.03		
Right Leg Kick to the Front	0.11		0.06		0.83			
Cross Arms in the Chest						1.00		
Basketball Shooting			0.06			0.22	0.72	
Both Arms Pointing to Both							0.17	0.78

Table 2. The homemade dataset. Av precision 84.37 %

	Sit	Drink	Pour	Gargle	Wear glass	Brush	Phone call	Stir
Sit	1.00			0.10	0.25			
Drink		0.90	0.05				0.15	0.10
Pour			0.65	0.05			0.05	
Gargle				0.80			0.05	
Wear glass			0.20		0.70			
Brush						0.95		
Phone				0.15		0.05	0.80	
Stir						0.05		0.95

Every sample is performed randomly and with different length. The body joints are captured at sample rate 30 Hz but down-sampled to 10 Hz.

After extracting the features in both datasets, we applied the Leave-one-out cross-validation is performed. Table 1, and Table 2 show results, respectively on MAD and on the homemade daily actions. The average precisions are resp. 89.93 % and 84.375 %, which comparable to state of the art results.

6   Conclusion and Future Work

In this paper, we presented latent force features based approach in solving human action recognition. The proposed features reach interesting results compared to existing works. Moreover, it allows more basic interpretation in relation with energetic and biomechanics aspects of human motion. In the near future, our aim is to include these two categories in the analysis to go deeper in human movement interpretation. Indeed, some preliminary tests showed the effectiveness of this coding to interpret some motor impairment such as trembling of limbs. The other point we want to address concerns segmentation. Our previous work will be combined to this one to provide a complete system.

Though the features are good to describe some actions, some actions are not described well by proposed features. A way to improve is to increase the number of latent forces. Unfortunately, this has a computational cost (inverting a large co-variance matrix) and a more adapted optimization technique should be investigated: we used a classical gradient descent while we have a quadratic form and local techniques should be faster avoiding the matrix inversion issue.

References

1. Shotton, J., Fitzgibbon, A., Cook, M., Sharp, T., Finocchio, M., Moore, R., Kipman, A., Blake, A.: Real-time human pose recognitionin parts from single depth images. In: CVPR (2011)
2. Alvarez, M.A., Luengo, D., Lawrence, N.D.: Latent force models. In: van Dyk, D., Welling, M. (eds.) Proceedings of 12th International Conference Artificial Intelligence and Statistics, pp. 9–16, April 2009
3. Wang, J., Liu, Z., Wu, Y., et al.: Mining actionlet ensemble for action recognition with depth cameras. In: 2012 IEEE Conference on Computer Vision and Pattern Recognition (CVPR), pp. 1290–1297. IEEE (2012)
4. Vemulapalli, R., Arrate, F., Chellappa, R.: Human action recognition by representing 3D skeletons as points in a lie group. In: Proceedings of the IEEE Conference on Computer Vision and Pattern Recognition, pp. 588–595 (2014)
5. Koppula, H.S., Gupta, R., Saxena, A.: Learning human activities and object affordances from RGB-D videos. Int. J. Robot. Res. **32**(8), 951–970 (2013)
6. Piyathilaka, L., Kodagoda, S.: Gaussian mixture based HMM for human daily activity recognition using 3D skeleton features. In: 2013 8th IEEE Conference on Industrial Electronics and Applications (ICIEA), pp. 567–572. IEEE (2013)
7. Sung, J., Ponce, C., Selman, B., et al.: Unstructured human activity detection from RGBD images. In: 2012 IEEE International Conference on Robotics and Automation (ICRA), pp. 842–849. IEEE (2012)
8. Zhang, H., Parker, L.E.: 4-dimensional local spatio-temporal features for human activity recognition. In: 2011 IEEE/RSJ International Conference on Intelligent Robots and Systems (IROS), pp. 2044–2049. IEEE (2011)
9. Cheng, G., Wan, Y., Saudagar, A.N., Namuduri, K., Buckles, B.P.: Advances in Human Action Recognition: A Survey. CoRR abs/1501.05964 (2015). http://arxiv.org/abs/1501.05964
10. Chaudhry, R., Ofli, F., Kurillo, G., Bajcsy, R., Vidal, R.: Bio-inspired dynamic 3D discriminative skeletal features for human action recognition. In: IEEE Conference on Computer Vision and Pattern Recognition Workshops (2013)
11. Slama, R., Wannous, H., Daoudi, M., Srivastava, A.: Accurate 3D action recognition using learning on the Grassmann manifold. Pattern Recogn. **48**(2), 556–567 (2015). Elsevier
12. Du, Y., Wang, W., Wang, L.: Hierarchical recurrent neural network for skeleton based action recognition. In: 2015 IEEE Conference on Computer Vision and Pattern Recognition (CVPR), Boston, MA, pp. 1110–1118 (2015). doi:10.1109/CVPR.2015.7298714
13. Bernier, E., Chellali, R., Thouvenin, I.M.: Human gesture segmentation based on change point model for efficient gesture interface. In: 2013 IEEE RO-MAN, South Corea, pp. 258–263 (2013)
14. Huang, D., Yao, S., Wang, Y., De La Torre, F.: Sequential max-margin event detectors. In: Fleet, D., Pajdla, T., Schiele, B., Tuytelaars, T. (eds.) ECCV 2014, Part III. LNCS, vol. 8691, pp. 410–424. Springer, Heidelberg (2014)
15. Chellali, R., Renna, I.: Emblematic gestures recognition. In: 2012 Proceedings of the ASME 11th Biennial Conference on Engineering Systems Design and Analysis (ESDA 2012). ASME ESDA 2012, vol. 2, pp. 755–753 (2012)

A Study on Classification of Food Texture with Recurrent Neural Network

Shuhei Okada, Hiroyuki Nakamoto[✉], Futoshi Kobayashi, and Fumio Kojima

Graduate School of System Informatics, Kobe University,
1-1 Rokkodai-cho, Nada-ku, Kobe 657-8501, Japan
shuhei.okada@kojimalab.com, nakamoto@panda.kobe-u.ac.jp
http://www.kojimalab.com/

Abstract. This study constructs a food texture evaluation system using a food texture sensor having sensor elements of 2 types. Characteristics of food are digitized by using the food texture sensor in imitation of the structure of the human tooth. Classification of foods is carried out by the recurrent neural network. The recurrent neural network receives the time-series outputs from the food texture sensor, and outputs classification signals. In the experiment, 3 kinds of food are classified by the recurrent neural network.

Keywords: Food texture · Recurrent neural network, RNN · Back propagation through time, BPTT · Classification of food texture

1 Introduction

Food texture greatly affects the taste that a human feels. In addition, it depends on the texture, whether it is the food which an infant and an elderly person are easy to take in. Methods to evaluate the food texture are mainly a sensory evaluation and an evaluation using the food texture measuring equipment now. The sensory evaluation is a method that large number of subjects eat food. However, this method includes individual difference and subject's preferences and needs much time and cost. In addition, it is difficult to quantify the food texture. On the other hand, Fig. 1 shows the food texture sensor which is generally used.

This sensor analyzes a time-series force in food structure. This evaluation method improves the problems of the sensory evaluation. However, the measured value is only the force, and it is difficult to express an index of the food texture in only it [2]. Therefore, a new method to quantify food texture is required. This study built food texture evaluation system which used a food texture sensor which has a sensor device with different kind. In the food texture evaluation system, firstly, the food texture sensor outputs the characteristics of the food as a numerical value. Secondly, the value is inputted to a network designed on a model of living nerve circuitry called the recurrent neural network. Finally, food is classified by the output of the recurrent neural network.

© Springer International Publishing Switzerland 2016
N. Kubota et al. (Eds.): ICIRA 2016, Part I, LNAI 9834, pp. 247–256, 2016.
DOI: 10.1007/978-3-319-43506-0_21

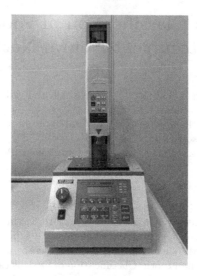

Fig. 1. Creep meter (JSV-H1000) [1].

2   Food Texture Measurement System

2.1   Human Receptors in Periodontal Membrane

There is periodontal membrane in the structure of the human tooth. It fixes a tooth and reduces load to depend on a tooth. Furthermore, it has perception receptors. These receptors have a slowly adapting and a quickly adapting. Therefore the food texture is perceived by the receptors in the periodontal membrane. When a human bites food, he senses a texture by letting a teeth vibrate. A tooth is displaced by stretching the periodontal membrane between a tooth and alveolar bones. The alveolar bone is a bone supporting the root of the tooth.

2.2   Measurement Principle

Figure 2 shows the structure of food texture sensor.

The food texture sensor used in this study has a plunger, a base, an elastomer layer. These correspond to the tooth, the alveolar bone, the periodontal membrane each. The plunger has a magnet. It is a columnar forms of 6 mm in diameter and the magnetizing force is 2070G/207mT. When the sensor touched food, a plunger is displaced. The board of the electronic circuit is glued on the lower part of the base. The board has GMR (Giant Magneto Resistive) elements and an inductor for coil element. The GMR elements and the inductor play a part in the slowly adapting and quickly adapting. The GMR element uses giant magneto resistive effect. When force increased to the plunger, the elastomer layer stretches and the plunger is displaced. The output of GMR elements changes by

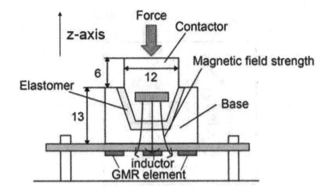

Fig. 2. Structure of food texture sensor.

increasing force for food. The output voltage of the inductor changes by changing a magnetic field when the magnet in the plunger is displaced. The inductor expresses the vibration when the human chewed food.

2.3 Prototype Sensor and Measurement System

Figures 3, 4 and 5 show the food texture sensor, the board of the electronic circuit using for the food texture sensor, and appliances used for the experiment in the measurement system for this study.

8 GMR elements are in a circle and at regular intervals. An inductor is the center of the circle. The texture sensor fractures food by moving the motorized

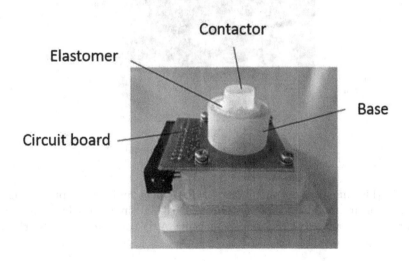

Fig. 3. Food texture sensor.

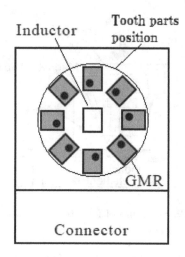

Fig. 4. Sensor circuit.

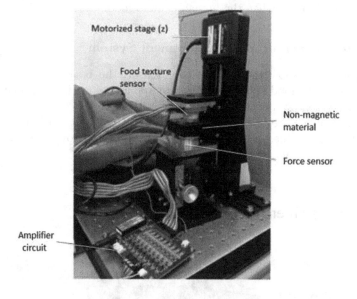

Fig. 5. Instrument for experiment.

stage. In this time, the output of the GMR element and the output voltage of the inductor are amplified it to 32 times and 10,000 times each by an amplifier circuit. After that, a PC obtains them using A/D converter. Finally, the PC calculates force based on the GMR's outputs.

3   Classification Using Recurrent Neural Network

3.1   Recurrent Neural Network

The recurrent neural network expands neural network in imitation of the structure of the human brain [3]. The recurrent neural network is easier to input time-series data than a neural network [4]. This study uses fully connected the recurrent neural network, which is a kind of the recurrent neural network (Fig. 6).

The way of learning is used back propagation through time (BPTT) which is a supervised learning [5]. At this time, it is assumed that the structure of fully connected recurrent neural networks becomes the laminar perceptron (Fig. 7) [6].

T is a maximum time.

Fully connected recurrent neural networks is defined as

$$s_i^{t+1} = \sum_{j=1}^{n} w_{ij} y_j^t, \tag{1}$$

$$y_i^t = f(s_i^t) \qquad (m+1 \le i \le n), \tag{2}$$

$$y_i^t = x_i^t \qquad (1 \le i \le m), \tag{3}$$

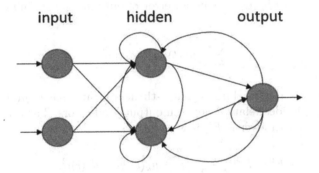

Fig. 6. Structure of recurrent neural networks

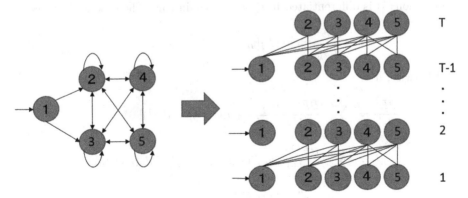

Fig. 7. Hierarchical structure

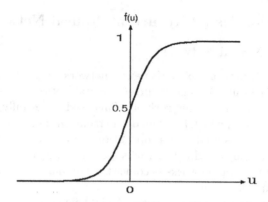

Fig. 8. Sigmoid function

where y_i^t the output of hidden layer and output layer, x_i^t is the input from outside, s_i^t is the internal state of the neuron, w_{ij} is a weight from the j-th neuron to the i-th neuron, n is the number of neurons, m is the number of input layer's neurons. The output function is used a sigmoid function (Fig. 8).

The BPTT learn by decreasing an error of outputs and teaching signals. It is defined as

$$\frac{1}{2} \sum_{k=m+1}^{n} \mu_k(t)[y_k^t - d_k(t)]^2, \tag{4}$$

where $d_k(t)$ is the teaching signal k-th neuron at time t, $\mu(t)$ is a mask function which decides relation between outputs and errors. The error from the T-th layer to the t-th layer is E^t, it is as follows:

$$E^t = E^{t+1} + \frac{1}{2} \sum_{k=m+1}^{n} \mu_k(t)[y_k^t - d_k(t)]^2. \tag{5}$$

Equation (5) is differentiated in s_k^t with a chain rule. The case of $T = t$ is

$$\frac{\partial E^T}{\partial s_k^T} = \mu_k(T)[y_k^T - d_k(T)]f'(s_k^T). \tag{6}$$

Otherwise, it is

$$\begin{aligned}
\frac{\partial E^t}{\partial s_k^t} &= \sum_{l=m+1}^{n} \frac{\partial E^{t+1}}{\partial s_l^{t+1}} \frac{\partial s_l^{t+1}}{\partial s_k^t} + \mu_k(t)[y_k^t - d_k(t)] \\
&= \sum_{l=m+1}^{n} \frac{\partial E^{t+1}}{\partial s_l^{t+1}} w_{lk} f'(s_k^t) + \mu_k(t)[y_k^t - d_k(t)]f'(s_k^t) \\
&= f'(s_k^t)\Big(\sum_{l=m+1}^{n} w_{lk} \frac{\partial E^{t+1}}{\partial s_l^t} + \mu_k(t)[y_k^t - d_k(t)]\Big).
\end{aligned} \tag{7}$$

Therefore, the weight is updated as follows:

$$\Delta w_{ij} = -\alpha \sum_{t=1}^{T} \frac{\partial E^t}{\partial w_{kl}}$$

$$= -\alpha \sum_{t=1}^{T} \sum_{k=m+1}^{n} \frac{\partial E^t}{\partial s_k^t} \frac{\partial s_k^t}{\partial w_{ij}}$$

$$= -\alpha \sum_{t=1}^{T} \frac{\partial E^t}{\partial s_i^t} \frac{\partial s_i^t}{\partial w_{ij}}$$

$$= -\alpha \sum_{t=1}^{T} \frac{\partial E^t}{\partial s_i^t} y_j^{t-1}, \tag{8}$$

$$w_{ij} = w_{ij} + \Delta w_{ij}. \tag{9}$$

Δw_{ij} is update amounts of the weight.

3.2 Classification

The classification of the food is carried out by the recurrent neural network. When a measured food texture data is input into the recurrent neural network, it learns the weight using 3 expressions. The teacher signal is set for each input. After learning, the weight is decided, and a new food texture data is input into recurrent neural networks using the weight. A food is classified by comparing the teacher signal with the output.

4 Experiment

4.1 Measurement of Food Textures

The sample to use for this experiment is cookies, gummy candies, corn snacks. These foods are put on the stage of the texture sensor, and the output force of GMR and the output voltage of the inductor of these foods is measured when they are fractured by the food texture sensor. They are measured for 6 s and the output voltage of the inductor is 10 KHz. They are measured by ten times per 1 food. Figures 9, 10 and 11 show the results of measurement of force and inducted voltage.

The characteristic of the biscuit is to significantly change the force from 3 to 6 s. The characteristic of the gummy candy isn't to change both the force and the inducted voltage from first to last. The characteristic of the corn snack is to significantly change the inducted voltage from 3 to 6 s.

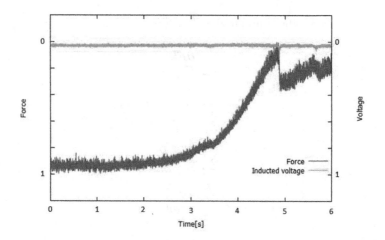

Fig. 9. Force and inducted voltage of biscuits.

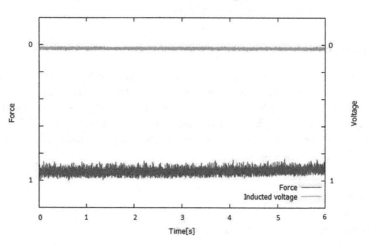

Fig. 10. Force and inducted voltage of gummy candy.

4.2 Classification

In the experiment, the training data is 1 and the test data are 9 out of measured data of 10 times. The test data are replaced with the training data, 10 times by classifications in total are carried out. The number of the inputs is 60000 because of one data in 6 s, but they are reduced to 1000 by 3 conditions, because neural network is easy to learn a little number of the inputs.

Condition 1, Measured data are averaged for every 60.
Condition 2, Measured data from 5 to 6 s are averaged for every 10.
Condition 3, The input is the data between 5.5 s and 5.6 s.

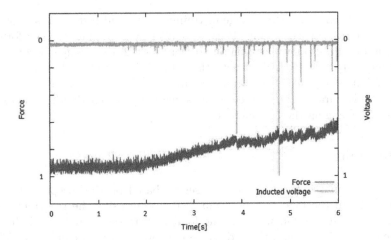

Fig. 11. Force and inducted voltage of corn snack.

Table 1. Parameters of the recurrent neural network

Time T	Number of learning	Number of neuron in hidden layer	Learning constant α
1000	10000	40	0.1

Table 2. Result of condition 1

	Biscuits	Gummy candy	Corn snack
Correct answer rate	70 %	62 %	54 %

Table 3. Result of condition 2

	Biscuits	Gummy candy	Corn snack
Correct answer rate	90 %	94 %	43 %

Table 4. Result of condition 3

	Biscuits	Gummy candy	Corn snack
Correct answer rate	100 %	96 %	76 %

The parameter of the recurrent neural network is set like Table 1. Tables 2, 3 and 4 show the result of the classification.

4.3 Discussion

Condition 3 was the best result in the three conditions. The reasons are as follows. First, the character of the food was reflected directly by not averaging measured data. Second, the change of training data and test data is about the same. In condition 1, the change became smaller by averaging it. Therefore, the classification of all foods was difficult. In condition 2, the change of corn snack of training data didn't get right with that of test data, so the classification was difficult. This study used the parameters in Table 1, but they aren't always the most suitable parameters. Therefore, it is necessary for conducting the experiment while changing the parameters.

5   Summary

This study proposed the food texture sensor based on the receptors that the human had and the food texture evaluation system using the recurrent neural networks. In the experiment, first, the food texture sensor measured the food texture of 3 foods. Second, those data were compressed by 3 conditions. Third, they were input into recurrent neural networks. Finally, the classification of the food texture was carried out by comparing the output of recurrent neural networks with the teacher signal. As a future work, this study will propose a method compressing measured data without reducing food characteristics and improve the correct answer rate of the classification.

References

1. Japan Instrumentation System Co., Ltd. http://www.jisc-jp.com/product/handy/jsv-hseries.html
2. Mitsuru, T., Takanori, H., Naoki, S.: Device for acoustic measurement of food texture using a piezoelectric sensor. Food Res. Int. **39**, 1099–1105 (2006)
3. Mikolov, T., Kombrink, S., Burget, L., Cernocky, J., Khudanpur, S.: Extensions of recurrent neural network language model. In: 2011 IEEE International Conference on Acoustics, Speech and Signal Processing, pp. 5528–5531 (2011)
4. Mikolov, T., Karafiat, M., Burget, L., Cernocky, J., Khudanpur, S.: Recurrent neural network based language model. In: INTERSPEECH, vol. 2, p. 3 (2010)
5. Heermann, P.D., Khazenie, Nahid: Classification of multispectral remote sensing data using a back-propagation neural network. IEEE Trans. Geosci. Remote Sens. **30**(1), 81–88 (1992)
6. Mirikitani, D.T.: Recursive bayesian recurrent neural networks for time-series modeling. IEEE Trans. Neural Netw. **21**, 262–274 (2009)

A Rotating Platform for Swift Acquisition of Dense 3D Point Clouds

Tobias Neumann, Enno Dülberg, Stefan Schiffer, and Alexander Ferrein[⊠]

Mobile Autonomous Systems and Cognitive Robotics (MASCOR) Institute,
FH Aachen University of Applied Sciences, Aachen, Germany
enno.duelberg@alumni.fh-aachen.de,
{t.neumann,s.schiffer,ferrein}@fh-aachen.de

Abstract. For mapping with mobile robots the fast acquisition of dense point clouds is important. Different sensor techniques and devices exist for different applications. In this paper, we present a novel platform for rotating 3D and 2D LiDAR sensors. It allows for swiftly capturing 3D scans that are densely populated and that almost cover a full sphere. While the platform design is generic and many common LRF can be mounted on it, in our setup we use a Velodyne VLP-16 PUCK LiDAR as well as a Hokuyo UTM-30LX-EW LRF to acquire distance measurements. We describe the hardware design as well as the control software. We further compare our system with other existing commercial and non-commercial designs, especially with the FARO Focus^{3D} X 130.

1 Introduction

In a number of domains *Mobile Mapping* is used for acquiring range data of the environment. The applications range from light detection and ranging (LiDAR) imaging to find out, say, if heavy-load trucks fit through a height-restricted area to services like Google Street View and other GIS for reconstructing exact street maps or complete urban areas by imaging from cars or aircrafts, respectively. Mobile Mapping in the field of robotics has further important applications. Mobile robot systems may use the acquired information for localisation, but they might also use the incoming sensor information for collision avoidance, terrain classification or path-planning. Outside, a GPS signal can be used for integrating a series of single sensor data into a consistent model of the environment; however, in indoor environments such information is not available. Here, good positioning information of the robot between consecutively recorded sensor readings is required. This information can be computed from overlapping sensor data, for instance, with Structure from Motion approaches in the case of 2D camera images, e.g. [4], or with the Iterative Closest Point (ICP) algorithm for 3D range data, e.g. [1,12]. For these methods to work, there must be sufficient overlap between consecutive images or scans. Pomerlau et al. [12] give a number of use cases for ICP-based point cloud registration, ranging from urban search & rescue data acquisition to autonomous driving applications.

For mobile robotic mapping applications, one can distinguish several different modes of how map data is acquired:

© Springer International Publishing Switzerland 2016
N. Kubota et al. (Eds.): ICIRA 2016, Part I, LNAI 9834, pp. 257–268, 2016.
DOI: 10.1007/978-3-319-43506-0_22

1. mapping of static vs. dynamic environments, i.e. whether there are dynamic objects in the environment such as pedestrians;
2. mobile vs. static data acquisition, i.e. the robot acquires data while standing still or while it is moving, respectively.

There is usually a trade-off between acquiring 3D data at a high frequency, on the one hand, and the accuracy of the data, on the other. Normally, slower devices yield higher accuracies in the scan data. There are several scenarios to be taken into account. In environments where many dynamic obstacles are around, one would usually prefer a high scan frequency, while in static environments long scanning times for a single 3D scan are not problematic. Acquiring 3D data in a dynamic environment with a device that takes about half a minute or more may lead to point clouds that are prone to motion blur and occlusions due to dynamic objects. Scanning the scene with a fast device may not yield the same accuracy and density for a single scan, but acquiring multiple scans (from the same position or from different view points) might allow the detection of dynamic obstacles in the data. Hence, the risk of integrating them into the final model is lower. A famous method for this is shown in [5], where rasterization of the data and using the probability on rasterizated cells is used. More methods are described in the literature. A different scenario is the acquisition with a moving device, for instance, if a scanning device is mounted onto a robot or a car. Then again, high frequencies per single 3D scan are advantageous.

In this paper, we present SWAP, a novel sensor platform equipped with a Velodyne VLP-16 PUCK[1] 3D LiDAR and a Hokuyo UTM-30LX-EW[2] 2D laser range finder (LRF). With our rotating platform we are aiming at getting the best from two worlds: it provides point clouds of the whole environment at reasonably high speeds with a high data rate. This is partly due to using a 3D instead of (only) a 2D sensor. In the next section, we give the technical description of our novel sensor platform's hardware and its control software. In Sect. 3, we compare our system with several other 3D scanning devices used for mobile mapping applications along with a set of system properties. Further, we compare point clouds taken with our platform with the commercial scanner FARO Focus[3D] X 130[3] in Sect. 4, before we conclude.

2   The SWAP Platform

In our own earlier work, we developed a tilting LiDAR device based on the Velodyne HDL-64[4] for acquiring dense 3D point clouds in mobile mapping applications [7]. While the device is very suitable for acquiring 3D dense point clouds (in [6] we used this device for mapping large-scale motorway tunnels), it has

---

[1] http://velodynelidar.com/vlp-16.html.

[2] http://www.hokuyo-aut.jp/02sensor/07scanner/utm_30lx_ew.html.

[3] http://www.faro.com/products/3d-surveying/laser-scanner-faro-focus-3d/overview.

[4] http://velodynelidar.com/hdl-64e.html.

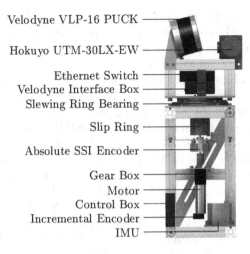

Velodyne VLP-16 PUCK

Hokuyo UTM-30LX-EW

Ethernet Switch
Velodyne Interface Box
Slewing Ring Bearing

Slip Ring

Absolute SSI Encoder

Gear Box
Motor
Control Box
Incremental Encoder
IMU

(a) Components of the platform.   (b) Photo of the platform.

Fig. 1. The components and a photo of our rotating sensor platform.

some drawbacks when it comes to the distribution of the range measurement in the point cloud. With a tilting scanner, the point clouds are particularly dense at the turning points of the device. A more even distribution of data is, in general, preferred. For this reason, we developed the SWAP (swift acquisition of point clouds) platform, a rotating 3D LiDAR based on the Velodyne VLP-16 PUCK LiDAR. With the novel system, we achieve a more even distribution of scan points in the point cloud while reducing the time needed for a whole 3D scan. This comes, however, at the cost of the density of the clouds. In the following, we give the technical description of the SWAP platform and its control software.

To be able to acquire the (nearly) complete sphere around the scanning device, we deploy a Velodyne VLP-16 PUCK LiDAR which is rotated perpendicular to its ranging plane. The LiDAR has a vertical field of view (VFoV) of 30° and a horizontal field of view (HFoV) of 360°. The VFoV is scanned with 16 individual laser beams; rotating the LiDAR horizontally yields the range data of the point cloud scanning a complete sphere around the device. The Velodyne is mounted with an inclination of 14° to the vertical rotation axis to maximise the area of high measurement density at the polar region above the platform. By deciding to use this 3D LiDAR we notably increase the number of points that we can register for a scan in a given time while maintaining an equal point distribution. E.g., a 2D LRF would need a frequency of about 300 Hz and capture 900 points in one revolution to acquire a point cloud with a similar resolution within the same time. As the minimum range of the Velodyne starts at about 0.9 m, we additionally mounted a Hokuyo UTM-30LX-EW on our scanning device for close-range data. The UTM-30-LX is a 2D multi-echo LRF with an angular resolution of 0.25°. Its range plane is mounted (almost) perpendicular to the rotating plane of our scanning platform. As a final ingredient, our SWAP platform is equipped with an inertial measurement unit (IMU) (μIMU from NG[5]) for providing the orientation of the platform w.r.t. the ground. Figure 1 shows a CAD drawing as well as a photo of the device.

The platform is separated into a base frame and the rotating sensor mount. Both parts are connected with a slip ring and a bearing. The base frame made of industrial aluminium profiles provides low weight and high stiffness which is needed to ensure that the transformation information between the IMU mounted in the lower part and the sensors in the upper part is consistent. The combination of motor and gear head provides us with 3 Nm of torque and allows for a maximum rotation speed of 2.6 Hz. However, a reasonable azimuth resolution can only be acquired with a scanning speed of up to 1.67 Hz while the full sphere point clouds are then captured with a half revolution which equals 3.34 Hz for this. Accurate information about the orientation of the rotating sensor platform is essential for correctly registering the raw data of the sensors. We deploy a 14 Bit industrial grade absolute Synchronous Serial Interface (SSI) encoder which is mounted on the drive shaft. The resolution provides a maximum error of 1.32′ or 0.022°. In a distance of 10 m, this corresponds to 3.8 mm. The second part of the platform is the rotating sensor mount. It houses a gigabit Ethernet switch, the interface box of the Velodyne VLP-16 PUCK and the Hokuyo UTM-30LX-EW, the power distribution for the sensors and several mounting rails for different sensors.

The raw data of the deployed Velodyne VLP-16 PUCK and the attached Hokuyo UTM-30LX-EW are registered making use of the SSI absolute encoder. Besides the absolute encoders, there is another incremental encoder attached to the motor shaft. Then, based on the readings of the absolute encoder, the raw data is collected and integrated into a point cloud for the device. This is done with a best-effort time-stamping on the data and where one UDP-package of the Velodyne VLP-16 PUCK is transformed all together. The time difference between the laser readings within one UDP-package are about 1.33 ms. For the rectification of the Hokuyo UTM-30LX-EW measurements the recording time for one sweep is taken into account. This setup yields a quite equal distribution of points in the sphere.

3   Comparing Existing 3D LiDAR Platforms

In the following, we compare our system with a number of other 3D LiDAR devices described in the literature. First, we review some of the devices deployed for 3D mobile mapping. Then, we define system properties of interest in order to compare different scanning devices such as resolution, range or scanning time, before we compare six different 3D scanning devices with our novel platform.

---

[5] http://www.northropgrumman.litef.com/en/products-services/
industrial-applications/product-overview/mems-imu/.

3.1   Acquiring Point Clouds

There are several ways for acquiring 3D point clouds with a mobile robot. One of the most straight-forward ways is to use two 2D LRF which are mounted orthogonally to each other. While one sensor is scanning along the XY-plane, the second LRF is mounted in the YZ-plane and measures wall and ceiling distances. However, this way only planar 3D maps [9] can be constructed. Similar setups are described in [3,19]. This sensor setup is suited only for dynamic acquisition, as due to the robot movement different scan lines along the YZ-plane are acquired. This also results in the sensor being unsuitable for collision avoidance.

Another common approach is to tilt one LRF to increase the field of view (FoV) of the ranging device shown, for instance, in [2,14,15]. Tilting, however, has several implications on the point distribution of the point cloud. A more even distribution is achieved by rotating a range device rather than tilting it. The differences between tilting, rolling and yawing an LRF are analysed in [10, 18], also w.r.t. the different point densities of the resulting point clouds. Some examples, which we investigate in this section, are the RTS/ScanDrive [18] and its improved version, the RTS/ScanDrive Duo [17] as well as a rotating Hokuyo UTM-30LX-EW range finder described in [13] (which we will call Rotating UTM-30LX or R-UTM-30LX, for short).

The method of an upwards facing LRF which rotates around the yaw-axis as described in [18] is also used in commercial architectural scanning devices where both rotations are already integrated in one system. We will have a closer look at the RIEGL VZ-400[6] and the FARO Focus^{3D} X 130. The former was, for instance, deployed in mobile robotics applications in [8], the latter in [16,18]. While both sensors have a horizontal FoV of 360°, the RIEGL VZ-400 has a vertical FoV of 100° with a maximum speed of 6 s for one scan; the FARO Focus^{3D} X 130 has a vertical FoV of 300°, here only the occlusion of the sensor platform itself limits the FoV. Other commercial systems such as the Velodyne LiDAR sensors follow a different principle. They use several laser-beams (between 16–64) to capture the VFoV and rotate them horizontally by 360°; they reach a scanning rate of up to 20 Hz. Their sensors have an VFoV between 26.8° and 40°. The Velodyne VLP-16 PUCK used for the 3D scanning device presented in this paper has 16 beams with a 2° vertical resolution.

3.2   Comparison Methodology

Mapping Use Case Scenarios. As pointed out in the introduction, there exist four different scenarios with different requirements for the acquisition sensor:

1. *Static environment:* The environment does not change during the acquisition of a scan. In particular, no dynamic objects occlude parts of the environment.
2. *Dynamic environment:* Dynamic obstacles exist and might cause occlusions or motion blur in the range scans.

---

[6] http://www.riegl.com/products/terrestrial-scanning/produktdetail/product/scanner/5/.

3. *Static data acquisition:* While the robot acquires a whole 3D scan (sphere around the robot), the robot stands still.
4. *Dynamic acquisition:* The robot moves while acquiring data; this adds another source of noise into the acquired 3D scan.

The different scenarios pose different requirements to the used 3D ranging device. It is, for example, advantageous if the device achieves a higher scan frequency when used in environments with dynamic obstacles. Consider a robot mapping a human-populated area. A ranging device with a low scanning rate will scan a pedestrian walking past the scanner multiple times, causing the point cloud to be blurred at these positions. Also, the same person occludes larger parts of the environment. On the other hand, if time for acquiring the scan is not of the essence, then the scan frequency is not important. A device then should enjoy a high point rate for providing accurate and dense point clouds of the static environment. This is often the case with architectural scanning devices.

System Properties. In the following we clarify which properties of the devices used in our overview will be compared. As, for instance, all devices have different azimuth resolutions, while higher resolutions can be reached with a higher scanning time, we use the fastest possible scanning time with an azimuth resolution of up to 2°.

Range: The range yields the maximum distance of the main LiDAR device used in the 3D scanner. The vendors of the devices state the maximal range at different reflectivity values in their data sheets. Therefore, this property just gives an indication of the distances up to which the device can be used.

Azimuth and Elevation Resolution: The azimuth resolution shows the horizontal resolution of the scanning device. Likewise, the elevation resolution states the vertical resolution.

Sphere Coverage: The sphere coverage sc indicates the theoretical coverage of a sphere given by the FoV of the sensor. We define it as $sc = \frac{\text{VFoV} \times \text{HFoV}}{360° \times 360°}$. What is also interesting for robotic mapping applications is whether the scanner has blind spots within this acquired sphere. As not all technical details of the scanning devices were available, we could not correlate the blind spots with the sphere coverage. Therefore, the blind spot indicates what is to be expected and covered within the text about each sensor.

3.3 System Comparison

In this section, we compare the following devices (also shown in Table 1) based on the properties described above.

– *RTS/ScanDrive (Duo)*. It is a rotating device with 2 SICK LMS 2D LRFs with a maximal resolution of 0.25°. In [17], the authors describe mapping experiments with this device where they use a resolution of 1°. This is the

Table 1. Specifications of reviewed systems. The vertical and horizontal resolutions shown correspond to the fastest scanning times of the respective device.

System	Range [m]	Resolution		Sphere coverage [%]
		Horizontal [deg]	Vertical [deg]	
RTS-Duo	30	2	1	100
R-UTM-30LX	30	2	0.25	75
VZ-400	350	0.5	0.288	27.77
Faro X130	130	0.035	0.07	83.33
T-HDL-64	120	0.09	∅0.015	32.44
SWAP	100	2	0.4	80.27

resolution we refer to in our comparison. The whole device has blind spots at the poles of the scanning sphere, as the two LRFs are mounted with an offset to the rotation centre.

- *R-UTM-30LX.* In [13], mobile mapping applications with a rotating Hokuyo UTM-30LX-EW are described. A single Hokuyo scanner is mounted perpendicular to the rotation plane. Therefore, the vertical resolution in Table 1 refers to the physical resolution of the Hokuyo device. The whole device is rotated horizontally. As the scanning time is not given in the paper, we calculated it based on the sensor data sheet for an azimuth resolution of 2°, compared to the RTS/ScanDrive. It covers 3/4 of a whole sphere with its scanning range.
- *Riegl VZ-400.* The Riegl VZ-400 is a commercial 3D range scanner. Its vertical resolution lies between 0.0024° and 0.288° and it has a range up to 600 m which, however, can be reduced to 350 m with a reduction in the scanning time. The values given in Table 1 show the value when the device is run in its high speed mode. It has a FoV of 100° × 360°. A complete scan requires 6 s; the device acquires up to 122 000 points/s.
- *FARO X130.* The other commercial scanner in the field is the FARO Focus^{3D} X 130. It has a vertical and horizontal resolution of 0.009°, respectively and a maximal range of about 130 m. It has a FoV of 300° × 360°. A complete scan takes up to 54 s, scanning up to 976 000 points/s.
- *Tilting HDL-64.* As described in [7], a Velodyne HDL-64 LiDAR is tilted between 0° and 90° which results in a VFoV of 118.6° but with a large blind spot which is produced by the robot base where the sensor is mounted on. It has a range of about 120 m. The data rate of the sensor is 1 330 000 points/s. A whole sweep takes about 25 s.

Figure 2a shows a comparison of the different systems w.r.t. the scanning times and the data rate of the respective 3D scanning devices. Note that both axes have a logarithmic scale. On the x-axis, the scanning rate in Hz is given. It shows the frequency with which a whole 3D scan (sphere or part of a sphere around the robot) can be acquired. On the y-axis, the point rate is given. It shows how many points the respective ranging sensor mounted on the device is acquiring per second. Devices such as the FARO or the T-HDL-64 have a high

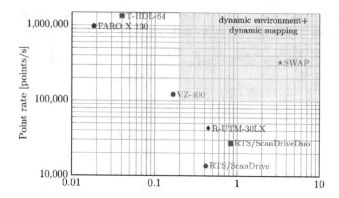

(a) A comparison of points/s and scans/s for typical mobile robotics sensors.

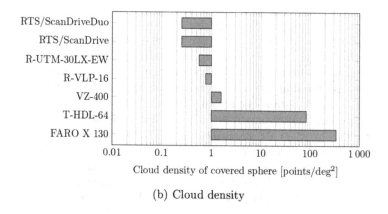

(b) Cloud density

Fig. 2. Comparison results.

data rate, but a whole sweep takes quite a long time (25 or 54 s, respectively). The other rotating devices such as the RTS or the R-UTM-30LX have a reasonable scanning time for a whole 3D sweep, but the data rate is not exceptionally high. Our novel SWAP platform has a really good trade-off between a reasonable scanning rate and a sufficiently high data rate. As shown in Fig. 2a, it is therefore the most suitable for dynamic mapping in dynamic environments.

Figure 2b compares the theoretical point cloud density that can be achieved by the compared devices with the settings given in Table 1. We compare the number of points per deg^2 which the respective device is able to scan. Note that this number is normalised w.r.t. the sphere coverage of the device. The densest point cloud, therefore, is generated by the FARO X 130, which scans over 300 points per deg^2. Our novel device produces point clouds which is close to a resolution of 1 deg^2, while the platforms with one or two rotating LRFs (RTS + R-UTM-30LX) acquire much sparser point clouds.

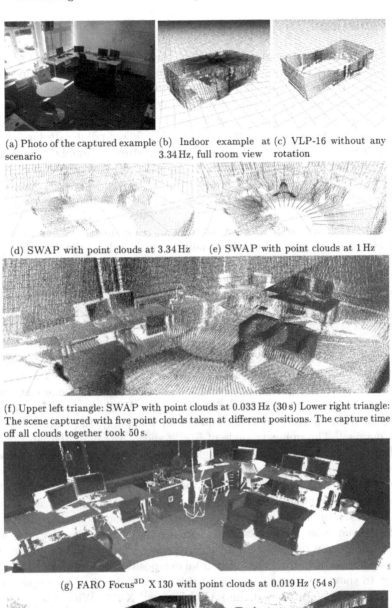

(a) Photo of the captured example (b) Indoor example at (c) VLP-16 without any scenario   3.34 Hz, full room view   rotation

(d) SWAP with point clouds at 3.34 Hz   (e) SWAP with point clouds at 1 Hz

(f) Upper left triangle: SWAP with point clouds at 0.033 Hz (30 s) Lower right triangle: The scene captured with five point clouds taken at different positions. The capture time off all clouds together took 50 s.

(g) FARO Focus³ᴰ X 130 with point clouds at 0.019 Hz (54 s)

(h) Outdoor scene captured by the SWAP platform in 30 s (left) and captured by the FARO Focus³ᴰ X 130 in 54 s (right) as a comparison.

Fig. 3. Example data of the SWAP platform captured with different rotation times and data from the FARO Focus³ᴰ X 130 as a comparison.

4   Discussion

As we have shown in the previous section, there are quite a few 3D scanning devices available. The properties of the scanners are quite different, some target at scanning architecture, some are rather suited for mobile mapping applications. The former devices usually have a high point rate but a very low scanning rate. The FARO Focus^{3D} X 130 is such a device. The scanners presented in our comparison that fall into the latter class trade a lower point rate for a higher scanning frequency. This makes them more interesting for mobile mapping. As we have shown in the scanner comparison, our rotating platform has a comparably high point rate (w.r.t. the other mobile mapping devices) and a high scanning rate. This makes it especially useful for dynamic mapping applications in dynamic environments.

In the rest of this section, we want to compare scanners not only based on their technical data sheet, but also qualitatively based on the resulting point clouds. In particular, we show point clouds from a scene in our laboratory environment and an outside scene of our university campus, both acquired with a FARO Focus^{3D} X 130 and the novel SWAP platform.

Figure 3 shows scans from our laboratory. Figure 3a shows a photograph of the scene to give an overview. Figure 3b shows a point cloud from the whole laboratory scanned with the SWAP platform in comparison of just deploying the Velodyne VLP-16 PUCK in the same room without rotating it (Fig. 3c). It becomes apparent why the Velodyne sensor needs to be rotated in order to acquire a full scanning sphere. Figure 3d to f show scans of the same scene taken at different rotation speeds of the SWAP device. While at 3.34 Hz the cloud is still somewhat sparse (but dense enough for mapping), the scan at 0.033 Hz is very detailed. The upper triangle of Fig. 3f shows a scan of the lab taken at 0.033 Hz; the cloud in the lower triangle shows an overlay of the whole scene captured from several view points. The total scan times of the different scans sum up to 50 s. This is nearly the time, which the FARO Focus^{3D} X 130 took from a single view point (Fig. 3g). One has to keep in mind that many point clouds are taken in mobile mapping applications to reconstruct the whole scene. Therefore, even more sparse scans are not a problem. On the contrary, while doing multiple scans taken with a higher frequency, shadows and occlusions that might occur when acquiring data only from one position can be avoided. Finally, we want to show two scans from an outside scene taken at our campus, again from the SWAP platform and the FARO Focus^{3D} X 130. While the scan from the FARO Focus^{3D} X 130 again is much more detailed, our rotating platform does not leave out major details of the scene. Note that the scanning time of our scanner compared to the FARO is cut in half. The distance to the campus buildings in the background is about 50 m.

5   Conclusion and Future Work

In this paper, we presented a novel platform for acquiring point clouds for robot mapping applications. The main idea is to continuously rotate a 3D LiDAR sensor around an axis that is perpendicular to its main ranging axis. In our case, we make use of a Velodyne VLP-16 PUCK together with a Hokuyo UTM-30LX-EW. With our setup we achieve more evenly distributed point clouds than with a tilting sensor while reaching nearly a full sphere coverage of the surrounding (only a cone of about 71° towards the sensor base cannot be acquired with the Velodyne VLP-16 PUCK, the occlusion of the Hokuyo UTM-30LX-EW is even lower). We have shown that our system is capable of capturing such full-sphere point clouds at a high speed of up to 3.34 Hz with a density that is still reasonably high enough for collision avoidance and generating 3D maps. This makes the SWAP platform the fastest 3D scanner with a near full-sphere coverage in the set of scanners that have been compared in this paper. Furthermore, we compared the results at slower revolutions with commercially available architecture scanner, yielding satisfyingly dense and accurate point clouds for mobile mapping applications.

The next steps for the system are to change from the best-effort time stamping to a real-time-system. The motor and encoders are currently controlled via USB, but they also support a serial connection, the data of the LiDAR sensors are delivered via Ethernet. The Velodyne VLP-16 PUCK gives a time-table for the time-offset of each laser within a UDP-package and the Hokuyo UTM-30LX-EW has a further synchronous output which sends a 1 ms long signal at a defined position of the laser.

Furthermore, the extrinsic calibration of the sensors needs to be measured precisely for correct range measurements. In [11] an automatic and fast method for this calibration is given which we are planing to adapt for our setup. With this automatic calibration we will measure the accuracy of the system compared to the architecture scanner FARO Focus^{3D} X 130 by aligning point clouds of the same area and calculating the mean point error.

Acknowledgments. This work was funded in part by the German Federal Ministry of Education and Research in the programme under grant 033R126C. We thank the anonymous reviewers for their helpful comments and Christoph Gollok for the simulation of the point distribution in Gazebo.

References

1. Besl, P.J., McKay, N.D.: Method for registration of 3-D shapes. IEEE Trans. Pattern Anal. Mach. Intell. **14**(2), 239–256 (1992)
2. Bohren, J., Rusu, R.B., Jones, E.G., Marder-Eppstein, E., Pantofaru, C., Wise, M., Mösenlechner, L., Meeussen, W., Holzer, S.: Towards autonomous robotic butlers: Lessons learned with the PR2. In: IEEE International Conference on Robotics and Automation (ICRA), May 2011
3. Früh, C., Zakhor, A.: 3D model generation for cities using aerial photographs and ground level laser scans. In: Proceedings of the Computer Society Conference on Computer Vision and Pattern Recognition (CVPR), vol. 2 (2001)
4. Hartley, R., Zisserman, A.: Multiple View Geometry in Computer Vision. Cambridge University Press (2003)

5. Hornung, A., Wurm, K.M., Bennewitz, M., Stachniss, C., Burgard, W.: OctoMap: an efficient probabilistic 3D mapping framework based on octrees. Auton. Robots **34**, 189–206 (2013)
6. Leingartner, M., Maurer, J., Ferrein, A., Steinbauer, G.: Evaluation of sensors and mapping approaches for disasters in tunnels. J. Field Robot. (2015)
7. Neumann, T., Ferrein, A., Kallweit, S., Scholl, I.: Towards a mobile mapping robot for underground mines. In: Proceedings of the 7th IEEE Robotics and Mechatronics Conference (2014)
8. Nüchter, A., Borrmann, D., Koch, P., Kühn, M., May, S.: A man-portable IMU-free mobile mapping system. In: ISPRS Annals of Photogrammetry, Remote Sensing and Spatial Information Sciences, vol. II-3/W5 (2015)
9. Nüchter, A., Lingemann, K., Hertzberg, J., Surmann, H.: 6D SLAM-3D mapping outdoor environments. J. Field Robot. **24**, 699–722 (2007)
10. Nüchter, A.: 3D Robotic Mapping. STAR, vol. 52. Springer, Heidelberg (2008)
11. Oberländer, J., Pfotzer, L., Roennau, A., Dillmann, R.: Fast calibration of rotating and swivelling 3-D laser scanners exploiting measurement redundancies. In: Proceedings of the International Conference on Intelligent Robots and Systems (IROS) (2015)
12. Pomerleau, F., Colas, F., Siegwart, R.: A review of point cloud registration algorithms for mobile robotics. Found. Trends Robot. **4**(1), 1–104 (2015)
13. Schadler, M., Stückler, J., Behnke, S.: Multi-resolution surfel mapping and real-time pose tracking using a continuously rotating 2D laser scanner. In: International Symposium on Safety, Security, and Rescue Robotics (SSRR) (2013)
14. Surmann, H., Lingemann, K., Nüchter, A., Hertzberg, J.: A 3D laser range finder for autonomous mobile robots. In: Proceedings of the 32nd International Symposium on Robotics (ISR) (2001)
15. Thrun, S., Thayer, S., Whittaker, W., Baker, C., Burgard, W., Ferguson, D., Hahnel, D., Montemerlo, D., Morris, A., Omohundro, Z., Reverte, C., Whittaker, W.: Autonomous exploration and mapping of abandoned mines. IEEE Robot. Autom. Mag. **11**(4), 79–91 (2004)
16. Wong, U., Morris, A., Lea, C., Lee, J., Whittaker, C., Garney, B., Whittaker, R.: Comparative evaluation of range sensing technologies for underground void modeling. In: Proceedings of the International Conference on Intelligent Robots and Systems (IROS) (2011)
17. Wulf, O., Nüchter, A., Hertzberg, J., Wagner, B.: Ground truth evaluation of large urban 6D SLAM. In: Proceedings of the International Conference on Intelligent Robots and Systems (IROS) (2007)
18. Wulf, O., Wagner, B.: Fast 3D scanning methods for laser measurement systems. In: Proceedings of the International Conference on Control Systems and Computer Science (2003)
19. Zhao, H., Shibasaki, R.: Reconstructing textured cad model of urban environmentusing vehicle-borne laser range scanners and line cameras. In: Second International Workshop on Computer Vision System (ICVS) (2001)

Kinematic Calibration and Vision-Based Object Grasping for Baxter Robot

Yanjiang Huang[1,2,3(✉)], Xunman Chen[1,2], and Xianmin Zhang[1,2]

[1] School of Mechanical and Automotive Engineering,
South China University of Technology, Guangzhou, China
mehuangyj@scut.edu.cn
[2] Guangdong Provincial Key Laboratory of Precision Equipment
and Manufacturing Technology, Guangzhou, China
[3] Research into Artifacts, Center for Engineering,
The University of Tokyo, Chiba, Japan

Abstract. In this paper, firstly, we analyze and compare four robot kinematic calibration methods for Baxter robot, including screw-axis measurement, extended kalman filtering, least squares method, integration of screw-axis measurement and kalman filtering. The performance of these four methods are evaluated through an experiment. Secondly, we analyze the monocular-vision and binocular-vision in object grasping, and propose a vision-based step-by-step object grasping strategy for the Baxter robot to grasp an object. Experiment results show that the object can be grasped based on the proposed method.

Keywords: Kinematic calibration · Vision-based object grasping · Baxter robot

1 Introduction

Since the dual arm robot can complete some tasks like human being, it has been focused by many researchers [1]. Many dual arm robots, such as SDA 10 [2], PR2 [3], Baxter [4], have been developed for varied applications. Bater robot is an industrial robot with two arms and two cameras on the arms, which has been applied in many applications, especial in picking tasks [4]. To complete the picking tasks, it is necessary to improve the positioning accuracy and grasp the object accurately.

There are many ways to improve the accuracy of positioning and object grasping for the robot. Calibration of the geometric parameters of the robot and sensor-based object grasping are the common methods, which received focuses by the researchers. Normally, the calibration of geometric parameters can be realized through the following four steps: modelling, measurement, parameter identification, and error compensation [5]. Homogeneous transformation matrix based D-H model is the usual model used for calibration [6]. However, the kinematic singularity occurs when using such D-H model. To overcome the kinematic singularity, some revised D-H model were proposed in [7–9]. Some studies realized the calibration of robot based on laser tracking [10] or CMM [11]. Such methods can obtain high accuracy, but with high economic cost. For parameter identification, Kalman filtering [12], and Maximum

© Springer International Publishing Switzerland 2016
N. Kubota et al. (Eds.): ICIRA 2016, Part I, LNAI 9834, pp. 269–278, 2016.
DOI: 10.1007/978-3-319-43506-0_23

likelihood method [13] are the common ways. Due to the advantages of vision sensors, the vision-based object grasping has been focused by many researchers [14–16]. In [14], a vision-based method was proposed to estimate the pose of the object. While in [15], a monocular vision-based 6D object localization method was proposed to realize the intelligent grasping. In [16], a fast algorithm for object detection using Scale Invariant Features Transform keypoint detector was proposed for a manipulator to detect and grasp an object. The methods for robot kinematic calibration and object grasping described above can used for some robots with high stiffness. However, they are difficult to be used for the robots with low stiffness, such as Baxter robot.

In this paper, we aim to propose methods to realize the kinematic calibration and object grasping for Baxter robot. We compare four methods in the kinematic calibration for Baxter robot, including the integration of screw-axis measurement and kalman filtering, screw-axis measurement, least square method, and extended kalman filtering. We also compare the monocular vision and the binocular vision, and propose a step by step object grasping method for the Baxter robot grasps an object. The proposed methods are evaluated through experiments with the laser tracker.

The problem is formulated in Sect. 2. The proposed method is described in Sect. 3. The experiments and discussions are presented in Sect. 4, and the conclusion is presented in Sect. 5.

2   Problem Formulation

In this paper, we take the Baxter robot as the research object. Baxter robot a dual-arm robot, which is developed based on the Robot Operating System (ROS) [4]. There are 7-DOF in each arm. Unified Robot Description Format (URDF) file is used to describe the kinematic model of the Baxter robot. The next frame is obtained by the transformation of previous frame in the three coordinate axis. The frame of the Baxter's arm is shown in Fig. 1. The kinematic model of Baxter robot can be described as follow:

$$
{}^{base}_{gripper}T = {}^{base}_{1}T\,{}^{1}_{2}T \cdots {}^{7}_{gripper}T \tag{1}
$$

According to the definition of URDF, the frame transformation between two adjacent frames can be described as follow:

$$
{}^{i}_{i+1}T = R_z(\theta_i)T_{xyz}(P_i)R_z(\alpha_{iz})R_y(\alpha_{iy})R_x(\alpha_{ix}) \tag{2}
$$

Here, R_a is the rotation matrix around a axis, T_a is the translation matrix along a axis. α_{ix}, α_{iy}, α_{iz} and P_i are the orientation and position of the link i, respectively. θ_i is the rotation angle of joint i.

There are many parameters in the frame transformation, it is difficult to identify these parameters. According to the specification of Baxter robot [4], to obtain the accurate kinematic model, we need calibrate 38 parameters including 14 orientation parameters and 24 position parameters. We compare four methods in the identification of parameters for Baxter robot.

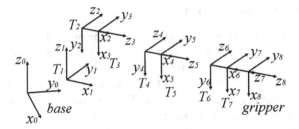

Fig. 1. The frames of Baxter's arm

To grasp an object accurately, the object should be positioned accurately. The eye-in-hand system is usually used in a robotic system to grasp an object due to its flexibility. Because a camera is attached on each arm for the Baxter robot, we can consider the system as an eye-in-hand system. We analyze the monocular vision and the binocular vision in the object grasping.

3 Proposed Method

3.1 Robot Kinematic Calibration

We compare the screw-axis measurement, extended kalman filtering, integration of screw-axis measurement and kalman filtering, and least square method. Following, we will describe these four methods simply.

Screw-axis measurement (SAM) is a way to calibrate the kinematic of robots [17]. When using screw-axis measurement, only one joint is moving, while other joints are motionless. The trajectory of a point on the moving joint is measured to obtain the position and orientation of the joint, which is used to calculate the kinematic parameters. The details of screw-axis measurement can be found in [17, 18].

Calman filtering (KF) is a way of linear quadratic estimation, which has been used to estimate the kinematic parameters of robots [19]. The extended calman filtering (EKF) is also used to estimate the kinematic parameters of robot [13, 20], which is considered as an optimization algorithm. The least square method (LSM) is a standard approach in regression analysis to the approximate solution of overdetermined systems, which has been used in the estimation of kinematic parameters for the robots [21]. To evaluate the previous system identification methods in the kinematic calibration of Baxter, we apply the EKF, LSM, and SAM in the kinematic calibration, including the position and orientation. We integrate the SAM and KF by using the SAM to estimate the orientation parameters and using the KF to estimate the position parameters. The details of KF can be referred to reference [19], and the details of EKF can be referred to reference [13].

3.2   Vision-Based Object Grasping

The coordinate frames for the monocular-vision and the binocular-vision are shown in Fig. 2(a) and (b), respectively. A, B, C are the robot base frame, end-effector frame, and camera frame, respectively. Normally, the grasped object is put on a flat. For the monocular-vision system, the robot should move to the upper of the flat, and make the optic axis of the camera be vertical to the flat. The positioning accuracy of monocular-vision depends on the verticality between the optic axis of the camera and the flat, and the calibration of the camera. When the working environment change, the calibration of the camera should be done once again. While using the binocular-vision, the flexibility and adaptability to the environment are improved, the matching points for these two cameras should be matched accurately.

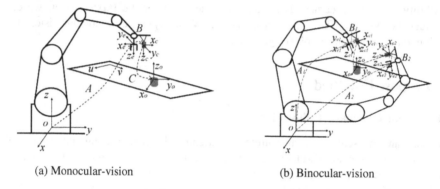

(a) Monocular-vision   (b) Binocular-vision

Fig. 2. Coordinate frames for eye-in-hand robot system

Whether using monocular-vision or binocular-vision, the positioning error exists in the real application. In this paper, we propose a step-by-step object grasping strategy. First, we move the robot to the approximate position based on binocular-vision, and then calculate the position error $\Delta X = (e_x, e_y, e_z, 0, 0, 0)^\mathrm{T}$ between the position of object and the current position of the end-effector. Because the position error is normally small, we can calculate the change of joints when compensating the position error based on the following equation:

$$\Delta\theta = J^{-1}\Delta X \tag{3}$$

Here, J is the Jaccobian matrix.

The implementation of step-by-step strategy in the object grasping is shown in Fig. 3. The object is grasped through two steps.

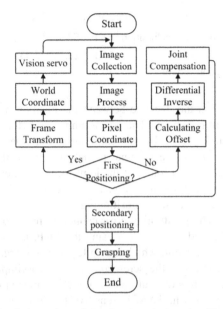

Fig. 3. Implementation of step-by-step strategy in object grasping

4 Experiment, Result and Discussion

4.1 Experiment for Kinematic Calibration

We evaluate the four robot kinematic calibration methods through an experiment with a laser tracker, as shown in Fig. 4. The measurement accuracy of the laser tracker is $(10 + 5L)$ um, L is the measurement distance. In this paper, we set L to 2 m. Therefore, the measurement accuracy of the laser tracker is 20 um. 150 points in the workspace of the robot are set to the measure points.

Fig. 4. Experiment setting in the evaluation of kinematic calibration methods

Table 1. The position error before and after calibration (unit mm)

Positioning error	No calibration	EKF	SAM	LSM	SAM + KF
Mean error	14.21	16.11	18.46	13.43	6.26
Standard deviation	5.39	10.78	6.21	4.96	3.22

The positioning error is the distance between the position measured by the laser tracker and the position of goal point. The positioning error before calibration and after calibration is shown in Table 1. Before calibration, i.e., the values of kinematic parameters are given based on the nominal model, the mean of positioning error is 14.21 mm. This is because the stiffness of the joint for the Baxter robot is low, which affect the positioning accuracy of the robot. By using the EKF, SAM, and LSM to calibrate the kinematic parameters, the positioning error is 16.11 mm, 18.46 mm, and 13.43 mm, respectively. The positioning error cannot be reduced after the calibration by using the EKF, SAM, and LSM. When using the EKF, it is difficult to adjust the covariance matrix of measurement, which may affect the convergence of the EKF. By using the SAM, the accuracy of the screw-axis equation is important to reduce the positioning error. However, it is difficult to obtain the accurate equation due to the low stiffness of the Baxter robot. The LSM is sensitive to the measure error, which will affect the effectiveness of the calibration. By using the SAM + KF, the positioning error is 6.26 mm, which is smaller than that based on the nominal model. This is because the SAM can calibrate the orientation parameter and the KF can calibrate the position parameter well.

4.2 Experiment for Object Grasping

In this experiment, we first analyze the measure error by using the monocular-vision and the binocular-vision, and then test the step-by-step strategy for object grasping. The experiment settings for monocular-vision and binocular-vision are shown

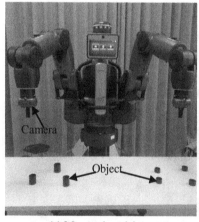
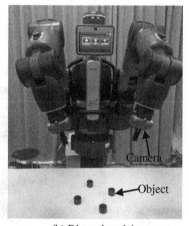

(a) Monocular-vision   (b) Binocular-vision

Fig. 5. Experiment setting

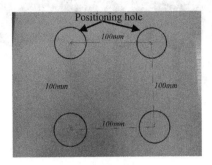

Fig. 6. Positioning plate

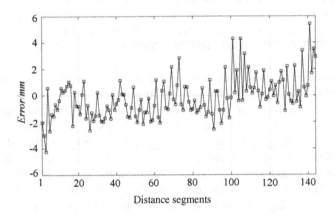

(a) Measure error derived by monocular-vision

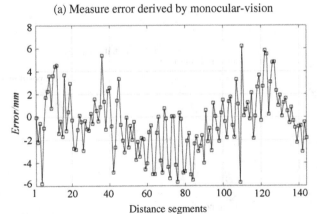

(b) Measure error derived by binocular-vision

Fig. 7. Measure error derived by monocular-vision and binocular-vision

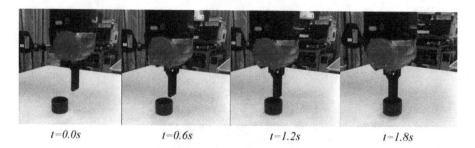

| $t=0.0s$ | $t=0.6s$ | $t=1.2s$ | $t=1.8s$ |

Fig. 8. Process of object grasping

in Fig. 5(a) and (b). We capture the image for a positioning plate, which has four holes on the plate. There are four distance segments on one plate. The length of each distance segment between the adjacent holes is 100 mm, as shown in Fig. 6. The measure error is the difference between the real length of the distance segment and the length measured by the camera. We put the plate in different positions on the table with different orientation, and test total measure 144 pieces of distance segments.

The measure errors derived by monocular-vision and binocular-vision are shown in Fig. 7(a) and (b). The mean of the measure error derived by the monocular-vision is 1.20 mm, while the mean of the measure error derived by the binocular-vision is 2.07 mm. The mean of the measure error derived by the monocular-vision is smaller than that derived by the binocular-vision. This is because the binocular-vision is affected by the accuracy of the matching point, the calibration of the camera, and so on.

To grasp the object successfully, we implement the step-by-step strategy in the experiment. After the robot is moved to the first positioning point, the difference between the positioning point and the goal point is calculated and compensated, the

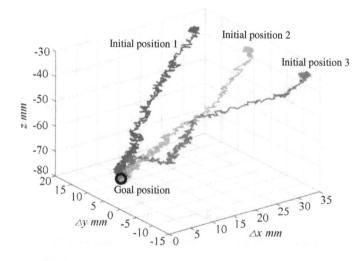

Fig. 9. End-effector positioning based on error compensation

process of object grasping is shown in Fig. 8. The object can be grasped by the end-effector based on the step-by-step object grasping strategy. To check the performance of the position error compensation, we set different position error and make the robot move to the goal position based on compensation, as shown in Fig. 9.

5   Conclusion

In this paper, we analyze and compare four robot kinematic calibration methods for Baxter robot. Due to the characteristic of the Baxter robot, the integration of screw-axis method and kalman filtering perform best in the kinematic calibration. We analyze the monocular-vision and binocular-vision in the object grasping, and propose a step-by-step object grasping strategy. The experiment results show that the proposed method can realize the object grasping for the Baxter robot.

In the future work, the calibration of non-geometrical parameters and the image processing algorithm will be taken into account.

Acknowledgment. This work was partially supported by the Scientific and technological project of Guangzhou (2015090330001), partially supported by the Fundamental Research Funds for the Central University (20152M004), and partially supported by the Natural Science Foundation of Guangdong Province (2015A030310239).

References

1. Smith, C., Karayiannidis, Y., Nalpantidis, L., Gratal, X., Qi, P., Dimarogonas, D.V., Kragic, D.: Dual arm manipulation – a survey. Robot. Auton. Syst. **60**, 1340–1353 (2012)
2. Bloss, R.: Robotics innovations at the 2009 assembly techonology expo. An Int. J. Industr. Robot **37**(5), 427–430 (2010)
3. Oyama, A., Konolige, K., Cousins, S., Chitta, S., Conley, K., Bradski, G.: Come on, our community is wide open for robotics research! In: The 27th Annual Conference of the Robotics Society of Japan, September 2009
4. Baxter robot, Rethink Robotics Company online (2016). http://www.rethinkrobotics.com/. Accessed 15 Mar 2016
5. Elatta, A.Y., Gen, L.P., Zhi, F.L.: An overview of robot calibration. J. Inf. Technol. **3**(1), 74–78 (2004)
6. Denavit, J.: A kinematic notation for lower-pair mechanisms based on matrices. ASME Trans. Appl. Mech. **22**, 215–221 (1955)
7. Park, I., Lee, B., Cho, S., Hong, Y., Kim, J.: Laser-based kinematic calibration of robot manipulator using differential kinematics. IEEE/ASME Trans. Mechatron. **17**(6), 1059–1067 (2012)
8. Zhuang, H., Roth, Z.: A note on singularities of the MCPC model. Robot. Comput. Integr. Manuf. **12**(2), 169–171 (1996)
9. Lou, Y., Chen, T., Wu, Y., Li, Z.: Improved and modified geometric formulation of POE based kinematic calibration of serial robots. In: IEEE/RSJ International Conference on Intelligent Robots and Systems, pp. 5261–5266 (2009)

10. Nubiola, A., Bonev, I.: Absolute calibration of an ABB IRB 1600 robot using a laser tracker. Robot. Comput. Integr. Manuf. **29**(1), 236–245 (2013)
11. Nubiola, A., Slamani, M., Joubair, A., Bonev, I.: Comparison of two calibration methods for a small industrial robot based on an optical CMM and a laser tracker. Robotica **32**(3), 447–466 (2014)
12. Renders, J., Rossignol, E., Becquet, M., Hanus, R.: Kinematic calibration and geometrical parameter identification for robots. IEEE Trans. Robot. Autom. **7**(6), 721–732 (1991)
13. Du, G., Zhang, P.: Online serial manipulator calibration based on multisensory process via extended kalman and particle fiters. IEEE Trans. Industr. Electron. **61**(12), 6852–6959 (2014)
14. Chen, C., Huang, H.: Pose estimation for autonomous grasping with a robotic arm system. J. Chin. Inst. Eng. **36**(5), 638–646 (2013)
15. Yang, Y., Cao, Q.: Monocular vision based 6D object localization for service robot's intelligent grasping. Comput. Math. Appl. **64**(5), 1235–1241 (2012)
16. Budiharto, W.: Robust vision-based detection and grasping object for manipulator using SIFT keypoint detector. In: International Conference on Advanced Mechatronics Systems, pp. 448–452 (2014)
17. Hollerbach, J., Giugovaz, L., Buehler, M., Xu, Y.: Screw axis measurement for kinematic calibration of the sarcos dexterous arm. In: IEEE/RSJ International Conference on Intelligent Robots and Systems, pp. 1617–1621 (1993)
18. Barker, L.: Vector-algebra approach to extract denavit-hartenberg parameters of assembled robot. NASA Technology Paper, NASA-TP-2191 (1983)
19. Lippiello, V., Siciliano, B., Villani, L.: Position and orientation estimation based on kalman filtering of stereo images. In: IEEE International Conference on Control Applications, pp. 702–707 (2001)
20. Nguyen, H., Zhou, J., Kang, H.: A calibration method for enhancing robot accuracy through integration of an extended kalman filter algorithm and an artificial neural network. Neurocomputing **151**, 996–1005 (2015)
21. Boroujeny, B.: Method of least-squares, Chap. 12. In: Adaptive Filters: Theory and Applications. Wiley, Chichester (2013)

Stiffness Estimation in Vision-Based Robotic Grasping Systems

Chi-Ying Lin[1(✉)], Wei-Ting Hung[2], and Ping-Jung Hsieh[1]

[1] Department of Mechanical Engineering,
National Taiwan University of Science and Technology, Taipei 106, Taiwan
chiying@mail.ntust.edu.tw
[2] Lite-On Technology Corporation, Taipei 114, Taiwan
Beach_81424@hotmail.com

Abstract. This paper presents an on-line estimation method which can find a mathematical expression of stiffness property of the objects grasped in vision-based robotic systems. A robot manipulator in conjunction with visual servo control is applied to autonomously grasp the object. To increase the accuracy of the object compression values associated with the used low-cost hardware, an extended Kalman filter is adopted to fuse the sensing data obtained from webcam and gripper encoder. The grasping forces are measured by a piezoresistive pressure sensor installed on the jaw of the manipulator. The force and position data are used to represent the stiffness property of the grasped objects. An on-line least square algorithm is applied to fit a stiffness equation with time-varying parameters. The experimental results verify the feasibility of the proposed method.

Keywords: Robotic grasping · Extended Kalman filter · Stiffness estimation · Sensor fusion

1 Introduction

Vision-based robotic grasping has played a central role in various industrial automated operations such as pick-and-place, packaging, and part assembly [1, 2]. During these operations, automatic inspection which aims to detect the defects of the grasped objects is also an imperative task to meet the increasingly high standards in the production line of factory automation. One efficient way to perform non-contract inspection is using the image sensing data from cameras to exclude both accidental failure and quality defects. However, inspection solely by using images can only detect surface defects [3]. In general cases, it is difficult to notice the defects such as stiffness change or texture damage just by its exterior, especially for some soft and delicate grasped objects (e.g., fresh food transportation). If the stiffness (i.e., impedance) of these objects is known in advance, appropriate grasping forces can be simply generated through the use of impedance control [4]. However, because the grasped objects may be damaged or changing its stiffness at any process from time to time, it is important to have the information of stiffness characterization in real-time [5] so as to maintain a high yield rate and perform accurate fault detection and diagnosis.

© Springer International Publishing Switzerland 2016
N. Kubota et al. (Eds.): ICIRA 2016, Part I, LNAI 9834, pp. 279–288, 2016.
DOI: 10.1007/978-3-319-43506-0_24

This study aims to develop an on-line method to measure object's stiffness using readily-available and low-cost components in vision-based robotic grasping systems. Instead of using extra displacement sensors and involving complicated manufacturing processes and system integration, the work directly applies a robotic gripper equipped with a piezoresistive force sensor to obtain the relationship between grasping forces and compression amount of the grasped objects. The goal is to monitor the stiffness change during the vision-guided autonomous grasping process and use this information to conduct more efficient automation tasks. For simplicity the object compression values are acquired from encoder feedback of the servo motor installed in the gripper. However, due to the limited encoder resolution, the position measurement is susceptible to quantization errors and subtle compression may be hardly detectable. Another issue is that unexpected deformation with the gripper may be occurred in grasping high stiffness objects by using such low-cost flexible gripper. Since image feedback is readily available in vision-based robotic grasping systems, this sensing information is also adopted to fuse the gripper encoder feedback data for estimation of object compression. Because image processing is widely vulnerable to environments with noises, an extended Kalman filter is applied to compensate the imperfect measurement from these two sensing data and obtain more robust stiffness estimation. In this presented system, a robot manipulator is first commanded to approach a static grasping target by using visual servo control technique. After contacting with the targeted object, the system then starts to estimate the object's stiffness during the grasping process. A least-square method is simultaneously applied to fit a time-varying stiffness equation for on-line evaluation. The experimental results demonstrate the feasibility of the proposed stiffness estimation method.

2   Experimental Setup

Figure 1 shows the photograph of the system setup used in this research, which is composed of three webcams, a robot manipulator, a force sensor, and an Arduino embedded board. Two downward-looking cameras are applied to build stereopsis to determine the grasping target's position and implement visual tracking control. Another camera (all made by Logitech C170) is installed in front of the manipulator to obtain the displacement information of the gripper. The robotic manipulator used in the experiments has 4 degree-of-freedom (DOF) and the servo motor installed in each joint is made from Robotis with model number AX-18A. Particularly, the servo model number used for the gripper is MX-28T and has an encoder with 4096 ppr. A FlexiForce sensor is equipped with the end of the gripper to measure the grasping forces. A low pass RC filter is applied to eliminate undesired high frequency electrical noises and the measured force values are sent to PC through the communication interface in Arduino. Figure 2 illustrates the schematic diagram of the proposed stiffness estimation system. The images of the target are first captured by using stereo cameras. After appropriate image processing the targeted object's center-position is given to drive the robot manipulator to approach the target by sending correct motor joint commands through visual servo control. During the entire grasping process, the encoder values associated with the gripper servo motor and the continuous images captured by the forward-looking camera are simultaneously integrated to estimate object compression values. The extended Kalman filter algorithm is implemented to reduce the influences of the error sources and noises from these two sensing data. Given measured grasping force feedback, a curve that illustrates the relationship between force and compression is continuously updated and fitted by using a least square method.

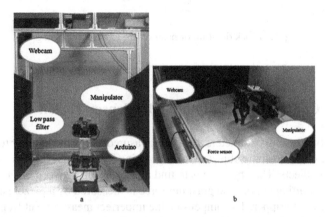

Fig. 1. Photograph of the experimental setup: a. front view; b. side view

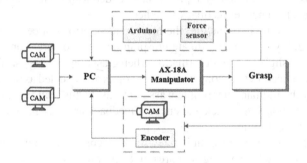

Fig. 2. Schematic diagram of the proposed stiffness estimation system

3   Visual Servoing

This study simply implemented a position-based visual servoing (PBVS) control system [6] for grasping and stiffness estimation experiments. The images captured from camera are first used to calculate the error between manipulator end-effector and grasping target. The motor position commands derived from classical resolved motion rate control are calculated to reduce this tracking error. Figure 3 depicts the applied PBVS control block diagram, where **g** represents the estimation of the goal position, **h** represents the calculation of manipulator end-effector with forward kinematic model K, and **J** is the robot Jacobian matrix. θ and $\dot{\theta}$ denote the joint angle and angular velocity of the robot manipulator, respectively. The control law for this visual servo system is given as

$$\dot{\theta} = \lambda J^{-1}(\mathbf{g} - \mathbf{h}) \tag{1}$$

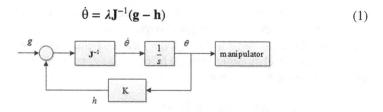

Fig. 3. Block diagram of position-based visual servoing

Note that λ is a proportional gain for tuning.

4   Object Compression Estimation by Sensor Fusion

In this study, two system parameters are to be measured/estimated in order to obtain the stiffness curve of the grasping objects. One is grasping force and the other is compression of the grasped objects. The objective is to find a relationship between this two physical properties and a fitting curve. The grasping force is measured by a piezoresistive force sensor at the end of gripper. To compensate the imperfect measurement from the available used low-cost hardware without adding extra displacement sensor, the amount of object compression is estimated by applying an extended Kalman filter [7] to fuse image feedback and gripper encoder feedback together.

As mentioned in the introduction, the measurement error sources by using a low-cost and low-rigidity gripper to estimate grasping object's compression include quantization errors and slight gripper deformation when grasping with high stiffness objects. These errors and uncertainties along with image noises are modeled as part of stochastic noises to satisfy the Kalman filter problem assumptions. In this section the measurement models associated image feedback and encoder feedback are presented to facilitate the Kalman filter problem setup.

For the purpose of visual recognition and tracking control, a red marker attached to the gripper end is treated as an image feature point. After following the image processing procedures mentioned in Sect. 3, the image coordinate of the gripper can be obtained and should be transformed to a world coordinate representation by using a camera model. The gripper movement is used to represent the object compression. The schematic diagram of the applied camera model is illustrated in Fig. 4 and the transformation formula can be represented as

$$p = \frac{f}{d}x + p_0 \tag{2}$$

where p denotes the gripper compression expressed in image plane with an unit in pixel. d is the distance between camera and gripper. f is the focal length of camera. x is the actual gripper compression presented in world coordinate. p_0 is the translation between pixel coordinate system and image coordinate system. For simplicity, in this study the compression value is expressed in camera coordinate system since this value is the same as the one expressed in world coordinate system.

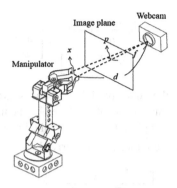

Fig. 4. Illustration of camera model associated with gripper compression

In addition to visual feedback, the encoder feedback available from the robotic gripper system is simultaneously applied to estimate object compression. As shown in a schematic diagram of gripper kinematic model (Fig. 5), the geometric relationship between gripper movement x and motor's rotation angle θ can be derived as

$$\theta = \cos^{-1}(\frac{x - L}{a}) - \alpha \tag{3}$$

where L is the half of the horizontal distance between the axis center of two motors. a is the vertical distance from the axis of rotation to the front-end of motor. α is an offset angle of θ.

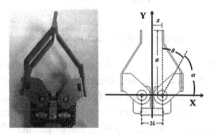

Fig. 5. Schematic diagram of the applied gripper for compression estimation

It is obvious in Eq. (3) that a nonlinear term $\cos^{-1}$ accompanies with the encoder measurement equation. This is the primary reason why an extended Kalman filter is applied to fuse both vision and encoder feedback information for compression estimation. Combining the above two measurement models with a static equation of motion using the fact that the gripper movement is slow, the system model and measurement model used in this study can be summarized as

$$x_t = x_{t-1} + u_t + w_t \tag{4}$$

$$
\begin{bmatrix} \theta_t \\ p_t \end{bmatrix} = \begin{bmatrix} \cos^{-1}(\dfrac{x_t - L}{a}) - \alpha \\ \dfrac{f}{d}x_t + p_0 \end{bmatrix} + \begin{bmatrix} v_{1,t} \\ v_{2,t} \end{bmatrix} \tag{5}
$$

where x_t, x_{t-1}, w_t and v_t share the same notations with the ones mentioned in the previous subsections. u_t represents the displacement produced by the motor command. Note that the constants α and p_0 can be simply offset to satisfy the linearity. To finish the problem setup the linearized measurement matrix $\mathbf{H}$ is derived as

$$
\mathbf{H} = \begin{bmatrix} \dfrac{-1}{a\sqrt{1 - (\dfrac{x_t - L}{a})^2}} \\ \dfrac{f}{d} \end{bmatrix} \tag{6}
$$

5 Experimental Results and Discussion

In order to evaluate the feasibility and performance of the proposed stiffness estimation method, a 4-DOF robot manipulator was adopted to conduct grasping experiments and the estimated stiffness parameters were updated on-line through a visual interface programmed by OpenGL libraries. This section presents object compression estimation with fused sensor data, and curve fitted results for on-line stiffness demonstration.

5.1 Grasping Object Compressing Estimation

The purpose of this experiment was to verify that the accuracy of object compression can be further improved by applying an extended Kalman filter technique integrating with the vision feedback and gripper encoder feedback obtained in the grasping process. Simulation and experiments were both conducted by sending motor commands to move the gripper to a desired position and comparing the differences between estimated and actual gripper position. In the simulation, the servo motors were commanded to rotate sequentially from 0° to 45° sampled with one degree, simulating the compression process when grasping the object. Gaussian noises were added to simulate the uncertainties and imperfect measurement due to the applied hardware in this study. Figure 6 shows the estimation errors of using three different sensing data: image feedback only, gripper encoder feedback only, and sensor fusion data. X-axis represents the motor command and Y-axis represents the estimated errors. Assume that the image noises have larger influences than the noises existed in the gripper system. The results by using fused sensor data (blue solid line in the plot) justify the improved accuracy of this proposed method.

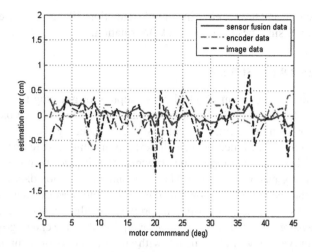

Fig. 6. Effectiveness of using different sensing data: simulation results (Color figure online)

In the experiment, the gripper was commanded to move from 10 cm to 0 cm and the data was recorded every 0.5 cm movement. Figure 7 shows the results by comparing the three sensing data with the values measured by a 1 mm resolution laser rangefinder. In Fig. 7, X-axis stands for the gripper movement, which is also treated as the amount of gripper compression. Obviously, the estimated errors in the proposed experimental system were dominated by image noises with an offset value. The regular triangular-like estimation errors by using encoder feedback only were primarily due to the quantization effects with limited encoder resolution. The estimation error distribution in experiments suggests that the normal distribution assumption may be not applicable to the cases using either encoder feedback or image feedback. However, the valid error

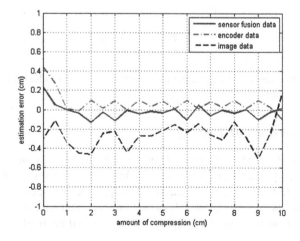

Fig. 7. Effectiveness of using different sensing data: experimental results

offset compensation and further error magnitude reduction still verifies the effectiveness of the method by fusing these two feedback information.

5.2   Measuring Object Stiffness

In this experiment, a sponge-made block was adopted for grasping tests and stiffness estimation. In order to establish a comparison basis, the stiffness of the block was first measured off-line using an electronic scale with 0.1 g accuracy. The block was placed on the scale and a drill installed on a small desktop machine was fed gradually to the block and provide different compression amounts. The displacement resolution of the scale on the drilling machine is 1 mm. The force and compression data were recorded and plotted as the red solid line shown in Fig. 8. The blue dashed line represents the estimated stiffness relationship by using the proposed method, where the compression forces were measured by a piezoresistive sensor equipped on the end of gripper. As can be seen, both curves show a close match with small amounts of compression and a deviation starts after the amount of compression is larger than 0.2 cm. Nevertheless, the overall trend of the proposed method is still similar to the one of off-line measurement results.

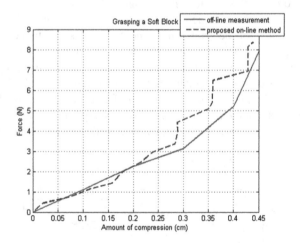

Fig. 8. Stiffness measurement using the proposed method and offline validation: grasping a soft block (Color figure online)

5.3   On-Line Stiffness Curve Fitting

To quantize the stiffness estimation results obtained in the previous subsection, a cubic polynomial was adopted as the stiffness equation in which its parameters were identified by using a standard least square method. The fitted parameter values were updated on-line in conjunction with the continuously-added measurement data. Figure 9 shows the curve fitting result with the whole compression process and the final fitted equation is represented as

$$y = 21.7263x^3 + 19.2486x^2 + 5.7681x + 0.1519 \tag{7}$$

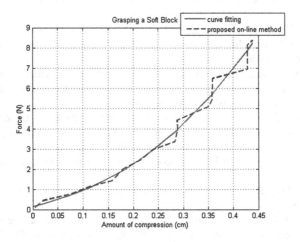

Fig. 9. Stiffness curve fitting results using least-square method: grasping a soft block

The coefficient of determination (R-squared) is 0.9857, which is very close to 1. Therefore, the fitting result is good enough for stiffness quantization and evaluation.

6   Conclusions

This paper presents an intelligent on-line stiffness method which identifies the time-varying stiffness property of objects grasped by vision-based robotic manipulators. One great feature in this proposed system is the exemption from using an extra displacement sensor for compression measurement by integrating with the image feedback and gripper encoder feedback information during grasping process. The experimental results justify the improved compression estimation with fused sensor data and the curve-fitted results demonstrate the feasibility of the proposed method. This technique may be applicable to many applications that require automated manipulation, monitoring, and stiffness inspection. Another application of this system is to further integrate the estimated stiffness with virtual reality technique to develop various haptic interfaces in games, surgical training, and military use. This study assumes that the measured noises and uncertainties are Gaussian noises and applies this assumption to facilitate the Kalman filtering problem formulation. To be more realistic as the true cases and obtain better estimation this strong assumption may be relaxed by applying other advanced algorithms that fit non-Gaussian noise distribution assumption such as particle filters [8].

References

1. Cheng, H., Zhang, Z., Li, W.: Dynamic error modeling and compensation in high speed delta robot pick-and-place process. In: IEEE International Conference on Cyber Technology in Automation, Control, and Intelligent System, pp. 36–41. IEEE Press, Shenyang (2015)

2. Feng, C., Xiao, Y., Willette, A., McGee, W., Kamat, V.R.: Vision guided autonomous robotic assembly and as-built scanning on unstructured construction sites. Autom. Constr. **59**, 128–138 (2015)

3. Zhou, A., Guo, J., Shao, W.: Automated detection of surface defects on sphere parts using laser and CCD measurements. In: Conference of IEEE Industrial Electronics Society, pp. 2666–2671. IEEE Press, Melburne (2011)

4. Li, M., Yin, H., Tahara, K., Billard, A.: Learning object-level impedance control for robust grasping and dexterous manipulation. In: IEEE International Conference on Robotics and Automation, pp. 6784–6791. IEEE Press, Hong Kong (2014)

5. Pedreno-Molina, J.L., Guerrero-Gonzalez, A., Calabozo-Moran, J., Lopez-Coronado, J., Gorce, P.: A neural tactile architecture applied to real-time stiffness estimation for a large scale of robotic grasping systems. J. Intell. Robot. Syst. **49**, 311–323 (2007)

6. Hashimoto, K.: A review on vision-based control of robot manipulators. Adv. Robot. **17**, 969–991 (2003)

7. Kalman, R.E., Bucy, R.S.: New results in linear filtering and prediction theory. J. Basic Eng. **83**, 95–108 (1961)

8. Arulampalam, M.S., Maskell, S., Gordan, N., Clapp, T.: A tutorial on particle filters for online nonlinear/non-Gaussian Bayesian tracking. IEEE Trans. Signal Process. **50**, 174–188 (2002)

Contour Based Shape Matching
for Object Recognition

Haoran Xu[1], Jianyu Yang[1(✉)], Zhanpeng Shao[2], Yazhe Tang[3],
and Youfu Li[4]

[1] School of Urban Rail Transportation, Soochow University, Suzhou, China
20144246017@stu.suda.edu.cn, jyyang@suda.edu.cn
[2] City University of Hong Kong Shenzhen Research Institute, Shenzhen, China
perry.shao@my.cityu.edu.hk
[3] Department of Precision Mechanical Engineering,
Shanghai University, Shanghai, China
yztang2008@yahoo.com
[4] Department of Mechanical and Biomedical Engineering,
City University of Hong Kong, Kowloon Tong, Hong Kong
meyfli@cityu.edu.hk

Abstract. To improve computational efficiency and solve the problem of low accuracy caused by geometric transformations and nonlinear deformations in the shape-based object recognition, a novel contour signature is proposed. This signature includes five types of invariants in different scales to obtain representative local and semi-global shape features. Then the Dynamic Programming algorithm is applied to shape matching to find the best correspondence between two shape contours. The experimental results validate that our methods is robust to rotation, scaling, occlusion, intra-class variations and articulated variations. Moreover, the superior shape matching and retrieval accuracy on benchmark datasets verifies the effectiveness of our method.

Keywords: Contour signature · Shape matching · Object recognition · Dynamic programming

1 Introduction

Object matching and recognition via shape matching is a basic and significant task in computer vision, and plays an important role in many applications, such as face recognition [1], biomedical image analysis [2], and robot navigation [3]. The shape of object contour has been widely studied in recent years for it provides a great deal of compact and meaningful information. According to the fruitful literatures in the relevant field [4, 5, 6, 7, 8], significant progress has been made in solving some challenging problems, like the influence of geometric transformations. However, there still are many difficult tasks including depressing the interference from intra-class variations and nonlinear deformations (articulated variations and occlusion). Hence a reliable shape description which can handle those variations is urgently needed.

In this paper, we propose a novel contour signature (CS) for shape matching and object recognition. This signature consists of five types of invariants in different scales to

© Springer International Publishing Switzerland 2016
N. Kubota et al. (Eds.): ICIRA 2016, Part I, LNAI 9834, pp. 289–299, 2016.
DOI: 10.1007/978-3-319-43506-0_25

get a rich representation of shape contour. The multiple scales guarantees capturing shape features including both local shape details and semi-global shape structure (object parts) in our method. Besides the powerful signature, our method also employs the Dynamic Programming (DP) [9] algorithm for shape matching. In the experimental section, our method is proved robust to translation, rotation, scaling, occlusion, intra-class variations and articulated variations. Moreover, the outstanding results of shape matching and retrieval on the benchmark datasets validate the good performance of our method.

2   Related Work

Among the existing literatures, there are a lot of descriptors proposed to describe shapes from different aspects. Shape context (SC) [10] is a representative work of shape descriptors. The SC descriptor of a certain point uses a set of segment vectors to represent the distribution of the remaining points relative to it. As a result, this method can offer a discriminative description for rigid shapes, but it is not robust enough for articulated deformations. Based on the SC method, Ling et al. proposed the inner distance shape context (IDSC) [4] method which replaces the Euclidean distance of shape context with inner distance to strengthen the invariant property of shape descriptor. The triangle area representation (TAR) [5, 11] method does not takes shape body parts into consideration which leads a less-than-ideal performance when it comes to the disturbance of noise.

Apart from the space structures of shape contour points, many shape descriptors generate powerful characterizations with local invariants. The typical representatives of this kind are the integral invariants [6] and differential invariants [12, 13]. These two methods are good at capturing features in small scales, whereas the disadvantage of dealing with intra-class differences and noise is obvious too. Mokhtarian et al. [14] proposed curvature scale space (CSS) method which keeps reducing the number of curvature zero-crossing points on shape contour until the whole shape is convex. This method uses CSS image as descriptions of shapes, so that it is unsuited to convex shapes. Multi-resolution polygonal method [15] captures shape features in multi-scales. It divides shape contour into several parts, and constructs a descriptor for each part. However, the one-type descriptor methods have their limits in capturing enough shape information for recognition tasks. In some related literatures [16, 17, 18], considering two or more one-type invariant descriptors can complement each other, it is feasible to mix various types of invariants together for building a novel hybrid descriptor.

3   Contour Signature

3.1   Definition of the Major Zone

To construct the contour signature, it is essential to make clear the definition of the major zone firstly. A closed shape contour consists of a sequence of contour points $S = \{p(i)|i \in [1, n]\}$, where n is the length of contour. Each point $p(i) = \{u(i), v(i)\}$ of the shape contour is described by its coordinates $u(i)$ and $v(i)$ on the image.

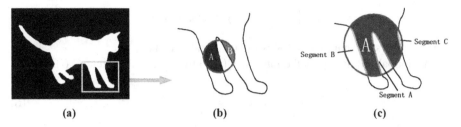

Fig. 1. The grey Zone A is the major zone within the circle, while the grey Zone B in (b) is not the major zone. The two segments B and C in (c) are not major segments. Only the blue Segment A crossing the circle center is used to calculate the arc length invariant. (Color figure online)

As shown in Fig. 1, given a circle $C_k(i)$ centered at $p(i)$ with radius r_k, one or more zones of shape S would be inside circle $C_k(i)$. Among these zones, the zone $Z_k(i)$ will be counted as the major zone if it coincides with the following: $p(i) \in Z_k(i)$, i.e., Zone A in the Fig. 1(b). The area of the major zone $Z_k(i)$, $s_k^*(i)$, can be calculated as follows:

$$s_k^*(i) = \int_{C_k(i)} B(Z_k(i), x) dx \tag{1}$$

where $B(Z_k(i), x)$ is defined by:

$$B(Z_k(i), x) = \begin{cases} 1, x \in Z_k(i) \\ 0, x \notin Z_k(i) \end{cases} \tag{2}$$

Similarly, the segment $L_k(i)$ of the shape contour inside the circle $C_k(i)$ which meets the condition: $p(i) \in L_k(i)$, is the major segment, i.e., Segment A in the Fig. 1(c). Denote the $l_k^*(i)$ arc length of the major segment $L_k(i)$:

$$l_k^*(i) = \int_{C_k(i)} B(L_k(i), x) dx \tag{3}$$

where $B(L_k(i), x)$ is defined by:

$$B(L_k(i), x) = \begin{cases} 1, x \in L_k(i) \\ 0, x \notin L_k(i) \end{cases} \tag{4}$$

3.2 Definition of the Contour Signature

The contour signature (CS) of shape contour is defined in terms of five types of invariants: normalized area s_k, changing rate of area s_k', normalized arc length l_k, changing rate of arc length l_k', and normalized central distance c_k, in the following form:

$$M(i) = \{s_k(i), s'_k(i), l_k(i), l'_k(i), c_k(i) | k \in [1, K], 1 \in [1, n]\} \tag{5}$$

where k is the scale number and K is the total scale number.

The normalized area s_k is the ratio of $s_k^*(i)$ to the area of the circle $C_k(i)$:

$$s_k(i) = \frac{s_k^*(i)}{(\pi r_k^2)} \tag{6}$$

As $s_k^*(i)$ could not be bigger than the area of circle $C_k(i)$, the value of $s_k(i)$ ranges from 0 to 1.

The normalized arc length $l_k(i)$ is the ratio of $l_k^*(i)$ to the circumference of circle $C_k(i)$:

$$l_k(i) = \frac{l_k^*(i)}{(2\pi r_k)} \tag{7}$$

The changing rate of area s'_k and the changing rate of arc length l'_k are defined as follows:

$$s'_k(i) = s_k(i) - s_k(i-1) \tag{8}$$

$$l'_k(i) = l_k(i) - l_k(i-1) \tag{9}$$

Denote $c_k^*(i)$ the central distance between $p(i)$ and $w_k(i)$, and the normalized central distance $c_k(i)$ is the ratio of $c_k^*(i)$ to r_k:

$$c_k(i) = \frac{c_k^*(i)}{r_k} = \frac{\|p(i) - w_k(i)\|}{r_k} \tag{10}$$

where $w_k(i)$ is the weighted center of the major zone $Z_k(i)$. The coordinate of $w_k(i)$ is computed by averaging the coordinates of all the pixels in the major zone. The value of $c_k(i)$ ranges from 0 to 1, because point $w_k(i)$ must be inside the major zone $Z_k(i)$.

So, for each contour point $p(i)$, there are $5 \times K$ invariants defined to construct the contour signature for extracting various kinds of shape features.

3.3   The Setting of Radius and Total Scale Number

The radius r_k of circle C_k for each scale is related with initial radius R. The detailed rule is:

$$r_k = \frac{R}{2^k} \tag{11}$$

That means, as the scale number k increases, the radius r_k decreases half by half, when the initial radius R is set according to the original shape:

$$R = \sqrt{area_S} \tag{12}$$

where $area_S$ is the area of the whole shape. A setting like this guarantees the capture of both local and semi-global shape features in different scales of CS.

The total scale number K also plays an important role in the performance of CS. It is not better when a bigger K is chosen at any time. Because bigger scale would not improve the discrimination of the contour signature from a certain value. On the contrary, bigger scales usually mean more computational cost and more sensitivity to noise. Generally speaking, bigger scales are necessary with higher complexity of the shape. About how to set K, our method follows the rule which meets a convergence condition: As the total scale K increases, the mean deviation of invariants between scales K and $K + 1$ is converge to a threshold. Once the mean deviation is smaller than the threshold, the current value of K is adopted and invariants in scale $K + 1$ should be abandoned.

4 Shape Matching

Powerful shape description approaches and efficient shape matching method are two key factors for shape-based object recognition. Before matching two shapes, it is important to assign appropriate correspondence of contour points from two shapes in advance. In this section, the Dynamic Programming (DP) [9] algorithm is presented to align two sequences of shape contour points. The similarity between two contours can be calculated as follows.

Given two shapes A and B, describe them with their contour point sequences: $A = \{p_1, p_2, \ldots, p_{n_A}\}$ and $B = \{q_1, q_2, \ldots, q_{n_A}\}$. Without loss of generality, assume $n_A \geq n_B$. The matching cost of two points p_i and q_j is defined as the Euclidean distance of their CS invariants:

$$d(p_i, q_j) = \sqrt{(s_k^p(i) - s_k^q(j))^2 + (s_k^{\prime p}(i) - s_k^{\prime q}(j))^2 + (l_k^p(i) - l_k^q(j))^2 + (l_k^{\prime p}(i) - l_k^{\prime q}(j))^2 + (c_k^p(i) - c_k^q(j))^2} \tag{13}$$

And a cost matrix $\mathbf{D}$ is generated to compute the total matching cost of two shapes:

$$D(A,B) = \begin{vmatrix} d(p_1, q_1) & d(p_1, q_2) & \cdots & d(p_1, q_{n_B}) \\ d(p_2, q_1) & d(p_2, q_2) & \cdots & d(p_2, q_{n_B}) \\ \vdots & \vdots & d(p_i, q_j) & \vdots \\ d(p_{n_A}, q_1) & d(p_{n_A}, q_2) & \cdots & d(p_{n_A}, q_{n_B}) \end{vmatrix} \tag{14}$$

Then, the DP algorithm is used to find the best correspondence π between shapes A and B with the lowest matching cost, where point p_i is mapped to point $q_{\pi(i)}$ uniquely. And the matching cost function $f_{A,B}(\pi)$ is defined as follows:

$$f_{A,B}(\pi) = \sum_{i=1}^{n_A} d(p_i, q_{\pi(i)}) \tag{15}$$

The similarity $sim(A, B)$ between two shapes is the minimum value of $f_{A,B}(\pi)$:

$$sim(A, B) = \min f_{A,B}(\pi) \tag{16}$$

5 Experiments

In this section, the robustness of the proposed contour signature is tested under various conditions, including: rotation, scaling, intra-class variations, partial occlusion and articulated deformation. The shape recognition accuracy of our method is evaluated by experiments on benchmark datasets, including MPEG-7 dataset, Kimia's datasets, and Articulated dataset. At the same time, the experimental results are compared with other important methods.

5.1 Robustness of the Contour Signature

Firstly, some experiments are conducted to validate the robustness of the contour signature to rotation, scale variations, and intra-class variations. In Fig. 2, the top two shapes in the first column with significant intra-class variations are selected from same class. Below them are the two shapes after spinning and scaling. Column 2–6 of Fig. 2 plot five invariants of each shape shown in the first column. By comparison, it is obvious that rotation, scaling transform and intra-class variations can barely affect the description results of contour signatures. Despite five types of invariants capture different kinds of shape features, they all have remarkable invariant properties for above-mentioned disturbances.

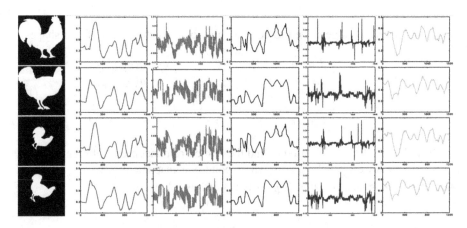

Fig. 2. Column 1 shows images with rotation, scaling, and intra-class variations. Column 2–6 are their CS functions of s, s', l, l' and c, respectively

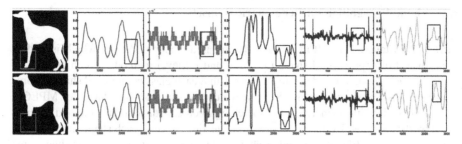

Fig. 3. The five CS invariants of occluded shape and original shape

Secondly, the contour signature is tested under partial occlusion. As the Fig. 3 shows, the sample shape of a dog with the plots of its contour signature are shown in the top row, and the bottom row is that of a dog whose forelegs are blocked. The blocked part are labeled by black boxes in the plots of CS invariants. Comparing each pair of CS invariants in the same column, only boxed functions are different while the rest parts are exactly same. It validates the contour signature can still perform well in shape description with existence of partial occlusion.

Besides, the articulated deformation, as another challenging variation in shape description, is also applied to test our method by experiment. We use two shapes of scissors to demonstrate the robustness of the contour signature against articulated deformations. As shown in Fig. 4, there are significant articulated variations between the scissors in the top row and the bottom row at the joint of two sharp blades. The CS invariants of two scissor shapes are plotted in Column 2−6 with the black circle marks. The black marks represent the red circle centers on respective joint of two scissors. It is not difficult to find the strong similarity between the corresponding plots of contour signatures, especially at the labeled parts.

In the above experiments, we only take the invariants of the contour signatures in the first scale into consideration. Because the circle $C_k(i)$ in the first scale has largest radius which get most information of shape. Therefore, the invariants in the first scale are most easily affected by various kinds of disturbance, like rotation, scaling, intra-class variations, occlusion and articulated deformation. As a result, if the CS invariant parameters are robust in the first scale, they are also robust in the higher scale.

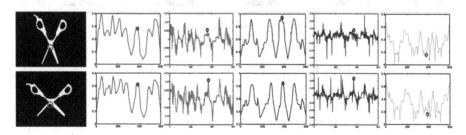

Fig. 4. The five CS invariants with articulated variations

5.2 Shape Recognition on Benchmark Datasets

MPEG-7 Dataset: The MPEG-7 [19] dataset is a standard dataset which is widely used for test of capability of shape matching and retrieval methods, which consists of 1400 silhouette images divides into 70 classes that each class contains 20 shapes. Some typical shapes are shown in Fig. 5, from which we can find that shapes of MPEG-7 have high variability. Generally, the matching accuracy of a certain method is measured by the so-called Bull's Eye score. In the testing process, each shape in the dataset is used as a query to match all other shapes. Among the 40 most similar shapes, the number of shapes from the same class of the query is counted. The Bull's Eye score is the ratio of the total number of shapes from the same class to the highest possible number (1400 × 20), and the best possible score is 100 %.

The Bull's Eye score of our method comparing with other methods are listed in Table 1, where our method achieves the best performance among the methods in this table. The comparison results validate our method is more representative and discriminative.

Kimia's Datasets: The Kimia's dataset is another widely used benchmark dataset for shape analysis, including Kimia's 25, Kimia's 99, and Kimia's 216 [21]. Considering that Kimia's 25 dataset is too small for test, our method is only tested on the Kimia's 99 and Kimia's 216 datasets, which contain 99 shapes belonging to 9 classes and 216 shapes belonging to 18 classes respectively. All the shapes of these two datasets are shown in Fig. 6. In this experiment, we use the same manner as that on the MPEG-7

Fig. 5. Typical shapes of the MPEG-7 dataset

Table 1. Accuracy (%) on the MPEG-7 dataset

Methods	Accuracy (%)
SC [10]	76.51
IDSC [4]	85.40
Contour Flexibility [20]	89.31
Shape Vocabulary [8]	90.41
Our Method	**91.25**

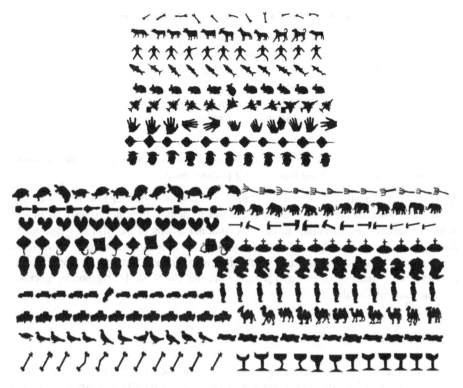

Fig. 6. All shapes of the Kimia's 99 and Kimia's 216 dataset

Table 2. Accuracy (%) on the Kimia's datasets

Method	Kimia's 99	Kimia's 216
SC [10]	76.36	76.14
CDPH + EMD [22]	81.61	87.12
Gen. Model [23]	89.39	N/A
Our Method	**94.46**	**90.81**

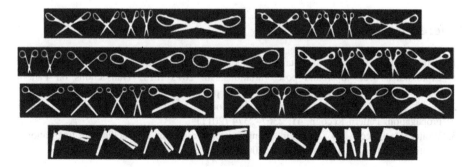

Fig. 7. All shapes of the articulated shape dataset

Table 3. Accuracy (%) on the articulated dataset

Method	Accuracy (%)
L2 (base line) [4]	38.75
SC [4]	36.25
MDS + SC [4]	58.76
IDSC [4]	85
Our Method	**89.38**

dataset to compute the accuracy of our method and other methods. The results are listed in Table 2, where our method has the best performance.

Articulated Dataset: The articulated dataset [4] contains 40 shapes from 8 different classes as shown in Fig. 7. We use this dataset to demonstrate the robustness of our method to articulated variations in the same manner as that of the MPEG-7 dataset. In the articulated dataset, shapes from the same class have serious articulated variations to each other. The experimental results are shown in Table 3, where our method outperforms other methods. It is worth pointing out that the IDSC [4] method is specially designed articulated shapes.

6   Conclusion

This paper presents a new contour signature for shape matching and object recognition. Five types of invariants capture shape features not only from different aspects, but also in different scales. Based on the contour signatures, the Dynamic Programming algorithm is also introduced for the calculation of shape similarity. The experimental results validates the robustness of the proposed method. At the same time, the recognition accuracy on benchmark datasets verifies the outstanding performance of our method for shape matching and retrieval.

Acknowledgements. This work was funded by research grants from the National Natural Science Foundation of China (NSFC No. 61305020 and No. 61273286), and the Natural Science Foundation of Jiangsu province, China (Grant No. BK20130316).

References

1. Drira, H., Ben Amor, B., Srivastava, A., Daoudi, M., Slama, R.: 3D face recognition under expressions, occlusions, and pose variations. In: 2013 IEEE Transactions on Pattern Analysis and Machine Intelligence, vol. 35(9), pp. 2270–2283 (2013)
2. Wang, J., Li, Y., Bai, X., et al.: Learning context-sensitive similarity by shortest path propagation. Pattern Recogn. **41**, 2367–2374 (2011)
3. Wolter, D., Latecki, L.J.: Shape matching for robot mapping. In: Pacific Rim International Conference on Artificial Intelligence (2004)

4. Ling, H., Jacobs, D.W.: Shape classification using the inner-distance. IEEE Trans. Pattern Anal. Mach. Intell. **29**(2), 286–299 (2007)
5. Alajlan, N., Rube, I.E., Kamel, M.S., Freeman, G.: Shape retrieval using triangle area representation and dynamic space warping. Pattern Recogn. **40**, 1911–1920 (2007)
6. Manay, S., Cremers, D., Hong, B.-W., Yezzi, A.J., Soatto, S.: Integral invariants for shape matching. IEEE Trans. Pattern Anal. Mach. Intell. **28**(10), 1602–1618 (2006)
7. Bai, X., Yang, X., Latecki, L.J., Liu, W., Tu, Z.: Learning context-sensitive shape similarity by graph transduction. IEEE Trans. Pattern Anal. Mach. Intell. **32**(5), 861–874 (2010)
8. Bai, X., Rao, C., Wang, X.: Shape vocabulary: a robust and efficient shape representation for shape matching. IEEE Trans. Image Process. **23**(9), 3935–3949 (2014)
9. Cormen, T.H., Leiserson, C.E., Rivest, R.L., Stein, C.: Introduction to Algorithms, 2nd edn. MIT Press, Cambridge (2001)
10. Belongie, S., Malik, J., Puzicha, J.: Shape matching and object recognition using shape contexts. IEEE Trans. Pattern Anal. Mach. Intell. **24**(4), 509–522 (2002)
11. Alajlan, N., Kamel, M.S., Freeman, G.H.: Geometry-based image retrieval in binary image databases. IEEE Trans. Pattern Anal. Mach. Intell. **30**(6), 1003–1013 (2008)
12. Sebastian, T.B., Klein, P.N., Kimia, B.B.: On aligning curves. IEEE Trans. Pattern Anal. Mach. Intell. **25**(1), 116–125 (2003)
13. Klassen, E., Srivastava, A., Mio, W., Joshi, S.H.: Analysis of planar shapes using geodesic paths on shape spaces. IEEE Trans. Pattern Anal. Mach. Intell. **26**(3), 372–383 (2004)
14. Mokhtarian, F., Abbasi, S., Kittler, J.: Efficient and robust retrieval by shape content through curvature scale space. Ser. Softw. Eng. Knowl. Eng. **8**, 51–58 (1997)
15. Attalla, E., Siy, P.: Robust shape similarity retrieval based on contour segmentation polygonal multiresolution and elastic matching. Pattern Recogn. **38**(12), 2229–2241 (2005)
16. Yang, J., Wang, H., Yuan, J., Li, Y.F., Liu, J.: Invariant multi-scale descriptor for shape representation, matching and retrieval. Comput. Vis. Image Underst. (CVIU) **145**, 43–58 (2016)
17. Yang, J., Xu, H.: Metric learning based object recognition and retrieval. Neurocomputing **190**, 70–81 (2016)
18. Yang, J., Yuan, J., Li, Y.F.: Parsing 3D motion trajectory for gesture recognition. J. Vis. Commun. Image Representation (JVCI) **38**, 627–640 (2016)
19. Latecki, L.J., Lakamper, R., Eckhardt, T.: Shape descriptors for non-rigid shapes with a single closed contour. In: IEEE Conference on Computer Vision and Pattern Recognition, Proceedings. vol. 1, pp. 424–429 (2000)
20. Xu, C., Liu, J., Tag, X.: 2D Shape matching by contour flexibility. IEEE Trans. Pattern Anal. Mach. Intell. **31**(1), 180–186 (2009)
21. Sebastian, T.B., Klein, P.N., Kimia, B.B.: Recognition of shapes by editing their shock graphs. IEEE Trans. Pattern Anal. Mach. Intell. **26**(5), 550–571 (2004)
22. Shu, X., Wu, X.: A novel contour descriptor for 2D shape matching and its application to image retrieval. Image Vis. Comput. **29**(4), 286–294 (2011)
23. Tu, Z., Yuille, A.: Shape matching and recognition-using generative models and informative features. In: Proceedings of European Conference Computer Vision (ECCV), Prague, Czech Republic, pp. 195–209 (2004)

Robust Gaze Estimation via Normalized Iris Center-Eye Corner Vector

Haibin Cai[1], Hui Yu[1], Xiaolong Zhou[2], and Honghai Liu[1(✉)]

[1] School of Computing, University of Portsmouth, Portsmouth, UK
honghai.liu@port.ac.uk
[2] Shenzhen Research Institute, City University of Hong Kong, Hong Kong, China

Abstract. Gaze estimation plays an important role in many practical scenarios such as human robot interaction. Although high accurate gaze estimation could be obtained in constrained settings with additional IR sources or depth sensors, single web-cam based gaze estimation still remains challenging. This paper propose a normalized iris center-eye corner (NIC-EC) vector based gaze estimation methods using a single, low cost web-cam. Firstly, reliable facial features and pupil centers are extracted. Then, the NIC-EC vector is proposed to enhance the robustness and accuracy for pupil center-eye corner vector based gaze estimations. Finally, an interpolation method is employed for the mapping between constructed vectors and points of regard. Experimental results showed that the proposed method has significantly improved the accuracy over the pupil center-eye corner vector based gaze estimation method with average accuracy of 1.66° under slight head movements.

Keywords: Gaze estimation · Eye tracking · Normalized iris center-eye corner vector · Interpolation

1 Introduction

Gaze estimation is a crucial technique in a wide spectrums of applications such as human machine interaction and cognitive processes. For example, Gaze can be applied in therapy for children with Autism Spectrum Disorders (ASD) [1]. Although accurate gaze estimation can be acquired using additional hardware such as IR sources and depth sensors, the complex setup hinders its application to daily interaction situations. The accuracy of traditional pupil center-corneal center based gazed estimation methods is sensitive to the head movements. In this paper, we explore robust and accurate gaze estimation method using a single, low cost web camera.

In the last three decades, active research has been carried out for eye tracking and gaze estimation [2]. In general, gaze estimation methods can be mainly categorized into two types: namely appearance based methods and features based methods. Appearance based gaze estimation methods take the whole eye image as an input and then construct mapping functions to gaze directions or the

© Springer International Publishing Switzerland 2016
N. Kubota et al. (Eds.): ICIRA 2016, Part I, LNAI 9834, pp. 300–309, 2016.
DOI: 10.1007/978-3-319-43506-0_26

points of regard. Lu et al. [3] proposed an adaptive linear regression method to sparsely select the training samples. The eye images are divided into many subregions and the summary of pixels intensities of each subregion is used to generalize a feature vector for mapping to the points on screen. Zhang et al. [4] used multimodal convolutional neural networks to estimate the gaze in the wild environment. A large training data from 15 participants using laptop over three months is collected for training. Funese et al. [5] fit a 3D morphable model to the depth data captured by a depth sensor, the eye images are then cropped to frontal face for gaze estimation. The final gaze is determined by combining frontal gaze and the head pose. Sugano et al. [6] proposed a multi-camera based system. The 3D shape of eye region is reconstructed and random regression forest is used for gaze estimation. However, appearance based gaze estimations either suffer from the large training data or the variations caused by head illumination, head pose and eye shapes.

Unlike appearance based methods, feature based methods estimate the gaze via extracting remarkable facial points such as eye corners, eye lids, pupil centers or corneal glints. The pupil center-corneal center (PCCR) based gaze estimation methods is one of the most popular feature based methods due to its simplicity and reasonable accuracy [7–10]. However, these methods require an additional IR source or strong visible light sources. Sesma et al. [11] propose to replace the corneal glints with eye corners thus the light sources can be removed. A PC-EC vector is calculated for gaze estimation where the pupil centers and eye corners are manually labeled to avoid image processing errors. Cheung et al. [12] further combine the PC-EC vector with estimated head pose to deal with the head movement.

Although the PC-EC vector can achieve reasonable accuracy when the head keeps still, it lacks the tolerance to deal with slightly head movement as stated in [11], which is an important ability when combining with estimated head pose for head pose free gaze estimation. In this paper we develop a normalized Iris center-eye corner (NIC-EC) vector to gain the robustness against slight head movements. Unlike the PC-EC vector in [11] where both outer and inner corners are used, the developed NIC-EC vector only uses the inner corners and is further normalized via the length of the inner corners and height of nose to overcome the slight head movements. Further more, this paper also shows that an acceptable gaze estimation accuracy can still be acquired by employing automatic facial feature localization and pupil center localization methods to track the features without manually labelling features.

In the proposed method, we firstly extract reliable facial features by using supervised decent method (SDM). Secondly, the accurate iris centers are localized by using the convolution based integer-differential eye localization method. Then the iris centers and inner eye corners are used to construct the iris center-eye corner vector. It is then further normalized according to the length of two inner eye corners and the hight of the nose. Finally, we use an interpolation method to map the NPC-EC to the points on the screen. Experimental results show that although there exist image processing errors in finding pupil centers and eye corners, the gaze estimation result is still acceptable. The rest of this paper is organized as follows. Section 2 describes the details of the gaze estimation method. Section 3 presents experiments including hardware setup, evaluation method and results of gaze estimation. The paper is concluded with discussions in Sect. 4.

2   Methodology

Many gaze estimation methods have been proposed in the recent years, among which, the conventional PCCR based gazed estimation has attracted much attention due to its easy implementation and acceptable accuracy [8]. However an additional IR source or visible light source is required in these methods. Sesma et al. [11] showed acceptable result can be obtained in a still head environment by replacing the glint with eye corner for PCCR gaze estimation method. In this paper, a NIC-EC vector based gaze estimation method is proposed by using the located facial features and iris centers. The flowchart of the gaze estimation system is shown in Fig. 1.

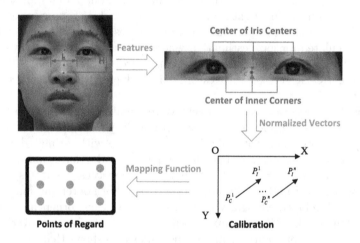

Fig. 1. Flowchart of gaze estimation algorithm

The algorithm firstly detects the facial features and eye centers. Then the length of two inner eye corners and the hight of the nose can be determined. The iris center-eye corner is constructed by the center of two inner corners and the iris centers. The final NIC-EC can be obtained after applying normalization for the IC-EC vector. Finally, a calibration procedure is performed to solve the parameters of the mapping function and the final gaze can be determined.

2.1 Facial Features and Iris Centers Localization

Many reliable facial features detection methods have been proposed in recent years. In this paper, we employ the SDM [13] to localize the inner eye corners and detect the length of nose. The SDM method can fast and accurately locate 48 facial features. For the iris center localization, we employ the convolution based integer-differential eye localization method [14]. This method mainly uses the large intensity difference of boundary between iris and sclera to fast locate the iris.

In practic, the boosted cascade face detector is firstly employed with default parameters to obtain the approximate location of the face [15]. The facial features detection and iris center localization are performed according to the face region.

2.2 Normalized Pupil Center-Eye Corner Vector

The conventional PC-CR methods use the pupil center and corneal glints vector to infer the point of regard (PoG) on screens. However it has the disadvantage of complex hardware setup. Recently research shows that the PC-EC vector has the potential to replace the PC-CR vector [11, 12]. However, this vector is sensitive to head movements. This paper proposes a NIC-EC vector to gain the robustness against slight head movements for gaze estimation. The NIC-EC vector is defined as follows:

$$
\begin{cases}
P_C = P_l^{corner} + P_r^{corner} \\
P_I = P_l^{iris} + P_r^{iris} \\
V_g = P_I - P_C \\
v_x = V_{gx}/L \\
v_y = V_{gy}/H
\end{cases}
\tag{1}
$$

where P_l^{corner} and P_r^{corner} stand for the located left eye inner corner and right eye inner corners, respectively. The center of the inner corners is represented by P_C. P_l^{iris} and P_r^{iris} stand for the located left and right iris centers, respectively. The center of two iris centers is represented by P_I. Then the iris center-eye corner vector V_g can be calculated by using P_I and P_C. To gain the robustness towards slight head movements, we further perform normalization for the vector. The vector V_g can be denoted as (V_{gx}, V_{gy}) in the image coordinate. We use the length of two inner corners to normalize the x direction of V_g. The y direction is normalized by dividing the height of the nose that can be calculated by the two of the located facial points.

2.3 Mapping Function

Many different types of mapping functions for PCCR have been proposed in the literature such as simple 2D linear interpolation [16] and polynomial model [7]. In this paper, we employ the mapping function proposed by [7]. The detail of the mapping function is as follows:

$$
\begin{cases}
u_x = a_0 + a_1 * v_x + a_2 * v_y + a_3 * v_x * v_y + a_4 * v_x^2 + a_5 * v_y^2 \\
u_y = b_0 + b_1 * v_x + b_2 * v_y + b_3 * v_x * v_y + b_4 * v_x^2 + b_5 * v_y^2
\end{cases}
\tag{2}
$$

where the coefficients $(a_0, a_1, a_2, a_3, a_4, a_5, b_0, b_1, b_2, b_3, b_4, b_5)$ are determined during a calibration stage. In the calibration stage, users are asked to look at several fixed points on the screen, thus the coefficients can be solved using a least square method.

3   Evaluation

In this section, we firstly introduce the hardware setup for the experiment. After explaining the evaluation method for gaze estimation accuracy, we then compare the results of our NIC-EC method with the PC-EC method proposed in [11].

3.1   Experiment Configuration

The experiment configuration includes a screen and a normal web camera as shown in Fig. 2. We use a 24 in. screen with a resolution of 1920 × 1080, the length and hight of the screen are around 52.1 cm and 29.3 cm, respectively. The adopted Logitech C910 camera has a resolution of 1920 × 1080 and a view angle of 78°. During the experiment, the users are asked to click 21 buttons on the screen to collect eye image data for gaze calibration and gaze estimation. When clicking buttons, the users are asked to keep their head roughly still without a chin rest. Furthermore, to test the robustness of the proposed NIC-EC against PC-EC method for slight head movements, the users are asked to have a rest and then go back to the roughly same head position to click buttons. By doing so, the testing data will have a slight different head movement compare to the calibration data.

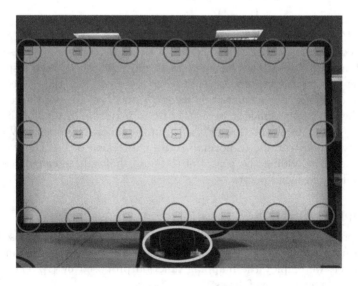

Fig. 2. Hardware setup of our system.

3.2   Accuracy Evaluation

To evaluate the accuracy of the proposed gaze estimation method, we use both of the pixel error e_{pixel} on the screen and the angular degree error e_{angel}.

$$\begin{cases} e_{pixel} = ||P_g - P||_2 \\ e_{angel} = arctan(Dg/Ds) \end{cases} \qquad (3)$$

where e_{pixel} and e_{angel} stand for the accuracy in pixels and degrees respectively. The estimated gaze point and real gaze point are represented as P_g and P respectively. Dg stands for the distance of real gaze points with estimated gaze points. Ds stands for the distance between human face and the screen.

3.3   Experimental Results

In the experiments, images data are captured by manually clicking the 21 buttons on the screen. The users are asked to keep the head still while clicking buttons. No chin rest is used during the clicking thus there might also be small head movement during the procedure. Then the facial features and iris center are localized using the method of [13] and [14] respectively.

Facial Feature Localization Result. Figure 3 shows the facial features localization result for all the 21 images when the distance between the subjects and the screen is 71 cm. The green points represent for the localized facial points. The red points are the iris centers. As shown in Fig. 3, the first column contains seven eye region images. They are corresponding to the first row buttons as shown in Fig. 2. The second and the third columns are corresponding to the second and the third rows buttons respectively. It should be noted that the low resolution of eye region in Fig. 3 is caused by the hardware. Although the selected web-camera

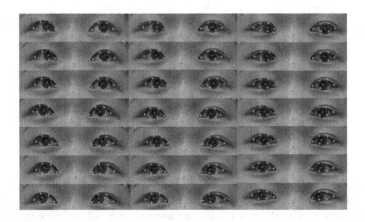

Fig. 3. Samples of facial feature localization results (Color figure online)

has a high resolution of 1920 ∗ 1080, its wide angle obstructs the way to obtain a high resolution in eye image region. As a result, the length of the located iris radius is 9 pixels.

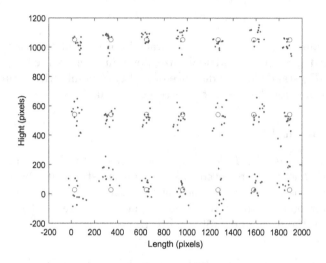

Fig. 4. Comparison of accuracy on calibration data. The result of the proposed NIC-EC method is marked in red points and result of PC-EC method is marked in blue points. (Color figure online)

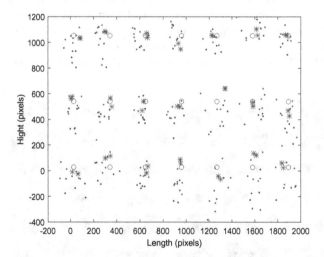

Fig. 5. Comparison of accuracy on testing data. The star points denote the estimated result on calibration data and the dot points denote the estimated result on testing data. The results of NIC-PC and PC-EC are marked with red and blue colors respectively (Color figure online)

Comparison with PC-EC Method. To show the good performance of the proposed NIC-EC method, we compare it with the PC-EC method proposed in [11]. We perform the comparison in two aspects, one is performed on calibration data and the other is on testing data. Both methods are tested on the same data.

Figure 4 shows the comparison result on calibration data where the red points are the results of NIC-EC method and the blue points are the results of PC-EC method. The calibration data consists of seven image sets where each set is collected at a different still head position. To compare the accuracy on slight head movements, we perform the calibration on one set of the data and use the rest six data sets for testing. The testing result is shown in Fig. 5 where the star points represent the estimated result on calibration data and the dot points represent the estimated result on testing data. The results of NIC-PC and PC-EC are marked with red and blue colors respectively. Figure 6 shows the quantitative comparison of two methods for the 21 points.

The average accuracy is shown in Table 1. The pixel error and degree error are calculated using Eq. 3. It can be observed that the calibration accuracy is roughly the same between the two methods. The proposed NIC-EC achieves an accuracy of 66.3 pixels for pixel error and 1.71° for degree error. The PC-EC method achieves an accuracy with 64.3 pixels for pixel error and 1.66° for degree error. For the slight head movements environment, the proposed NIC-EC method can achieve around 1° better performance than the PC-EC method. The accuracy of NIC-EC method is 103.7 pixels for pixel error and 2.68° for degree error. The accuracy of PC-EC methods is 141.4 pixels for pixel error and 3.65° for degree error.

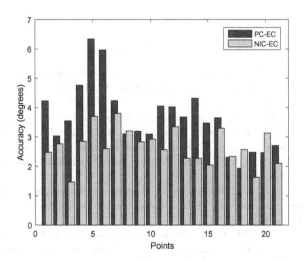

Fig. 6. Comparison of accuracy on testing data for 21 different points. The results of NIC-PC and PC-EC are marked with yellow and blue colors respectively (Color figure online)

Table 1. The average processing time of one eye region.

Method	Calibration		Testing	
	Pixel error	Degree error	Pixel error	Degree error
Sesma et al. [11] (PC-EC)	64.3 pixels	1.66°	141.4 pixels	3.65°
Proposed method (NIC-EC)	66.3 pixels	1.71°	103.7 pixels	2.68°

4 Conclusion

This paper proposes a single web-cam base gaze estimation method by mapping the NIC-PC vector to the points of regard. The proposed method has a significant benefit in that it is more robust and accurate than the PC-EC based gaze estimation method. The experimental results also show that the proposed method is robust against slight head movement which is an important requirement when integrating with head pose information to acquire head pose-free gaze estimation.

Future work will be targeted to explore head pose-free and calibration-free gaze estimation method for humanrobot interaction and humanrobot skill transfer [17,18] by incorporating effective visual tracking methods [19,20].

Acknowledgments. This work was supported by EU Seventh Framework Programme (611391, Development of Robot-Enhanced therapy for children with AutisM spectrum disorders (DREAM)) and National Natural Science Foundation of China (61403342,U1509207,61325019,61273286) and China Scholarship Council (201408330184).

References

1. Cai, H., et al.: Gaze estimation driven solution for interacting children with ASD. International Symposium on Micro-Nano Mechatronics and Human Science (MHS). IEEE (2015)
2. Hansen, D.W., Ji, Q.: In the eye of the beholder: a survey of models for eyes and gaze. IEEE Trans. Pattern Anal. Mach. Intell. **32**(3), 478–500 (2010)
3. Lu, F., Sugano, Y., Okabe, T., et al.: Adaptive linear regression for appearance-based gaze estimation. IEEE Trans. Pattern Anal. Mach. Intell. **36**(10), 2033–2046 (2014)
4. Zhang, X., Sugano, Y., Fritz, M., et al.: Appearance-based gaze estimation in the wild. In: Proceedings of the IEEE Conference on Computer Vision and Pattern Recognition, pp. 4511–4520 (2015)
5. Mora, K., Odobez, J.M.: Geometric generative gaze estimation (G3E) for remote RGB-D cameras. In: Proceedings of the IEEE Conference on Computer Vision and Pattern Recognition, pp. 1773–1780 (2014)
6. Sugano, Y., Matsushita, Y., Sato, Y.: Learning-by-synthesis for appearance-based 3d gaze estimation. In: Proceedings of the IEEE Conference on Computer Vision and Pattern Recognition, pp. 1821–1828 (2014)

7. Morimoto, C.H., Mimica, M.R.M.: Eye gaze tracking techniques for interactive applications. Comput. Vis. Image Underst. **98**(1), 4–24 (2005)
8. Topal, C., Gunal, S., Kodeviren, O., et al.: A low-computational approach on gaze estimation with eye touch system. IEEE Trans. Cybern. **44**(2), 228–239 (2014)
9. Sigut, J., Sidha, S.A.: Iris center corneal reflection method for gaze tracking using visible light. IEEE Trans. Biomed. Eng. **58**(2), 411–419 (2011)
10. Cho, D.C., Kim, W.Y.: Long-range gaze tracking system for large movements. IEEE Trans. Biomed. Eng. **60**(12), 3432–3440 (2013)
11. Sesma, L., Villanueva, A., Cabeza, R.: Evaluation of pupil center-eye corner vector for gaze estimation using a web cam. In: Proceedings of the Symposium on Eye Tracking Research and Applications, pp. 217–220. ACM (2012)
12. Cheung, Y., Peng, Q.: Eye gaze tracking with a web camera in a desktop environment. IEEE Trans. Hum. Mach. Syst. **45**(4), 419–430 (2015)
13. Xiong, X., De la Torre, F.: Supervised descent method for solving nonlinear least squares problems in computer vision. arXiv preprint arXiv:1405.0601 (2014)
14. Cai, H., Liu, B., Zhang, J., et al.: Visual Focus of Attention Estimation Using Eye Center Localization
15. Viola, P., Jones, M.J.: Robust real-time face detection. Int. J. Comput. Vision **57**(2), 137–154 (2004)
16. Zhu, J., Yang, J.: Subpixel eye gaze tracking. In: Fifth IEEE International Conference on Automatic Face and Gesture Recognition, 2002, Proceedings. IEEE (2002)
17. Liu, H.: Exploring human hand capabilities into embedded multifingered object manipulation. IEEE Trans. Industr. Inf. **7**(3), 389–398 (2011)
18. Ju, Z., Liu, H.: Human hand motion analysis with multisensory information. IEEE/ASME Trans. Mechatron. **19**(2), 456–466 (2014)
19. Zhou, X., Yu, H., Liu, H., et al.: Tracking multiple video targets with an improved GM-PHD tracker. Sensors **15**(12), 30240–30260 (2015)
20. Zhou, X., Li, Y., He, B., et al.: GM-PHD-based multi-target visual tracking using entropy distribution and game theory. IEEE Trans. Industr. Inf. **10**(2), 1064–1076 (2014)

Planning, Localization, and Mapping

Large Scale Indoor 3D Mapping Using RGB-D Sensor

Xiaoxiao Zhu[1(✉)], Qixin Cao[1], Hiroshi Yokoi[2], and Yinlai Jiang[2]

[1] State Key Lab of Mechanical Systems and Vibration, Research Institute of Robotics,
Shanghai Jiao Tong University, Shanghai, China
{ttl,qxcao}@sjtu.edu.cn
[2] The University of Electro-Communications, Tokyo, Japan
{yokoi,jiang}@hi.mce.uec.ac.jp

Abstract. 3D Mapping using RBG-D sensor is a hot topic in the robotic field. This paper proposes a sub-map stitching method to build map in the large scale indoor environment. We design a special landmark, and place it in the environment. Every sub-map contains those landmarks, and then can be stitched by BA optimization. The result shows that the map error is blow 1 % in a room with the dimensions of 13 m × 8 m.

Keywords: Large scale indoor mapping · RGB-D sensor · Map stitching

1 Introduction

There are many ways to build a 3D model of the indoor environment using RGB-D sensor, such as RGBDSLAM [1, 2] SLAM6D from 3Dtk [3] or KinectFusion [4]. KinectFusion has the highest accuracy and real-time performance of those methods. RGBDSLAM usually runs at 2 FPS and will become slower when more data is captured, while SLAM6D is an offline method. To build a hole-less map, we have to capture many views and also need the real-time feedback of the current model, so the real-time performance is very important. The KinectFusion method runs at 15 fps, so we can building the map by just handing Kinect and scanning all the environment with a real-time feedback. However, the biggest disadvantage of KinectFusion is that, it can only build limited size of map, because this method is memory consuming and especially because it uses a GPU memory. Although simply reducing the resolution of the map can enlarge the mapping area, this will decrease accuracy and make the mapping process unstable. To resolved this limitation, paper [5, 6] propose a GPU-Octree data structure to replace the 3-dims array of the Truncated Signed Distance Function (TSDF) model [7], so it can reduce the demands of GPU memory and keep the accuracy. Paper [8] proposal the Kinectous algorithm, which use the GPU as a ring buffer so can recycle use the GPU. However, those methods all finish the mapping process in one try. And also requires a lot of time to complete the entire map creation process, so reduce the flexibility of the map created in the indoor complex cases difficult to practice.

In this paper, we resolved this problem by a sub-map stitching method. Since the KinectFusion algorithm suit to create the small scale map, so we just use the KinectFusion algorithm to build a sub-map with dimensions of 6 × 3 × 3 m, and the

© Springer International Publishing Switzerland 2016
N. Kubota et al. (Eds.): ICIRA 2016, Part I, LNAI 9834, pp. 313–321, 2016.
DOI: 10.1007/978-3-319-43506-0_27

sub-maps were then stitched into a large map. To help the stitching process, a kind of special landmark will be proposed, which is easy to be detected and localized in the sub-map. The experiment result shows good precision.

2 Procedure of Mapping

2.1 The Main Idea and the Landmark

Since the KinectFusion algorithm will produces a point cloud style sub-map, so the stitching target is the point cloud. The traditional point cloud stitching or alignment problem is solved by ICP [9] method, which is not suitable for this mapping case for three reasons:

1. The traditional ICP method need a good initial for converging. But in the mapping process, the RGB-D sensor is holding by human, so there is no odometer information, then a good initial position cannot be provided for ICP; and
2. The general ICP method is a 3D matching method, but notice that a consistent ground plane is especially important in the indoor environment; and
3. The ICP need the overlap between the two point clouds, and more overlap the more coverage the more precision. However, in the mapping case, more overlap between the sup-maps means more number of sub-map, and will cost more mapping time.

Therefore, we proposed a ground plane consistent sub-map stitching using 3D landmark. As shown in Fig. 1, the 3D landmark is made with balls because the spherical shape is easy to be detected in the point cloud and a small plane for computing the localization data. The landmarks are distinguishable by radius and height.

Fig. 1. Four ball landmarks

2.2 Layout of the Landmarks

Before the mapping process, every sub-map will be placed with four ball landmarks and every two sub-maps have two corresponding ball landmarks, as shown in Fig. 2. For least overlap between sub-maps, the landmark is place near the edge of the sub-map. And to make the recognition of the corresponding relationship easy, beside the landmark's identification, the distance between two landmarks is also used to determine the correspondence of two sub-maps. For example, in sub-map1 and sub-map2, the distance between landmark1 and landmark2 should be distinctive, so landmark3 and landmark4 are the corresponding landmarks.

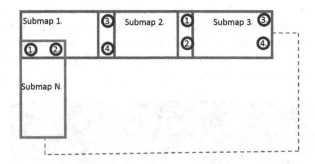

Fig. 2. Illustration of the alignment of the landmarks

2.3 Sub-map Building Using Advanced KinectFusion Algorithm

As mentioned before, the original KinectFusion method is not very stable, especially in some simple environment. So we use an advanced KinectFusion algorithm proposed in our previous work [10]. This advanced KinetFusion has two improvements of KinectFusion algorithm. On one hand, the edge feature points in the environment are matched to improve the robustness, on the other hand, a ground plane point cloud in the model is preset to improve the accuracy. The improved algorithm decreases the modeling error from 4.5 % to 1.5 % in a room of 6 m ×3 m ×3 m. Although the efficiency is influenced, the running speed of the algorithm is still very high, and the user experience during modeling is good.

An example of sub-map is shown in Fig. 3.

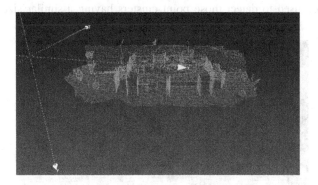

Fig. 3. An example of a sub-map

2.4   Ground Plane Extraction

The ground plane is extracted using the Random sample consensus (RANSAC) method [11], which is an iterative method to estimate parameters of a mathematical model from a set of observed data which contains outliers, and the threshold value is set as 2 cm. A temporary ground plane coordinate system is defined and the sub-map is transformed to the ground coordinate system. Although the ground plane coordinates of the sub-maps are not yet consistent, by doing this the dimension of the sub-map stitching problem is decreased to 2D. The transformation results are shown in Fig. 4, in which the x-y plane is coincident with the ground plane.

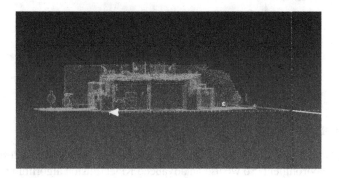

Fig. 4. The result of ground plane extraction and point cloud transformation

2.5   Ball Landmark Extraction and Corresponding Computation

Extracting the balls from the sub-map also uses the RANSAC method, but directly running the RANSAC for the sub-map will fail because the ball is small compared to the sub-map, meaning that the number of outliers is far larger than the inliers. Therefore, the ground plane is first extracted and deleted from the sub-map and the Euclidean cluster method is then used to detect those point clusters having a similar size to the ball;

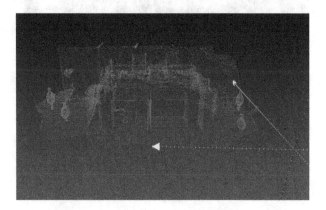

Fig. 5. The result of ball extraction

RANSAC is then run to find balls in every cluster. The result is shown in Fig. 5. After ball extraction, determining the correspondences by ball radius and height and ball-ball distance is straightforward.

2.6 Computation of the Pairwise Transformation

In this step, we intend to compute the transformation between two adjacent sub-maps. There are two correspondence points (the ball centers) and a 2D transformation must be computed. The general computation method for an optimized transformation from two sets of corresponding points is reviewed by [12]. In our case, only two corresponding points are used to compute a 2D transformation, so a simpler way of solving the problem was found. First, the ball points were projected to the x-y plane as in Fig. 6(b) (where c1 and c2 are the centers of the points), then the optimized transformation is the one that makes the two centers coincide and the four points lie in one line (as shown in Fig. 6(c)). This result concerns with the result of the SVD based transform optimized method [13].

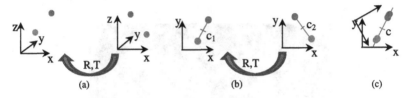

(a)   (b)   (c)

Fig. 6. (a) Illustration of the original problem (b) Projection to the x-y plane decreases the problem to 2D (c) The optimized 2D transformation

2.7 Loop Closure and BA Optimization

When building the sub-maps, the end map and the first map must be connected, giving a loop closure. Based on a good initial value from the pairwise transformation result, a BA algorithm [14] is used to optimize the transformation between those sub-maps.

The origin BA algorithm assume there are n 3D points projected into m pictures, let x_{ij} to be the projection of point i in the image j, v_{ij} to be a binary variable, $v_{ij} = 1$ if point i can be seen in image j, otherwise $v_{ij} = 0$. And let a_j to the camera internal and external parameters, let b_i represent the 3D point i in the world. BA algorithm minims the projection error and compute the optimal coordinate of those 3D points and also the transformations between those cameras.

$$\min_{a_j, b_i} \sum_{i=1}^{n} \sum_{j=1}^{m} v_{ij} d(Q(a_j, b_i), x_{ij})^2 \qquad (1)$$

In which the Q is the prediction of the projection of the point i in image j, and $d(x, y)$ represent the Euclidean distance between two 2D points in the image.

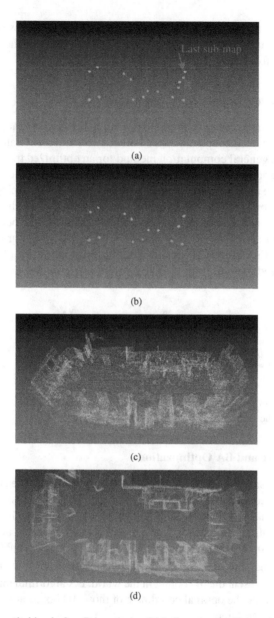

Fig. 7. (a) sub-map stitching before BA optimize (b) balls map after BA optimize (c) Final map after BA optimize (d) Final map without the landmarks

Here, in our case, the b_i is the real 2D coordinates on the ground of the ith land-mark(x_i, y_i), and a_j is j[th] sub-map's 2d pose information include the rotation angle Theta and the transformation component T, v_{ij} can determined by the neighborhood relation-ship, $Q(a_j, b_i)$ represent the coordinates of landmark i respect to the sub-map j.

The Fig. 7(a) show the error of initial stitching result using the result of the previous section, show as Fig. 7(b), after the optimization, the ball landmarks in each sub-map are in good agreement. Figure 7(c) and (d) show the final map with and without the landmarks.

3 Experiment

This experiment mainly test the accuracy of the map built by our method. The test lab environment is show as Fig. 8(a), in which several test points were selected shown as the red points and the distance between each other were measured manually. The built 3D map is shown in Fig. 8(b), we import the map into a tool software named Meshlab and also measured the corresponding distance. Then the two types of measured distances are compared with each other, and the result is shown in Table 1. The d_{AB} and d_{BC} show the accuracy inside a sub-map, and the others show the accuracy of the sub-map stitching method. The result shows that map error is 1 % in the room with dimensions of 13 m × 8 m.

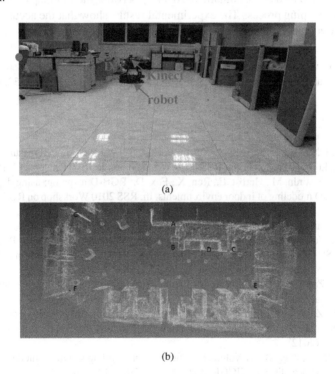

(a)

(b)

Fig. 8. (a) The test scene (b) Some points used for distance comparison

Table 1. Distance comparison result

	Real distance (cm)	Map distance (cm)	Absolute error (cm)
d_{AB}	136.5	137	0.5
d_{BC}	278	279.2	1.2
d_{DE}	400.252	394.049	6.20283
d_{DF}	811.758	812.607	0.8493
d_{DG}	822.771	816.12	6.65138
d_{EF}	1089.81	1080.58	9.22867
d_{EG}	1193.04	1181.14	11.9036

4 Conclusion

In this paper, a sub-map stitching based mapping method for building a large scale 3D indoor environment model was presented, with emphasis on the consistency of the ground plane. And by using landmark, the overlap between the sub-maps can be reduced, so make the mapping process The experimental results show that the accuracy the final 3D map is less than 1 % that in a room with dimensions of 13 m × 8 m.

Acknowledgment. We thank the support of China Postdoctoral Science Foundation, No. 2015M571561 and the National Natural Science Foundation of China, No. 61273331.

References

1. Engelhard, N., Endres, F., Hess, J., Sturm, J., Burgard, W.: Real-time 3D visual SLAM with a hand-held RGB-D camera. In: RSS 2010 Workshop on RGB-D Cameras (2010)
2. Henry, P., Krainin, M., Herbst, E., Ren, X., Fox, D.: RGB-D mapping: using depth cameras for dense 3D modeling of indoor environments. In: RSS 2010 Workshop on RGB-D Cameras (2010)
3. Slam6d. Slam6d toolkit. http://slam6d.sourceforge.net/index.html
4. Izadi, S., Newcombe, R.A., Kim, D., Hilliges, O., Molyneaux, D., Kohli, P., Shotton, J., Hodges, S., Freeman, D., Davison, A., Fitzgibbon, A.: Kinectfusion: real-time dynamic 3D surface reconstruction and interaction. In: ACM SIGGRAPH 2011 (2011)
5. Zeng, M., Zhao, F., Zheng, J., Liu, X.: Octree-based fusion for realtime 3D reconstruction. In: Graphical Models (2012)
6. Zeng, M., Zhao, F., Zheng, J., Liu, X.: A memory-efficient KinectFusion using octree. In: Hu, S.-M., Martin, R.R. (eds.) CVM 2012. LNCS, vol. 7633, pp. 234–241. Springer, Heidelberg (2012)
7. Curless, B., Levoy, M.: A volumetric method for building complex models from range images. In: Proceedings of SIGGRAPH 1996, pp. 303–312 (1996)
8. Whelan, T., Kaess, M., Fallon, M., Johannsson, H., Leonard, J., McDonald, J.: Kintinuous: spatially extended kinectfusion (2012)
9. Besl, P.J., McKay, N.D.: A Method for Registration of 3-D Shapes. IEEE Trans. Pattern Anal. Mach. Intell. **14**(2), 239–256 (1992)

10. Zhu, X., Cao, Q., Yang, Y., et al.: An improved kinect fusion 3D reconstruction algorithm. Robot **36**(2), 129–136 (2014)
11. Fischler, M.A., Bolles, R.C.: Random sample consensus: a paradigm for model fitting with applications to image analysis and automated cartography. Commun. ACM **24**(6), 381–395 (1981)
12. Eggert, D.W., Lorusso, A., Fisher, R.B.: Estimating 3-D rigid body transformations: a comparison of four major algorithms. Mach. Vis. Appl. J. **9**(5), 272–290 (1997)
13. Arun, K.S., Huang, T.S., Blostein, S.D.: Least-squares fitting of two 3-D point sets. IEEE Trans. Pattern Anal. Mach. Intell. J. **5**, 698–700 (1987)
14. Triggs, B., McLauchlan, P.F., Hartley, R.I., Fitzgibbon, A.W.: Bundle adjustment – A modern synthesis. In: Triggs, B., Zisserman, A., Szeliski, R. (eds.) ICCV-WS 1999. LNCS, vol. 1883, pp. 298–372. Springer, Heidelberg (2000)

Performance Comparison of Probabilistic Methods Based Correction Algorithms for Localization of Autonomous Guided Vehicle

Hyunhak Cho[1], Eun Kyeong Kim[2], Eunseok Jang[2], and Sungshin Kim[3(✉)]

[1] Department of Interdisciplinary Cooperative Course: Robot,
Pusan National University, Busan, South Korea
darkruby1004@pusan.ac.kr
[2] Department of Electrical and Computer Engineering,
Pusan National University, Busan, South Korea
{kimeunkyeong, esjang}@pusan.ac.kr
[3] School of Electrical and Computer Engineering, Pusan National University,
Busan, South Korea
sskim@pusan.ac.kr

Abstract. This paper presents performance comparison of probabilistic methods based correction algorithms for localization of AGV (Autonomous Guided Vehicle). Wireless guidance systems among the various guidance systems guides the AGV using position information from localization sensors. Laser navigation is mostly used to the AGV of a wireless type, however the performance of the laser navigation is influenced by a slow response time, big error of rotation driving and a disturbance with light and reflection. Therefore, the localization error of the laser navigation by the above-mentioned weakness has a great effect on the performance of the AGV. There are many different methods to correct the localization error, such as a method using a fuzzy inference system, a method with probabilistic method and so on. Bayes filter based estimation algorithms (Kalman Filter, Extended Kalman Filter, Unscented Kalman Filter and Particle Filter) are mostly used to correct the localization error of the AGV. This paper analyses performance of estimation algorithms with probabilistic method at localization of the AGV. Algorithms for comparison are Extended Kalman Filter, Unscented Kalman Filter and Particle Filter. Kalman Filter is excluded to the comparison, because Kalman Filter is applied only to a linear system. For the performance comparison, a fork-type AGV is used to the experiments. Variables of algorithms is set experiments based heuristic values, and then variables of same functions on algorithms is set same values.

Keywords: Extended Kalman Filter · Unscented Kalman Filter · Particle Filter · Performance comparison · Localization

© Springer International Publishing Switzerland 2016
N. Kubota et al. (Eds.): ICIRA 2016, Part I, LNAI 9834, pp. 322–333, 2016.
DOI: 10.1007/978-3-319-43506-0_28

1 Introduction

Guidance system technique among various AGV (Autonomous Guided Vehicle) technique is separated to a wire guidance system and a wireless guidance system. The technique guided AGV to destination using the physical properties and information of the logical. AGV of the wire guidance type is guided by underground guide line (magnet, induction line). Wire guidance system has advantages such as high safety, low cost of a sensor, however disadvantages of the system are high expense of laying guidelines and the difficulties of change of work paths [1, 2].

Wireless guidance systems complements the mentioned problems of wire guidance systems using the calculated position by installed landmarks in an environment. Localization technique of the wireless type AGV is comprised of global localization and local localization. The technique of the local localization calculates the relative position from the previous position using acceleration and angular speed. Sensors of the local localization has advantages fast response time, fast calculation time, low cost and so on. However disadvantages are a sensor error, a cumulative error and a bias error [3, 4].

The technique of the global localization calculates the global position in the environment over known landmarks. For global localization, matching methods between measured features and known landmarks are used for the technique of the global localization. However disadvantages the method are long calculation time, a sensor error and inaccurate calculated position over the wrong matching.

To complement the above-mentioned problems of the localization, the position of AGV is complementarily calculated by the local localization and the global localization. Correction methods are various, and then probabilistic methods based correction algorithm among various algorithms are used a lot in AGV. Methods are Kalman Filter, Extended Kalman Filter [5, 6], Unscented Kalman Filter [7, 8] and Particle Filter [9, 10]. Extensive research of correction algorithms has been done. Mentioned algorithms are estimated a state of a system (non-linear system or non-Gaussian system) including inputs with much noises, and then there have a high efficiency to the localization.

For the performance comparison of algorithms (Extended Kalman Filter, Unscented Kalman Filter and Particle Filter) in this paper, there are implemented using the local localization and the global localization, and then performance of the applied algorithms to AGV are analyzed by experiments.

In this paper, Sect. 2 describes the used AGV, and Sect. 3 describes localization method using probabilistic methods. Section 4 explains experiments and results and Sect. 5 describes conclusion.

2 System Configuration

2.1 Automatic Guided Vehicle – Fork Type

For experiments of this paper, Fork-type AGV is used to analysis. The used fork-type AGV was remodeled the manual fork-lift production, and then that has an axle drive unit with a full-electric power steering.

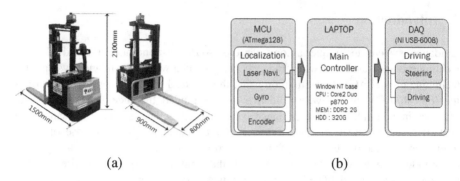

(a)   (b)

Fig. 1. Autonomous Guided Vehicle; (a) Fork-type AGV, (b) System configuration

Encoders and a gyro for a linear velocity and an angular velocity for local localization are installed to AGV, and then a laser navigation is used to global localization of AGV. The laser navigation is installed to the top of AGV for minimizing effect of disturbance by surroundings objects, and then encoders are installed to road wheels (auxiliary wheels) under forks of AGV. The gyro is installed above of the axle drive of AGV. System of the used AGV consists of a localization part, a control part and a driving part.

To calculate positions of AGV, information (x, y, t of the laser navigation, linear velocities of encoders and an angular velocity of the gyro) of a localization part are transmitted to the control part over a micro controller unit every 100 ms. The control part calculates AGV's position and controls the driving part using the calculated position. Figure 1 shows the system configuration of AGV.

2.2   Localization Sensors

Used sensors are relative localization sensors (encoders and a gyro) and global localization sensor (laser navigation).

To local localization, the used encoder is E40H-12-1000-3-V-5 (1000 pulse resolution), and then the gyro is myGyro300-SPI ($\pm300°$/s sensitivity). Tables 1 and 2 show the specification of sensors. To global localization, the used sensor is NAV200 (laser navigation).

Table 1. Specification of encoder (E40-H12-1000-3-V-5)

Model	Specification	
	Supply Voltage	12v
	Resolution	1000 pulse
	Output	Voltage
	Output Phase	3 Phase (A, B, Z)
	Shaft Diameter	Φ12mm

Table 2. Specification of gyro (myGyro300-SPI)

Model	Specification	
	Supply Voltage	5 V
	Output	SPI Communication
	Measurement Angle Range	±300 °/s
	Sensitivitiy	0.2439 °/LSB

That calculates a position over comparison of positions between measured reflector and known reflectors, therefore NAV200 system needs to know positions of reflectors in an environment.

More specifically, the header of NAV200 transmits infrared light over 360 rotation and receives reflected infrared light from installed reflections in the environment. NAV200 calculates distances and angles over measured information between the header and reflectos, and then the system calculates now position and now angle using a matching method. Table 3 shows the specification of NAV200.

Table 3. Specification of laser navigation (NAV200)

Model	Specification	
	Supply Voltage	24 V
	Field of View	360
	Output	RS-232 communication
	Operating Range	1.2~28.5 m
	Position Accuracy	±25 mm
	Angular Accuracy	0.1 °

3 Localization Method

Localization of AGV is composed local localization and global localiztion. To local localization, this paper used kinematics of AGV with the axle-drive type and used sensors are 2 encoders and 1 gyro. Global localization uses bayes filter based correction method such as EKF (Extended Kalaman Filter), UKF (Unscented Kalman Filter) and PF (Particle Filter). The prediction step of above-mentioned filters uses kinematics of AGV and the estimation step uses the correction method of each filters. Subsection 3.1 explains the kinematics for local localization and Subsect. 3.2 explains correction localization method using probabilistic method with global localization and local localization.

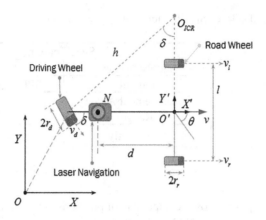

Fig. 2. Kinematics

3.1   Kinematics

The used AGV has the axle-drive model, and kinematics of the used system is same Fig. 2.

In Fig. 2, *OICR* is the central axis of rotation in the axle-drive system. Angular velocities (w_l and w_r)of road wheels are measured by encoders, linear velocities are calculated by Eq. (1). In Eq. (1) r_l and r_r are radiuses of each road wheels.

$$v_l = w_l 2 r_l$$
$$v_r = w_r 2 r_r$$
(1)

The linear velocity (v) of AGV and the angular velocity (w) of AGV are calculated by Eq. (2), and then l of Eq. (2) is the length between road wheels.

$$v = \frac{v_r + v_l}{2}$$
$$w = \tan^{-1}\left(\frac{v_r - v_l}{l}\right)$$
(2)

The angular velocity using the gyro is calculated by Eq. (3), in Eq. (3) $Gyro_{Center}$ is the mean value of 1000 data on the stop state and $Gyro_{ADC}$ is the output value of the gyro. Sensitivity in Eq. (3) is the value of the specification.

$$w = (Gyro_{Center} - Gyro_{ADC})Sensitivity$$
(3)

The angular velocity of AGV is the mean value (angular velocity using encoders and angular velocity using the gyro). The local localization is calculated by 3 method (Euler Integration – Eq. (4), 2nd order Runge-Kutta Integration – Eq. (5) and Exact Integration – Eq. (6)).

Used localization method is exact integration because of the most efficient performance based on experiments.

$$x_{k+1} = x_k + v \cos \theta_k$$
$$y_{k+1} = y_k + v sin \theta_k \quad (4)$$
$$\theta_{k+1} = \theta_k + w$$

$$x_{k+1} = x_k + v \cos \left(\theta_k + \frac{w}{2} \right)$$
$$y_{k+1} = y_k + v sin \left(\theta_k + \frac{w}{2} \right) \quad (5)$$
$$\theta_{k+1} = \theta_k + w$$

$$x_{k+1} = x_k + \frac{v}{w} (\sin(\theta_{k+1}) - \sin(\theta_k))$$
$$y_{k+1} = y_k + \frac{v}{w} (\cos(\theta_{k+1}) - \cos(\theta_k)) \quad (6)$$
$$\theta_{k+1} = \theta_k + w$$

3.2 Correction of Localization Method Using Probabilistic Algorithm

Probabilistic based correction methods of the localization of AGV are EKF, UKF and PF. Above-mentioned filters has high effectiveness to a dynamic system, non-linear system and non-gaussian system including noise inputs. Base algorithm of above-mentioned filters is bayes filter, and that recursively estimates the state of a system using measurement values of sensors. Figure 3 shows the estimation process of the bayes filter.

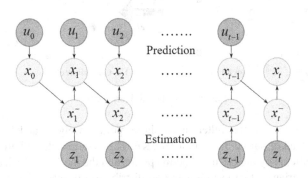

Fig. 3. Bayes filter

In Fig. 3, x_t^- is the prediction state of the time t, x_t is the estimation state of the time t. u is the input of system model, z is the measurement value.

Bayes filter based EKF is consisted like Fig. 4.

EKF is the designed algorithm to apply Kalman Filter at a non-linear system, and that used linearlization over partial differential of the system model and the

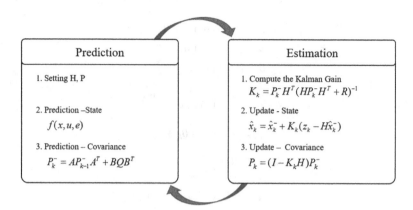

Fig. 4. Extended Kalman Filter

measurement model in the estimation model system. Prediction step of EKF predicts the system state and the P covariance. The system state and the P covariance are corrected by Kalman Gain in the estimation step. Covariances (Q of predtion step and R of estimation step) are selected by a user over experiments.

In Fig. 4, x is the position of AGV, u is values of sensors for the local localization and e is noise values of sensors. P is covariance of the system, A and B are model using partial differential. z_k is measurement values.

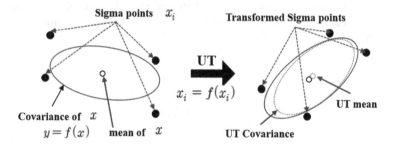

Fig. 5. Unscented Transform

UKF (Unscented Kaman Filter) used sigma points, weights and UT (Unscented Transform) instead of partial differential in EKF. The system state and error covariance in UKF are predicted by the UT using sigma points and the model. UT calculates a mean value and a covariance, that used a gaussian distribution instead of partial differential using jacobian. Figure 5 shows the UT process and Fig. 6 shows the process of UKF.

PF (Particle Filter) is non-parametric filter that has not parameters unlike EKF and UKF. The method repeatedly estimates optimal states using particles. Particles has the system state each differently. PF estimates the system state using information particles over properly proposed probability distributions based randomly particles. However, initial state of particles, resampling step and so on are selected by a user on experiments. Figure 7 shows the process of PF.

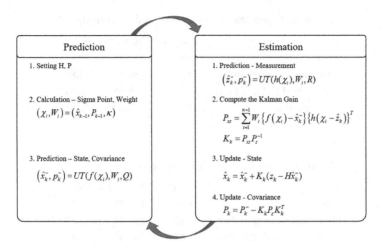

Fig. 6. Unscented Kalman Filter

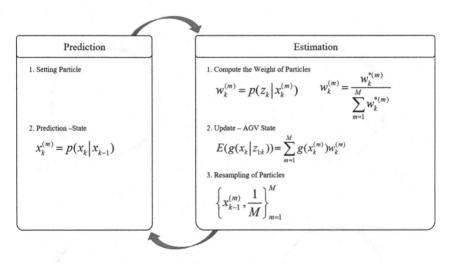

Fig. 7. Particle Filter

In this paper, setting variables (covariance, function) by the user are selected the most effective value by experiments.

4 Experiments and Results

For analysis of localization performance, the experiment environment is 8400 × 2100 mm. Reflectors cylindrical for the laser navigation were installed to the surface of a wall, and the length and the radius are 600 mm and 60 mm.

Table 4. Experiment results of straight driving (unit:mm)

	Mean error	Max error	Std
Non	88.76	264.65	80.82
Extended Kalman Filter	17.68	39.11	13.83
Unscented Kalman Filter	**7.32**	**29.66**	**6.91**
Particle Filter	20.06	36.64	12.81

Table 5. Experiment results of rotation driving (unit:mm)

	Mean error	Max error	Std
Non	121.04	146.83	23.21
Extended Kalman Filter	94.82	**126.06**	12.73
Unscented Kalman Filter	**71.67**	184.04	30.86
Particle Filter	119.07	135.60	**3.70**

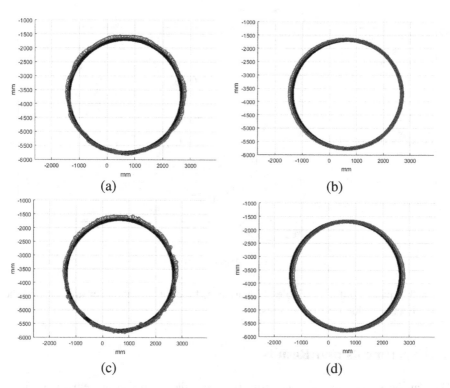

(a)

(b)

(c)

(d)

Fig. 8. Experiment result of rotation driving; (a) Non Filter, (b) Extended Kalman Filter, (c) Unscented Kalman Filter, (d) Particle Filter

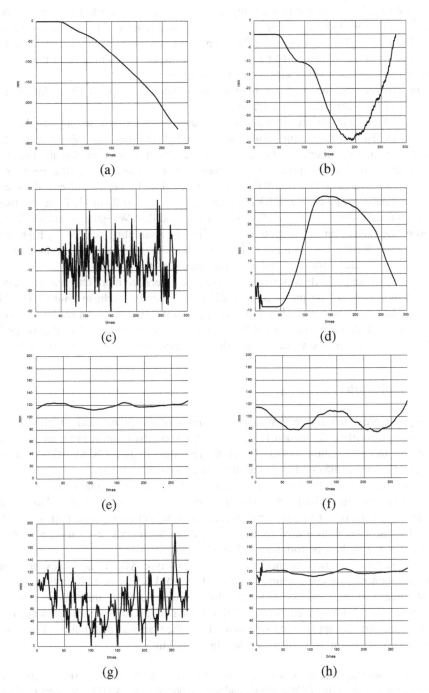

Fig. 9. Localization Error; (a) Non Filter of straight driving, (b) EKF of straight driving, (c) UKF of straight driving, (d) PF of straight driving, (e) Non Filter of rotation driving, (f) EKF of rotation driving, (g) UKF of rotation driving, (h) PF of rotation driving

Straight driving and rotation driving are used to performance test for performance analysis of filter algorithms. The speed of straight driving is 24 cm/s, and then the speed and the angle are 24 cm/s and 30°. Number of experiments are each 10 times. For comparison of algorithms performance, results of filters were compared to the driving line of the simulation using kinematics without noise.

In experiments, the used kinematics model among 3 type is Exact Integration because of high efficiency at experiments.

EKF, UKF and PF are implemented by experiments based setting variables of algorithms. Table 4 shows the result value of straight driving and Table 5 shows the result value of rotation driving. Figure 8 shows results of the rotation driving. Figure 9 shows errors of driving experiments, and then there expressed one of all experiments.

System covariance (Q) of EKF changes every time by outputs of encoders, and measurement covariance (R) of EKF and covariance (Q and R) of UKF are the fixed constant matrix. Number of particles to PF is 200, and then Weighted Mean Method is used to the estimation of the system state at PF. In resampling step at PF, high rank 20 % reuses for the next estimation step, and then the others are redistributed to the arbitrary positions (±100 mm and ±10°) of the estimated position.

To experiments, UKF performance versus the others are more efficient. However, the position may not be predicted to UKF model, because kalman gain of UKF is greatly influenced by noise. That only can apply to the system model of the Gaussian distribution. Parameters of correction algorithms are important that the properly selected, and then performance of algorithms is determined by the parameters.

5 Conclusion

This paper presented the performance comparison of the probabilistic method based correction algorithms (Extended Kalman Filter, Unscented Kalman Filter and Particle Filter) of the localization method of AGV. The algorithms apply to AGV, and the performance are compared by straight driving and rotation driving.

For localization method in this paper, the local position is calculated by kinematics method (Exact Integration) with over encoders and the gyro. The used kinematics model among 3 type is Exact Integration because of high efficiency at experiments. The global localization is measured by laser navigation. Setting variables from a user are selected to the highest performance value by experiments.

To comparison of performance, the designed fork-type AGV is used to experiments (each straight driving and rotation driving 10 times). UKF among algorithms are the most efficient to the correction of localization method. However, the position may not be predicted to UKF model, because kalman gain of UKF is greatly influenced by noise. That only can apply to the system model of the Gaussian distribution. Therefore, future work is optimization of parameters in algorithms.

Acknowledgments. This work was supported by BK21PLUS, Creative Human Resource Development Program for IT Convergence and was supported by the MOTIE (Ministry of Trade, Industry & Energy), Korea, under the Industry Convergence Liaison Robotics Creative Graduates Education Program supervised by the KIAT (N0001126).

References

1. Vis, I.F.A.: Survey of research in the design and control of automated guided vehicle systems. Eur. J. Oper. Res. **170**(3), 677–709 (2006)
2. Xiang, G., Tao, Z.: Robust RGB-D simultaneous localization and mapping using planar point feature. Robot. Auton. Syst. **72**, 1–14 (2015)
3. Spandan, R., Sambhunath, N., Ranjit, R., Sankar, N.S.: Robust path tracking control of nonholonomic wheeled mobile robot experimental validation. Int. J. Control Autom. Syst. **13**, 897–905 (2015)
4. Christof, R., Daniel, H., Christopher, K., Frank, K.: Localization of an omnidirectional transport robot using IEEE 802.15.4a ranging and laser range finder. In: IEEE/RSJ International Conference on Intelligent Robots and Systems, pp. 3798–3803 (2010)
5. Miguel, P., Antonio, P.M., Anibal, M.: Localization of mobile robots using an Extended Kalman Filter in a LEGO NXT. IEEE Trans. Educ. **55**(1), 135–144 (2012)
6. Jakub, S., Michal, R., Vladimir, K.: Evaluation of the EKF-based estimation architectures for data fusion in mobile robots. IEEE/ASME Trans. Mechatron. **20**(2), 985–990 (2015)
7. Maral, P., Guangjun, L.: An adaptive unscented Kalman filtering approach for online estimation of model parameters and state-of-charge of lithium-ion batteries for autonomous mobile robots. IEEE Trans. Control Syst. Technol. **23**(1), 357–363 (2015)
8. Ramazan, H., Hamid, D.T., Mohammad, A.N., Mohammad, T.: A square root unscented fast SLAM with improved proposal distribution and resampling. IEEE Trans. Ind. Electron. **61** (5), 2334–2345 (2014)
9. Bin, Y., Yongsheng, D., Yaochu, J., Kuangrong, H.: Self-organized swarm robot for target search and trapping inspired by bacterial chemotaxis. Robot. Auton. Syst. **71**, 83–92 (2015)
10. Gerasimous, G.R.: Nonlinear Kalman filters and particle filters for integrated navigation of unmanned aerial vehicles. Robot. Auton. Syst. **60**, 978–995 (2012)

Driving Control by Based upon the Slip Pattern of the Ball Robot

Howon Lee, Moonjeon Hwan, Dongju Park, Seon-ho Park, and Jangmyung Lee[(✉)]

Department of Electronics Engineering, Pusan National University, Busan 609-735, Korea
{hollove,jeonghwan1696,djpark1696,seonho1696,jmlee}@pusan.ac.kr

Abstract. This paper proposes a driving course plans using the slip patterns of an Mecanum wheel Ball Robot. Slip of Mecanum wheel Ball Robot causes uncertainty on the driving. This solution is necessary in order to reduce the uncertainty of the robot. Analyzing the slip pattern according on the driving angle change of the robot, which was shown in the graph for slip pattern. On the basis of a formal slip pattern can establish the moving performance of the Mecanum wheel Ball Robot and the optimal path planning. Using a self-produced robots, the experiment on the optimal path planning using a slip pattern and the performance was evaluated.

Keywords: Mecanum wheels · Slip pattern · Inverted ball robot

1 Introduction

Recently, the interest towards the human robot interaction through such service robots as nursing robots and guiding robots has been increased. Also, a number of robot engineers have studied the cooperation between humans and mobile robots or the path planning in the indoor. Such mobile robots are moved in various inner environments. The stably-moving four-wheeled robot is considered to be a general kind of mobile robot. However, in recent years, such two-wheel robots as the Segway and one-wheel robots have been actively studied [1, 2]. The robots with such structures absolutely require rotations when every moving direction is changed. As the radius of each rotation becomes smaller, the level of accuracy for the destination of the robot is degraded by the slip on the ground. Therefore, the biggest possible radius is required for each rotation. Such a point generally brings a lot of limitations for the execution of tasks in a narrow indoor environment. Therefore, the omni-directional robot that does not require any rotation when the moving direction is changed provides a great advantage in the indoor environment.

The omni-directional robot which is to be addressed in this paper is the ball robot shown in Fig. 1. Regarding such a ball robot, the four-wheel structure using a general wheel or a roller and the three-wheel structure using an omni-wheel have been studied [3, 4]. The three-wheel structure shows the arrangement of motors at the angle of 45° along the vertical axis and the horizontal axis is located at the center of the ball. In case of the driving process towards the directions other than those at the angles of 0, 120 and 240°, a considerable amount of slip occurs due to the structural problem [5, 6]. However, the four-wheel structure shows almost no slip which is caused by the structural problem. Notice that the points of action between the wheel and the ball towards the orthogonal

© Springer International Publishing Switzerland 2016
N. Kubota et al. (Eds.): ICIRA 2016, Part I, LNAI 9834, pp. 334–342, 2016.
DOI: 10.1007/978-3-319-43506-0_29

Fig. 1. Types of the omni-directional ball robot

direction of each axis are symmetric when the four wheels are arranged vertical to each other [7, 8]. However, in the driving process towards non-orthogonal direction of each axis, the direction of torque of the motor and the point of action of the ball are different, which causes the occurrence of traction problems and non-linear slip.

In this paper, the patterns for the slip occurring based on the structure of the ball robot are analyzed to drive an optimal way which reduces the loss of torque caused by slip and increases the level of accuracy. Such a method is used to measure and utilize the direction for the occurrence of slip through the differences between the torque-delivering and the driving directions of the robot. In order to verify the functions of the suggested driving-controlling method, the time and the positional accuracy to the destination are compared for three basic driving methods.

2   Structure

An inverse ball driving robot with Mecanum wheels is implemented by installing four Mecanum wheels and by using the ball casters between a main body and a bowling ball as shown in Fig. 2.

The specifications of the robot used in the experiment are as follows. The main body is 25 kg, cylinder of 470 mm tall with a radius of 150 mm. A common bowling ball with the weight of 3.63 kg and the radius of 112 mm is used. The inverse ball drive robot is equipped with one AHRS sensor, one 22.2 V 4,500 mA battery, four 24 V DC motors and one main controller. The main controller which consists of one DSP (TMS320F28335), the Bluetooth network module, four encoders and four DC motor drivers.

Fig. 2. Macanum wheel ball robot

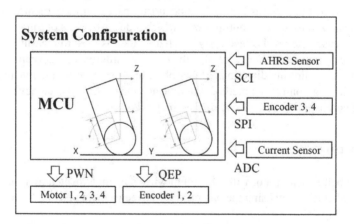

Fig. 3. The configuration of the control system

Figure 3 shows the control board such as system configuration. In the system configuration, it is input SCI, SPI, QEP and ADC value. The output of the control board has a PWM control for the motor.

3   Driving Principles of the Ball Robot

Such a ball robot is basically driven by rolling a ball through the operation of the motors equipped for four wheels. The method of operating the ball with four motors in order to drive the robot towards a direction of each axis is shown in Fig. 4.

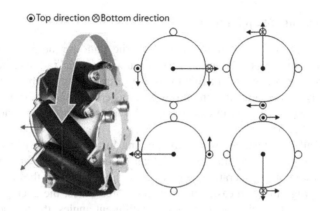

Fig. 4. Driving method for each axis

Also, the ball robot can be regarded as an omni-directional robot which can be driven not only towards the direction of each axis but also towards any direction. The method of operating the motor for the movement towards a particular direction is shown in Fig. 5.

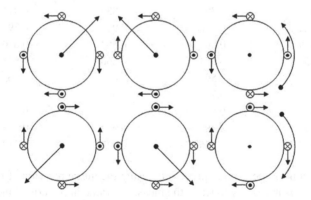

Fig. 5. Operating of the suggested ball robot to various directions

4   Slip Measurements for on the Driving Angle of the Ball Robot

4.1   Definition of the Slip Problems of the Robot

Slip covered in this research is not a slip between the ground and Ball Robot. To the structural problem of the robot is to pattern the problem for the slip coupled that occurs when moving the ball and the driving wheels at a specific angle. In Fig. 7 robot structure, there is seen between the wheel and constant contact pressure of four wheels.

Ball Robot can drive in all directions. So, the change in the wheel slip is also according to the variation of the angle of the direction of the Ball Robot and the direction of driving.

4.2 Measurement of Slip Pattern

It was known that every direction would show the same torque-delivering rate. In order to figure out the occurring slip, the torque-delivering rate can be measured through an actual experiment. The robot was overturned before the ball was installed. Then, by supplying the same amount of power for a certain period of time, the torque-delivering rate was measured in case of the driving process towards each direction with the interval of 5°.

When the ball was operated to move for the same moving distance by modifying the value with the interval of 5°, the torque-delivering rate occurring was measured by changing the value of the operating velocity three times based on the structure of the robot. By carrying out the measurement process three times, the average value was obtained. Based on the values of each axis for different angles, the torque-delivering rate for the measured angle was arranged in Table 1.

Table 1. The measured slip for the same moving distance in each driving direction

Angel	Measured value1	Measured value2	Measured value3	Measured value4
0	1	1	1	1
5	1.004	0.890	0.839	0.911
10	0.722	0.740	0.911	0.791
15	0.679	0.664	0.667	0.670
20	0.650	0.583	0.666	0.633
25	0.759	0.592	0.638	0.663
30	0.737	0.751	0.615	0.701
35	0.743	0.896	0.752	0.797
40	0.897	0.886	0.932	0.905
45	0.883	0.926	0.927	0.912

In order to control the driving process by using the measured value for the torque-delivering rate, it is necessary to show the measured value in patterns. Since the measured value has a symmetrical characteristic based on each axis or the starting point, the average value was shown in patterns by using such a characteristic. The graph in patterns is same as the one shown in Fig. 6. In reality, the data values in patterns are based on the repetition of the range from 0 to $\pi/4$, so the values up to the subject range are saved and the remaining values are shown in patterns based on the symmetry and the reversal of the saved values.

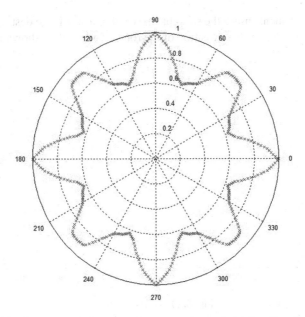

Fig. 6. Torque-delivery rate in patterns for each driving direction

5 Experiment

The experiment was driving in three different cases
    Case 1 is a straight line driving (0° of slip pattern) at the target point.
    Case 2 is a straight line driving (22.5° of slip pattern) at the target point.
    Case 3 is diagonal line driving (45° of slip pattern) and straight line driving (0° of slip pattern) at the parget point.
    It shows a driving method for the case 1~3 in Fig. 7.

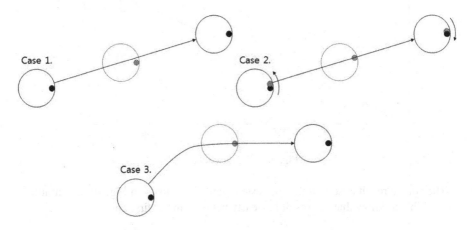

Fig. 7. Execution methods of the experiment

The first experiment shows the straight-line driving towards the destination without any change of the route or the direction of the robot. The result is shown in Fig. 8.

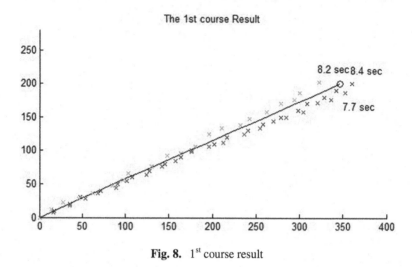

Fig. 8. 1ˢᵗ course result

The second experiment shows the case of backlashing towards the original direction of the robot after the robot is rotated towards the driving direction in order to match the direction of the robot with the driving direction. As shown in Fig. 9, due to the first rotation and the backlashing at the destination, a lot of time is consumed.

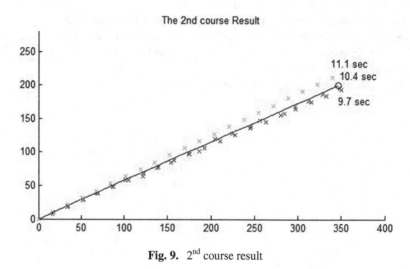

Fig. 9. 2ⁿᵈ course result

The course result based on the suggested method is shown in Fig. 10. By intuition, it is possible to know that the result is clearly better than the first one.

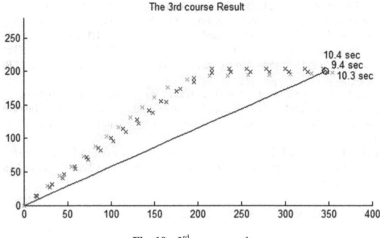

Fig. 10. 3rd course result

6   Conclusion

This paper focused on the possible algorithm with precise driving by utilizing the route with the least number of slip when a ball robot, the omni-directional robot suitable for the execution of tasks in the inner environment, was driven. For the research, the ball robot which uses the mecanum wheel that can minimize the number of structural slip was designed, while the slip patter was experimentally analyzed. Then, by utilizing the measured slip pattern, the algorithm setting the possible precise driving was suggested. By utilizing the route-establishing method suggested through an experiment, it was experimentally confirmed that the ball robot could reach the destination more precisely than those following other route-establishing methods when it was moved to the destination. In the future, the correlation between such a precise route and the one using the least amount of energy will be analyzed and the route-establishing algorithm which optimizes two factors at the same time will be studied.

Acknowledgment.   This research is supported by the MOTIE (Ministry of Trade, Industry & Energy), Korea, under the Industry Convergence Liaison Robotics Creative Graduates Education Program supervised by the KIAT (N0001126).

References

1. Howon, L., Youngkuk, K., Jangmyung, L.: Hill climbing of an inverted pendulum robot using an attitude reference system. Int. J. Robot. Autom. **27**(3), 255–262 (2012)
2. HongZhe, J., JongMyung, H., JangMyung, L.: A balancing control strategy for a one wheel pendulum robot based on dynamic model decomposition: simulations and experiments. IEEE/ASME Trans. Mechatron. **16**(4), 763–768 (2011)
3. Endo, T., Nakamura, Y.: An omnidirectional vehicle on a basketball. In: Proceedings of the ICAR 2005, pp. 573–578, July 2005

4. Lauwers, T.B., Kantor, G.A., Hollis, R.L.: A dynamically stable single-wheeled mobile robot with inverse mouse-ball drive. In: Proceedings of the ICRA 2006, pp. 2884–2889, May 2006
5. Fankhauser, P., Gwerder, C.: Modeling and Control of a Ballbot. ETH Zurich (2010)
6. Hertig, L., Schindler, D., Bloesch, M., Remy, C.D., Siegwart, R.: Unified state estimation for a ballbot. In: 2013 IEEE International Conference on Robotics and Automation (ICRA), pp. 2471–2476, May 2013
7. Nagarajan, U., Mampetta, A., Kantor, G.A., Hollis, R.L.: State transition, balancing, station keeping, and yaw control for a dynamically stable single spherical wheel mobile robot. In: Proceedings of the IEEE International Conference on Robotics and Automation, pp. 998–1003 (2009)
8. Liao, C.W., Tsai, C.C., Li, Y.Y., Chan, C.-K.: Dynamic modeling and sliding-mode control of a ball robot with inverse mouse-ball drive. In: Proceedings of SICE 2008, Tokyo, Japan, pp. 2951–2955, August 2008

Straight Driving Improvement of Mobile Robot Using Fuzzy/Current Sensor

Ha-Neul Yoon$^{(\boxtimes)}$, Dong-Eon Kim$^{(\boxtimes)}$, Byeong-Chan Choi$^{(\boxtimes)}$, Min-Chul Lee$^{(\boxtimes)}$, and Jang-Myung Lee$^{(\boxtimes)}$

Department of Electronic Engineering, Pusan National University, Jangjeon-Dong, Geumjeong-Gu, Busan, South Korea
{haneul1696,dongeon1696,byeongchan1696, mclee,jmlee}@pusan.ac.kr

Abstract. In this paper, the research was done focus on the driving efficiency improvement of mobile robot in outdoor environments. The slip is occurred during driving because dynamic characteristic of mobile robot and external environmental factors have an effect on the driving efficiency. For reducing the slip, researches have been done such as Optimal Slip Ratio Control and Model Following Control. But, reducing a slip has many difficulties such as disturbance, cumulative error of sensor, measurement imprecision. So, this paper proposed a robust ASS (Anti-Slip System) focus on the outdoor mobile robot. For reducing a slip, current sensor and encoder was used because current sensing and encoder has not cumulative error. Using the current sensing, designed the FSC (Fuzzy Slip Control), to complete the ASS (Anti Slip System) by combining PI. To demonstrate the control performance, real experiments are performed using the mobile robot in outdoor.

Keywords: Slip control · Straight driving performance · PID · Fuzzy · Current sensor · Mobile robot

1 Introduction

Mobile robot has been widely used in various fields such as services, military and industrial. In order to control smoothly mobile robot according to the user's purpose, the precise technique to control the position and movement of the robot is needed. In this regard, there are many progressing studies at domestic and abroad about driving efficiency. In this paper, we conducted a study focusing on the improving driving efficiency of a mobile robot in an outdoor environment.

Driving efficiency of the mobile robot is strongly influenced by robot's dynamic characteristics and external environmental factors [1, 2]. The outdoor environment is different from indoor flat environment. Also, it includes uncertainty such a sloping terrain, Irregular surface. The slip is occurred because of changing a friction coefficient in irregular surface. Especially, primary factor changing a driving efficiency is friction-slip characteristic between robot's wheel and floor surface. So, driving safety is may be decreased because of unplanned wheel's slip. If the slip is serious, the robot may be break away from the planned path and becomes uncontrolled situation.

© Springer International Publishing Switzerland 2016
N. Kubota et al. (Eds.): ICIRA 2016, Part I, LNAI 9834, pp. 343–350, 2016.
DOI: 10.1007/978-3-319-43506-0_30

For reducing a slip, there are many researches like Optimal Slip Ratio Control that maintains the optimal slip ratio and Model Following Control that controls the wheel slip based on the identified information by adaptive observer, driving efficiency improvement by using sensors [3–6]. However, most of the studies are confined to the actual car and there are many difficulty to analyze the mobile robot's dynamics because of disturbance (=wind, road surface condition, etc.) and internal factors (=tire's dynamics, friction factor, etc.). Also it`s hard to obtain accurate mobile robot speed due to inaccuracy of the sensor measurements. When the error of the mobile robot speed is increased, the control performance is fall in unstable situation.

In this paper, we suppose the current sensing and encoder that have no cumulative error, slip control algorithm based on non-model Fuzzy control focused on a mobile robot [7–10]. Proposed slip control is robust algorithm to cumulative error and it is easy to apply real-time systems. Also, it is strong on modeling error to use non-model based Fuzzy control. In the Sect. 2, we analyze the current characteristic according to road surface condition using current sensor. In the Sect. 3, we describe the method about design the Fuzzy logic based on the current characteristic. And in the Sect. 4, we investigate the result of the slip control algorithm through an experiment, and finally we derive the conclusion of the study in the Sect. 5.

2   Current Characteristic According to the Road's Surface Condition

Above all, analysis of the road's surface condition should be done for improving driving efficiency. It is need to compensate breakaway from the planned path by the slip that made by ground's friction change and irregular road condition. For estimating current characteristic according to road condition, the system is made using a PI controller (Fig. 1).

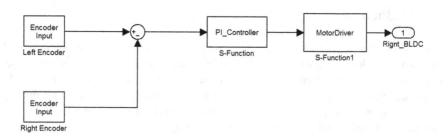

Fig. 1. PI controller for current characteristic experiment

First, the desired velocity was set on the left wheel for experiment. And then, PI controller controls right wheel's velocity to compensate velocity error between left wheel and right wheel. Current characteristic experiments were conducted to establish the Test-bed by using an artificial turf in Fig. 2.

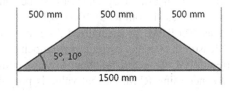

Fig. 2. Slope structure of test-bed

When the mobile robot drives the slope in the test-bed, the electric currents are flowing through both sides motor. So, interrelationship of Slip with electric current can be estimated by measuring electric currents at both sides motor. Left and right motor's

Table 1. Average current per velocity

RPM	Velocity (m/s)	Average current (mA)
1000	0.1170	150
1500	0.1675	211
2000	0.2233	270
2500	0.2792	325
3000	0.3350	378
3500	0.3908	427

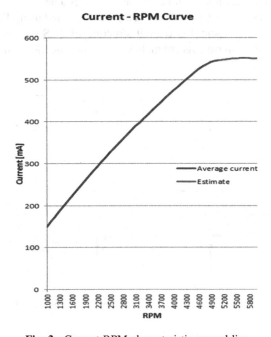

Fig. 3. Current-RPM characteristic curved line

current were measured by changing slope range from 5° and 10°. Driving velocity range is increased from RPM 1000 (0.11 m/s) to RPM 5000 (0.56 m/s) by incremental value as 500 RPM for specialized time interval.

Measured average current per velocity are arranged in Table 1 and Fig. 3 is Current-RPM characteristic about average current.

At increasing RPM, current value is proportionate to the ground's friction. But at more than 4800 RPM, current increasing ratio is very small. The friction is increased according to increasing actuating force. When driving in velocity higher than the ground's friction, slip was occurred because friction isn't increased. When robot is driving in more than 4800 RPM, it is not assured the ground's friction. So, ASS (Anti-Slip System) is designed to drive up to 4600 RPM.

3 Robust Anti-Slip System

In this paper, the difference between the actual driving distance and the desired driving distance of mobile robot was defined as a slip. When starting or running or stopping the mobile robot, a slip occurs because the road surface friction is reduced. In the research on the slip compensation, typically there are an optimal slip rate control method for limiting the torque of the motor using the sensor measurements and MFC (Model following control). MFC has a sensitive characteristic about changing the status of the mobile robot because robot's dynamics are seriously affected by changing external and internal factors. Also, it has difficulty to obtain an accurate mobile robot's speed due to the cumulative error and inaccuracy of measurement sensors. If error of the mobile robot's speed is increased, mobile robot's control is unstable.

In this paper, robust ASS (Anti-slip system) was designed using Fuzzy control and current sensor, encoder. Figure 4 is overall structure of ASS proposed in this paper. ASS is a robust Slip control algorithm based on the encoder and current sensing

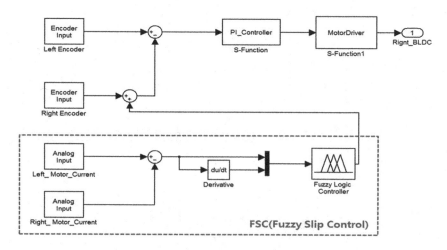

Fig. 4. Overall structure of ASS (Anti-Slip System)

without accumulated error. ASS is composed of a PI controller and Fuzzy Slip Control. PI controller has a function to maintain the target speed and correct the speed difference between the left and right motor. Fuzzy Slip Control reduces the slip generated from the irregular ground using measured current information.

Equation 1 is input of the Fuzzy control, and Eq. 2 is output of the Fuzzy control. Input variables of Fuzzy Slip Control are defined as two inputs about the current difference between left and right wheel, a differential of error. In Eq. 1, I_L and I_R is a current value of the left and right wheel.

$$e(t) = I_L(t) - I_R(t), \quad \frac{de(t)}{dt} = \frac{e(t) - e(t-1)}{T} \tag{1}$$

$$du(t) = e(t) \times \frac{de(t)}{dt} \tag{2}$$

Figure 5 shows the Membership function. It was designed using triangular form and the weight was set from 0 to 9.

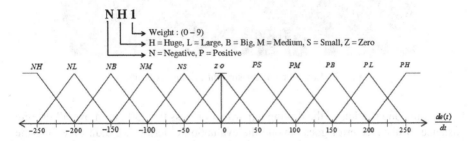

Fig. 5. Membership function

		\multicolumn{11}{c}{Error}
		NH	NL	NB	NM	NS	ZO	PS	PM	PB	PL	PH
Error variation	NH	NH0	NL5	NB5	NM5	NS5	Z00	Z05	PS5	PM5	PB5	PL5
	NL	NH0	NL4	NB4	NM4	NS4	Z00	Z06	PS6	PM6	PB6	PL6
	NB	NH0	NL3	NB3	NM3	NS3	Z00	Z07	PS7	PM7	PB7	PL7
	NM	NH0	NL2	NB2	NM2	NS2	Z00	Z08	PS8	PM8	PB8	PL8
	NS	NH0	NL1	NB1	NM1	NS1	Z00	Z09	PS9	PM9	PB9	PL9
	ZO	NH0	NL0	NB0	NM0	NS0	Z00	PS0	PM0	PB0	PL0	PH0
	PS	NL9	NB9	NM9	NS9	Z09	Z00	PS1	PM1	PB1	PL1	PH0
	PM	NL8	NB8	NM8	NS8	Z08	Z00	PS2	PM2	PB2	PL2	PH0
	PB	NL7	NB7	NM7	NS7	Z07	Z00	PS3	PM3	PB3	PL3	PH0
	PL	NL6	NB6	NM6	NS6	Z06	Z00	PS4	PM4	PB4	PL4	PH0
	PH	NL5	NB5	NM5	NS5	Z05	Z00	PS5	PM5	PB5	PL5	PH0

Fig. 6. Fuzzy rule table

Fuzzy inference was used by Min-Max method and the de-fuzzification method was used by Mamdani COG (Center Of Gravity). The rule table used Fuzzy Slip Control are summarized in the Fig. 6.

4 Experimental Result

Driving experiment was performed in front of grass field in Pusan National University. The overall driving distance is 20 m. We measured distance error between actual driving distance and the desired driving distance each 4 m (Fig. 7).

Fig. 7. Environment for straight driving experiment

Evaluation criterion of the driving experiment is less than 50 cm in overall driving distance. If distance error goes beyond the evaluation criterion, we didn't measure the experiment result. The experiment was conducted Open loop 6 times, PI 6 times, ASS 12 times. Table 2 shows distance error at the goal line.

Table 2. Distance error on the ground

No	Open loop control	PI control	ASS control
1	31 cm	30 cm	6.5 cm
2	33 cm	50 cm	−17.5 cm
3	−48 cm	−42 cm	10.5 cm
4	87 cm	−28 cm	17.5 cm
5	−52 cm	32 cm	6 cm
6	52 cm	38 cm	15 cm
Average	50.5 cm	36.6 cm	12.2 cm

Open-loop control's average error is 50.5 cm. So it was regarded as unstable control. When using only PI control, it shows poor stability because of the second, third result. The ASS control shows a distance error of up to 21.5 cm. Also it was measured only one time during 12 times experiments. It is show more high stability than the Open-Loop and PI control.

5 Conclusion

In this paper, ASS was designed by using an FSC and PI controller. FSC is a robust Slip control algorithm using a Fuzzy based on non-model, current sensing without accumulated error. FSC was designed using a current sensor, and then ASS was developed by combining PI with FSC. To demonstrate the control performance, real experiments are performed using the mobile robot in outdoor. The ASS was reduced by 60 % compared with Open-Loop and PI control. So, ASS is robust algorithm compare to the Open-Loop and PI control. Slip control technique has been doing much research in electric vehicle area. So, ASS is very useful algorithm on vehicle industry based on the electric motor. Also, the current sensing skill can be utilized in diagnosing fault using a variation in the load. Therefore, it can be widely used in vehicle and machinery industrial, etc.

In further research, we're planning to conduct experiment according to s-curve and circular motion and design stable driving control of mobile robot on the Slope. Also, we will apply Adaptive control to the mobile robot for driving improvement.

Acknowledgments. This research is supported by the MOTIE (Ministry of Trade, Industry & Energy), Korea, under the Industry Convergence Liaison Robotics Creative Graduates Education Program supervised by the KIAT (N0001126).

References

1. Bekker, M.G.: Introduction to Terrain-Vehicle Systems. University of Michigan Press, MI (1969)
2. Wong, J.Y.: Theory of Ground Vehicle. Wiley, New York (1976)
3. Lee, T.Y.: A study on electric vehicle slip control using model following control. Korean Soc. Mech. Technol. **12**(4), 33–39 (2010)
4. Kim, I.T., Kazuki, N., Hong, S.K.: Odometry error compensation for mobile robot navigation using gyroscope. Korean Inst. Electr. Eng. **2004**(7), 2206–2208 (2004)
5. Shin, H.S., Hong, S.K., Chwa, D.K.: A study in movement of wheeled mobile robot via sensor fusion. Korean Inst. Electr. Eng. **2005**(10), 584–586 (1997)
6. Song, J.B., Hong, D.W.: HWILS implementation of slip control based on TCS engine control approach. Korean Soc. Mech. Eng. **2**(1), 852–857 (1996)
7. Yeo, H.J., Sung, M.H.: Fuzzy control for the obstacle avoidance of remote control mobile robot. Inst. Electron. Eng. Korea – Syst. Control **48**(1), 47–54 (2011)
8. Lee, J.H., Lee, W.C.: Position calibration and navigation of mobile robot using inertial sensor and fuzzy rules. J. Korean Inst. Inf. Technol. **11**(5), 23–31 (2013)
9. Shin, J.H.: Robust adaptive fuzzy backstepping control for trajectory tracking of an electrically driven nonholonomic mobile robot with uncertainties. J. Inst. Control, Robot. Syst. **18**(10), 902–911 (2012)

10. Choi, B.J., Kim, S.: Design of simple-structured fuzzy logic system based driving controller for mobile robot. Korean Inst. Intell. Syst. **22**(1), 1–6 (2012)
11. Jung, S.W., Jo, Y.G., Lee, W.S., Son, J.W., Han, S.H.: Conference of KSMTE the Korean Society of Manufacturing Technology Engineers, p. 138, April 2012
12. Park, K.W.: Fuzzy logic based auto navigation system using dual rule evaluation structure for improving driving ability of a mobile robot. J. KOREA Multimedia Soc. **18**(3), 387–400 (2015)
13. Bae, K.H., Choi, Y.K.: A formation control scheme for mobile robots using a fuzzy compensated PID controller. Korea Inst. Inf. Commun. Eng. **19**(1), 26–34 (2015)
14. Kim, H.J., Kang, G.T., Lee, W.C.: Indoor location estimation and navigation of mobile robots based on wireless sensor network and fuzzy modeling. Korean Inst. Intell. Syst. **18**(2), 163–168 (2015)
15. Park, J.H., Baek, S.J., Chong, K.T.: Design of adaptive fuzzy controller to overcome a slope of a mobile robot for driving. J. Korea Acad.-Ind. Cooperation Soc. **13**(12), 6034–6039 (2012)

Efficiency of Dynamic Local Area Strategy for Frontier-Based Exploration in Indoor Environments

Serkan Akagunduz[1(✉)], Nuri Ozalp[1], and Sirma Yavuz[2]

[1] Informatics and Information Security Research Center, The Scientific and Technological Research Council of Turkey, Gebze, Kocaeli, Turkey
{serkan.akagunduz,nuri.ozalp}@tubitak.gov.tr
[2] Department of Computer Engineering, Yildiz Technical University, Itanbul, Turkey
sirma@ce.yildiz.edu.tr
http://www.bilgem.tubitak.gov.tr

Abstract. Exploration in unknown environments is a fundamental problem for autonomous robotic systems. The most of the existing exploration algorithms are proposed for indoor environments and aim to minimize the overall exploration time and total travelled distance. In this paper, a modified version of frontier-based exploration approach is presented to decrease exploration time and total distance. This approach introduces two more parameters to the Exploration Transform (ET) algorithm, which evaluates all detected frontiers to select next target point. On the other hand, the proposed approach considers locally observed frontiers, which might be changed with the parameter of dynamic distance. The proposed algorithm is individually tested and compared to ET algorithm with five random starting points in three different environments. Experimental results show that the proposed algorithm provides superior performance over conventional ET algorithm.

Keywords: Frontier-based · Exploration · Dynamic distance

1 Introduction

Exploration and mapping are necessary operations for autonomous mobile robots navigating in unknown indoor environments. Traditional exploration algorithms are based upon the concept of frontier that indicates a boundary region between the known (mapped) and unknown (unexplored) part of an environment. The exploration problem within this framework can be defined by choosing the optimum way to extend the map and to plan a path from the current robot position to the selected frontier by avoiding obstacles. In general, frontier-based exploration algorithms work efficiently on grid maps. These maps distinguish explored and unexplored area unlike geometric feature maps [1, 2].

The crucial problems in search and rescue are how to access the victim as soon as possible and how to use the robots energy in an efficient manner during

© Springer International Publishing Switzerland 2016
N. Kubota et al. (Eds.): ICIRA 2016, Part I, LNAI 9834, pp. 351–361, 2016.
DOI: 10.1007/978-3-319-43506-0_31

exploration. Exploration algorithms used in search and rescue environments are of great importance in the searching strategy. These algorithms concern many different strategies in terms of navigation of mobile robots. The purpose of these strategies is to determine the safest and shortest way simultaneously during exploration and discovery. The advantages and disadvantages of these strategies are discussed in [3,4].

Frontier-based exploration strategies include the approaches of the nearest frontier, the safest frontier, and the farthest frontier [5]. These approaches consider the cost minimization and the gain maximization problem. The cost is defined as the total distance and time spent to reach a frontier, and the gain is defined as the expected coverage by reaching this frontier [6–8].

In this paper, a modified Exploration Transform (ET) algorithm is proposed in order to provide an alternative solution approach to the exploration problem for Unmanned Ground Vehicles (UGVs). This proposed algorithm is called as Dynamic Distance Exploration (DDE) and integrates two new parameters, which are dynamic distance and beta, to ET algorithm. Dynamic distance parameter determines the distance from robot to frontiers and the distance between destination and robot. Beta parameter is used to select appropriate path for UGV. The goal of the modified ET algorithm is to minimize the total exploration time and travelled distance of UGV.

The rest of the paper is structured as follows. In the following section, related works focusing on the frontier-based exploration problem are given. The proposed algorithm, the problem formulation and contributions are introduced in Sect. 3. The performance of the different scenarios is compared, and the results are presented in Sect. 4. Finally, the conclusion is given in the last section.

2   Related Work

Exploring an unknown environment with single autonomous mobile robot requires an efficient exploration strategy to minimize the total duration of exploration and total travelled distance by the robot during exploration. Recent studies have been focused on the solution of exploration problem using frontier selection methods such as the nearest frontier, the safest frontier and the farthest frontier strategies [9,10]. In [6], Amna, Lakmal and Jorge presented a frontier selection strategy. This strategy considers minimizing total travelled distance and maximizing the map coverage of the environment during exploration. It maps the environment geometrically and stores the map using Dynamic Triangulation Tree structure (DTT). The geometrical information embedded in the DTT is used in the frontier selection step to reduce the exploration time by combining both the cost and gain values. Their experimental results are limited and have not been compared to any well-known exploration strategies.

Gonzalez-Banos and Latombe proposed an algorithm that selects candidate goal points among all frontiers which help to extend the map optimally and to provide safer navigation [11]. This algorithm aims to determine convenient frontiers where can be obtained the optimal sensor data. For each candidate goal point, visible area and maximum range is calculated to enhance of coverage area. However this approach does not consider total exploration time and total travelled distance during exploration.

Kim and co-authors proposed another frontier-based exploration approach called Sensor-based Random Tree [1]. An improvement on it is the frontier-based SRT (FB-SRT) method that provides a robot exploring more efficiently by limiting the randomized generation of position to the boundary between explored and unknown regions called the frontier. But if there are no frontier areas during exploration, a robot has to go back to previous positions until finding a new frontier. In some cases, it can be a long detour until the robot reaches a position which has unexplored frontier.

2.1 Exploration Transformation Algorithm

Wirth and Pellenz proposed a frontier-based exploration method to get new information about environment and search for victims [3]. They use hector-slam to obtain a map. With the algorithm that they suggest in order to extend the exploration area, they consider the safe of the path instead of distance among frontiers. If a frontier among all frontiers is closer but not safe for path planning, this frontier is not chosen for a goal. On the other hand if an another frontier is farther but it is safer compared with all other frontier, it is chosen to path planning as a goal. However, this approach might cause an increase in overall exploration time and the total travelled distance by the robot during exploration. Therefore, our exploration methodology outperforms this method in the manner of total exploration time and travelled distance. This approach finds solutions to the exploration problem by combining the Yamauchi's exploration and Zelinsky's path transformation methods in [3]. Yamauchi's frontier-based exploration method proposes an approach to find the next exploration goal depending on the closest target. Occupancy-Grid map is used as an input data. According to this algorithm, If more than one potential goal is detected in map, then the closest goal is selected [12]. Zelinsky's path transformation considers path problem on occupancy-grid map using distance transformation and obstacle transformation. The distance transformation calculates the cost for each free cell in order to arrive the goal cell. They calculate the cost between current position of robot and goal position. The obstacle transformation, Ω, figure out for each cell distance to the nearest obstacle [13,14].

The path transformation (Φ) of any starting cell c to reach the target frontier cell c_g is defined as follows:

$$\Phi[c, c_g] = \min_{c \in X_c^{c_g}} [l(C) + \alpha \sum_{c_i \in C} C_{danger}(c_i)] \tag{1}$$

In formula (1), $X_c^{c_g}$ is the cluster of entire likely paths from c to c_g, $l(C)$ is the length of the trajectory path C, $C_{danger}(c_i)$ is the cost function for "discomfort" of the entering cell c_i, and α is a weighting factor ≥ 0.

The path $l(C)$ is between two cell can be calculated as follows:

$$l(C) = l(c_0, \ldots\ldots c_n) = \sum_{i=0}^{n-1} d(c_i, c_{i+1}) \qquad (2)$$

In Eq. (2) $d(c_i, c_{i+1})$ indicates the distance between subsequent cells in the path. The cost function for the entering cell (c_{danger}) of a cell can be calculated via obstacle transformation of this cell [13,14] as provided in (3).

$$C_{danger}(c) = \begin{cases} \infty, & \text{if } d < d_{min} \\ (d_{opt} - d)^2, & else \end{cases} \qquad (3)$$

In the case statement (3), robot is forced to travel a certain distance from the obstacles. Because if robot stays away from any obstacle, it can not detect any of them due to the limitation of sensors. Moreover, robot should see obstacles to obtain new information and to determine location itself. So the distance between robot and obstacles has to be optimum at the distance d_{opt} to get new information and should not be closer than d_{min} to avoid obstacles.

Transformation of an elegant solution can be obtained for the exploration problem: The path transformation should consider not only the cost of a path to a goal frontier, but also the cost of a path to a close frontier. The path may not have to be the shortest, and the frontier does not have to be the closest, because the path cost is determined with the distance and the safe metrics [3]. Exploration Transform formula is given in (4). According to this formula, F represents all frontier cells which are observed in all global map. Algorithm considers these frontier cells to generate a safe path by selecting a goal frontier.

$$\Psi(c) = \min_{c_g \in F} \left(\min_{C \in X_c{}^{c_g}} [l(C) + \alpha \sum_{c_i \in C} C_{danger}(c_i)] \right) \qquad (4)$$

3   Dynamic Distance Exploration

According to our exploration method, detected frontiers at dynamic distance area are continually evaluated in terms of cost and gain during the robot's operation. The cost of a given frontier is simply the required travel distance to reach it while gain is the expected information gain if this frontier was selected as the next candidate position [6]. Consideration of all the frontiers from the beginning of exploration could increase both the complexity of the algorithm and the calculation time. The problem with evaluating all frontiers identified to address a limited space instead of closer to the current location of the robot has been shown to produce more effective results of the evaluation of the detected frontiers. Determined frontiers according to a distance parameter can be dynamically changed. Dynamic distance parameter is reduced when a certain number of the frontiers are found, otherwise it is increased when a certain number of frontiers are not found. Thus, the assessment of appropriate frontiers for the robot can be done more quickly and effectively.

3.1   Modified Exploration Transform Based on Dynamic Distance

In this study, we propose two parameters to conventional exploration transform algorithm in order to reduce the total exploration time together with the travelled distance. In this context, we have modified the exploration transformation algorithm, introduced by Wirth and Pellenz, by choosing the next frontier from a Bounding Area (BA) instead of choosing from the list of all available frontiers. The size of the bounding area is determined dynamically by looking at the number of frontiers in this region. BA is changed with a parameter called dynamic distance (DD). If there are not enough frontiers in the BA, the size of bounding area is increased, otherwise it is decreased. Using the DD, the robot selects the most appropriate frontier close to itself and could create a safe path within the bounding area. We have named the proposed exploration methodology as Dynamic Distance Exploration (DDE) algorithm.

$$DD_{next} = \begin{cases} (DD_{pre} + i_{rate}) & \text{if } LF \leq f_{min} \\ (DD_{pre} - d_{rate}) & else \end{cases} \tag{5}$$

DD parameter is described in case statement (5). DD_{next} is the next dynamic distance, LF represents local frontiers in bounding area, DD_{pre} stands for previous dynamic distance, i_{rate} is the rate of increase of dynamic distance, d_{rate} is the rate of decrease of dynamic distance, f_{min} is the minimum frontier number in bounding area.

Algorithm 1. Pseudo-code of DDE Algorithm

Data: frontiers
Result: trajectory path
1 initialization of DD, rate of increase, rate of decrease, minimum number of frontiers and β parameters;
2 **while** *DD value is less than a predifined threshold value* **do**
3   find frontiers in BA depending on DD;
4   **if** *there are not enough frontiers to find goal* **then**
5   increase DD with a rate of increase;
6   explore for new frontiers;
7   select a frontier as a goal in BA;
8   calculate a safe path to be followed
9   **else**
10   decrease DD with a rate of decrease;
11   select a frontier as a goal in BA;
12   calculate a safe path to be followed
13   **end**
14 **end**

According to pseudo-code of DDE Algorithm, DD, rate of increase, rate of decrease, minimum number of frontiers and beta parameter values are initiated. The initial values of the DD, increase rate, decrease rate and minimum number of frontiers can vary depending on the complexity and size of the test environment.

$$\Psi(c) = \min_{c_g \in LF} \left(\min_{C \in X_c{}^{c_g}} [l(C) * \beta + \alpha \sum_{c_i \in C} C_{danger}(c_i)] \right) \qquad (6)$$

Formula of DDE Algorithm is given in (6). According to this formula, LF represents the set of local frontiers in bounding area. If there are not enough determined frontiers in BA, DD is increased with rate of increase. Minimum number of frontiers is a predifined threshold value in BA. β is weight factor of trajectory working better if β is $\geq$ two, $C_{danger}(c_i)$ the cost function for the discomfort of entering cell c_i, and α a weighting factor≥ 0. DDE Algorithm considers local frontiers in BA to generate a safe path by selecting a goal frontier.

4   Simulation Results

As shown in Fig. 1, ET algorithm selects a frontier among all frontiers in the explored area. Additionally this algorithm does not look at local optimum. It can be inferred from the figure, ET methodology might select a frontier which is located far away. It causes an increase in the total time and total travelling distance. On the other hand, as shown in Fig. 2, DDE algorithm selects local frontiers within the BA which is determined by DD. This figure shows that DDE algorithm looks at local optimum using DD parameter. It considers local frontiers in order to plan a path. With this approach, it is shown that the total exploration time and the total travelled distance are reduced.

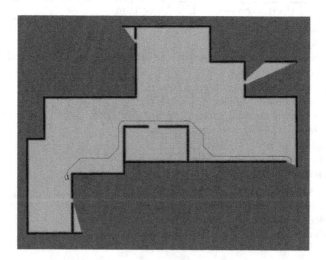

Fig. 1. Illustration of frontier selection for ET algorithm. In this figure the robot is modelled yellow square. Red lines inside the rooms display frontiers. Red line between the furthest frontier and robot is represent path generated by ET algorithm. (Color figure online)

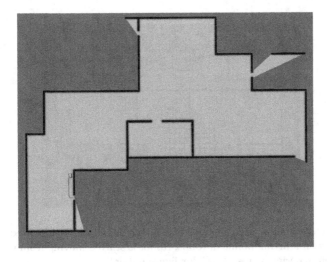

Fig. 2. Illustration of frontier selection for *DDE* algorithm. Red line between the nearest frontier and robot is represent path generated by *DDE* algorithm. (Color figure online)

To illustrate the capability of the proposed approach about the frontier-based exploration problem for autonomous mobile robot, simulations are performed for three different indoor environments. *DDE* algorithm is tested under dynamic local distances which has initial values as 20, 25, 30. Additionally, tests are performed in the same three indoor environment using *ET* algorithm. Algorithms are executed from five different starting point in each environment. Firstly *DDE* algorithm is run four times for all dynamic distance values from each starting point. Then *ET* algorithm is run four times from same starting points. Performance results for *ET* and *DDE* algorithms are obtained by simulating with these scenarios four times.

To compare the performance of the proposed *DDE* and *ET* algorithms, the total travelled distance by the UGV throughout the exploration of all indoor environment, the amount of covered area that completed in time and the total time to exploring whole environment were collected for each environment. The proposed *DDE* algorithm is run with three different dynamic distance (*DD*) parameter settings which controls the application performance rate of the proposed algorithm.

First, second and third environments are shown in Figs. 3, 4 and 5 respectively. UGV started moving from five different starting points for each individual test. *DDE* algorithm is run using dynamic distance which are 20,25,30 for each starting point. For all environments, initial values are given as $\alpha = 0.5$, $d_{opt} = 20$, $d_{min} = 0$, $increaserate = 3$, $decreaserate = 3$. Accoording to these values, *ET* and *DDE* algorithm results are seen in Tables 1, 2 and 3. As shown in Tables 1, 2 and 3 total travelled distance and total duration of exploration in DDE algorithm is significantly smaller than ET algorithm.

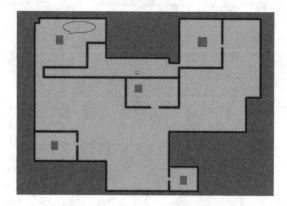

Fig. 3. First environment used in tests of *ET* and *DDE* algorithm. Red squares are represent different starting points of robot. Robot is started moving from these starting points for each experiment result. (Color figure online)

Table 1. Simulation results for first environment

		Total travelled distance (m)	Total duration of exploration (s)
DDE	DD = 20	308.936	447.2
	DD = 25	324.652	480.95
	DD = 30	335.148	490.6
ET		350.802	509.35

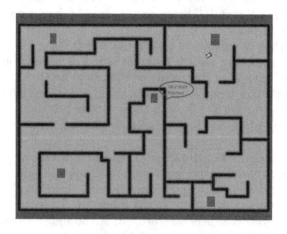

Fig. 4. Second environment used in tests of *ET* and *DDE* algorithm. Red squares are represent different starting points of robot. Robot is started moving from these starting points for each experiment result. (Color figure online)

Table 2. Simulation results for second environment

		Total travelled distance (m)	Total duration of exploration (s)
DDE	DD = 20	504,4	747,4
	DD = 25	472,05	695,18
	DD = 30	482,65	716,65
ET		524,16	765,825

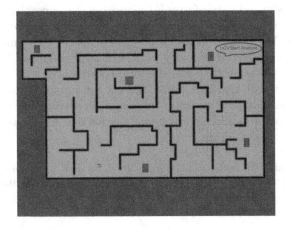

Fig. 5. Third environment used in tests of *ET* and *DDE* algorithm. Red squares are represent different starting points of robot. Robot is started moving from these starting points for each experiment result. (Color figure online)

Table 3. Simulation results for third environment

		Total travelled distance (m)	Total duration of exploration (s)
DDE	DD = 20	466,815	709,4
	DD = 25	477,231	705,9
	DD = 30	476,01	718,6
ET		494,364	747,25

According to simulation results, it can be clearly inferred that *DDE* algorithm outperforms *ET* algorithm. Additionally, efficiency of the algorithm is tested under a variety of dynamic distance values and, at the end, it was observed that the proposed algorithm performs well for $DD = 20$, $DD = 25$ and $DD = 30$ in the manner of total travelled distance, total exploration time and the total covered area. Performance illustrations for total travelled distance and total exploration time are given in Tables 1, 2 and 3 respectively.

5 Conclusion

In this paper a modified frontier-based exploration algorithm for autonomous mobile robots has been presented. The traditional algorithms have limited performance capacity in terms of cost and gain. We studied different search strategies based on dynamic distance local area parameter on Exploration Transform (ET) algorithm.

The proposed method, Dynamic Distance Exploration (DDE) Algorithm, is seen to be successful and outperforms ET algorithm. To assess the effectiveness of the proposed DDE algorithm, we have performed simulations for three different environments with five random initial position of UGV. The simulation results demonstrated that DDE methodology works well in the manner of total travelled distance, total exploration time and the total covered area.

As a future study, it will be better to test the proposed exploration algorithm under more complex environment cases. Moreover, investigation of the achievement of the proposed algorithm among different exploration methodologies would be a valuable study.

Acknowledgment. Research is supported by The Scientific and Technological Research Council of Turkey (TUBITAK BILGEM), conducted within the UAVs-Research Lab- project (project number 3920-S513000), which is part of the Avionics and Air Defense Systems research program.

References

1. Kim, J., Seong, K.J., Kim, H.J.: An efficient backtracking strategy for frontier method in sensor-based random tree. In: 2012 12th International Conference on Control, Automation and Systems. ICC, Jeju Island, Korea, 17–21 October 2012
2. Amigoni, F., Behnke, S.: Experimental evaluation of some exploration strategies for mobile robots. In: IEEE International Conference on Robotics and Automation, Pasadena, CA, USA, 19–23 May 2008
3. Wirth, S., Pellenz, J.: Exploration transform: a stable exploring algorithm for robots in rescue environments. In: IEEE International Workshop on Safety, Security and Rescue Robotics, SSRR 2007, Rome, Italy, September 2007
4. Holz, D., Basilico, N., Amigoni, F., Behnke, S.: A comparative evaluation of exploration strategies and heuristics to improve them. In: Proceedings of European Conference on Mobile Robotics(ECMR), Oerebro, Sweden (2011)
5. Basilico, N., Amigoni, F.: Exploration strategies based on multi-criteria decision making for searching environments in rescue operations. Auton. Robots **31**(4), 401–417 (2011)
6. AlDahak, A., Seneviratne, L., Dias, J.: Frontier-based exploration for unknown environments using incremental triangulation. In: 2013 IEEE International Symposium on Safety, Security, and Rescue Robotics (SSRR). ICC, Linkoping, Sweden, 21–26 October 2013
7. Wettach, J., Berns, K.: Combining dynamic frontier based and ground plan based ex-ploration: a hybrid approach. In: 41st International Symposium on Robotics ISR/Robotik 2014, Munich, Germany, 2–3 June 2014

8. Holz, D., Basilico, N., Amigoni, F., Behnke, S.: Evaluating the efficiency of frontier-based exploration strategies. In: Proceedings of Joint 41th International Symposium on Robotics and 6th German Conference on Robotics, Munich, June 2010

9. Wettach, J., Berns, K.: Dynamic frontier based exploration with a mobile indoor robot. In: 41st International Symposium on Robotics (ISR), Munich, Germany, 7–9 June 2010

10. Meger, D., Rekleitis, I., Dudek, G.: Heuristic search planning to reduce exploration uncertainty. In: IEEE/RSJ International Conference on Intelligent Robots and Systems (IROS), Nice, France, 22–26 September 2008

11. Gonzales-Banos, H.H., Latombe, J.-C.: Navigation strategies for exploring indoor environments. Int. J. Robot. Res. **21**(10–11), 829–848 (2002)

12. Yamauchi, B.: A frontier-based approach for autonomous exploration. In: 1997 IEEE International Symposium on Computational Intelligence in Robotics and Automation, Monterey, CA, p. 146, 10–11 June 1997

13. Zelinsky, A.: Using path transforms to guide the search for findpath in 2D. I. J. Robotic Res. **13**(4), 315–325 (1994)

14. Zelinsky, A.: Environment exploration and path planning algorithms for a mobile robot using sonar. Ph.D. thesis. Wollongong University, Australia (1991)

Optimization of a Proportional-Summation-Difference Controller for a Line-Tracing Robot Using Bacterial Memetic Algorithm

Brandon Zahn[1,2(✉)], Ivan Ucherdzhiev[1,3], Julia Szeles[1], Janos Botzheim[1], and Naoyuki Kubota[1]

[1] Graduate School of System Design, Tokyo Metropolitan University, Hachioji, Japan
`brandonktm11@gmail.com`, `ivan_ucherdzhiev@smartcom.bg`,
`{perecka,botzheim,kubota}@tmu.ac.jp`
[2] University of Newcastle, Callaghan, Australia
[3] Technical University of Sofia, Sofia, Bulgaria

Abstract. The smart home of the future will require a universal and dynamic mapping system of a complex labyrinth that devices can use for a wide range of tasks, including item delivery, monitoring and streamlining transportation. In this paper, the core concept of this system is proposed and is demonstrated by a line-tracing robot that has self optimizing properties using an evolutionary algorithm. Tests were performed to find the best controller for the robot and a proportional-summation-difference (PSD) controller was found to be best suited for the robot's hardware specifications. An evolutionary computing algorithm called Bacterial Memetic Algorithm (BMA) was then implemented to optimize the PSD controller's properties to suite the conditions of a track. After several thousand candidate solutions of evolutionary computing and simulated tests, the robot was able to complete a lap in real time with a significant decrease in both lap time and average error.

Keywords: Line-tracing · Proportional-discrete-summation · Evolutionary algorithm · Smart home

1 Introduction

Smart homes of the future will become more complicated and an effective, small-scale, internal transportation system will be required to meet the demands of an efficient top-tier smart home, factory or office. The goal of this paper is to propose a mapping system that can meet these needs of the future by building the foundation for a line-tracing robot that can be used in a modern, internet of things, oriented smart home. The robot will display the basic features of the system by following a single, complex track using a PSD controller that has self optimizing properties.

The robot was designed to follow a target line using an array of six infrared reflectance sensors which served as a digital input to the PSD controller that

N. Kubota et al. (Eds.): ICIRA 2016, Part I, LNAI 9834, pp. 362–372, 2016.
DOI: 10.1007/978-3-319-43506-0_32

drove the robot's motors. Several other controllers were considered such as fuzzy logic and crisp logic but due to the digital input of the sensor array, the PSD controller was superior in performance. An evolutionary algorithm called Bacterial Memetic Algorithm (BMA) was used to optimize the properties of the controller to minimize the lap time and average error of the robot around a track. A software simulation was developed to simulate the robot on any given track that the user required. Several thousand evaluations were performed using this algorithm together with the simulator to generate the most optimal coefficients for the PSD controller which best suited the conditions of the track. Once the simulation was complete, the most efficient coefficients were found, they were updated and the robot was able to complete a lap with approximately 10 % greater speed and an increase in line-tracing accuracy than when the initial PSD controller's coefficients were used.

Future work is planned to further develop the mapping system. The robot should be able to perform tasks such as monitoring using a camera, reflectance sensors, a gyroscope and other peripherals, while constantly being connected to a base station via a Wi-Fi connection. Some more examples of planned work include a manual control override from a user and to incorporate an intelligent control system so that the robot can efficiently find its way through a maze-like labyrinth.

There is a lot of research that is investigating possible ways on how to make smart homes more efficient and comfortable through the use of many varying techniques. Inhabitant and object tracking in smart homes is a large topic in most research that was investigated. The existence and location of an inhabitant and the number of inhabitants existing at the same place are major challenges that are being discovered for the advancement of smart home research. As a result, many different solutions have been proposed to accurately monitor and track inhabitants of a smart home in a noninvasive way [7].

Currently, there are various techniques for monitoring and sensing in smart homes. Some smart homes use RFID chips to easily detect various objects as they do not have to be within line of sight of a sensor. RFID chips are very cheap and can be easily integrated into an existing system. There is a requirement that the object being detected must have a RFID chip installed or attached to it. This causes a problem when an object does not have a RFID chip attached as it cannot be detected. There has been some development using this technique, but there are many other possible ways for monitoring [4,12]. Objects can also be detected with voice or sound sensors. There are some objects which cannot be detected by voice because they do not have one and sound sensors may be prone to large amounts of interference, making it difficult to monitor a target object. Another alternative is using temperature sensors. Objects which have similar or near equivalent temperatures or environments that have a small temperature range can make monitoring via temperature very problematic.

Detecting inhabitants and various objects in smart homes can also be achieved through the use of cameras. To monitor the majority of a smart home's contents, a lot of individual cameras are required. The cameras need to be placed in the right position and at the right angle for maximum efficiency, making the installation process tedious [14]. If too many cameras are in operation, it can often cause distress to the inhabitants as it is essentially an invasion of privacy even though they would be only used to better their living conditions and add more functionality to a smart home [10]. In this example, a line-tracer using the mapping system being proposed can fulfill the requirements of smart home monitoring [13]. It will only require one camera and can follow a given labyrinth to detect inhabitants or objects so it is not necessary to install many cameras in the home. The line-tracer can be equipped with various sensors to detect more than just the actual objects, but gain information on the environment's condition as well due to it's powerful hardware and the numerous peripherals available for use. It is also possible to control the robot from an external source for more accurate control [8].

A controlling method of the behavior of a mobile robot in a dynamic environment has also been researched that is useful for the advancement of the mapping system. In order for a robot to have sufficient control in a changing environment, it should control the computational cost according to the state of the environment, and it must take a feasible behavioral alteration within a finite time [9]. Another method of controlling the behavior of a mobile robot that can be implemented into the system is perception-based robotics such as logical inference and fuzzy computing [6].

This paper covers concept introduction, the specifications of the robot used, a proposed solution method including the controllers considered and the testing method, the experimental results and analysis and finally, conclusions and planned future work.

2   Robot Specifications

The robot was made with the following parts: STM32 Nucleo F466RE development board [11], Zumo reflectance infrared sensor array and Zumo robot V1.2. The pin configuration of these devices can be seen in Fig. 1a and an image of the complete robot can be seen in Fig. 1b. The STM32 Nucleo F466RE development board was chosen because of it's powerful hardware and numerous peripheral pins available. This development board was also designed to be connected with Arduino devices so it could be attached to the Zumo robot.

Some features of the development board include a 32-bit STM32 microcontroller with a clock frequency up to 180 MHz and 225 DMIPS, 512 kB of flash memory and 128 kB RAM, up to 17 internal timers, up to 114 I/O ports with interrupt capability, three audio to digital converters and up to 20 communication interfaces. This board gives the ability for many additional features to be integrated into the robot.

The operation of the robot in real time is as follows: the digital input from the six IR sensors is used to determine the current location and orientation of the robot with respect to the black target line. Any sensor that does not detect a reflection of IR light, will be set to HIGH, meaning the sensor is over a black surface. This input is then stored in a char variable in the program. Please refer

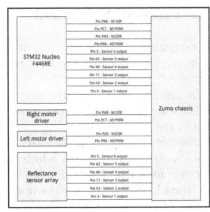 

(a) Block diagram of the pin config   (b) An image of the complete robot

Fig. 1. Robot details

Table 1. PSD lookup table

Binary input	Hexadecimal input	Output
0000 0001	0×01	error = 5
0000 0011	0×03	error = 4
0000 0010	0×02	error = 3
0000 0110	0×06	error = 2
0000 0100	0×04	error = 1
0000 1100	$0 \times 0C$	error = 0
0000 1000	0×08	error = -1
0001 1000	0×18	error = -2
0001 0000	0×10	error = -3
0011 0000	0×30	error = -4
0010 0000	0×20	error = -5
0011 1111	$0 \times 3F$	interrupt flag

to Table 1 for the lookup table that shows the valid inputs used by the PSD controller. If, for example, the error input was -5, the PSD controller would make a difference between the left and right tracks to be 100 %, meaning the left motor would have 0 % velocity of the maximum speed, while the right motor would have 100 % of the maximum speed. This would result in an immediate 90 degree turn to the left. When the robot is removed from the track, an interrupt flag is called and will then stop the motors and timers of the robot. The program will wait for an input from the user via a serial connection and then output the lap time and average error around the track to a terminal. Finally, the robot's program is terminated.

3   Proposed Solution Method to Problem Statement

The problem statement is as follows: The target objective for experiment and analysis is to optimize the PSD controller in order to minimize the lap time of the robot and increase line-tracing accuracy.

In order to meet the requirements of the problem statement, the proposed solution method consists of developing a software simulation of the robot to simulate the tests that would be performed in real time. Initial coefficients are selected and the robot is tested in the simulator. Once this preliminary test is completed, the lap time is sent from the simulator to the BMA which then performs calculations to find the best coefficients for a minimization in lap time. Once the optimal solution has been found, tests in real time with the optimal coefficients are performed to check the validity of the simulation results.

3.1   The Controllers and Logic of Line-Tracing

Fuzzy logic is a problem solving control system methodology used in a wide range of applications for data acquisition and system control and provides a simple way to arrive at a definite conclusion based upon vague, ambiguous or noisy analog inputs. Fuzzy logic mimics the thought process of humans and how a person would make a decision. For this reason, it is considered to be one of the three main branches of computational intelligence, along with evolutionary computation and neural networks. Fuzzy logic incorporates a simple rule based IF X AND Y THEN Z approach to solving a problem rather than attempting to model a system mathematically [5]. Fuzzy logic uses a membership function that gives a variable a certain membership value within a set. This means an input can be in several independent sets but will have varying membership values for each.

Fuzzy logic and crisp logic were tested with the robot's hardware setup and several problems were encountered with the logic operating at an acceptable level. This is due to the input of these systems being of a discrete type. Fuzzy logic is best applied when an analog input is being used and a decision is made on how to indirectly control a target output. In the case of the robot's hardware configuration, discrete inputs from the sensor array were being transformed into analog outputs to the motor drivers. For this reason, a PSD controller was the most suitable controller for the task.

The PSD controller was derived from the proportional-integral-derivative (PID) controller which is a control loop feedback mechanism that continuously calculates an error value as the difference between a measured variable and a target set point. The controller attempts to minimize the error over time by adjustment of a control variable to a new value determined by the formula

$$u(t) = K_P e(t) + K_I \int_0^t e(\tau)d\tau + K_D \frac{de(t)}{dt}. \tag{1}$$

The coefficients K_P, K_I, K_D are non negative and influence the behavior of the PID controller [3]. Since the PID controller needed to be discrete to work with the robot's hardware specifications, it was converted into a PSD controller represented by the formula

$$u[n] = K_P e[n] + K_S \sum_{n=1}^{\infty} e[n] + K_D \frac{e[n] - e[n-1]}{n - (n-1)}. \tag{2}$$

Where $e[n]$ and $e[n-1]$ inputs represent the error and previous error respectively from the target set point, which are the two center sensors in the sensor array. A dozen preliminary tests were performed to find suitable initial values for the coefficients K_P, K_S, K_D of the PSD controller. It was then used to drive the left and right motors of the robot to guide it along the target line by changing the difference in velocities of the motors.

3.2 Bacterial Memetic Algorithm

Bacterial Memetic Algorithm (BMA) [1] is applied for finding the optimal parameter setting of the controller's K_P, K_S, K_D parameters. BMA is a population based stochastic optimization technique which effectively combines global and local search in order to find a good quasi-optimal solution for the given problem. In the global search, BMA applies the bacterial operators, the bacterial mutation and the gene transfer operation. The role of the bacterial mutation is the optimization of the bacteria's chromosome. The gene transfer allows the information's transfer in the population among the different bacteria. As a local search technique, the Levenberg-Marquardt method is applied by a certain probability for each individual [1].

In the case of evolutionary and memetic algorithms, the encoding method and the evaluation of the individuals, or bacteria, have to be discussed. The bacterium consists of the three controller parameters to be optimized, K_P, K_S, K_D. The evaluation of the bacteria is calculated by the simulator program explained in Sect. 3.3.

The operation of the BMA starts with the generation of a random initial population containing N_{ind} individuals. Next, until a stopping criterion is fulfilled, which is usually the number of generations N_{gen}, we apply the bacterial mutation, local search, and gene transfer operators. Bacterial mutation creates N_{clones} number of clones of an individual, which are then subjected to random changes in their genes. The number of genes that are modified with this mutation is a parameter of the algorithm by the name l_{bm}. After the bacterial mutation, for each individual the Levenberg-Marquardt algorithm is used by a certain probability LM_{prob}, until a certain number of iterations LM_{iter}. The two other parameters of this step are the initial bravery factor γ_{init} and the parameter for the terminal condition τ. The last operator in a generation is the horizontal gene transfer. It means copying genes from better individuals to worse individuals. For this reason, the population is split into two halves, according to the cost values. The number of gene transfers in one generation N_{inf}, as well as the number of genes l_{gt}, that get transferred with each operation, are determined by the parameters of the algorithm. In the experiment, the parameter settings for BMA is presented in Table 2.

Bacterial memetic algorithm has been successfully applied to a wide range of problems. More details about the algorithm can be found in [1,2].

3.3   Robot Simulator

As the BMA creates several thousand candidate solutions of the robot's PSD controller, a software simulation of the robot was required to perform thousands of evaluations within a reasonable time frame. The simulator was developed in the C# environment using Visual Studio. In the simulation, the user can load a top-view image of the actual track as a .jpg or .png file. In order to make it easier to see, the loaded image is resized to fit the screen of the program. The track contrast is enhanced by coloring all the pixels that have a RGB value less than a predefined threshold pure black, as they are considered to be the line of the track. All the pixels in the image that have a RGB value greater than a predefined threshold are colored pure white and then the whole image is redrawn to get a clear track with no interference.

The line-tracer is placed in a picture box and the user can place the starting line and the line-tracer anywhere on the track. Then the user can define the orientation of the robot related to the track. The scale of the robot related to the size of the track is then calculated and corrected by measuring the size of the robot in pixels and the width of the track line in pixels to a specific ratio that matches the real ratio between the track line and the robot's size. The line-tracing animation is done by using a forward kinematics model. The image of the track is a static frame and the image of the robot is a dynamic frame. The frame details are shown in Fig. 2.

$$
T_i = \begin{bmatrix} \cos(\theta) & \sin(\theta) & 0 & X_{Ti} \\ -\sin(\theta) & \cos(\theta) & 0 & Y_{Ti} \\ 0 & 0 & 1 & 0 \\ 0 & 0 & 0 & 1 \end{bmatrix} \begin{bmatrix} X_{Tr} \\ Y_{Tr} \\ 0 \\ 1 \end{bmatrix} \tag{3}
$$

A rotation matrix shown in Eq. 3 is used to rotate the robot's frame with reference to the track's frame where X_{Ti} is the x coordinate of the center point of the robot in the track frame, Y_{Ti} is the y coordinate of the center point of the robot in the track frame, θ is the rotation angle between the track's frame and the robot's frame, X_{Tr} is the x coordinate of a point in robot's frame, Y_{Tr} is the y coordinate of a point in robot's frame and T_i is the result vector which contains the coordinates of the robot's frame point with respect to the track's frame. With this kinematic model, the simulation software recalculates the robot sensor's coordinates every millisecond. The PSD controller checks the color of the pixels under the sensors and recalculates the linear and angular velocity of the robot depending on the position and orientation of the robot with respect

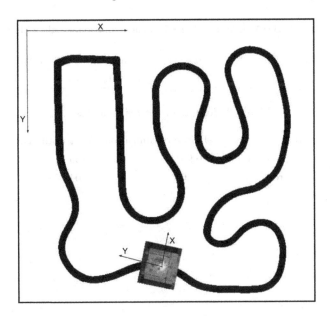

Fig. 2. A diagram of the robot simulation software setup

to the line. After every calculation of the velocities, the simulation software updates the position of the robot image and starts the whole process again from the beginning. Once the robot completes a full valid lap of the track, the lap time is then sent to the BMA to begin a new evaluation. If the robot looses the line, an invalid lap time is sent to the BMA.

4   Experimental Results and Analysis

The track used in both the simulation and real time tests had a total length of 4.8 meters, a thickness of 2 cm and included curves to test all lookup cases for the PSD controller. A comparison can be seen in Fig. 4a between the real world test conditions and Fig. 4b showing the simulator configuration. During the simulated tests, a decrease in lap time from 13.2 s to 10.4 s was achieved over 20 generations in the BMA, which is approximately a 20 % decrease from the initial generation's lap time. Table 2 shows the parameters that were used in the BMA for the optimization process. 20 generations, 4 clones and 10 individuals were used in the simulation. Figure 3 shows how the target error has been minimized.

Real time tests before and after applying the optimized PSD controller show a similar result in Table 3 with an approximate decrease in lap time of 9.7 % and showed a 0.856 % increase in line-tracing accuracy. There are several factors that may have affected the tests in real time, such as friction between the tracks of the robot and the surface of the track and interference like dust and dirt blurring the track for the robot. Also due to an inconsistent power supply from four 1.5 V batteries, the robot's maximum speed has likely been reduced. Inaccuracies in the software simulation could also be reason for slightly differing results.

Table 2. BMA parameters used in the evolutionary algorithm

Parameter	N_{gen}	N_{ind}	N_{clones}	l_{bm}	N_{inf}	l_{gt}	LM_{prob}	LM_{iter}	γ_{init}	τ
Value	20	10	4	1	3	1	10%	8	1.0	0.0001

Table 3. Results of the real time test

Parameter	Before optimization	After optimization
Average Lap Time (sec)	10.951	9.896
Average Error (%)	8.56	7.704

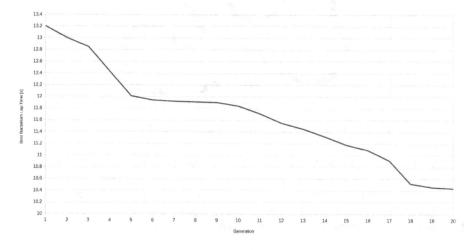

Fig. 3. A line graph showing the bacterium lap time over number of generations

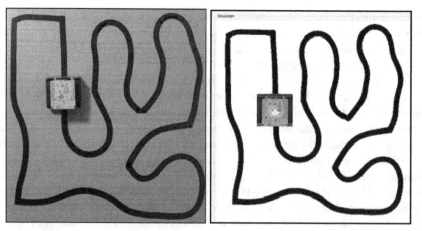

(a) Image of the real time tests   (b) Image of the simulation tests

Fig. 4. A comparison between the real time tests and the simulated tests

5 Conclusion

From the results, the conclusion can be made that the optimization of the PSD controller has been successful. This is a good foundation to start developing more features of the robot and increase the number of tasks that the robot can perform. By doing this, the mapping system that the robot will be using will also be further developed to include many important features that a universal, dynamic mapping system requires to be successful. There are many tasks that require completion to fully realize the proposed mapping system.

Avoiding obstacles blocking the pathway is a priority task and is an important capability for real world applications so the ability to follow a line with obstacle avoidance capabilities will be investigated. The robot should also be able to navigate more complex tracks with broken lines seen in a labyrinth that consists of intermittent broken lines and interference. The current method for optimizing the controller's coefficients requires an external software simulation and evolutionary algorithm but it should internally optimize the controlling software in real time without requiring external inputs so the ability to optimize the controller's properties internally and automatically in real time will be investigated. A manual controller will also be developed to allow a user to manually control the robot from a computer or smart phone through a Wi-Fi connection. It will also be possible to allow an external source, such as a smart home, to control the robot. Line mapping and memorizing is a key feature for a universal mapping system. Another key task is to memorize and store the location of a line and it's features. If it is a maze-like labyrinth, the robot will know the route to reach a destination after it has analyzed and memorized the whole course. If multiple routes to a target destination are present, then an algorithm to find the fastest route will be required so an intelligent control system that can efficiently find the best route through a labyrinth will be investigated. Instead of using a discrete input, the robot will be modified to allow for an analog input to be used so that fuzzy logic may be more easily used in controlling the robot, which will increase the accuracy of the robot's control at the cost of a more resource demanding software. Finally, through a Wi-Fi connection, the robot should be able to stream a video feed from a camera attached to the robot to a user or the smart home.

Due to the powerful hardware used, there are many opportunities available for expansion of active peripherals and sensors. This will give the robot the flexibility and capability to fulfill many different tasks and play an integral role in the smart home of the future.

References

1. Botzheim, J., Cabrita, C., Kóczy, L.T., Ruano, A.E.: Fuzzy rule extraction by bacterial memetic algorithms. Int. J. Intell. Syst. **24**(3), 312–339 (2009)
2. Botzheim, J., Toda, Y., Kubota, N.: Bacterial memetic algorithm for offline path planning of mobile robots. Memet. Comput. **4**(1), 73–86 (2012)

3. Hagglund, T.: PID controllers: theory, design and tuning. In: ISA: The Instrumentation, Systems, and Automation Society, Upper Saddle River, New Jersey (1995)
4. Jin, P.Y., Bann, L.L., Kit, J.L.W.: RFID-enabled elderly movement tracking system in smart homes (2014)
5. Klir, G.J., Yuan, B.: Fuzzy Sets and Fuzzy Logic. Theory and Applications. Prentice Hall, Upper Saddle River (1995)
6. Kubota, N.: Perception-based robotics based on perceiving-acting cycle with modular neural networks. In: Proceedings of the 2002 International Joint Conference on Neural Networks, pp. 477–482, May 2002
7. Lyman, G.: Human tracking methods comparsion for smart house (2014)
8. Mowad, M., Fathy, A., Hafez, A.: Smart home automated control system using android application and microcontroller (2014)
9. Nojima, Y., Kubota, N., Kojima, F., Fukuda, T.: Control of behavior dimension for mobile robots. In: Proceedings of the Fourth Asian Fuzzy System Symposium, pp. 652–657, May 2000
10. Robels, R.J., Kim, T.H.: Systems and methods in smart home technology: a review (2010)
11. STMicroelectronics: Nucleo-F446RE Data Sheet (2015)
12. Tjiharjadi, S.: Design of an integrated smart home control (2013)
13. Valtonen, M., Vuorela, T.: Capacitive user tracking methods for smart environments (2012)
14. West, G., Newman, C., Greenhill, S.: Using a camera to implement virtual sensors in a smart house (2003)

Navigation and Control for an Unmanned Aerial Vehicle

Jiahao Fang[1], Xin Ye[1], Wei Dong[1,2], Xinjun Sheng[1(✉)], and Xiangyang Zhu[1]

[1] State Key Laboratory of Mechanical System and Vibration, School of Mechanical Engineering, Shanghai Jiaotong University, Shanghai 200240, China
{fangjiahao,yexin5120209375,chengquess,xjsheng,mexyzhu}@sjtu.edu.cn
[2] State Key Laboratory of Fluid Power and Mechatronic Systems, Zhejiang University, Hangzhou 310058, China

Abstract. Recent advancement in micro-electronic technology makes the development of quadrotor fully possible. This paper elaborates on the complete designing process of a quadrotor control platform. Firstly a hardware structure with a microcontroller and embedded sensors is reported. Then through the implementation of a linear complementary filter, the real-time attitude estimation of the quadrotor is realized. Furthermore, aiming at the outdoor environment navigation, a barometer and a Global Positioning System (GPS) are also integrated within this platform. A multisensory data fusion algorithm is then proposed to combine these sensors for the outdoor navigation. Finally, a group of PID controllers are designed to stabilize the attitude and position. The experiment results demonstrate the capability of this platform to navigate autonomously through user-defined waypoints.

Keywords: Quadrotor control platform · Multisensory data fusion · PID controller

1 Introduction

In recent decades, the application of unmanned aerial vehicle (UAV) has been widely reported in both military and civil fields. Application of UAV is usually considered in the dangerous environments. For instance, with additional sensors, a UAV is reported to support wildfire observations [1]. After the earthquake in Fukushima, a T-Hawk Micro Aerial Vehicle equipped with radiation sensors accomplishes the missions to help operators locate the nuclear fuel debris [2]. In this way, the use of UAV could promote the success rate of the missions and ensure the safety of the human lives.

Among all these UAVs, quadrotor stands out due to its flexibility and potential capabilities. The designing of the control algorithm as well as the real-time state estimation algorithm for the quadrotor is challenging due to the complex dynamic model and the unreliable measurement from the sensors on the UAV. During the past decade, many researchers involve themselves into the designing

© Springer International Publishing Switzerland 2016
N. Kubota et al. (Eds.): ICIRA 2016, Part I, LNAI 9834, pp. 373–383, 2016.
DOI: 10.1007/978-3-319-43506-0_33

of the quadrotor controller. In their works, some mathematic models are built to understand the dynamics of the quadrotor [3,4]. Several control algorithms, both linear and nonlinear, are also proposed to stabilize the quadrotor. Examples include a non-linear controller derived using back-stepping approaches [5], a PD controller presented by V. Kumar and performs well in their papers [6]. Apart from the controller designing, work has also been done to do the flight state estimation [7,8]. However, Most of these studies are verified in a simulation environment or carried out in some commercial platforms which provide little experience in the practical development of the flight controller. In practical experiments, the accuracy of attitude and position estimation would be influenced by the noise of the embedded sensor. And the performance of these control algorithms depend on the accuracy of the real-time state estimation.

Meanwhile, success has been reported from many research groups or commercial companies in constructing their own quadrotors. From the commercial companies, the DJI Wookong, Parrot ARDrone, Asctec Humming bird have been introduced. Nevertheless, these controllers are based on the closed-source projects. There is little information we could get. As for the open-source project, such as the STARMAC [9], the pixhawk [10], X4-flyer [11] and so on, the published articles just report parts of their work.

For these reasons, a complete development process of the flight controller is presented in this paper. Firstly, this paper proposes a hardware platform which functions as the flight controller. Inertial sensor, barometer and GPS are integrated into the platform to estimate the real-time state of the quadrotor. Taking the noise of the sensors into consideration, a pose estimation algorithm based on a linear complementary filter and a group of PID controllers are implemented on this platform. To verify their validity and stability, flight tests are carried out.

The remainder of this paper is organized as follows. Section 2 develops the dynamic model of the quadrotor. Section 3 describes the hardware structure of this platform. Section 4 elaborates on the multisensory data fusion algorithm for attitude and position estimation, along with the corresponding attitude and position control algorithm. Outdoor flight test results are given in Sect. 5 and a conclusion is reached in Sect. 6.

2 Modeling

In order to develop the dynamic model of the quadrotor, the rotation matrix between two coordinate systems, the brushless motor dynamics and the rigid body dynamics are studied in this section.

2.1 Rotation Matrix

Firstly, coordinate systems including an inertial frame and a body fixed frame are defined. All of the following equations are derived in these two coordinate systems.

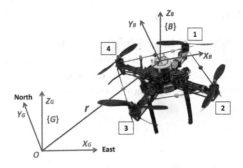

Fig. 1. Inertial frame and body fixed frame

The body fixed frame $\{B\}$ is defined in Fig. 1. The X axis points to the front of the quadrotor. The Y axis points to the left. Then the Z axis is defined by the right hand rule.

For the inertial frame $\{G\}$, the origin point O would be set at the takeoff point of the quadrotor. Then an east-north-up orthogonal frame is established by the right hand rule.

Three attitude angles are defined in Fig. 2. ϕ is the attitude angle of roll which represents the quadrotor rotates around the X axis. θ is the attitude angle of pitch which represents the quadrotor rotates around the Y axis. ψ is the attitude angle of yaw which represents the quadrotor rotates around the Z axis.

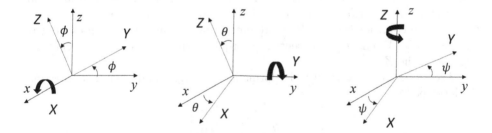

Fig. 2. Attitude definition

With these attitude angles and a rotation order of X-Y-Z, a rotation matrix $\{R\}$ from the inertial frame to the body fixed frame is defined as

$$R(\phi, \theta, \psi) = \begin{bmatrix} c\psi c\theta & c\psi s\theta s\phi - s\psi c\phi & c\psi s\theta c\phi + s\psi s\phi \\ s\psi c\theta & s\psi s\theta s\phi + c\psi c\phi & s\psi s\theta c\phi - s\phi c\psi \\ -s\theta & c\theta s\phi & c\theta c\phi \end{bmatrix} \quad (1)$$

where s stands for sin, c stands for cos.

2.2 Brushless Motor Dynamics

The sequence of the motors is defined in Fig. 1 with a clock-wise direction. The thrusts F and the momentums M of the motors are all defined in the Z direction of the body fixed frame. In the steady state, the thrust and the momentum from a single motor would be proportional to the square of this motor's angular speed [12].

$$
\begin{cases} F_i = k\omega_i^2 \\ M_i = b\omega_i^2 \end{cases}
\tag{2}
$$

where i is from 1 to 4, the coefficient k and b would be decided by the shape of the blade, and ω is the motor's angular speed.

2.3 Rigid Body Dynamics

Rigid body dynamics governs the motion of the quadrotor. In this subsection, the Newton-Euler equations for the quadrotor would be developed. Firstly, four control variables are defined as

$$
\begin{cases} U_1 = F_1 + F_2 + F_3 + F_4 \\ U_2 = (F_4 + F_3 - F_2 - F_1)\frac{L}{\sqrt{2}} \\ U_3 = (F_1 - F_2 - F_3 + F_4)\frac{L}{\sqrt{2}} \\ U_4 = M_1 - M_2 + M_3 - M_4 \end{cases}
\tag{3}
$$

where F_1, F_2, F_3, F_4 are the thrusts, M_1, M_2, M_3, M_4 are the momentums, and L is the distance between the rotor and the center of the quadrotor.

In these control variables, U_1 represents the total thrusts. U_2 represents the torque on the Y axis. U_3 represents the torque on the X axis. U_4 represents the torque on the Z axis.

According to the Newton-Euler formalism, the rigid body dynamic equation is developed as

$$
\begin{cases} m\ddot{r} = R\begin{bmatrix} 0 \\ 0 \\ U_1 \end{bmatrix} - \begin{bmatrix} 0 \\ 0 \\ mg \end{bmatrix} \\ I\ddot{q} = \begin{bmatrix} U_2 \\ U_3 \\ U_4 \end{bmatrix} - \dot{q} \times I\dot{q} \end{cases}
\tag{4}
$$

where m stands for the mass of the quadrotor, r stands for the position in the inertial frame, g stands for the gravity constant, q stands for the attitude in the body fixed frame, and I stands for the moment of inertia.

3 Hardware Design

This section describes the hardware structure of this platform including the Microcontroller Unit (MCU), the embedded sensors and the other related peripherals. The computation power comes from a 48 MHz 32 bit ARM RISC MCU with 64 KB SRAM and 256 KB Flash. This MCU is a low cost MCU without the Float Computation Unit (FPU) and does not support Digital Signal Processing Instructions. The poor computation performance of this MCU gives a high requirement for the simplicity of the algorithm.

The Inertial Measurement Unit (IMU) MPU-9150 consists of an integrated 3-axis Accelerometer, a 3-axis Gyroscope and a 3-axis Magnetometer. With these inertial sensors, the attitude estimation is achieved. Regarding of the position estimation, a barometer MS5611 for measuring the altitude and a GPS receiver U-Blox M8N for measuring the horizontal position are also integrated in the platform.

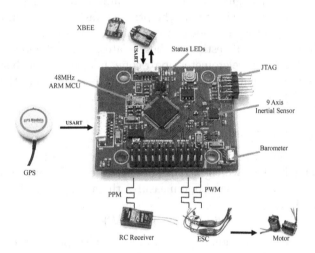

Fig. 3. Hardware structure

As for the other peripherals, the XBEE has been chosen as the wireless communication system between this platform and the ground station. The signal from the remote control (RC) is transformed into a PPM signal by the RC receiver then processed by the MCU. Besides, this platform has 4 PWM outputs to driver the corresponding DC motors. The status LEDs and an external buzzer are also used to indicate the real-time working status of this platform.

With all these sensors and peripherals, the input and the output data streams within this platform is summarized in Fig. 4.

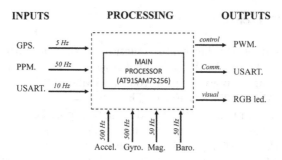

Fig. 4. Input and output data streams of the system

4   State Estimating and Control

To control a quadrotor, the current states including the local position, linear velocity, attitude and angular velocity are required [13]. Among these states, the attitude and angular velocity are the primary variables used in the attitude control of the quadrotor. The local position and the linear velocity are used in the position control. To estimate these states, many different sensors are required including the accelerometer, gyroscope, magnetometer, barometer and the GPS. In some specific application, lidar, VICON system and the Kinect are also integrated into the estimating process.

4.1   Estimating Attitude

In this paper, a linear complementary filter is implemented to combine the sensors data from MPU9150 which includes the accelerometer, gyroscope and magnetometer. Generally, the gyroscope measures the angular velocity in the body fixed frame $\{B\}$

$$\dot{\theta}_u = \dot{\theta} + b + \eta \in \{B\} \tag{5}$$

where $\dot{\theta}_u$ is the measured angular velocity of the gyroscope. $\dot{\theta}$ is the true value of the angular velocity. b is the slow time-varying gyroscope bias. η is the additive measurement noise.

The major defect of the gyroscope is the slow time-varying bias which makes the method of integrating these measurements to estimate the attitude impossible. For these reasons, we need to combine the measurements of accelerometer which is far more stable in the long term but susceptible in a short time to compensate the bias of the gyroscope.

Let $\dot{\theta}_u$ be the measurement angular velocity of the gyroscope and $\dot{\hat{\theta}}$ the compensated angular velocity which is used for the integration. $\hat{\theta}$ denotes the attitude estimation and θ_a is the attitude measured by the accelerometer. $\hat{b}$ is the time-varying bias estimation of the gyroscope (Fig. 5).

$$\begin{cases} \dot{\hat{\theta}} = \dot{\theta}_u - \hat{b} + K_p(\theta_a - \hat{\theta}) \\ \dot{\hat{b}} = K_i(\theta_a - \hat{\theta}) \end{cases} \tag{6}$$

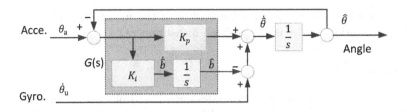

Fig. 5. Complementary filter for attitude estimation

4.2 Estimating Position

To estimate the local position of the quadrotor, the primary sensor would be the accelerometer. The mathematic model of the accelerometer is defined as

$$a_u = R^T(a + g\vec{z}) + b_a + \eta_a \in \{B\} \tag{7}$$

where a_u is the acceleration measurement. R is the rotation matrix defined in Sect. 2. a is the true motion acceleration in the inertial frame. $g\vec{z}$ is the gravity. b_a is the bias of the accelerometer. η_a is the additive measurement noise (Fig. 6).

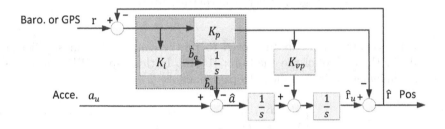

Fig. 6. Position estimation

However, due to the steady error and the stochastic noise listed in the formula, there is no effective way to use the accelerometer solely. With careful calibration, these defects could be relieved but not completely eliminated [14]. For these reasons, in this paper the platform would include a barometer to get the altitude information and a GPS to provide the position information in the horizontal plane. And a PI controller is designed to fuse the data from the accelerometer, barometer and the GPS. The key point of the PI controller is to correct the steady error of the accelerometer with the measurement from barometer and the GPS.

The position estimation algorithm is defined as

$$\begin{cases} \hat{a} = a_u - \hat{b}_a \\ \dot{\hat{b}}_a = K_i(r - \hat{r}) \\ \hat{r} = \hat{r}_u - K_p(r - \hat{r}) \end{cases} \tag{8}$$

where a_u is the acceleration measurement. $\hat{a}$ is the modified acceleration. $\hat{r}$ is the position estimation. r is the position measurement.

4.3 Control

In this section, the controllers will be designed based on the dynamic model in Sect. 2 and the states estimation algorithm in Subsects. 4.1 and 4.2. The block diagram of this controller is shown in Fig. 7, which includes a position loop(**P**), a velocity loop(**V**), an attitude loop(**A**)and a rate loop(**R**).

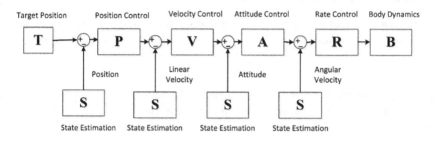

Fig. 7. Structure of control system

The PID controller has been proved to be the simplest but most efficient methods for the control of the quadrotor [15, 16]. In this paper, PID controllers are also implemented in these four loops defined as

$$
\begin{cases}
P = K_{pp} \\
V = K_{vp}(1 + \frac{1}{T_{vi}s} + T_{vd}s) \\
A = K_{ap} \\
R = K_{rp}(1 + \frac{1}{T_{ri}s} + T_{rd}s)
\end{cases}
\tag{9}
$$

where P controller is adopted in the position loop and the attitude loop. PID controller is implemented in the velocity loop and the rate loop.

5 Flight Test

In order to verify the effectiveness of this platform including the hardware and the algorithm, flight test has been carried out in the outdoor environment. Position controller and Velocity controller are running at 250 Hz. Meanwhile, Attitude controller and Rate controller are running at 500 Hz. Thanks to the wireless communication system XBEE which is shown in the Fig. 3, all of the states in the quadrotor during the experiments are collected and sent back to the ground station. The altitude control experiment and horizontal position control experiment are carried out separately.

5.1 Altitude Control

For the altitude control, the quadrotor is ordered to hover at a specific height. The test result is shown in Fig. 8. Z_p is the current altitude estimation and Z_{cmd} is the corresponding target altitude.

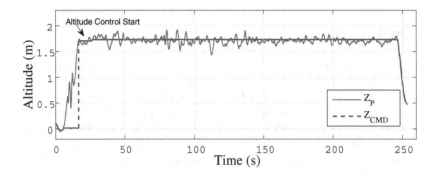

Fig. 8. Results altitude control

The experiment result shows that the quadrotor well responds to the command and the steady error is within 0.3 m. Comparing with the reported result [16], the performance is good enough.

5.2 Position Control

For the horizontal position control, the experiment is carried out at a specific height of 1.7 m and the response of the quadrotor has been tested with the hovering and ramp command in the horizontal plane. The test result is shown in Fig. 9.

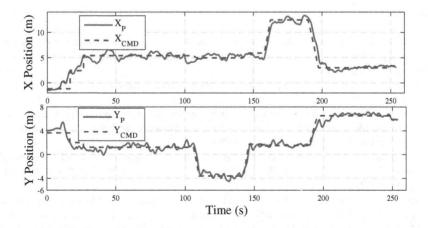

Fig. 9. Results position control

In the X axis of the inertial frame, the quadrotor firstly hovers for 150 s, then it follows several ramp commands. In the Y axis of the inertial axis, the quadrotor firstly hovers for 100 s, then it also follows several ramp commands. The experiment result shows that the quadrotor well responds to the command. The steady error is within 1 m and the delays are within 2 s. Comparing with the reported result [17], the performance is good enough.

6   Conclusions

This paper elaborates on the development of a quadrotor platform which is able to navigate through the user-defined waypoints in the outdoor environment. The hardware structure is reported in Sect. 3 with the necessary sensors. In Sect. 4, a simple linear complementary filter and a PI-based multisensory data fusion algorithm are proposed to estimate the attitude and the local position of the quadrotor. In the outdoor flight experiment, this platform has demonstrated good performance in the attitude stability and position control ability through the PID controllers which are discussed in Sect. 4. The test result shows the steady error is within 0.3 m in the altitude direction. As for the horizontal direction, the steady error is within 1 m and the delay is within 2 s. Future work would include some additional functions such as the path-planning and collision avoidance. To realize these functions, some external sensors such as the laser rangefinder and sonar would be added to this platform. These sensors would improve the navigation ability of this platform in the exploration of the unknown environment.

Acknowledgments. This work was funded by the special development fund of Shanghai Zhangjiang Hi-Tech Industrial Development Zone (No.201411-PD-JQ-B108-009) and Open Foundation of the State Key Laboratory of Fluid Power Transmission and Control (No. GZKF-201510).

References

1. Ambrosia, V., Buechel, S., Wegener, S., et al.: Unmanned airborne systems supporting disaster observations: near-real-time data needs. Int. Soc. Photogramm. Remote Sens. **144**, 1–4 (2011)
2. Jzsef, L.: Exploring the capacities of airborne technology for the disaster assessment
3. Guo, W., Horn, J.: Modeling and simulation for the development of a quad-rotor UAV capable of indoor flight. In: Modeling and Simulation Technologies Conference and Exhibit. Keystone, Colorado (2006)
4. Bouabdallah, S., Siegwart, R.: Full control of a quadrotor. In: IEEE/RSJ International Conference on Intelligent Robots and Systems, IROS 2007, pp. 153–158. IEEE (2007)
5. Mian, A.A., Daobo, W.: Modeling and backstepping-based nonlinear control strategy for a 6 DOF quadrotor helicopter. Chin. J. Aeronaut. **21**(3), 261–268 (2008)
6. Michael, N., Mellinger, D., Lindsey, Q., et al.: The GRASP multiple micro-UAV testbed. IEEE Robot. Autom. Mag. **17**(3), 56–65 (2010)

7. Earl, M.G., D'Andrea, R.: Real-time attitude estimation techniques applied to a four rotor helicopter. In: Proceedings of 43rd IEEE Conference on Decision and Control, Atlantis, Paradise Island, Bahamas, 14–17 December, pp. 3956–3961 (2004)
8. Hong, S.K.: Fuzzy logic based closed-loop strapdown attitude system for unmanned aerial vehicle (UAV). Sens. Actuators A Phys. **107**, 109–118 (2003)
9. Hoffmann, G., Rajnarayan, D., Waslander, S., Dostal, D., Jang, J., Tomlin, C.: The stanford testbed of autonomous rotorcraft for multi agent control (STARMAC). In: Proceedings of the Digital Avionics Systems Conference, vol. 2, pp. 12.E.4–12.E.121.10 (2004)
10. Meier, L., Tanskanen, P., Fraundorfer, F., Pollefeys, M.: PIXHAWK: a system for autonomous flight using onboard computer vision. In: Proceedings of the ICRA, pp. 2992–2997, May 2011
11. Guenard, N., Hamel, T., Mahony, R.: A practical visual servo control for an unmanned aerial vehicle. IEEE Trans. Robot. **24**(2), 331–340 (2008)
12. Zhang, Q., Chen, J., Yang, L., Dong, W., Sheng, X., Zhu, X.: Structure optimization and implementation of a lightweight sandwiched quadcopter. In: Liu, H., Kubota, N., Zhu, X., Dillmann, R. (eds.) ICIRA 2015, Part III. LNCS, vol. 9246, pp. 220–229. Springer, Heidelberg (2015)
13. Mahony, R., Kumar, V., Corke, P.: Multirotor aerial vehicles: modeling, estimation, and control of quadrotor. IEEE Robot. Autom. Mag. **19**(3), 20–32 (2012)
14. Lim, C.H., Lim, T.S., Koo, V.C.: A MEMS based, low cost GPS-aided INS for UAV motion sensing. In: 2014 IEEE/ASME International Conference on Advanced Intelligent Mechatronics (AIM), pp. 576–581. IEEE (2014)
15. Salih, A.L., Moghavvemi, M., Mohamed, H.A.F., et al.: Modelling and PID controller design for a quadrotor unmanned air vehicle. In: 2010 IEEE International Conference on Automation Quality and Testing Robotics (AQTR), vol. 1, pp. 1–5. IEEE (2010)
16. Dong, W., Gu, G.Y., Zhu, X., et al.: Modeling and control of a quadrotor UAV with aerodynamic concepts. World Acad. Sci. Eng. Technol. **7**, 377–382 (2013)
17. Remes, B.D.W., Esden-Tempski, P., Van Tienen, F., et al.: Lisa-s 2.8 g autopilot for GPS-based flight of MAVS. In: IMAV 2014: International Micro Air Vehicle Conference and Competition 2014, Delft, The Netherlands. Delft University of Technology, 12–15 August 2014

Robot Path Control with Rational-Order Calculus

Adrian Łęgowski$^{(\boxtimes)}$, Michał Niezabitowski, and Tomasz Grzejszczak

Institute of Automatic Control, Silesian University of Technology,
16 Akademicka St., 44-100 Gliwice, Poland
{adrian.legowski,michal.niezabitowski,tomasz.grzejszczak}@polsl.pl

Abstract. In this paper an application of the fractional calculus to path
control is studied. The integer-order derivative and integral are replaced
with the fractional-order ones in order to solve the inverse kinematics
problem. The proposed algorithm is a modification of the existing one.
In order to maintain the accuracy and to lower the memory requirements
a history limit and a combination of fractional and rational derivation are
proposed. After reaching assumed accuracy or iteration limit the algorithm switches to integer order derivative and stops after few additional
iterations. This approach allows to reduce the positional error and maintain the repeatability of fractional calculus approach. The simulated path
in task space have been designed in a way that causes the instability of
standard *Closed Loop Pseudoinverse* algorithm. Our study proves that
use of fractional calculus may improve the joint paths.

Keywords: Fractional calculus · Path planning · Motion planning ·
Inverse kinematics · Manipulator

1 Introduction

A problem of the *inverse kinematics* (IK) is well known and deeply studied.
There are many algorithms that find the set of joint values for desired task
space coordinates.

The old and well known algorithms are presented in [1–5] however, one has
to recognize that the topic is still of great importance. Many researchers develop
new approaches. For the problem of uncertainty of joint lengths the efficient
method is proposed in [6]. In papers [7,8] authors prove that for well known
structures it is possible to utilize the new neural network based approach. For
manipulators with active spherical ball joints the solution is presented in [9].
There are manipulators for which finding the *inverse kinematics* solution as
closed form formulas may be hard or even impossible and therefore, numerical
algorithms for few of them are proposed in [10]. Paper [6] suggests an application
of the *Interval Newton* method to solving the IK problem. In order to address
the time-efficiency combining the numerical and analytical approach is proposed
in paper [11].

© Springer International Publishing Switzerland 2016
N. Kubota et al. (Eds.): ICIRA 2016, Part I, LNAI 9834, pp. 384–395, 2016.
DOI: 10.1007/978-3-319-43506-0_34

Manipulators are much more than simple industrial robots. Many papers proves that methods from industrial robotics can be utilized in modeling human arm [12–14].

The inverse kinematics algorithm should be predictable, repeatable and accurate. The approach presented in this paper allows to meet these requirements with limited memory consumption which is important for implementation in robots' controllers. Altering the existing methods results in similar solutions with deterministic memory consumption and very high positional accuracy.

Our paper is organized as follows. The second section presents three definitions of fractional derivative and shows the discrete approximation of the Grunwald-Letnikov operator with use of the *short memory principle*. Third section introduces the standard Jacobian-based inverse kinematics algorithm with integer-order derivative and suggests use of the *Moore-Penrose Pseudoinverse* matrix. The next part proposes replacing integer-order calculus with fractional one and modification of the existing method proposed in [15] by varying the derivative order. Section five introduces the simulation settings, defines the task and initial conditions. Part six is dedicated to the simulation results and positional accuracy. The last section summarizes obtained results and underlines the advantages as well as disadvantages of the proposed approach.

2   Fractional Calculus

The idea of non-integer order derivative and integral is nearly as old as well known integer-order calculus. It goes back to the 1695 and Leibniz's letter to L'Hospital [16]. There are many definitions of fractional derivative and integral. The three popular ones are presented in this paper.

The Riemann-Liouville (RL) derivative is defined as follows [17]:

$$_0D_t^\alpha y(t) = \frac{1}{\Gamma(1-\alpha)} \frac{d}{dt} \int_0^t (t-\tau)^{-\alpha} y(\tau) d\tau. \tag{1}$$

The Caputo's definition of derivative [18]:

$$_0D_t^\alpha y(t) = \frac{1}{\Gamma(1-\alpha)} \int_0^t \frac{y'(\tau)}{(t-\tau)^\alpha} d\tau. \tag{2}$$

The Grunwald-Letnikov definition is given by the equation [17,18]:

$$_aD_t^\alpha y(t) = \lim_{\Delta t \to 0} \left[\frac{1}{(\Delta t)^\alpha} \sum_{k=0}^\infty y(t - k\Delta t)\gamma(\alpha, k) \right], \tag{3}$$

$$\gamma(\alpha, k) = (-1)^k \frac{\Gamma(\alpha+1)}{\Gamma(k+1)\Gamma(\alpha-k+1)}. \tag{4}$$

The function $\Gamma()$ is the Gamma function. For every definition we assumed that derivative order $\alpha \leq 1$ and $\alpha > 0$.

Often, the approximation of expression (3) is implemented with use of the short-memory principle [19] by formula (5):

$$D^\alpha y(t) = \frac{1}{(\Delta t)^\alpha} \sum_{k=0}^{N} y(t - k\Delta t)\gamma(\alpha, k), \tag{5}$$

where Δt is sampling time and N is the truncation order [15].

The last definition seems to be well suited for software implementation. With use of *look up table* (LUT) there is no need for computing the value of γ in every iteration.

Currently researchers are looking for new applications of fractional calculus (FC) in various branches of science. Many researchers proved that the FC can be applied in control theory in order to design new type of controllers [20,21]. In paper [22] fractional continuous models have been studied. Implementation of fractional models requires accurate methods of approximation. Many of them are based on approximating the s^α in Laplace domain. They have been studied in [17,23].

These and many others applications prove the usability of fractional calculus and therefore, need for finding new applications in order to improve solutions for well known problems. We believe that one of these applications would be signal processing like in [24].

3   Integer-Order Inverse Kinematics

Inverse kinematics solution can be found in various ways e.g. by finding the analytical formulas [25]. In this paper we focus on differential IK solution. The common approach suggests using the following equation:

$$\frac{dx}{dt} = J(q)\frac{dq}{dt}, \tag{6}$$

where x is the vector of coordinates in task space, q is the vector of coordinates in joint space and $J(q)$ is the Jacobian matrix for given q. For given desired manipulator position (X_{ref}) we can rewrite the Eq. (6) as follows:

$$J^{-1}(q_{i-1})\Delta x_i = \Delta q_i, \tag{7}$$

where

$$\Delta x_i = X_{ref} - x_{i-1},$$
$$q_i = \Delta q_i + q_{i-1},$$
$$i = 1, 2, 3, \dots .$$

The process of finding solution is iterative and accuracy may be sensitive to number of iterations i. Trajectory realization requires computing the IK solution for every given X_{ref}. In every iteration the algorithm takes into account demanded position (X_{ref}) and previous values of task and joint space coordinates. This fact can be written as follows:

$$q_{i-1} + J^{-1}(q_{i-1})(X_{ref}(j\Delta t) - x_{i-1}) = q_i \tag{8}$$

where

$$j = 1, 2, 3, \dots .$$

It is worth noting that for given $X_{ref}(j\Delta t)$ we use the value of x computed by solving the forward kinematics problem in previous iteration. It is important to reach desired accuracy or number of iterations before the new value of X_{ref} is demanded. The Jacobian matrix J has to be invertible in order to derive the value of Δq. Many researchers suggest using *Moore-Penrose pseudoinverse* of matrix. This approach allows to use the expression (8) for even redundant structures. Considering this we can write:

$$J^{\#}(q_{i-1})\Delta x(i) = \Delta q(i) \tag{9}$$

which can be rewritten as follows:

$$J^{\#}(q_{i-1})(X_{ref} - x_{i-1}) = q_i - q_{i-1}, \tag{10}$$

$$q_{i-1} + J^{\#}(q_{i-1})(X_{ref} - x_{i-1}) = q_i \tag{11}$$

and finally:

$$q_{i-1} + J^{\#}(q_{i-1})(X_{ref}(j\Delta t) - x_{i-1}) = q_i, \tag{12}$$

where $J^{\#}$ is the *Moore-Penrose pseudoinverse* of matrix J.

It has been proven that this simple approach often called *closed loop pseudoinverse* (CLP) leads to an aperiodic joints motion for certain cyclic end-effector trajectories [15]. This problem mostly concerns redundant manipulators.

4   Inverse Kinematics Based on Fractional Calculus

In paper [15] authors study the application of fractional calculus (FC) to closed loop pseudoinverse (CLP) method. The fractional-order derivative is the global operator that has a memory of all past events. This property may force the periodic motion for desired cyclic end-effector trajectories. Considering the approximation of Grunwald-Letnikov derivative, in $j - th$ time moment, we can rewrite the formula (10) as follows:

$$J^{\#}(q_{i-1})\Delta^{\alpha}x(i) = \Delta^{\alpha}q(i), \tag{13}$$

$$q_i + \sum_{k=1}^{N} \gamma(\alpha, k)q_{i-k} = J^{\#}(q_{i-1}) \left[X_{ref}(j\Delta t) + \sum_{k=1}^{N} \gamma(\alpha, k)x_{i-k} \right], \tag{14}$$

where N is the truncation order and

$$\gamma(\alpha, k) = (-1)^k \frac{\Gamma(\alpha + 1)}{\Gamma(k + 1)\Gamma(\alpha - k + 1)}. \tag{15}$$

Having formula (14) we can compute:

$$q_i = J^\#(q_{i-1}) \left[X_{ref}(j\Delta t) + \sum_{k=1}^{N} \gamma(\alpha, k)x_{i-k} \right] - \sum_{k=1}^{N} \gamma(\alpha, k)q_{i-k}. \tag{16}$$

The Eq. (16) allows to design an iterative procedure for finding the solution of IK problem. For given X_{ref} at the time $j\Delta t$ we compute the value of q_i. The only variable that is the function of time is the reference position. This leads to a conclusion that for very large number of iterations and small truncation order N, the algorithm would take into account only the values of x_i and q_i between two nearest reference positions X_{ref}.

In paper [15] authors suggest that lowering the differential order α lowers the positioning accuracy. We can confirm that for most studied trajectories. To address this issue we propose a variable order α. It has been proven that integer order derivation maintains high positional accuracy. Having that in mind we can write that derivation order is given by the expression (17). This is based on method proposed in [26].

$$\alpha(c) = \begin{cases} 1 & \text{if } c \geq I - d, \\ \alpha_s & \text{if } c < I - d, \end{cases} \tag{17}$$

where α_s is the initial order of derivative, I is the maximal number of iterations, d defines the number of iterations with rational order derivation and c is the iteration number between $X_{ref}((j-1)\Delta t)$ and $X_{ref}(j\Delta t)$. With that in mind we can write:

$$q_i = J^\#(q_{i-1}) \left[X_{ref}(j\Delta t) + \sum_{k=1}^{N} \gamma(\alpha(c), k)x_{i-k} \right] - \sum_{k=1}^{N} \gamma(\alpha(c), k)q_{i-k}. \tag{18}$$

Proposed approach may improve the accuracy. It is worth noting that this method may be more time-efficient since few last iterations considers only one set of previous values.

5   Simulated Path and Settings

In simulation we consider 4-DOF manipulator. The task is to follow the given trajectory. The path is generated by periodic functions. The path is chosen in a way that allows for periodic motion of joints.

We have omitted the lower and upper bounds for joints in order to ease the task for the algorithm. It is clear that in practical application these constraints

Table 1. Denavit-Hartenberg (D-H) parameters [27]

i	λ_i[cm]	l_i[cm]	$\alpha_i[°]$	$\theta_i[°]$
1	0	0	0	θ_1
2	λ_2	-3	0	-90
3	0	-3	0	θ_3
4	0	-1	0	θ_4

have to be included during the path planning. Considering the presented D-H parameters (Table 1) the Jacobian matrix takes the form:

$$J = \begin{bmatrix} -3C_{13} - 3C_1 - C_{134} & 0 & -3C_{13} - C_{134} & -C_{134} \\ -3S_{13} - 3S_1 - S_{134} & 0 & -3S_{13} - S_{134} & -S_{134} \\ 0 & 1 & 0 & 0 \end{bmatrix}, \quad (19)$$

where $C_{lm} = cos(\theta_l + \theta_m)$, $S_{lm} = sin(\theta_l + \theta_m)$, $S_{lmn} = sin(\theta_l + \theta_m + \theta_n)$, $C_{lmn} = cos(\theta_l + \theta_m + \theta_n)$.

The path in task space is given as a location described in Cartesian space:

$$X_{ref}(j\Delta t) = \begin{bmatrix} x_{ref}(j\Delta t) \\ y_{ref}(j\Delta t) \\ z_{ref}(j\Delta t) \end{bmatrix} = \begin{bmatrix} 2.5sin(\omega_1 j\Delta t) + 1 \\ 2.5cos(\omega_2 j\Delta t) + 2.3 \\ 10cos(\omega_3 j\Delta t) \end{bmatrix}, \quad (20)$$

where

$$\Delta t - timestep,$$
$$\omega_1, \omega_2, \omega_3 - angular\,frequencies,$$
$$0 \le j\Delta t \le 50,$$
$$j = 0, 1, 2, \dots .$$

The chosen path is designed to cause specific issues. The integer order derivation may result in joints' rotation angles larger than $2\pi[rad]$. The initial time is $t = 0$. The angular frequencies vary in experiments and therefore, it is specified in every figure or table. The simulation maximal length is set to $t_{max} = 30[s]$. The stop criterion is defined as a maximal number of iterations $I = 250$. This value has been chosen experimentally.

For every trajectory the initial position is as follows:

$$q_0 = [6.2024, 10, 5.8199, 3.7168]^T .$$

6   Results

In the first experiment we study the influence of memory length. Figures 1-3 present the trajectories for manipulators' joints. We decided to omit the second joint since its trajectory seem to remain untouched by the changes.

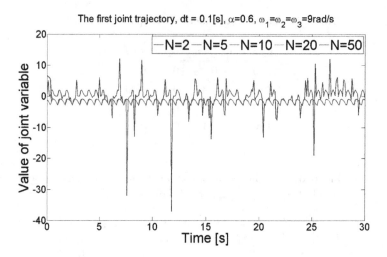

Fig. 1. First join trajectory for various N [own source]

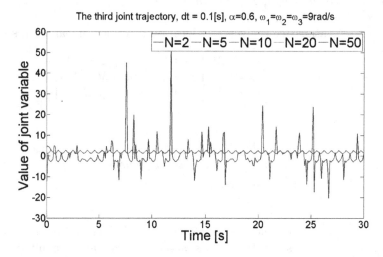

Fig. 2. Third join trajectory for various N [own source]

These figures (Figs. 1, 2 and 3) reveal small differences at the few first seconds of trajectory for $N > 2$. For $N = 2$ one can observe poor quality of motion. In this test we require a periodic motion for every joint with extreme values as close to zero as possible. Considering this criterion we conclude that $N = 5$ may be sufficient for our test trajectory. The small number of N meets our other requirement which is small memory consumption. A proper truncation of the sums in (18) allows for relatively low-cost implementation. With use of the *lookup table* there should not be any performance issues during the computation of γ.

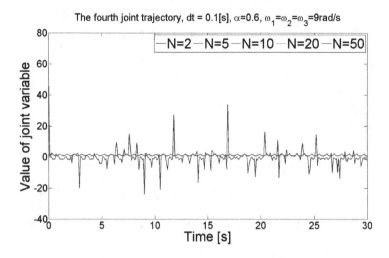

Fig. 3. Fourth join trajectory for various N [own source]

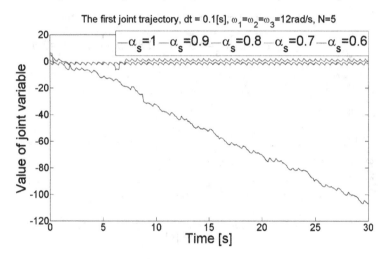

Fig. 4. First join trajectory for $\omega_1 = \omega_2 = \omega_3 = 12$ [rad/s] [own source]

Since for $N = 5$ we obtain relatively good results, we decided to use this truncation in other simulations. We conclude that for $\alpha \geq 0.6$ this parameter would have rather poor impact. One must notice that function γ decays slower for higher derivative order (we consider $0 \leq \alpha \leq 1$).

Figures 4 and 5 present the joint trajectories for various order α. In these figures one can observe that integer-order approach causes the multiple rotations for the first joint. We conclude that for specific cases lowering the order may improve the motion performance which can be observed in the presented figures for $\alpha < 0.9$.

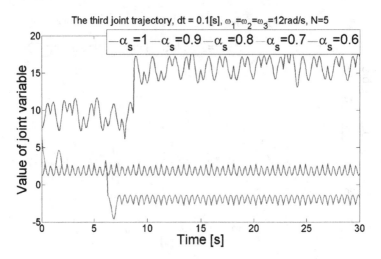

Fig. 5. Third join trajectory for $\omega_1 = \omega_2 = \omega_3 = 12$ [rad/s] [own source]

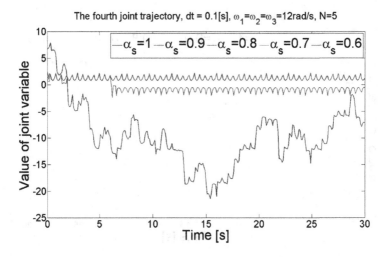

Fig. 6. Fourth join trajectory for $\omega_1 = \omega_2 = \omega_3 = 12$ [rad/s] [own source]

Table 2 presents the accuracy for various α_s, constant $\alpha = 0.6$ and $\omega = \omega_1 = \omega_2 = \omega_3$. The accuracy is defined as follows:

$$X_d(j) = X_{ref}(j\Delta t) - X_c(j\Delta t),$$

$$P_{err} = \frac{\sum_{j=0}^{J-1} \sqrt{X_d(j)^T X_d(j)}}{J}, \tag{21}$$

where J is the number of points, $X_c(j\Delta t)$ is a direct kinematics result for computed joint variables at $j - th$ point. Results contained in Table 2 reveal that

Table 2. The positional error

$\omega[\text{rad/s}]$	$\alpha_s = 1$ $\times 10^{-15}$	$\alpha_s = 0.8$ $\times 10^{-16}$	$\alpha_s = 0.7$ $\times 10^{-16}$	$\alpha_s = 0.6$ $\times 10^{-16}$	$\alpha = 0.8$	$\alpha = 0.6$
1	1.9295	6.3962	4.9055	4.5553	0.8029	1.7497
2	1.8982	5.1492	4.4086	4.3252	0.7397	1.7433
4	2.7414	4.6763	4.4075	4.3544	0.7064	1.7399
7	5.1422	4.6163	4.4146	4.3647	0.6929	1.7419
10	9.6118	4.3995	4.5751	4.4561	0.6862	1.7424
12	7.0679	4.2633	4.0103	4.4104	0.7010	1.7410

proposed approach allows to ignore the accuracy issues mentioned in [15] with maintaining the benefit which is periodic motion for every joint.

7   Conclusions

In this paper we studied the possible application of fractional calculus in solving the inverse kinematics problem and therefore, to manipulator path control. In this paper we presented method based on the existing algorithm. Our experiments confirm that there are benefits of the memory effect. We conclude that lowering the derivation order may improve the repeatability of joint trajectories (Figs. 4, 5, 6) for specified end-effector trajectory but it may increase the positional error which can be seen it Table 2. Our modification allows to maintain the accuracy of well known *CLP* algorithm without loosing the cyclic joints motion. The number of iterations have to be chosen with respect to processing power (maximal available number). Our experiments reveal that it should be at least ten or more iterations with integer-order derivation.

The disadvantage of the proposed method is the ignorance of the joint limitations. Properly designed algorithm has to meet these constraints. Moreover, this simple approach does not affect the singular configurations since the Jacobi matrix is computed without any adjustments. This is the reason why presented method may only be the base for more advanced algorithms.

In our case the fractional calculus proved to be the good solution. We believe that the fractional calculus may be used in processes where simple PID controller is not enough like in nuclear fusion plasma control [28,29].

Acknowledgments. The research presented here was done by first author as part of the project funded by the National Science Centre in Poland granted according to decision DEC-2014/13/B/ST7/00755. Moreover, research of the second author was supported by Polish Ministry for Science and Higher Education under internal grant BK-213/RAu1/2016 for Institute of Automatic Control, Silesian University of Technology, Gliwice, Poland. The calculations were performed with the use of IT infrastructure of GeCONiI Upper Silesian Centre for Computational Science and Engineering (NCBiR grant no POIG.02.03.01-24-099/13).

References

1. Wang, L.T., Chen, C.: A combined optimization method for solving the inverse kinematics problems of mechanical manipulators. IEEE Trans. Robot. Autom. **7**(4), 489–499 (1991)
2. Manocha, D., Canny, J.F.: Efficient inverse kinematics for general 6R manipulators. IEEE Trans. Robot. Autom. **10**(5), 648–657 (1994)
3. Szkodny, T.: Forward and inverse kinematics of IRb-6 manipulator. Mechanism Mach. Theory **30**(7), 1039–1056 (1995)
4. Goldenberg, A.A., Benhabib, B., Fenton, R.G.: A complete generalized solution to the inverse kinematics of robots. IEEE J. Robot. Autom. **1**(1), 14–20 (1985)
5. Szkodny, T.: Basic Component of Computational Intelligence for IRB-1400 Robots. In: Cyran, K.A., Kozielski, S., Peters, J.F., Stańczyk, U., Wakulicz-Deja, A. (eds.) Man-Machine Interactions. AISC, vol. 59, pp. 637–646. Springer, Heidelberg (2009)
6. Kumar, V., Sen, S., Shome, S.N., Roy, S.S.: Inverse kinematics of redundant serial manipulators using interval method in handling uncertainties. In: Proceedings of the 2015 Conference on Advances in Robotics, AIR 2015, Goa, India, 2–4 July 2015, pp. 1:1–1:6 (2015)
7. Son, N.N., Anh, H.P.H., Chau, T.D.: Inverse kinematics solution for robot manipulator based on adaptive MIMO neural network model optimized by hybrid differential evolution algorithm. In: 2014 IEEE International Conference on Robotics and Biomimetics, ROBIO 2014, Bali, Indonesia, 5–10 December 2014, pp. 2019–2024 (2014)
8. Köker, R.: A genetic algorithm approach to a neural-network-based inverse kinematics solution of robotic manipulators based on error minimization. Inf. Sci. **222**, 528–543 (2013)
9. Dong, H., Fan, T., Du, Z., Chirikjian, G.S.: Inverse kinematics of active rotation ball joint manipulators using workspaces density functions. In: Advances in Reconfigurable Mechanisms and Robots II, pp. 633–644. Springer International Publishing, Cham (2016)
10. Kucuk, S., Bingul, Z.: Inverse kinematics solutions for industrial robot manipulators with offset wrists. Appl. Math. Model. **38**(78), 1983–1999 (2014)
11. Ananthanarayanan, H., Ordnez, R.: Real-time inverse kinematics of $(2n + 1)$ DOF hyper-redundant manipulator arm via a combined numerical and analytical approach. Mech. Mach. Theory **91**, 209–226 (2015)
12. Babiarz, A., Klamka, J., Zawiski, R., Niezabitowski, M.: An approach to observability analysis and estimation of human arm model. In: 11th IEEE International Conference on Control Automation, pp. 947–952, June 2014
13. Babiarz, A.: On mathematical modelling of the human arm using switched linear system. AIP Conf. Proc. **1637**, 47–54 (2014)
14. Babiarz, A.: On control of human arm switched dynamics. In: Man-Machine Interactions 4: 4th International Conference on Man-Machine Interactions, ICMMI 2015 Kocierz Pass, Poland, 6–9 October 2015, pp. 151–160. Springer International Publishing, Cham (2016)
15. Duarte, F.B.M., Machado, J.A.T.: Pseudoinverse trajectory control of redundant manipulators: a fractional calculus perspective. In: Proceedings of the 2002 IEEE International Conference on Robotics and Automation, ICRA, Washington, DC, USA, pp. 2406–2411, 11–15 May 2002 (2002)

16. Ross, B.: A brief history and exposition of the fundamental theory of fractional calculus. In: Fractional Calculus and Its Applications: Proceedings of the International Conference Held at the University of New Haven, pp. 1–36, June 1974. Springer, Berlin (1975)
17. Vinagre, B., Podlubny, I., Hernandez, A., Feliu, V.: Some approximations of fractional order operators used in control theory and applications. Fract. Calculus Appl. Anal. **3**(3), 231–248 (2000)
18. Garrappa, R.: A Grünwald-Letnikov scheme for fractional operators of Havriliak-Negami type. Recent Adv. Appl. Modell. Simul. **34**, 70–76 (2014)
19. Podlubny, I.: Fractional differential equations: an introduction to fractional derivatives, fractional differential equations, to methods of their solution and some of their applications, vol. 198. Academic press (1998)
20. Cao, J.Y., Cao, B.G.: Design of fractional order controllers based on particle swarm optimization. In: 2006 1ST IEEE Conference on Industrial Electronics and Applications, pp. 1–6, May 2006
21. Mackowski, M., Grzejszczak, T., Lęgowski, A.: An approach to control of human leg switched dynamics. In: 2015 20th International Conference on Control Systems and Computer Science (CSCS), pp. 133–140, May 2015
22. Aoun, M., Malti, R., Levron, F., Oustaloup, A.: Numerical simulations of fractional systems: an overview of existing methods and improvements. Nonlinear Dyn. **38**(1), 117–131 (2004)
23. Barbosa, R.S., Machado, J.T.: Implementation of discrete-time fractional-order controllers based on is approximations. Acta Polytechnica Hungarica **3**(4), 5–22 (2006)
24. Binias, B., Palus, H.: Feature selection for EEG-based discrimination between imagination of left and right hand movements. Measur. Autom. Monit. **61**(4), 94–97 (2015)
25. Lęgowski, A.: The global inverse kinematics solution in the adept six 300 manipulator with singularities robustness. In: 2015 20th International Conference on Control Systems and Computer Science, pp. 90–97, May 2015
26. Lęgowski, A., Niezabitowski, M.: Manipulator path control with variable order fractional calculus. In: Proceedings of International Conference on Methods and Models in Automation and Robotics (MMAR2016), 29 Aug–1 Sept 2016
27. Hartenberg, R.S., Denavit, J.: Kinematic Synthesis of Linkages. McGraw-Hill, New York (1964)
28. Garrido, I., Garrido, A.J., Romero, J.A., Carrascal, E., Sevillano-Berasategui, G., Barambones, O.: Low effort nuclear fusion plasma control using model predictive control laws. Math. Probl. Eng. **2015**, 1–8 (2015). Article ID 527420
29. Garrido, I., Garrido, A.J., Sevillano, M.G., Romero, J.A.: Robust sliding mode control for tokamaks. Math. Probl. Eng. **2012**, 1–14 (2012). Article ID 341405

Path Planning Based on Direct Perception
for Unknown Object Grasping

Yasuto Tamura[1(✉)], Hiroyuki Masuta[2], and Hun-ok Lim[1]

[1] Department of Mechanical Engineering,
Kanagawa University, Yokohama, Japan
{tamura,holim}@kanagawa-u.ac.jp
[2] Department of Intelligent Systems Design Engineering,
Toyama Prefectural University, Imizu, Japan
masuta@pu-toyama.ac.jp

Abstract. This paper discusses a robot path planning based on the sensation of grasping for a service robot. The previous method should recognize the accurate physical parameters to grasp an unknown object. Hence, we propose the path planning by the sensation of grasping for a decreasing of computational costs. The sensation of grasping affords the possibility of action to a robot directly without inference from physical information. The sensation of grasping is explained by an inertia tensor of three-dimensional point cloud and a fuzzy inference. The proposed method involves an unknown object detection by a depth sensor, and the path planning based on the sensation of object grasping that is determined by the characteristic of the robot and the state of the object. As experimental results, we show that the robot can grasp the unknown object which the robot arm cannot reach on the table due to the movement to the position suitable for the object grasping.

Keywords: Unknown object grasping · Depth sensor · Robot vision · Path planning

1 Introduction

Recently, various types of intelligent robots have been developed and studied for supporting humankind. These intelligent robots are expected to work not only in environments such as factories but also in homes and commercial and public facilities [1]. In a real environment, an intelligent robot should remain in action in order to successfully perform specific tasks, even in an unknown environment. Therefore, an intelligent robot should perceive the environment flexibly, just as humans do. Many previous studies focus on the perception of unknown objects in real environments [2–4]. In fact, research on vision-based approaches is extremely active. One of the most recognized vision-based approaches for the detection of unknown objects is template matching. Template matching can detect a specific object accurately. However, templates need to be provided by an operator or be acquired in advance [5], but it is difficult to prepare templates of objects that need to be grasped, in advance. Consequently, it is difficult to detect an unknown object immediately in a human living environment.

© Springer International Publishing Switzerland 2016
N. Kubota et al. (Eds.): ICIRA 2016, Part I, LNAI 9834, pp. 396–406, 2016.
DOI: 10.1007/978-3-319-43506-0_35

Further, recent well-known object detection approaches detect object shapes by three-dimensional (3D) point processing. Several 3D point processing methods are used for object detection by a search of primitive shapes or visual features. 3D point cloud data for object detection are obtained using feature matching methods. A feature matching method can detect a specific object robustly. Therefore, feature-based or shape-based approaches have been proposed for 3D reconstruction tasks [6–8]. However, these methods require template features in advance. It is difficult to detect unknown objects in the framework of feature matching. Iterative methods such as RANSAC can find imperfect instances of objects within a certain category of shapes. However, these methods require a high computational cost to obtain accurate results. Therefore, a method to detect unknown objects at a low computational cost is required for an intelligent service robot system. In this study, we developed an intelligent service robot system for clearing dishes from a table. This intelligent robot is equipped with various types of sensors for perceiving an environment. We realized the clearing tasks through human interactions using this robot system [9]. However, a robot cannot operate appropriately in an unexpected situation that has unknown objects or unde-tectable objects because it is based on a premise that a robot is given the object information in advance. 3D information of an object is required for object grasping. Therefore, a robot should perceive an unknown object quickly through 3D information without previous object templates in order to execute a flexible action. In order to solve these problems, we focus on plane detection based on the geometrical changes in a 3D point cloud for the discrimination between unknown objects. Further, we proposed an online processable unknown object extraction method based on 3D plane detection [10]. The proposed method consists of plane detection that applies particle swarm optimization (PSO) with region growing (RG) and plane unification based on geo-metrical structures.

Our research target is to develop an object extraction method that can execute a flexible robot action in real time and has high extraction precision without object templates. For the unknown object extraction by a service robot, we intend to achieve a high-speed object extraction based on the detection of object's small planes which has difficulty in detection by conventional methods.

The rest of this paper is organized as follows: Sect. 2 explains our system for environmental measurement. Section 3 explains the unknown object detection and robot path planning based on the sensation of grasping. Section 4 shows experimental results. Section 5 concludes this paper.

2 System for Environmental Measurement

2.1 Outline of Service Robot System

Figure 1 shows an overview of an intelligent robot system for clearing a table. This intelligent robot system consists of a service robot and an interactive robot. The interactive robot recognizes human orders from hand gestures and spoken commands by using a RGB-D vision system and a voice recognition system [10]. The service robot has an arm, and the mobile robot collects an object on the basis of the vision information obtained using the 3D depth sensor and the laser range sensor (LRF). The robot arm called Gamma is small and lightweight, and is similar in size to the human arm. Gamma has a seven degree-of-freedom (7DOF) structure and eight motors, including a gripper. The service robot detects the object position and then, collects the objects from the table.

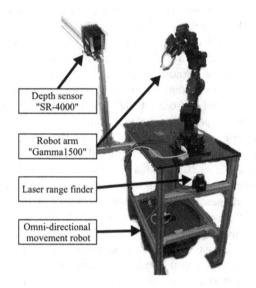

Fig. 1. Service robot

2.2   3D Depth Sensor for Measuring Environment

We use SR-4000 as the 3D depth sensor to measure the environment. SR-4000 is a time-of-flight depth image sensor made by MESA Imaging AG. It can measure 3D distance and two-dimensional (2D) luminance data [11]. The depth range of this sensor is a maximum of 7.5 m, and the output signal is in the quarter common intermediate format (QCIF: 176 × 144 pixels). Therefore, this sensor can measure 25,344-directional distance information simultaneously. SR-4000 is located at a height of 60 cm from the base of the robot arm to fit certain target objects in the sensor range. The maximum frame rate of this sensor is 50 fps, but we set a fixed frame rate of 20 fps, considering the sensing stability and real-time sensing.

Therefore, an intelligent robot can measure significant volumes of data simultaneously and thus reduce the computational cost.

Figure 2 shows a snapshot of the distance and amplitude image. We define the measuring point of the image pixel as the w–h axis of the 2D image. The left output image is left rotated by 90° according to the camera specifications. Moreover, the measuring point of the real world is defined as the x–y–z axis of the robot coordinates, where the origin is the base of the robot arm. The cup is recognized on the basis of 3D distance data by observation. However, the distance data of the cup surface are not complete. A part of the cup is not measured because of the sensor noise and occlusion. The surface boundary is smoothed with a smoothing filter. Further, the edge of the measured field is curved because of lens distortion. Therefore, all the aforementioned conditions should be considered in order to perceive unknown situations. To detect an object to grasp on the basis of these sensing data, we focus on the plane detection method because we consider a cluster of certain planes to be an object.

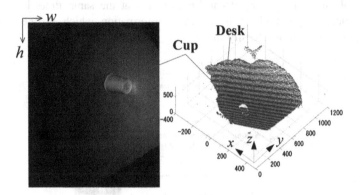

Fig. 2. Snapshot of SR-4000

3 Robot Path Planning by Sensation of Grasping

This section explains the unknown object detection method, the sensation of grasping for object grasping, and the path planning by the estimation of suitable grasping position.

3.1 Unknown Object Detection Method

We have proposed the Simplified Plane Detection based SEgmentation of a point cloud into object's clusters (SPD-SE). Figure 3 shows the flow chart of the proposed method with SPD-SE. The SPD-SE consists of the simplified plane detection (SPD) with region growing (RG), the particle swarm optimization (PSO) based seeded RG and the plane integrating object extraction [12, 13]. The PSO based seeded RG is the first step to detect planes which is composed of object from point cloud data. Figure 3(a), (b) and (c) shows a snapshot of SR-3000 image, distribution of particles and a result of simplified plane detection, respectively. The particles of PSO are updated to gather around detected small planes [14]. However, it is difficult to detect a steady plane because of sensing noise. We applied the estimation accuracy parameter to consider the changing time series. As a result, a stable plane detection result is obtained, similar to Fig. 3(d). But, there are many planes on the object position in the case of a round object and multi-object situation. Therefore, the planes are integrated to one plane set as an object based on geometric invariance of each plane, like Fig. 3(e) [14]. The largest detected plane as a table plane is eliminated from the detected plane.

Figure 3(e) shows the object extraction results of SPD-SE. The point cloud set corresponding to the cup is extracted accurately. The SPD-SE was shown that the extraction accuracy of point cloud on an object is higher and the computational cost is less than a point cloud library [15]. Moreover, the calculation time of SPD-SE is less 50 [ms]. Therefore, the SPD-SE may have applicability to robot perception at real time. The extracted object of SPD-SE has not only point group of a box, but also posture and combination of composing plane. So, there are advantages that the point group and properties of an unknown object can be extracted at the same time. However, the detected object of point cloud is imperfect information which is one side view. Therefore, it is difficult to identify the specific object size and posture.

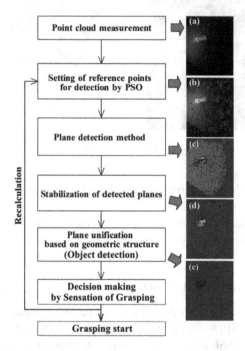

Fig. 3. Flow of the proposed method with SPD-SE

3.2   Sensation of Grasping Based on Characteristic of Robot Arm

A gravity position of the extracted plane set is calculated from set of point cloud data. However, it is difficult to recognize accurate object position, posture, shape and size without predefined knowledge. Because, there are no point cloud data of behind the side of the box and the edges of planes are not sharp by smoothing filter. Hence, we propose a sensation based perception which perceives exclusively the relevant information to the behavior of the robot.

The accurate object size, posture and shape are not important if a person decides to grasp an object. There is a direct perception concept in ecological psychology [16, 17]. A direct perception insists that relevant information for action is perceived directly without integrating some physical information such as size, posture and so on. Therefore, we propose the sensation of grasping based on ecological psychology as following equation.

$$
\mathbf{I}_t = \begin{bmatrix} I_{x'} & I_{x'y'} & I_{x'z'} \\ I_{y'x'} & I_{y'} & I_{y'z'} \\ I_{z'x'} & I_{z'y'} & I_{z'} \end{bmatrix} = \begin{bmatrix} \frac{m}{N}\sum\limits_{n=1}^{N}\left(y_n'^2+z_n'^2\right) & \frac{m}{N}\sum\limits_{n=1}^{N}\left(-y_n'x_n'\right) & \frac{m}{N}\sum\limits_{n=1}^{N}\left(-z_n'x_n'\right) \\ \frac{m}{N}\sum\limits_{n=1}^{N}\left(-x_n'y_n'\right) & \frac{m}{N}\sum\limits_{n=1}^{N}\left(x_n'^2+z_n'^2\right) & \frac{m}{N}\sum\limits_{n=1}^{N}\left(-z_n'y_n'\right) \\ \frac{m}{N}\sum\limits_{n=1}^{N}\left(-x_n'z_n'\right) & \frac{m}{N}\sum\limits_{n=1}^{N}\left(-y_n'z_n'\right) & \frac{m}{N}\sum\limits_{n=1}^{N}\left(x_n'^2+y_n'^2\right) \end{bmatrix}
$$
(1)

where $\mathbf{I}_t$ is an inertia tensor. x'_n, y'_n and z'_n are distance points transformed into an object gravity center coordinate shown by Fig. 4. m is unit weight, N is the number of the point cloud on an object. The value of principal moment of inertia indicates the size of an extracted object. For example, the value of principal moment of inertia around the y-axis means the sense of a width size in robot view. Especially, the inertia moment around y' and z' axis should be smaller than the object size by grasping of robot hand, because the robot arm can only reach from the front side. Therefore, the combination of principal moments of inertia is explained the sensation of grasping. The part of sensation of grasping in relation to an inertia tensor is calculated as follows based on simplified fuzzy inference [18, 19].

$$
\mu_r = \min\{A_{1,r}(I_{x'}), \quad A_{2,r}(I_{y'}), \quad A_{3,r}(I_{z'})\}
$$
(2)

$$
F = \frac{\sum\limits_{r=1}^{R}\mu_r \cdot y_r}{\sum\limits_{r=1}^{R}\mu_r}
$$
(3)

where μ_r is the fitness value of the r-th rule applying membership functions A of 3 values of principal moment of inertia, R is the number of rules, y_r is output of r-th rule. In our fuzzy inference, x_1, x_2, and x_3 of inputs are $I_{x'}$, $I_{y'}$, $I_{z'}$, respectively. F is the output which means the graspability by the value of principal moment of inertia. The membership function shows Fig. 5, SM, MM and LM are shown small, middle and large of inertia moment, respectively. The size of membership function is decided in consideration of the embodiment of the robot. Table 1 shows the rules of fuzzy inference. The consequent part function yf means the sensation of grasping given by singleton. There are 20 rules, 1.0 means a most likely graspability situation. In this case, the rule is decided by heuristics. The rules consist of the principal moment of inertia, because the movement of the robot arm can not change the coordinate of x'_n, y'_n and z'_n.

A robot can decide to grasp an object when the sensation of grasping has high value. Therefore, the sensation of grasping affords the possibility of action to a robot directly without inference from physical information such as size, posture and shape.

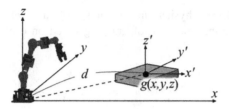

Fig. 4. The coordination for the sensation of grasping

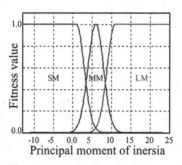

Fig. 5. Membership function for sensation of grasping

Table 1. Fuzzy rules for sensation of grasping

Ix'	SM								
Iy'	SM			MM			LM		
Iz'	SM	MM	LM	SM	MM	LM	SM	MM	LM
yf	0			0.9			0		0

Ix'	MM								
Iy'	SM			MM			LM		
Iz'	SM	MM	LM	SM	MM	LM	SM	MM	LM
yf	0.4	0.4	0.1	0.5	1	0.9	0	0	0

Ix'	LM								
Iy'	SM			MM			LM		
Iz'	SM	MM	LM	SM	MM	LM	SM	MM	LM
yf		0.2	0.1		0.7	0.9	0	0	0

Moreover, the distance between robot and object has an effect on the sensation of grasping. The part of the sensation of grasping in relation to the distance between robot and object is calculated as follows.

$$D = \frac{1}{1 + \exp\{q_1 * (d - q_2)\}} * \frac{1}{1 + \exp\{q_3 * (d - q_4)\}} \tag{4}$$

where q_1, q_2, q_3, and q_4 are parameters for adjustment of perspective sensibility and d is the distance between robot and object. d is restricted by the shape around the table. Therefore, d is a function of an angle θ around the object on the table. Furthermore, the result of fuzzy inference is the function of an inertia tensor. Then the sensation of grasping is calculated as follows based on Eqs. (2), (3) and (4).

$$S(\mathbf{I}_t, \theta) = F(\mathbf{I}_t) * D(\theta) \tag{5}$$

$$\theta_{suitable} = \arg\max_{\theta \in [-\pi, \pi]} \{S(\mathbf{R}_z(\theta) \cdot \mathbf{I}_t \cdot \mathbf{R}_z^{-1}(\theta))\} \tag{6}$$

where S is the sensation of grasping, $\theta_{suitable}$ is the angle to maximize S from -180 to 180 degrees. An inertia tensor $\mathbf{I}_t$ can change a principal axis of inertia by multiplying rotation matrix $\mathbf{R_z}(\theta)$. Therefore, a robot can simulate the higher graspability direction without an accurate 3D information data.

3.3   Path Planning by the Estimation of Suitable Grasping Position

The service robot can detect a relative self-position between the table and the robot by using LRF. The robot detects an outer edge of a table by RANSAC and 2-dimensional point cloud from LRF. Then the robot can move around a table while maintaining the distance of 100 [mm] between the table and the robot. The shape information excluding the position of the table is assumed to be known. Figure 6 shows the result of the table position detection in the robot initial state, where small blue points are 2D depth data, red lines are results of line detection by RANSAC, and green points are detected endpoints based on the space continuity of depth points. The depth points on the horizontal red line toward the robot is estimated to be the table edge. Then the robot coordinates the self-position so that the robot front parallels the table edge in the initial state. Subsequently, the robot estimates the suitable grasping position by a higher graspability direction and a robot movement track around a table. There is a case of several high graspability directions by Eq. (6) such as the position A and D in Fig. 7.

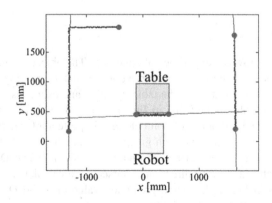

Fig. 6. Table position detection by LRF in robot initial state (Color figure online)

The blue dotted line in Fig. 7 is a track which the robot can move around a table. The robot moves to the position with higher graspability value based on Eq. (6) in consideration of change of $d(\theta)$.

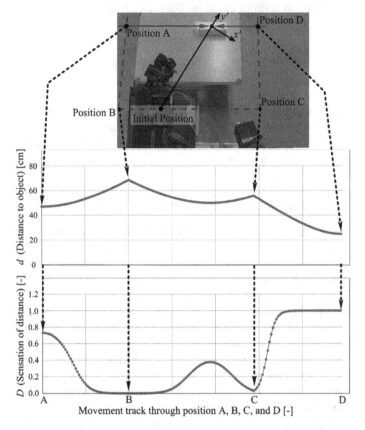

Fig. 7. Estimation of suitable grasping position (Color figure online)

4 Experiment

Top figure in Fig. 7 shows an experimental condition. The object is outside the reach of the robot arm. Therefore, the robot must move to a suitable grasping position.

The robot detects the posture of the table by LRF, and estimates a track which the robot can move. The blue dotted line in Fig. 7 is a track which the robot can move around a table. The detected object of point cloud is imperfect information which is one side view. Hence, two symmetrical high graspability directions are estimated. Despite the view from position D is imperfect information, Position A and D where the blue dotted line and high graspability directions intersects are calculated.

Bottom figures in Fig. 7 shows the prediction value of d and D by the estimated track with Position A, B, C, and D. In the experiment, parameters q_1, q_2, q_3, and q_4 in

Eq. (4) are 2.0, 9.0, −10.0, and 0.5, respectively. The robot moves to the high graspability position D. There is the interval which is high value of D between the initial position and Position C. Despite this, the robot moved to the position D through this interval due to a low value of F by fuzzy inference. The higher graspability position is predicted by simple calculation using tensor expression and robot itself physicality without complicated 3D model.

5 Conclusion

This paper discusses the direct perception and the path planning for unknown object grasping using depth sensor. To perceive an unknown object, we have proposed the plane detection based approach of the SPD-SE.

The SPD-SE is the PSO-based plane detection method with RG and the object extraction method based on geometric structure. However, it is difficult to recognize accurate object position, posture, shape and size without predefined knowledge. Because, there are no point cloud data of behind the side of the box and the edges of planes are not sharp by smoothing filter. Hence, we propose a sensation based perception which perceives exclusively the relevant information to the behavior of the robot. The sensation of grasping is explained by an inertia tensor, a fuzzy inference, and an evaluation function of distance to object. The sensation of grasping affords the possibility of action to a robot directly without inference from physical information such as size, posture and shape.

As experimental results, we show that the robot can detect a relevant information of a grasping behavior directly. And, the sensation of grasping presents own state and environmental state together by one parameter.

As future work, we will verify the sensation of grasping during the operation of the robot.

References

1. Mitsunaga, N., Miyashita, Z., Shinozawa, K., Miyashita, T., Ishiguro, H., Hagita, N.: What makes people accept a robot in a social environment. In: International Conference on Intelligent Robots and Systems, pp. 3336–3343 (2008)
2. Liu, Z., Kamogawa, H., Ota, J.: Fast grasping of unknown objects through automatic determination of the required number of mobile robots. J. Robot. Soc. Jpn. **27**(6), 445–458 (2013)
3. Yamazaki, K.: A method of grasp point selection from an item of clothing using hem element relations. J. Robot. Soc. Jpn. **29**(1), 13–24 (2015)
4. Dube, D., Zell, A.: Real-time plane extraction from depth images with the randomized hough transform. In: 2011 IEEE International Conference on Computer Vision Workshops, pp. 1084–1091 (2011)
5. Opromolla, R., Fasano, G., Rufino, G., Grassi, M.: A model-based 3D template matching technique for pose acquisition of an uncooperative space object. IEEE Sens. **15**(3), 6360–6382 (2015)

6. Segundo, M.P., Gomes, L., Bellon, O.R.P., Silva, L.: Automating 3D reconstruction pipeline by surf-based alignment. In: 19th IEEE International Conference on Image Processing (ICIP) 2012, pp. 1761–1764 (2012)
7. Toshev, A., Taskar, B., Daniilidis, K.: Shape-based object detection via boundary structure segmentation. Int. J. Comput. Vision 99(2), 123–146 (2012)
8. Belongie, S., Malik, J., Puzicha, J.: Shape matching and object recognition using shape contexts. IEEE Trans. Pattern Anal. Mach. Intell. 24(24), 509–522 (2012)
9. Masuta, H., Hiwada, E., Kubota, N.: Control architecture for human friendly robots based on interacting with human. In: Jeschke, S., Liu, H., Schilberg, D. (eds.) ICIRA 2011, Part II. LNCS(LNAI), vol. 7102, pp. 210–219. Springer, Heidelberg (2011)
10. Masuta, H., Makino, S., Lim, H., Motoyoshi, T., Koyanagi, K., Oshima, T.: Unknown object extraction based on plane detection in 3D space. In: 2014 IEEE Symposium on Robotic Intelligence In Informationally Structured Space (RiiSS), pp. 135–141, December 2014
11. Dellen, B., Alenyà, G., Foix, S., Torras, C.: 3D object reconstruction from Swissranger sensor data using a spring-mass model. In: International Conference on Computer Vision Theory and Applications (VISAPP), pp. 368–372 (2015)
12. Adams, R., Bischof, L.: Seeded region growing. IEEE Trans. Pattern Anal. Mach. Intell. 16, 641–647 (1994)
13. Kennedy, J., Eberhart, R.C.: Particle swarm optimization. In: Proceedings of the 1995 IEEE International Conference on Neural Networks, vol. 4, pp. 1942–1948 (1995)
14. Masuta, H., Makino, S., Lim, H.: 3D plane detection for robot perception applying particle swarm optimization. In: World Automation Congress (WAC), pp. 549–554 (2014)
15. Point Cloud Library. http://pointclouds.org/
16. Gibson, J.J.: The Ecological Approach to Visual Perception. Lawrence Erlbaum Associates, Hillsdale (1979)
17. Turvey, M.T., Shaw, R.E.: Ecological foundations of cognition. Int. J. Conscious. Stud. 6, 95–110 (1999)
18. Takagi, T., Sugeno, M.: Fuzzy identification of systems and its applications to modeling and control. IEEE Trans. Syst. Man Cybern. 15(1), 116–132 (1985)
19. Masuta, H., Lim, H.: Direct perception and action system for unknown object grasping. In: Proceedings of the 24th IEEE International Symposium on Robot and Human Interactive Communication, pp. 313–318 (2015)

An Efficient RFID-Based Localization Method with Mobile Robot

Haibing Wu, Zeyu Gong, Bo Tao$^{(\boxtimes)}$, and Zhouping Yin

State Key Laboratory of Digital Manufacturing Equipment Technology,
Huazhong University of Science and Technology,
Wuhan 430074, Hubei, The People's Republic of China
taobo@hust.edu.cn

Abstract. A RFID-based localization method, using the phase data and its gradient information of radio signal between the RFID reader and tag, is proposed in this paper. The unknown RFID tags location can be calculated through calibrating phase drift and structuring count matrix and difference matrix. A mobile robot equipped with odometer sensors and a RFID reader is employed herein to form the synthetic aperture. Simulations are performed, and the results show that the proposed method can achieve a satisfied localization accuracy at shorter synthetic aperture length, and has the potential in warehouse management.

Keywords: RFID · Localization method · Mobile robot · Localization · Synthetic aperture · Phase gradient

1 Introduction

In recent years, mobile robot has been widely applied to management and transportation of goods. The capability of locating goods automatically will significantly improve the efficiency and automation level of the mobile robots. Radio Frequency Identification (RFID) has been applied to warehouse logistics, and enjoyed the reputation of highly efficient in remote reading and identification. Embedding the RFID technology into mobile robot becomes a promising solution of goods localization. Therefore, RFID-based location awareness technology attaches growing researcher's attention, and reveals satisfactory locating accuracy with a relatively low cost [1–5]. Generally, the RFID reader is installed on the robot, and RFID tag on the goods.

Received signal strength indicator (RSSI) and phase are two main information outputted by RFID reader which is related to orientation and distance between reader and tag [6]. In many localization method, RSSI is used as a visual indication for distance. However, RSSI is not sensitive to distance, and is easily interfered by the circumstance. In addition, RSSI may vary with the change of the relative angle between reader antenna and tag, resulting in severe degrading of localization accuracy [7, 8]. In contrast, phase has a high range resolution that means it can perceive small distance variety [3, 7]. Meanwhile, it is direction-independent and only dependent on distance. For these reasons, phase information is preferable to develop the localization method in this paper. Despite its advantages, phase has its defects, such as full-cycle ambiguity

© Springer International Publishing Switzerland 2016
N. Kubota et al. (Eds.): ICIRA 2016, Part I, LNAI 9834, pp. 407–417, 2016.
DOI: 10.1007/978-3-319-43506-0_36

(different distance, but same phase) and phase drift. These are the challenges to be coped with in phase applications.

Some published papers employ phase information to establish models with the sum of unit vectors [9–11]. To achieve a precise localization result, a long synthetic aperture to sample a considerable number of location points to overcome the ambiguity of phase. However, the motion distance of mobile antenna may be restricted in certain scenarios. So, developing precision in limited motion distance is the main purpose of this paper.

In this paper, an efficient RFID-based localization method, in which the direct relationship between distance and phase is used, which can exclude most of impossible points, instead of doubting all of points. Besides, the change scope of phase is different in the same movement distance of antenna, call it phase gradient, it is believed that this information is positive to improve the localization quality.

The rest of this paper is organized as follows. The phase feature is analyzed, and localization scene and problems are described in Sect. 2. The localization algorithm is built in Sect. 3. Simulation experiments and results are given in Sect. 4. Finally, the conclusions are given in Sect. 5.

2 Localization Problems

In this section, the features of phase, such as full-cycle ambiguity, direction-independence and phase gradient, are discussed in detail. Then, the localization scene and pretreatment of this localization method are described.

2.1 Phase Feature

Phase. In this paper, ultrahigh frequency (UHF) RFID device and passive tag are employed. Tag obtains energy and information from the reader's RF signal, meanwhile, it feedbacks RF signal to the reader through back-reflection. The reader measure the signal's phase value. The relationship between phase and distance can be expressed as:

$$\theta = \left(\frac{2\pi}{\lambda} \times 2d + \varphi + \varepsilon \right) \bmod 2\pi \tag{1}$$

Where θ is the output phase value. λ is the wavelength of RF signal, it can be calculated by $\lambda = c/f$, c is the speed of light, f is the frequency of RF signal, in china, it is $920.625 \sim 924.375$ MHz. d is the distance from reader to tag. φ represents the phase drift, which is related to the hardware characteristics of reader and tag. ε is the measurement noise, $\varepsilon \in N(\mu, \sigma^2)$. Reader can't acquire the numbers of full period, only the remainder of 2π, so phase is full-cycle ambiguity. Figure 1 describes the ideal distribution of phase when φ is 2, in which reader antenna is located in the center point (0, 0). From this figure, it can be observed that phase is completely the same at same radius, it is direction-independent. Besides, phase is highly sensitive to distance, approximately 0.45 mm/°.

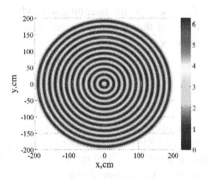

Fig. 1. The ideal distribution of phase   **Fig. 2.** Schematic diagram of tag movement

Phase Gradient. As mentioned above, phase is a troublesome but exact physical parameter. In this paper, the information of phase's gradient difference is employed to realize localization. Figure 2 describes that a tag moves from point $A1$ to point $A2$ in the achievable reading region of reader antenna O. Suppose $A1\left(x_1^T, y_1^T\right), A2\left(x_2^T, y_2^T\right)$, and $O(x^R, y^R)$. Tag's movement direction along x-axis is α, and its movement distance is L. In point $A1$, the distance between tag and reader antenna is d_1, point $A2$ is d_2. Phase gradient can be defined as

$$G = \frac{\Delta\theta}{L} \tag{2}$$

Where $\Delta\theta$ is the phase difference value of $A1$ and $A2$, namely,

$$\Delta\theta = \theta_1 - \theta_2 = \left(\frac{4\pi}{\lambda} d_1 \bmod 2\pi - \frac{4\pi}{\lambda} d_2 \bmod 2\pi\right) \bmod 2\pi = \left(\frac{4\pi}{\lambda} d_1 - \frac{4\pi}{\lambda} d_2\right) \bmod 2\pi$$

$$= \left(\frac{4\pi}{\lambda} \left(\sqrt{\left(x_1^T - x^R\right)^2 + \left(y_1^T - y^R\right)^2} - \sqrt{\left(x_2^T - x^R\right)^2 + \left(y_2^T - y^R\right)^2}\right)\right) \bmod 2\pi$$

Because,

$$\begin{cases} x_2^T = x_1^T + L\cos\alpha \\ y_2^T = y_1^T + L\sin\alpha \end{cases}$$

So, G can be expressed as:

$$G = \frac{\Delta\theta}{L} \triangleq g(x_1^T, y_1^T, \alpha, L) \tag{3}$$

From Eq. 3, phase gradient G varies with the position of starting point $A1$, movement direction α, and movement distance L. Based on this phenomenon, when a tag is moving, the position of tag can be distinguished through phase gradient difference. In this paper, tag's motion carrier is mobile robot.

2.2 Localization Scene and Pretreatment

Localization Scene. Figure 3 describes one possible scene of this localization method. A mobile robot equipped odometer sensors and RFID reader moves along with the black solid line, a lot of RFID tags (unknown location) distribute in the side of movement track. This goal of localization method is to localize these tags through moving robot. The motion path of robot can be seen as a synthetic aperture. This paper only considers and analyzes the situation of straight synthetic aperture.

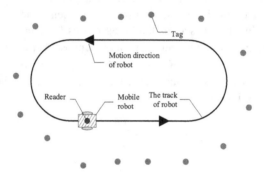

Fig. 3. The localization scene

In general, the trajectory of mobile robot can be reconstructed by its odometer sensors. Suppose the mobile robot is a differential drive kinematic platform with two drive wheels and one universal wheel. $\mathbf{x}_{r,t} = (x_t, y_t, \phi_t)$ represents robot's pose at time t with position (x_t, y_t) and orientation ϕ_t. Robot's kinematic model can be built as below [12]:

$$\mathbf{x}_{r,t+1} = \begin{bmatrix} x_t + \Delta x_t \\ y_t + \Delta y_t \\ \phi_t + \Delta \phi_t \end{bmatrix} = \begin{bmatrix} x_t + \frac{\mu_{R,t} + \mu_{L,t}}{2} \cos \frac{\mu_{R,t} - \mu_{L,t}}{d} \\ y_t + \frac{\mu_{R,t} + \mu_{L,t}}{2} \sin \frac{\mu_{R,t} - \mu_{L,t}}{d} \\ \phi_t + \frac{\mu_{R,t} - \mu_{L,t}}{d} \end{bmatrix} \triangleq f(\mathbf{x}_{r,t}, \mu_{R,t}, \mu_{L,t}) \tag{4}$$

Where $\mu_{R,t}$ and $\mu_{L,t}$ is movement distance of the right and left drive wheel at time interval $(t, t+1)$. This model can realize relative location of robot with recursive calculation.

Actually, because robot's wheel may slip and its odometer information contains noise, with the accumulation of time, the trajectory built by robot's kinematic model may be fail to reflect the real trajectory. Moreover, due to the limited read range of RFID reader, a tag may be irradiated by antenna only in a small scope. So it is very important to achieve satisfied accuracy at shorter robot's trajectory (synthetic aperture length), i.e. the shorter the better. In this localization scene, the localization error of 5 cm can reach required specification of accuracy, which is adequate in warehouse management.

Localization Pretreatment. Before introducing the whole localization algorithm, the localization pretreatment (the conversion of relative coordinate and relative motion) is given, as shown in Fig. 4. In this localization scene, there are two frames, the robot frame UV and the absolute frame XY. In robot frame, suppose tag is in $T'\left(x_0', y_0'\right)$. If the starting positioning of robot is known, the robot's position $R(x, y)$ and motion velocity $\overrightarrow{V}$ in absolute frame can be clearly known through its kinematic model. Suppose $T(x_0, y_0)$, if vector $\overrightarrow{RT'} = \overrightarrow{OT}$, tag T in absolute frame has same coordinate with tag T' in robot frame. Using the similar method, if robot (install antenna) moves at velocity $\overrightarrow{V}$, then it is assumed that tag T moves at velocity $\overrightarrow{V'}$, which their size is equal and direction is opposite. The purpose of doing this is to transform a synthetic aperture (mobile antenna and static tag) to an inverse synthetic aperture (static antenna and mobile tag), namely, solving the position of static tag T' in robot frame is the same as solving the starting position of moving tag T in absolute frame. This conversion would be beneficial for understanding and handling localization method.

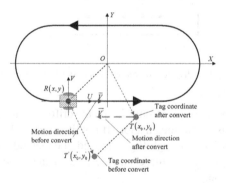

Fig. 4. The conversion of relative coordinate and relative motion

3   Localization Algorithm

The flowchart of this localization method's algorithm is provided in Fig. 8. It is distributed into two stages, offline process and online process. The former is to prepare for the later. The detail position solving process is given as follows.

3.1   Offline Process

The Fingerprint Database of Phase. As shown in Fig. 5, the region of reader's accessible reading is divided into a lot of grids H, whose side length is S. These vertices of grids can be seen as virtual tags. For accelerating the position solving, an offline fingerprint database of phase value θ is prepared through Eq. 1, denoted as:

$$\theta = \begin{pmatrix} \theta_{1,1} & \cdots & \theta_{1,N} \\ \vdots & \vdots & \vdots \\ \theta_{M,1} & \cdots & \theta_{M,N} \end{pmatrix} \tag{5}$$

Where M and N is the number of these grids's row and column. In this paper, reader's position is assumed at the original point O of the absolute frame XY as mentioned above.

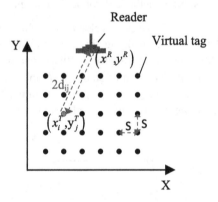

Fig. 5. The division of reader's accessible reading region

The Phase Drift. Especially point out, the phase drift φ of RFID device must been calibrated before it is used into localization. If not, it is impossible to directly calculate potential area of tag position through Eq. 1. Indeed, the method can exclude most area, then only minority area may be the candidate area of tag coordinate, which is very beneficial for localization.

3.2   Online Process

Now, suppose the mobile robot is moving along with its track, and the RFID device always outputs the phase value. Two matrixes are built, one count matrix I, one difference matrix Q, and they are initialized to zero. Their dimension is equal to grids H. The following is a detailed description about online process.

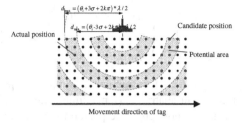

Fig. 6. Regional division. Given a phase value, antenna's accessible reading region can be divided into potential area and impossible area of tag position, all of grids H in potential area are candidate positions of tag's position, and those in impossible area can be ruled out.

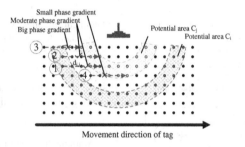

Fig. 7. The application principle of phase gradient. Combining two potential areas C_i, C_j with robot's movement distance $d_{i,j}$, the size of candidate position's phase gradient can be divided into immoderate(big-tab 3, small-tab 4) and moderate(tab 1 and tab 2). Count matrix I would reflect this difference between immoderation and moderation.

- STEP 1: Obtain current phase value θ_i, and calculate the potential area C_i that tag may be in through Eq. 1. Considering Gaussian measurement noise ε, $C_i = \{H|(|\boldsymbol{\theta} - \theta_i| \bmod \pi) \leq 3\sigma\}$, as shown in Fig. 6.

- STEP 2: After robot moves, its movement distance $d_{i,j}$ is calculated according to robot's kinematic model, simultaneously, obtain its phase value θ_j, and calculate the potential area C_j.

- STEP 3: Make use of the application principle of phase gradient, as shown in Fig. 7, if its phase gradient is moderate, the corresponding element in count matrix I would plus one, else plus zero. The form of count matrix can be expressed as:

$$I = \begin{pmatrix} I_{1,1} & \cdots & I_{1,N} \\ \vdots & \vdots & \vdots \\ I_{W,1} & \cdots & I_{W,N} \end{pmatrix} + \begin{pmatrix} \chi_{1,1} & \cdots & \chi_{1,N} \\ \vdots & \vdots & \vdots \\ \chi_{W,1} & \cdots & \chi_{W,N} \end{pmatrix}, \chi_{i,j} = 0 \quad or \quad 1 \quad (6)$$

- STEP 4: In STEP 3, the phase gradient of tab 1 and tab 2 are both moderate, the phase difference value between the measured phase value and its ideal value stored in fingerprint database of phase $\boldsymbol{\theta}$ is calculated to distinguish these two tabs. Then, a difference matrix Q is maintained, it can be denotes as:

$$Q = \begin{pmatrix} Q_{1,1} & \cdots & Q_{1,N} \\ \vdots & \vdots & \vdots \\ Q_{M,1} & \cdots & Q_{M,N} \end{pmatrix} + \begin{pmatrix} q_{1,1} & \cdots & q_{1,N} \\ \vdots & \vdots & \vdots \\ q_{M,1} & \cdots & q_{M,N} \end{pmatrix}, q_{i,j} = \left| \boldsymbol{\theta} - \theta_j \right| \bmod \pi \quad (7)$$

- STEP 5: Repeat STEP 1—STEP 4, extract the most probable point of tag's position where the element of count matrix I is maximum, and difference matrix Q is minimum. This point is the most probable position that the tag is in.

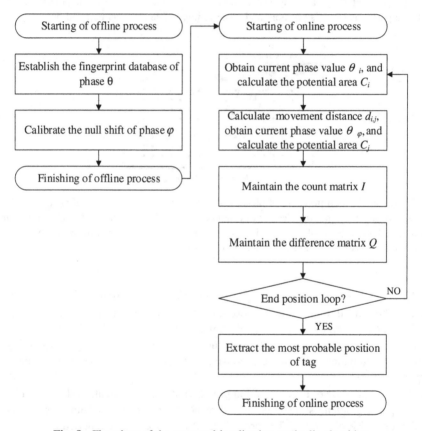

Fig. 8. Flowchart of the proposed localization method's algorithm

4   Simulations and Discussions

A simulation experiment is carried out to compare, analyze and verify this localization method's performance. In simulation, only one antenna is considered, the side length S of these grids is 0.5 cm, the speed of mobile robot is about 20 cm/s, and the sample interval is about 0.05 s.

4.1   Synthetic Aperture Length

The localization method needs to make use of the synthetic aperture to realize position calculation. As mentioned above, the trajectory (synthetic aperture) calculated by the robot kinematic model may be not very precise when the trajectory is lengthen, and the length of synthetic aperture is limited due to the limited read range of RFID reader. So the length of synthetic aperture that this method can reach a satisfied accuracy is an important manifestation of the effectiveness of the algorithm. In this paper, it is carried out to analyze the relationship between localization accuracy and synthetic aperture length. The algorithm given in ref. [11] is used to compare with this paper's algorithm in the case of same phase measurement noise. These experiments are repeated 50 times, and calculate their averages of localization accuracy. The result is shown in Fig. 9. From it, it can be seen that these two methods's localization accuracy can be improved with the growth of synthetic aperture length, finally, it can achieve stable performance. This paper's algorithm requires shorter synthetic aperture length at localization accuracy of 5 cm compared with the algorithm given in ref. [11]. In fact, considering the phase drift is calibrated, many positions that their phase gradient are similar with true tag position are directly excluded, meanwhile, the combination of count matrix and difference matrix are also beneficial for localization.

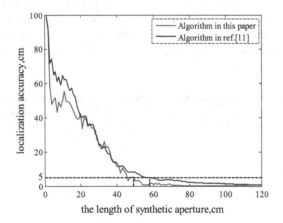

Fig. 9. Localization accuracy with the length of synthetic aperture

4.2   Phase Measurement Noise

Phase measurement noise is Gaussian distribution, $\varepsilon \in N(\mu, \sigma^2)$, where μ is average value, σ is standard deviation. Suppose $\mu = 0$ radian, and $\sigma = 0.1 - 0.2$ radian, which matches the practical case. During the simulation, it is noticed that the standard deviation has important influence on the performance of localization method, the smaller the better, as shown in Fig. 10. But, although at the biggest noise ($\sigma = 0.2$ radian), this localization method is still relatively stable, and can reach a satisfied accuracy, of course, it needs a longer synthetic aperture length.

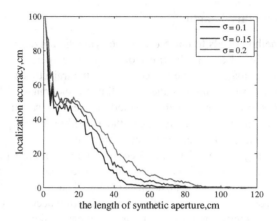

Fig. 10. The influence of phase measurement noise

5 Conclusions

This paper fully exploits the available information of phase, phase gradient and synthetic aperture formed by mobile robot kinematic model, and realizes localization of the unknown tags. In the aspect of localization algorithm, the calibration of phase drift and the structure of count matrix and difference matrix are both conducive to localization performance. The simulation results indicate this method can reach a satisfied localization accuracy at shorter synthetic aperture length. This method is expected to apply to warehouse management.

Acknowledgments. This work is supported by the National Science Foundation of China under Grant 51575215 and 51535004, the National Science and Technology Major Project of China under Grant 2014ZX04014101 and the Guangdong Innovative Research Team Program under Grant 2011G006.

References

1. Ko, C.-H.: RFID 3D location sensing algorithms. Autom. Constr. **19**(5), 588–595 (2010)
2. Nan, L., Becerik-Gerber, B.: Performance-based evaluation of RFID-based indoor location sensing solutions for the built environment. Adv. Eng. Inform. **25**(3), 535–546 (2011)
3. Yang, L., Chen, Y., Li, X.-Y., Xiao, C., Li, M., Liu, Y.: Tagoram: real-time tracking of mobile RFID tags to high precision using COTS devices. In: 20th ACM Annual International Conference on Mobile Computing and Networking, MobiCom 2014, 7-11 September 2014, pp. 237–248. Association for Computing Machinery, Maui (2014)
4. Zhou, J., Shi, J.: Localisation of stationary objects using passive RFID technology. Int. J. Comput. Integr. Manuf. **22**(7), 717–726 (2009)

5. Azzouzi, S., Cremer, M., Dettmar, U., Knie, T., Kronberger, R.: Improved AoA based localization of UHF RFID tags using spatial diversity. In: 2011 IEEE International Conference on RFID-Technologies and Applications (RFID-TA 2011), 15-16 September 2011, pp. 174-80. IEEE, Piscataway (2011)
6. Nikitin, P.V., Martinez, R., Ramamurthy, S., Leland, H., Spiess, G., Rao, K.V.S.: Phase Based Spatial Identification of UHF RFID Tags. In: 2010 IEEE International Conference on RFID (IEEE RFID 2010), 14–16 April 2010, pp. 102-109. IEEE, Piscataway (2010)
7. Martinelli, F.: A robot localization system combining RSSI and phase shift in UHF-RFID signals. IEEE Trans. Control Syst. Technol. **23**(5), 1782–1796 (2015)
8. Tianci, L., Lei, Y., Qiongzheng, L., Yi, G., Yunhao, L.: Anchor-free backscatter positioning for RFID tags with high accuracy. In: IEEE INFOCOM 2014 - IEEE Conference on Computer Communications, 27 April-2 May 2014, pp. 379–387. IEEE, Piscataway (2014)
9. Miesen, R., Kirsch, F., Vossiek, M.: Holographic localization of passive UHF RFID transponders. In: 2011 IEEE International Conference on RFID (IEEE RFID 2011), 12-14 April 2011, pp. 32-37. IEEE, Piscataway (2011)
10. Miesen, R., Kirsch, F., Vossiek, M.: UHF RFID localization based on synthetic apertures. IEEE Trans. Autom. Sci. Eng. **10**(3), 807–815 (2013)
11. Parr, A., Miesen, R., Vossiek, M.: Inverse SAR approach for localization of moving RFID tags. In: 2013 IEEE International Conference on RFID (IEEE RFID 2013), 30 April-2 May 2013, pp. 104–109. IEEE, Piscataway (2013)
12. Cook, G.: Mobile Robots: Navigation, Control and Remote Sensing. John Wiley & Sons, Inc., Hoboken (2011)

Vector Maps: A Lightweight and Accurate Map Format for Multi-robot Systems

Khelifa Baizid[1,2](✉), Guillaume Lozenguez[1,2], Luc Fabresse[1,2], and Noury Bouraqadi[1,2]

[1] Mines Douai, IA, 59508 Douai, France
baizid.khelifa@gmail.com
[2] Univ. Lille, 59000 Lille, France
http://car.mines-douai.fr

Abstract. SLAM algorithms produce accurate maps that allow localization with typically centimetric precision. However, such a map is materialized as a large *Occupancy Grid*. Beside the high memory footprint, *Occupancy Grid Maps* lead to high CPU consumption for path planning. The situation is even worse in the context of multi-robot exploration. Indeed, to achieve coordination, robots have to share their local maps and merge ones provided by their teammates. These drawbacks of *Occupancy Grid Maps* can be mitigated by the use of topological maps. However, topological maps do not allow accurate obstacle delimitations needed for autonomous robots exploration. So, robots still have to handle with *Occupancy Grid Maps*. We argue that *Vector-based Maps* which materialize obstacles using collections of vectors is a more interesting alternative. *Vector Maps* both provide accurate metric information likewise *Occupancy Grid Maps*, and represent data as a graph that can be processed for path planning and maps merging as efficiently as with topological maps. Conclusions are backed by several metrics computed with several terrains that differ in size, form factor, and obstacle density.

1 Introduction

Metric SLAM attempts to model a given environment typically in terms of *Occupancy Grids* [1] They discretise the environment as a collection of cells of fixed size. Their size grows with the size of the mapped environments and the precision of range sensors. Beside memory overhead, metric maps are also resource intensive both in terms of CPU, and network bandwidth. Path planning based on metric maps is computationally intensive, even in environments with few obstacles. The larger is the map the more CPU is required for computing a path.

Additionally, more drawbacks of metric maps, based on *Occupancy Grids*, appears in the context of Multi-Robot Systems (MRS) that collaboratively explore unknown environments.

This work was supported by Région Nord Pas-de-Calais as part of the SUCRé project (Human-robots cooperation in hostile environment).

N. Kubota et al. (Eds.): ICIRA 2016, Part I, LNAI 9834, pp. 418–429, 2016.
DOI: 10.1007/978-3-319-43506-0_37

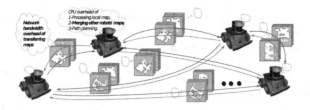

Fig. 1. Multi-robot exploration requires exchanging and merging maps.

Collaboration requires robots to exchange local maps and merge them to build a global map faster Fig. 1). However, the larger is the map, the larger is the network traffic to transfer it. Given that $required\_bandwidth(kbps) = transferred\_data\_size(kb)/time(s)$, in case of many large maps transferred simultaneously, the required bandwidth may exceed the network capacity.

Besides, map fusion is also computationally expensive. The larger is the robotic fleet, the more maps will be merged by each robot. As a result the amount of required CPU for map fusion can quickly become significantly high.

An alternative mapping approach relies on topological maps [2,3]. A topological map can be modeled as a graph, where nodes denote locations in the environment, while edges connect nodes that are adjacent locations. Such a graph requires much less memory than an *Occupancy Grid*. It also requires less CPU for path planning and map fusion when a fleet of robots performs a complex coordination [4]. Indeed, the size of a topological map is independent from the size of the terrain. However, pure topological maps are unsuitable for localization or navigation because of their limited metric information. Thus, topological maps cannot always replace metric maps in heterogeneous multi-robot fleets context and they include metrical information.

In the context of multi-robot exploration, an ideal map should:

- provide metric information that accurately model the environment;
- minimize the CPU overhead for path planning;
- minimize the CPU overhead for map fusion;
- minimize the memory footprint;
- minimize the network bandwidth overhead for map sharing.

We focus on 2D maps produced out of laser range sensor scans. Our long-term goal is to build a SLAM algorithm that directly produces an ideal map from laser scans. But, before developing such an algorithm, we need first to identify a map format that meets requirements listed above.

In this paper, we investigate the use of *Vector-based Maps*[1] as an ideal format for maps. We have chosen the vector maps because they exhibit the advantages of both topological and metric maps. A *Vector Map* materializes obstacles as a collection of line segments. Such a map occupies significantly less memory (typically Kilobytes) than a metric map (typically Megabytes). As a result, transmitting a

---

[1] We use *Vector-based Map* and *Vector Map* interchangeably throughout the paper.

Vector Map over the network consumes less bandwidth. A map with smaller size and organized as a graph structure also drastically reduces CPU consumption for both path planning and map fusion.

In the following, we first give a brief overview of state of the art (Sect. 2). Then, we introduce the vector-based map format (Sect. 3). We show how to convert a *Vector Map* from and to a metric map. Next, we introduce a selection of metrics to compare vector and metric maps (Sect. 4). We then use these metrics to evaluate *Vector Maps* built for several terrains that differ in size, form factor, and obstacle density (Sect. 5). Our *Vector Maps* are built out of *Occupancy Grids* maps generated with the Karto SLAM algorithm [5]. Last, we draw conclusions and sketch some future works (Sect. 6).

2   State of the Art

The most used metric map consist in *Occupancy Grids* [6], where the environment is represented in terms of occupied cells for obstacles and unoccupied cells for free spaces. However, they are extremely expensive in terms of memory size. Moreover, they are unsuitable for data associations which implies a huge CPU load [7]. CPU overhead is even worse in multi-robot exploration missions, since robots have to merge maps. This makes it very expensive to compute complex coordination. Deliberative approaches are limited to sequential frontier attribution that requires efficient communications.

It is possible to reduce the size of the occupancy memory of metric maps by proposing an approach of compression using RANSAC map matching and sparse coding as building blocks [8]. Authors claim that the proposed approach performs well in terms of compression ratio, speed and retrieval performance of compressed/decompressed maps. While this compression reduces network overhead, the path planning and map fusion still require dealing with the full size *Occupancy Grid*. A relevant example of reducing the map size, applied in 3D environment, the one given by Armin et all [9]. The approach was based on Octrees tree data structure, where the space is partitioned by recursively subdividing it into eight octants. In fact, the structure is a tree of 3D spaces (e.g., parallelepiped). This approach proves to be efficient and it is used most in 3D graphics and 3D game engines, however, it explicitly represents free volumes in the tree. Moreover, it is expected to have computational complexity when the map precision is high and/or the environment is unstructured. In this paper, we compare vector maps to compressed occupancy grid maps (sparse matrices). This gives insights on similar other approaches such as Octree maps (in 2D space) and the RANSAC approaches.

Topological approaches [2,3] in map building have been considered as a great alternative to the metric approaches to reduce memory footprint. Moreover, they help to speed up path planning [10] and allow a fleet of robots to handle complex coordination as task allocation in punctual distributed path planning phases [4]. However, metric environment representation is usually mandatory for modeling the navigable free space and for localization. This leads to a new trend in SLAM algorithms to have both representations (i.e. topological and metric) [3,11].

Hybrid metric-topological approaches take advantage from topological relations allowing to use algorithms based on graph theory and metric representation, which come with euclidean mathematics. A generic interface toward hybrid metric-topological SLAM was proposed by Blanco et al. [10]. The algorithm builds a graph of *Occupancy Grid Maps*. *Occupancy Grids* are nodes modeling local areas and edges materialize as topological transformation between *Occupancy Grids*. Such a representation is well suited for large-scale environments mapping and exploration. Similarly, Jongwoo et al. [12] proposed a vision-based mapping of large-scale environments. The approach builds a hybrid representation of the environment of topological and metric maps. The approach relies on optimizing the global map (topological), while it maintains the metric property of the local map. Authors claim that loop closures issues tackled in [10] are handled very efficiently, by benefiting from topological representation. Other alternatives hybrid metric-topological are proposed using collection of range scans [7] or vector-based features [13] but there are not deeply investigated. These approaches permit to model the surrounding environment using a graph structuring and positioned objects with relations between them. In [13] objects are frames of vector-based obstacle delimitation with distance and uncertainty relations. It implements the Extended Kalman Filter (EKF) to estimate the maps and robot poses.

While we appreciate benefits from these approaches, we believe that they rely on a heavy structure generally based on one large or several *Occupancy Grids* with the addition of a graph structure.

To speed up global path planning and help to achieve faster navigation and localization several researchers have attempted to extract topological map models from grid maps [11,14–17]. Fabrizi and Saffiotti [15] proposed an approach based on fuzzy grid map to deal with sensors uncertainties and benefits from image processing to define free and occupied space regions. While, Myunget et al. [14] extract virtual doors by overlapping the Generalized Voronoi Diagram (GVD) and a configuration space eroded by the half size of the door. Kwangro et al. [17] have attempted to improve results given in [14] by building their approach on an algorithms given in [18] to detect edges and obstacle curvature. The resulting map is optimized by genetic algorithms to merge nodes and reduce edges. While we are aware that this approach is dedicated to path planning of a robotic cleaner, we argue that the quality of the generated map (e.g., obstacle counters) is rather poor.

Many small areas (e.g., non free space) were not represented. This decrease of map precision is not suitable for a multi-robot systems that explores and builds a shared map in a coordinated way. Indeed errors on a local map accumulate for the whole team.

There is in the literature a work that builds a map based on geometric primitives [19]. Vandorpe et al. used lines and circles to represent a given environment. Other work [20] presents a technic to create a map of the surroundings of a mobile robot by converting the raw data of a scanning sensor to a map composed of polygonal curves. Nevertheless, there exist no evaluation or comparison of these approaches with *Occupancy Grids*.

3   Vector-Based Map

A *Vector Map* is a graph that delimits the navigable space (i.e. free space) and the un-navigable one (i.e. obstacles). The graph is defined by vertices and edges where each vertex $v \in V$ is a particular delimitation position (e.g., corner) and each edge $e \in E$ is a continuous delimitation. Assuming a two dimensional *Vector Map*, a vertex v defines a position at the coordinates $(x_v, y_v) \in \mathbb{R}^2$ and an undirected edge $e(v_e, v_e')$ defines a line-segment obstacle delimitation. Each point p_e of an edge e is at the frontier between navigable space and obstacle ($p_e = v_e + k \cdot \overrightarrow{v_e v_e'}$ with $k \in [0, 1]$).

Using a range sensor or *Occupancy Grid Map*, obstacle delimitation is first obtained in a more or less structured points cloud format. Obstacle delimitations are defined as a finite set of points $P \subset \mathbb{R}^2$. A valid *Vector Map* $\langle V, E \rangle$ has to respect a constraint of ϵ-error threshold and a constraint of continuity (example Fig. 2e). The constraint of ϵ-error threshold defines the resolution of the generated map. Whatever a point $p \in P$ in the initial obstacle delimitation, there is at least one segment $e \in E$ in the *Vector Map* where the distance between the point p and the segment e is less than ϵ:

$$\forall p \in P, \quad \exists e \in E, \quad distance(p, e) < \epsilon.$$

The constraint of ϵ-error threshold is mainly responsible for the growing on the number of edges in the *Vector Map*. The smaller is ϵ, the less is the number of edges that are necessary to approximate the initial points cloud. The continuous constraint ensures that *Vector Map* edges do not introduce obstacles that are considered as free in the initial points cloud. For each position v_e'' of any edge segment e, there is at least a position p in the initial points cloud P the distance between v_e'' and p is shortest than ϵ: $\forall e \in E, \quad \forall v'' \in e, \quad \exists p \in P, \quad distance(p, v'') < \epsilon$.

An optimal *Vector Map* is a valid map that minimizes first the number of edges and second, distances between the initial points cloud and edge segments. The number of edges responsible for the required resources for map processes (map transfer, map fusion, path planning, etc.) and it is the main criterion to be optimized in our approach. While minimizing the distances allows to minimize errors due to the vectorization. Thus, given the minimal number of edges, an optimal *Vector Map* minimizes the following sum: $\sum_{p \in P} (\min_{e \in E} distance(p, e))$. A *Vector Map* is similar to k-means clustering method of vector quantization, where means are replaced by segments. This may imply that an optimal *Vector Map* is hard to compute. Therefore, we use a naïve heuristic method to generate the vector map from a given initial points cloud.

From an initial complete *Vector Map*, we use a naïve greedy algorithm to minimize the number of edges (cf. Fig. 2). The initial *Vector Map* involves all point in points cloud P and a edge each times the distance between two points is smaller than ϵ. The algorithm improves the map by detecting and removing shortcuts. A shortcut is defined by three successive vertices v_1, v_2 and v_3 ($\exists e_{12} = (v_1, v_2)$ and $\exists e_{23} = (v_2, v_3)$) if the distance between v_2 and the segment (v_1, v_3)

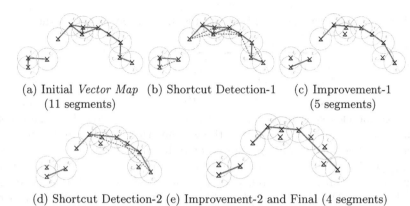

(a) Initial *Vector Map* (b) Shortcut Detection-1   (c) Improvement-1
     (11 segments)   (5 segments)

(d) Shortcut Detection-2 (e) Improvement-2 and Final (4 segments)

Fig. 2. Generation of a *Vector Map* from points cloud with a fixed ϵ distance.

is less than ϵ. In such a case, the edges e_{12} and e_{23} are reduced to one edge linking v_1 and v_3 and vertex v_2 is removed. This process is repeated until no more shortcut is found.

4 Benchmarking Metrics

To compare maps, we used several metrics described in [21–23]. However, we've modified some of them and express them as percentage to ease understanding. We also consider two additional metrics which are the unoccupied picture distance [21] and the memory footprint as the size of the map.

Map Score (MS) [24]: It compares two maps cell-by-cell. It starts from the value of 0, and then it increases by a ratio of one divided by the total cells number, if the chosen cells are similar. It can be described by the following equation: $MS = \frac{100}{n} \cdot \sum_{i=1}^{n} R_i - G_i$, where R_i and G_i are the robot generated and the ground-truth maps, respectively, and n is the number of pixels in the ground-truth map.

Cross Correlation (CC) [25]: It consists of calculating the coefficient of similarity between two maps. A fitness measure of the robot-generated map is calculated using Barons *cross correlation coefficient*. It is normalized between 0 and 1 and it is based on the averaging cell values: $CC = \frac{\overline{R_i \cdot G_i} - \overline{R_i} \cdot \overline{G_i}}{\sigma(R) \cdot \sigma(G)} \cdot 100$, where, $\overline{R_i \cdot G_i}$ is the mean of $R_i \cdot G_i$ and σ is the standard deviation.

Pearson's Correlation (PC) [26]: It gives the measure of how likely it is possible to infer a map from another and it is applied only for occupied space. According to [21] this metric suffers from two drawbacks, one because it requires similar occupied pixels between maps, and the other is that it is perturbed by the outliers (e.g., [27]): $PC = \dfrac{\sum\limits_{i=1}^{n}(R_i-\overline{R_i})(G_i-\overline{G_i})}{\sqrt{\sum\limits_{i=1}^{n}(R_i-\overline{R_i})^2}\cdot\sqrt{\sum\limits_{i=1}^{n}(G_i-\overline{G_i})^2}} \cdot 100.$

Occupied Picture-Distance-Function (OPDF) [21,22]: It calculates the ratio of the sum of the closest Manhattan-distance for each occupied cell of the robot-generated map to an occupied cell in the ground-truth divided by the number of occupied pixels of the robot-generated map. It comes with the advantages of removing the inherent problem of pixel-to-pixel comparison. It is worth to note that we make a limit in the search space (e.g., a rectangle of width *wscpace* and height *hsspace*): $OPDF = (1 - \frac{1}{no\cdot r}\sum_{i=1}^{no} d_i)\cdot 100$, where *no* is the number of occupied cells, d_i is the Manhattan-distance of each occupied cell of the robot-generated map to the closest occupied cell on the ground-truth map. In case no closest cell was found we set $d_i = r$, where r is the maximum search space distance $r = \sqrt{(wsspace^2 + hsspace^2)}$.

Unoccupied Picture-Distance-Function (UPDF) : It has the same concept of the OPDF however it is applied to unoccupied pixels: $UPDF = (1 - \frac{1}{nu\cdot r}\sum_{i=1}^{nu} d_i)\cdot 100$, where *nu* is the number of unoccupied cells, d_i is the Manhattan-distance of each unoccupied cell of the robot-generated map to the closest unoccupied cell on the ground-truth map. In case no closest cell was found we set $d_i = r$.

Memory Occupancy Size : This metric evaluates the size of the map.

5   Experiments

5.1   Setup

We conducted multiple experiments to evaluate and compare *Vector Maps* and *Occupancy Grid* maps using the previously defined metrics. Our objective is to cover as much as we can in terms of diversity, complexity and scalability. Therefore, we used two different axes of evaluation: terrain topology and terrain size as shown in Fig. 3. We used three different terrain topologies: maze like, office like and unstructured like, to assess *Vector Maps* on different shapes. We also used three different sizes for each terrain topology: small ($10\,m \times 10\,m$), medium ($40\,m \times 40\,m$) and large ($80\,m \times 80\,m$).

Those parameters lead to 9 experiments that we have performed using the MORSE [28] robotics simulator. We created the 2D ground-truth maps and we generated the 3D MORSE map files out of them. All our experiments use the Pioneer3dx robot model. The control architecture of the robot has been developed using ROS[2] packages such as Karto SLAM for the construction of the *Occupancy Grid* map. We chose Karto SLAM because it provides better results compared to other SLAM algorithms implementations [22]. During each experiment, the simulated robot was automatically driven through the same predefined trajectory according to the current terrain.

---

[2] http://www.ros.org.

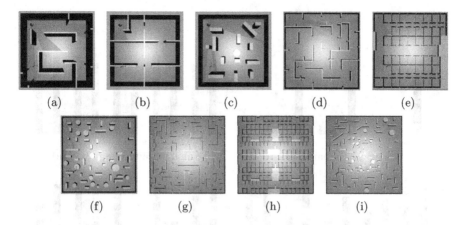

Fig. 3. Snapshots of the experiment terrain taken from MORSE simulator: (a) maze 10 × 10 m, (b) office 10 × 10 m, (c) unstructured 10 × 10 m, (d) maze 40 × 40 m, (e) office 40 × 40 m, (f) unstructured 40 × 40 m, (g) maze 80 × 80 m, (h) office 80 × 80 m and (i) unstructured 80 × 80 m.

The output of each experiment is one *Occupancy Grid* map. Based on each occupancy grid map, we generated the corresponding *Vector Map*. Indeed, we don't have a SLAM algorithm that directly builds a vector map from laser scans since we first need to assess whether the use of vectors are a good alternative to occupancy grids. The points cloud matches occupied cell centers and comes with regular discrete coordinates. The value ϵ is initialized according to the *Occupancy Grid* map resolution ϵ' (*i.e.* the size of a cell in the real word). ϵ is the minimal distance to potentially connect the eight neighbors of an occupied cell ($\epsilon = \sqrt{2} \times \epsilon'$).

Figure 4(a) shows the size of the Karto map (grid cell), zipped Karto map, *Vector Map* and zipped *Vector Map*, where each value is an average of the three terrain types of the same size (P (.PGM), PZ (Zipped .PGM) V (.Vector Map), VZ (Zipped .Vector Map)). Figures 4(b) to (j) present our experimental results through comparisons between maps generated by Karto SLAM (left bar with blue color), *Vector Maps* based on Karto (middle bar with green color) and *Vector Maps* based on ground-truth (right bars with brown color) for the three different terrain sizes and the three different types of terrain.

5.2   Discussion

As one might expect, a vector-based map has a much smaller memory footprint (39.2 KB in average for a 80 × 80 m terrain) compared to the one of an *Occupancy Grid* map produced by Karto (2532.4 KB in average). Indeed, a single edge in the *Vector Map* can replace many points of the *Occupancy Grid*. Moreover, the number of edges in a vector increases only if there are more obstacle corners. In contrast, the memory footprint of an *Occupancy Grid* map increases with the size of the modeled terrain.

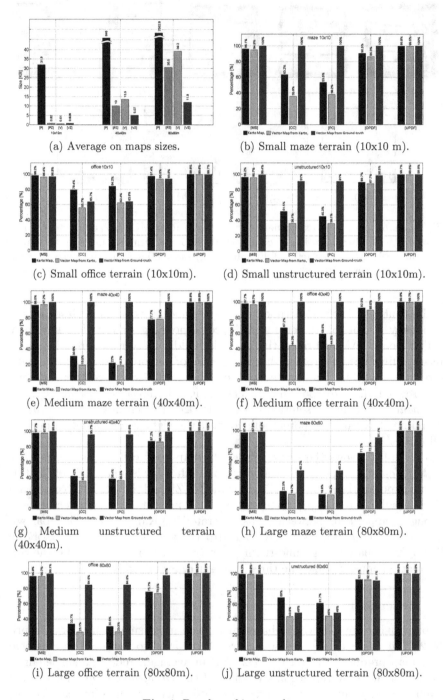

(a) Average on maps sizes.

(b) Small maze terrain (10x10 m).

(c) Small office terrain (10x10m).

(d) Small unstructured terrain (10x10m).

(e) Medium maze terrain (40x40m).

(f) Medium office terrain (40x40m).

(g) Medium unstructured terrain (40x40m).

(h) Large maze terrain (80x80m).

(i) Large office terrain (80x80m).

(j) Large unstructured terrain (80x80m).

Fig. 4. Benchmarking results

When it comes to sending a map over the network, which is a critical issue in Multi-Robotics exploration, a large file might be compressed to save the bandwidth. Interestingly, the non-compressed *Vector Map* requires an amount of memory that is close to what is required for a compressed *Occupancy Grid*. Thus, *Vector Map* makes it possible to send the uncompressed map to avoid the computational overhead of the compression/decompression.

The evaluation of *Vector Maps* includes the analysis of the quality. Our goal was to highlight potential degradations of the map quality due to the approximation by line segments. Therefore, we have chosen five different metrics: Map Score (MS), Cross Correlation (CC), Pearson's Correlation (PC), Occupied Picture-Distance-Function (OPDF) and Unoccupied Picture-Distance-Function (UPDF). We performed a normalization to provide results as percentages with best value being 100 %. So we can compare them for different terrain sizes.

For the three metrics MS, OPDF, and UPDF, the quality is always over 91 %. The quality of the ground-truth vectorized map is often 100 % as one might expect, while the vectorized Karto maps obtain values that always equal or very close to scores of the actual Karto map (max 1 % difference). For the two remaining metrics (CC and PC), the quality of the regenerated map may decrease in some terrains. Surprisingly, this is even true for ground-truth vectorized map. CC and PC reach the value of 100 % in three terrains, over 63 % in other three terrains and around 49 % in the two remaining terrains. However, these two metrics are criticized as suffering from some drawbacks given in [21], which seem to be reinforced by behaviors drown in this paper. We suspect that line segments approximation is one possible source for these results.

However, we believe that reaching 100 % quality for all setups is also a matter of trade-off with the memory footprint. A map of higher quality is likely to require more edges, and hence occupy more memory. Still, since *Vector Maps* have a small memory footprint, scarifying some more memory in favor of an increase in quality is definitely acceptable.

6 Conclusion and Future Work

In this paper, we show that *Vector Maps* are an interesting alternative to occupancy grids for representing the environment. Indeed, *Vector Maps* which are based on line segments defined in metric coordinates can be processed based on graph theory likewise topological maps. They are also lightweight and accurate when generated from either a map produced by a SLAM algorithm (i.e. Karto) or the ground truth. While this is already interesting in the context of a single robot, it is even more critical for multi-robot exploration. Indeed, to achieve coordination robots have to share their local maps and merge ones provided by their teammates. Vector maps outperform occupancy grids by requiring less network bandwidth for the transfer and are promising less maps merging CPU consumption.

We used several metrics to compare maps produced by Karto, the best laser SLAM algorithm available for ROS [22] and our *Vector Maps*. As expected,

Vector Maps have drastically reduced memory footprint. We also compared quality using 5 metrics. For 3 of them, we can conclude that the quality of the *Vector Map* is similar to the *Occupancy Grid* map. However, the 2 remaining quality metrics show a loss that can be important under some conditions. Considering the overall quality, the memory reduction and the possibility to manipulate maps as graphs, we can conclude that the use of *Vector Maps* is an interesting alternative to *Occupancy Grid* maps. However, the 2 remaining quality metrics show a loss that can be important under some conditions. Considering the overall quality, the memory reduction and the possibility to manipulate maps as graphs, we can conclude that the use of *Vector Maps* is an interesting alternative to *Occupancy Grid* maps, that can be used in many robotics settings.

As for future work, we aim at evaluating *Vector Maps* in merging and path planning process during multi-robot exploration missions. In the long term, we plan to investigate a SLAM algorithm that directly produces *Vector Maps* out of laser scans. By suppressing the intermediate step of building *Occupancy Grid* maps, we expect an improvement of the final *Vector Map* quality.

References

1. Stachniss, C. (ed.): Robotic Mapping and Exploration. STAR, vol. 55. Springer, Heidelberg (2009)
2. Choset, H., Nagatani, K.: Topological simultaneous localization and mapping (SLAM): toward exact localization without explicit localization. IEEE Trans. Robot. Autom. **17**(2), 125–137 (2001)
3. Savelli, F., Kuipers, B.: Loop-closing and planarity in topological map-building. In: IEEE/RSJ International Conference on Intelligent Robots and Systems (IROS), vol. 2, pp. 1511–1517, September 2004
4. Lozenguez, G., Adouane, L., Beynier, A., Mouaddib, A.-I., Martinet, P.: Punctual versus continuous auction coordination for multi-robot andmulti-task topological navigation. Auton. Robots, 1–15 (2015)
5. Gerkey, B.: Karto slam, July 2015. http://www.ros.org/wiki/karto
6. Moravec, H., Elfes, A.: High resolution maps from wide angle sonar. Rob. Autom. **2**, 116–121 (1985)
7. Lu, F., Milios, E.: Globally consistent range scan alignment for environment mapping. Auton. Robots **4**, 333–349 (1997)
8. Tomomi, N., Kanji, T.: Dictionary-based map compression for sparse feature maps, In: IEEE International Conference on Robotics and Automation (ICRA), pp. 2329–2336, May 2011
9. Hornung, A., Wurm, K.M., Bennewitz, M., Stachniss, C., Burgard, W.: OctoMap: an efficient probabilistic 3D mapping framework based on octrees. Auton. Robots **34**, 1–17 (2013)
10. Blanco, J.-L., Fernandez-Madrigal, J.-A., Gonzalez, J.: A new approach for large-scale localization and mapping: hybrid metric-topological slam. In: 2007 IEEE International Conference on Robotics and Automation, pp. 2061–2067, April 2007
11. Thrun, S.: Learning metric-topological maps for indoor mobile robot navigation. Artif. Intell. **99**, 21–71 (1985)
12. Jongwoo Lim, M.P., Frahm, J.-M.: Online environment mapping using metric-topological maps. Int. J. Robot. Res., 1–15, November 2007

13. Moutarlier, P., Chatila, R.: Stochastic multisensory data fusion for mobile robot location and environment modeling. In: International Symposium on Robotics Research (ISRR), pp. 207–216 (1989)
14. Myung, H., Jeon, H., Jeong, W.: Virtual door algorithm for coverage path planning of mobile robot. In: IEEE International Symposium on Industrial Electronics (ISIE), pp. 658–663, July 2009
15. Fabrizi, E., Saffiotti, A.: Extracting topology-based maps from gridmaps. In: IEEE International Conference on Robotics and Automation (ICRA), vol. 3, pp. 2972–2978 (2000)
16. Choi, J., Choi, M., Chung, W.K.: Incremental topological modeling using sonar gridmap in home environment. In: IEEE/RSJ International Conference on Intelligent Robots and Systems, pp. 3582–3587, October 2009
17. Joo, K., Lee, T.K., Baek, S., Oh, S.Y.: Generating topological map from occupancy grid-map using virtual door detection. In: IEEE Congress on Evolutionary Computation (CEC), pp. 1–6, July 2010
18. Nunez, P., Vazquez-Martin, R., del Toro, J., Bandera, A., Sandoval, F.: Feature extraction from laser scan data based on curvature estimation for mobile robotics. In: IEEE International Conference on Robotics and Automation (ICRA), pp. 1167–1172, May 2006
19. Vandorpe, H.X.J., Van Brussel, H.: Exact dynamic map building for a mobile robot using geometrical primitives produced by a 2D range finder. In: IEEE International Conference on Robotics and Automation (ICRA), pp. 901–908, April 2011
20. Lakämper, R., Latecki, L.J., Sun, X., Wolter, D.: Geometric robot mapping. In: Andrès, É., Damiand, G., Lienhardt, P. (eds.) DGCI 2005. LNCS, vol. 3429, pp. 11–22. Springer, Heidelberg (2005)
21. Balaguer, B., Balakirsky, S., Carpin, S., Visser, A.: Evaluating maps produced by urban search and rescue robots: lessons learned from robocup. Auton. Robots **27**, 449–464 (2009). Springer
22. Santos, J.M., Portugal, D., Rocha, R.: An evaluation of 2D SLAM techniques available in robot operating system. In: IEEE International Symposium on Safety, Security (SSRR), pp. 1–6, October 2013
23. Yan, Z., Fabresse, L., Laval, J., Bouraqadi, N.: Metrics for performance benchmarking of multi-robot exploration. In: IEEE/RSJ International Conference on Intelligent Robots and Systems (IROS), October 2015
24. Martin, M.C., Moravec, H.P.: Robot evidence grids. In: Technical report CMU-RI-TR-96-06, Robotics Institute-Carnegie Mellon University (1996)
25. O'Sullivan, S.: An empirical evaluation of map building methodologies in mobile robotics using the feature prediction sonar noise filter and metric grip map benchmarking suite. Master Thesis, University of Limerick (2003)
26. Guyon, I., Gunn, S., Nikravesh, M., Zadeh, L.A.: Feature Extraction: Foundations and Applications. StudFuzz, vol. 207. Springer-Verlag New York, Inc., Secaucus (2006)
27. Miranda Neto, A., Correa Victorino, A., Fantoni, I., Zampieri, D., Ferreira, J., Lima, D.: Image processing using pearson's correlation coefficient: applications on autonomous robotics. In: International Conference on Autonomous Robot Systems (Robotica), pp. 1–6, April 2013
28. Echeverria, G., Lassabe, N., Degroote, A., Lemaignan, S.: Modular open robots simulation engine: MORSE. In: IEEE International Conference on Robotics and Automation, Shanghai, China, pp. 46–51, May 2011

Interactive Intelligence

Information Intelligence

Optimal Viewpoint Selection Based on Aesthetic Composition Evaluation Using Kullback-Leibler Divergence

Kai Lan[✉] and Kosuke Sekiyama

Department of Micro-Nano System Engineering, Nagoya University,
Furo-cho, Chikusa-ku, Nagoya 464-8601, Japan
lan@robo.mein.nagoya-u.ac.jp

Abstract. In this paper, we construct a robot photographic system to search for the optimal viewpoint of a scene. Based on some known composition rules in the field of photography, we propose a novel aesthetic composition evaluation method by the use of Kullback-Leilber divergence. For viewpoint selection, we put forward a method called Composition-map, which can estimate the aesthetic value of scenes for each candidate viewpoint around the target group. At last, the effectiveness of our robot photographic system is confirmed with practical experiments.

Keywords: Robot photographic system · Aesthetic composition evaluation · Kullback-Leibler divergence · Composition-map

1 Introduction

With the popularization of digital cameras, people are surrounded by photographs with a huge number in their daily life. Digitized photographs can be easily edited and composed by raster graphics editors if they are not satisfactory based on the aesthetic sensibility of human. In recent years, attempts have been tried to value aesthetic appreciation of photos by transforming humans' preferences into computable factors [1]. This kind of researches becomes a new field called *Computational Aesthetics* [2] and aims to achieve the understanding of human's aesthetic sensibility for future robots.

Composition is the way of placing or arranging visual components within the picture frame [3]. In the field of photography, it is the most basic technique to shoot a better looking photo. Several principles help to compose natural and aesthetic compositions. For example, *Rule of thirds* and *Diagonal composition* are two of the most well-known compositions which can be frequently found in photographic works from both professionals and amateurs.

Conventional researches succeeded in digitalizing humans' aesthetic evaluations on photos. As region size and visual balance of targets show obvious influence on aesthetic appreciation, Liu et al. [4] and Zhang et al. [5] proposed an evaluation function by transforming them into computational factors based

© Springer International Publishing Switzerland 2016
N. Kubota et al. (Eds.): ICIRA 2016, Part I, LNAI 9834, pp. 433–443, 2016.
DOI: 10.1007/978-3-319-43506-0_38

on *Rule of thirds*. Lan et al. [6] showed an expansion version by constructing other evaluation functions for not only *Rule of thirds* but also *Diagonal composition* and *Triangle composition*. The evaluation functions were verified to have a high consistency with human evaluation. However, as the evaluation functions for different composition rules varies with each other, it is a complicated process to construct a new evaluation fucntion for a new composition rule. There exists no common function which can be applied to all kinds of compositions.

The approach for aesthetic property enhancement of a scene can be classified into two typical methods generally. One is to modify the composition of targets by image after-processing. [4,5] constructed a crop-and-retarget operator by trimming the size of photos and replacing foreground targets based on the composition evaluation scores. Although they realized to enhance aesthetics of photos, recomposing positions of targets forcibly without considering the semantic relations of them cannot reflect objective reality of actual scenes. Furthermore, Byers et al. [7,8] tried to combine aesthetic composition evaluation with viewpoint selection for aesthetic property enhancement. Exploiting the mobile capability of monitoring robots, pleasing compositions can be found based on composition principles. However, due to the lack of a well-planed viewpoint searching process, the success ratio for their system to obtain the satisfying compositions is only 29 %.

In this paper, we advance a novel system for optimal viewpoint selection based on aesthetic composition evaluation. We put forward a novel composition evaluation method by using Kullback-Leibler divergence which can be easily expanded with various compositions. For optimal viewpoint selection, we propose a stable selection method called Composition-map, which can estimate the aesthetic value of scenes for each candidate viewpoint around the target group. The effectiveness of our robot photographic system is confirmed with practical experiments.

2   Robot Photographic System

We propose an robot photographic system to search for the optimal viewpoint of targets. This system consists of five modules: *Image Processing, Aesthetic Evaluation, Optimal Viewpoint Selection, Path Planning* and *Robot Motion*. Figure 1 shows a schematics of the proposed system.

In *Image Processing* module, color image and cloud image of the original scene obtained by a RGB-D camera are input to the system. Using the cloud image, all targets can be detected in the color image. The background and floor of the scene will be subtracted by a segmentation method [9], then the information of objects in the pixel coordinate of images can be estimated. We prepare an interface for users so that they can select any object as a composition target from the operational screen. Those, which are not selected as targets, will be totally ignored.

In *Aesthetic Evaluation* module, we propose a novel aesthetic composition evaluation method by using Kullback-Leibler divergence. We make a description

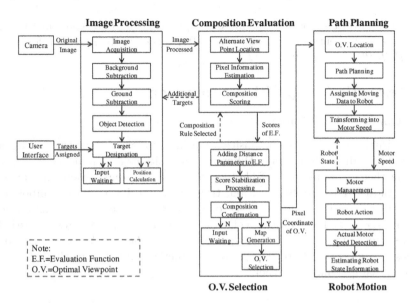

Fig. 1. Robot Photographic System.

of compositions using the Gaussian mixture distribution. As the best position and size of targets can be confirmed, we define different models showing the best arrangements for different composition rules respectively. Then differences between distributions of the present composition and a model can be judged by the Kullback-Leibler divergence from the vision of information theory. The minimal value of the Kullback-Leibler divergence implies the best looking composition for the present targets.

In *Optimal Viewpoint Selection* module, we set some viewpoint candidates around the target group. The composition of targets observed from each viewpoint can be presumed by the position information of targets and viewpoints. After evaluating the aesthetic properties for each composition, the optimal viewpoint can be confirmed with considering a distance factor. The distance factor is a reflection of energy and time efficiency, since the robot is requested to find the optimal viewpoint within a limited time and power consumption. The distance factor allows a monitoring robot to select viewpoints near its present location.

In *Path Planning* module, we estimate the actual position of the optimal viewpoint in the robot coordinate system. In order to realize a smooth and continuous motion of the robot, we design a straight moving path for it. Although the path will update every three seconds approximately, we apply a proportional feedback control to the robot before the path updates. Moving speed of the robot is decided by the distance to the goal position while moving direction depends on the feedback control. For the omni-directional robot, rotation speed and orientation are decided by a visual feedback control in a real time.

In *Robot Motion* module, the micro-computer of the robot controls the rotation speed and orientation for each wheel after recieving the moving information from *Path Planning* module.

3   Aesthetic Composition Evaluation

3.1   Composition Elements

In the visual arts field, the composition means the arrangement or organization of visual ingredients in a scene. Different compositions are employed to express different themes or emotions of photographers. These composition rules include rule of thirds, diagonal composition, triangle composition, curve composition, symmetric composition and so on. As composition mainly considers positions and sizes of targets, we focus on these elements and summarize features of aesthetci compositions using them. In this section, we take *Rule of thirds* and *Diagonal composition* for example to explain our aesthetic evaluation method. Figure 2 shows the sample images of them.

(a) Rule of thirds   (b) Diagonal composition

Fig. 2. Typical compositions of photographs

3.2   Composition Distribution

In our theory, the existence probability of all objects in the pixel image equals 1. We suppose the existence probability density of an object is described as a two-dimensional Gaussian distribution. The coordinate of the target center corresponds to the mean of the Gaussian distribution while width and height of a target relate to variance of this distribution in x and y axes. The arrangement of plural targets in the scene can be described using a Gaussian mixture distribution with peaks whose number equals the number of targets. We define that existence probability of the region inside the detected boundary of the target is 1 if there is only one object in the pixel image. Referring to the two-dimensional Gaussian distribution, the formulation for existence probability density $f(x, y)$ of a target is as follows:

$$f(\mathbf{x}^{\mathrm{T}}) = \frac{1}{2\pi|\mathbf{\Sigma}|^{\frac{1}{2}}} \exp\left(-\frac{1}{2}(\mathbf{x} - \mathbf{u})^{\mathrm{T}}\mathbf{\Sigma}^{-1}(\mathbf{x} - \mathbf{u})\right) \tag{1}$$

Where $\mathbf{x}^{\mathrm{T}} = (x, y)$. $\mathbf{u}^{\mathrm{T}}$ is the coordinate of the target center. Σ is the variance-covariance matrix, determined by the width and height of the target and the correlation coefficient ρ. Correlation coefficient ρ, showing the linear correlation between two variables x and y, can reflect the directional information of targets in the scene.

If the number of targets in a scene is n and each target has a different importance as constituent elements for a composition, the existence probability of target $T_i (1 \leq i \leq n)$ can as expressed by its weight k_i. Existence probability distribution of all targets $p(x, y)$ can be described as follows:

$$p(x, y) = \sum_{i=1}^{n} k_i f_i(x, y). \tag{2}$$

Where $\sum_{i=1}^{n} k_i = 1$.

(a) Sample scene of two targets.   (b) Gaussian mixture with three dimensions.   (c) Top view of Gaussian mixture.

Fig. 3. Composition distribution of a scene.

Figure 3(a) is a sample scene with two targets detected. According to our assumption, the Gaussian mixture distribution of the sample scene can be generated as shown in Fig. 3(b). For convenience, in the following discussions we will use a top view of Gaussian Mixture distribution Fig. 3(c) to represent the composition of a scene since a pixel image is in a two-dimensional coordinate.

3.3   Model Distribution

In rule of thirds, targets are expected to be deposited on the crossing points. And the region size of all targets are desired to account for 12 % of the image frame [4]. So a model composition can be confirmed with the information of target position and size [6]. The model composition in rule of thirds for Fig. 3(a) is shown in Fig. 4(a). In the same way, model compositions in diagonal composition for this sample can be determined as well, shown as Fig. 4(b).

Formulation of the existence probability for model compositions $q(x, y)$ is similar to Eq. (1) by adjusting means $\mathbf{u}^{\mathrm{T}}$ and the variance-covariance matrix Σ for each Gaussian distribution.

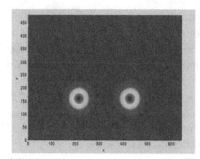 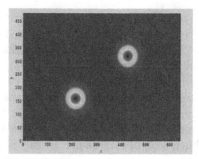

(a) Model composition for rule of thirds.   (b) Model composition for diagonal composition.

Fig. 4. Model compositions.

3.4 Kullback-Leibler Divergence for Composition

As Kullback-Leibler Divergence has a function of measuring the difference between two probability distributions [10], we apply this concept to measure the difference between the present composition and the corresponding model composition. A small value of Kullback-Leibler divergence implies a good composition. It is defined to be a double integral as follows:

$$D = \iint_H p(x, y) \log \frac{p(x, y)}{q(x, y)} \, dx dy. \tag{3}$$

Where H denotes the whole region of a pixel image.

As a small value of Kullback-Leibler divergence can reflect the aesthetic property of a scene, we define the aesthetic evaluation functions as follows:

$$E = e^{-tD}. \tag{4}$$

$t = \frac{1}{3}$ from experimental results. E varies from 0 to 1.

4   Optimal Viewpoint Selection

We develop a viewpoint selection algorithm for robots. By estimating the relationship of targets in different observation position based on coordinates and angles at present in the real world, we can estimate the composition of a viewpoint before the robot reaches the position.

Figure 5 shows geometric relations to calculate pixel information in different observation positions. Pixel coordinates (x'_i, y'_i) and area $M'(S_i)$ of Target i are estimated as follows:

$$x'_i = \frac{2}{W} \frac{\tan \theta_i}{\tan \varphi}, \tag{5a}$$

$$y'_i = \frac{2}{V} \frac{\tan \gamma_i}{\tan \varphi_2}, \tag{5b}$$

$$M'(S_i) = M(S_i) \left(\frac{d}{d_i}\right)^2.$$ (5c)

Where W and V are pixel width and height of the photo frame. d is the distance between Target i and the robot at the present position. d_i is the distance from Target i to the expected observation position.

With x'_i, y'_i and $M'(S_i)$, the aesthetic properties can be evaluated with the proposed functions. As a result, we can estimate the aesthetic property of all the candidate viewpoints. The optimal viewpoint is determined with the highest score according to our aesthetic evaluation scores and a distance factor mentioned in Sect. 2. Figure 6 is a viewpoint map from a vertical view. Different evaluations scores are represented in different colors. Higher scores are in blue while lower scores are in red. S shows the position where the monitoring robot is at present, while G is the optimal viewpoint estimated.

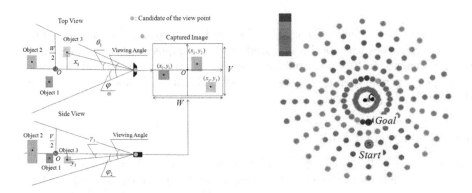

Fig. 5. Top view and side view of targets.   **Fig. 6.** Composition map. (Color figure online)

5 Experiment

5.1 Experimental Condition

Figure 7 shows the experimental condition. The monitoring robot is a omnidirectional mobile robot with a PC and a RGB-D camera(Xtion Pro Live). Motors of the robot are controlled by a Arduino microcomputer. We prepared a user interface, allowing users to select any object to be a target at will. Composition rules can also be designated according to users' favours although we only prepared two selections at present.

5.2 Experimental Process

In this experiment, we use two kinds of composition rules for aesthetic evaluation, *Rule of thirds* and *Diagonal composition*. As mentioned, targets selected through

Fig. 7. Experimental condition.

the user interface should be motionless, otherwise the optimal viewpoint will be not converged to a fixed position. Based on each composition rule, we repeated the experiment for 10 times and for each time the robot was put in the same initial position and orientation. We recorded the Kullback-Leilber divergence for the initial and optimal viewpoints.

5.3   Experiment Results

Figure 8 are snapshots taken by an RGB-D camera during the viewpoint searching process. Pictures in the left-side are evaluated by rule of thirds while that in the right are valued by diagonal composition. We picked snapshots of frame = 1, 21, 30 to show the variation of scenes. With the decrescence of Kullback-Leibler divergence, better appreciations can be obtained from the scenes.

Figure 9 shows the variation of Kullback-Leibler divergence for all frames. Although the curves fluctuate during the viewpoint searching process, we can clearly read the decreasing trend of Kullback-Leibler divergence from the charts. From Figs. 8 and 9, we conclude that aesthetic compositions can be obtained by Kullback-Leibler divergence minimization.

As mentioned above, we repeated the viewpoint selection experiment 10 times for each composition rule and recorded the Kullback-Leilber divergence for the initial and optimal viewpoint as Fig. 10. Although values of the Kullback-Leilber divergence for the initial scenes vary violently based on different intial viewpoints, the Kullback-Leilber divergence for the optimal viewpoints converge to a low level with little variance. Comparing with the initial compositions, Kullback-Leilber divergence of scenes in the optimal viewpoint selected by our system decreased by 81 % and 76 %.

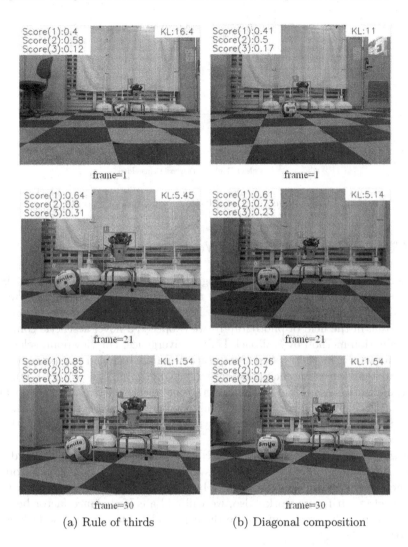

Fig. 8. Snapshots during viewpoint selection process.

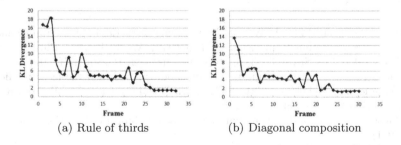

Fig. 9. KL divergence for composition (D).

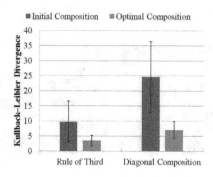

Fig. 10. KL divergence of initial and optimal viewpoints.

6   Conclusions and Future Works

In this paper, we proposed a robot photographic system to search for the optimal viewpoint of a scene. We demonstrated that different composition rules in art field can be computationally evaluated. Based on *Rule of thirds* and *Diagonal composition* in the field of photography, we proposed a novel aesthetic composition evaluation method by Kullback-Leilber divergence. For viewpoint selection, we put forward a method called Composition-map, which can estimate the aesthetic value of scenes for each candidate viewpoint around the target group. At last, the effectiveness of our optimal viewpoint selection system was confirmed with practical experiments. As the result showed, the aesthetic evaluation of the optimal viewpoints selected by our system was improved compared with the initial compositions.

However, we only dealt with the relative position of the targets and did not take the background of photographs into consideration. Since backgrounds influence aesthetic values sometimes, a background evaluation element will be introduced in our future work. Also, we will add a color balance factor because in photography the artist can choose the color palette to express the feelings of the works.

References

1. Datta, R., Joshi, D., Li, J., Wang, J.Z.: Studying aesthetics in photographic images using a computational approach. In: Leonardis, A., Bischof, H., Pinz, A. (eds.) ECCV 2006. LNCS, vol. 3953, pp. 288–301. Springer, Heidelberg (2006)
2. Wu, Y., Bauckhage, C., Thurau, C.: The good, the bad, and the ugly: predicting aesthetic image labels. In: 20th IEEE International Conference on Pattern Recognition (ICPR), pp. 1586–1589 (2010)
3. Dunstan, B.: Composing Your Paintings. Watson-Guptill Publications, New York (1971)
4. Liu, L., Chen, R., Wolf, L., Cohen-Or, D.: Optimizing photo composition. Comput. Graph. Forum **29**, 469–478 (2010)

5. Zhang, F., Wang, M., Hu, S.: Aesthetic image enhancement by dependence-aware object recomposition. IEEE Trans. Multimedia **15**, 1480–1490 (2013)
6. Lan, K., Sekiyama, K.: Autonomous viewpoint selection of robots based on aesthetic composition evaluation of a photo. In: The 2015 IEEE Symposium Series on Computational Intelligence, pp. 295–300 (2015)
7. Byers, Z., Dixon, M., Goodier, K., Grimm, C., Smart, W.D.: An autonomous robot photographer. In: The 2003 IEEE/RSJ International Conference on Intelligent Robots and Systems, pp. 2636–2641 (2003)
8. Byers, Z., Dixon, M., Smart, W.D., Grimm, C.: Say cheese! Experiences with a robot photographer. AI Mag. **25**, 37–46 (2004)
9. Filliat, D., et al.: RGBD object recognition and visual texture classification for indoor semantic mapping. In: The 2012 IEEE International Conference on Technologies for Practical Robot Applications (TePRA), pp. 127–132 (2012)
10. Kullback, S., Leibler, R.A.: On information and sufficiency. Ann. Math. Stat. **22**, 79–86 (1951)

Guidance of Robot Swarm by Phase Gradient in 3D Space

Keita Horayama[1], Daisuke Kurabayashi[1(✉)], Syarif Ahmad[2],
Ayaka Hashimoto[3], Takuro Moriyama[4], and Tatsuki Choh[5]

[1] Tokyo Institute of Technology, 2-12-1 Ookayama,
Meguro-ku, Tokyo 152-8552, Japan
dkura@ctrl.titech.ac.jp

[2] Gemalto Pte Ltd, 12 Ayer Rajah Crescent, Singapore 139941, Singapore

[3] Yaskawa Electric, 480 Fujisawa, Iruma-shi, Saitama 358-8555, Japan

[4] Toshiba, 1 Komukaitoshiba-cho, Saiwai-ku, Kawasaki-shi,
Kanagawa 212-8582, Japan

[5] Toshiba, 1 Toshiba-cho, Fuchu-shi, Tokyo 183-8511, Japan

Abstract. In this paper, we proposed a simple system that could control
the direction of a swarm of mobile robots by using phase gradient. We
implemented a system of autonomous mobile robots, in which each robot
was equipped an oscillator with local interaction ability. By introducing
relative direction among robots, we realized a particular phase distribution so that we could guide a swarm of robots in 2D or 3D spaces. We
verified the proposed system by experiments and simulations.

Keywords: Swarm robots · Oscillator network · Phase gradient

1 Introduction

In this paper, we proposed a system to guide a set of distributed autonomous
mobile robots by using phase gradients emerged over the robots. We supposed
that a swarm was a set of anonymous autonomous robots, which had no ability
to identify others. We didn't assume a global communication system for distributed robots. Instead, we employed oscillators that had local interactions only,
and implemented a guidance system for a swarm. Like a true slimemold [1], the
network of oscillators, in which a node is a mobile robot, emerges phase gradient
over the robots. We can manipulate the state of the phase gradient by changing
parameters of oscillators and their interactions. By introducing relative orientation among robots into local interactions, we successfully realized guidance of a
swarm in 2D or 3D spaces.

2 Problem Settlement

2.1 Mobile Robots and Basic Control Law

Let a mobile robot be a circular-shape with radius R. Suppose that it can measure distance between objects within it's visible region, and it has the dynamics

© Springer International Publishing Switzerland 2016
N. Kubota et al. (Eds.): ICIRA 2016, Part I, LNAI 9834, pp. 444–451, 2016.
DOI: 10.1007/978-3-319-43506-0_39

(1) proposed in [2], where $\boldsymbol{p}_i = [x_i, y_i]^T$ and $\boldsymbol{n}_i$ ($|\boldsymbol{n}_i| = 1$) represent the position and the attitude of a robot, respectively. $\boldsymbol{p}_O$ notes the nearest point of a detected object, and $\alpha \sim \gamma$, τ, d_r and d_O are constant parameters.

$$m\ddot{\boldsymbol{p}}_i = \alpha \sum_j \boldsymbol{f}_{ij} + \beta \boldsymbol{n}_i - \gamma \dot{\boldsymbol{p}}_i + \delta \boldsymbol{g}_i \tag{1}$$

$$\boldsymbol{f}_{ij} = \left(|d_{ij}|^{-3} - |d_{ij}|^{-2}\right) e^{-|d_{ij}|} d_{ij} \qquad \tau \dot{\boldsymbol{n}}_i = \boldsymbol{n}_i \times \frac{\dot{\boldsymbol{p}}_i}{|\dot{\boldsymbol{p}}_i|} \times \boldsymbol{n}_i$$

$$d_{ij} = \frac{\boldsymbol{p}_j - \boldsymbol{p}_i}{d_r} \qquad \boldsymbol{g}_i = \begin{cases} (d_o - |\boldsymbol{p}_i - \boldsymbol{p}_o|) \frac{\boldsymbol{p}_i - \boldsymbol{p}_o}{|\boldsymbol{p}_i - \boldsymbol{p}_o|} & (|\boldsymbol{p}_i - \boldsymbol{p}_o| < d_o) \\ 0 & (|\boldsymbol{p}_i - \boldsymbol{p}_o| \geq d_o) \end{cases}$$

2.2 Oscillators and Phase Gradient

We employed phase oscillators [3] to generate a phase gradient over robots. We assumed that every robot was equipped with the same oscillator system.

$$\dot{\phi}_i = \Omega + \frac{\kappa}{|N_i|} \sum_{j \in N_i} (\phi_j - \phi_i) \tag{2}$$

In Eq. (2), Ω indicates the common natural frequency of oscillators, κ is a common constant gain for interactions, ϕ_i shows the phase of oscillator i, and N_i means a set of neighbor oscillators around i.

A single oscillator, $\dot{\phi}_i = \Omega$ (supposing $\kappa = 0$), can be considered an one-hand clock. As in Fig. 1(a), clocks may have differences (phase gaps). When we apply (2) with $\kappa > 0$, the clocks (oscillators) will be synchronized with no phase gap [4] (Fig. 1(b)).

When we set the natural frequency of oscillator T as $\Omega_T = \Omega + \Delta\Omega$, the oscillators can be sill synchronized (yet $|\Delta\Omega|$ is limited). However, the gaps among phases will not be zero, and we will see a phase gradient toward T (Fig. 1(c)). By changing $\Delta\Omega = 0$, $\Delta\Omega > 0$, or $\Delta\Omega < 0$, we can change the direction of 1D phase gradient over the oscillators.

2.3 Problem Settlement

Goal of this paper is to realize a control system to guide a swarm in 3DOF space. Suppose that each robot is equipped with an oscillator formulated in (2). As we reported in [5], we can guide robots in 1 direction by installing motion law to climb up a phase gradient, where a robot T, which has Ω_T, selects appropriate phase gradient.

In this paper, our objective is to extend the guidance system into 2DOF or 3DOF control. For the purpose, we introduced relative directions among robots so that robots could emerge phase gradients in radial directions of those by previous system in [5].

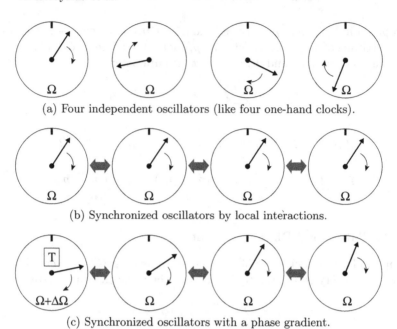

(a) Four independent oscillators (like four one-hand clocks).

(b) Synchronized oscillators by local interactions.

(c) Synchronized oscillators with a phase gradient.

Fig. 1. Synchronization of coupled oscillators.

3   Emergence of Phase Gradients in 2D or 3D Space

In order to add one more DOF, we proposed a directional interaction for oscillators as (3), where ε_1 and ε_2 are constant parameters, θ and ψ denoted relative orientation of robot j in a sight of i, as indicated in Fig. 2. As the same of the case of Ω_T, we select robot T and apply (3) to robots. Then, we can generate a phase gradient around robot T.

$$\dot{\phi}_i = \Omega + \frac{\kappa}{|N_i|} \sum_{j \in N_i} \left(\hat{\phi}_j - \phi_i \right) \qquad (3)$$

$$\hat{\phi}_j = \begin{cases} \phi_j - \varepsilon_1 \theta_{ji} - \varepsilon_2 \psi_{ji} & (j = T) \\ \phi_j & (j \neq T) \end{cases}$$

Let ϕ_T be the phase of robot T, and $\xi_i := \phi_i - \phi_T$. Then, $\phi_j - \phi_i = \xi_j - \xi_i$. According to (2), (4) holds.

$$\dot{\xi}_i = \dot{\phi}_i - \dot{\phi}_T = \frac{\kappa}{|N_i|} \sum_{j \in N_i} (\xi_j - \xi_i) - \Delta\Omega \qquad (4)$$

Suppose an adjacency matrix $A = (a_{ij})$ that represents connections among robots, where $a_{ij} = 1$ if robot i and j are neighbors (otherwise, $a_{ij} = 0$). To simplify the problem, we assume that A is symmetric. Because (5) holds, (4) can be transformed to (6).

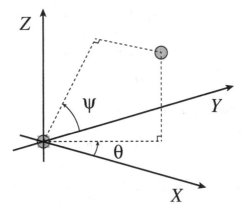

Fig. 2. Relative angles among two robots.

$$\sum_{j \in N_i} (\xi_j - \xi_i) = \sum_{j=1}^{n} a_{ij} (\xi_j - \xi_i) \tag{5}$$

$$\dot{\xi}_i = -\frac{\kappa}{|N_i|} \xi_i \sum_{j=1}^{n} a_{ij} + \frac{\kappa}{|N_i|} \sum_{j=1}^{n} a_{ij} \xi_j - \Delta \Omega$$

$$= -\kappa \xi_i + \frac{\kappa}{|N_i|} \sum_{j=1}^{n} a_{ij} \xi_j - \Delta \Omega \tag{6}$$

Let $\boldsymbol{z} := [\xi_1, \xi_2, \cdots, \xi_n]^T$, D be an order matrix of A, and $L := D - A$ be a graph laplacian of A. Then, (6) is summarized as (7), where $\boldsymbol{1}_n :=$ $[1, 1, \cdots, 1]^T$.

(a) Overview   (b) LEDs and sensors

Fig. 3. Implemented robot system.

$$\dot{z} = -\kappa z + \kappa \begin{bmatrix} \frac{1}{|N_i|} \sum_{j=1}^n a_{1j}\xi_j \\ \vdots \\ \frac{1}{|N_i|} \sum_{j=1}^n a_{nj}\xi_j \end{bmatrix} - \Delta\Omega 1_n$$

$$= \kappa D^{-1}Lz - \Delta\Omega 1_n \tag{7}$$

Here, let us introduce $\hat{\phi}_j$ to (7).

$$\dot{\xi}_i = \frac{\kappa}{|N_i|} \sum_{j \in N_i} (\xi_j - \xi_i - (\varepsilon_1\theta_{ji} + \varepsilon_2\psi_{ji})) - \Delta\Omega$$

$$= -\kappa\xi_i + \frac{\kappa}{|N_i|} \sum_{j=1}^n a_{ij}\xi_j - \Delta\Omega - \frac{\kappa}{|N_i|} \sum_{j=1}^n a_{ij}(\varepsilon_1\theta_{ji} + \varepsilon_2\psi_{ji}) \tag{8}$$

Initial state

(a) $\Omega_T > \Omega$ (b) $\Omega_T < \Omega$

(c) $\varepsilon_1 > 0$ (d) $\varepsilon_1 < 0$

Fig. 4. 2D experiments with mobile robots.

Then, (9) holds.

$$\dot{z} = -\kappa D^{-1}Lz + \Delta\Omega 1_n - \kappa D^{-1}\Theta \tag{9}$$

$$\Theta := \left[\sum_{j=1}^{n} a_{1j}(\varepsilon_1\theta_{j1} + \varepsilon_2\psi_{j1}), \quad \cdots \quad \sum_{j=1}^{n} a_{nj}(\varepsilon_1\theta_{jn} + \varepsilon_2\psi_{jn}) \right]^T$$

Let $M := -\kappa D^{-1}L$, $m_{\Delta\Omega} := \Delta\Omega 1_n$, and $m_\varepsilon := -\kappa D^{-1}\Theta$. Because the equilibrium of (9), z_e, is (10), we can see that we can modify the shape of the phase gradient by parameters $\Delta\Omega$, ε_1, and ε_2.

$$z_e = -M^{-1}(m_{\Delta\Omega} + m_\varepsilon) \tag{10}$$

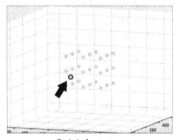

Initial state

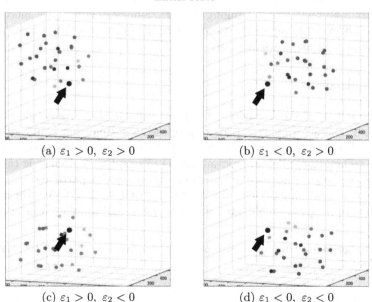

(a) $\varepsilon_1 > 0,\ \varepsilon_2 > 0$ (b) $\varepsilon_1 < 0,\ \varepsilon_2 > 0$

(c) $\varepsilon_1 > 0,\ \varepsilon_2 < 0$ (d) $\varepsilon_1 < 0,\ \varepsilon_2 < 0$

Fig. 5. Simulation of 28 robots in 3D.

4   Experiments

4.1   2D Guidance of Mobile Robots

We implemented the system to e-puck [6], a small mobile robot (Fig. 3). We added photo-sensors and a LASER distance sensor [7] on the top of it. We programmed a motion algorithm (11), by which a robot climbed up a phase gradient. Each robot included a programmed oscillator and it illuminated the LEDs when the phase of the oscillator acrossed zero.

$$\tau \dot{\boldsymbol{n}}_i = \begin{cases} \boldsymbol{n}_i \times \frac{\boldsymbol{p}_k - \boldsymbol{p}_i}{|\boldsymbol{p}_k - \boldsymbol{p}_i|} \times \boldsymbol{n}_i \ \text{(phase gap exists)} \\ \boldsymbol{n}_i \times \frac{\dot{\boldsymbol{p}}_i}{|\dot{\boldsymbol{p}}_i|} \times \boldsymbol{n}_i \quad \text{(others)} \end{cases} \tag{11}$$

Since we implemented mobile robots on 2D plane, we set $\varepsilon_2 = 0$. Figure 4 shows the motions exhibited by the robots. The robot at the left-end of the photos was T. When we set $\varepsilon_1 = 0$, we could control the robots attracted to or got away from the robot T. When we gave positive or negative ε_1, the robots moved up or down in the photos. The results shown that we successfully guided the robot swarm in 2 directions.

4.2   3D Guidance by Simulation

We have demonstrated the guidance algorithm in 3D space by simulations. Figure 5 illustrated behaviors of 28 robots. In the results, a robot indicated with arrowmark was robot T. By changing the ratio among ε_1 and ε_2, we successfully controlled the robot group in 3D space. Notice that a robot selected its neighbors according to Delaunay tesselations.

5   Conclusions

Swarming is one of the behaviors shown by animals that has no global communication systems. In this paper, we proposed a system that could guide the direction of a swarm of robots in two or three dimensions based on the local interactions among oscillators programmed in robots. By manipulating the parameters of oscillators and their interactions, we realized a particular phase distribution that enabled guidance of a swarm in 2DOF or 3DOF. We verified the system by implementing mobile robots on 2D plane and simulation in 3D space.

Acknowledgement. This work was partially supported by KAKENHI 25420212, MEXT Japan.

References

1. Takamatsu, A., et al.: Controlling the geometry and the coupling strength of the oscillator system in plasmodium of Physarum polycephalum by microfabricated structure. Protoplasma **210**, 164–171 (2000)

2. Shimoyama, N., et al.: Collective motion in a system of motile elements. Phys. Rev. Lett. **76**(20), 3870–3873 (1996)
3. Kuramoto, Y.: Chemical Oscillation, Waves, and Turbulance. Springer, Heidelberg (1984)
4. Jadbabaie, A., et al. On the stability of the kuramoto model of coupled nonlinear oscillators. In: Proceeding of the 2004 American Control Conference, pp. 4296–4301 (2004)
5. Kurabayashi, D., et al.: Adaptive formation transition of a swarm of mobile robots based on phase gradient. J. Robot. Mech. **22**(4), 467–474 (2010)
6. Mondada, F., et al.: The e-puck a robot designed for education in engineering. In: Proceeding of the 9th Conference on Autonomous Robot Systems and Competitions, vol. 1(1), pp. 59–65 (2009)
7. Kawata, H., et al.: Development of ultra-small lightweight optical range sensor systems. In: Proceeding of IEEE/RSJ International Conference on Intelligent Robots and Systems, pp. 3277–3282 (2005)

Collective Construction by Cooperation of Simple Robots and Intelligent Blocks

Ken Sugawara[✉] and Yohei Doi

Department of Information Science, Tohoku Gakuin University,
2-1-1, Tenjinzawa, Izumi-ku, Sendai 981-3193, Japan
sugaken@mail.tohoku-gakuin.ac.jp

Abstract. Collective construction is one interesting topic for swarm robots. We proposed a novel method for it introducing simple robots and relatively intelligent blocks. The blocks are assumed to have a processor, memory and communication device, and assemble the designed structure by controlling the growth direction of which the robot places the next block. The robots just load a block stochastically, and they are conducted to unload the block by the block that forms the structure. We first explain the mechanism of our proposal, and show some fundamental characteristics achieved by computer simulation. Next we explain a robot and blocks which enable to construct the small scale structure in real world.

Keywords: Collective construction · Swarm robots · Intelligent block

1 Introduction

Collective construction is one of the most attractive topic in swarm robotics. There are some approaches for realizing collective construction by swarm robots. One general approach is centralized method: the structure to be constructed is clearly designed in detail, whole process for construction is determined in advance, and concrete tasks are scheduled and assigned to each robot. The other approach is distributed method: there is no central system which controls whole process for construction, and each robot works by distributed method.

Collective construction by distributed method is challenging problem, which we still have open question how to establish it. In this sense, biological system is suggestive because we often find some clues for well-organized distributed mechanism in it. We know some animals such as ants, termites build large nests. The nests constructed by some kind of termite could be five meters or more, and it is hundreds times as large as their own body size.

Distributed algorithm for construction is exciting topic for scientists as well as engineers. From the viewpoint of algorithm, there are some interesting reports which were derived from observations of social insects. Denebourg et al. [1], for example, proposed simple probabilistic algorithm which enables the agents to collect the distributed matter and to form a circular cluster. Theraulaz and Bonabeau proposed asynchronous automata model with some micro-rule sets. They showed the system could generate various types

© Springer International Publishing Switzerland 2016
N. Kubota et al. (Eds.): ICIRA 2016, Part I, LNAI 9834, pp. 452–461, 2016.
DOI: 10.1007/978-3-319-43506-0_40

of structure and some of them resemble to real nests which are constructed by social insects [2, 3].

In robotics field, many papers have discussed the construction by swarm robots. Beckers et al., for example, assembled a real robot system which confirm the feasibility of Deneubourg's algorithm [4]. Parker et al. proposed simple bulldozing algorithm which enables robots to make a circular area by pushing the matters [5]. Some papers succeeded in construction of more complicated structures introducing a beacon system to conduct robots to form the structure [6, 7]. Another series of research which dealt with two dimensional and three dimensional structure construction has also reported [8, 9]. They are inspired by termite's nest construction. Target structure is represented by pile of blocks, and a proper process for construction is designed automatically. Each robot picks up a block at the entrance of the structure, moves along with it, and places the block based on the pre-calculated design.

2 Objective

Major objective of this paper is to confirm the feasibility of our proposal for robotic construction by an experiment of real robot system. We have proposed a method for robotic construction introducing simple dynamics for swarm robots and semi-active blocks which forms the structure [10]. The blocks as well as the robots have a processor and form the designed structure by their cooperation (Fig. 1).

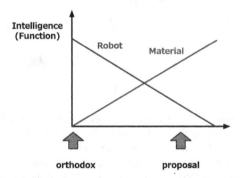

Fig. 1. In orthodox method, materials are passive and the robots are intelligence enough to construct a structure. Our proposal is to share the intelligence between the robots and materials and construct the desired structure by their interaction.

The point of our proposal was the balance of intelligence between the robot and the block. When we consider robotic construction problem, it is general approach that each robot has sufficient information and places the block (materials of the structure) in proper position. In other word, the robots are active and the materials are completely passive. In this approach, we sometimes encounter some difficulties for constructing a complex structure. One simple idea for overcoming the difficulty is to give processing capability and communication methods to the block, and reduce the burden of the robots [11].

In our proposal, the information for assembly is provided to the blocks, and the blocks conduct the robots to place next blocks in position. Defining a set of rules, we showed the desired structure could be constructed. We also showed the system has interesting secondary effect: the structure could have the characteristic of dynamic equilibrium because of simple algorithm for the robots.

Our proposal is concretely explained in next section. In Sect. 4, some fundamental characteristics of the proposed method is demonstrated. In Sect. 5, we show the result of small scale experiment by real robot's system, and conclude this paper.

3   Proposal

As described in previous section, the block is assumed to have processing capability and communication method. The assembled blocks recognize its role by communicate with neighboring block and conduct the robots to place another block in proper position. It means that each robot just loads/unloads the block (Fig. 2).

(a)   (b)

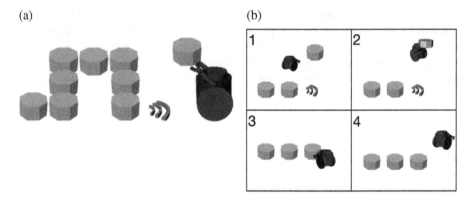

Fig. 2. (a) Conceptual image of the proposal. The block conducts the robot to place next block in proper position. (b) Sequence of the block and the robot action. (1) Each robot moves randomly searching for the block. (2) When the robot encounters a block, it picks up the block and moves randomly again. (3) When the robot detects the signal from the other block, it places the block there, and (4) searches for other blocks again. The placed block starts to send signal to desired direction.

3.1   Algorithm for a Robot

One of the simplest algorithms for clustering the distributed materials was proposed by Deneubourg et al. [1]. The robots just pick up or deposit the materials based on simple stochastic dynamics. In our proposal, we modify this algorithm for semi-active blocks (ALGORITHM 1).

ALGORITHM 1

```
LOOP
  DO random walk UNTIL find a block

  IF holding no block THEN
      Pick up the block
  ENDIF
  ELSE IF holding a block THEN
      IF detect signal from the found block THEN
            Drop off the holding block at the signal source
      ENDIF
  ENDIF
```

In short, the robot holding no block picks up the encountered block, and the robot holding a block puts down the block when it detects the signal from the other blocks.

3.2 Dynamics of a Block

In order to share the information for structure construction between the blocks, each block has a processor, small memory and devices to transmit a signal for the robots and the blocks. In the memory, it has a set of rules and a counter value, and has short range transceivers. Based on the rule set, placed block starts to send signal to proper side on which next block should be placed.

The rule set is composed of initial value, trigger values, growth directions, and a new counter value as an option. For simplicity, we assume the grid world, in which a series of blocks has seven growth directions. Following symbols indicate the side on which the next block should be placed.

S: straight from the former block
DL: diagonally forward left
L: left
DR: diagonally forward right
R: right
DBL: diagonally backward left
DBR: diagonally backward right

Note that the side such as right or left is relatively determined by its growing direction.

Two or more directions could be indicated in parallel by transmitting the signal to two or more sides at once; we can say the growing structure branches out.

In order to control the growth limit, a counter value is introduced. Basically it is used for controlling total length. Initial value 10, for example, means total length of the structure is ten. However, it could also be used for determining the length of the branch by combination of optional value.

Description of the rule set is as follow:

```
{initial count value: (trigger value, direction, (optional
value))}.
```

The following examples generate "L" shape, "Y" shape and "q" shape, respectively.

(1)   A rule set {6:(3,L)}
(2)   A rule set {7:(3,DL), (3,DR)}
(3)   A rule set {5:(1,L,2)}

Each block also has the following rules.

– Single block does not transmit a signal. Even if a block works as a part of the structure, it stops transmission and clears its counter value when it is removed from the structure.
– The block in the structure does not change its communicating side. Even when another branch grows and makes contact with a part of the structure, the contacted block does not accept the signal from the newcomer (for example, the block of which internal counter value is 3 in Fig. 3(ex3) can be 3 (based on the value of the block on the left-hand side) or 1 (based on the value on the upper side). In this case, the counter value of this block is 3 because the block on the upper side is newcomer for this block.

(1)   (2)   (3)

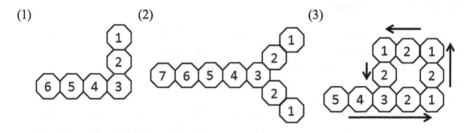

Fig. 3. Constructed structures described by rule set above. The numbers show the counter value in each block.

4   Fundamental Characteristics

We confirmed the behavior of our proposal by computer simulation. Table 1 shows the condition.

Table 1. Simulation condition

- Simulation field:	Discrete (lattice) space
- Field size:	300 x 300 pixels
- Neighborhood:	Moore neighborhood
	(Eight surrounding cells are available for robots and blocks)
- The number of robots:	5 to 300.
- The number of blocks:	100, 400, 500

Figure 4 shows a snapshot of two examples of construction. White dots are single blocks and red dots are the blocks in the structure. Robots are not displayed in these snapshot.

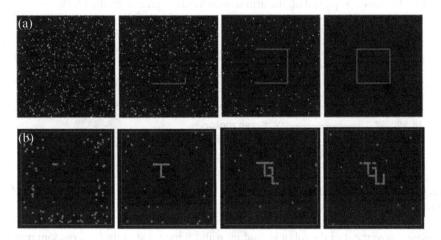

Fig. 4. Examples of construction of structure. (a) rectangle (b) asymmetric structure (Our university's logo)

5 Experiment

We constructed a small scale robot system and made experiments for our proposal. The block is shown in Fig. 5. Its shape is octagonal column, has eight pairs of IR LED and IR module for communication between the blocks, and has another IR LED on the top for controlling the robots to be held or released. Steel plate covers the face of the block.

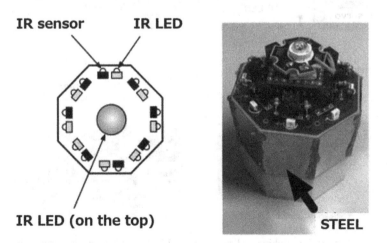

Fig. 5. (left) Schematic top view of the block. It has eight pairs of IR-LED and IR-module for communication between the blocks. (right) A picture of the block.

The robot is illustrated in Fig. 6. It has a pair of motors for moving and single arm with a magnet for carrying the block. Five IR sensors are placed in front of the robot for avoiding collision. Two IR sensors on the rear surface are used for detecting the block. Another IR sensor is placed on the arm to receive the signal from the block.

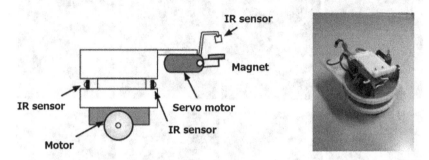

Fig. 6. (left) Schematic side view of the robot. It has a single arm for carrying the block. (right) a picture of the robot.

Basic movement of the robot is random walk: it moves straight for a random period and chooses random direction. When the front sensors detect wall or the block (or other robots) it turns about 180-degree. If the detected object is the block, the IR sensor on the arm should detects the IR signal from the IR-LED on the top of the block (Fig. 7(A)). In this case, the robot presumes the block is caught by the magnet and starts to move randomly again with the block. When the carried block detects a signal for construction

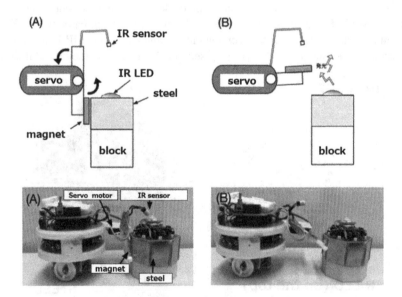

Fig. 7. A mechanism for catching and releasing the block. (A) The block is caught by magnet automatically. (B) When the robot receives a signal from the block by IR sensor on the arm, it rotates the arm and detaches the block mechanically.

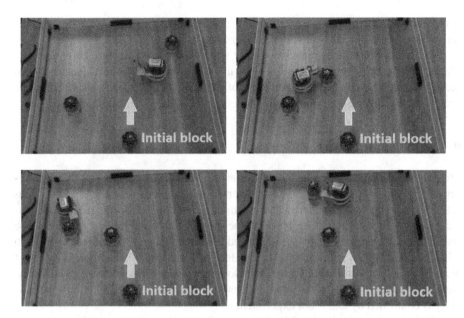

Fig. 8. A snapshot for constructing a line structure. The rule is described as {3:(null)}.

Fig. 9. A snapshot for constructing a line structure. The rule is described as {3:(2,DBL)}.

from other blocks, it transmits a signal to the robot by the IR LED on the top, and the robot releases the block by moving the magnet on its arm (Fig. 7(B)).

We carried out to construct two simple structures: a line structure (Fig. 8) and a triangular structure (Fig. 9). The initial block in the bottom is fixed. The distance between the blocks can be controlled by the strength of IR signal. In this experiment, the distance becomes longer as we set the strength of IR signal relatively high.

6 Conclusion

We propose a method for collective construction, which is accomplished by the interaction between simple robots and advanced blocks. The block is not passive but semi-active; it has a processor, memory and communication device, and the blocks communicate their counter value. The block at the growth point determines the side to be communicated by reference to the rule set and its counter value. On the other hand, each robot works using an extension of the fundamental clustering algorithm, in which it picks up or drops off the block according to local conditions. In spite of their simplicity, the blocks and the robots cooperatively construct structures as we designed. Its feasibility is confirmed by real robot experiment.

We do not describe in detail in this paper, but we have reported this system has another remarkable aspect. The constructed structure is stable globally, but the blocks are locally attached or detached repeatedly. As robots' behavior is too simple, they do not hesitate to remove the block which already forms a part of structure. We can say that the structure is under dynamic equilibrium. As for artificial construction problem, few researches focus on this characteristic, but we consider the introduction of "dynamic equilibrium" is advantageous to adaptability to environmental changes. This characteristic is now under investigation.

This work was partially supported by a Grant-in-Aid for Scientific Research on Innovative Areas "Molecular Robotics" (No. 24104005) of The Ministry of Education, Culture, Sports, Science, and Technology, Japan.

References

1. Deneubourg, J.L., et al.: The dynamics of collective sorting robot-like ants and ant-like robots. In: Animals to Animats, pp. 356–363 (1990)
2. Theraulaz, G., Bonabeau, E.: Coordination in distributed building. Science **269**, 686–688 (1995)
3. Theraulaz, G., Bonabeau, E.: Modelling the collective building of complex architectures in social insects with lattice swarms. J. Theor. Biol. **177**, 381–400 (1995)
4. Beckers, R., Holland, O.E., Deneubourg, J.L.: From local actions to global tasks: stigmergy and collective robotics. In: Artificial Life IV, pp. 181–189. MIT Press (1994)
5. Parker, C., Zhang, H.: Robot Collective Construction by Blind Bulldozing. In: IEEE International Conference on Systems, Man and Cybernetics, vol. 2, pp. 59–63 (2002)
6. Melhuish, C., Holland, O., Hoddell, S.: Collective sorting and segregation in robots with minimal sensing. In: Proceedings of the Fifth International Conference on Simulation of Adaptive Behavior, pp. 465–470 (1998)

7. Stewart, R.L., Russell, R.A.: A distributed feedback mechanism to regulate wall construction by a robotic swarm. Adapt. Behav. **14**(1), 21–51 (2006)
8. Werfel, J.: Building blocks for multi-robot construction. In: Alami, R., Chatila, R., Asama, H. (eds.) Distributed Autonomous Robotic System 6, pp. 285–294. Springer, Heidelberg (2007)
9. Werfel, J., Petersen, K., Nagpal, R.: Distributed multi-robot algorithms for the TERMES 3D collective construction system. In: Proceedings of the IEEE/RSJ International Conference on Intelligent Robots and Systems (2011)
10. Sugawara, K., Doi, Y.: Collective construction of dynamic structure initiated by semi-active blocks. In: Proceedings of the IEEE/RSJ International Conference on Intelligent Robots and System, pp. 428–433 (2015)
11. Werfel, J., Yaneer, B-Y., Rus, D., Nagpal, R.: Distributed construction by mobile robots with enhanced building blocks. In: Proceedings of the IEEE International Conference on Robotics and Automation, pp. 2787–2794 (2006)

Can Haptic Feedback Improve Gesture Recognition in 3D Handwriting Systems?

Dennis Babu$^{(\boxtimes)}$, Seonghwan Kim, Hikaru Nagano, Masashi Konyo,
and Satoshi Tadokoro

Human Robot Informatics Lab, Graduate School of Information Sciences,
Tohoku University, Sendai 980-8579, Japan
{dennis,nagano,konyo,tadokoro}@rm.is.tohoku.ac.jp

Abstract. Current gesture interfaces accept relatively simple postures and motions for reliable inputs, which are still far from natural and intuitive experiences for the users. This paper suggests a unique idea that haptic feedback has a potential to improve not only user experiences but also gesture recognition performances. We expect haptic feedback to provide users cues for natural writing movement and accordingly generate movement easily recognized by the system. We developed a writing gesture recognition system using the K-Means clustering algorithm for writing state estimation and a haptic feedback system which involved frictional sensation during writing and impulsive sensation at the beginning and ending of writing. The experiments on five participants showed an approximately 5 % and 4 % improvement in the true positive and the false negative gesture recognition rate with visual-haptic feedback compared to visual feedback alone. We confirmed that the improvement was due to changes in hand motion by haptic feedback, which led to a higher correlation between reference waveform and performed motion and an increase of the finger stopping time at the end of the writing. We also confirmed the positive effects of our haptic feedback on the user experiences.

Keywords: Gesture recognition · Haptic feedback · Mid air interaction

1 Introduction

Recent advances in consumer markerless motion tracking systems have opened up new user experiences on mid-air interaction. For example, the Leap Motion controller realizes a multi-finger interaction with sub-millimeter accuracy [1]. Microsoft's Kinect sensor provides full-body 3D motion capture. Current gesture recognition, however, accepts relatively simple postures and motions for reliable inputs, which are still far from natural and intuitive experiences for the users. The limitations of gesture recognitions come not only from the unsatisfactory performance of recognition algorithms but also the incomplete user's behaviors in the air.

© Springer International Publishing Switzerland 2016
N. Kubota et al. (Eds.): ICIRA 2016, Part I, LNAI 9834, pp. 462–471, 2016.
DOI: 10.1007/978-3-319-43506-0_41

This paper suggests a novel approach to improving the system recognition robustness. A potential issue for the mid-air tracking system is that users cannot perform their intended motion precisely due to lack of physical contact. The authors expect that haptic feedback to represent virtual contact information will provide the users a cue to control their movement and consequently improve the system robustness even with the same recognition algorithm. Figure 1 shows our main idea, which proposes a positive loop to improve the system performance of the gesture recognition by haptic feedback. The target task in this paper is handwriting in the air. The 3D tracking data of the user's hand in mid-air for gesture recognition system. The gesture recognition system estimates motion states of the handwriting to generate haptic feedback. The haptic feedback system provides vibrotactile stimuli at the user's wrists depending on the estimated states.

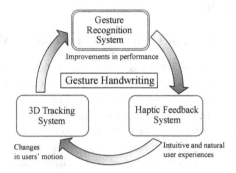

Fig. 1. Proposed gesture handwriting approach.

Although a potential technology to stimulate a hand freely in the air with an airborne ultrasound tactile display has been proposed recently [2], most of the conventional studies have applied small vibrators attached on the hand for providing vibrotactile stimuli during gesture motion. Targets of the midair interaction systems can be broadly classified into a human-computer interaction and a motion guidance. Relevant studies related to haptic feedback assisted human-computer interaction systems includes an investigation of the effects of vibrotactile feedback in mid-air gesture interaction [3], comparison between visual, aural and haptic feedback types in a simple remote pointing task [4,5], and tactile cueing navigation coupled with visual feedback in mid-air gesture interaction [6]. Vibrotactile feedback can also guide human motion for static pose correction [7], dynamic motion apart from desired trajectories [8]. The comparative performances were also reported on single and multi-sensory cues [9]. Conventional studies targeted on relatively large and rough motion of gestures.

The novelty of this study is to investigate the potential of haptic feedback on the system recognition performance, i.e., not only for a user but a system. The contributions of this paper are as follows:

– Providing initial proofs to show the potential of haptic feedback to improve gesture recognition performance in terms of the recognition rates and reduction of retrials in handwriting tasks.

- Investigating the effects of the haptic feedback on hand motion to discuss why the recognition performance has improved for specific users.
- Confirming positive effects of our haptic feedback on the user experiences, including intuitiveness, user satisfaction, and learnability by subjective rating scale questions.

First we discuss a typical writing motion profile in a real screen. Then, we develop a gesture recognition algorithm and a vibrotactile stimulation method to represent the writing and transition state based on the gesture recognition. In the experiments, we confirm the system recognition rates and discuss the changes in motion by the haptic feedback.

2   System Overview

2.1   Observation of Motion Characteristics During Writing

An experimental study for gaining insight into the handwriting recognition algorithm was conducted. We measured the hand motion characteristics in an ideal 3D handwriting, in which a real screen was placed in front of the subjects thereby receiving natural haptic feedback using a motion capture system. The screen size, height from the ground and the relative distance to the user was fixed to be 300, 600 and 450 [mm] respectively. For simplifying the analysis the number of characters to be written in each trial are fixed to be three combinations, namely '0 1 2', '3 4 5' and '6 7 8'. As a result, we found a typical absolute velocity pattern at each transition of the writing states.

2.2   Gesture Recognition System

The gesture recognition system consists of a five state machine which uses the feature vectors extracted from motion parameters for time series clustering. A realistic haptic feedback system should give the feeling of a real screen while writing with no feedback in other states. In a 3D handwriting system, there are frequent transitions from write_state to transition_state with entry_state

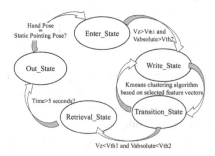

Fig. 2. State diagram for midair handwriting state recognition

and retrieval_state occurring in the initial and final phases of writing. Thus, the primary goal of the gesture recognition system is to distinguish accurately between write_state and transition_state for realistic feedback generation. While a K-Means clustering algorithm using the motion parameters determines the traversal from write_state to transition_state and vice versa, static and dynamic hand poses are used to instigate traversal through out_state, entry_state and retrieval_state during real-time 3D handwriting as shown in Fig. 2.

The Euclidean distance between the real-time motion frame array with that of a typical velocity minimum array vector obtained from the ideal motion characteristics is the primary feature vector to detect start or end points of writing phase. The other feature vectors employed are directions of V_y and V_x vectors. The above feature vectors are fed to a K-Means clustering algorithm which clusters the motion data into two clusters: minimum velocity points and other points. The continuous length of the velocity minimum cluster elements, henceforth mentioned in this paper as finger stopping time is used to distinguish motion states into write_state and transition_state.

2.3 Vibrotactile Feedback

We use a linear resonant type actuator (Force Reactor, Alps Electric Corp.,) as a vibrator at the user's wrist as shown in Fig. 4. The LRA actuator was placed in a 3D printed fitting box of size 55×25 [mm] and attached to users hand using an adjustable wristband with the overall weight of the device being 35 g. The primary reason for the selection of wrist-based haptic feedback compared to a more intuitive fingertip based feedback is the commercial availability of such a system.

Two types of haptic feedback were generated in response to the changes in the estimated states by the gesture recognition system. One is the friction sensation in the writing phase (Write_State). The friction display method using vibrotactile stimuli was proposed by the authors [10]. This method represents the stick-slip transitions between the skin and a surface by amplitude modulations of high-frequency vibrations (380 Hz) depending on hand exploration speed and pressing force, which are reflected in a physical vibration system model. Details are shown in [10].

The other is impulsive stimuli to represent the timing of initial and ending of contact in handwriting. Damping sinusoidal signals, which are known to produce contact sensation in an impact situation, were generated in the beginning and end of each write_state. We use the following waveform $Q(t)$ as the impulsive stimulation, which was used in [11].

$$Q(t) = A(v)e^{-Bt} \sin(2\pi ft) \qquad (1)$$

where A is the initial amplitude, B is the damping coefficient, and f is the vibration frequency. We used different stimulation at the beginning and end point by changing the damping coefficients; $B = 10$ for the beginning point and $B = 40$ for the end point. Vibration frequency was constant at $f = 240$ Hz for

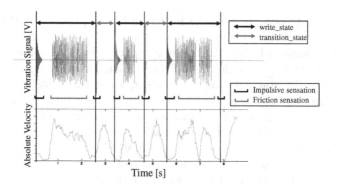

Fig. 3. Vibrotactile feedback during writing of '0 1 2'.

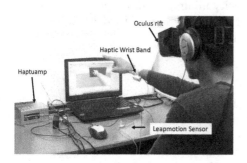

Fig. 4. Experimental setup.

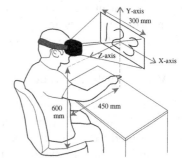

Fig. 5. Pictorial representation of experimental setup.

the both. Figure 3 shows the haptic signal excitation during a typical writing scenario of '0 1 2'. The absolute velocity at the bottom represents the hand motion speed.

3 Experimental Setup

3.1 System

The experimental set up as shown in Fig. 4 consists of dual infrared camera (Leap motion) based real time hand motion tracking system which extracts the motion vectors and a Head Mounted Display (Oculus Rift) for real time visual rendering of the written characters. The visual feedback of the written characters is rendered in realtime by the head mount display during writing based on the gesture recognition algorithm. A virtual skeletal hand model is also rendered from the tracked hand model along with black screen is devised to replicate an ideal writing scenario.

3.2 Participants and Tasks

Five adult subjects well aware of haptic feedback systems in the age range of 21–25 with an average age of 22.5 was selected for evaluating the devised 3D handwriting system. The experimental conditions required the participants to write characters with and without haptic feedback while shunting external noises by pink noise generation. During all the experiments an initial briefing of the experimental tests along with training on the system for a predefined time duration was conducted.

The experimental task required the subjects to write the characters '0 1 2', '3 4 5' and '6 7 8' three characters each with and without haptic feedback. The writing was repeated four times to mitigate occasional variations and statistical analysis. The order of writing characters was changed after each experiment so as to avoid the effect of haptic feedback in subsequent experiments and vice versa. After each experiment, an interval of 3 min was allowed to avoid fatigue due to stress. The subjects were required to use the entire writing area of the screen with no other specific requirements of writing pattern. After the completion of all the experiments, each subject was asked to rate the system on a scale of 0–5 with and without haptic feedback based on a predefined subjective rating scale questionnaire.

4 Results

4.1 Gesture Recognition Rate

Table 1 shows the results of gesture recognition rates with and without the haptic feedback system. Each subject wrote characters of '0 1 2', '3 4 5' and '6 7 8' four times on each condition with and without haptic feedback, thus, the total number of write_state and transition_state was 180 and 120 times, respectively. The individual count of the write_state and the transition_state were 36 and 24, respectively. The overall recognition rates of write_state or True Positive rate (TP) increased from 92.77 % with visual feedback alone to 98.21 % with visual and haptic feedback whereas the recognition rate of transition_state or False Negative rate (FP) increased from 94.64 % to 98.21 %.

While four subjects had improved recognition rates with visual-haptic feedback compared to visual feedback alone, the improvement in the recognition rate of subject 2 was significant as shown in Table 2. The recognition rate with False Negative (FN) is higher compared to True Positive Rate (TP) for every subject irrespective of the presence of haptic feedback.

4.2 Number of Retrials for Task Completion

When the gesture recognition is wrong, the absence of visual and haptic feedback during writing prompted the subjects to try again for successful writing. Thus, a number of retrials became a significant parameter along with gesture recognition rate for evaluation of haptic feedback in the current system. The overall number

Table 1. Gesture recognition rate.

	Visual feedback		Visual+Haptic feedback	
	write_state	transition_state	write_state	transition_state
write_state	92.77 % (TP)	7.33 %	97.02 % (TP)	2.98 %
transition_state	5.36 %	94.64 % (FN)	1.79 %	98.21 % (FN)

Table 2. Individual gesture recognition rate.

	Visual feedback		Visual+Haptic feedback	
Subjects	TP	FN	TP	FN
Subject1	93.33 %	95 %	100 %	100 %
Subject2	70 %	96.67 %	85 %	100 %
Subject3	100 %	93.75 %	100 %	100 %
Subject4	96.67 %	100 %	100 %	100 %
Subject5	100 %	100 %	93.75 %	100 %

of retrials to execute the specified writing task decreased from 41 to 16 and that in the starting phase decreased from 30 to 9 with the addition of haptic feedback along with visual feedback as shown in Fig. 6. While the number of retrials for subjects 1, 2, 3 and 4 either decreased or remained same with haptic feedback, the decrease of number of retrials by subjects 1 and 4 were significant.

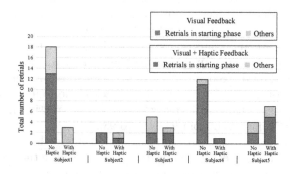

Fig. 6. Number of retrials with and without haptic feedback for each subject.

4.3   Euclidean Distance Score and Finger Stopping Time

Figures 7 and 8 shows the box plot of Euclidean distance score and finger stopping time of the subjects after writing each character with and without haptic feedback respectively. The Euclidean distance score is significantly reduced in 4 subjects whereas the reduction in subject 2 is marginal. Figure 8 shows that

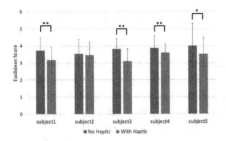

Fig. 7. Euclidean distance score

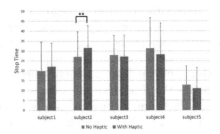

Fig. 8. Stopping time modulation

Table 3. Subjective rating scale

Parameters	Visual feedback	Visual+Haptic feedback
Ease of use	2.8 ± 0.8	3.6 ± 0.6
Interactive	2.6 ± 1.0	3.6 ± 0.6
Intuitive	2.2 ± 0.8	3.4 ± 0.5
User satisfaction	3.2 ± 0.75	4.4 ± 0.6
Learnability	2.4 ± 1.0	3.6 ± 0.6

haptic feedback significantly modulates the finger stopping time of the subjects after writing each character. While subjects 1 and 2 tends to increase the finger stopping time, other subjects tend to shorten the stopping time. The subject 2 had significant change in the stopping time but that of other subjects remained marginal.

4.4   Questions on User Experiences

Table 3 shows the average value of 5 user experience parameters of all the subjects under study. As can be seen from the results the subjects had an apparent leniency towards visual-haptic feedback compared to visual feedback only. While all the users asserted that haptic feedback improved all the five user experience parameters, the user satisfaction and intuitiveness fared the best of all.

5   Discussions

As the haptic and visual feedback during writing and transition phases depends on the gesture recognition results, an improvement in the recognition rate and/or the number of retrials can significantly improve the user satisfaction with the devised writing interface. From the results, it is evident that there is the improvement in overall gesture recognition rates as well as the number of retrials of the subjects with visual and haptic feedback compared to visual feedback alone. The Euclidean distance score and the finger stopping time had significant modulation due to haptic feedback thereby increasing the gesture recognition rate.

The gesture recognition algorithm depends mainly on two parameters, the Euclidean distance score and the finger stopping time. Any improvements in these parameters due to haptic feedback can improve the system recognition rate for the same algorithm. The haptic feedback modulates the minimum velocity curve thereby making the curve more correlated to the reference curve and thus decreases the Euclidean distance score. This result makes the input to the K-Means clustering algorithm more smooth with reduced fluctuations. Moreover, the haptic impulse feedback at the starting and beginning of writing causes the subjects to change their finger stopping time. These changes in the finger motion pattern improve the gesture recognition rates of the subjects along with improvement in user satisfaction as seen from the subjective rating scale.

While Subjects 1 to 4 had improved gesture recognition results, the decrease in Euclidean score of Subjects 1, 3 and 4 and the increase in stopping time of subject 2 was very significant. Though Subject 5 had a significant decrease in the Euclidean score, the finger stopping time was marginally lower compared to other subjects in both cases thus leading to reduced gesture recognition rates.

The varying effect of the haptic feedback on different users can be attributed to the virtual and real haptic stimulation positions, i.e., while the subjects see their finger touching the screen, the stimulation occurs on the wrist. Out of the five subjects, four subjects clearly chose visual-haptic feedback over the other, whereas a subject found visual-haptic feedback confusing and less interactive compared to its visual counterpart.

6   Conclusions

This paper investigates the potential of vibrotactile feedback to improve the system recognition performance in a gesture handwriting task. We provided initial proofs to show the potential of improvements by developing a gesture recognition system to provide the two types of haptic feedback; friction sensation during writing and impulsive stimulation at the beginning and ending of each letter writing.

The results with visual-haptic feedback compared to visual feedback alone showed that the haptic feedback improved the gesture recognition rate. The total number of retrials for the task completion was also reduced or remained constant with haptic feedback. The improvement in the gesture recognition rate was estimated to be due to modulation of hand motion during writing phase with haptic feedback. The modulation of the Euclidean distance score and the finger stopping time led to smooth feature vectors for the K-Means clustering algorithm to distinguish the transitions of the estimated states.

Presented concept of the positive loop shown in Fig. 1 suggests us a possibility of further improvement in recognition performance by optimizing the gesture recognition algorithm in taking account of the changes in user's motion induced by haptic feedback. In the future research efforts, the authors will try to improve the gesture recognition algorithm by including more motion features to adapt the motion changes.

Acknowledgment. This work was supported in part from ImPACT Program (Tough Robotics Challenge).

References

1. Weichert, F., Bachmann, D., Rudak, B., Fisseler, D.: Analysis of the accuracy and robustness of the leap motion controller. Sensors **13**, 6380–6393 (2013)
2. Monnai, Y., Hasegawa, K., Fujiwara, M., Yoshino, K., Inoue, S., Shinoda, H.: HaptoMime: mid-air haptic interaction with a floating virtual Screen. In: Proceedings of ACM UIST, pp. 663–667 (2014)
3. Adams, R.J., Olowin, A.B., Hannaford, B., Sands, O.S.: Tactile data entry for extravehicular activity. In: 2011 IEEE World Haptics Conference (WHC), pp. 305–310 (2011)
4. Krol, L.R., Aliakseyeu, D., Subramanian, S.: Haptic feedback in remote pointing. In: Proceedings of CHI 2009 Extended Abstracts on Human Factors in Computing Systems (CHI EA 2009), pp. 3763–3768 (2009)
5. Burke, J.L., Prewett, M.S., Gray, A.A., Yang, L., Stilson, F.R.B., Coovert, M.D., Elliot, L.R., Redden, E.: Comparing the effects of visual-auditory and visual-tactile feedback on user performance: a meta-analysis. In: Proceedings of the 8th International Conference on Multimodal Interfaces (ICMI 2006), pp. 108–117 (2006)
6. Lehtinen, V., Oulasvirta, A., Salovaara, A., Nurmi, P.: Dynamic tactile guidance for visual search tasks. In: Proceedings of the 25th Annual ACM Symposium on User Interface Software and Technology, pp. 445–452 (2012)
7. Rotella, M.F., Guerin, K., Xingchi, H., Okamura, A.M.: HAPI Bands: a haptic augmented posture interface. In: 2012 IEEE Haptics Symposium (HAPTICS), pp. 163–170 (2012)
8. Bark, K., Hyman, E., Tan, F., Cha, E., Jax, S.A., Buxbaum, L.J., Kuchenbecker, K.J.: Effects of vibrotactile feedback on human learning of arm motions. IEEE Trans. Neural Syst. Rehabil. Eng. **23**(1), 51–63 (2015)
9. Xu, M., Wang, D., Zhang, Y., Song, J., Wu, D.: 2015 IEEE Performance of simultaneous motion and respiration control under guidance of audio-haptic cues. In: World Haptics Conference (WHC), pp. 421–427 (2015)
10. Tsuchiya, S., Konyo, M., Yamada, H., Yamauchi, T., Okamoto, S., Tadokoro, S.: Virtual Active Touch II: vibrotactile representation of friction and a new approach to surface shape display. In: IEEE/RSJ International Conference on Intelligent Robots and Systems (IROS), pp. 3184–3189 (2009)
11. Okamura, A.M., Cutkosky, M.R., Dennerlein, J.T.: Reality-based models for vibration feedback in virtual environments. IEEE/ASME Trans. Mechatron. **6**(3), 245–252 (2001)

Meta-module Self-configuration Strategy for Modular Robotic System

Zhen Yang, Zhuang Fu[✉], Enguang Guan, Jiannan Xu, and Hui Zheng

State Key Laboratory of Mechanical System and Vibration, Shanghai Jiao Tong University,
Shanghai 200240, People's Republic of China
zhfu@sjtu.edu.cn

Abstract. Modular robotic system (MRS) consisting of several identical modules is able to adapt to situations by configuring to the morphology which best suits the environment. Utilizing this characteristic, this paper presents a self-configuration strategy for the M-Lattice MRS. Based on this strategy, the construction task under complex environment is converted to a self-configuration task accomplished by modular robots. Meta-module method is utilized to avoid system cracking. Meanwhile, a gradient greedy method in a virtual gravity field is selected to indicate the locomotion of meta-modules. Computational simulations demonstrate the feasibility and scalability of the proposed strategy.

Keywords: Self-configuration · Meta-module · Modular robot · M-Lattice

1 Introduction

Over recent years, applications of MRS have increasingly drawn the attention of the robotics and control communities. MRS made up of a given number of modules can attain different configurations based on their relative coupling. Using this property, this paper introduces a self-configuration strategy for the M-Lattice MRS. Under this self-configuration strategy, the traditional construction task under complex environment is converted to a self-configuration task for MRS.

Modular robots can be geometrically classified into three types: chain-type, lattice-type and hybrid-type. And besides compact modular robots, there are several metamorphic modular robots. Chirikjian [1] develops a metamorphic MRS. The metamorphic module consists of six-bar linkages. And two linkages are connected by a rotation joint. The whole system is an aggregation of several hexagon modules. The metamorphic property makes the system have better performance in self-reconfiguration than compact modular systems. Inspired by the design of metamorphic module and the unit-compressible modular robot, Crystal [2], the system grid of M-Lattice is designed to be a hexagon which consists of six unit-compressible modules.

Using a MRS to solve the construction task is first proposed by Terada and Murata [3]. In their work, the construction task is accomplished by heterogeneous modules, the cubic structure modules and the assembler robots. In our work, a homogeneous modular robot is chosen to finish the construction task.

© Springer International Publishing Switzerland 2016
N. Kubota et al. (Eds.): ICIRA 2016, Part I, LNAI 9834, pp. 472–484, 2016.
DOI: 10.1007/978-3-319-43506-0_42

The control strategies for self-reconfiguration have referential meanings for self-configuration. In a typical centralized control strategy such as the simulated annealing method [4], current configuration and target configuration are compared in real time. The controller drives specific modules into several target positions. As the centralized control strategy suffers from the system scale, distributed control strategy attracts more attention. Golestan [5] uses a graph signature method to analyze relationships between module connections and target configurations. Zhu [6] simplifies the cellular automata method in self-reconfiguration task, using UBot. In our self-configuration strategy, modules moves along the system boundary without presetting their final positions. And the global message is unnecessary. Enlightened by Park [7] and Arney [8], a finite state machine is used to change modules from movable status to stable status.

Meta-module can be defined as a group of modules which act as one unit. During the locomotion, the modules that compose a meta-module keep the connection with each other. Compared with single module, meta-module gets more flexible mobility. Several studies have been reported in literature on the meta-module applications. One primary motion mode for meta-modules is crawling along the system boundary like a tiny chain MRS, such as Telecube [9] and ATRON [10]. Dewey [11] introduces a control strategy that has a good performance in flexible meta-module systems. In our work, meta-module is designed to be capable of crawling along the system boundary and then entering into the optimal position. Owing to the feature of self-configuration task, a distributed control strategy is designed to indicate the locomotion of meta-modules. And inspired by Christensen [12], a virtual gravity gradient field is used to evaluate the possible target positions for meta-modules.

2   Design of M-Lattice Modular Robot

The most challenging task in self-configuration is how to deal with the motion constrains. In a MRS, the motion constrains can be classified into two types: local constrains and global constrains.

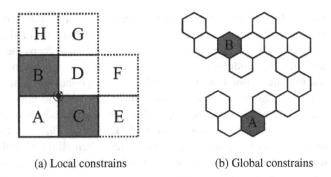

(a) Local constrains   (b) Global constrains

Fig. 1. The motion constrains. (a) Local constrains. If a square module wants to move from C to B around module A, the positions of D, E, F, G and H must be empty; (b) Global constrains. During the configuration process, the hexagon module A and B can't move. Otherwise, the system will break apart.

As shown in Fig. 1, local constrains is mainly determined by the module structure and the motion mode. Global constrains is a critical problem in the development of path planning algorithms. Thus at the stage of module structure design, besides of the efficient module transportation and reliable connection, there is an extra essential requirement that the module structure should release the local constrains as much as possible.

2.1   Basic Module

M-Lattice modular robot is designed to consist of two parts: a centre prism frame and three identical mechanical arms. As shown in Fig. 2, three mechanical arms are located on the symmetrical sides of the prism frame. Each mechanical arm contains two DOF which are the rotation DOF and expansion/contraction DOF. The rotation DOF drives one module connect with different modules and then changes module's position. The expansion/contraction DOF can change the rotation radius which could release the local constrains significantly. In addition, one module can keep the connection with three adjacent modules simultaneously [13].

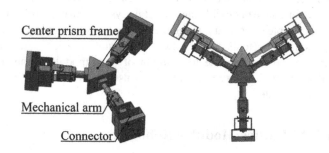

Fig. 2. M-Lattice modular robot (TFR-ISO view and expansion/contraction DOF)

2.2   Basic Motion

The locomotion of M-Lattice module is realized through the attach/detach operations and the rotations of mechanical arms within a sequence of connected modules. During the locomotion, one module in the module sequence called *Base* keeps the connection with system while the rest modules of sequence called *Assistants* are disconnected with system. Thus the module locomotion can be marked as a single-base and multi-assistants (SBMA) locomotion. The SBMA locomotion can be described in detail as follows. (1) Based on the initial position and target position, the module sequence can be determined. One specific module is chosen as *Base*. (2) Except for *Base*, all the modules within the sequence break the connections with system, then all the modules contract their mechanical arms. Based on the number of modules within the sequence, mechanical arms rotate a certain angle. Then module sequence forms a closed polygon. (3) One module breaks its former connection within the sequence. All the mechanical arms rotate back to their initial status. Then modules expand their mechanical arms and reconnect themselves with system.

An example of SBMA locomotion is shown in Fig. 3. The number of modules within the sequence is three and only one module is ordered to change its position. It can be found that owing to the expansion/contraction DOF, whether the positions adjacent to the module sequence are occupied or not, there is no collision risk.

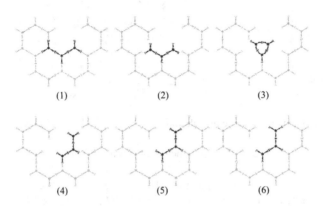

Fig. 3. An example of SBMA locomotion with three modules

As mentioned before, mechanical arms need to rotate a certain angle to form a closed polygon. According to the geometric relationship, every module can be treated as three vertices in the closed polygon. The number of edges can be found as $E = 3n$. To compose a closed polygon, the rotation angle changes with the number of modules within the sequence as follows:

$$\frac{2\pi}{3}n + (\pi - \theta) \times 2n = \pi \times (E - 2) \Rightarrow \theta = \pi \times \frac{6 - n}{6n}$$

where n is the number of modules within the sequence; E is the number of edges of the closed polygon; Θ is the rotation angle of mechanical arms.

2.3   System Structure

Six interconnected modules compose a hexagon grid and many hexagon grids connected together will form a huge mesh plane. In order to express modules' positions and connection status clearly in the following sections, the module is described in its topological form, as shown in Fig. 4(a). In our work, the position of each module is marked under a special X-Y index frame, as shown in Fig. 4(b). Under this index frame, each module's position is represented by an ordered pair of integers (x, y), where x represents the column number of module's location and y represents the row number. In order to analyze the self-configuration task, some definitions are given firstly.

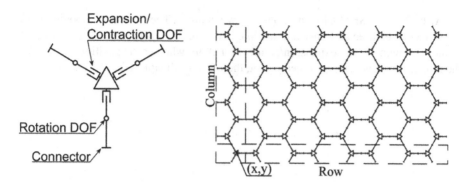

(a) Topological form of M-Lattice module frame   (b) The X-Y index

Fig. 4. System structure of M-Lattice MRS

Definition 1. The configuration space of M-Lattice MRS is represented as a vertex index matrix, marked as *VI*. The form of the matrix *VI* is shown as follows:

$$VI = \begin{bmatrix} 1 & \cdots & X \\ \vdots & \ddots & \vdots \\ 0 & \cdots & 1 \end{bmatrix}$$

where each element *VI* (x, y) represents a position that can be occupied by a module and (x, y) is the index representation introduced in Fig. 4(b).

VI $(x, y) = 1$: the current vertex is occupied by a module;
VI $(x, y) = 0$: the current vertex is empty;
VI $(x, y) = X$: the current vertex can't be occupied by a module. This could be caused by the reserve for special shape buildings.

Definition 2. The connection status of M-Lattice MRS in the configuration space is represented by the adjacent edge matrix, marked as *AE*. If the position of module is (x, y), the connection status of this module is marked as follows:

$$AE = \begin{bmatrix} x & y+1 & 1 \\ x & y-1 & 0 \\ x \pm 1 & y & X \end{bmatrix}$$

where each row represents a connection status for module (x, y).

AE $(i, 3) = 1$: the i-th adjacent position is occupied by a module;
AE $(i, 3) = 0$: the i-th adjacent position is empty;
AE $(i, 3) = X$: the i-th adjacent position can't be occupied by a module. This could also be caused by the reserve for special shape buildings.

Under the X-Y index frame of M-Lattice MRS, modules have two possible gestures. The only difference between these two gestures is the relative location of the adjacent module which has the same index value along Y-axis as module (x, y). Thus in the matrix AE, if the position of module is (x, y), the first row represents the connection status of upper adjacent position $(x, y+1)$, the second row represents the connection status of under adjacent position $(x, y-1)$ and the third row represents the connection status of left/right adjacent position whose index representation is $(x-1, y)$ or $(x+1, y)$ determined by the gesture of module (x, y).

Based on these definitions, the network composed by M-Lattice modules can be represented as an undirected weighted graph $G = (VI, AE, w)$, where VI is the vertex index matrix which means a set of modular robots, AE is the adjacent edge matrix representing how the modules are connected, and w is a weight function. Thus the self-configuration task is now transformed to an abstract problem that is how to convert the vertex index matrix VI to a nonzero matrix via the locomotion of modules. Nonzero matrix VI means all the positions in the configuration space are occupied by modules or reserved for special shape buildings.

3 Self-configuration Strategy Based on Meta-module Locomotion

During the SBMA locomotion, the disconnection between *Assistant* and system is critically dangerous for maintaining the system connectivity. This problem can be solved as follows. It can be found that the primary problem is that it's hard to make sure whether *Assistant* is the only module connected with two local systems. But if the locomotion of single module does not involve any existing system module, system connectivity can be guaranteed with no doubt. Although the SBMA locomotion determined by the module structure and connection manner is unchangeable, we can give a new definition for 'single module'. That is the meta-module method.

Meta-module is a unit which consists of several basic modules. Although meta-module increases the granularity of the system, it has a significantly better mobility than the single module. Utilizing a meta-module method, the SBMA locomotion may not involve any existing system module and a single meta-module is able to move itself from one position to another. In our work, meta-module is designed to consist of four M-Lattice modules. During the locomotion of meta-module, the connection relations of meta-module are not fixed. The locomotion of meta-module along the system boundary is a succession of cycles of sense-think-act. These three parts can also be called as path searching, path planning and polymorphic locomotion.

3.1 Path Searching

As a special module unit, the locomotion of meta-module is realized through connecting with different modules. Path searching is executed by the meta-module members connected with system. In the M-Lattice MRS, each grid is a hexagon containing six modules. In order to search all the possible positions, the depth of path searching is set as four. All the possible positions are saved as follows:

$$Path = \begin{bmatrix} TgX_1 & TgY_1 & SysX_1 & SysY_1 & MetaX_1 & MetaY_1 \\ \vdots & \vdots & \vdots & \vdots & \vdots & \vdots \\ TgX_n & TgY_n & SysX_n & SysY_n & MetaX_n & MetaY_n \end{bmatrix}$$

where $(TgXi, TgYi)$ is the location of the possible target position; $(SysXi, SysYi)$ is the location of the system module which is connected with the searching meta-module member; $(MetaXi, MetaYi)$ is the location of the searching meta-module member.

During the path searching, each meta-module member sends a message to its adjacent system module. If there is an empty place adjacent to it, the system module will send a message back to the meta-module member. If there exists another adjacent system module, it sends a message to the subsequent adjacent module and asks for empty place information. The search depth is four, and then meta-module gets all the empty places within this range.

3.2   Path Planning

From the grid geometry, it can be found that a hexagon grid is confirmed by three connected modules while two adjacent grids is confirmed by two connected modules. For each row of data in **Path**, the target position $(TgXi, TgYi)$ is filtered whether it is contained in the grid which is confirmed by the searching meta-module member and its adjacent system module. As each meta-module consists of four M-Lattice modules, the target position contained by a grid which also contains a meta-module member at least is reachable for the meta-module in theory.

A gradient greedy method in a virtual gravity field is utilized to choose the optimal target position from all the reachable positions. In the configuration space, the direction of virtual gravity is set along the X-axis and the position which has the largest gradient value is selected as the optimal target position. In each locomotion round, the meta-module member with the smallest index representation is set to stand for meta-module. This module determines a possible target position vectors $\overrightarrow{V_i}$ from its position to the possible target position. The projection gradient $grad(i)$ along the X-axis vector $\overrightarrow{V_m}$ for each possible target position vector is calculated. If the projection gradient is positive, the probability P of taking $grad(i)$ as the target gradient is given by the Boltzmann's law. And the modules with negative projection gradients are recorded with a small probability.

$$grad(i) = \left\langle \vec{V_i}, \frac{\vec{V_m}}{\left|\vec{V_m}\right|} \right\rangle \qquad P(gradT = grad(i)) = \begin{cases} \dfrac{e^{grad(i)/\tau}}{\sum\limits_{i=1}^{n} e^{grad(i)/\tau}} & grad(i) > 0 \\[4mm] \varepsilon & grad(i) < 0 \end{cases}$$

where $grad(i)$ is the projection gradient of position i; τ is a positive temperature parameter in the Boltzmann distribution; ε is a small greedy probability for negative gradient; $gradT$ is the target gradient.

If several target positions have the same probability, the one with largest index representation along Y-axis is chosen as the optimal target position. This choice guarantees that the subsequent incoming meta-modules will have further search space and increases the success rate of self-configuration.

3.3   The Polymorphic Locomotion of Meta-module

The polymorphic locomotion of meta-module is a sequence of module's motions to connect the meta-module with target position, absolutely based on meta-module members. In order to adapt to various target positions, the structure of meta-module is polymorphic just like a water-flow. In order to explain the locomotion clearly, some definitions are given firstly.

Definition 3.   Target grid is the grid which contains the searching meta-module member, its adjacent system module and the target position.

Definition 4.   Tail module is a meta-module member. It belongs to the target grid, and it is the tail of the meta-module sequence which means that Tail's adjacent position should be empty.

Definition 5.   Transitional grid can be confirmed by the tail module, its adjacent empty position and its adjacent position which does not belong to the target grid.
The relations of target grid, tail module and transitional grid are shown in Fig. 5.

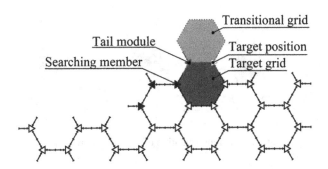

Fig. 5.   The presentation of target grid, tail module and transitional grid

The polymorphic locomotion of meta-module is designed as a succession of stable grid transitions where meta-module members are driven into the target grid until the target position is occupied by a meta-module member. The polymorphic locomotion consists of two stages: combination to the transitional grid and separation to the target grid.

In the combination stage, all the meta-module members which do not belong to the target grid are relocated into the transitional grid. During the combination process, if a meta-module member is outside of both target grid and transitional grid, it will be relocated into the transitional grid under the SBMA locomotion. One representation of combination is shown in Fig. 6.

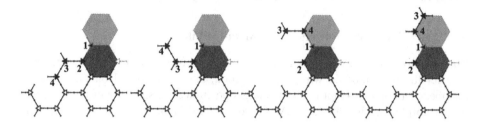

Fig. 6. One representation of combination process

In the separation stage, meta-module members in the transitional grid are relocated into the target grid. If the target position is occupied by a meta-module member, the polymorphic locomotion ends. The separation process is similar with the combination while the main difference is that, in the separation, the amount of meta-module members which should move into the target gird is determined by the occupied status of target grid. At the beginning of the separation, target grid is assumed as a connected half-occupied grid and the occupied amount can be obtained. Then meta-module members in the transitional grid start to move into target grid from the empty position adjacent to the tail module. One representation of separation is shown in Fig. 7.

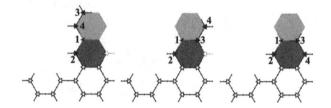

Fig. 7. One representation of separation process

4   Simulations of Self-configuration

A simulator to evaluate the properties of meta-module self-configuration strategy is built on MATLAB, running on a desktop with CPU Pentium E8400, RAM 4 GB.

The simulation scene is shown in Fig. 8. M-Lattice modular robot is represented as a node consisting of a filled circular and three bold lines for simplicity. For arbitrary configuration structure, the initial configuration space is assumed as rectangle and the system is initialized under the presentation of an undirected weighted graph $G = (VI, AE, w)$. Module supply area is on the left. Meta-module consisting of four modules is taken into configuration space from the module supply area and then crawls along the system boundary step by step. After a meta-module is terminated, the next meta-module starts to move. If the entrance is blocked by system modules, one simulation round is finished.

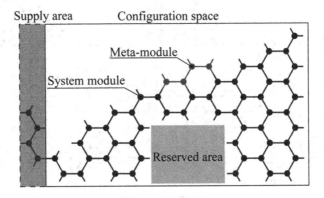

Fig. 8. Simulation scene

In order to test the scalability of self-configuration strategy, the system is set with different scales from 10 × 10 to 60 × 60. And two structures are chosen as the target configurations which are rectangle and L-shaped geometric structure. During the self-configuration process, path searching, path planning and polymorphic locomotion are operated by the meta-module in a distributed manner. If a collision happens, the simulation would be shut down. If system cracks because of SBMA locomotion, the self-configuration task is considered as a failure, and the simulation is also terminated. The self-configuration process of a configuration space with 400 (20 × 20) modules is shown in Fig. 9.

The simulations are repeated twenty times for every system scale. For each experiment, time consumption is recorded. As shown in Fig. 10, time consumption grows approximately exponentially as the number of modules increases. There are some reasons. First, that is because in the simulation, the local and global constrains are monitored by the simulator. And this operation consumes extra time which is not pertinent to the self-configuration process. Second, combination and separation process cost extra time because of the transitional grid. We are working on a modified locomotion manner where the transitional grid is unnecessary. Third, in a huge configuration space, it is possible to realize parallel self-configuration while multiple meta-modules entry into the configuration space at regular intervals and self-configure at the same time. Parallel process shortens the time consumption and is achievable because of the distributed property of self-configuration strategy where the involved modules of path searching, path planning and polymorphic locomotion are limited.

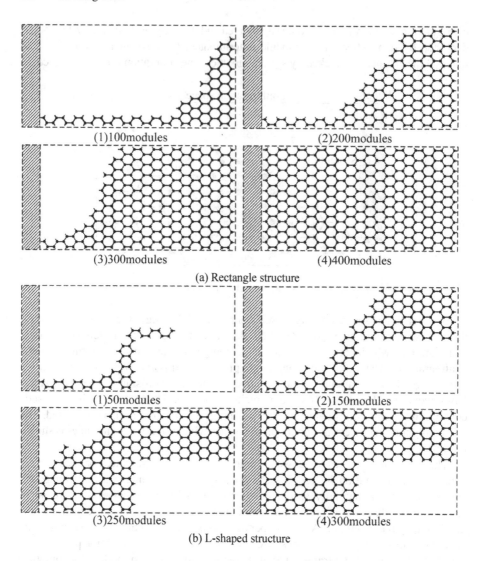

(1)100modules

(2)200modules

(3)300modules

(4)400modules

(a) Rectangle structure

(1)50modules

(2)150modules

(3)250modules

(4)300modules

(b) L-shaped structure

Fig. 9. Self-configuration processes of rectangle and L-shaped geometric structure

In all of the simulation rounds, the success rate of self-configuration always maintains 100 %. Furthermore, through analysing the message load, the behaviours of meta-modules in the self-configuration process can be approximately evaluated. And this can be considered as a characterization factor of system scalability. It can be found that the message load grows approximately linearly as the number of modules increases. The scalability of self-configuration is advisable. And the property of scalability needs further research.

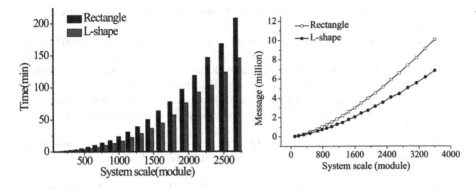

Fig. 10. Time consumption and message load of self-configuration process

5   Conclusions

This paper proposes a novel self-configuration strategy for the M-Lattice MRS. The validity and scalability of this strategy for M-Lattice MRS have been presented. The module design and topological features of the M-Lattice MRS have been described. The construction task under complex environment is converted to a self-configuration task accomplished by modular robots. A meta-module method is designed to realize the self-configuration process. Formed by four M-Lattice modules, each meta-module changes its position through connecting with different modules. While the connection status is changeable within the meta-module members, the connections within the system do not change during the self-configuration process, which reduces the global constrains successfully. Under our self-configuration strategy, single meta-module is able to finish path searching, path planning and polymorphic locomotion to form a specific target structure. Results from simulations have shown that the self-configuration strategy for the M-Lattice MRS has an advisable scalability for the regular target structures. However, it still needs further research about it. And it may be applied in the domain of scalable MRS.

Acknowledgments. This work is partially supported by the National Natural Science Foundation of China (Grant No. 61473192, U1401240, 61075086) and National Basic Research Program of China (2014CB046302).

References

1. Chirikjian, G.S.: Kinematics of a metamorphic robotic system. In: IEEE International Conference on Robotics and Automation, pp. 449–455 (1994)
2. Rus, D., Vona, M.: Crystalline robots: self-reconfiguration with compressible unit modules. Autonom. Robots **10**(1), 107–124 (2001)
3. Terada, Y., Murata, S.: Automatic modular assembly system and its distributed control. Int. J. Robot. Res. **27**(3), 445–462 (2008)

4. Chiang, C.J., Chirikjian, G.S.: Modular robot motion planning using similarity metrics. Autonom. Robots **10**(1), 91–106 (2001)
5. Golestan, K., Asadpour, M., Moradi, H.: A new graph signature calculation method based on power centrality for modular robots. In: Martinoli, A., Mondada, F., Correll, N., Mermoud, G., Egerstedt, M., Hsieh, M., Parker, L.E., Støy, K. (eds.) Distributed Autonomous Robotic Systems. STAR, vol. 83, pp. 505–516. Springer, Heidelberg (2013)
6. Zhu, Y., Bie, D., Iqbal, S., Wang, X., Gao, Y., Zhao, J.: A simplified approach to realize cellular automata for UBot modular self-reconfigurable robots. J. Intell. Robot. Syst. **79**(1), 1–18 (2014)
7. Park, H.W., Ramezani, A., Grizzle, J.W.: A finite-state machine for accommodating unexpected large ground-height variations in bipedal robot walking. IEEE Trans. Robot. **29**(2), 331–345 (2013)
8. Arney, D., Fischmeister, S., Lee, I., Takashima, Y., Yim, M.: Model-based programming of modular robots. In: IEEE International Symposium on Object/Component/Service-Oriented Real-Time Distributed Computing, pp. 66–74 (2010)
9. Aloupis, G., Collette, S., Damian, M., Demaine, E.D., El-Khechen, D., Flatland, R., Langerman, S., O'Rourke, J., Pinciu, V., Ramaswami, S., Sacristán, V., Wuhrer, S.: Realistic reconfiguration of crystalline (and telecube) robots. In: Chirikjian, Gregory S., Choset, H., Morales, M., Murphey, T. (eds.) Algorithmic Foundation of Robotics VIII. STAR, vol. 57, pp. 433–447. Springer, Heidelberg (2009)
10. Shokri, A., Masehian, E.: A meta-module approach for cluster flow locomotion of modular robots. In: Proceeding of the 2015 3rd RSI International Conference on Robotics and Mechatronics, pp. 425–431 (2015)
11. Dewey, D.J., Ashley-Rollman, M.P., Michael, D.R., et al.: Generalizing metamodules to simplify planning in modular robotic systems. In: IEEE/RSJ International Conference on Intelligent Robots and Systems, pp. 1338–1345 (2008)
12. Christensen, D.J., Østergaard, E.H., Lund, H.H.: Metamodule control for the atron self-reconfigurable robotic system. In: The 8th Conference on Intelligent Autonomous Systems, pp. 685–692 (2004)
13. Guan, EG., Yan, WX., Jiang, DS., Fu, Z., Zhao, YH.: A novel design for the self-reconfigurable robot module and connection mechanism. In: 3rd International Conference on Intelligent Robotics and Applications, pp. 400–408 (2010)

Quantized Consensus of Multi-agent Systems with Additive Noise

Jiayu Chen and Qiang Ling[✉]

Department of Automation,
University of Science and Technology of China, Hefei, China
jiayuc@mail.ustc.edu.cn, qling@ustc.edu.cn

Abstract. This paper investigates the average consensus problem of multiple discrete-time integrator agents under communication constraints and additive noise. In real applications, both quantization error and additive noise are often unavoidable and may terribly degrade the consensus performance. To handle quantization, we adopt a distributed dynamic encoding and decoding policy, under which the resolution of quantizers can change over time to tightly catch up the states and provide more accurate information. Moreover, bounded additive noise is considered. By generalizing the original noise-free protocol in [14], we propose a modified protocol with a new scaling function and prove that under our protocol, one can achieve approximate consensus even with 1 bit per channel use under the perturbation of additive noise. Furthermore we set up a quantitative relationship between the consensus performance, measured by the ultimate consensus error bound, and the number of available bits per channel use.

1 Introduction

In recent years more and more attention has been paid to the distributed coordination of networked multi-agent systems due to its wide application areas, such as coordination control of Unmanned Aerial Vehicle (UAV) [2], flocking [3], rendezvous [4] and formation control [5]. (More details can be found in [1]). When it comes to the consensus problem, each agent has to implement a distributed protocol based on the limited information about itself and its neighboring agents.

Traditional consensus problems have been well solved [6,7] when all agents can continuously exchange state information with infinite information precision. However, constraints on sensor cost, communication bandwidth, and energy budget dictate that information transmitted between agents has to be quantized, which yields unavoidable quantization error. For the single-agent systems, the bit rate conditions to stabilize systems are derived in [8–10]. Some results of the single-agent systems are adopted to multi-agent systems, [11–13] where the quantized variables are allowed to be any integer, i.e., their quantization ranges are the whole integer set and an infinite bandwidth is needed to transmit such quantized variables, which is difficult to implement in reality.

© Springer International Publishing Switzerland 2016
N. Kubota et al. (Eds.): ICIRA 2016, Part I, LNAI 9834, pp. 485–497, 2016.
DOI: 10.1007/978-3-319-43506-0_43

To overcome the above unrealistic infinite bandwidth requirement, the Minimum Data Rate(MDR) problem is investigated in the field of quantized consensus of multi-agent system. [14] proposes a distributed protocol based on dynamic encoding and decoding and prove that under its protocol, average consensus can be achieved for a connected network with merely one bit information exchange between adjacent agents. More general systems were considered in [15,16]. But all these results do not consider process noise. If we simply implement them, the desired consensus could be broken due to the additive process noise.

In this paper, we generalize the existing results to handle additive process noise. We first design a new scaling function, which takes the effects of the additive noise into account. Then we propose a modified consensus protocol based on the new scaling function and show that a uniform quantizer utilizing this protocol will never be saturated and the system will achieve approximate consensus with bounded consensus error. Furthermore, we provide a quantitative relationship between the number of available bits per channel use and the consensus performance, measured by the ultimate consensus error.

The rest of this paper is organized as follows. Section 2 presents the models of the communication network and agents, and the basic consensus protocol. In Sect. 3, we propose the modified consensus protocol and prove that approximate average consensus can be achieved under our protocol. Section 4 examines the achieved theoretical results through extensive simulations. Some concluding remarks and future research topics are presented in Sect. 5. To improve readability, technical proofs are placed in the Appendix.

Notations: I denotes an proper dimensional identity matrix and the subscript is omitted for short. $\mathbf{1} = [1, ..., 1]^T$ is a proper dimensional vector with all elements equal to 1 and $J_N = \frac{1}{N}\mathbf{11}^T$. $\|\cdot\|$ and $\|\cdot\|_\infty$ respectively represent the Euclidean and infinity norms on vectors or their induced norms on matrices.

2   Mathematical Models

2.1   Communication Graph

An undirected graph $\mathcal{G} = \{\mathcal{V}, \mathcal{E}, \mathcal{A}\}$ is used to represent the communication topology of N agents, where $\mathcal{V} = \{1, 2, ..., N\}$ is the index set of N agents with i representing the ith agent, $\mathcal{E} \subseteq \mathcal{V} \times \mathcal{V}$ is the edge set of paired agents and $\mathcal{A} = [a_{ij}] \in \mathbb{R}^{N \times N}$ with nonnegative elements $a_{ij} \in \{0, 1\}$ is the weighted adjacency matrix of $\mathcal{G}$. Note that $\mathcal{A}$ is a symmetric matrix. An edge $(j, i) \in \mathcal{E}$ if and only if $a_{ij} = 1$, which means that agent j can send information to agent i. The neighborhood of the ith agent is denoted by $N_i = \{j \in \mathcal{V} | (i, j) \in \mathcal{E}\}$. $D_i = |N_i|$ is called the degree of agent i and $D^\star = \max_i D_i$ is called the degree of $\mathcal{G}$. A sequence of edges $(i_1, i_2), (i_2, i_3), ..., (i_{k-1}, i_k)$ is called a path from agent i_1 to agent i_k. The graph $\mathcal{G}$ is called a connected graph if for any $i, j \in \mathcal{V}$, there exists at least one path from i to j. Denote $\mathcal{D} = diag(D_1, ..., D_N)$ and the Laplacian matrix of $\mathcal{G}$ by $\mathcal{L} = \mathcal{D} - \mathcal{A}$. The eigenvalues of $\mathcal{L}$ in an ascending order are denoted by $0 = \lambda_1 \leq \lambda_2 \leq ... \leq \lambda_N$, where λ_N is the spectral radius of $\mathcal{L}$ and λ_2 is called the algebraic connectivity of $\mathcal{G}$.

2.2 Dynamic Encoding and Decoding Algorithms

In a digital communication network, the jth agent possesses an encoder and D_j decoders, each of which serves a neighbor of agent j. The encoder quantizes the state of agent j and encodes it into a bit sequence. The decoders receives the encoded state information of D_j neighbors of agent j and decodes them to generate estimates of the states of neighboring agents.

Encoders implement a $(2L)-$level uniform quantizer $q(\cdot) : \mathbb{R} \to \Gamma$, which is a map from $\mathbb{R}$ to the set of quantized levels $\Gamma = \{\pm i, i = 1, 2, ..., L\}$ and mathematically expressed as

$$q(y) = \begin{cases} i - 1/2, \ i - 1 \leq y < i, \quad i = 1, ..., L \\ L - 1/2, \ y \geq L \\ -q(-y), \ y < 0 \end{cases} \tag{1}$$

As [14], we adopt a difference encoder with scaling. More specifically, the encoder Φ_i of the ith agent is composed of an estimator and a quantizer, which are given below.

Estimator:

$$\begin{cases} \xi_{ii}(0) = 0, \\ \xi_{ii}(k) = g(k - 1)m_i(k) + \xi_{ii}(k - 1), \quad k = 1, 2, ... \end{cases} \tag{2}$$

Quantizer:

$$m_i(k) = q\left[\frac{x_i(k) - \xi_{ii}(k - 1)}{g(k - 1)}\right], \quad k = 1, 2, ... \tag{3}$$

where k is the time step, x_i is the state of the ith agent, $\xi_{ii}(k)$ is the internal state of Φ_i, and $m_i(k)$ is the information sent to the neighbors of the ith agent. $g(k)$ is the scaling function to be designed. The decoder of the ith agent's neighbours can decode the information received from agent i and get an estimate of $x_i(k)$, $\xi_{ji}(k)$, according to

$$\begin{cases} \xi_{ji}(0) = 0, \\ \xi_{ji}(k) = g(k - 1)m_i(k) + \xi_{ji}(k - 1), \quad k = 1, 2, ... \end{cases} \tag{4}$$

2.3 System Model

In this paper, the agents are governed by the following integrator dynamics

$$x_i(k + 1) = x_i(k) + u_i(k) + \omega_i(k) \quad i = 1, ..., N \quad k = 0, 1, ... \tag{5}$$

where $x_i(k) \in \mathbb{R}$ represents the state of the ith agent, $u_i(k) \in \mathbb{R}$ is the input of the ith agent, $\omega_i(k) \in \mathbb{R}$ is the additive process noise. In this paper, we assume that the exact states of the neighbours are not available because of quantization. According to (2) and (4), we see that both the ith agent's encoder and its neighbours' decoders can get estimates of state x_i, and their estimates are exactly the same when there is no delay in the communication channel.

Therefore we denote such estimate of x_i as $\hat{x}_i (= \xi_{ii} = \xi_{ji})$. Thus we adopt the following quantized protocol,

$$u_i(k) = h \sum_{j \in N_i} a_{ij} \left[\hat{x}_j(k) - \hat{x}_i(k) \right], \quad i = 1, 2, ..., N \qquad (6)$$

Denote $X(k) = [x_1(k), ..., x_N(k)]^T$, $\hat{X}(k) = [\hat{x}_1(k), ..., \hat{x}_N(k)]^T$ and $\omega(k) = [\omega_1(k), ..., \omega_N(k)]^T$. Define the *quantization error* vector

$$\eta(k) = X(k) - \hat{X}(k) \qquad (7)$$

and the *deviation* vector

$$\delta(k) = X(k) - J_N X(k) \qquad (8)$$

By the above notation, we can combine the closed-loop system equation and the coding algorithm together into a matrix form,

$$\begin{cases} X(k+1) = (I - h\mathcal{L})X(k) + h\mathcal{L}\eta(k) + \omega(k) \\ \hat{X}(k+1) = g(k)q \left[\frac{X(k+1) - \hat{X}(k)}{g(k)} \right] + \hat{X}(k) \end{cases}, \qquad (9)$$

where $q(\cdot)$ is the component-wise quantizer, i.e., $q([y_1, ..., y_N]) = [q(y_1), ..., q(y_N)]$ and $\mathcal{L}$ is the Laplacian matrix of the network.

The following assumptions will be adopted in the subsequent analysis:

(A1) $||\delta(0)||_\infty \leq ||x(0)||_\infty \leq B_x$
(A2) $||\omega(k)||_\infty \leq M$, $k = 1, 2, ...$

where B_x and M are some nonnegative constants.

Due to the inclusion of additive process noise, we cannot guarantee the traditional average consensus, i.e., $\lim_{k \to \infty} \delta(k) = 0$. Instead, we pursue the following approximate consensus,

$$\lim_{K \to \infty} \sup_{k \geq K} ||\delta(k)|| < \infty. \qquad (10)$$

3   Quantized Consensus Protocol

3.1   Scaling Function Design

The design of the scaling function $g(k)$ is the key point of our quantized consensus protocol, which involves the topology of the communication network and the additive process noise. That design needs the following preliminary result.

Lemma 1. [14] *If the topology of network is connected and $h < 2/\lambda_N$, then $\rho_h < 1$, where*

$$\rho_h = \max_{2 \leq i \leq N} |1 - h\lambda_i|. \qquad (11)$$

Furthermore, if $h < 2/(\lambda_2 + \lambda_N)$, then $\rho_h = 1 - h\lambda_2$.

Remark 1. The maximum eigenvalue of $(I - h\mathcal{L})$ is 1 and the others are inside the unit circle. In the following we will show that the convergence rate of the discrete-time system mainly depends on ρ_h. To make sure that the quantizer will never saturate and always provide meaningful quantization result, it is necessary to require that the scaling function converge slower than ρ_h.

Here we propose a scaling function

$$\begin{cases} g(0) = g_0 \\ g(k+1) = \gamma g(k) + M \end{cases} \tag{12}$$

where the convergence factor $\gamma \in [\rho_h, 1)$ with $h \in (0, 2/\lambda_N)$. Note that M is the norm bound of the additive noise. On the one hand, $g(k)$ still decreases as [14], especially under large g_0. On the other hand, there exists an additive "overmeasure" each step so that the additive noise will not result into the saturation of the quantizers. To make sure the scaling function $g(k)$ is monotonically decreasing with respect to k, we choose $g_0 > \frac{M}{1-\gamma}$, which will simplify our analysis without loss of generality.

3.2 Quantized Consensus Protocol

Before presenting the quantized consensus protocol, we first analyze the dynamic models of the quantization error and the deviation error. Note that $\mathcal{L}J_N = J_N \mathcal{L} = 0$, therefore $h\mathcal{L}\delta(k) = h\mathcal{L}[X(k) - J_N X(k)] = h\mathcal{L}X(k)$.

By the system model (9), we define an *innovation* vector as

$$e(k) = X(k+1) - \widehat{X}(k) = (I + h\mathcal{L})\eta(k) - h\mathcal{L}\delta(k) + \omega(k). \tag{13}$$

We can derive

$$\begin{cases} \delta(k+1) = (I - h\mathcal{L})\delta(k) + h\mathcal{L}\eta(k) + (I - J_N)\omega(k) \\ \eta(k+1) = e(k) - g(k)q\left(\frac{1}{g(k)}e(k)\right) \end{cases} \tag{14}$$

Let $\acute{\delta}(k) = \frac{\delta(k)}{g(k)}$, $\acute{\eta}(k) = \frac{\eta(k)}{g(k)}$ and $\acute{\omega}(k) = \frac{\omega(k)}{g(k)}$. By (12), (14), we get

$$\begin{cases} \acute{\delta}(k+1) = G(k)(I - h\mathcal{L})\acute{\delta}(k) + G(k)h\mathcal{L}\acute{\eta}(k) + G(k)(I - J_N)\acute{\omega}(k) \\ \acute{\eta}(k+1) = G(k)\left[\acute{e}(k) - \mathcal{Q}(\acute{e}(k))\right] \end{cases} \tag{15}$$

where $G(k) = g(k)/g(k+1)$ and

$$\acute{e}(k) = \frac{e(k)}{g(k)} = (I + h\mathcal{L})\acute{\eta}(k) - h\mathcal{L}\acute{\delta}(k) + \acute{\omega}(k). \tag{16}$$

Note that $\acute{e}(k)$ is exactly the information to be quantized. Now we are ready to present our first consensus result in Theorem 1, whose proof is moved into the Appendix to improve readability.

Theorem 1. *Suppose Assumptions (A1) − (A2) hold. For any given $h \in (0, 2/\lambda_N)$ and $\gamma \in [\rho_h, 1)$, let*

$$L \geq L(h, \gamma) \tag{17}$$

$$L = h\lambda_N \left[\frac{\sqrt{N}(B_x + g_0)}{g_0} + \frac{\sqrt{N}h\lambda_N M}{2(1-\gamma)} + \frac{\sqrt{N}h\lambda_N[(1-\gamma)g_0 - M]}{2M\gamma e} \right]$$
$$+ \frac{1 + 2hD^\star + 2\gamma(1-\gamma)}{2\gamma} \tag{18}$$

The protocol given in (2)–(4) and (6) is implemented with 2L-level uniform quantizers (1), the scaling function (12) whose initial value satisfies

$$g_0 > \max \left\{ \frac{M}{1-\gamma}, B_x + M \right\} \tag{19}$$

Then the closed-loop system (9) achieves approximate consensus with the following bounded consensus error

$$\lim_{k \to \infty} ||\delta(k)|| \leq \frac{\sqrt{N}M[h\lambda_N + 2(1-\gamma)]}{2(1-\gamma)^2} \tag{20}$$

Remark 2. By (18), we see that the L to achieve the desired approximate consensus is determined by h and γ. When h is small enough and γ is close to 1, L can be as low as 1, i.e., 1 bit per channel use is enough to guarantee that the ultimate consensus error is bounded. Of course, $L = 1$ may yield very large consensus error bound in (20). The balance between the required L and the achievable consensus error is critical and will be discussed further in Sect. 3.3.

Remark 3. According to (20), the topology of the communication network, particularly λ_N, is an important factor of the consensus error. If we choose $h < 2/(\lambda_2 + \lambda_N)$ and $\gamma = \rho_h$, then $1 - \gamma = h\lambda_2$ and

$$\lim_{k \to \infty} ||\delta(k)|| \leq \frac{\sqrt{N}M}{2h\lambda_2} \left[\frac{\lambda_N}{\lambda_2} + 2 \right] \tag{21}$$

λ_2 is defined as the *algebraic connectivity of graph* in [6] and λ_2/λ_N is referred to as the *eigenratio* of an undirected graph [17], which has an upper bound

$$\frac{\lambda_2}{\lambda_N} \leq \frac{\min_i D_i}{\max_i D_i} \tag{22}$$

where $\min_i D_i$ and $\max_i D_i$ represent the minimum and maximum degrees of agents in the communication network, respectively.

When all degrees of agents in the network are relatively large and the difference between the maximum and minimum degrees is relatively small, the system can achieve relatively small consensus error, i.e., more balanced and connected networks yield more accurate consensus.

3.3 Consensus Protocol Control Gain Design

In this subsection, we propose a design method of the control gain when there exists limitations on the communication resource and the quantized consensus performance.

In Theorem 1 we present a quantized consensus protocol to assure the system to achieve approximate consensus utilizing finite bit communication. One can see from the result (18) that the larger the control gain h we choose, the more quantization levels the protocol need. However in practical scenario the communication resource is always limited, therefore the control gain has an upper bound. We show this result in the following lemma and for computation convenience, we choose the convergence factor of the scaling function $\gamma = \rho_h$.

Lemma 2. *Suppose Assumptions $(A1) - (A2)$ hold and set $\gamma = \rho_h$. For any given $L_1 \in \mathbb{N}^+$, under the quantized consensus protocol in Theorem 1 with the $2L_1-$level uniform quantizers, there always exist a control gain $h \in (0, h(L_1))$ that guarantees the closed-loop system (9) to achieve approximate consensus.*

Proof. The proof is omitted due to the space limitation.

From equality (21) we know that to achieve accurate approximate consensus, the control gain h should be sufficiently large. So when we set a consensus performance requirement δ_1 that $\lim_{k \to \infty} \|\delta(k)\| \leq \delta_1$, the choice of h has an lower bound $h(\delta_1)$. But when there exists some limitations on the use of communication resource, from the result in Lemma 2 we know that $h(\delta_1)$ should be smaller than $h(L_1)$. Therefore we have the following lemma.

Lemma 3. *Suppose Assumptions $(A1) - (A2)$ hold and set $\gamma = \rho_h$. For any given $\delta_1 \in [\underline{\delta}, \infty)$, under the quantized consensus protocol in Theorem 1 with $2L_1-$level uniform quantizers, there always exist a control gain $h \in \left(h(\delta_1), \bar{h}\right)$, where $\bar{h} = \min\left\{h(L_1), \frac{2}{\lambda_N}\right\}$ that guarantees the closed-loop system (9) to achieve approximate consensus with consensus error no more that δ_1, where $\underline{\delta}$ and $h(\delta_1)$ satisfy the following equations*
(a) When $h \in (0, \frac{2}{\lambda_2 + \lambda_N}), \gamma = \rho_h = 1 - h\lambda_2,$

$$h(\delta_1) = \frac{\sqrt{N}M(\lambda_N + 2\lambda_2)}{2\delta_1 \lambda_2^2}, \qquad \underline{\delta} = \frac{\sqrt{N}M(\lambda_N + 2\lambda_2)}{2\bar{h}\lambda_2^2}$$

(b) When $h \in (\frac{2}{\lambda_2 + \lambda_N}, \frac{2}{\lambda_N})$, $\gamma = \rho_h = h\lambda_N - 1$, $h(\delta_1)$ is the root of equation

$$2\delta_1 \lambda_N h^2 x + (\sqrt{N}M\lambda - 8\delta_1 \lambda_N)h + (8\delta_1 - 4\sqrt{N}M) = 0, \quad \underline{\delta} = \frac{\sqrt{N}M(4 - \bar{h}\lambda_N)}{2(2 - \bar{h}\lambda_N)^2}$$

Consider an complete network of eight agents, i.e. each agent has the access to all the other agents' information in the network, as an example. From Fig. 1 one can see that as the restricted quantization level L_1 increases from 1 to 50, the lower bound of the consensus performance $\underline{\delta}(L_1)$ decreases. The shaded area presents the feasible choice of control gain.

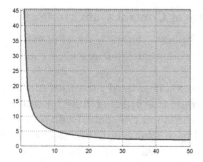

Fig. 1. Curves of the restricted quantization level L_1 and the lower bound $\underline{\delta}(L_1)$

4 Simulations

The validity of the quantized consensus protocol is demonstrated in this section. Consider a multi-agent system consisting of 8 agents with integrator dynamics (9). The interconnection topology is described by a regular undirected graph of degree 4 where the eigenvalues of the Laplacian matrix $\lambda_2 = 4, \lambda_N = 8$.

Set the initial states equal to $X(0) = [5, 10, 15, 20, 25, 30, 35, 40]^T$ and the noise bound $M = 0.5$. Choose the control gain $h = 0.0045$ and consequently according to (17)–(19) we have $\gamma = \rho_h = 0.982$, $g(0) = 40.5$, $L(h, \gamma) = 0.799$ and the bit rate $L = 1$. Figure 2 illustrates the trajectories of states, the deviation vector's Euclidean norm and the norm bound we estimate. One can see that all the agents converge to approximate consensus after several steps and the consensus error remain bounded thereafter which verifies our protocol. In this way we sacrifice the system performance to save the communication resource, which is very important in practical scenario.

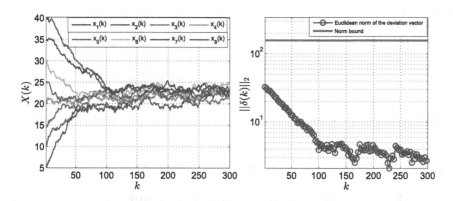

Fig. 2. Trajectories of states and Euclidean norm of deviation vector under Network topology one with one bit communication

5   Conclusion

In this paper we show that under the protocol we designed, a group of dynamic agents can achieve approximate consensus even when there exists additive noise. It is shown that the consensus error depends on the scaling function, the consensus gain and the communication topology. Furthermore we propose a way to design the consensus protocol when there exists constrictions on the communication resource and system performance. However achievement of accurate consensus is still a tough mission. Quantizers designed with various structures can only weaken the impact of the additive noise on current state but can not eliminate their cumulative effect in the future.

For future research, robustness with respect to packet-loss, link failures and time-delay may be considered and the results may be extended to high-ordered system.

6   Appendix: Proofs

6.1   Proof of Theorem 1

The proof is consist of two parts. In part 1, we assume that each communication channel between agents possesses adequate communication resources, and therefore we focus on the interruption that the quantization operation and additive noise bring into the system. In part 2, we work on how many data bit is enough on earth for the system to achieve approximate consensus when applied to the quantized consensus protocol.

Part 1. Since matrix $\mathcal{L}$ is symmetric, we can diagonalize it with the unitary matrix $T = [(1/\sqrt{N})\mathbf{1}, \phi_2, ..., \phi_N]$ defined by $\phi_i^T \mathcal{L} = \lambda_i \phi_i^T, i = 2, ..., N$. Therefore we have $I - h\mathcal{L} = \begin{bmatrix} \frac{1}{\sqrt{N}} & \phi \end{bmatrix} \begin{bmatrix} 1 & \mathbf{0} \\ \mathbf{0} & \Lambda \end{bmatrix} \begin{bmatrix} \frac{\mathbf{1}^T}{\sqrt{N}} \\ \phi^T \end{bmatrix} = \phi \Lambda \phi^T + J_N$, where $\phi = [\phi_2, ..., \phi_N]$, $\Lambda = diag(1 - h\lambda_2, ..., 1 - h\lambda_N)$ and $\mathbf{0}$ is the zero matrix of proper dimension. Then the recursion of the deviation vector (15) can be transformed into

$$\acute{\delta}(k+1) = G(k)(\phi \Lambda \phi^T + J_N)\acute{\delta}(k) + G(k)h\mathcal{L}\acute{\eta}(k) + G(n)(I - J_N)\acute{\omega}(k)$$
$$= G(k)\phi \Lambda \phi^T \acute{\delta}(k) + G(n)h\mathcal{L}\acute{\eta}(k) + G(k)(I - J_N)\acute{\omega}(k) \qquad (23)$$

where $J_N \acute{\delta}(k) = J_N(I - J_N)X(k)/g(k) = 0$. Let $G_0^k = \prod_{i=0}^{k} G(i) = \frac{g(0)}{g(k+1)}$ and it follows that

$$\|\acute{\delta}(k+1)\| \leq \|G_0^k \phi \Lambda^{k+1} \phi^T \acute{\delta}(0)\| + \|h \sum_{j=0}^{k} G_{k-j}^k (\phi \Lambda \phi^T)^j \mathcal{L}\acute{\eta}(k-j)\|$$

$$+ \|\sum_{j=0}^{k} G_{k-j}^k (\phi \Lambda \phi^T)^j (I - J_N)\acute{\omega}(k-j)\| \qquad (24)$$

where $(\phi \Lambda \phi^T)^0 = I$.

We estimate the three terms on the right hand of (24) separately.

Note that $\rho_h \leq \gamma$, $||\phi|| = ||\phi^T|| = 1$ and $||x||_\infty \leq ||x|| \leq \sqrt{N}||x||_\infty$ for any N dimensional vector x. For the first term,

$$\left\| G_0^k \phi \Lambda^{k+1} \phi^T \acute{\delta}(0) \right\|_2 \leq G_0^k (\rho_h)^{k+1} \frac{\sqrt{N}||\delta(0)||_\infty}{g(0)} \leq \frac{\sqrt{N}B_x}{g(k+1)} \gamma^{k+1} \qquad (25)$$

Since we assumed that each quantizer possesses adequate quantization levels, at each recursion step the quantizers will not be saturated, i.e. the quantization error are no more than $1/2$. Then together with (15) one can see that

$$||\acute{\eta}(k)||_\infty \leq \frac{1}{2}G(k-1), \quad k = 1, 2, \ldots \qquad (26)$$

Note that $||\mathcal{L}|| = \lambda_N$. For the second term, together with (26), we have

$$\left\| h\sum_{j=0}^{k} G_{k-j}^k (\phi\Lambda\phi^T)^j \mathcal{L}\acute{\eta}(k-j) \right\|_2 \leq \sqrt{N}h\lambda_N \left[\sum_{j=0}^{k} \frac{g(k-j)}{g(k+1)} (\rho_h)^j \frac{g(k-j-1)}{2g(k-j)} \right]$$

$$\leq \frac{\sqrt{N}h\lambda_N}{2g(k+1)} \left[(k+1)g_0\gamma^{k-1} + M\sum_{j=0}^{k}\sum_{i=0}^{k-j-2} \gamma^i\gamma^j \right]$$

$$= \frac{\sqrt{N}h\lambda_N}{2g(k+1)} \left(g_0 - \frac{M}{1-\gamma} \right)(k+1)\gamma^{k-1} + \frac{\sqrt{N}h\lambda_N M}{2g(k+1)} \frac{1-\gamma^{k+1}}{(1-\gamma)^2} \qquad (27)$$

Similarly for the last term, with the Assumption $(A2)$ and note that $||I - J_N|| = 1$, we can get

$$\left\| \sum_{j=0}^{k} G_{k-j}^k (\phi\Lambda\phi^T)^j (I - J_N)\acute{w}(k-j) \right\|_2 \leq \frac{\sqrt{N}M}{g(k+1)} \frac{1-\gamma^{k+1}}{1-\gamma}$$

As proved above we know that the scaled deviation vector $\acute{\delta}(k)$ can be bounded. Moreover, when step k tends to infinity, together with (25), (27) and (28) we have

$$\lim_{k\to\infty} ||\delta(k+1)|| = \lim_{k\to\infty} g(k+1)||\acute{\delta}(k+1)||$$

$$\leq \lim_{k\to\infty} \left[\sqrt{N}B_x\gamma^{k+1} + \frac{\sqrt{N}h\lambda_N}{2} \left(g_0 - \frac{M}{1-\gamma} \right)(k+1)\gamma^{k-1} \right.$$

$$\left. + \frac{\sqrt{N}h\lambda_N M}{2} \frac{1-\gamma^{k+1}}{(1-\gamma)^2} + \sqrt{N}M\frac{1-\gamma^{k+1}}{1-\gamma} \right]$$

$$\leq \frac{\sqrt{N}h\lambda_N M}{2(1-\gamma)^2} + \frac{\sqrt{N}M}{1-\gamma} \qquad (28)$$

Therefore the closed-loop system can achieve approximate consensus with bounded consensus error.

Part 2. Under a quantized consensus protocol, it is necessary and important that the quantizers will never saturate. Otherwise, what one agent broadcasts could be the saturated value of the state instead of its real value and that saturated information may mislead its neighbours. We prove here that the uniform quantizers (1) with $2L$ levels will never be saturated when $L \geq L(h, \gamma)$ with the help of mathematical induction.

Note that $\acute{e}(k)$, $k = 1, 2, \ldots$ is the information to be quantized. At the initial time, we know that $\widehat{X}(0) = 0$. According to Assumption (A1) and (19) we have

$$||\acute{e}(0)||_\infty = ||(I + h\mathcal{L})\acute{\eta}(0) - h\mathcal{L}\acute{\delta}(0) + \omega(0)||_\infty \leq \frac{B_x + M}{g_0} < L(h, \gamma) < L \quad (29)$$

Hence, when $k = 0$ the quantizers are unsaturated.

For any given nonnegative integer n, suppose that when $k = 0, 1, \ldots, n$ the quantizer are not saturated, i.e. the quantization error are less than $1/2$. Then at the $n + 1$ step,

$$||\acute{e}(n + 1)||_\infty \leq ||(I + h\mathcal{L})||_\infty ||\acute{\eta}(n + 1)||_\infty + h||\mathcal{L}||||\acute{\delta}(n + 1)|| + ||\acute{\omega}(n + 1)||_\infty$$

$$\leq \frac{1 + 2hD^\star}{2} G(n) + h\lambda_N ||\acute{\delta}(n + 1)|| + \frac{||\omega(n + 1)||_\infty}{g(n + 1)}$$

$$\leq \frac{g(n)(1 + 2hD^\star)}{2[g(n)\gamma + M]} + h\lambda_N ||\acute{\delta}(n + 1)|| + \frac{M(1 - \gamma)}{M}$$

$$\leq \frac{1 + 2hD^\star}{2\gamma} + h\lambda_N ||\acute{\delta}(n + 1)|| + (1 - \gamma) \quad (30)$$

The norm bound of $\acute{\delta}(n + 1)$ can be estimated similarly as in the proof of Theroem 1. From (25)–(28), we have

$$||\acute{\delta}(n + 1)|| \leq \frac{\sqrt{N} B_x \gamma^{n+1}}{g(n + 1)} + \frac{\sqrt{N} h\lambda_N M}{2g(n + 1)} \frac{1 - \gamma^{n+1}}{(1 - \gamma)^2} + \frac{\sqrt{N} M}{g(n + 1)} \frac{1 - \gamma^{n+1}}{1 - \gamma}$$

$$+ \frac{\sqrt{N} h\lambda_N}{2g(n + 1)\gamma^2} \left(g_0 - \frac{M}{1 - \gamma} \right) (n + 1)\gamma^{n+1} \quad (31)$$

Function $f(x) = x\gamma^x$, $x \in \mathbb{R}$, $\gamma \in (0, 1)$ has an upper bound $f(x) \leq \frac{1}{-e \log(\gamma)}$ when $x = \frac{1}{-\log(\gamma)}$, where e is the base of the natural logarithm. And since $\gamma \in (0, 1)$ we know that $\log(\gamma) \geq 1/\gamma$, then the term $(n + 1)\gamma^{n+1} \leq \frac{\gamma}{e}$.

Note that $g(k) > g(\infty) = \frac{M}{1-\gamma}$, we can get

$$||\acute{\delta}(n + 1)|| \leq \frac{\sqrt{N} B_x}{g_0} + \frac{\sqrt{N} h\lambda_N(1 - \gamma)}{2M\gamma e} g_0 - \frac{\sqrt{N} h\lambda_N}{2\gamma e} + \frac{\sqrt{N} h\lambda_N M}{2(1 - \gamma)} + \sqrt{N} \quad (32)$$

Therefore

$$\|\acute{e}(n+1)\|_\infty \leq h\lambda_N \left[\frac{\sqrt{N}(B_x + g_0)}{g_0} + \frac{\sqrt{N}h\lambda_N M}{2(1-\gamma)} + \frac{\sqrt{N}h\lambda_N[(1-\gamma)g_0 - M]}{2M\gamma e} \right]$$
$$+ \frac{1 + 2hD^\star}{2\gamma} + (1 - \gamma)$$
$$= L(h, \gamma) \leq L \tag{33}$$

So at time $k = n+1$ the quantizers are still unsaturated. Therefore, by induction, we conclude that if a set of $(2L)$-level uniform quantizers with $L \geq L(h, \gamma)$ are applied to the system, they will never be saturated.

According to the result in part 6.1, under the quantized consensus protocol the agent system can achieve approximate consensus with bounded consensus error.

Acknowledgement. This work was partially supported by the National Natural Science Foundation of China under Grant 61273112.

References

1. Cao, Y., Yu, W., Ren, W., et al.: An overview of recent progress in the study of distributed multi-agent coordination. IEEE Trans. Ind. Inform. **9**(1), 427–438 (2013)
2. Alarcon Herrera, J.L., Chen, X.: Consensus algorithms in a multi-agent framework to Solve PTZ camera reconfiguration in UAVs. In: Su, C.-Y., Rakheja, S., Liu, H. (eds.) ICIRA 2012, Part I. LNCS, vol. 7506, pp. 331–340. Springer, Heidelberg (2012)
3. Saber, R.O.: Flocking for multi-agent dynamic systems: algorithms and theory. IEEE Trans. Autom. Control **51**(3), 401–420 (2006)
4. Lin, J., Morse, A.S., Anderson, B.D.O.: The multi-agent rendezvous problem - the asynchronous case. SIAM J. Control Optim. **46**(6), 2120–2147 (2007)
5. Fax, J.A., Murray, R.M.: Information flow and cooperative control of vehicle formations. IEEE Trans. Autom. Control **49**(9), 1465–1476 (2004)
6. Saber, R.O., Murray, R.M.: Consensus problems in networks of agents with switching topology and time-delays. IEEE Trans. Autom. Control **49**(9), 1520–1533 (2004)
7. Saber, R.O., Fax, J.A., Murray, R.M.: Consensus and cooperation in networked multi-agent systems. Proc. IEEE **95**(1), 215–233 (2007)
8. Brockett, R.W., Liberzon, D.: Quantized feedback stabilization of linear systems. IEEE Trans. Autom. Control **4**(7), 1279–1289 (2000)
9. Elia, N., Mitter, S.K.: Stabilization of linear systems with limited information. IEEE Trans. Autom. Control **46**(9), 1384–1400 (2001)
10. Tatikonda, S., Mitter, S.: Control under communication constraints. IEEE Trans. Autom. Control **49**(7), 1056–1068 (2004)
11. Kashyap, A., Basar, T., Srikant, R.: Quantized consensus. Automatica **43**(7), 1192–1203 (2007)

12. Cai, K., Ishii, H.: Gossip consensus and averaging algorithms with quantizaiton. In: Proceedings of American Control Conference, pp. 6306–6311 (2010)
13. Lavaei, J., Murray, R.M.: Quantized consensus by means of gossip algorithm. IEEE Trans. Autom. Control **57**(1), 19–32 (2012)
14. Li, T., Fu, M., Xie, L., Zhang, J.: Distributed consensus with limited communication data rate. IEEE Trans. Autom. Control **56**(2), 279–292 (2011)
15. Li, T., Xie, L.: Distributed coordination of multi-agent systems with quantized-observer based encoding-decoding. IEEE Trans. Autom. Control **57**(12), 3023–3037 (2012)
16. Qiu, Z., Hong, Y., Xie, L.: Quantized leaderless and leader-following consensus of high-order multi-agent systems with limited data rate. In: IEEE Conference on Decision and Control, pp. 6759–6764 (2013)
17. You, K., Xie, L.: Network topology and communication data rate for consensusability of discrete-time multi-agent systems. IEEE Trans. Autom. Control **56**(10), 2262–2275 (2011)

Design and Development of a Multi-rotor Unmanned Aerial Vehicle System for Bridge Inspection

Jie Chen[1], Junjie Wu[1], Gang Chen[1], Wei Dong[1,2],
and Xinjun Sheng[1(✉)]

[1] State Key Laboratory of Mechanical System and Vibration,
School of Mechanical Engineering, Shanghai Jiaotong University,
Shanghai 200240, China
{cjl19921229,jesse0120,chg947089399,chengquess,
xjsheng}@sjtu.edu.cn
[2] State Key Laboratory of Fluid Power and Mechatronic Systems,
Zhejiang University, Hangzhou 310058, Zhejiang, China

Abstract. To prevent the occurrence of bridge structural failure, regular bridge inspections are required to find the defects of bridges. Traditional detection methods are mainly carried out manually, which are cumbersome and unsafe. The objective of this study is to develop an unmanned aerial vehicle (UAV) for bridge inspection in a fully autonomous manner. In view of the characteristics of the bridge environment and detection principle, a hexarotor frame with an upward camera gimbal is specially designed. Complete control system, sensor system, and image processing system are integrated into the system on the Robot Operation System (ROS). In addition, position estimation and obstacle detection with multi-sensor fusion technique are proposed for obstacle avoidance, human-friendly control flying under the bridge environment without Global Position System (GPS). Compared with traditional bridge inspection method, UAVs will not be limited to space, which will simplify the inspection process, improve the inspection efficiency, guarantee personnel safety, and reduce the incidence of high-risk job accidents.

Keywords: Unmanned Aerial Vehicle · Industrial inspection · Bridge inspection · Obstacles avoidance · Multi-sensor fusion

1 Introduction

In recent years, with the development of materials, micro-electromechanical systems (MEMS), micro inertial measurement unit (MIMU) and flight control technology, micro UAVs are growing rapidly and drawing more and more attentions. Particularly, multi-copter stands out because of its compact size, simple mechanical structure, good agility, and capability of vertical landing and hovering [1]. In addition to aerial photography, multi-rotor UAVs have many potential applications in industry, agriculture, public security, scientific research and even military [2–4]. Especially in industrial field, compared to the traditional stationary industrial robots, micro UAVs, as one kind of new

© Springer International Publishing Switzerland 2016
N. Kubota et al. (Eds.): ICIRA 2016, Part I, LNAI 9834, pp. 498–510, 2016.
DOI: 10.1007/978-3-319-43506-0_44

mobile robots, can work in three-dimensional space and have a wider range of applications [5].

Bridge inspection is one such scenario that could well demonstrate the superiorities of the multicopters. In order to prevent the occurrence of bridge structural failures, regular bridge inspections are required to find the defects of the bridges, especially in China, bridge-building continue to develop during the last decades. And current detection methods are mainly performed manually, which are cumbersome and unsafe. New bridge detection methods are urgently demanded.

Motivated by the requirements for autonomous inspection, some research institution also tried to use UAVs to do bridge inspections. Central Laboratory for Roads and Bridges in Paris tried to use a micro helicopter for bridge inspection [6], a vision system was installed for local position estimation and a front camera utilized for image capturing. The process of inspection was done manually. Such system was not so human-friendly for non-professional operators to control a helicopter without security protection. Some companies like AIBOTIX from Germany also offer solutions for bridge, wind turbine, and power line inspections with UAV [7]. One example from AIBOTIX is a hexarotor equipped with a protected frame and an upward camera, operated manually with the help of live video. However, without sensors to detect the environment, such system cannot be close enough to the bridge to take photos and also need to be operated manually under the bridge without GPS. Thus localization is a great challenge for autonomous flight under the bridge, and some related work in other fields can also provide reference for bridge inspection like the application for culverts inspection by UAV system from MIT [8], which used a Light Detection and Ranging sensor (LIDAR) for localization without GPS. Another research from the University of Pennsylvania can also inspire us [9–11], they fused multi-sensors including IMU, GPS, laser scanner, and vision system to estimate the position of the drone for indoor and outdoor autonomous flight.

To draw a conclusion from the current research and application of bridge inspection with the UAV, some challenges still need to be overcome. First is to find a precise localization method in bridge environment for position control of the UAV. Second is the integration of a high-performance autonomous flight system with automatic obstacle-avoid algorithm especially for bridge inspection. Last but not least is to make the flight controlling system more human-friendly.

In light of this, a hexarotor UAV system is designed and developed in this study, which is mainly used for bridge inspection. According to the characteristics of bridge environment and detection principle, a hexarotor frame with an upward camera gimbal is specially designed. Control system, sensor system, and image system are integrated into the system on the ROS. Moreover, position estimation and obstacle detection by fusion of multi-sensors is done for obstacle avoidance, human-friendly control flying under the bridge environment without GPS. Test results are shown in the result section and limitation are discussed in the conclusion section.

2 Design and Integration of the System

2.1 Modeling

Mathematic model and dynamic model of the hexarotor UAV are studied in this part.

Mathematic Model. For the inertial coordinate, a north-east-down (NED) orthogonal coordinate system is established by the right hand rule. For the body fixed coordinate system, the head direction is the chosen as X axis, the right as Y axis, and down as Z axis.

The conversion between the inertial coordinate system and the body fixed frame is described via Euler-Angles. The position of the drone in the inertial frame and the attitude vector is defined as

$$r = \begin{bmatrix} x \\ y \\ z \end{bmatrix}, \quad q = \begin{bmatrix} \varphi \\ \theta \\ \psi \end{bmatrix}, \tag{1}$$

where r is the position, q is the attitude, φ is the rotation angle around X axis anti-clockwise, θ around Y axis, and ψ around Z axis.

With the attitude angle defined, the rotation matrix to transform a vector from the inertial frame to the body fixed frame is

$$R = \begin{bmatrix} cos\psi cos\theta & cos\theta sin\varphi & -sin\theta \\ sin\varphi sin\theta cos\psi - cos\varphi sin\psi & sin\varphi sin\theta sin\psi + cos\varphi cos\psi & sin\psi cos\theta \\ cos\varphi sin\theta cos\psi + sin\varphi sin\psi & cos\varphi sin\theta sin\psi - sin\varphi cos\psi & cos\theta cos\varphi \end{bmatrix} \tag{2}$$

Dynamic Model. Referring to the modeling of quadrotor in [12], define the input control variable as

$$\begin{cases} U_1 = F_1 + F_2 + F_3 + F_4 + F_5 + F_6 \\ U_2 = \dfrac{\sqrt{3}}{2}(F_2 + F_3)L - \dfrac{\sqrt{3}}{2}(F_5 + F_6)L \\ U_3 = \left(F_1 + \dfrac{1}{2}F_2 + \dfrac{1}{2}F_6\right)L - \left(F_4 + \dfrac{1}{2}F_3 + \dfrac{1}{2}F_5\right)L \\ U_4 = M_1 + M_3 + M_5 - M_2 - M_4 - M_6 \end{cases} \tag{3}$$

where F_1, F_2, F_3, F_4, F_5 and F_6 are the thrusts, and M_1, M_2, M_3, M_4, M_5 and M_6 are the momentums, and L is the diagonal wheelbase of the frame.

Then the rigid body dynamics can be derived as

$$\begin{cases} m\ddot{r} = R \begin{bmatrix} 0 \\ 0 \\ U1 \end{bmatrix} - \begin{bmatrix} 0 \\ 0 \\ mg \end{bmatrix} - \dot{q} \times m\dot{r} \\ \\ I\ddot{q} = \begin{bmatrix} U3 \\ U2 \\ U4 \end{bmatrix} - \dot{q} \times I\dot{r} \end{cases} \tag{4}$$

where m is the mass, I is the inertia, g is the gravity constant, R is the rotation matrix, q and r are defined as the attitude and the position to the initial coordinate system.

2.2 Structure Design and Analysis

The frame consists of several parts, including the body, arms, landing gear, protection parts, upward camera gimbal, and laser scanner gimbal. Based on the mathematic and dynamic model of hexarotor aerial vehicle, a framework is designed as shown in Fig. 1.

Compared with quadrotor, hexarotor has two more rotors to provide a redundant power for the vehicle, which can keep the vehicle flying safely even when one of the rotor is broken down. The diagonal wheelbase of the framework is designed to 950 mm which is capable for 18 inch's propeller. Such size can guarantee a stable flight performance with or without wind. Considering the size and portability, the framework is designed to be foldable to make it convenient to be carried and transported.

The materials in consideration for the design include aluminum alloy, and carbon fiber reinforced polymer (CFRP) material. The carbon fiber tube and plate frame is used because it is strong and light and can be easily manufactured.

The upward two-axis camera gimbal is designed to reduce the vibration of the body and stabilize the camera to capture a clear image. The camera gimbal is mounted on the top of the vehicle in order to get the images of both bottom and side of the bridge.

The laser scanner gimbal is designed to keep the attitude of the laser scanner to get point cloud data of the environment in a same level. What is more, it can also rotate the laser scanner to get three dimensional point cloud data.

(a)   (b)

(c)   (d)

Fig. 1. The structure of designed hexarotor framework. (a) Overview of the framework. (b) Folding design of the structure. (c) Upward gimbal for camera. (d) Gimbal for laser scanner

2.3   System Design

The system consists mainly four parts: mechanical system, control system, sensor system, and image system as shown in Fig. 2.

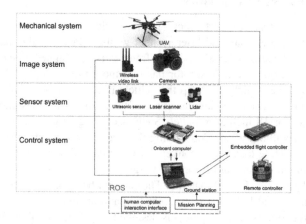

Fig. 2. Composition of the system

Control System. The control system is the core of the whole system. It plays a role in processing data of the sensors, mission planning, and flight control. It includes four parts: an embedded flight controller, an onboard computer, a ground station, and a remote controller.

The embedded flight controller is used for flight control. We choose an open source autopilot named Pixhawk to take up this position. The accelerometer, gyroscope and magnetometer are integrated in Pixhawk to estimate the attitude of the vehicle, while the barometer and GPS are for position estimation. It ensures a basic flight function of the vehicle.

The onboard computer is mainly for processing data and sending command to the flight controller. We use the Raspberry Pi 2 whose computing performance can meet the requirement. An Ubuntu server system is installed, with the Robot Operating System (ROS) running on it.

The ground station performs as a human computer interaction interface as shown in Fig. 3. Status information of the vehicle can be acquired from it such as attitude, altitude, and velocity and so on. Images from the camera can also be seen from the ground station to monitor the defects of the bridge in real time. Moreover, mission planning can also be done on it to make the vehicle fly autonomously.

Sensor System. Besides of the onboard sensors including IMU, barometer and GPS for flight control, some additional sensors are also integrated into the system for environment detection, obstacle avoidance and position estimation in no GPS signal environment. Additional sensors include a laser scanner, ultrasonic sensors, a single-point LIDAR.

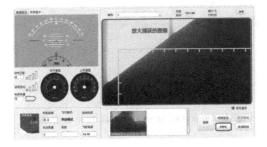

Fig. 3. Ground station for bridge inspection

The laser scanner can detect the environments of the bridge, we can use the scan data to realize obstacle avoidance, localization and mapping. UTM-30LX from Hokuyo is chosen, which has 30 m and 270° scanning range and can be used in outdoor environments.

Ultrasonic sensors are also used to detect the obstacles where the laser scanner cannot reach in order to guarantee the safety of the vehicle.

An upward mounted single-point LIDAR is used to detect the distance between the drone and the bottom of the bridge. With this data, we can avoid the drone crashing upward to the bridge. The data can also be fused to estimate the altitude of the vehicle.

Image System. The image system includes a camera and a wireless video link. The image system is for online monitoring and offline image processing.

For online monitoring, the wireless video link can transmit the compressed images of the camera to the ground station in real time so that the operator can determine whether there is a defect in time. There is an infrared LED on the drone to control the camera remotely. The uncompressed pictures can be saved on camera for offline image processing including image mosaic and defects recognition.

3 Fusion of Multi-sensors for Bridge Environment

Fusion of multi-sensors will be introduced in this section to estimate the status of the vehicle. With the status information, obstacle avoidance algorithm and human-friendly flying control algorithm are designed especially for bridge environment.

3.1 Position Estimation

The accelerometer, barometer and LIDAR are fused to estimate the altitude of the vehicle. The accelerometer can provide high dynamic acceleration data in short term, however its error will accumulate over time, while barometer and LIDAR will not. However, the barometer is very sensitive to the pressure change especially under the bridge where the wind is strong. The LIDAR can provide precise distance data, however, it will be disturbed by the step change of the surface. So we use complementary filter to estimate the altitude and vertical velocity of the vehicle as shown in Fig. 4.

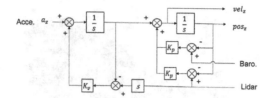

Fig. 4. Fusion of sensors for altitude estimation

The accelerometer, laser scanner and GPS are fused to estimate the horizontal position of the vehicle. When the GPS signal is good, the GPS and accelerometer is enough to estimate the position. When under the bridge where GPS is poor, laser scanner and accelerometer will be fused to estimate the position as shown in Fig. 5.

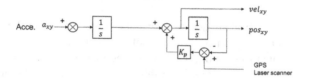

Fig. 5. Fusion of sensors for position estimation

As for laser scanner, an incremental laser scan matcher [13] can be used for indoor environment to estimate position, however this algorithm fails to work under the bridge, because few features can be detected except for the bridge piers. So here we make use of such special feature to find one or two of the piers, and make an average of the point cloud to represent the center of the pier, and then calculate the relative position of the vehicle to the pier, as shown in Fig. 6. And then we use this data to fuse with the accelerometer.

One thing need to pay attention to is the coordinate transform. The inertial coordinate is a NED system, meanwhile the position estimated by the laser scanner is under the bridge coordinate where X axis is the connection direction of the two pier, so when the system is initialized, head the drone at the direction perpendicular to the connection direction of the two piers, so that the rotation matrix can be determined between the bridge coordinate and the inertial coordinate be measured.

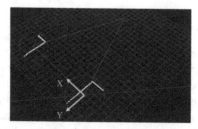

Fig. 6. Schematic for position estimation of the laser scanner under the bridge

3.2   Obstacle Avoidance

Obstacle avoidance algorithm will ensure the flight safety under the bridge. The laser scanner and the ultrasonic sensor can detect the obstacles in the environment. Once obstacle is detected, an artificial potential force field will be generated around the obstacle to push away the drone as shown in Fig. 7.

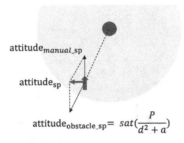

Fig. 7. Schematic of obstacle avoidance

Set a safe distance d_{safe}, when the distance of the obstacle d is less than d_safe, it generates an artificial potential field

$$F = \frac{P}{d^2 + \alpha} \tag{5}$$

where d is the distance of the obstacle, P is coefficient corresponding to the attitude adjustment, α is a constant.

In order to avoid a violent change of attitude, make a saturation to the attitude set point produced by the obstacle, and add this value to the original attitude set point by vector sum

$$\overline{Att}_{.sp} = \overline{Att}_{.manual} + \overline{sat}\left(\frac{P}{d^2 + \alpha}\right) \tag{6}$$

In the vertical direction, safe distance is d_{h_safe}, and the distance of the bridge is d_h, when $d_h \leq d_{h\_safe}$, disable the manual input of upward motion, and set the vertical position setpoint as

$$z_{sp} = z - (d_h - d_{h\_safe}) \tag{7}$$

where z_{sp} the setpoint of the altitude, z is the current altitude.

3.3   Human-Friendly Control

In order to make the it easy to operate, some human-friendly control algorithms are designed to the bridge inspection.

For bridge inspection, the piers are required to be inspected. In order to get good picture of the piers, the drone need to keep a constant distance with the pier to take photos for image mosaic and defects recognition. So we make use of the characteristic of the pier to control the drone with the laser scanner.

As shown in Fig. 8, The laser scanner gets the point cloud of the environment, set a point cloud filter range so that we can get the point cloud of the pier, then calculate the center of the pier, so we can get a relative position between the drone and the pier.

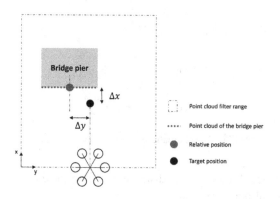

Fig. 8. Flying of constant distance to the pier

The position control system includes a position controller, a velocity controller, an attitude controller and an attitude rate controller as shown in Fig. 9. Such control model is a most common method to control the position of a drone [13].

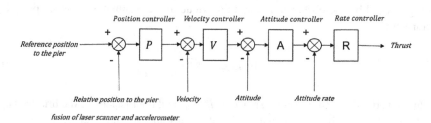

Fig. 9. Control diagram

To keep the vehicle flying with a constant distance to the pier, the input of the position controller is the reference position to the pier, and the relative position is the feedback, which is fused of relative position calculated by laser scanner and the integral of the acceleration from Sect. 3.1.

4 Result

The hexarotor aircraft is manufactured and assembled as shown in Fig. 10. Then we carried out some tests on the hexarotor aircraft, including weight measurement, flight endurance testing, and obstacle avoidance testing, to verify the performance of the vehicle. We also tested the system at the Nanpu Bridge and Yangpu Bridge in Shanghai to verify the capability of the inspection system.

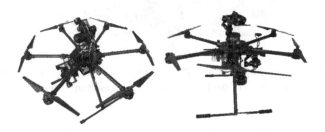

Fig. 10. Prototype of the hexarotor

4.1 Flight Performance

The weight of the developed hexarotor frame is about 4.7 kg. And its size is 940 mm. With a 10000 mAh battery, it can fly about 25 min without payload. With sensors, gimbals and the camera integrated to the system, its weight is 7.4 kg, which can fly about 14 min full load with 10000 mAh battery. And with 16000 mAh battery, its flight time is 20 min with 8.1 kg take-off weight.

The width of the bridge is about 36 m, and the length between two bridge pier is about 50 m, the average height of the pier is about 40 m. To guarantee the safety, we suppose the velocity of the drone is 3 m/s and it stops about 2 s to take a photo every 3 m. To finish inspection task of such region, it will take about 12.7 min. So the flight time of our system is enough to complete the inspection of the region enclosed by four piers. However, if another region need to inspected, we will replace with another battery.

4.2 Image Processing

We take some photos from Yangpu Bridge by our UAV system, and make some simple process on these photos.

The result of image mosaic of the bridge pier and bottom using the SURF feature matching algorithm are shown in Fig. 11.

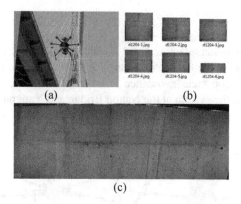

Fig. 11. Flight test and image mosaic of the bridge. (a) The drone is conducting bridge inspection. (b) Several pictures of the bridge pier. (c) The result of image mosaic.

4.3   Comparison

As shown in Table 1, compared with some similar products, most of the multicopter like DJI S1000 + are only for aerial photograph, so they cannot get the upward view of the bridge bottom and there is no additional sensor for no-GPS flight. Some products like Aibotics X6 are especially for industrial inspection, but without sensors to detect the environment, such system cannot be close enough to the bridge to take photos and also need to be operated manually under the bridge without GPS. So our developed UAV will be more capable to do bridge inspection work.

Table 1. Comparison with other products

Products	Developed UAV	DJI S1000+	Aibotics X6
Structure	Hexacopter	Octocopter	Hexacopter
Size(mm)	940	1045	1050
Mounted camera	Upwardview Forward view	Downward view Forward view	On-Top-Camera Downward -Camera
Take-off weight	6kg-12kg	6kg-11kg	4.6-6.6kg
Flight time	14min(10000mAh&7.4kg); 20min(16000mAh&8.1kg)	15min(15000mAh & 9.5Kg)	30min(10000mAh)
Additional sensors	Laser scanner, ultrasonic sensor, and Lidar for environment detection	none	Ultrasonic sensor

5 Conclusion

In this paper, a hexarotor UAV system is designed and developed for bridge inspection. With fusion of multi-sensors, the drone can detect the environment and estimate current status to guarantee the safety under the bridge. Human-friendly control algorithm makes it easier for operators to carry out the inspection process. Compared with the traditional bridge inspection method, such UAV bridge inspection system can partly replace human work, which will simplify the inspection process, improve the inspection efficiency, and guarantee personnel safety.

However, position estimation using laser scanner can only work in the specific bridge environment, it is still a problem without GPS when the environment is much more complex. To fly in a more complex environment, other localization method need to be integrate to the system to get more precise position to control the flight of the drone. Only with precise and reliable position estimation and control, further work including path planning and fully autonomous flight can be carried forward.

Acknowledgments. This work was funded by the special development fund of Shanghai Zhangjiang Hi-Tech Industrial Development Zone (No.201411-PD-JQ-B108-009) and Open Foundation of the State Key Laboratory of Fluid Power Transmission and Control (No. GZKF-201510).

References

1. Kumar, V., Michael, N.: Opportunities and challenges with autonomous micro aerial vehicles. Int. J. Robot. Res. **31**(11), 1279–1291 (2012)
2. Williamson, W.R., Abdel-Hafez, M.F., Rhee, I., et al.: An instrumentation system applied to formation flight. IEEE Trans. Control Syst. Technol. **15**(1), 75–85 (2007)
3. Baker, R.E.: Combining micro technologies and unmanned systems to support public safety and homeland security. J. Civ. Eng. Archit. **6**(10), 1399 (2012)
4. Ambrosia, V., et al.: Unmanned airborne systems supporting disaster observations: near-real-time data needs. Int. Soc. Photogram. Remote Sens. **144**, 1–4 (2011)
5. Mahony, R., Kumar, V., Corke, P.: Multirotor aerial vehicles: modeling, estimation, and control of quadrotor. IEEE Robot. Autom. Mag. **19**(3), 20–32 (2012)
6. Metni, N., Hamel, T.: A UAV for bridge inspection: visual servoing control law with orientation limits. Autom. Constr. **17**(1), 3–10 (2007)
7. AIBOTIX. Discover the possibilities of uav drones. http://www.aibotix.com/applications.html
8. Serrano, N.E.: Autonomous quadrotor unmanned aerial vehicle for culvert inspection. Diss. Massachusetts Institute of Technology (2011)
9. Shen, S., et al.: Multi-sensor fusion for robust autonomous flight in indoor and outdoor environments with a rotorcraft MAV. In: 2014 IEEE International Conference on Robotics and Automation (ICRA), IEEE (2014)
10. Shen, S., Michael, N., Kumar, V.: Autonomous multi-floor indoor navigation with a computationally constrained MAV. In: 2011 IEEE international conference on Robotics and automation (ICRA), IEEE (2011)

11. Shen, S., Michael, N., Kumar, V.: Autonomous indoor 3D exploration with a micro-aerial vehicle. In: 2012 IEEE International Conference on Robotics and Automation (ICRA), IEEE (2012)
12. Dong, W., et al.: Modeling and control of a quadrotor UAV with aerodynamic concepts. World Acad. Sci. Eng. Technol. **7**, 377–382 (2013)
13. Censi, A.: An ICP variant using a point-to-line metric. In: 2008 IEEE International Conference on Robotics and Automation, ICRA 2008, IEEE (2008)

Cognitive Robotics

Evaluating Trust in Multi-Agents System Through Temporal Difference Leaning

Rishwaraj Gengarajoo[1], Ponnambalam S.G.[1(✉)], and Chu Kiong Loo[2]

[1] Advanced Engineering Platform, School of Engineering, Monash University Malaysia,
Jalan Lagoon Selatan, 47500 Bandar Sunway, Selangor, Malaysia
sgponnambalam@monash.edu
[2] Faculty of Computer Science and Information Technology, University of Malaya,
50603 Kuala Lumpur, Malaysia

Abstract. In any mission that requires cooperation and teamwork of multiple agents, it is vital that each agent is able to trust one another to accomplish the mission successfully. This work looks into incorporating the concept of trust in a multi agent environment, allowing agents to compute trust they have on each other. The proposed trust evaluation model called TD Trust model is developed by adapting temporal difference (TD) learning algorithm into its evaluation framework. The proposed trust model evaluates the trust of an agent based on experience gained from interaction among agents. The proposed model is then tested using simulation experiments and its performance is compared against the Secure Trust model, which is a comprehensive model reported in literature.

Keywords: Reinforcement learning · Trust estimation · Multi-agent

1 Introduction

Multi-agents systems (MAS) is widely acknowledged as an effective framework for a broad range of application such as robotics, decision support systems, data mining and computing [1–6]. In any MAS applications, it is critical that the agents are able to cooperate effectively with each other in order to complete the given tasks successfully [7, 8]. As a part of the collaboration effort, the agents might have to share information and resources among themselves in order to optimize their cooperation efforts [9, 10].

Agents in MAS are often required to work in a highly uncertain and dynamic environment, especially when the agents are working in real world scenarios, such as factory robots, search and rescue robots and so on. Under such working conditions, misinterpretation of the environment and sharing of false data is highly likely albeit unwillingly by the agents since their sensor readings are highly probable to noise and uncertainty. Under such circumstances, it is important to have a measure of trust in agents that are sharing sensor readings or any other data among the agents in the system.

This work looks into integrating the concept of trust in MAS as a means to improve cooperation among agents by allowing each agent to evaluate the trustworthiness of other agents in the system. Some of the approaches studied to incorporate trust into a system are direct trust evaluation model, indirect/reputation-based trust evaluation

N. Kubota et al. (Eds.): ICIRA 2016, Part I, LNAI 9834, pp. 513–524, 2016.
DOI: 10.1007/978-3-319-43506-0_45

models, socio-cognitive trust evaluation models, organizational trust evaluation model, conceptual model and models based on various information sources [1, 11].

This study focuses on the development of a trust evaluation model using reinforcement learning as a means of direct trust evaluation. The developed model is capable of objectively evaluating trust in agents in MAS, and at the same time exhibits a more subjective approach in its trust evaluation. The contributions of this paper are:

1. Development of a trust evaluation model using reinforcement learning.
2. Simulation experiments to analyze the performance of the developed model.
3. Highlights the potential application of the trust model in real world MAS.

1.1 Related Work

The literature review highlight several approaches employed to model trust in MAS. Two of the commonly used approaches in trust evaluation are individual learning and social learning [12–14]. Individual learning is where an agent evaluates the trustworthiness of other agents through direct interaction while social learning is when an agent uses any information available from its social network to estimate trust value. One of the most frequently used method for trust estimation in both individual learning and social learning is mathematical modeling.

Mohammed et al. [9] applied a hyperbox accuracy formulation in bucket brigade algorithm in their trust evaluation system. Basheer et al. [7] proposed a confidence model which incorporates certainty and evidence as an evaluation agent. Another method for modeling a trust estimation model is using a probabilistic evaluation incorporating certainty and evidence based modeling [15]. One of such method is investigated by Wang and Singh [16], where they studied and developed an algorithm that could mathematically formulate the bijection between trust and evidence in trust modeling. An improved model is later presented using additional parameters: certainty and referral from third party trust estimation [17].

Reputation based trust is another branch of social learning where the trustworthiness of an agent is evaluated from the trust and reputation information obtained through interactions of other agents with the target agent. In this branch of study, the research integrates interaction trust, role-based trust, witness reputation and certified reputation to empirically evaluate trust [14]. Mantel and Clark [18] developed a reputation management system consisting of a distributed trust approximation model, where the agents are capable of evaluating the trustworthiness of other agents via a trust estimation model. Jøsang and Haller [19] employed a multinomial probability distribution function to evaluate trust in a reputation system.

The literature study shows that the idea of experience based learning is not thoroughly explored in trust modeling. In this paper, a trust estimation framework incorporating reinforcement learning is presented where trust is empirically evaluated through experience obtained via direct interaction among agents.

2 Problem Statement

To understand the study presented, a simple ball collecting example can be used. Consider an environment with two agents, A_1 and A_2, set to randomly explore with the objectives of acquiring different colored balls scattered in the environment. The balls shall be referred to as the goals in future sections. Each agent is expected to collect a specific colored ball: A_1 is tasked to collect red ball and A_2 is tasked to collect blue ball. The agents can independently explore the environment to collect their balls but if they encounter a ball that belong to another agent i.e. A_1 encounters a blue ball, that agent shares the location information with the other agent i.e. A_1 shares the information with agent A_2.

However, the information shared by the agents may be compromised or false due to several reasons:

1. Faulty or damaged sensors resulting in false readings.
2. High degree of uncertainty in the environment creating high level of noise.
3. A highly dynamic environment that changed from initial observation.

When agent A_1 shares this information with A_2, agent A_2 would use its time and resources to arrive at target location. If the information shared is false, no ball is observed. This would be a tremendous waste of precious time and resources which may lead delay in completing the mission.

This is the running example used in this paper. The proposed algorithm should be able to:

1. Empirically evaluate the trust value of an agent.
2. Iteratively update the trust value through experience obtained from interaction.
3. Determine non-value added robots to the system i.e. robots with high probability of sharing false information.

3 Proposed Framework

The proposed model is developed using temporal difference (TD) learning method, a branch of reinforcement learning technique. Temporal difference learning has the combining characteristics of dynamic programming and Monte Carlo method [20]. Using this technique, the previous trust value is iteratively bootstrapped with the experience gained via observing outcomes from an interaction with a target agent. The underlying trust evaluation is calculated by means of Beta Reputation System (BRS) [21]:

$$T_{i,j} = \left(\frac{r+1}{r+s+2} \right) \tag{1}$$

In Eq. (1), $T_{i,j}$ denotes the trust value agent i has on agent j. During agent i's interaction with agent j, one of two possible outcomes can be observed: 'r' or 's'. 'r' represents positive observation (the goal present) while 's' is represents negative observation. Through the interactions, the resulting observations serve as sample experience or '*f(exp)*'. The sample experience is a function of reinforcement reward and observational trust.

With each interaction, a successful observation results in positive experience gain and vice versa. In the proposed sample experience evaluation model, the BRS is used to compute the trust value related directly to the observed interaction. This value is then added to the internal reward received by the agent based on the observed outcome. The sample experience, '*f(exp)*' model as:

$$f(exp) = r + \gamma \left(\frac{r+1}{r+s+2} \right) \qquad (2)$$

The gamma in Eq. (2) is the discount factor which signifies the immediate importance of the experience gained which in this case is '1'. The values of 'r' and 's' are established after receiving observations from interactions. The observations can be classified as positive or negative [17]:

- Positive observation, 'r': the evidence obtained or observation after taking an action supports the data or information received from robots. The value of 'r' during positive outcome is '1' while the value of 's' is '0'.
- Negative observation, 's': the evidence obtained or the observation after taking an action does not support the data or information received from other robots. The value of 's' during negative outcome is '1' while the value of 'r' is '0'.

Based on the observations, the internal reward mechanism either rewards the agent for trusting a trustworthy agent or punishes the agent for trusting a lying agent. The complete '*f(exp)*' is given as:

$$f(exp) = \begin{cases} (0.333) + 1\left(\dfrac{r+1}{r+s+2} \right), & \text{if } r = 1 \text{ and } s = 0 \\ (-0.333) + 1\left(\dfrac{r+1}{r+s+2} \right), & \text{if } r = 0 \text{ and } s = 1 \end{cases} \qquad (3)$$

The 'r' and 's' values are discrete values. After the outcome of each interaction is obtained, the 'r' and 's' are set as '1' or '0' accordingly i.e. if positive outcome, 'r' is '1' while 's' is '0' and vice versa. The '*f(exp)*' value from Eq. (3) is then bootstrapped to the previous trust value to compute the updated trust on an agent. The proposed trust evaluation model combining the sample experience and the bootstrapping process is:

$$T_{i,j} \leftarrow T'_{i,j} + \alpha \left(f(exp) - T'_{i,j} \right) \qquad (4)$$

Based on experiments conducted, the internal reward values are selected. In Eq. (4), 'α' denotes the learning rate which factors the value or the importance of the new experience during the iteration and '$T_{i,j}$' is the previous trust value. The proposed model in Eq. (4) is developed using temporal difference learning and is called TD Trust model. The outcome of this equation characterizes a probabilistic approach of evaluating the trust value where the trust falls between [0, 1].

4 Simulation

Simulation experiments are conducted in Visual Studio 2013 C++ platform to investigation and analysis the performance of the developed trust evaluation model. The simulation experiments are similar to the ball (interchangeable term with goal) collecting exercise described in Sect. 2.

The environment is set up as a fifty by fifty grid cell world with three agents (A_1, A_2 and A_3) starting from the same initial position with a neutral trust value estimate on each other, '0.5'. The agents are required to explore the environment in search of ten randomly scattered goals. While exploring the environment to achieve their objective of finding their respective goals, the agents could work together to expedite the completion time. When one agent encounters the goal of another agent (for example, A_1 found a goal that belongs to A_3), that agent shares the location information with the respective agent (A_1 shares goal location information with A_3). The agent receiving the information then proceed to explore the target location to verify the goal color (A_3 explores the location shared by A_1).

If the observation shows that there is goal present and meets the objectives of the agent, the interaction results in a positive outcome with 'r' is '1' while 's' is '0' and vice versa. With the outcome observation in hand, the agent then updates its trust on the agent that shared the information using Eq. (4). If the trust evaluated on an agent drops to below minimum value, that agent is considered untrustworthy. Any further information shared by this agent is considered false and is not trusted or followed. In these experiments, the minimum trust value is set as '0.3' which corresponds to 40 % below the neutral trust value, '0.5'. If trust on an agent sharing information is less than the minimum value, the other agent executes its own exploration strategy. The exploration strategy of the agents is a simple algorithm that always explores the least explored grid cell first.

The objective of the simulation is to collect all the goals where each agent is required to find ten goals scattered in the environment. The simulation terminates once all the goals have been collected. During the simulation experiment, the agents are subjected to various probability of lying. The simulation experiments are conducted under these three conditions:

1. A_1, A_2 and A_3 has 0 %, 20 % and 40 % of lying respectively.
2. A_1, A_2 and A_3 has 0 %, 40 % and 60 % of lying respectively.
3. A_1, A_2 and A_3 has 0 %, 60 % and 80 % of lying respectively.

Each of these conditions are executed in thirty simulation experiments using different environment maps where the goal locations vary in each of the environment. The thirty simulation experiments ensure that an average performance of the model is obtained.

The performance of the proposed model is compared against another trust evaluation model found in the literature, Secure Trust [4]. Secure Trust model is chosen for comparison study for two main reasons. First, Secure Trust evaluates trust through interactions outcomes, similar to the proposed trust model, therefore this model is justified for comparison study since both models have same evaluation principles.

Second, the Secure Trust model is one the most recently published trust evaluation model at the start of this research.

The performance parameters studied in this simulation are:

1. Time (time steps) taken to complete objective.
2. Fluctuation in the trust values.

4.1 Performance – Time Step

One of the important parameter to be considered in the simulation study is the total time taken to complete the given objectives. However, the varying hardware processing capabilities among computers results in the simulation executing at different speed on different computers. Therefore, a more generalized approach to analyze the completion time is considered in this study to evaluate the performance of the models. The proposed approach is by examining the time steps taken.

Since the simulation experiment is formalized as a grid world environment, each time the agent moves from one grid location to the next is considered taking one step. Each step taken by the agent corresponds to one-time step. When an agent travels to a potential goal location shared by another agent, those time steps are also added to the total time steps. If a goal is not observed at the target location, then those time steps becomes non-value added which in turn, increases the total time steps taken to complete the objectives. Having a certain degree of trust in the information provider ensures that an agent is not constantly misled to wrong location and the time steps taken are always value added. Efficiency of the MAS is denoted by the total time steps taken to complete the given objectives i.e. the smaller the total time steps, the faster the objectives are completed.

Figure 1 illustrates the time steps performance comparison when agents A_1, A_2 and A_3 are subjected to 0 %, 20 % and 40 % probability of lying respectively. Under these circumstances, the proposed TD Trust model performs averagely 33.13 % faster than Secure Trust model in completing the given objectives. When the probabilities are increased for A_2 and A_3 to 40 % and 60 % respectively, the TD Trust model continuously

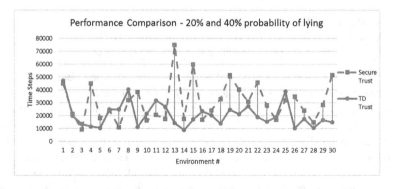

Fig. 1. Comparing the time steps performance between models at 20 % and 40 % probability of sending false information

performs better than Secure Trust model, by completing the objectives averagely 33.55 % faster. Details of the performance comparison in these conditions is shown in Fig. 2. Further increase in the lying probability of A_2 and A_3 to 60 % and 80 % respectively shows that TD Trust model still performed faster than Secure Trust model by 18.98 %. Figure 3 shows the performance details when the agents are subjected to 60 % and 80 %.

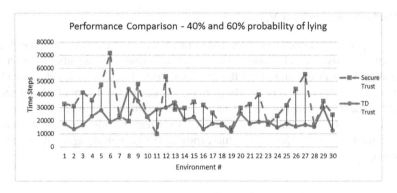

Fig. 2. Comparing the time steps performance between models 40 % and 60 % probability of sending false information

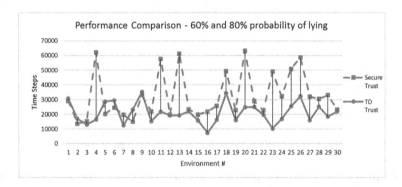

Fig. 3. Comparing the time steps performance between models 60 % and 80 % probability of sending false information.

Table 1 shows the average time steps of taken by the two models to complete simulation objectives under the different lying conditions. The trust evaluation model is useful in identifying potentially lying and untrustworthy agents based the interactions outcomes. By disregarding any information from untrustworthy agents, an agent is able to improve the time steps counts since only value added time steps are included to the total time taken to complete the given objectives. When comparing the proposed TD Trust model and Secure Trust model, agents using TD Trust model are more effective in identifying and following information from trustworthy agents only.

Table 1. The summary of time steps taken to complete achieve simulation objectives.

	% Sending false information	Secure trust	TD trust
Time steps	20–40 %	30779	*20582*
	40–60 %	32298	*21462*
	60–80 %	25904	*20987*

4.2 Performance – Trust Value

Another important parameter investigated in this simulation experiment is the fluctuation patterns in trust value. Following an interaction between two agents, the experience gained should not drastically affect the previous trust placed in an agent. Instead, the trust in an agent should be objectively increased or decreased in a gradual manner. An example of the trust value fluctuation pattern observed in the 5th simulation experiment is shown in Fig. 4. The figure shows agent A_1 evaluating the trust on agent A_3 who in a series of interactions. If the trust value drops to 0.3 or below, agent A_3 is deemed untrustworthy.

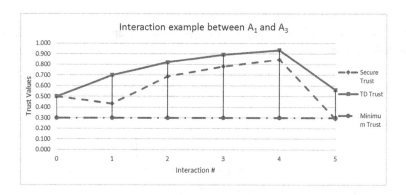

Fig. 4. Trust values fluctuation example in an environment

Interactions 1 to 4 shows a positive outcome with goal observed at the location indicated by A_3 while interaction 5 shows a negative outcome where the goal is not observed by A_1 at the location indicated by A_3. Analyzing the TD Trust values, a more gradual increase in trust is noticeable from interactions 1 through 4 while the Secure Trust values show a more dramatic increase in trust value. For example, from interaction 1 to 2, Secure Trust calculated 58.42 % increase in trust while TD Trust evaluated 17.14 % in trust. At the first instance lying (interaction 5), both TD Trust and Secure Trust calculated a drop in trust. However, Secure Trust estimated a drastic drop in trust, by 65.49 % while TD Trust models made a cautious drop in trust value, by 40 %. A drastic drop in trust after the first failed interaction may lead to classifying wrong agents as untrustworthy. In this regard, the TD Trust model exhibit a more gradual and preferred estimation patterns of the trust values than the Secure Trust model. Table 2 highlights the changes of fluctuation in the trust estimation by both models.

Table 2. Change in trust value recorded from the 5th simulation experiment.

Interaction #	Change in trust values	
	Secure trust	TD trust
1	13.30 %	40.00 %
2	58.42 %	17.14 %
3	13.68 %	8.78 %
4	8.32 %	4.84 %
5	−65.49 %	−40.00 %

The Fig. 5 illustrates another example of trust fluctuation with higher number of negative outcomes in a series of five interactions. Out of the five interactions, interaction number 2, 4 and 5 resulted in negative outcomes i.e. no goals present after arriving at target locations. When the first negative outcome interaction is encountered, the trust values evaluated by both the trust models showed a drop in trust. Out of the two models, Secure Trust model calculated a drop in trust value almost reaching to the minimum value. When the next interaction results in a positive outcome, both trust models showed an increase in trust. However, Secure Trust model showed a higher increase (82.57 % increase) in trust value when compared to TD Trust model (55.24 % increase). The next interactions 4 and 5 both show a negative outcome, resulting in decrease in trust value. During interaction 4, the trust value calculated by Secure Trust plunged to the minimum value line (trust value dropped by 46.39 %). TD Trust model on the other hand, made a subjective drop in trust (by 40 %), giving the agent a benefit of doubt. Only in the next interaction TD Trust model calculated a drop in trust value past the minimum trust limit.

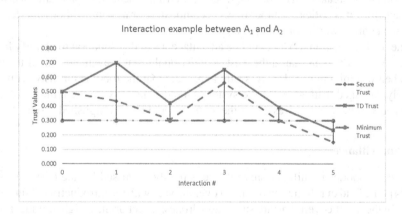

Fig. 5. Trust fluctuation with high number for negative interactions

Important criteria in any trust evaluation model is the requirement for a subjective evaluation of trust with a sense of benefit of the doubt. The model should be able to improve the efficiency of the system while correctly identifying trustworthy agents. With regards to this, the proposed TD Trust model meet the criteria. Besides exhibiting a gradual increase in trust, the model does not deem an agent untrustworthy with just one negative observation (see Table 2).

5 Applications of the Proposed Approach

The proposed trust evaluation model has potential application in any MAS field that requires multiple agents to cooperatively work together to achieve common mission objectives. This section highlights some of the application of the proposed trust model in the robotics field. Many robotics application requires multiple robots to work in the same environment and share resources to achieve common objectives. Some examples of such applications are surveillance and search and rescue. The proposed trust model could be incorporated into the robots to ensure only trustworthy robots are cooperating together to ensure higher success rate.

5.1 Search and Rescue

Robots could prove to be an incredible resource in search and rescue operations. The use of robots are especially critical when the search and rescue involves dangerous environment such as in unstable building structures, structures under fires, radioactive area or any environment that is dangerous for human to operate [22, 23]. The robots could be used to gather information vital to the search and rescue missions such as identifying trapped victims, spot potential hazardous zones, scout for radioactive leaks, build a map of the environment and etc.

The robots could then share this information among each other and with the human search and rescue teams for further actions. However, operating under such hazardous condition could be potentially damage robots resulting in faulty sensors, misreading of environment or breakdown of the robot. As a search and rescue mission is almost always time sensitive, it is critical that the robots accomplish their mission without interference from such untrustworthy or faulty robots.

The trust estimation model could be advantageous in this situation to identify faulty robots. With every data shared the trust on individual robot is evaluated and if the trust drops below a minimum value, it is safe to assume that the robot is untrustworthy and possibly damaged. Efficiency of the system should improve with the removal of non-value added robots and the rescue of the victims could be accomplished faster.

5.2 Surveillance

The use of robotics in military surveillance is highly discussed for the past few years [22, 24]. The field of robotics has evolved over the year with the introduction of futuristic robots in the art of covert warfare such nano drones, insect drones, high altitude drones and more. These drones are often operating remotely in order to ensure secrecy of the mission. In a surveillance mission, the drones are used for various purpose such as tracking a moving target, surveying a target location or building and mostly for intelligence gathering. In some of the cases, multiple drones are used in concert to ensure a thorough information gathering.

Working under such secrecy and distance, it is difficult to ascertain if the drones are damaged or faulty due to unpredicted reason. The drones could be sharing false information such as incorrect sensor reading, wrong map layout, different target tracked and so on. Any follow up mission based on this information could result in failure or worst, lost in human or soldier's life.

Therefore, a trust mechanism could be used in the surveillance system to ensure the drones are working in perfect order. Each time a drone shares information with each other, the drone is accessed for its accuracy and quality of transmitted information. If the drone often sends a false information, the drone is deemed untrustworthy for the operation and need to be recalled for maintenance.

6   Conclusion and Future Work

This paper details the development and investigation of a trust evaluation model for a MAS framework. The proposed trust model termed TD Trust model, uses an adaptation of temporal difference learning technique to evaluate trust in agents of a MAS. The model empirically evaluates trust on agents by adding experience gained through interactions to the previous trust placed on them. The proposed TD Trust model is tested using simulation experiments to investigate the performance of the proposed model against a comprehensive model found in the literature. The simulation outcomes indicate that the proposed model provides an objective evaluation of trust and also a more subjective evaluation of trust resulting in a better performance when compared to the model from literature. The TD Trust model has potential application is various field that requires multiple agents to cooperate and work together such as surveillance, navigation, search and rescue. Future work for this study includes investigating the application of the proposed model in a real world system.

Acknowledgement.   This research is funded by e-Science grant provided by Ministry of Science, Technology and Innovation (MOSTI), Malaysia, Project Number: 03-02-10-SF0200.

References

1. Han, Y., Zhiqi, S., Leung, C., Chunyan, M., Lesser, V.R.: A survey of multi-agent trust management systems. IEEE Access **1**, 35–50 (2013)
2. Busoniu, L., Babuska, R., De Schutter, B.: A comprehensive survey of multiagent reinforcement learning. IEEE Trans. Syst. Man Cybern. Part C Appl. Rev. **38**, 156–172 (2008)
3. Aref, A., Tran, T.: Using fuzzy logic and Q-learning for trust modeling in multi-agent systems. In: 2014 Federated Conference on Computer Science and Information Systems (FedCSIS), pp. 59–66 (2014)
4. Das, A., Islam, M.M.: SecuredTrust: a dynamic trust computation model for secured communication in multiagent systems. IEEE Trans. Dependable Secur. Comput. **9**, 261–274 (2012)
5. Cavalcante, R.C., Bittencourt, I.I., da Silva, A.P., Silva, M., Costa, E., Santos, R.: A survey of security in multi-agent systems. Expert Syst. Appl. **39**, 4835–4846 (2012)
6. Luo, L., Chakraborty, N., Sycara, K.: Distributed algorithms for multirobot task assignment with task deadline constraints. IEEE Trans. Autom. Sci. Eng. **12**, 876–888 (2015)

7. Basheer, G.S., Ahmad, M.S., Tang, A.Y., Graf, S.: Certainty, trust and evidence: towards an integrative model of confidence in multi-agent systems. Comput. Hum. Behav. **45**, 307–315 (2015)
8. Yugang, L., Nejat, G., Vilela, J.: Learning to cooperate together: a semi-autonomous control architecture for multi-robot teams in urban search and rescue. In: 2013 IEEE International Symposium on Safety, Security, and Rescue Robotics (SSRR), pp. 1–6 (2013)
9. Mohammed, M., Lim, C., Quteishat, A.: A novel trust measurement method based on certified belief in strength for a multi-agent classifier system. Neural Comput. Appl. **24**, 421–429 (2014)
10. Fullam, K.K., Klos, T.B., Muller, G., Sabater, J., Schlosser, A., Topol, Z., Barber, K.S., Rosenschein, J.S., Vercouter, L., Voss, M.: A specification of the agent reputation and trust (ART) testbed: experimentation and competition for trust in agent societies. In: Proceedings of the Fourth International Joint Conference on Autonomous Agents and Multiagent Systems, pp. 512–518 (2005)
11. Sabater, J., Sierra, C.: Review on computational trust and reputation models. Artif. Intell. Rev. **24**, 33–60 (2005)
12. Vanderelst, D., Ahn, R.M., Barakova, E.I.: Simulated trust-towards robust social learning. Artif. Life, 632–639 (2008)
13. Vanderelst, D., Ahn, R.M., Barakova, E.I.: Simulated trust: a cheap social learning strategy. Theor. Popul. Biol. **76**, 189–196 (2009)
14. Huynh, T.D., Jennings, N.R., Shadbolt, N.R.: An integrated trust and reputation model for open multi-agent systems. Auton. Agent Multi-Agent Syst. **13**, 119–154 (2006)
15. Wang, Y., Singh, M.P.: Formal trust model for multiagent systems. In: Proceedings of the 20th International Joint Conference on Artifical Intelligence, pp. 1551–1556 (2007)
16. Wang, Y., Singh, M.P.: Evidence-based trust: a mathematical model geared for multiagent systems. ACM Trans. Auton. Adapt. Syst. **5**, 1–28 (2010)
17. Wang, Y., Hang, C.-W., Singh, M.P.: A probabilistic approach for maintaining trust.pdf. J. Artif. Intell. Res. **40**, 47 (2011)
18. Mantel, K.T., Clark, C.M.: Trust networks in multi-robot communities. In: 2012 IEEE International Conference on Robotics and Biomimetics (ROBIO), pp. 2114–2119 (2012)
19. Jøsang, A., Haller, J.: Dirichlet reputation systems. In: The Second International Conference on Availability, Reliability and Security, ARES 2007, pp. 112–119 (2007)
20. Sutton, R.S., Barto, A.G.: Reinforcement Learning: An Introduction. MIT Press, Cambridge (1998)
21. Jsang, A., Ismail, R.: The beta reputation system. In: Proceedings of the 15th Bled Electronic Commerce Conference, vol. 5, pp. 2502–2511 (2002)
22. Gautam, A., Mohan, S.: A review of research in multi-robot systems. In: 2012 7th IEEE International Conference on Industrial and Information Systems (ICIIS), pp. 1–5 (2012)
23. Nagatani, K., Kiribayashi, S., Okada, Y., Otake, K., Yoshida, K., Tadokoro, S., Nishimura, T., Yoshida, T., Koyanagi, E., Fukushima, M., Kawatsuma, S.: Emergency response to the nuclear accident at the Fukushima Daiichi Nuclear Power Plants using mobile rescue robots. J. Field Robot. **30**, 44–63 (2013)
24. Burdakov, O., Doherty, P., Holmberg, K., Kvarnstrom, J., Olsson, P.M.: Relay positioning for unmanned aerial vehicle surveillance. Int. J. Robot. Res. **29**, 1069–1087 (2010)

Controlling Logistics Robots
with the Action-Based Language YAGI

Alexander Ferrein[1]($\boxtimes$), Christopher Maier[2], Clemens Mühlbacher[2],
Tim Niemueller[3], Gerald Steinbauer[2], and Stavros Vassos[4]

[1] Mobile Autonomous Systems and Cognitive Robotics Institute,
FH Aachen University of Applied Sciences, Aachen, Germany
`ferrein@fh-aachen.de`
[2] Institute for Software Technology, Graz University of Technology, Graz, Austria
`{muehlbacher,steinbauer,maier}@ist.tugraz.at`
[3] Knowledge-Based Systems Group, RWTH Aachen University of Technology,
Aachen, Germany
`niemueller@kbsg.rwth-aachen.de`
[4] Department of Computer, Control, and Management Engineering,
Sapienza University of Rome, Rome, Italy
`vassos@dis.uniroma1.it`

Abstract. To achieve any meaningful tasks, a robot needs some form
of task-level executive which acquires knowledge, reasons or plans, and
performs and monitors actions. A formal approach for such agent programming is the GOLOG agent programming language. GOLOG is based
on a first-order logic representation, and a drawback of common implementations is that in order to program agents, also knowledge of Prolog
functionality is typically needed. In this paper, we present a prototype
implementation of YAGI, a language rooted in GOLOG that offers a practical subset of the rich GOLOG framework in a more familiar syntax.
Bridging imperative-style programming with an action-based specification, YAGI is more accessible to developers and provides a better ground
for robot task-level executives. Moreover, we developed bindings for popular robotics frameworks such as ROS and Fawkes. As a proof of concept
we present a YAGI-based agent for the RoboCup Logistics League which
shows the expressiveness and the possibility to easily embed YAGI into
robot applications.

1 Introduction

For a mobile robot to fulfill its tasks, some high-level decision making strategy is needed. There is a variety of paradigms for encoding and handling the
high-level tasks of a robot. For instance, state machines or decision trees are
often used to decide which action to take depending on sensor inputs, e.g., [13],
and Belief-Desire-Intention architectures are used as the basis for high-level control languages such as 3APL [8] and PRS [9]. Another possibility to realize
high-level control is classical task planning using standardized description languages such as the planning domain definition language (PDDL) [5]. Similarly,

© Springer International Publishing Switzerland 2016
N. Kubota et al. (Eds.): ICIRA 2016, Part I, LNAI 9834, pp. 525–537, 2016.
DOI: 10.1007/978-3-319-43506-0_46

approaches based on hierarchical tasks networks (HTNs) have been used [22]. Another expressive approach for high-level control that is based on a rich logical framework is GOLOG [11], along with its many descendants. GOLOG has a formal semantics based on the situation calculus [21] and has shown its usefulness in prototype applications ranging from educational robotics [12] and robot soccer [2] to domestic service robots [3].

While the GOLOG family languages have many advantages, e.g., advanced expressiveness, a formal underlying semantics, and the capability to combine imperative-style programming along with logical reasoning and planning, nonetheless the existing implementations have some drawbacks: (1) nearly all implementations are based on Prolog, which makes the integration of a GOLOG interpreter into a typical robot or agent architecture a less straightforward task; (2) GOLOG interpreters are often bounded by features of the underlying Prolog interpreter; (3) when writing agent programs, the distinction between GOLOG features and Prolog functionality is often not quite clear for most GOLOG implementations. Moreover, we see that Prolog is not the first choice for a roboticist making it less familiar than other programming environments. There are approaches to address these shortcomings, for example an interpreter based on Lua [4], and YAGI (Yet Another Golog Interpreter) [1]. YAGI is a language specification for a subset of GOLOG that was designed to reach out to a larger user group appealing to a more familiar imperative-style programming syntax. The state reported in [1] only concerned a first version of a user-friendly syntax and a semantic bridge to the situation calculus and GOLOG. A fully implemented YAGI interpreter and interfaces to robotics systems though was not available and the evaluation of YAGI (and GOLOG) in wider robotics scenarios was not possible.

In this paper, we present a fully-functional interpreter for the YAGI language that has been further extended with concepts necessary to control a robot system like sensing and exogenous events. Moreover, we show how robotics frameworks like ROS [20] and Fawkes [16] can be hooked up easily through a plug-in system. As a proof of concept we programmed YAGI agents for the RoboCup Logistics League [15,19] in simulation [23], where a team of robots has to fulfill logistics tasks in a virtual factory environment. YAGI overcomes the tight and confusing coupling between syntax and semantics of previous Prolog implementations of GOLOG and even allows to use different back-end algorithms for interpreting the language. We are convinced that tools like YAGI are a good step to motivate roboticist to use deliberative concepts like action-based programming to control their robots.

2   The YAGI Framework

2.1   Situation Calculus and Golog Prerequisites

YAGI (Yet Another Golog Interpreter) is an action-based robot and agent programming language, has its roots in the GOLOG language family [11] and its formal foundations in the situation calculus [21]. The situation calculus is a first-order logic language with equality (and some limited second-order features) which allows for reasoning about actions and their effects [21]. The major concepts are *situations*, *fluents*, and *primitive actions*. Situations reflect the evolution of the world and are represented as sequence of executed actions rooted in an initial situation S_0. Properties of the world are represented by fluents, which are situation-dependent predicates. Actions have preconditions, which are logical formulas describing if an action is executable in a situation. Successor state axioms define whether a fluent holds after the execution of an action. The precondition and successor state axioms together with some domain-independent foundational axioms form the so-called basic action theory (BAT) describing a domain.

GOLOG [11] is an action-based agent programming language based on the situation calculus. Besides primitive actions it also provides *complex actions* that can be defined through programming constructs for control (e.g. while loops) and constructs for non-deterministic choice (e.g. non-deterministic choice between two actions). This allows to combine iterative and declarative programming easily. The program execution is based on the semantics of the situation calculus and querying the basic action theory whenever the GOLOG program requires the evaluation of fluents.

2.2 Design Aims

Our work to provide a GOLOG-like interpreter with YAGI aims at overcoming some of the shortcomings of existing Prolog-based implementations, in order to make the action-based programming concepts used in GOLOG accessible to a larger community. A major problem of the existing implementations and also a reason for limited outreach is that it is hard for non-expert users to distinguish GOLOG from Prolog concepts. As a basic action theory and GOLOG are essentially defined in first-order logic, there is in fact no syntax specification for the language used in the implemented systems which typically use a mix of Prolog functionality with GOLOG-defined constructs. The YAGI syntax on the other hand, aims at a specification that is independent of the underlying implementation, and using concepts that are similar to familiar programming languages in order to make it easy to use. The YAGI syntax and semantics should follow the action-based framework of GOLOG, also providing constructs for realizing other important tasks for robotics, such as sensing or exogenous events. Finally, a YAGI interpreter should allow for an easy integration into other programming languages or robotics frameworks using clear and lean interfaces.

2.3 System Architecture

In order to achieve the above design aims, the YAGI framework uses the 3-tier architecture. The architecture comprises (1) a front-end, (2) a back-end and (3) a system interface. (Please refer to [14] for a detailed description.) This architecture allows a clear separation between the YAGI language and its syntax (see Sect. 2.4) and the implementation of the interpreter. The front-end provides the user-interface allowing queries and invocation of actions and programs and checks the syntactical correctness of queries, statements or programs. Currently the front-end is realized by an interactive console that allows to query fluents and to execute single statements or entire programs. The interface to the back-end is realized using an abstract syntax tree (AST) and a string-based callback. Therefore, in contrast to Prolog-based implementations the user has not to be aware of the concrete implementation of the system and different implementations of the back-end are easily possible. The back-end interprets YAGI statements and maintains a representation of the YAGI domain theory, the program to execute, and the state of the world. Moreover, it is able to execute imperative parts of programs as well as to plan over declarative sections. The back-end also incorporates information coming from the system interface such as result to sensing actions or exogenous events. The communication to the system interface is realized through an open string-based signal mechanism. The system interface provides the coupling of the back-end and the remaining robot system. It is responsible for the execution of actions triggered by the back-end and the collection and interpretation of feedback from the robot system. Using this system interface allows to hook-up different robot frameworks easily.

2.4   The YAGI Language

The aim of the definition of YAGI's syntax was to provide a clear definition of the language that allows to specify domain theories and control programs in same language while staying with the well-elaborated concepts of the situation calculus and GOLOG, but providing a more common style. The language definition is given in BNF and is therefore independent of any implementation. The full language definition can be found in [14].

The language provides statements to build up a domain theory with fluents, actions, and procedures. *Fluents* are used to represent the state of the world. While in the situation calculus fluents are first-order predicates that depend on a situation, in YAGI a fluent is represented as a set of tuples. The structure of the tuple defines the signature of a fluent. The interpretation of a fluent is that a fluent holds for a parameter tuple if the related tuple is in the set. Currently, the possible domain for parameters is a finite set of strings. For instance, the fluent $is\_at(p, l)$ represents a puck p being at a location l. Listing 1.1 depicts the fluent declaration with finite domains for pucks and locations. The definition is similar to well-known associative arrays. The set-based representation of fluents allows for an intuitive syntax for manipulation, queries and iteration over fluents. The assignment `is_at += <"Pk2","M3">`, which describes the effect on a fluent, looks similar to statements in common programming languages and express that the fluent additionally holds for the pair `"Pk2"` and `"M3"`. Similar statements exist for deleting pairs, assigning fluents, and more advanced assignments such as iterative or conditional assignments. *Query formulas* may include logical connectives and quantifiers similar to first-order formulas. For example, a statement such as `exists <$x,"M2"> in   is_at` – where `$x` denotes a variable named x – asks if any puck `$x` is at machine `"M2"`.

```
1 fluent is_at[{"Pk1",...,"Pk18"}]
2              [{"D1",...,"D6","M1",...,"M24","R1","R2"}];
```

Listing 1.1. Definition of a binary fluent representing the location of a puck.

```
1 action goto($place) external ($status)
2 effect:
3   skill_status = {<$status>};
4 signal:
5   "skill-exec-wait ppgoto{place='"+ $place +"'}";
6 end action
```

Listing 1.2. Action that calls the Behavior Engine to move to a specified place.

YAGI further provides the concept of primitive *actions* that mostly follows the situation calculus definition by Reiter [21]. Listings 1.2 and 1.3 show two examples of action definition. The action **read_light** consists of four parts. The head defines the action's name plus a list of internal and external variables (l. 1). The internal variables are the regular parameters to the action. If external variables are defined, the action becomes a setting action. Such actions are a way to integrate sensing. We follow here the pragmatic way of directly set a fluent based on sensing result rather than reasoning about incomplete knowledge and alternative models [2]. Besides exogenous events, setting actions are extensions made to YAGI during this work in order to be able to reflect the behavior of a robot system. External variables are bound by the system interface after the action execution. The precondition (ll. 2–3) is a YAGI formula and denotes if an action is actually executable. In contrast to Reiter where successor state axioms are used to describe the effect of actions to fluents, in YAGI the effect on fluents (ll. 4–5) is directly expressed using fluent assignments. Please note that effects can comprise a sequence of simple as well as complex statements such as iterative and conditional assignments. Finally, the signal block (ll. 6–7) represents the communication to the system interface. Once the action is triggered for execution, the related string is sent to the system interface. Variables will be bound before transmission.

Statements to define *procedures* are also provided by YAGI. Procedures work the same way as in GOLOG. Listing 1.5 depicts a procedure definition comprising a head with the procedure's name and parameters and a body. The body of a procedure comprises primitive and complex actions. Complex actions are

```
1 action read_light() external ($red, $yellow, $green)
2 precondition:
3   blackboard_connected == {<"true">};
4 effect:
5   light_state = {<$red, $yellow, $green>};
6 signal:
7   "bb-get RobotinoLightInterface::/machine-signal";
8 end action
```

Listing 1.3. Action that reads the light signal in front of the robot and stores the outcome into the fluent **light_state**.

```
1  proc production()
2    while not (exists <$M> in machines) do
3      sleep_ms("1000");
4    end while
5    while phase == {<"PRODUCTION">} do
6      get_raw_material("Ins1");
7      pick <$M,"T5"> from machine_types such
8        perform_production_at($M,"T5");
9      end pick
10     deliver("deliver1");
11   end while
12 end proc
```

Listing 1.4. Procedure that implements the production main control loop.

```
1  proc exploration()
2    while not (exists <$M> in expl_machines) do
3      sleep_ms("1000");
4    end while
5    while phase == {<"EXPLORATION">} do
6      if (exists <$M> in expl_machines) then
7        pick <$M> from expl_machines such
8          explore($M);
9        end pick
10     end if
11   end while
12 end proc
13
14 proc explore($M)
15   goto($M);
16   read_light();
17   exploration_report($M);
18   mark_explored($M);
19 end proc
```

Listing 1.5. Procedure that implements the exploration main control loop.

for and while loops, conditionals and non-deterministic statements for picking arguments or statements. For example, the procedure in Listing 1.5 uses a **pick** statement to non-deterministically choose a tuple where a fluent holds. Please note that pattern matching with variables can be used. For instance **<$p,"R1"> in is_at** would select a puck $p which is at recycling machine "R1". Also, a non-deterministic selection between actions is possible. For instance, the statement **choose A1() or A2()** chooses non-deterministically between the actions A1 and A2. These non-deterministic statements play a major role if *planning* is used in YAGI. By default, YAGI performs on-line execution where the next best action is derived and executed immediately. No backtracking is done and the program is aborted if the selected action cannot be executed because the precondition does not hold. YAGI provides a **search** statement where parts of a program are executed off-line analogous to the original GOLOG semantics. If this statement is used, first a complete valid action sequence through the program is sought and once one is found, the sequence is executed. Therefore, YAGI allows to combine iterative programming and planning.

2.5   The Database Back-End

The two major tasks of the back-end are the representation of the YAGI domain theory including definitions of all fluents, actions, and procedures as well as the program execution.

For the first task, internal data structures are maintained and updated if the AST of the YAGI program contains domain-related statements like action or procedure definitions. The second task is tackled in the current implementation using a database representation for fluents [7]. Fluents are represented as tables in a relational database. Rows in a table represent the tuples the fluent holds for while the columns represent the fluent's parameter. If a fluent is defined, the according table is added to the database. Using a relational database allows for quick updates of fluents as well as efficient queries. The realization of the back-end using a database is the main reason for finite domains concerning fluents. It also imposes a closed-world assumption. The current implementation uses SQLite. YAGI formulas are evaluated using these tables and the usual SQL semantics.

For the execution of a YAGI program, its AST is traversed according to the YAGI semantics. The on-line execution of YAGI programs (basically the main procedure) follows the on-line transition semantics defined for INDIGOLOG [6]. This semantics uses the predicates *Trans* and *Final*. The predicate $Trans(y, d, y', d')$ denotes if a YAGI program y with the database d legally can lead to a remaining program y' and an updated database d'. The predicate $Final(y, d)$ denotes if a YAGI program y with a database d can terminate legally. In the on-line execution, once a legal transition is present this transition is executed. If the transition is final the program is terminated. During on-line semantics non-deterministic statements like pick and choose are random.

For the search statement with an off-line execution semantics, a complete trace through the AST is determined before any actions get executed. The current implementation uses a simple but complete breadth-first search algorithm. Once a legal transition within the AST is found, the remaining program is considered a valid state in the search space. For each state a database is kept which is updated according to the transitions leading to that respective state. The search is continued until a state with legal termination is found. Then, the actions along the path from the root node of the AST are executed in sequence.

The back-end is also responsible for the interaction with the system interface. If the back-end selected an action for execution, the according string is sent to the system interface including the proper binding of variables. Once the action was executed, the back-end updates the database according to the action definition's effects. In the case of a setting action, the external binding of variables by the system interface is considered. Please note that setting actions are not allowed within a search statement as the actions are not executed during the search and therefore no results from the system interface are available. Moreover, the back-end handles exogenous event in an asynchronous way. Exogenous events reported by the system interface are stored in a queue. After a transition step is finished all stored events are processed in a first-come-first-serve fashion. For each exogenous event, the variables are bound and the database is updated according to the statements in the event definition. An event definition is similar to a setting action definition, but has no precondition section.

2.6   System Interface

The system interface is realized as a C++ plug-in system allowing for an easy connection to different robot systems. Only two functions have to be reimplemented. The first function is called once an action (primitive or setting) is triggered. The function only needs to call the appropriate action within the robot system and to report the result. The second function handles the exogenous events reported by the robot system and adds them to the event queue of the back-end. Currently, implementations for ROS [20] and Fawkes [16] exists and are available as open source projects, see http://yagi.ist.tugraz.at. The Fawkes system interface used in this work is detailed in Sect. 4.

3   The RoboCup Logistics League in Simulation

RoboCup [10] is an international initiative to foster research in the field of robotics and artificial intelligence. It serves as a common testbed for comparing research results in the robotics field. RoboCup is particularly well-known for its various soccer leagues. In the past few years application-oriented leagues received increasing attention. In 2012, the new industry-oriented Logistics League Sponsored by Festo (LLSF) – renamed to RoboCup Logistics League (RCLL) in 2015 – was founded to tackle the problem of production logistics. Groups of three robots have to plan, execute, and optimize the material flow in a smart factory scenario and deliver products according to dynamic orders. Therefore, the challenge consists of creating and adjusting a production plan and coordinate the group of robots [15].

In 2014, the *LLSF competition* took place on a field of 11.2 m × 5.6 m surrounded by walls (Fig. 1). Two teams are playing at the same time competing for points, (travel) space and time. Each team has an exclusive input storage (blue areas) and delivery zone (green area). Machines are represented by RFID readers with signal lights on top indicating the machine state. The lights indicate the current status of a machine, such as "ready", "producing" and "out-of-order". There are three delivery gates, one recycling machine, and twelve production machines per team. Material is represented by orange pucks with an RFID tag. In the beginning, all pucks (representing the products) are in raw material state and located in the input storage (blue zones).

The game is controlled by the *referee box (refbox)*, a software component which keeps track of puck states, instructs the light signals, and posts orders to the teams. After the game is started, no manual interference is allowed, robots receive information only from the refbox.

The game is split into two major phases. During the *exploration phase* the machines will show a light pattern to indicate a specific type. At game start,

Fig. 1. Top: LLSF finals at RoboCup 2014. Bottom: The simulation of the LLSF in Gazebo. (Color figure online)

the refbox randomly determines the machine types, light pattern, and team assignment and announces it to the robots. Then, the robots have to explore the machines and report them back to the referee box. After four minutes, the refbox switches to the *production phase* during which teams win points for delivering ordered products, producing complex products, and recycling. The initial raw materials must be refined through several stages to final products using the production machines. Finished products must then be taken to the active gate in the delivery zone. Machines can be unavailable ("broken") for a limited time.

We use a *simulation of the LLSF* based on Gazebo [23] shown in Fig. 1 for development and testing. The simulation provides a 3D environment with optional physics engines and a variety of already supported sensors and actuators. The integrated system and many of the components are available as open source (http://carologistics.org). A noteworthy property of the LLSF is that the referee box provides *accountable agency to the environment*. The simulation uses the same game controller as in real games. Additionally, the simulation allows to provide different *levels of abstraction*. Here, we use mixed higher-level (signal lights directly from simulation instead of image processing) and lower-level abstractions (laser data for self-localization and navigation).

4   Implementing a Logistics Agent with YAGI

As we build on the Carologistics system, we use the Fawkes [16] robot software framework for integration. It provides the necessary functional components and integration with the simulation. We have developed a Fawkes-specific YAGI system interface which provides generic access to the Fawkes middleware as well as communication with the refbox. The basic skills, e.g. to grab a product from the input storage, are provided through the Lua-based Behavior Engine [17]. The skills perform limited execution monitoring and recover from local failures if possible. Failures that require a change of strategy or deliberation are reported to the YAGI program to deal with it. YAGI interprets the high-level control program and executes actions by calling skills in Fawkes. Information like the light signal states are read via setting actions, i.e. data is read explicitly at certain points in the program (upper left arrows). The communication with the referee box is provided through signal and exogenous events (upper right arrows). Note that there is no direct interaction between YAGI system interface and the simulation. Perception and actuation is performed through Fawkes only. This allows for a simple deployment on a real robot running the Fawkes framework as the interfaces remain the same. Integrating YAGI with additional frameworks only requires an appropriate system interface.

We have implemented a local, incremental, and in principal distributed agent program [19]. It is local in that its scope is a single robot, and not the group as a whole. Because it does not plan ahead the whole game but commits to a course of actions at certain points in time, we call the agent incremental. In principal, the agent could coordinate with other robots for a more efficient production without a centralized instance, making it a distributed system. However, at this time we limit our effort to the single robot case.

The `exploration` program (Listing 1.5) reads the machine mapping from the referee box (ll. 2–4), and then explores the machines by picking one machines after another calling the `explore` procedure for each (ll. 6–10). This procedure drives to the machine, reads the light signal, and reports it (ll. 14–19). Listing 1.2 shows the action declaration that calls the Behavior Engine to move to a specified machine in a blocking fashion which returns only after the skill has been finished (ll. 4–5). It then binds the status value to the `skill_status` fluent (ll. 1–3) to indicate success or failure of the skill execution. Listing 1.3 shows a setting action which gathers data from the blackboard. If a blackboard connection has been established (ll. 3–4), it calls to the system interface (ll. 6–7), data is returned in the external variables (l. 1), and stored in a fluent (ll. 4–5). The program consistently scores about half of the possible 48 points in simulation. To improve this, the pick could be modified to choose the closest unexplored machine. To achieve more points, an efficient use of multiple robots is necessary.

The `production` program (Listing 1.4) repeatedly completes the production chain for P_3 products, which require a single refinement step at a specific machine. Once machine information has been received (ll. 2–4), it retrieves a raw material from the input store (l. 6) and picks an appropriate machine to perform the production (ll. 7–9). After this production step, the raw material will have been converted to a P_3 product and can be delivered (l. 10). The score in the production phase has a greater variance because of the greater influence of travel distances and posted orders. The top score achieved was 55 points (out of possible 60 with P_3 products alone).

5 Discussion

The integration of the YAGI framework with the existing Fawkes-based system and the coding of an agent control program was achieved in about a week. The Fawkes system interface exploits middleware introspection to provide a generic integration. Other systems lacking this capability, like ROS, may not be able to offer a generic integration, but require a platform and domain specific one. The restriction to finite domains for fluents prevent the use of arbitrary numerics or user pointers as opaque data structures for closer C++ system integration. This, however, is not a principal issue, but can be improved in later revisions. Compared to a more elaborated CLIPS-based agent system [18], on-line YAGI emphasizes imperative constructs, while an event or rule-based system allows for easier revising a robot's goal. Integrating INDIGOLOG's interrupts or using YAGI's search capabilities could remedy this disadvantage. The particular benefit of YAGI is that it combines the strengths of two worlds: for one, it is based on the well-studied and well-understood foundations of GOLOG. This provides a sound semantics to the language and its interpretation. For another, it overcomes concerns prevalent in the robotics community by providing a more familiar and easy to use syntax for robot task-level executives. Our experiments implementing an agent program for the RoboCup Logistics League have shown that already the current prototype is a viable platform for such medium complex domains. Some of the current limits will be addressed in YAGI's vivid development process shortly. Finally, due to the fact that YAGI is based on GOLOG's formal logical semantics over programs and their executions, it is possible to perform much more advanced reasoning tasks, gaining all the benefits that come with the research that is being done in the situation calculus community. For example, it is not difficult to imagine future versions of a YAGI interpreter to be able to evaluate whether properties hold over future executions of programs or verify programs over different domains and scenarios, following current state-of-the-art research in verifying properties for GOLOG programs.

Acknowledgments. T. Niemueller was supported by the German National Science Foundation (DFG) research unit *FOR 1513* on Hybrid Reasoning for Intelligent Systems (http://www.hybrid-reasoning.org). C. Mühlbacher was supported by the Austrian Research Promotion Agency (FFG) with the grant Guaranteeing Service Robot Dependability During the Entire Life Cycle (GUARD). We thank the anonymous reviewers for their helpful comments.

References

1. Ferrein, A., Steinbauer, G., Vassos, S.: Action-based imperative programming with YAGI. In: Cognitive Robtics Workshop. AAAI Press (2012)
2. Ferrein, A., Lakemeyer, G.: Logic-based robot control in highly dynamic domains. J. Robot. Auton. Syst. **56**(11), 980–991 (2008)

3. Ferrein, A., Niemueller, T., Schiffer, S., Lakemeyer, G.: Lessons learnt from developing the embodied AI platform CAESAR for domestic service robotics. In: Proceedings of AAAI Spring Symposium. AAAI Technical Report, vol. SS-13-04. AAAI (2013)

4. Ferrein, A., Steinbauer, G.: On the way to high-level programming for resource-limited embedded systems with Golog. In: Ando, N., Balakirsky, S., Hemker, T., Reggiani, M., von Stryk, O. (eds.) SIMPAR 2010. LNCS, vol. 6472, pp. 229–240. Springer, Heidelberg (2010)

5. Fox, M., Long, D.: PDDL2. 1: an extension to PDDL for expressing temporal planning domains. J. Artif. Intell. Res. (JAIR) 20, 61–124 (2003)

6. Giacomo, G.D., Lespérance, Y., Levesque, H.J., Sardina, S.: IndiGolog: a high-level programming language for embedded reasoning agents. In: Multi-Agent Programming: Languages, Tools and Applications. Springer, US (2009)

7. Giacomo, G.D., Palatta, F.: Exploiting a relational DBMS for reasoning about actions. In: Cognitive Robotics Workshop (2000)

8. Hindriks, K., de Boer, F., van der Hoek, W., Meyer, J.J.: Agent programming in 3APL. Auton. Agent. Multi-Agent Syst. 4(2), 357–401 (1999)

9. Ingrand, F., Chatila, R., Alami, R., Robert, F.: PRS: a high level supervision and control language for autonomous mobile robots. In: IEEE International Conference on Robotics and Automation (ICRA), vol. 1 (1996)

10. Kitano, H., Asada, M., Kuniyoshi, Y., Noda, I., Osawa, E.: RoboCup: the robot world cup initiative. In: Proceedings of 1st International Conference on Autonomous Agents (1997)

11. Levesque, H.J., Reiter, R., Lespérance, Y., Lin, F., Scherl, R.B.: Golog: a logic programming language for dynamic domains. J. Logic Program. 31(1–3), 59–83 (1997)

12. Levesque, H., Pagnucco, M.: LeGolog: inexpensive experiments in cognitive robotics. In: Cognitive Robotics Workshop (2000)

13. Loetzsch, M., Risler, M., Jungel, M.: XABSL - a pragmatic approach to behavior engineering. In: IEEE/RSJ International Conference on Intelligent Robots and Systems (2006)

14. Maier, C.: YAGI - An Easy and Light-Weighted Action-Programming Language for Education and Research in Artificial Intelligence and Robotics. Master's thesis, Faculty of Computer Science, Graz University of Technology (2015)

15. Niemueller, T., Ewert, D., Reuter, S., Ferrein, A., Jeschke, S., Lakemeyer, G.: RoboCup logistics league sponsored by festo: a competitive factory automation testbed. In: Behnke, S., Veloso, M., Visser, A., Xiong, R. (eds.) RoboCup 2013. LNCS, vol. 8371, pp. 336–347. Springer, Heidelberg (2014)

16. Niemueller, T., Ferrein, A., Beck, D., Lakemeyer, G.: Design principles of the component-based robot software framework fawkes. In: Ando, N., Balakirsky, S., Hemker, T., Reggiani, M., von Stryk, O. (eds.) SIMPAR 2010. LNCS, vol. 6472, pp. 300–311. Springer, Heidelberg (2010)

17. Niemüller, T., Ferrein, A., Lakemeyer, G.: A lua-based behavior engine for controlling the humanoid robot nao. In: Baltes, J., Lagoudakis, M.G., Naruse, T., Ghidary, S.S. (eds.) RoboCup 2009. LNCS, vol. 5949, pp. 240–251. Springer, Heidelberg (2010)

18. Niemueller, T., Lakemeyer, G., Ferrein, A.: Incremental task-level reasoning in a competitive factory automation scenario. In: Proceedings of AAAI Spring Symposium 2013 - Designing Intelligent Robots: Reintegrating AI (2013)

19. Niemueller, T., Lakemeyer, G., Ferrein, A.: The RoboCup logistics league as a benchmark for planning in robotics. In: Workshop on Planning and Robotics (PlanRob) at ICAPS-15 Scheduling (ICAPS) (2015)
20. Quigley, M., Conley, K., Gerkey, B.P., Faust, J., Foote, T., Leibs, J., Wheeler, R., Ng, A.Y.: ROS: an open-source Robot Operating System. In: ICRA Workshop on Open Source Software (2009)
21. Reiter, R.: Knowledge in Action. Logical Foundations for Specifying and Implementing Dynamical Systems. MIT Press, Cambridge (2001)
22. Sacerdoti, E.: Planning in a hierarchy of abstraction spaces. Artif. Intell. $5(2)$, 115–135 (1974)
23. Zwilling, F., Niemueller, T., Lakemeyer, G.: Simulation for the RoboCup logistics league with real-world environment agency and multi-level abstraction. In: Bianchi, R.A.C., Akin, H.L., Ramamoorthy, S., Sugiura, K. (eds.) RoboCup 2014. LNCS, vol. 8992, pp. 220–232. Springer, Heidelberg (2015)

Robots that Refuse to Admit Losing – A Case Study in Game Playing Using Self-Serving Bias in the Humanoid Robot MARC

Mriganka Biswas[✉] and John Murray

University of Lincoln, Lincoln, UK
mrbiswas@lincoln.ac.uk

Abstract. The research presented in this paper is part of a wider study investigating the role cognitive bias plays in developing long-term companionship between a robot and human. In this paper we discuss how the self-serving cognitive bias can play a role in robot-human interaction. One of the robots used in this study called MARC (See Fig. 1) was given a series of *self-serving trait* behaviours such as denying own faults for failures, blaming on others and bragging. Such fallible behaviours were compared to the robot's non-biased friendly behaviours. In the current paper, we present comparisons of two case studies using the self-serving bias and a non-biased algorithm. It is hoped that such humanlike fallible characteristics can help in developing a more natural and believable companionship between Robots and Humans. The results of the current experiments show that the participants initially warmed to the robot with the self-serving traits.

Keywords: Human-robot interaction · Cognitive bias in robot · Imperfect robot · Human-robot long-term companionship

1   Introduction

The study presented in this paper seeks to influence robot-human interaction and with the selected 'cognitive biases' to provide a more humanlike interaction. Existing robot interactions are mainly based on a set of well-ordered and structured rules, which can repeat regardless of the person or social situation (Parker 1996). This can tend to provide an unrealistic interaction, which might make it hard for humans to relate with the robot after a number of interactions. In the previous research studies (Biswas et al. 2014, 2015) *misattribution* and *empathy gap* biases were tested on some of our robots. These biases provided interesting results and elicited interesting responses from the participants, as such we proceeded to test further biases and imperfectness in robot behavior. In this paper we test the attributes of the *self-serving* bias with humanoid robot MARC (Multi-Actuated Robotic Companion, see Fig. 1).

People make internal attributions for desired outcomes and external attributions for undesired outcomes (Shepperd 2008). This means, people tends to credit themselves for a desired result but blames onto others for failure or undesired results – which is basic self-serving trait. In the current experiments we developed a 'self-serving biased

© Springer International Publishing Switzerland 2016
N. Kubota et al. (Eds.): ICIRA 2016, Part I, LNAI 9834, pp. 538–548, 2016.
DOI: 10.1007/978-3-319-43506-0_47

algorithm' for the MARC to interact with the participants in a game (rock-paper-scissors) playing scenario. In such interactions, the robot shows biased behaviours such as denial, blaming others for losing and brags for winning. We compare this biased algorithm with another baseline algorithm which is 'friendly' and without the self-serving bias's components. At the end, we compare the results from both interactions to find out the participant's preferences of likeness.

In this paper, the term 'companionship' is defined by the five stages of the interpersonal relationship (Levinger 1983). Such stages are acquaintance, build up, continuation, deterioration and the end. The main goal of the use of these biases is to make such 'continuation' stage longer and delay the deterioration and termination stages.

In the current experiments, three interactions were performed between the participant and the robot spanning a two-month time period in order to provide 'long-term' effects. In this paper we introduce a model demonstrating self-serving biased behaviours in MARC and how this influences the interactions with the participant. In the experiments, we compare participant's responses between the robot with a self-serving biased behaviour and the robot without the bias, showing the impact on long-term human-robot interaction caused by the cognitive bias.

2 The Model: Imperfect Robot Using Cognitive Bias

Social robots take their persona from humans, as robots become much more popular in society, there is a need to make them much more sociable. Current social robots are able to anthropomorphize and mimic human actions but their actions are limited and may not provide sufficient reasons for people to create and maintain social relationships (Baxter 2011). But in human interactions, people usually meet with others and are able to form different relationships. From that we raise a simple question, 'What happens in human-human interaction which is lacking in robot-human interaction that prevents a social relationship between the robot and human?'

Cognitive biases among individuals plays an important role to make human interactions unique, natural and human-like (Wilke 2012). In existing social robotics, the robots are imitating human's social queues for example: eye-gazing, talking and body movements etc. But it is the human's behavioural variety which includes faults, unintentional mistakes, task imperfections etc. which is absent in these social robots. Sometimes a robot's social behaviours lacks that of a human's common characteristics such as idiocy, humour and common mistakes. Many robots are able to present social behaviours in human-robot interactions but unable to show human-like cognitively imperfect behaviours. Or 'human-like behaviours' are presented in such manner that participants cannot relate themselves with the robot. Studies suggest that various cognitive biases have a reasonable amount of influences on human thinking process to make misjudgments, mistakes and fallible activities (Baron, 2007). Such misjudgments, mistakes and fallible activities creates an individual's social behavior, making it human-like and cognitively imperfect (Tomasello 1999). Bless et al. (2004) suggested that cognitive biases can influence a human's behaviours in both positive and negative ways. Biases have effects on individual's decision making, characteristics behaviours and social beliefs. Robots, on the other hand, currently lack human-like cognitive characteristics in their robot-human interaction and that might prevent the interaction becoming human-like. We therefore expect that such 'neutrality issue' in human-robot interaction can be solved by using humanlike cognitive imperfections (e.g. making mistakes, forgetfulness and fallible activities) in a robot's mode of interactions. The study presented in this paper describes a long-term experiment using a humanoid robot MARC and a self-serving cognitive bias. We hope that human-like cognitive imperfections can result in a model which will allow for a stronger preference towards the robot from the human participant.

There has been much research which shows the potential behind the idea of implementing human-like cognitive abilities in robots. Reeves and Nass have demonstrated with several experiments that users are naturally biased to ascribe certain personality traits to machines, PCs, and other types of equipment (Reeves 2000). There is similar research shown by Park et al. where the researchers developed emotions generation model based on personality and mood (PME model) (Park 2010). At the Samsung Research Lab, researchers developed TAME model which is based on traits, attitudes, mood and emotions 'intended to promote effective human-robot interaction by conveying the robot's affective state to user in easy-to-interpret manner' (Moshkina 2009).

The above studies show that researchers have used a variety of methods in human-robot interaction, such as emotion expression, mimicking human actions, anthropomorphic behaviours, assigning specific personality traits etc., but developing cognitive biases in the context presented in this paper is still a new area. Bless et al. (2004) suggested that cognitive biases can have a positive and negative affect on human behaviour and on an individual's decision making, characteristic behaviours and social constructs. Cognitive biases affect an individual's general communicative behaviour in a human-human interaction and this makes the interaction human-like and natural. Robots in the other hand, lack this cognitive characteristic in interaction and this might prevent the interaction from becoming human-like. Our previous studies (Biswas M, 14, 15) show that biases such as misattribution and empathy gap can be helpful to develop long-term human-robot interactions. Inspired by the previous research we carry out the research using much complex biases such as self-serving effects.

3   Methodology & Experiments

The experiments presented in this paper show how a robot with a self-serving cognitive bias can influence long-term robot-human interaction. In the experiments the robot plays the popular 'paper-rock-scissors' (roshambo) game with the participants and expresses different behaviours for the biased and unbiased (baseline) algorithms.

3.1   The Robot MARC

A 3D printed humanoid robot called MARC has been used for the experiments. MARC was inspired by the open project InMoov (2015). The reason behind the use of a humanoid robot in these experiments as opposed to our previous work (Biswas M, 14,15) is that as research suggests, a human-like body in a robot helps users to understand the robot's gestures intuitively (Kanda 2005). MARC can move its hands, arms and body, tilt its head and look around. Figure 1 shows the robot MARC and, an experiment situation where MARC interacting with the participant.

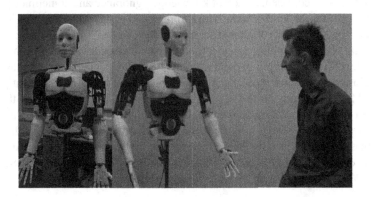

Fig. 1. MARC the humanoid robot, and participant interacting with MARC

3.2 Selection of Cognitive Bias

As discussed earlier, in the current experiment self-serving bias was developed in the robot MARC. The robot exhibited its biased behaviours whilst playing roshambo with participants. The self-serving bias is important for an individual to maintain and enhance their self-esteem, but in the process an individual tends to ascribe success to their own

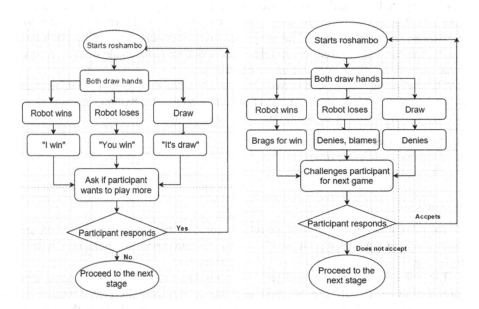

Fig. 2. The game playing algorithm for both algorithms. Left side image represents the baseline, and the right side image represent the self-serving biased algorithm.

abilities and efforts, but fails to take into account external factors (Campbell and Sedikides 1999). Individual's motivational processes and cognitive processes influence the self-serving bias (Shepperd 2008). The general characteristics of such biased behaviours could be over confidence, lack of knowledge, ignorance and sometimes perusing. For example, in exams, students attribute earning good grades to themselves but blame poor grades on the teacher's poor teaching ability or other external causes. Studies suggests that self-serving bias can affect interpersonal closeness and relationships in a social context. When working in pairs to complete interdependent outcome tasks, relationally close pairs did not show a self-serving bias while relationally distant pairs did (Campbell 2000). A study on self-serving bias in relational context suggests this is due to the idea that close relationships place limits on an individual's self enhancement tendencies (Sedikides 1998). In our experiments, the robot personality with the self-serving bias algorithm, usually takes credit for winning games against the participants, but blames external causes for losing – which is basic self-serving bias effects. See Fig. 2.

3.3   Experiment Algorithms

The current experiments presented in this paper compare the responses of the participants to the robot's self-serving biased behaviours versus the behaviours from the unbiased robot. We call such 'non-biased' behaviour the baseline. The 'baseline' algorithm was developed without the effects of the self-serving cognitive bias. For example, in baseline behaviours, if the robot loss a game it simply says "You win" or "I lose", but if the self-serving bias effect triggered, then the robot should not acknowledge its own lose, and instead blames its loss on others. Therefore, in the self-serving algorithm, MARC tends to blame failure on the external factors and responses such as "I was not ready" or "You are cheating" are used. In our experiment, the self-serving biased robot even denies any acceptance of a draw in a game, and it blames this also on the participants and accuses them of cheating. Such differences in dialogues are made in all conversational parts of the interactions during the experiments.

The interaction was divided in five stages. Such stages were there for making clear differences between baseline and biased algorithms, so that the baseline algorithm can be compared with the biased algorithm. The five stages were:

  i.   Meet and greet the participant
 ii.   Explaining the game rules
iii.   Game playing
iv.   Game result
 v.   Farewell

The robot may need to explain the rules to some of the participants. Based on the participant's group and algorithm, there can be differences in dialogues in rule explaining. This stage is only for the 1st interactions.

The outcomes of a single hand game-play are three: robot wins, robot loses, and draw. Based on such outcomes the robot responds differently in different algorithms.

The 'game result' is a state where the robot calculates and declares the winner. However, MARC's dialogues would be different at this stage based on the particular algorithm in operation, i.e. biased or unbiased. For the self-serving algorithm, the robot will praise itself, brag about winning, but it blames others for losing. The robot motivates itself if it loses in all games of an interaction, and similarly, it influences its self-esteem if it wins all the game hands in an interaction.

The game hands were drawn in random order, therefore the outcomes could not be fixed. However, the experiments were designed to get the reactions from the participants in different situations of interactions. Therefore, the robot could lose in all games in all three interactions, or win it all, but finding out the preferences of participants to an algorithm is the goal.

The core differences between the baseline and self-serving algorithms are in the construction of the bias based conversations, so that robot's responses could be biased. As seen in Fig. 3, the baseline conversation structure is straightforward which starts with the robot saying something or asking a question, then participant's response and ends with another statement by the robot. In between two dialogues from the robot the participant can respond only one time. The 2nd dialogue from the robot usually comes as 'Okay', 'I understand' or a compliment, so that there is no open end for that particular conversation part, and the robot moves to the next dialogue. On the other hand, the biased dialogues are structured to take responses from the participant and to state the robot's own opinions. As discussed earlier, in the self-serving bias, the robot blames external causes for losing a game hand. In our case, such external causes included the robot stating it was not ready, the robot was 'looking somewhere else', or something went wrong with the robots control. If the participant doesn't agree with the robot, the robot tries to convince the participant and challenges to play again. In such cases, the robot sometimes blames the participants of cheating in the games. Figure 3 shows the differences in the two algorithms in each of the stages of the interactions.

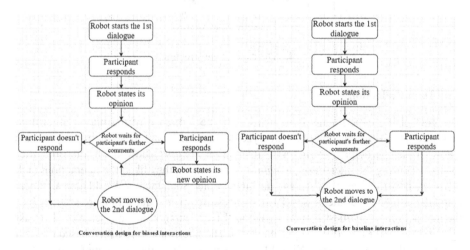

Fig. 3. Conversation designs for baseline and biased algorithms

4   Experimental Procedures

Participants were invited for three human-robot interactions by advertisements. 30 participants were selected to interact with any of the individual algorithms. Therefore, for each algorithm there are 15 participants. Although, participants in both groups were selected randomly, the male-female ratio was balanced in both groups. Participants were different ages, but age range was between 15 to 48.

The experiment consisted a total of three interactions for both algorithms maintaining at least a week interval between two interactions. The groups of 15 participants interacted in such interactions should tell us the effects of each individual algorithm in long period of time. Figure 4 shows the general experiment structure. All the interactions were on a one to one basis, where each participant interacted with MARC individually for at least 8 to 10 min.

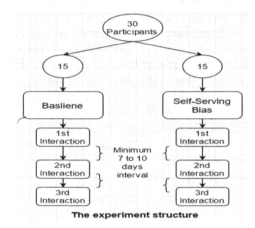

Fig. 4. The experiment structure

4.1   Data Collection

The goal of the experiments is to investigate the influence of the biased algorithm, therefore, we chose four factors to analyze the data with, and these included, *pleasure* – how pleased participants were for the interaction, *comfort* – how much comfort participants felt during the interaction, *likeability* – how much they liked the interactions and, *rapport* – how involved they were in the interactions. Such factors should help to understand the influences of the biases. Participants were given a questionnaire after each of the interactions. The questionnaires were in 'Likert' method using a scale of '1' (least agreeablenesses) to '7' (most agreeableness). The questions were asked based on such factors of pleasure (8 items) (Kim 2014), experience likability (5 items) (Hone et al. 2000), comfort (6 items) (Hassenzahl 2004) and rapport (20 items) (Multu 2006). At the end of the final experiment, we took an interview of each participants to know their experiences (winning, losing games etc.).

4.2 Statistical Analysis

A mixed (4 × 2) ANOVA was carried out on the dimensions (4) and algorithms (2). Figure 7 shows a descriptive table of each dimensions from all interactions. It shows that the Means of each dimensions are higher for the biased algorithms than the baseline. Among all the chosen factors, the self-serving biased algorithm scored higher than the baseline. There were stable positive increments in the ratings for each of the dimensions in all biased algorithms over the baseline algorithm. The Sig (p value) came out as < 0.05 which indicate the significance our collected data over large population. There was a statistically significant difference between means and therefore, we can reject the null hypothesis and accept the alternative hypothesis. Figure 5 shows plotting four dimensions of two algorithms. As clearly seen in the graph the participants rated much higher for biased interactions than the baseline interaction. Figure 6 shows the Average of Means ratings of the participants in the all 3 interactions. Figure 8 shows the overall Means plots from each algorithms in all the three experiments. The X-axis represents the algorithms and the Y-axis represents the marginal Means of two algorithms. As seen in the graph, the overall Means from baseline in much less than the self-serving. This graph can be called as the 'influence on participant' graph, as the graph represents the Mean ratings from all factors. The graph was generated in the repeated measure test in SPSS using post-Hoc analysis.

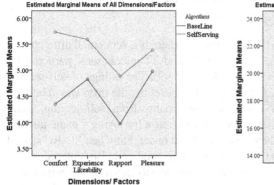 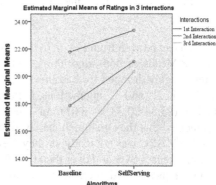

Fig. 5. The means of 4 dimensions in two algorithms

Fig. 6. Means for 3 interactions in both algorithms

5 Discussion

Overall statistical analysis shows very a positive influence of the self-serving bias. In Fig. 5 we can see that in all the factors the biased robot scored higher than the baseline. The differences between Means of four factors (Fig. 7) are: comfort 1.38, experience likeability 0.77, rapport 0.97 and pleasure 0.41. Self-serving bias scored high in all factors, but as seen in the pleasure factor the difference is much less than others. To measure the 'pleasure' dimension, we added eight items in the questionnaire some of these are: "Playing the game and having conversation with the robot is pleasurable to me", "Playing the game and having conversation with the robot is satisfying to me", "Playing the game and having conversation with the robot is enjoyable to me", "Playing the game and having conversation with the robot is entertaining to me" and similar. In their rating sheet, participants from the baseline interactions rated higher for first two questions (higher in 'pleased' and 'satisfaction') but lower for the other twos (lower in 'enjoyable' and 'entertainment'). On the other hand, the participants from self-serving interaction rated much higher for the last two questions (highly 'enjoyable' and 'entertainment'). In the comment section, some of the participants commented that it was very entertaining when the robot denied that it lost the game.

Algorithms		Means
Total Comfort Ratings Average	Baseline	4.35
	Self-Serving Bias	5.73
Total Experiences Likeability Ratings Average	Baseline	4.82
	Self-Serving Bias	5.59
Total Rapport Ratings Average	Baseline	3.97
	Self-Serving Bias	4.89
Total Pleasure Ratings Average	Baseline	4.97
	Self-Serving Bias	5.38

Fig. 7. Means of four dimensions

Fig. 8. Influence graph – based on the total ratings

As it can be seen from Fig. 6, in the 1st interaction there is a very small difference in average Means of both algorithms. But in the 2nd and 3rd interactions, participant's ratings hugely dropped for baseline (21.75-14.76 = 6.99). The ratings of self-serving dropped in 2nd and 3rd interactions, but compared to baseline, the drop was relatively smaller (23.34-20.33 = 3.01). In interview, participants from self-serving group commented that they genuinely believed the robot's excuse for losing a game initially was that MARC was not ready, or they drew their hand faster, but when MARC started making excuses over and over then they found it 'interesting' and also 'entertaining'.

In the 2nd and 3rd interactions, self-serving biased MARC accuses the participants of cheating whenever it loses in a game. In the participant's opinion they found it highly entertaining and liked it very much. To measure such bias affect, we added a 'comfort' factor which had six questions in the questionnaire. Surprisingly, participants in biased interactions did not find MARC's such behaviours as uncomfortable, uneasy or very difficult for them. As they commented, they were surprised to be accused of 'cheating' from a robot. Participants also mentioned, it is very common human behaviour not to accept losing, and the robot acting same like their friends. Moreover, they found that MARC's 'bragging' behaviours after winning a game was hilarious. However, the participants from baseline interactions did not find any such humanlike behaviours from MARC, and to them its behaviours were 'as common as a robot'. In their interviews and comments, they pointed out that playing game with a robot was enjoyable but it became less enjoyable after a few times. The participants found MARC's responses are 'stereotype', 'very mechanical' and 'common as robot-like' in the baseline.

Figure 8 shows an overall Means difference between baseline and self-serving algorithms. As it can be seen the baseline Mean is much smaller than the self-serving. From this graph it can be said that participants in biased interactions were more influenced by the robot's biased and imperfect behaviours, and rated higher than the baseline. However, to the baseline group the robot's behaviours were mechanical and as usual like a robot. In the 1st interactions, both groups participants enjoyed the game and rated high, but in the later interactions, MARC continued to show biased humanlike behaviours in biased interactions, so that the participants found it interesting and rated very similar as the first interaction. But, the robot with baseline algorithm failed to show such humanlike behaviours in later interactions, and so participants found their interactions as mechanical and, the ratings dropped higher than the biased interactions.

Therefore, it can be concluded that self-serving biased behaviours in robot were able to develop better interactions than a robot without such behaviours. All three biased interactions received more popularity and gained more positive responses from the participants. The participants liked the robot's behaviours in different situations in games, such as winning, losing and draw – that the robot brags about a win but blames on the participants or the external causes for losing and make draw, but despite of that the robot behaves very friendly – greets them, bid them farewell and requested for coming next times. Such kind of behaviours are very common in people, between close friends. In friendships, close friends could be very competitive in game playing and do not want to lose easily. Such types of behaviours are common human nature which we do and see in our daily life. When participants found out the same behaviours from our imperfect robot MARC they might found it easy to relate with it, and that might be the reason for biased and imperfect algorithm getting higher ratings than baseline. On the other hand, baseline MARC did not show any humanlike common behaviour rather than very generic impressions - which might be expected from a robot to our participants, and that could be the reason of the differences in ratings between biased and baseline algorithms. However, from the experiments and analysis of collected data it can be concluded that the humanlike biases and imperfectness in robot's interactive behaviours can enhance its abilities of companionship with its users over a robot without biases.

6   Conclusion

The experiment show that MARC with bias algorithms is more likely to keep the participants interested over time, therefore become more influential regarding interactions. Participants enjoyed playing rock-paper-scissors (roshambo) with the robot and expressed their feedback in favor of the biased algorithm. From the overall feedback it can be said that participants in self-serving group rated higher for their experiences which indicates that the self-serving biased algorithm made better interaction than baseline. We can see that the robot with baseline algorithms, developed a preliminary attachment with the participants, but robot with biased algorithms made the interactions more interesting to the participants. The participants felt more personable with the robot when the robot bragged and blames, denies of losing, accusing others and showed related human behaviours during interactions.

We realize that robot's imperfect biased behaviours also relies in part on the robot itself, i.e., the way MARC communicates. So questions may rise about the effects and relations between the robot's abilities and imperfect behaviours shown by it. To answer this question, previously experiments were done with the robots ERWIN and MyKeepon (Biswas 2015) where ERWIN was machine-like and MyKeepon is toy-like robot. But in both cases participants preferred the biased (misattribution and empathy Gap) interactions. In the current experiments self-serving bias was tested during a game playing interactions with humanoid robot MARC. In all cases, participants liked the interactions using selected biases more than the robot without bias. From these different experiments with different robots it can be said that the biases chosen in experiments influenced the participants to accept the biased interactions significantly, which helped to develop the attachment between the participants and robot which can help to make long-term robot-human companionship.

References

Baxter, P., et al.: Long-term human-robot interaction with young users. In: Proceedings of the IEEE/ACM HRI-2011, Lausanne (2011)

Bless, H., Fiedler, K., Strack, F.: Social Cognition: How Individuals Construct Social Reality. Psychology Press, Hove, New York (2004)

Biswas, M., Murray, J.: Effect of cognitive biases on human-robot interaction: A case study of a robot's misattribution. In: 2014 RO-MAN, pp. 1024–1029 (2014)

Campbell, W.K., Sedikides, C.: Self-threat magnifies the self-serving bias: A meta-analytic integration. Rev. Gen. Psychol. **3**, 23–43 (1999)

Campbell, W., et al.: Among friends? An examination of friendship and the self-serving bias. Br. J. Soc. Psychol. **39**(2), 229–239 (2000)

Park, J.W., et al.: Artificial emotion generation based on personality, mood and emotions for life-like facial expressions of robots. In: Forbrig, P., Paternó, F., Pejtersen, A.M. (eds.). IFIP, vol. 332, pp. 223–233Springer, Heidelberg (2010)

Moshkina, L., Arkin, R.C., Lee, J.K., Jung, H.: Time-varying affective response for humanoid robots. In: Kim, J.-H., Ge, S.S., Vadakkepat, P., Jesse, N., Al Manum, A., Puthusserypady K, S., Rückert, U., Sitte, J., Witkowski, U., Nakatsu, R., Braunl, T., Baltes, J., Anderson, J., Wong, C.-C., Verner, I., Ahlgren, D. (eds.) Progress in Robotics. CCIS, vol. 44, pp. 1–9. Springer, Heidelberg (2009)

Reeves, B., Clifford, N.: Perceptual Bandwidth. ACM, vol. 43(3), March 2000

Tomasello, M.: The human adaption of culture. Ann. Rev. Anthropol. **28**, 509–529. doi:10.1146/annurev.anthro.28.1.509

Shepperd, J., Malone, W., Sweeny, K.: Exploring causes of the self-serving bias. Soc. Pers. Psychol. Compass **2**(2), 895–908 (2008)

Wilke, A., Mata, R.: Cognitive bias. In: The Encyclopedia of Human Behavior, vol. 1, pp. 531–535. Academic Press (2012)

http://www.instructables.com/id/Make-Robot-voices-using-free-software/. Retrieved 10 December 2015

Cognitive Architecture for Adaptive Social Robotics

Seng-Beng Ho[✉]

Institute of High Performance Computing, A*STAR,
1 Fusionopolis Way #16-16 Connexis, Singapore 138632, Singapore
hosengbeng@gmail.com

Abstract. We describe a general adaptive cognitive architecture that is applicable to a wide variety of situations from video surveillance to robotics. A cognitive system must have a set of built-in motivations or primary goals to drive its behavior. In addition, the system must have an adequate model of the motivations and primary goals of humans to be able to interact effectively with humans. Rapid causal learning is used to learn about the causalities in the world, which in turn form the grounded knowledge necessary for problem solving to achieve these primary goals as well as secondary goals derived from them. A general adaptive cognitive system using this architecture has the ability to understand human motivation and emotion, and observe and learn from the environment and human behavior. This architecture is an ideal platform for adaptive social robotics.

1 Introduction

A general adaptive cognitive system must be able to observe and learn from the environment and other agents' behavior, including the behavior of humans [1]. The cognitive system has its own set of built-in motivations or primary goals to drive its behavior. In addition, to be able to interact effectively with humans, the system must have a model of human motivations and primary goals as well. To be an effective social robot, the backbone of cognitive processing consists primarily of executing problem solving to achieve these primary goals of the robot and the humans it is in social contact with. A series of secondary goals are created in the process. For the cognitive system to be maximally adaptive, the causal connections between certain actions taken and their attendant effects must be learned from observing and learning from the environment, including learning about the causal determinants of the behavior of other agents such as humans. These connections or causal rules can then be used to achieve the various necessary goals. Rapid causal learning [2–4] is used to learn these causalities.

Figure 1 depicts the basic components of the cognitive system [1]. The primary processing backbone is shown on the left as consisting of the application of causal rules or scripts (a series of actions [5, 6]) to perform actions to satisfy some primary or secondary goals. The goals compete for priorities of being considered. Perception and conception perform the service function of providing the system with the observed properties of the external world, allowing it to learn the causal rules.

© Springer International Publishing Switzerland 2016
N. Kubota et al. (Eds.): ICIRA 2016, Part I, LNAI 9834, pp. 549–562, 2016.
DOI: 10.1007/978-3-319-43506-0_48

Fig. 1. Cognitive architecture for a general adaptive intelligent system [1].

If the cognitive system involved is one that is meant to model an agent like that of a human, then the motivations and needs of a human must form the primary goals situated at the apex of the cognitive architecture of Fig. 1 [1]. If the agent involved is a robot, then a corresponding robot's version of its internal motivations and needs would form the primary goals. In Fig. 2(a), we show the Maslow hierarchy of needs that we adopt to encode the primary goals of a human [1, 7]. Maslow [7] proposes that the various types of human needs can be thought of as forming a hierarchy, with the physiological needs such as the needs for food, water, sex, etc. being at the lowest level. The next levels up would be the needs for safety, companionship, and so on. Only when the lower level needs are satisfied would the next higher levels be addressed. The Maslow hierarchy of needs is not without controversy in terms of whether the strict adherence to a level-by-level attempts to satisfy needs is supported by real-world data [8], but it suffices as a first approximation for the purpose of social robotics.

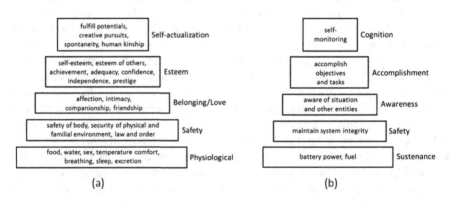

Fig. 2. Hierarchy of needs. (a) For humans [1, 7]. (b) For robots [9, 10].

For a robotic system, a needs hierarchy tends to be relatively simple. Quek [9, 10] used an artificial system's needs hierarchy shown in Fig. 2(b) to drive a robotic system to achieve survivability. There are 5 levels of needs that interact with each other, which mirrors the Maslow hierarchy of Fig. 2(a), but at each level, there are relatively fewer needs to be satisfied compared to that of the human. However, for a robot to be able to understand humans in social situations and interact with them effectively, it must have a model of the humans' needs hierarchy. Thus a social robot has a dual task of satisfying its own internal needs and that of humans, using a different set of primary goals in each case (See Sect. 3).

2   The Adaptive Social Robot

As mentioned above, whether it is for humans or robots, there is a set of primary goals to be achieved [1]. In the process of satisfying the primary goals, a series of secondary goals may need to be satisfied first. The actions needed to satisfy the goals, primary or secondary, depend on certain causalities between the actions taken and the consequences of those actions. In humans, the knowledge of some of these causalities is built-in for survival purposes and it is conceivable that the builder of a robot would build-in certain rules to ensure the survival of the robot as well. For example, humans are born with the ability to avoid certain noxious odors, and a robot builder may build-in the ability for a robot to seek electrical outlets to charge itself.

However, whether for humans or robots, having the ability to continue to learn about causalities will ensure continual survival of the agent involved [1]. For example, humans are not born to know that money can be used to buy food and other necessary survival items. This is something that has to be learned. A robot may not know that taking certain actions will create happiness or unhappiness in the humans it interacts with, and this has to be learned. There are always new things and new ways available in the environment to satisfy needs, and hence to achieve maximal adaptability, the ability to learn these causalities is of paramount importance.

In the following discussion, we first consider problem solving scripts used to satisfy needs followed by a discussion on the learning of scripts.

2.1   Problem Solving Script for Need Satisfaction

As mentioned above, the process of need satisfaction begins with a need to be satisfied or a primary goal to be achieved. In the process, a series of secondary goals may be created. Figure 3 shows an example using the satisfaction of the need to alleviate hunger or consume food to illustrate the process [1].

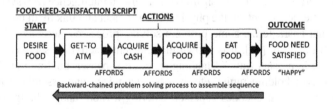

Fig. 3. A learned problem solving solution in the form of a FOOD-NEED-SATISFACTION SCRIPT to address the need for food or hunger alleviation [1].

One sequence of actions for this need to be satisfied is to first withdraw some cash from an ATM (Automatic Teller Machine) - an ATM affords the acquisition of cash. This is followed by using the cash to buy food – cash affords the acquisition of food. This is then followed by eating or ingestion of the food. The ingestion of the food causes the internal need of food consumption/hunger alleviation to be satisfied which is the achievement of a primary goal. The acquisition of cash and food, and the ingestion of food are all secondary goals. Thus what Fig. 3 illustrates is a chain of affordances and causalities, which represents a problem solving process encoded in a "script," that allows a primary goal to be achieved. The cognitive system achieves a state of "happiness" when a need is satisfied. The script is divided into three portions: the START, ACTIONS, and OUTCOME portions as shown [1, 3].

When a need is satisfied, the system will temporarily suspend the effort toward satisfying the need involved. The need, say for food, will arise again when the internal energy level of the system is low again. Exactly what problem solving process or what script is activated depends on the starting situation when the need arises. If food is available/accessible to the system, the system begins eating the food, which activates only a subpart of the script of Fig. 3. If food is not available but cash is, then the system would seek to acquire food with the cash. If cash is not available, then the system would seek an ATM, a bank, or to borrow from someone, and so on. There could be more than one way to acquire the item/carry out the action in the script of Fig. 3 in each step of the script, thus in general the script is in the form of an AND-OR graph [11–14].

When a need arises, if a script is available that can exploit the current state of the system and the environment, it will be immediately executed. Otherwise, a backward chained problem solving process can begin to assemble earlier learned, known individual steps (which are basically "causal rules" such as "if you pass some cash to a food-seller she will pass you some food") to solve the problem. If there is no rule available to achieve some of the steps, then learning would have to take place. This learning process can involve direct observation of the environment or being informed by another agent, say in the form of natural language communication. In the next section, we discuss the learning of problem solving scripts through observation.

2.2   Learning of Problem Solving Scripts

The state of the art in computer vision allows an artificial intelligence (AI) system to learn about the activities and events in the physical environment and link them up into causally linked units that characterize some problem solving processes (scripts) [12, 13]. In Si et al. [12] and Pei et al. [13], it is shown that elemental activities such as "standing at door," "walking in passageway," "standing at water dispenser, holding mug," "bending at water dispenser, holding mug," etc. can be recognized and assembled into, say, a sequence of actions that encodes a person entering a room and walking to the water dispenser to obtain water to satisfy her need to quench thirst. However, in these earlier works, there is no principled division of goals into primary and secondary goals, neither is there an attempt to characterize the needs of a cognitive system with something akin to the full complexity of the Maslow hierarchy. To address the issues faced by social robotics, there is also a need to formulate the interactions between scripts of problem solving executed by robots as well as humans.

In this paper, we build on the "learning through visual input" mechanisms that have been explored in these earlier papers [12, 13] and enhance the system with a computationally sophisticated characterization of human needs based on the Maslow hierarchy [1, 7] to address the issues of social robotics. Issues on script interaction are also important in characterizing social interactions between robots and humans which are not explored in the earlier papers. In this section, we describe the learning of scripts that will form the foundation for problem solving and social interaction in social robots.

Figure 4 shows an example of learning a script through visual observation. The script learned is "ACQUIRE-CASH-FROM-ATM." This is a sub-script of the FOOD-NEED-SATISFACTION SCRIPT of Fig. 3. The process begins with observing a person walking toward an ATM. This is followed by the person stopping next to the ATM, pressing the buttons on the ATM, the ATM dispensing some cash, the person picking up the cash, and the person walking away from the ATM. All these activities and the causal sequences in which they are embedded are recognizable by state of the art computer vision technologies [11–14]. With this sequence of activities identified, a script of "ACQUIRE-CASH-FROM-ATM" can be encoded as shown in Fig. 5. Ho and Liausvia [3] investigated the learning of script in a different domain using rapid causal learning [1, 2] which is also applicable to the current example.

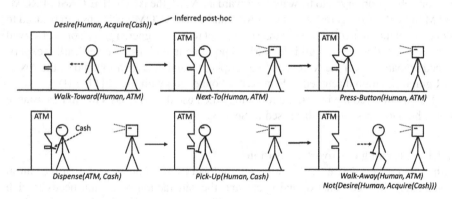

Fig. 4. The learning of ACQUIRE-CASH-FROM-ATM SCRIPT – observing activities.

ACQUIRE-CASH-FROM-ATM SCRIPT

START	ACTIONS	OUTCOME
Desire(Human, Acquire(Cash))	-Walk-Toward(Human, ATM) until Next-To(Human, ATM) -Press-Button(Human, ATM) → Dispense(ATM, Cash) -Pick-Up(Human, Cash) → Acquired(Human, Cash) -Walk-Away(Human, ATM)	Acquired(Human, Cash) Not(Desire(Human, Acquire(Cash))) Synchronic causal condition Diachronic causal condition Synchronic causal condition

Fig. 5. The ACQUIRE-CASH-FROM-ATM SCRIPT learned from the events in Fig. 4.

The statement in the START portion of the script, *"Desire(Human, Acquire(Cash))"* in Fig. 5 is inferred post-hoc from the activities, as is the *"Not(Desire(Human, Acquire(Cash))"* statement in the OUTCOME part. These constitute the covert part of the observation while the visually identifiable activities constitute the overt part.

There is a number of ways *Desire(Human, Acquire(Cash))* can be inferred from the overtly observed portion of the process. Between the boundary activities of *Walk-Toward(Human, ATM)* and *Walk-Away(Human, ATM)*, if a person acquires/holds on to something extra at the end of the activities compared to at the beginning, it is construed that the item is the desired item. Two other methods of inferring covert desires will be discussed in connection with Fig. 8.

In the script of Fig. 5, we also distinguish diachronic vs synchronic causal preconditions for the activities involved, as articulated in Ho [1, 2]. A diachronic causal precondition is the event that must be present and that precedes the event that is its causal consequence. Therefore, the buttons on the ATM must be pressed (and in a particular manner) before cash is dispensed. A synchronic condition is the state of the system that must be present before an action can be taken – e.g., the person must be next to the ATM before she can press the buttons.

Other than being used to solve problems – i.e., to arrive at a certain desired OUTCOME given a certain START state, a system can carry out the ACTIONS to achieve the outcome – a script can also be used to infer intention. For example, if a person/robot is observed to be walking toward an ATM, the ACQUIRE-CASH-FROM-ATM SCRIPT is triggered, and the START and/or OUTCOME portions can be used to infer the intention involved. More than one script may be triggered given some observed behavior, and the system will infer various possible intentions based on earlier experienced probabilities of the corresponding scripts' instantiation. (This would be the AND-OR graph situation discussed in Sect. 2.1 in connection with Fig. 3 [11–14]).

Having inferred the intent, the robot may then use that as a basis to offer assistance to the human. This will be discussed in Sect. 3.

2.2.1 Learning of Covert Causation

Other than covert desire that must be inferred as discussed above, some of the causal connections between an external event and the satisfaction of certain needs (which represent primary goals) may also be covert. Figure 6 shows the situations under which overt or covert causal connections between an external event and the satisfaction of an internal need may be derived.

Ho and Liausvia [15] outlines how a causal learning process allows the causation of an external event and the change of some internal state to be learned – basically through a temporal correlation process. For the physiological level needs, humans take advantage of this kind of learning to associate many external events and objects to the change in their internal states. For example, putting something in the mouth (an external event) that gives rise to a pleasant taste immediately and a subsequent alleviation of hunger (internal events) allows one to learn a new kind of food. Walking into a sunlit location and immediately experiencing a rise in body temperature allows one to learn the causation between sunlight and warmth. How would a robot learn these so that it can understand what pleases or displeases humans?

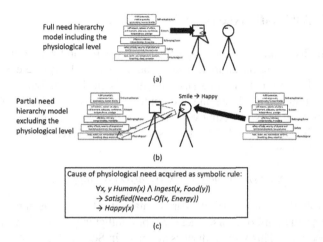

Fig. 6. Learning of causation between an external event and the satisfaction of an internal need. Robot is represented by the agent with the square head and human, round head. (a) The robot has a full built-in hierarchy of needs similar to that of the humans. The causations involved are overt. (b) The robot has a partial human need hierarchy excluding the physiological needs. Some of the causations involved are covert. (c) Learning from symbolic knowledge.

Figure 6(a) shows one scenario in which the causation between an external event and the satisfaction of an internal need of a human can be learned by a robot. This is if the full model of the Maslow hierarchy of needs for humans is built into the robot. This method faces one difficulty at the physiological level. To replicate the physiological responses of a human as a consequence of external events – food, water, temperature, etc. – would require complex biological sensors and processes that replicate the entire complexity of that of the human. The needs at the cognitive levels – the Safety level and above – would also be a challenge to model but possibly, some reasonably complex computational models can handle it.

Figure 6(b) shows another approach to learn the causation between external events and the satisfaction of internal needs of humans, especially for those at the physiological level. This is by observing the external signs of human behavior as a consequence of some external events. E.g., if the robot observes that when a human ingests something, it is followed by the emergence of a smile on the human's face, it can infer that the ingestion process must have satisfied some needs of the human involved. However, heuristics would be needed to map the external event to a particular need – e.g., one can assume that if it is an event that involves the ingestion of something, it is most likely satisfying a physiological need of food consumption, and not, say, the sleep need or some higher level cognitively oriented needs. Of course, at some levels of functioning of the robot, this exact mapping may not really matter, if the purpose of the robot is just to understand what makes a person happy and use that knowledge to good effect.

Figure 6(c) shows yet another way for a robot to acquire the knowledge of the causation between an external event and a need, which is simply to learn it in the form of symbolic commonsense knowledge.

3   Robot's Altruistic Needs

It is not usually listed among the needs in the Maslow's hierarchy but *altruism* is listed as a reward for a typical social animal, of which human is a kind, say by the neuroscientist Edmund Roll [16]. If robots are to assist humans in their daily activities, they need to understand what human needs are and attempt to help humans satisfy them.

In the same vein as Isaac Asimov's three laws of robotics, which specify the top level control priorities for a robot [17], we specify the top level control priorities of a social robot as follows:

$$\forall x, y \, Robot(x) \wedge Human(y) \wedge Happy(x) \wedge Happy(y) \tag{1}$$

That is, a robot's top level priorities are to make itself as well as humans optimally happy. In some situations, of course, there will be conflict in these two priorities and the robot must resolve it. This will be discussed below in Sect. 3.2. The rules concerning the needs or primary goals are as follows:

$$\forall x, y \, Robot(x) \wedge Satisfied(Need\text{-}Of(x, y)) \rightarrow Happy(x) \tag{2}$$

$$\forall x, y \, Human(x) \wedge Satisfied(Need\text{-}Of(x, y)) \rightarrow Happy(x) \tag{3}$$

The rules concerning secondary goals that involve acquiring some objects are as follows:

$$\forall x, y \, Robot(x) \wedge Desire(x, Acquire(y), t_1) \wedge Acquired(x, y, t_2 > t_1) \wedge Not\big(Desire\big(x, Acquire(y), t_2\big)\big) \rightarrow Happy(x) \tag{4}$$

$$\forall x, y \, Human(x) \wedge Desire(x, Acquire(y), t_1) \wedge Acquired(x, y, t_2 > t_1) \wedge Not\big(Desire\big(x, Acquire(y), t_2\big)\big) \rightarrow Happy(x) \tag{5}$$

The above rules together with the rules that encode the causations between external events and the satisfaction of internal (covert) needs could be used in a backward-chained process to generate scripts that can solve the problems of needs satisfaction.

3.1   Script Interaction

To satisfy the altruistic need, a human or robot needs to provide assistance to other humans or robots to achieve their goals. This scenario is portrayed in Fig. 7.

The scenario begins with a Human 1 pursuing a course of actions to satisfy one of her internal needs, in this case the food consumption need. An affordance/causal chain like that shown in Fig. 3 is established (from a relevant script if it is available, and if not, through a backward-chained problem solving process) to solve the problem. Now, suppose a robot or another Human 2 were to intervene in the process in a positive manner, such as to provide the items involved or assist in the actions sought after by Human 1 in the affordance chain, it will facilitate Human 1 in achieving her goal. Robot/Human 2 may concoct another script to provide the assistance – in Fig. 7 we show a "SUPPLY-FOOD SCRIPT" being invoked to serve this end. The provision of food obviates the

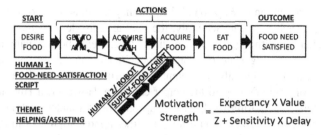

Fig. 7. Theme of HELPING/ASSISTING built from FOOD-NEED-SATISFACTION SCRIPT and SUPPLY-FOOD SCRIPT. See text for explanation of the Motivation Strength formula.

need for the earlier steps for Human 1– looking for an ATM and acquiring cash from it – and hence accelerates the solution execution of Human 1. It will also help conserve the energy expanded by Human 1 to achieve her goal. The interaction between the two scripts in Fig. 7 constitutes a "theme" of HELPING/ASSISTING.

This kind of script-script interaction to involve other agents' goals in the plans of an agent has been investigated by Abelson [18]. Others themes investigated by Abelson include "admiration," "devotion," "appreciation," "cooperation," "alienation," "betrayal," etc. These could be similarly applied to the area of social robotics.

In Fig. 7 we also show a formula derived from psychological research [19] for the computation of Motivation Strength. The formula shows that the strength of motivation is proportional to Expectancy - the expected probability of success of the task (in this case, when applied to the motivation of executing SUPPLY-FOOD SCRIPT say, it would be the probability of successfully supplying the food) - and the Value of the task (how much satisfaction the robot receives from the act), and is inversely proportional to the Sensitivity of the agent involved to the Delay in executing the task and the Delay itself. Z is a constant in case the Delay is negligible.

With this formula, if the robot is equally able to intervene in any of the steps, it can compute which intervention is more favored. For example, supplying food is better than supplying cash, as it will save Human 1 more effort, so Value is higher for supplying food. However, if the Expectancy of supplying food is lower (based on past experiences), then there will be a trade-off. Given a certain Sensitivity level, the task with the smaller Delay is favored. This Delay value can be derived from the script involved – the total time taken to execute the script is the sum of all the times taken for individual steps. When all the values are plugged into the Motivation Strength formula, a most favored intervention will emerge.

A partial representation of the scenario of Fig. 7 in logical form is as follows:

$$\forall x, y, z \, Human(x) \land Robot(y) \land Desire(x, Acquire(z)) \land Assist(y, Acquire(x, z)) \land Acquired(x, z) \rightarrow Happy(x) \land Happy(y)$$

$$(6)$$

The reason why this representation is partial is that there are other internal state changes that are not captured in the representation. Firstly, the helping agent – the robot or Human 2 – can gain a sense of social competence in providing the assistance. This touches on the Esteem level of the Maslow hierarchy of needs (Fig. 2(a)). This is a higher resolution characterization of *Happy(y)* – i.e., the happiness is derived from this particular primary goal achievement. Secondly, the helping episode can create a number of internal state changes with regards to needs in the agent being helped, namely Human 1. Depending on the starting state of Competence in Human 1, the consequences are different. If Human 1 is originally competent in acquiring food, the assistance is appreciated and a state of gratitude toward the helper can arise. However, if Human 1 is originally incompetent in acquiring food, depending on other variables such as her personality, the rendering of assistance could lead to either appreciation or a further sense of incompetence. For the case of a competent Human 1, too much offering of assistance can also create in her a sense of being harassed.

In the next sub-section we consider the issue of "competence motivation competition" which addresses the issue of conflict of priorities that we mentioned at the beginning of this section (Sect. 3).

3.2   Need Competition

In Fig. 8 we show a sequence of situations which is based on Fig. 3 which involves a person carrying out a sequence of actions that leads to the satisfaction of her food consumption need. A robot, observing some of the actions emitted by the human, infers her intentions at various stages of the problem solving process using scripts and offer possibly relevant acts of assistance. (The mechanisms of using scripts to infer intentions was discussed at the end of Sect. 2.2). More than one possible intention may be present at a given time and the robot would either offer a few alternative acts of assistance, or wait for more observation to take place to be more certain of her intention and hence offer a more certain act of assistance.

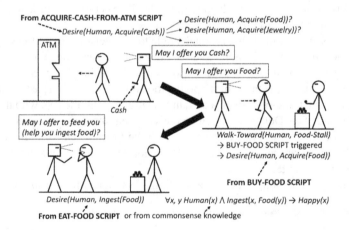

Fig. 8.   Robot offering assistance in various situations.

Other than the discussion in connection with Fig. 5 in which we described how covert desires can be inferred from actions in the scene for script construction, Fig. 8 also shows two other situations in which covert desires may be inferred. In the BUY-FOOD

SCRIPT, there is a change of the item possessed by the human from the beginning to the end of the script. A heuristic dictates that the final item possessed by the human is the desired item. In the EAT-FOOD SCRIPT, the processes discussed in connection with Fig. 6 are used to infer the covert desire.

However, as mentioned above in the previous section, over-offering of assistance can result in the intended recipient of the assistance feeling uncomfortable. The gain of a kind of competence in the offering agent – in this case, "social competence" – can mean the loss of competence in the recipient agent in terms of her ability to complete certain tasks. Figure 9 illustrates the process of "competence motivation competition."

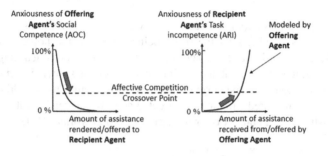

Fig. 9. Motivation competition.

The competition is formulated in terms of "anxiousness competition" in the same vein as discussed in Ho [1], in which an agent in a battle situation has to balance its need for charging up its energy on the one hand and dealing with the enemy on the other. The anxiousness of having a low energy level competes with the anxiousness of being threatened by the enemy and when one exceeds the other, the goal is switched to address the higher priority need first. In the case here, the competition is between the Anxiousness of the Assistance Offering Agent's Social Competence (AOC) and the Anxiousness of the Assistance Recipient Agent's Task Incompetence (ARI). In the left graph in Fig. 9, the AOC decreases as an exponential function of the amount of assistance offered/rendered to others – i.e., the agent satisfies its altruistic need. In the right graph, the ARI increases as an exponential function of the amount of assistance offered by/received from other agents – i.e., the agent feels incompetent if it is always offered help or is always needing and receiving help.

An altruistic social robot, in addition to satisfying its immediate altruistic need, must also take into consideration of the assistance recipient's need for feeling competent, and must have the model as shown in the right graph of Fig. 9 to guide its behavior. As shown in Fig. 9, as more assistance is offered/rendered, the robot's AOC is reduced, but the recipient's ARI is increased. When the ARI is higher than the AOC, which means an "Affective Competition Crossover Point" is reached [1], the robot must change goal and hold back its offering/rendering of assistance. Only then a balance can be struck to satisfy Eq. (1) – making both robot and human optimally happy.

Anxiousness enters the Motivation Strength formula of Fig. 7 in the form of the Sensitivity parameter. If a robot is anxious about completing a certain task, it will not tolerate a script that requires a long delay to execute.

The exact shape of both graphs in Fig. 9 can be changed as a result of experience. For example, given a certain amount of assistance offered or received, the amount of the corresponding anxiousness can be up or down-adjusted. This is the process of affective learning [1]. If the recipient agent finds that there is no reason to feel incompetent about receiving assistance as a result of some other events – e.g., realizing that being helped actually results in some better outcomes in other aspects such as safety - then the entire graph will be down-shifted.

4   A More Complex Scenario

Figure 10 depicts a more complex scenario in which a social robot assists a person at home. The process begins with the robot being placed in the home to observe the behavior of the human in her daily activities. After a rapid causal learning process as discussed above, it learns a number of scripts that encode the knowledge of the human's desires and the actions that she normally takes to fulfil them. This should only require the human having repeated similar goal-directed activities a small number of times, as has been illustrated in a number of prior computer vision works [12, 13].

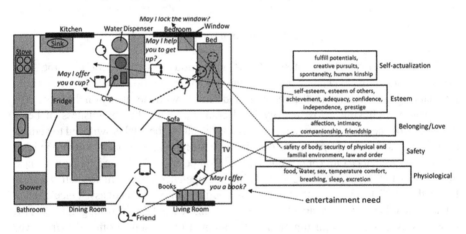

Fig. 10. A more complex scenario for a social robot – observing and learning how to assist a human at home to satisfy various levels of need. Square head is the robot, round head, the human.

In Fig. 10, it is shown that the robot offers the human a cup as the human seems to have the intention of obtaining water from the water dispenser but she does not have a cup at hand. This intention is inferred from her actions of placing herself next to the water dispenser and looking at the dispenser, which are the actions embedded in an ACQUIRE-WATER-FROM-DISPENSER script. The human normally satisfies her entertainment need (this need is normally not included in the Maslow hierarchy of needs [7] but should be added to it) by watching TV. This was observed by the robot in its first round of observation. However, suppose later some friends/relatives brought in a shelf of books and the robot then observed that books could also be used to alleviate boredom (satisfy the entertainment need) from the behavior of the human. Now, when it sees that the human looks bored (an expression recognizable by a computer vision system), it infers that entertainment need is to be satisfied and hence it offers a book to the human. Figure 10 also shows the companion need being observed (a friend visiting) and the safety need being addressed (offering to lock the window).

As with the case of Fig. 8, the human could begin to feel incompetent if assistance is always offered. Based on the competence motivation competition process of Fig. 9, the robot would modulate its assistance-offering behavior accordingly. The activities in Fig. 10 thus involve the first 4 levels of the Maslow hierarchy (Fig. 2) – physiological, safety, companionship, and esteem needs.

5 Conclusion

We have demonstrated how a problem solving oriented adaptive cognitive architecture installed with a Maslow-type hierarchy of needs endows a robot with the abilities to understand human needs and interact with humans effectively, offering assistance when necessary. Building on state of the art computer vision systems [11–14] and taking advantage of the rapid causal learning methods we and others investigated earlier [1–4], we show how scripts that encode sequences of relevant actions to achieve a goal (such as the satisfaction of a need) can be learned from observing and interacting with the environment, thus providing the mechanisms for robots to be adaptive. Scripts can also be used to infer intention, and that forms the basis for robot interaction with humans to achieve the overall goals of "happiness" for everyone, robots and humans alike. We also demonstrated the mechanisms of goal competition and how its resolution can lead to overall optimally happy states for robots and humans. Our architecture is thus useful for adaptive social robotics.

Acknowledgement. This research is supported by the Social Technologies + Programme funded by A*STAR Joint Council Office. The author wishes to thank Dr. Han Lin for providing information on the Motivation Strength formula of Fig. 7.

References

1. Ho, S.-B.: Principles of Noology: Toward a Theory and Science of Intelligence. Springer, Switzerland (in press, 2016)
2. Ho, S.-B.: On effective causal learning. In: Goertzel, B., Orseau, L., Snaider, J. (eds.) AGI 2014. LNCS, vol. 8598, pp. 43–52. Springer, Heidelberg (2014)
3. Ho, S.-B., Liausvia, F.: Rapid learning and problem solving. In: Proceedings of the IEEE Symposium Series on Computational Intelligence for Human-like Intelligence, Orlando, Florida, pp. 110–117. IEEE Press, Piscataway (2014)
4. Fire, A., Zhu, S.-C.: Learning perceptual causality from video. ACM Trans. Intell. Syst. Technol. **7**(2), 23 (2016)
5. Schank, R., Abelson, R.: Scripts, Plans, Goals and Understanding. Lawrence Erlbaum Associates, New Jersey (1977)

6. Regneri, M., Koller, A., and Pinkal, M.: Learning script knowledge with web experiments. In: Proceedings of the 48th Annual Meeting of the Association for Computational Linguistics, Uppsala, Sweden, pp. 979–988. Association for Computational Linguistics, Stroudsburg (2010)

7. Maslow, A.H.: Motivation and Personality. Harper & Row, New York (1954)

8. Reeve, J.: Understanding Motivation and Emotion. John Wiley & Sons, New Jersey (2009)

9. Quek, B.K.: Attaining operational survivability in an autonomous unmanned ground surveillance vehicle. In: Proceedings of the 32nd Annual Conference of the IEEE Industrial Electronics Society, Paris, France, pp. 3969–3974. IEEE Press, Piscataway (2006)

10. Quek, B.K.: A Survivability Framework for Autonomous Systems. Ph.D. thesis, National University of Singapore (2008)

11. Gupta, A., Srinivasan, P., Shi, J., Davis, L.S.: Understanding videos, constructing plots: learning a visually grounded storyline model from annotated videos. In: 2009 IEEE Conference on Computer Vision and Pattern Recognition, Miami, Florida, pp. 2012–2019. IEEE Press, Piscataway (2009)

12. Si, Z., Pei, M., Yao, B., Zhu, S.-C.: Unsupervised learning of event AND-OR grammar and semantics from video. In: Proceedings of the 13th International Conference on Computer Vision, Barcelona, Spain, pp. 41–48. IEEE Press, Piscataway (2011)

13. Pei, M., Jia, Y., Zhu, S.-C.: Parsing video events with goal inference and intent prediction. In: Proceedings of the 13th International Conference on Computer Vision, Barcelona, Spain, pp. 487–494. IEEE Press, Piscataway (2011)

14. Tu, K., Meng, M., Lee, M.W., Choe, T.E., Zhu, S.-C.: Joint video and text parsing for understanding events and answering queries. IEEE Multimedia $21(2)$, 42–70 (2014)

15. Ho, S.-B., Liausvia, F.: Knowledge representation, learning, and problem solving for general intelligence. In: Kühnberger, K.-U., Rudolph, S., Wang, P. (eds.) AGI 2013. LNCS, vol. 7999, pp. 60–69. Springer, Heidelberg (2013)

16. Rolls, E.: Memory, Attention, and Decision-Making. Oxford University Press, Oxford (2008)

17. Asimov, I.: I, Robot. Gnome Press, New York (1950)

18. Abelson, R.P.: The structure of belief systems. In: Schank, R.C., Colby, K.N. (eds.) Computer Models of Thought and Language. W.H. Freeman, San Francisso (1973)

19. Steel, P., Konig, C.J.: Integrating theories of motivation. Acad. Manag. Rev. $31(4)$, 889–913 (2006)

Accelerating Humanoid Robot Learning from Human Action Skills Using Context-Aware Middleware

Charles C. Phiri[✉], Zhaojie Ju[✉], and Honghai Liu

School of Computing, University of Portsmouth, Portsmouth, UK
{charles.phiri,zhaojie.ju,honghai.liu}@port.ac.uk

Abstract. In this paper we propose the creation of context-aware middleware to solve the challenge of integrating disparate incompatible systems involved in the teaching of human action skills to robots. Context-aware middleware provides the solution to retrofitting capabilities onto existing robots (agents) and bridges the technology differences between systems. The experimental results demonstrate a framework for handling situational and contextual data for robot Learning from Demonstration.

Keywords: Context-aware middleware · JSON-LD · SLAM

1 Introduction

Smart environments apply sensor technology to constantly monitor the spaces for variable controlling interests and conditions. A robot operating in this smart environment could adjust its belief-set and modify its decisions in response to the additional stimulus. While accessing the computable data is challenging, making online decisions in a timely manner is even more difficult. On the other hand, if the access link between the robot and the smart data source is compromised, the robot could be left exposed. In this paper, we propose a framework for deriving posterior beliefs from smart environments to enable adaptable context-sensitive behavior in robots. We explore how to add online adaptive behavior to enable robot Learning from Demonstration, LfD, allowing end-users to intuitively teach robots new tasks without programming knowledge.

Increased availability of ultra-low-power sensor devices is continually driving down the cost and power budget required to embed connected sensing technology within human environments [17,20] creating observable smart environments. On the other hand, increasingly robots are being introduced to share human spaces. The ability to take advantage of the shared capabilities with humans has made humanoid robots among the most common types of robots being introduced in human environments. Humanoid robots can access the same tools and facilities as humans and are considered expendable, which makes them useful in disaster recovery and situations or places which are not compatible with human

© Springer International Publishing Switzerland 2016
N. Kubota et al. (Eds.): ICIRA 2016, Part I, LNAI 9834, pp. 563–574, 2016.
DOI: 10.1007/978-3-319-43506-0_49

comfort and life [12]. Programming robots to exhibit emergent intelligent behavior in a stochastic environment requires specialist skills.

Unlike record and replay methodology, LfD, focuses on learning and generalization of observed demonstrations in order to perform the task [4]. By learning what it means to perform the task, the robot can make the correct decision at the right time; this is generally agreed as emergent intelligent behavior. Learning the meaning of a task involves the interpretation of the context in which the task is performed. Rich contextual data enables the robot to make decisions about situations and events at the right time improving the learning rate, acquisition of new tasks and their variations. The perceptual and physical equivalence problems are abstracted by providing the context data in a universally accessible format allowing the robot to make decisions based on its own embodiment and sensory perception above. This allows a symbiotic collaborative learning environment in which the observable state-space is richly mapped and computational localization data for the robot, objects and obstacles is highly available [5].

Conditional discrimination and disambiguation of context allows the context to be refocused and directed to the observable problem space. On the other hand, by classifying the robot's perception data as immediately required or not, the robot could distribute the computation of fringe details necessary for deriving posteriors to edge systems which can then provide the feedback as actionable data with the appropriate weighting. This allows the robot to form part of the interactive ecosystem with the smart spaces [5]. The architecture presented in this research does not hinder the application of the necessary security concerns. Although the experiments in this paper use the Aldebaran NAO H25 robot platform [2] the concepts explored are largely language and platform agnostic.

The robot uses its stereo cameras to estimate its location on a map that may be provided as prior information or estimated online. The motion planning (including collision avoidance) algorithms are determined using localization and mapping data estimated and calculated using onboard belief-set and reference frame. Using the smart environment, the robot location, landmarks, and map of the environment can be estimated using a global reference frame. The main challenge of motion planning can then be restated as a spartiotemporal correspondence problem. The robot is this work uses Active Online SLAM for motion planning. Figure 1 shows the Active Online SLAM algorithm summary based on [22,23].

A fixed Microsoft Kinect multisensor device is used to provide context data. Active Online implies that when Kinect data is not available, the robot may apply the same strategy using the onboard sensors with an increased computation burden and reduced field of vision. The mapping and localization data from the robot is also published allowing other robots and agents in the smart space to implement cooperative strategies with a high degree of complexity in reasonable timescales [12] based on the context data. Presence information plays a key part in determining the utility of the context data. The method proposed lets heterogeneous robots and smart environments to cooperate on tasks while permitting a high degree of autonomy for the robot [1]. We propose a context-aware

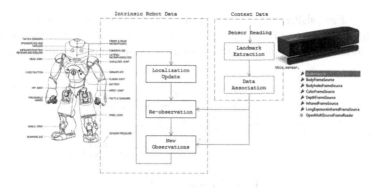

Fig. 1. Leveraging distributed common context

middleware that allows events and actionable inputs from external agents to be applied to the online control policy determination, including the reduction of the latency between context acquisition to the derivation of actionable control policy. We decompose LfD into a motion planning problem in this environment. Both the robot and human instructor are observable in a shared state space.

1.1   Related Work

The Beecham sector map shows segmentation and availability of data sources constituting smart spaces [20]. Mesh networks have been proposed by researchers [16] as a way of alleviating the communication challenge of collaborative agents. 5G networks are set to be the core M2M communication technology for IoT enabling immersive software defined networks [15]. OpenFlow and SDN offer methods of separating the control plane from the service delivery [18]. The proposed methodology in this research does not compete with these transport infrastructure-heavy technologies but rather adds a complementary cross-domain service for sharing contextual knowledge between heterogeneous controllers in a common configuration space.

Kim, Cho [11] propose a context-aware middleware service for networked robots based on Common Object Request Broker Architecture, CORBA, technology. Context between agents is shared using the distributed object paradigm. A system that recognizes situations automatically and uses various sensors to observe and adjust its current knowledge is termed context-aware [3]. Babu and Sivakumar [3] propose an adaptive and autonomous context-aware middleware for multi-agents with Type 2 Fuzzy rough context ontology. In their work, they provide a model that allows context collection, context processing and application reaction to significant context changes.

Smart spaces are a product of the paradigm of Ambient Intelligence [13]. Context-awareness breaks the isolated clusters of computable knowledge and distributes it among agents and controllers thereby allowing collaborative learning [21]. Li, Eckert [14] present a survey of context-aware middleware between 2009 and 2015. In their work they show the graph in Fig. 2 which breaks down the process of acquisition of context through to distribution to the participating agents [2]. Li, Eckert [14] also present a very good summary of the prevalent context modeling techniques, their advantages, disadvantages and applicability.

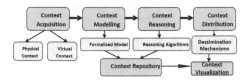

Fig. 2. Context-modeling [8]

Most recent and noteworthy in this domain is Googlés project Tango [8] which has an express desire to use computer vision to give devices the ability to understand their position relative to the world around them. Project Tango uses machine learning algorithms to understand the Motion Tracking, Area Learning, and Depth Perception. Area Learning uses an, Area Description File, ADF, to store the area map data and devices within it. The ADF can either be extracted during exploration or prepared beforehand and shared accordingly. At Google I/O 2015, Google in partnership with Infineon Technologies demonstrated a 60 GHz radar-based sensor that allows highly sensitive decomposition of human action skills [7]. Technology explored in the presentation would greatly improve accessibility of human action skills in humanoid Learning from Demonstration.

2   System Architecture and Overview of Concept

Figure 3 shows the deployment view of the multi-sensory system model implementing the concept with the Kinect as an example data source. Other suitable sensors can equally be used [20]. The middleware receives the data, shapes the data and distributes messages to all involved parties in the topic using the publisher-subscriber pattern. The messages are formulated as JSON-LD objects.

The topics provide a grouping concept for the context, a bounding condition. The bounding data define the boundaries of the search space according to the context defined. This may be used to limit the search space or even apply business rules, as may be applicable, enabling the robot to determine relevancy of both the source and the context data to the task in hand.

Looking at communication as a bounding condition, for example, the robot may, consider the quality of the signal based on the RSSI of the air interface to predict the expected QoS and from which it can then determine whether it should outsource computation of non-critical fringe data without breaking soft real time constraints of the task in hand. If the link is acceptable, a second level constraint may be based on availability of mapping and localization data from the Kinect Data source and the latency of sourcing this data, for instance. Based on the context data the robot probabilistically determines to employ a computationally less expensive $(O(n))$ Spike landmark detection algorithm using the onboard stereo cameras to check for obstacles in its immediate environment thereby reducing both the search space for optimal solutions and in turn decreasing the contributing power budget. Spike landmarks are recognized by detecting pixels that protrude from their surrounding pixels. The context data includes labeled originators so that the agents can selectively ignore the irrelevant sources [22, 23]

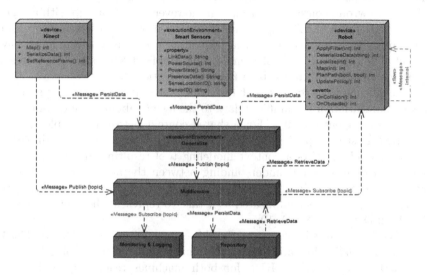

Fig. 3. Architecture deployment view

The context data is the generalized data from both the smart environment and the participating robots. The artifacts propagated to the robot or persisted to the repository are formatted to be immediately available for computation of the control policy. The system may also add bounding context data before the computation of the policy [22]. The proposed middleware is completely auditable allowing all contributors to be appraised and monitored via a non-intrusive monitoring channel Fig. 3. The robot can decide the levels of exploration and exploitation to employ depending on availability of the relevant computational model and the distribution of resources [22, 23].

2.1 Context-Aware Middleware

Context-awareness enables service discovery and adaptation of computing devices based on the availability of ambient intelligence. It is generally agreed that context-aware system should be responsive to multi-agents, covering a large number of devices, assisting a large number of system actors and serving a large number of purposes [3]. Context-aware computing is a style of computing in which situational and environmental information is used to anticipate immediate needs and proactively offer enriched, situation-aware and usable content, functions and experiences [19]. Context-aware systems can sense their physical environment, and adapt their behavior accordingly. Knowledge representation of features extracted from the environment involves prioritization of what is needed. An autonomous robot has to make these determinations in real time. By providing computable context the robot, given prioritized data within a bounded configuration-space, can find a more directed solution in the in a shorter period of time. The wrong context or estimation, however, could end up with a highly biased greedy robot [11,22,23]. In this study, by providing the repository, contextual memory is enhanced allowing exploitation of previously mapped environment. The repository is implemented using Hopfield Network which forms an associative memory controlled by a Lypunov Function [9].

The challenge with context is that it is recursive. It represents the situation and communication of the ambient data to the agents about the environment in which the said agent operates. For a shared context between multiple systems, the context must be expressed in such a way that it is understood by all parties. Ontology modeling offers a context modeling tool resulting in rich descriptive models providing a shared understanding between the agents [10,18–20]. The task starts by determining the ontology of the context. In this framework, the task only needs to be as complicated as it take to represent the context as W3C JSON Link Data, JSON-LD [10].

JSON-LD format is built on top of and is compatible with JSON (RFC7624). This allows serialization and deserialization of W3C RDF ontology data to JSON. JSON-LD simplifies RDF for both machines readability and human inspection [10]. The JSON-LD messages are published using ZeroMQs, ZMQ, extended publisher-subscriber pattern, XPUB-XSUB. XPUB-XSUB pattern allows multiple subscribers with very minimal overheads. The entire architecture takes advantage of the ZMQs lightweight scalability to share data across domains [6]. JSON-LD forms a well-defined technology neutral interface between participating applications removing the need of deep integration between components which may be heterogeneous. Messages can be decomposed and forwarded to the target without regard to implementation of either the source of the messages or the sink. As only messages are shared, the solution is platform, language and OS agnostic.

Unlike standard JSON messages, JSON-LD has defined primitives which allows a common understanding of the self-describing JSON document.

A critical fundamental system requirements guiding the framework sets that the computation data is delivered in soft real time and that the data is reliable, the system is highly observable and auditable. Figure 4 shows the component view of the architecture focusing on the communication pattern. This illustrations the realization of scalable middleware using ZMQ which has very low latency and handles concurrency and self-discovery. In this configuration, the added advantage is that subscribers and publishers may join or leave without affecting the other. ZMQ also proposes strategies for dealing with slow

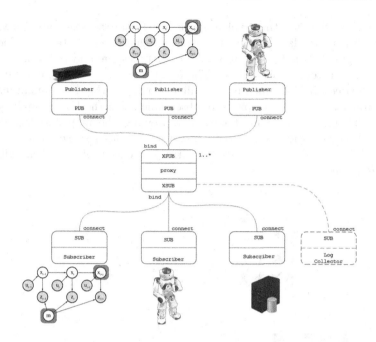

Fig. 4. Scaling with ZMQ

subscribers such as encouraging them to drop the connection. Using ZMQ, the framework can handle multiple publishers and subscribers segregating them using topics. Although Fig. 4 shows a homogeneous ZMQ solution, ZMQ is designed to allow distributed multi-protocol integration and deployment.

3   Preliminary Results Using SLAM

The use case employs the probabilistic EKF-SLAM algorithm to show applicability of the framework in LfD. Simultaneous Localization and Mapping, SLAM, is an example of a context problem that involves use of the best view of the world given the resources and information available to map stochastic environments. SLAM formally is concerned with the problem of building a map of an unknown environment by a mobile robot while at the same time navigating the environment using the map. SLAM consists of multiple parts; Landmark extraction, data association, state estimation, state update and landmark update. There are numerous solutions to the SLAM problem but in this paper we focus on the implementation of a software solution to aid sharing information to improve the observation of the state space to bias the (Active) Online SLAM problem towards a localization or mapping problem by taking advantage of collaborative learning [22, 23].

Figure 5 shows feature extraction and classification from both the robot and the human instructor. Feature extraction in LfD is a high dimension problem.

PCA is applied to reduce the problem space. The feature selection applies weighting to the feature to classify topical relevance. Relevance is contextual. The generalized output data is framed into a JSON-LD message and published via the middleware to listening subscribers. The robot uses the context information to adjust its trajectories obtained using Online Active SLAM. The human instructor, the robot and the state space are observable using the Kinect multi-sensor device which build. The Kinect currently provides the map and localization data based on the vision system only, however, this other sensors could equally be applied. The challenge is to solve the correspondence problem when switching reference frames.

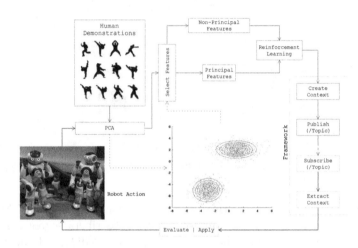

Fig. 5. Feature extraction using principal component analysis, PCA and reinforcement learning to build context for distribution.

The output of the work is the architecture of a message-based generic context-aware framework and how to implement it. For completeness, however, we briefly describe the notions behind SLAM to illustrate how it fits into the framework. *Given*

The Robot's Controls:	$u_{1:T} = \{u_1, u_2, u_3, , u_T\}$
Observations:	$z_{1:T} = \{z_1, z_2, z_3, , z_T\}$

Wanted

Map of the environment:	m
Path of the Robot:	$x_{1:T} = \{x_1, x_2, x_3, , x_T\}$

Full SLAM represents an estimate of the entire path and map expressed as shown in Eq. (1):

$$p(x_{1t}, m | z_{1t}, u_{1t}) \tag{1}$$

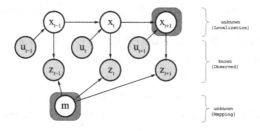

Fig. 6. Online SLAM

Online SLAM, on the other hand, only considers the most recent pose and map. Online SLAM is expressed in terms of Eq. (2) and has the graph model shown in Fig. 6:

$$p(x_t, m|z_{1t}, u_{1t}) = \int \int ... \int p(x_{1t}, m|z_{1t}, u_{1t})dx_1 dx_2...dx_{t-1} \qquad (2)$$

Recursive based filter is applied to the robots estimated perception data to adjust the map. Extended Kalman Filter, EKF, is then applied to improve the noisy sensory data to obtain a more accurate model. Thrun et al. [22,23] have shown how SLAM techniques scale to real world challenges including the application of the Victoria Park Data Set in vehicular self-navigation tasks [22]. In this study, the focus is on LfD. The pose estimates of the human instructor are serialized and made available to the robot to reproduce using its own joint coordinate frame. The objects representing the human and humanoid robot joint coordinates are serialized to the same JSON-LD structure opening up a future possibility to automate the scoring of the learning effort as the data can be accordingly scaled and the directly compared. Figure 7 shows Kinect joint and vector data objects. The data can be represented in the Kinect Depth or Camera Space.

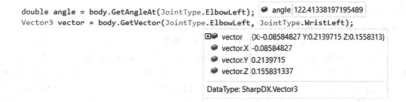

Fig. 7. Kinect joint angles and bone vector data using kinectex extensions.

Looking at NAO H25 Kinematics frame, the data in Table 1 shows the limits of the right ankle joint to avoid collisions with its own body frame. Taking the details of the first row (in degrees) in Table 1, the data is formulated into RDF

Table 1. NAO H25 right leg joints anti-collision limitation [2]

RAnklePitch (degrees)	RAnkleRoll Min	RAnkleRoll Max	RAnklePitch (radians)	RAnkleRoll Min	RAnkleRoll Max
−68.15	−4.3	2.86	−1.189442	−0.075049	0.049916
−48.13	−9.74	10.31	−0.840027	−0.169995	0.179943
−40.11	−12.61	22.8	−0.700051	−0.220086	0.397935
−25.78	−44.06	22.8	−0.449946	−0.768992	0.397935
5.73	−44.06	22.8	0.100007	−0.768992	0.397935
20.05	−31.54	22.8	0.349938	−0.550477	0.397935
52.87	−2.86	0	0.922755	−0.049916	0

```
{
    "@context": {
        "name": "urn:{e1fd1d97-00b4-4be7-b759-589d7edf71e8}/name",
        "description": "urn:{e1fd1d97-00b4-4be7-b759-589d7edf71e8}/description",
        "rAnklePitch": {
            "@id": "urn:{e1fd1d97-00b4-4be7-b759-589d7edf71e8}}/RAnklePitch",
            "@type": "xsd:float"
        },
        "rAnkleRoll": "urn:{e1fd1d97-00b4-4be7-b759-589d7edf71e8}}/RAnkleRoll",
        "min": {
            "@id": "urn:{e1fd1d97-00b4-4be7-b759-589d7edf71e8}/min",
            "@type": "xsd:float"
        },
        "max": {
            "@id": "urn:{e1fd1d97-00b4-4be7-b759-589d7edf71e8}}/max",
            "@type": "xsd:float"
        },
        "xsd": "http://www.w3.org/2001/XMLSchema#"
    },
    "name": "RightLeg",
    "description": "NAO H25 Right Leg Joints Anti-collision Limitation",
    "rAnklePitch": -68.15,
    "rAnkleRoll": {
        "min": -4.30,
        "max": 2.86
    }
}
```

Fig. 8. JSON-LD representation of NAO H25 right leg joint limitation

ontology using JSON-LD [10] as shown in Fig. 8. The object data represented as JSON-LD shows a directed graph of the linked objects. JSON-LD, being JSON, is self-descriptive.

Using the W3C JSON-LD 1.01 format, Fig. 8, the @context symbol encapsulates metadata carries forward knowledge that the consumer uses to understand the data types and purpose of the tuples. The resource name attached to the GUID (urn:e1fd1d97-00b4-4be7-b759-589d7edf71e8) allows bounding of the context to a particular NAO robot data set. The @type forward declares the data as W3C XML Schema float data type which, in turn, defines the boundaries of expected data. As the standard requires that the JSON-LD be valid JSON, it also follows that the data can be consumed without knowledge of either RDF or indeed JSON-LD [10], at the risk of losing most of the intrinsic benefits of the formats.

4   Conclusion

In this work, we shown the concept and architecture of a context-aware middleware to improve on the availability and accessibility of the simultaneous localization and mapping data for robot learning from demonstration, LfD. We have demonstrated a cooperative strategy using context data to derive more accurate posterior beliefs based leveraging the global view provided by smart environment to map the state space and locate self within the state space as well as locating obstacles and landmarks. For the Active Online SLAM implementation the robot is able to switch between onboard sensors and the environment sensors and can offer redundant data to improve it confidence. Associative memory allows the robot to choose between exploitative or exploratory schedule [22,23]. By using standard protocols, the knowledge-sharing is kept transparent to the transport mechanisms used such that only topical subscribers can take advantage of the published information regardless of the connectivity infrastructure [6,10]. Using low-latency transfers and caching, the robot can off-board some other non-critical computation without compromising on intelligent behavior. Context-aware middleware provides the solution to not only retrofitting capabilities onto existing robots (agents) but also bridges the technology debt within future and existing system architectures. Preliminary results using LfD show that these features help accelerate the learning effort for humanoid robots.

In future work, we would consider using BSON format for the messages over the wire. For a truly immersive experience, smart environment must be able to take advantage of the robots' closed world knowledge as a starting point. Future work will explore using FGMM for pattern recognition leveraging on previous work by the authors [24,25].

Acknowledgments. The authors would like to acknowledge support from DREAM project of EU FP7-ICT (grant no. 611391), Research Project of State Key Laboratory of Mechanical System and Vibration China (grant no. MSV201508), and National Natural Science Foundation of China (grant no. 51575412).

References

1. Ahmad, A., et al.: Cooperative robot localization and target tracking based on least squares minimization. In: 2013 IEEE International Conference on Robotics and Automation (ICRA). IEEE (2013)
2. Aldebaran. NAO Documentation (2016). http://doc.aldebaran.com/. Accessed 09 Mar 2016
3. Babu, A.K.A., Sivakumar, R.: Development of type 2 fuzzy rough ontology-based middleware for context processing in ambient smart environment. In: Mandal, D., Kar, R., Das, S., Panigrahi, B.K. (eds.) Intelligent Computing and Applications. Advances in Intelligent Systems and Computing, vol. 343, pp. 137–143. Springer, India (2015)
4. Billard, A., Grollman, D.: Robot learning by demonstration. Scholarpedia **8**(12), 24–38 (2013)

5. Cheng, L., et al.: Design and implementation of human-robot interactive demonstration system based on Kinect. In: 2012 24th Chinese Control and Decision Conference (CCDC). IEEE (2012)

6. Corporation, i. MQ - The Guide -MQ - The Guide (2014). http://zguide.zeromq.org/. Accessed 03 Apr 2016

7. Developer, G. Google I/O 2015 - A little badass. Beautiful. Tech and human. Work and love, ATAP (2015)

8. Google. Google's Project Tango (2016). https://www.google.com/intl/en_us/atap/project-tango/. Accessed 04 Apr 2016

9. Hopfield, J.J.: Hopfield network. Scholarpedia 2(5), 1977 (2007)

10. JSON-LD. JSON for Linking Data (2016). http://json-ld.org/. Accessed 03 Apr 2016

11. Kim, H., et al.: A middleware supporting context-aware services for network-based robots. In: IEEE Workshop on Advanced Robotics and its Social Impacts. IEEE (2005)

12. Kohlbrecher, S., et al.: Humanrobot teaming for rescue missions: team ViGIR's approach to the 2013 DARPA robotics challenge trials. J. Field Robot. 32(3), 352–377 (2015)

13. Li, R., et al.: Towards ROS based multi-robot architecture for ambient assisted living. In: 2013 IEEE International Conference on Systems, Man, and Cybernetics (SMC). IEEE (2013)

14. Li, X., et al.: Context aware middleware architectures: survey and challenges. Sensors 15(8), 20570–20607 (2015). (Basel)

15. Lin, B.-S.P., et al.: The roles of 5G mobile broadband in the development of IoT, big data, cloud and SDN. Commun. Netw. 8(01), 9 (2016)

16. Murray, D., et al.: An experimental comparison of routing protocols in multi hop ad hoc networks. In: 2010 Australasian Telecommunication Networks and Applications Conference (ATNAC). IEEE (2010)

17. Newmeyer, N.: Changing the future of cyber-situational awareness. Warfare 14, 32–41 (2015)

18. Opennetworking.org. Software-Defined Networking (SDN) Definition - Open Networking Foundation (2016). https://www.opennetworking.org. Accessed 01 Apr 2016

19. Perera, C., et al.: Context aware computing for the internet of things: a survey. IEEE Commun. Surv. tutorials 16(1), 414–454 (2014)

20. Research, B.: M2M world of connected services: Internet of Things (2011)

21. Rutishauser, S., et al.: Collaborative coverage using a swarm of networked miniature robots. Robot. Auton. Syst. 57(5), 517–525 (2009)

22. Stachniss, C.: Introduction to robot mapping. SLAM-Course (2014)

23. Thrun, S., et al.: Probabilistic Robotics. MIT press, Cambridge (2005)

24. Ju, Z., Liu, H.: Fuzzy Gaussian mixture models. Pattern Recogn. 45(3), 114–115 (2012)

25. Ju, Z., Li, J.,Liu, H.: An integrative framework for human hand gesture segmentation in RGB-D Data. IEEE Syst. J. (in press). doi:10.1109/JSYST.2015.2468231

Bio-inspired Robotics

Optimization of Throwing Motion by 2-DOF Variable Viscoelastic Joint Manipulator

Hiroki Tomori[1(✉)], Hikaru Ishihara[2], Takahiro Nagayama[2], and Taro Nakamura[2]

[1] Department of Mechanical Engineering, Graduate School of Science and Engineering,
Yamagata University, 4-3-16 Jonan, Yonezawa, Yamagata, Japan
tomori@yz.yamagata-u.ac.jp
[2] Department of Precision Mechanics, Faculty of Science and Engineering, Chuo University,
1-13-27 Kasuga, Bunkyo-ku, Tokyo, Japan
{h_ishihara,t_nagayama}@bio.mech.chuo-u.ac.jp,
nakamura@mech.chuo-u.ac.jp

Abstract. This paper focuses on control of variable viscoelasticity joint manipulator. Each joint consists of pneumatic rubber artificial muscle and magneto-rheological fluid. And the joint can generate instantaneous force by accumulating potential energy in artificial muscle. Using instantaneous force appropriately, robots can perform dynamic motion such as jumping and throwing like a human. These motions are expected to contribute to the efficient transport of objects and improve the robot's mobility. And also, elasticity and viscosity of joints are needed to control appropriately to achieve target task. Therefore, we proposed a method to control variable viscoelasticity of joint. We set throwing motion as a target task. And elasticity and viscosity are decided by simulation. In addition, simulation result is optimized by Particle Swarm Optimization (PSO) algorithm. Finally, we conducted throwing experiment to reproduce the simulation result. As a result, simulation result showed that elasticity and viscosity changed to accelerate end effector. However, experimental result showed deviations from simulation result because of model error.

Keywords: Manipulator · Pneumatic · Artificial muscle · MR fluid

1 Introduction

In modern society, various kinds of robot are working for human. Especially, robots for human assistance such as rehabilitation, medical care, nursing care, etc. has been required and studied. Almost of all these robots need to be safe for human while they are working nearby human. Therefore, soft actuators are focused for these robots because their flexibility can absorb impact of collision between robot and human. In addition, high compliance of a soft actuator has advantages in collaboration work with human.

Therefore, we focused on a pneumatic artificial muscle [1–4]. It is a soft actuator made from rubber and fiber, and actuated by air pressure. An advantage of this actuator is structural high compliance from rubber and air. By this characteristic, it can absorb impact of unprovided collision without requiring feedback control like a motor. However, pneumatic artificial muscle is hard to control because of their nonlinearity,

© Springer International Publishing Switzerland 2016
N. Kubota et al. (Eds.): ICIRA 2016, Part I, LNAI 9834, pp. 577–588, 2016.
DOI: 10.1007/978-3-319-43506-0_50

and its flexibility causes a vibration. To this problem, we have suggested approaches from a mechanical equilibrium model [5, 6] of the artificial muscle. By this study, artificial muscles can be controlled stably by feedforward controller with the model [6].

And we also suggested to use MR (Magnetorheological) fluid to suppress the vibration. MR fluid is functional fluid that can change own apparent viscosity quickly by magnetic field [7, 8]. And the reason for using MR fluid is that MR fluid has higher yield stress comparison with other functional fluid like ER (Electrorheological) fluid [9]. This advantage makes us easy to develop compact MR fluid device. Furthermore, it needs low voltage and current to work.

Then, we developed a variable viscoelastic joint manipulator using pneumatic artificial muscles and a MR fluid brake. The joint of the manipulator has variable viscoelasticity like a human. Therefore, we expects this manipulator to perform various tasks (position control, low vibration motion, impact absorption, motion with instantaneous force, etc.). Before this research, we have controlled arm position and suppressed vibration successfully [10]. And we also suggested method for generating instantaneous force using elasticity of this joint [11]. In previous studies on generating instantaneous force, Hondo developed a spring–motor coupling system [12], and Watari developed a MR cylinder [13]. However, the elastic coefficient of the actuator used was constant, and the instantaneous force and position could not be controlled independently. Furthermore, viscosity control was not used to control the output.

Furthermore, using instantaneous force appropriately, robots can perform dynamic motion such as jumping and throwing like a human. These motions are expected to contribute to the efficient transport of objects and improve the robot's mobility. And also, elasticity and viscosity of joints are needed to control appropriately to achieve target task. However, it is hard to calculate desired elasticity and viscosity from target task because of high complexity of motion increasing with increasing degree of freedom of manipulator.

Therefore, we proposed a method to control variable viscoelasticity of joint in this paper. We set throwing motion as a target task. And elasticity and viscosity are decided by simulation. In addition, simulation result is optimized by Particle Swarm Optimization (PSO) algorithm that is type of metaheuristics. Finally, we conducted throwing experiment to reproduce the simulation result.

2   Variable Viscoelastic Joint Manipulator

2.1   Straight-Fiber-Type Artificial Muscle

The straight-fiber-type artificial muscle developed in our laboratory has a tubular shape and is made of natural rubber latex. Figure 1 shows a schematic of the artificial muscle. The muscle structure includes a lengthwise carbon fiber layer. Consequently, when air pressure is applied to the muscle, it expands in the radial direction and contracts lengthwise. The contractile force is then used as an actuator. The artificial muscle is flexible and lightweight and provides high output. Furthermore, its elasticity changes with the applied air pressure.

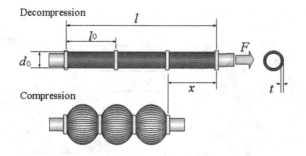

Fig. 1. Schematic of fiber-double-layer-type artificial muscle

2.2 MR Fluid Brake

To generate and control instantaneous force, variable viscosity and friction are required. Therefore, we focused on the MR fluid brake. By applying a magnetic field, the apparent viscosity of the MR fluid can be changed in several milliseconds. In addition, this device has a friction damper function, which can be used for the accumulation of elastic energy in pneumatic artificial muscles. In addition, it can be used as a variable viscosity damper to control the arm's vibration by controlling its apparent viscosity.

In this study, we used an MRB-2107-3 MR fluid brake (LORD Co., USA) (MR brake A) and a compact MR brake (ER tec Ltd., Japan) (MR brake B). Figure 2 illustrates MRB-2107-3 as basic mechanism of MR brake. The MR fluid in this device generates high yield stress by the magnetic field. Thus, this mechanism can halt the arm's rotation.

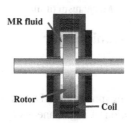

Fig. 2. Configuration of MR brake

2.3 DOF Variable Viscoelastic Joint Manipulator

Figures 3 and 4 shows the developed 2DOF artificial muscle manipulator. In each joint, two artificial muscles are arranged antagonistically through a pulley. Then the pulley transmits the contractive force of the artificial muscles to the rotation axis. By this mechanism, each joint can be independently controlled angle of joint and joint stiffness. And the MR brake is fixed to the first joint through a gear. As a result, it is possible to apply a brake to the rotation axis.

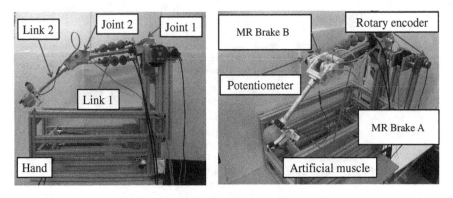

Fig. 3. 2-DOF variable viscoelasticity joint manipulator

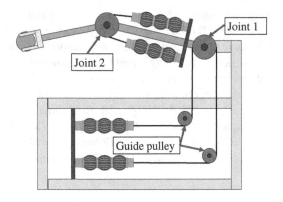

Fig. 4. Mechanism of manipulator

3   Throwing Motion by Manipulator

3.1   Throwing Motion

In this section, we describe throwing motion by 2-DOF variable viscoelastic joint manipulator. First, the method for generation of instantaneous force is shown in Fig. 5. In Fig. 5, (1): MR brake halt the joint. Then, (2): air pressure is applied to both artificial muscles. Here, air pressure is calculated from desired angle of joint and desired joint stiffness. In this phase, potential energy is accumulated in artificial muscle. Finally, (3): by releasing the brake, potential energy is converted into kinetic energy in short time. Using this method, potential energy accumulated in artificial muscle depends on angle of joint and joint stiffness.

Next, Fig. 6 shows Throwing motion by 2-DOF manipulator. In this study, joints start to move in order of joint 1 and joint 2. Then, the hand of manipulator releases ball. In addition, joint stiffness and viscosity of both joints can be changed in each phase (1), (2) and (3). Therefore, we should decide these values by simulation: release timing of joint 2 after joint 1, release timing of the hand, joint stiffness, viscosity of MR brake and desired angle of joint.

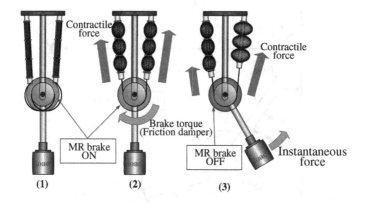

Fig. 5. Generation of instantaneous force

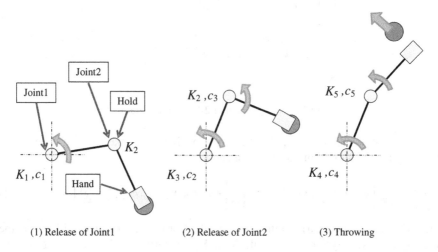

(1) Release of Joint1   (2) Release of Joint2   (3) Throwing

Fig. 6. Throwing by 2-DOF variable viscoelastic joint manipulator

3.2 Spring Model of 2-DOF Manipulator

Our study aims at the optimization of throwing motion using simulation. Therefore, it is preferable that the dynamic model requires only few calculations, thus speeding up the simulation. Therefore, we propose that the motion of the manipulator, driven by a method for generating instantaneous force, be regarded as the free oscillation or damped oscillation of the arm with a spring shown in Fig. 7. In this spring model, the artificial muscles are represented as springs that can change their natural lengths and spring constants. Thus, the spring joint can be controlled own equilibrium angle (desired angle) and joint stiffness K_j. And the joint can also change its viscosity coefficient c. In the figure, I and m represent inertia and load, respectively.

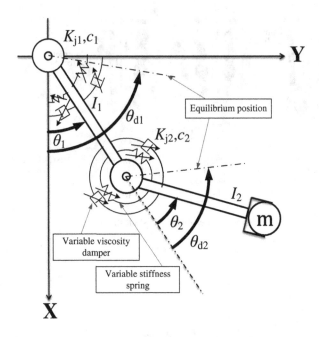

Fig. 7. Spring model of 2-DOF manipulator

By this approach, our manipulator can be modeled as the mass-spring system with damper. And equation of motion is presented as follows:

$$\mathbf{H}(\theta_1, \theta_2)\begin{bmatrix} \ddot{\theta}_1 \\ \ddot{\theta}_2 \end{bmatrix} + \mathbf{h}(\dot{\theta}_1, \dot{\theta}_2, \theta_1, \theta_2) + \mathbf{D}(E_1, E_2)\begin{bmatrix} \dot{\theta}_1 \\ \dot{\theta}_2 \end{bmatrix} + \mathbf{K}_j(P_{1j1}, P_{2j1}, P_{1j2}, P_{2j2})\begin{bmatrix} \theta_1 - \theta_{d1} \\ \theta_2 - \theta_{d2} \end{bmatrix} = 0 \qquad (1)$$

where $\mathbf{H}$ is the inertia matrix, $\mathbf{h}$ the matrix of the Coriolis and centrifugal force, $\mathbf{D}$ the matrix of the viscosity coefficient of the joint, $\mathbf{K}_j$ the matrix of joint stiffness, E_1 and E_2 the voltage applied to the MR brakes, P_{1j1} and P_{2j1} the air pressure of the artificial muscles at the first joint, and P_{1j2} and P_{2j2} the air pressure of the artificial muscles at the second joint.

4   Simulation for Throwing

4.1   Evaluation Function

To evaluate throwing motion by the manipulator, evaluation function was set. In this study, we search values expressed in Sect. 3.1 to obtain better (smaller) score from this evaluation function.

Here, Eq. (2) shows evaluation function. It has 6 evaluation items and 6 weight coefficients described in Table 1. I_{e1} is about throwing ability. I_{e2} and I_{e3} relate to throwing motion. Large change of viscoelasticity can be used to accelerate arm. On the other hand, if we throw an object gently, large change of viscoelasticity isn't needed.

Then, I_{e4} and I_{e5} are about suppression of long swing of arm. They are useful when the robot works at narrow space or around people. Finally, I_{e6} is ratio of potential energy accumulated in artificial muscles to kinetic energy of the ball. Even though the manipulator throws the ball far, I_{e6} will decrease if large potential energy is accumulated for throwing. By this evaluation function, throwing motion is evaluated not only from throwing distance but also from other points depending on situation. In this paper, we set same values as weight coefficients B_1 to B_6 on a trial basis.

$$I_e = B_1 I_{e1}^2 + B_2 I_{e2}^2 + B_3 I_{e3}^2 + B_4 I_{e4}^2 + B_5 I_{e5}^2 + B_6 I_{e6}^2 \tag{2}$$

Table 1. Items of evaluation function

I_{e1}	Reciprocal of throwing distance
I_{e2}	Amount of change (Joint stiffness)
I_{e3}	Amount of change (Viscosity coefficient)
I_{e4}	Overshoot of joint 1
I_{e5}	Overshoot of joint 2
I_{e6}	Reciprocal of energy efficiency
$B_1 \sim B_6$	Weight coefficient (=1)

4.2 Optimization of Throwing Motion by Simulation

In this section, we searched values on throwing by simulation. To search these values, we used Particle Swarm Optimization (PSO) algorithm. It can obtain practical solution instead of exact solution. However, it doesn't need differential calculation of model. Therefore, PSO algorithm can be applied to complex problem solution like our study.

Basically, optimization move ahead as follows: (1) Spring model (Sect. 3.2) is applied random initial values (Sect. 3.1) and simulates throwing. (2) Simulation result is evaluated by evaluation function (Sect. 4.1). (3) PSO algorithm decide next values from score of evaluation function. (4) Simulation is conducted and go to (2). After prescribed number of simulation, PSO algorithm shows values when best score is obtained.

Table 2. Simulation conditions

	Joint1	Joint2
Initial angle [deg]	0	−90
Desirable angle [deg]	160	90
Joint stiffness (Max.) [Nm/deg]	0.040	0.100
Joint stiffness (Min.) [Nm/deg]	0.022	0.040
Viscosity coefficient (Max.) [Nm.s/deg]	0.100	0.100
Viscosity coefficient (Min.) [Nm.s/deg]	0.000	0.000

Here, simulation condition is shown in Table 2. Viscoelasticity is searched range of this table. And manipulator does overhand throw. Then, weight of the ball is 20 g. Next, joint 1 is placed 1 m above from ground, and throwing distance is horizontal distance from the hand to the ball on the ground. Finally, number of simulation is 1500 times.

4.3   Simnulation Result

First, optimization results are shown in Table 3. And response of each joint in Fig. 8. Release timing of joint 2 in Table 3 is the angle of joint 1, and throwing timing is the angle of the joint 2. Then, motion of the arm calculated from Fig. 8 is shown in Fig. 9. The arm moves in a counterclockwise direction from fourth quadrant. In addition, response of joint stiffness and viscosity coefficient of joints are shown in Fig. 10.

Table 3. Optimization results

Release timing of joint 2 [deg]	129.4
Throwing timing [deg]	−26.1
Throwing distance [m]	1.56
Overshoot of joint 1 [deg]	0
Overshoot of joint 2 [deg]	5.39
Energy efficiency [%]	60.3

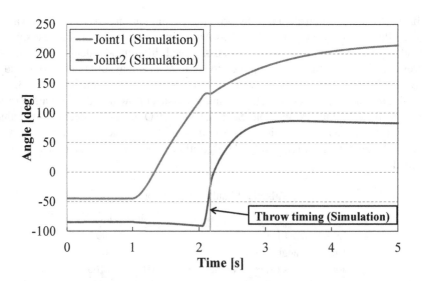

Fig. 8. Angles of the manipulator (Simulation result)

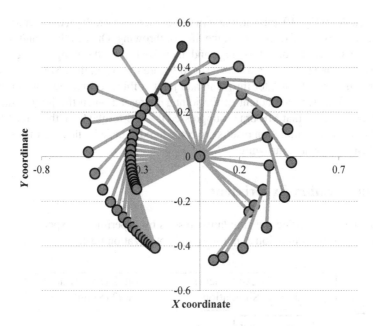

Fig. 9. Motion of the manipulator (Simulation result)

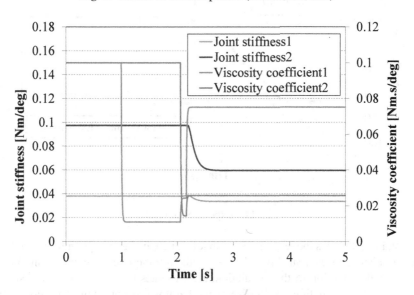

Fig. 10. Joint stiffness and viscosity coefficient (Simulation result)

From Table 3 and Fig. 8, we can see that overshoot of each joint is suppressed, although there is large steady state error on joint 1. Next, from Fig. 9, the manipulator kept the joint 2 bent right up until releasing the ball. Thereby, we guess that the moment of inertia of the manipulator was kept low, it contributed to increase rotational speed of joint 1. Then, from Fig. 10, joint stiffness in the initial phase was high in order to increase the potential energy, and decreased at the time of throwing. On the other hand, viscosity coefficient was low before throwing, and increased after throwing. We guess that viscoelasticity changed to increase arm speed before throwing, then MR brake worked for suppressing overshoot after throwing and joint stiffness decreased not to inhibit it. Furthermore, viscosity coefficient of the joint 1 increased when the joint 2 started to move. We considered that MR brake suppressed the anti-torque from the joint 1 to joint 2. For these reasons, simulation results are considered to be led reflecting the evaluation function, and reasonable throwing.

5   Demonstration Experiment

In this chapter, we verified the simulation results by experiment. Experimental conditions were based on simulation conditions and optimization results.

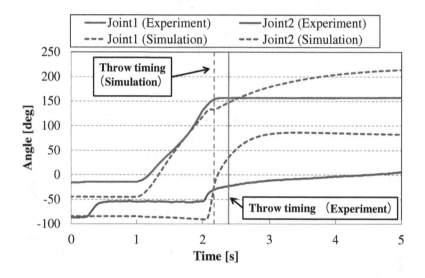

Fig. 11. Experimental result

First, Fig. 11 shows the response of each joint angle in the experimental result. Here, throwing distance was 1 m. From Fig. 11, although the experimental result showed similar trend of motion to the simulation, steady states of joint angles were smaller compared to the simulation results. We guess that it was occurred by wires connecting the artificial muscle and a pulley. Spring model is not considered a tensile by wire, the spring of joint is constant within the movable range. However, actual joints are pulled by the wire, thus wires tensed stronger at the edge of the movable range. Therefore, it affected the movement of the joint. Further, joint 2 started to move a little despite it was halt by MR brake. It is because a margin of brake torque halting the joint 2 was not enough. As a result, part of the potential energy was lost that had been accumulated in the joint 2. Since these differences between simulation and experiment were occurred, throwing distance in experiment was shortened than simulation. In addition, friction and viscosity on joints need to be considered. Although these of MR brakes has been considered, effects by bearings and gears are considerable. It is solved by more accurate assembling and simple mechanism that rotary shaft driving smoothly.

6   Conclusion

In this paper, we aimed to design the throwing operation by the 2-DOF variable viscoelastic joint manipulator. Then, release timing of joint 2 after joint 1, release timing of the hand, joint stiffness, viscosity of MR brake and desired angle of joint were set as the design parameters. And to determine these values, a spring model constructed. In addition, the evaluation function of throwing operation was produced that included throwing distance, amount of viscoelastic change, response of the joint and energy efficiency. This is because we expects that variable viscoelastic joint manipulator has high versatile for working in varied circumstances. Next, we optimized throwing motion by simulation with PSO algorithm. As a result, joint stiffness worked for increasing the potential energy, and viscosity coefficient tended to work for increasing velocity of hand and suppressing overshoot. It is considered that optimization results reflect the evaluation function, and throwing motion is reasonable. Finally we conducted experiment to reproduce the simulation result. The experimental result showed similar tendency to the simulation results. However, differences between these results occurred, and throwing distance of experiment was shorter than one of simulation. We guess that the spring model does not consider effects of wire enough, and MR brake needs a margin of brake torque for halting joint. These problems can be solved by improving the models and actual machine. And, parameters determined by the search is seems to be reasonable. Therefore, we conclude that motion design using the search of parameters by model and metaheuristics is useful for variable viscoelastic joint manipulator.

References

1. Daerden, F., Lefeber, D.: Pneumatic artificial muscles: actuators for robotics and automation. Eur. J. Mech. Environ. Eng. **47**(1), 10–21 (2002)
2. Nakamura, T., Saga, N., Yaegashi, K.: Development of pneumatic artificial muscle based on biomechanical characteristics. In: Proceedings of 2003 IEEE International Conference on Industrial Technology (ICIT 2003), pp. 729–734 (2003)
3. Nakamura, T.: Experimental comparisons between McKibben Type artificial muscles and Straight Fibers Type artificial muscles. In: Proceedings of SPIE International Conference on Smart Structures, Devices, and Systems III, p. 641424 (2006)
4. Tomori, H., Nakamura, T.: Theoretical comparison of McKibben-Type artificial muscle and novel Straight-Fiber-Type artificial muscle. Int. J. Autom. Technol. (IJAT) **5**(4), 544–550 (2011)
5. Ferraresi, C., Franco, W., Bertetto, A.M.: Flexible pneumatic actuators: a comparison between the McKibben and the Straight Fibres muscles. J. Robot. Mechatron. **13**(1), 56–63 (2001)

6. Nakamura, T., Shinohara, H.: Position and force control based on mathematical models of pneumatic artificial muscles reinforced by straight glass fibers. In: Proceedings of 2007 IEEE International Conference on Robotics and Automation (ICRA 2007), pp. 4361–4366 (2007)
7. Park, B.J., Park, C.W., Yang, S.W., Kim, H.M., Choi, H.J.: Core-shell typed polymer coated-carbonyl iron suspension and their magnetorheology. Phys.: Conf. Ser. (11th Conf. Electrorheological Fluids Magnetorheological Suspensions) **149**(1), 012078 (2009)
8. Li, W., Nakano, M., Tian, T., Totsuka, A., Sato, C.: Viscoelastic properties of MR shear thickening fluids. J. Fluid Sci. Technol. **9**(2), 172–173 (2014)
9. Liu, J.: Magnetorheological fluids: from basic physics to application. JSME Int. J. Ser. B Fluids Therm. Eng. (Spec. Issue Adv. Fluid Inf.) **45**(1), 55–60 (2002)
10. Nagai, S., Tomori, H., Midorikawa, Y., Nakamura, T.: The position and vibration control of the artificial muscle manipulator by variable viscosity coefficient using MR brake. In: 37th Annual Conference of IEEE Industrial Electronics Society, vol. 37, no. 1, pp. 307–312 (2011)
11. Tomori, H., Nagai, S., Majima, T., Nakamura, T.: Variable impedance control with an artificial muscle manipulator using instantaneous force and MR brake. In: 2013 IEEE/RSJ International Conference on Intelligent Robots and Systems (IROS), pp. 5396–5403 (2013)
12. Hondo, T., Mizuuchi, I.: Design and modal analysis of feedback excitation control system for vertical series elastic manipulator. In: 2013 IEEE/RSJ International Conference on Intelligent Robots and Systems (IROS), pp. 2888–2893 (2013)
13. Watari, E., Tsukagoshi, H., Kitagawa, A., Kitagawa, Y.: Control of magnetic brake cylinder and its application to pneumatically driven robots. In: JSME Robotics and Mechatronics Conference 2008, p. 1A1-B20 (2008)

Development of the Attachment for the Cable of Peristaltic Crawling Robot to Reduce Friction in Elbow Pipe

Ryutaro Ishikawa, Takeru Tomita[✉], Yasuyuki Yamada[✉], and Taro Nakamura[✉]

Department of Precision Mechanics, Faculty of Science and Engineering, Chuo University, 1-13-27 Kasuga, Bunkyo-ku, Tokyo 112-8551, Japan
{r_ishikawa,t_tomita}@bio.mech.chuo-u.ac.jp,
yamada156@2009.jukuin.keio.ac.jp,
nakamura@mech.chuo-u.ac.jp

Abstract. It is difficult to research small diameter pipes using conventional devices. In order to solve this challenge, this paper focuses on the locomotion of an earthworm that is capable of moving stably in a narrow space. A peristaltic crawling robot was developed which was capable of traveling 100 m through a 100A specification pipe using a peristaltic movement. However, especially in long distance inspection, and in elbow pipe, friction force between cable and pipes are too larger to ignore. Therefore, the friction force makes the robot impossible to inspect for long distance. And the friction force injure the cable and air tube equipped in the robot. In this paper, we developed an attachment to reduce the friction force. Using this device, we conducted experiment to measure friction in elbow pipe, and confirmed its effectiveness.

Keywords: Peristaltic crawling robot · Sewer pipe inspection · Artificial muscle · Reduce friction force

1 Introduction

Sewer pipes are useful in cities around the world, but recently aging sewer pipes (pipes which are over 50 years old) are causing road subsidence accidents [1]. To prevent this accidents, it is needed to research sewer pipes and conduct maintenances.

Presently, pressure feed pipes are often used as sewer pipes. There are two reasons for this. Primarily, pressure feed pipes are highly flexible and can twist in a horizontal or vertical direction, and therefore free layout irrespective of the terrain is possible. Secondarily, pressure feed pipes are inexpensive, durable, and require a short construction period. However, pressure pipes have the disadvantage of increased pipeline complexity.

There is a sewer pipe inspection method of visually confirming the damaged site by placing a camera into the sewer pipe. Operating a camera inside a sewer pipe is mainly achieved with the use of an endoscope or a camera-carrying robot [2–4]. The former has a camera as the endoscope and is pushed into the pipe. The latter is a robot that is

© Springer International Publishing Switzerland 2016
N. Kubota et al. (Eds.): ICIRA 2016, Part I, LNAI 9834, pp. 589–595, 2016.
DOI: 10.1007/978-3-319-43506-0_51

equipped with a camera which moves freely along the pipe. However, there are drawbacks in each of these inspection methods. In the first method, due to a complex conduit path constructed using pressure pipes, friction between the endoscope and pipe increases with camera feed distance thus limiting the pipe inspection distance to 10–30 m. The second method is limited to pipe diameters larger than 150 mm due to the size of the robot. In order to output a pulling force required to drive the power and camera cables necessary for driving the robot, a certain size of the robot is required to house the mechanism to drive the robot along the pipe with all the cables. From the above drawbacks, long-distance inspection inside small-diameter (diameter 150 mm or less) and complex pipes is challenging.

To solve this problem, focus was put on the peristaltic movement of an earthworm that is capable of stable locomotion in a narrow space. It is capable of stable loco-motion in a narrow space, because it has large ground surface area. And more, this movement demand narrow space for move, because earth worm expand and contract only back and forth. Using this movement, a peristaltic crawling robot was developed that is able to move through a 100A specification sewer pipe which has an inner diameter of 108 mm. To emulate the peristaltic crawling motion of an earthworm, the robot is equipped with pneumatically powered artificial muscles [5]. This robot aimed to achieve long-distance complex pipe inspections of more than 100 m, while traveling at a target speed of 50 mm/s to improve the efficiency of sewer pipe inspection.

Previously, the authors developed a robot that can operate in 100A pipe as shown in Fig. 4. The robot consists of a seven unit section, six joints, and a head unit equipped with a camera and cable which supply electricity and air pressure. It was confirmed through validation tests that locomotion in horizontal vertical pipes is possible. In horizontal pipes, the robot obtained a traveling speed of 51.2 mm/s. When we conduct experiment, we confirmed the phenomenon of the friction are larger in bent pipe. It is assumed that when cable are pulled, the cable are pushed in corner of inner elbow pipe by pulling force of the robot. When consider long distance inspection, the friction between cable and pipe are too large to ignore the friction. So we developed device attached to the cable of Peristaltic Crawling Robot to reduce friction in elbow pipe. Using this device, we conducted experiment to measure friction force in elbow pipe, and confirmed its effectiveness. The effectiveness of the new friction force reduce device was evaluated by conducting measure friction force experiments in elbow pipe.

2 Peristaltic Crawling Motion of the Earthworm

Figure 1 shows the structure of an earthworm's body. An earthworm is composed of between 110–200 body segments, where each section has a two layer muscle of lon-gitudinal and circular muscles. Their mode of locomotion is through extending and retracting the segments one by one.

Figure 2 demonstrates locomotion of the earthworm by peristaltic crawling. Earthworms deflate the body segments of the head at the beginning. While this con-traction is in turn propagated to the rear body segments, it extends the body segments of the head. At this time, the contracted body segments create friction with the ground,

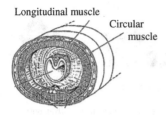

Fig. 1. The structure of an earthworm body

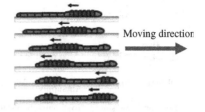

Fig. 2. Earthworm locomotion

thereby allowing the extended body segments to obtain a reaction force to extend forward. By repeating this contraction and expansion, earthworms can move forward.

3 Peristaltic Crawling Robot

3.1 Peristaltic Crawling Robot

The appearance of the peristaltic crawling robot is as shown in Fig. 3. The robot consists of 7 joints, 6 unit sections, and a search camera attached to the head section.

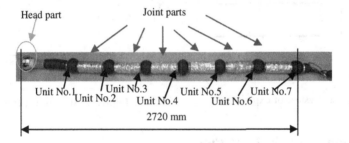

Fig. 3. Peristaltic crawling robot

3.2 Unit Sections

Figure 4 shows the structure of the unit which is used for the robot, and Fig. 6 shows the appearance of the unit section. The unit makes use of pneumatically powered artificial muscles. Therefore, the unit performs contraction and expansion by the supply and discharge of air. This characteristic enables the robot to perform movement through peristalsis. Each unit is equipped with a solenoid valve for supplying and discharging air directly to the unit. This prevents the delay of air transfer from the compressor when the robot travels distances of 100 m or more.

Figure 5 shows the appearance of the unit section in elongation and contraction states. In addition to the artificial muscle, this unit is composed of bellows and two flanges.

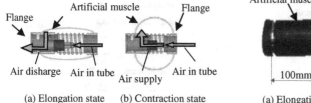
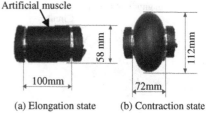

(a) Elongation state (b) Contraction state	(a) Elongation state (b) Contraction state
Fig. 4. Cross section of unit section	**Fig. 5.** Structure of unit section

3.3 Pulling Force of the Robot

Experiments were conducted to measure the pulling force of the robot. Figure 6 shows a system of this experiment, where a load cell was attached to the end of the robot, which was inserted into a straight pipe. The experiment to measure pulling force was recorded as shown in Fig. 7. From the result, maximum pulling force was 215 N.

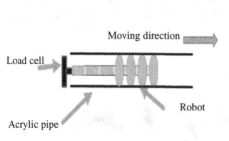
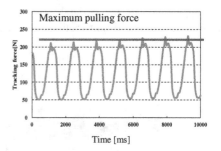

Fig. 6. System of experiment	**Fig. 7.** System of experiment

4 Friction Force in Bent Pipe

4.1 Measure Friction Force in Bent Pipe

We conduct experiment to measure friction force in bent pipe, and developed device to reduce friction force. As shown in Fig. 8, using bent pipe and robot`s cable and spring scale, we measured friction force. To reproduce the friction force in former part of the pipe, the load (attached 850 g road to gain friction force in straight acritic pipe as 10 N) are attached to the end of the cable. The result of the experiment is 100 N. This was concluded that the cable was pulled and pushed in the corner of inner elbow by contraction force and increased friction force in the point. It is difficult to ignore the friction force. Because in long distance driving and more elbow pipe, it is estimated that friction force is larger than pulling force of the robot and it is difficult to drive against friction force. And more, pressing force with cable and the corner can injure the cable.

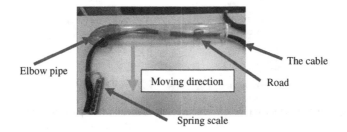

Fig. 8. The system of experiment

4.2   Reduce Friction Force in Elbow Pipe

To reduce the friction force in elbow pipe, we developed cable attachment as shown in Fig. 9(a) and (b). Figure 9(c) shows the cross section of the device. At first, a groove of device are caught in corner of bent pipe and left behind at inner corner of elbow pipe. Second, the cable can pass smoothly through the hole of the attachment along curved surface with smaller friction force.

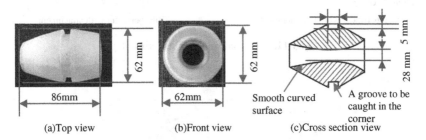

Fig. 9. The appearance of attachment

Using this attachment, we conducted experiment to confirm the usefulness of the attachment. Using elbow pipe and robot's cable and spring scale, we measured friction force. As shown in Fig. 10, a groove of attachment are caught in a corner of inner elbow pipe, and cable was able to pass the corner. The friction force with attachment is 73 N. That is concluded that the curved surface in the attachment left behind is smoothness, thus the cable was able to pass corner of inner elbow pipe with smaller friction force. Figure 10 shows the appearance of cable at the corner of inner bent pipe. In Fig. 11(a), the cable are pushed to the corner by pulling force, so the friction force are large. On the other hand, Fig. 11(b) shows state of cable with attachment in the corner. When we look at Fig. 11(b), groove of device are caught at corner of bent pipe and left behind at inner corner of elbow pipe. Therefore, the cable can pass smoothly through the hole of the attachment along curved surface with smaller friction force.

So we confirmed the usefulness of the attachment. Figure 12 shows the comparison of the friction force without attachment and with attachment.

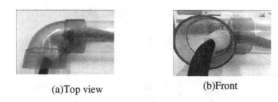

(a)Top view   (b)Front

Fig. 10. The appearance of cable and attachment when passing bent pipe

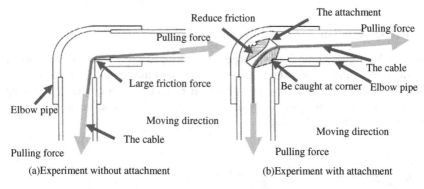

(a)Experiment without attachment   (b)Experiment with attachment

Fig. 11. The comparison of experiment

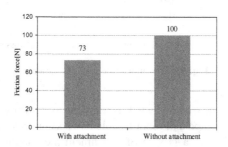

Fig. 12. The comparison of the result with attachment version and without attachment version

From Fig. 12, friction force without attachment is 100[N], while with attachment is 73[N]. We were able to reduce friction force by 23 %. It is concluded that a groove of attachment are caught at corner of bent pipe and left behind at corner of inner elbow pipe, and the curved surface in the device is smoothness, thus the cable was able to pass corner of inner elbow pipe with smaller friction force. So we confirmed effectiveness of the attachment by experience. However, actual pipe has more than one elbow pipe. So, in the future, we will develop the attachment that uses in more than one elbow pipe.

5   Conclusions

A new attachment was added to cable to reduce its friction force. We measured friction force in bent pipe.

And we confirmed effectiveness of new attachment by experience. The new attachment is recommended in order to reduce friction force in bent pipes to lengthen driving distance and protect cable.

In the future, research in this area should focus on the following:

- Target further reduce of friction force in bent pipe.
- Conduct driving experiment using attachment with peristaltic crawling robot.
- Develop attachment expand by air pressure by find elbow pipe passively to cope with pipe line that has more than one elbow pipe.

References

1. Uchida, H., Ishii, K.: Basic Research on Crack Detection for Sewer Pipe Inspection Robot Using Image Processing. In: Proceedings of the 2009 JSME Conference on Robotics and Mechanics, No. 09–4 (2009)
2. Li, P., Ma, S., Li, B., Wang, Y.: Development of an adaptive mobile robot for in-pipe inspection task. In: Proceedings of the IEEE International Conference on Mechatronics and Automation, pp. 3622–3627 (2007)
3. Okada, T., Sanemori, T.: MOGER: A Vehicle Study and Realization for In-pipe Inspection Tasks. IEEE J. Rob. Autom **RA-3**(6), 573–582 (1987)
4. Heidari, A.H., Mehrandezh, M., Paranjape, R., Najjaran, H.: Dynamic Analysis and Human Analogous Control of a Pipe Crawling Robot. In: Proceedings of the IEEE/RSJ International Conference on Intelligent Robots and Systems, pp. 733–740 (2009)
5. Nakamura, T., Shinohara, H.: Position and force control based on mathematical models of pneumatic artificial muscles reinforced by straight glass fibers. In: Proceedings of the IEEE International Conference on Robotics and Automation, pp. 4361–4366 (2007)

Multimodal Recurrent Neural Network (MRNN) Based Self Balancing System: Applied into Two-Wheeled Robot

Azhar Aulia Saputra[1,2(✉)], Indra Adji Sulistijono[2], and Naoyuki Kubota[1]

[1] Graduate School of System Design, Tokyo Metropolitan University,
6-6 Asahigaoka, Hino, Tokyo 191-0065, Japan
{azhar,kubota}@tmu.ac.jp
[2] Graduate School of Engineering Technology,
Politeknik Elektronika Negeri Surabaya, Kampus PENS,
Jalan Raya ITS, Sukolilo, Surabaya, East Java 60111, Indonesia
indra@pens.ac.id

Abstract. Biologically inspired control system is necessary to be increased. This paper proposed the new design of multimodal neural network inspired from human learning system which takes different action in different condition. The multimodal neural network consists of some recurrent neural networks (RNNs) those are separated into different condition. There is selector system that decides certain RNN system depending the current condition of the robot. In this paper, we implemented this system in pendulum mobile robot as the basic object of study. Several certain number of RNNs are implemented into certain different condition of tilt robot. RNN works alternately depending on the condition of robot. In order to prove the effectiveness of the proposed model, we simulated in the computer simulation Open Dynamic Engine (ODE) and compared with ordinary RNN. The proposed neural model successfully stabilize the applied robot (2-wheeled robot). This model is developed for implemented into humanoid balancing learning system as the final object of study.

Keywords: MRNN · RNN · Two-wheeled robot

1 Introduction

There has been a significant increase in the development of robotic stability systems. Most researchers implemented physical method for the stabilization. In our previous research, we also implemented physical approach. We applied inverted pendulum model and zero moment point in humanoid robot locomotion [1,2]. Other researcher implemented biological approach learning system for stability system or control system in any object of studies. In biological approach, recurrent neural networks are suitable for control system and have been applied in several cases. Lin et al. proposed a robust recurrent neural network

© Springer International Publishing Switzerland 2016
N. Kubota et al. (Eds.): ICIRA 2016, Part I, LNAI 9834, pp. 596–608, 2016.
DOI: 10.1007/978-3-319-43506-0_52

(RRNN) for a biaxial motion mechanism in CNC machine. The aim of their proposed method is for improving the motion tracking performance [3]. In 2008, Zhang et al. implemented RNN for model predictive control. Zhang et al. shows the effectiveness of their proposed method in simulation [4]. Other researchers implemented diagonal RNN that is known for dynamic mapping and for nonlinear dynamical systems. It also can be combined with conventional PID controller for tracking a straight line under the trapezoidal velocity planning [5].

In a specific case, especially in humanoid robot locomotion, there are many researcher also implemented RNN in humanoid robot that will be final object of study in our proposed research. In 2014, Tran et al. developed central pattern generation based the biped robot locomotion and applied RNN controlling the walking mechanism and walking stabilization in their applied robot [6]. In the same year, Jeong et al. applied RNN for generating the suitable behavior sequence for autonomous robot [7]. In 2015, Lin et al. implemented RNN with Elman model to achieve satisfactory control without performance degradation for humanoid biped robot locomotion with unknown uncertainties and faults [8]. RNN also can be combined with evolutionary algorithm for controlling the stability of humanoid robot [9].

In our previous research, we implemented single RNN Elman model with back propagation through time (BPPT) for stability system in humanoid robot locomotion [10]. In this proposed research, we will improve the stability level in robotics cases. This research are inspired from the human learning system. Humans give different responses in different condition. Therefore, we applied multimodal recurrent neural network composed by some RNNs system which one RNN represent one certain condition. In this proposed method we used RNN Elman model [11]. In modular neuron structure studies, Yamaguchi et al. proposed a new modular neural network architecture used simple feedback controller for training the neural networks in each module. The proposed research was applied for a mobile robot controller [12]. In 2000, modular learning system has been proposed by Farooq. Different with our proposed model which considers previous condition, he implemented ordinary artificial neural network in each module without considered previous condition [13]. Multimodal neural system has been proposed by several researcher. Kiros et al. proposed multimodal neural language model for image processing [14]. Other researchers also implemented multimodal neural learning system for image processing [15]. Modular Neural Network is also implemented for adaptive control in Arm robot manipulator. It's involving Support Vector Machines (SVM) and an adaptive unsupervised neural network. Karras also implemented it for forming the trajectory mapping and tracking control [16].

We are going to implement this proposed learning model in stability system on humanoid robot in order to build a self-adaptation system. Before implementing in a humanoid robot as the final object of study and in order to prove the effectiveness of the proposed learning model, we implemented this learning model in two-wheeled mobile robot as the first object of study because its simplicity. In experimental result, we will compare our proposed method with conventional

RNN which considers the previous condition to be processed. The contribution of this proposed model is that, we design multimodal learning system for stability which implement combined RNNs work alternately depending on the current condition. Therefore, by using this method robot can respond different action in different condition.

This paper is organized as follows. Section 2 explains the design of applied robot and how the applied robot can represent the humanoid robot. In Sect. 3, the applied RNN model are explained. Section 4 discusses the proposed MRNN model applied for stability system. In Sect. 5, we show the experimental result for proving the effectiveness of the proposed model.

2   Overview Robot Design

In this section, we explain the robot model that we used to prove the effectiveness of the proposed model. We used two-wheeled mobile robot that is inspired from inverted pendulum model that is depicted in Fig. 1. This mobile robot model is often implemented for Segway, which is used for helping human move from one place to other place. This robot design is the simple implementation of the proposed model that can represent the balancing system in humanoid robot. We design the applied robot in the Open Dynamic Engine simulation. We installed 2 kinds of sensor in this robot, tilt sensor and angular velocity sensor. Those are represented the angle value of robot and angle speed value of robot from vertical direction, respectively.

In the future, we would like to implement the proposed model in the humanoid biped robot. Humanoid biped robot design represented by the applied robot design can be seen in Fig. 2.

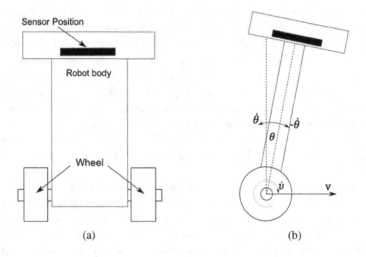

Fig. 1. (a) Robot design from front side. (b) Robot design from side

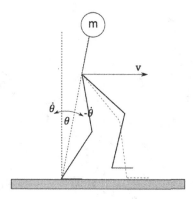

Fig. 2. Humanoid biped robot design representation of proposed robot design. The body speed is controlled by controlling the walking step behavior in further model

Our proposed robot design in this paper has two wheels as the response actuators of the stabilization system. The wheels movement results the speed v which is computed in Eq. (1). The speed value is depending the value of the output of certain yth RNN ($O_k^y(t)$) where it represents the speed acceleration of wheels. Each RNN module result different speed acceleration of wheels depending on the sensors condition. In Eq. 1, a_{min} and a_{max} are the minimum and maximum value of the speed acceleration, respectively. Where, θ is the tilt angle of the robot from vertical direction and $\dot{\theta}$ is the angular velocity or the robot. In humanoid biped robot design depicted in Fig. 2, movement speed v is resulted by step length of walking robot. However the additional modeling is required in order to model the walking behavior so that can control the speed of the robot walking. In the previous research, we have built the walking speed control [17] that will be combination with proposed research.

$$v(t) = v(t-1) + \big(O_k^y(t)\big(a_{max} - a_{min}\big) + a_{min}\big) \tag{1}$$

3   RNN Model

In this section, we explain how MRNN's system implemented for two-wheeled mobile robot. Many researcher have used RNN model for control system. In the previous research, we used single RNN in order to build the stability system in humanoid robot [10]. In this proposed model, MRNN composed more than one RNN system which works alternately depending on the certain condition in the certain time. RNNs process their previous condition in current process, therefore the previous condition will influence and become input in the feed forward process.

In this paper the RNN system is divided into four layers which are input layer, hidden layer, context layer, and output layer. The number of hidden and context layer are same since context layer represents the previous condition of

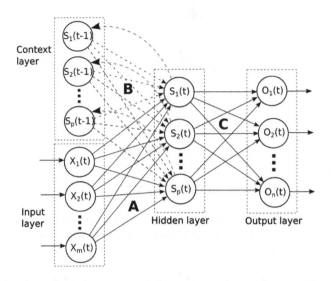

Fig. 3. Diagram of RNN model

hidden layer. The mathematical model of feed forward process of RNN can be seen in Eqs. (3) and (4). In Eqs. (3) and (4), $X_i(t)$ is the input value in certain time from the sensor feedback from the ith input neuron, $S'_j(t)$ and $S_j(t)$ are the inputs and outputs of the jth hidden neuron, and $O'_k(t)$ and $O_k(t)$ are the inputs and outputs of the kth output neuron. The number of neurons in the input layer, in the hidden layer, and in the output layer are denoted by m, p, and n, respectively. Parameter y shows the current RNN activated. The RNN model can be seen in Fig. 3.

$$e = \begin{cases} 0 & \text{if } \alpha_t < (\alpha_{t-1} - \beta) \\ \text{sgn}(d_k - g(O_k(t)))(\alpha_t - (\alpha_t - \beta)) & \text{otherwise} \end{cases} \quad (2)$$

In this case, the desired output is the condition where the robot maintains the angular velocity at zero value. δ_k is the error propagation in the output node and δ_j is the error propagation in the hidden node computed in Eqs. (5) and (6), e is the error value calculated in Eq. 2, where $\alpha = \text{abs}(d_k - g(O_k(t)))$, d_k is the desired output, and the output function $g(x)$ is the output of the sensor.

$$S^y_j(t) = f(S'_j(t)) = f\left(\sum_i^m X_i(t)A^y_{ij}(t) + \sum_h^p S_h(t-1)B^y_{hj}(t)\right) \quad (3)$$

$$O^y_k(t) = f(O'_k(t)) = f\left(\sum_j^p S_j(t)C^y_{jk}(t)\right) \quad (4)$$

$$\delta^y_k = (e_t)f'(O'_k(t)) \quad (5)$$

$$\delta_j^y = \sum_k^n \delta_k^y C_{jk}^y(t) f'\left(S_j^{'y}(t)\right) \tag{6}$$

$$\mathbf{C}^y(t+1) = \mathbf{C}^y(t) + \eta \mathbf{S}^y(t)\delta_k^y \tag{7}$$

$$\mathbf{B}^y(t+1) = \mathbf{B}^y(t) + \eta \mathbf{X}^y(t)\delta_j^y \tag{8}$$

$$\mathbf{A}^y(t+1) = \mathbf{A}^y(t) + \eta \mathbf{S}^y(t-1)\delta_j^y \tag{9}$$

In Eq. (3), the activation function for the hidden layer, $f(x)$ uses sigmoid function. In Eqs. (7), (8), and (9), the weight parameters of the neurons are presented by $\mathbf{A}, \mathbf{B},$ and $\mathbf{C}$ matrices acquired by a learning process. In BPTT, the error propagation is done recursively, where η is the learning rate of the weight of synapse between the motoric and the sensoric neuron. Those weight parameters can be dynamically regenerated depending on the environmental condition.

In this RNN model, 2 neurons are used in input layer as the tilt or angular velocity angle and the previous speed acceleration of the wheels. Hidden layer and context layer consist 10 neurons, while in output layer consist 1 neuron represent the speed acceleration of the robot.

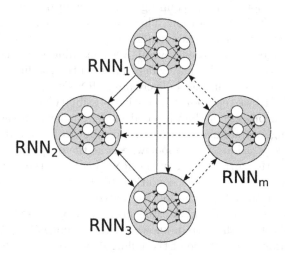

Fig. 4. Illustration model of MRNN

4   MRNN Model

In this section, new neuron based stabilization model is explained. In the previous research we used single model of learning system that assume in one certain condition represented all cases in the robot. Here, we used multimodal learning system that can assume different condition. We assume that robot will result different response in different condition. In the humanoid robot case, if the robot get a small disturbance such as the push or uneven terrain, then the robot only give hands response for protecting its stable condition. When robot get

---

Algorithm 1. Learning process in MRNN

Data: tilt angle or angular velocity of the robot
Result: angular velocity of the motor
initialization;
selection process;
response action → delay time sampling;
while *not at end of the process* **do**
    read current condition → evaluate previous state;
    if *Robot falling down* **then**
      | set the initial condition;
    else
      | selection process;
      | response action → delay time sampling;
    end
end

---

a high disturbance, then robot will response different action by positioning its footsteps. If the robot is often acquiring certain condition, then the robot has many experiences in that condition. Therefore, the robot has a good response when get experience in that certain condition.

Our MRNN model composed several RNN learning system. The learning system in this model work alternately depending on the condition of robot. The MRNN model can be seen in Fig. 4. There are m RNN's composed the structure of MRNN which arrows represent the moving possibility between each RNN.

The algorithm process of the proposed model is shown in the Algorithm 1. There are tilt angle or angular velocity of the robot as the input data, and the current speed acceleration of the robot wheels as the output data. First of all, the system select the current state of the RNN based on the current condition of the robot. Based on the current condition, the robot's action is activated and given delay time response. In the looping process, the robot read the current condition of the robot and evaluate the previous RNN based on the current condition. In this state, the system knows whether robot falling down or not. If the robot condition is in outside of the falling down condition then the system select the appropriate state of RNN. Based on the current condition, the system activate the response action, and time delay response is given. If the robot is falling down, then robot starts its initial condition.

The selection process is explained in Algorithm 2. Based on the current θ or $\dot{\theta}$ condition, system evaluate the weight parameters of y_{t-1}-th RNN. After that, the current id of RNN (y_t) is selected based on the Eq. 10. Where parameter m_{rnn} is the number of RNNs. There are 2 method for placing the range of each RNN depicted in Fig. 6, first placing the desire angle point inside the RNN range. Second, placing the desire angle point in the RNN barrier.

$$y_t = int\left(\frac{(\theta + \theta_{min})}{\theta_{max} - \theta_{min}} m_{rnn}\right) \tag{10}$$

Algorithm 2. Selection process

Selection process procedure
θ or $\dot{\theta}$ are the current condition of the robot;
updating the weight parameters of y_{t-1}-th RNN in previous process;
y_t = the state id of RNN depending on the θ & $\dot{\theta}$;
if *selection parameter is n* **then**
| calculating process in n-th RNN;
end
response action;

An example MRNN process can be seen in Fig. 5 where the desire point is 0.5. There are 5 RNN, each RNN has supported range where when the condition is inside the certain range of RNN then that RNN is activated. In this process illustration the first condition of the robot is 0.9 there for the selector state select RNN 1, after that RNN 1 send the output to the robot. After 1 time sampling, the result condition is 8.0. this value is used for evaluation process of RNN 1 and the RNN selection of next sampling. Based on the illustration example, the running process flows from RNN $1 \rightarrow 2 \rightarrow 2 \rightarrow 4 \rightarrow 3 \rightarrow 4 \rightarrow 3 \rightarrow 3$.

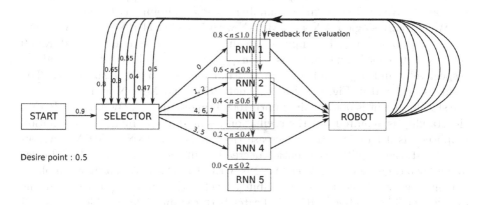

Fig. 5. Illustration of MRNN process

5   Experimental Result

In these experimental results, we would proof the proposed stabilization model into 2-wheeled mobile robot. There are 2 manipulated output parameter in these parameter, which are acceleration or the speed as the output parameter of the wheels. In each manipulation, we compared the single RNN and MRNN with different number of RNN model.

In the first experiment we analyze the influence of number of RNNs toward the stability with setting the desire point not in the RNN barrier which its

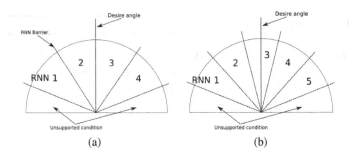

Fig. 6. The illustration RNN barrier (a) desire angle in the RNN barrier (b) desire angle inside the RNN range

example can be depicted in Fig. 6b. In this experiment, we used 3 neurons in input layer, 20 neurons in hidden layer, 20 neuron in context layer, and 1 neuron in output layer. Where the 3 input neurons are required from previous neuron output ($O_k^y(t-1)$), Angular velocity of the robot's tilt ($\dot{\theta}$), and the previous wheel speed $v(t)$. The scenario of this experiment is, the proposed system should stabilize the robot and keep the desired tilt angle of the robot. We used Open Dynamic Engine in order to applied and analyze the proposed system in the simulation. We set the maximum number of time sampling as 10000.

We compare single RNN with MRNN with 5 RNNs, 11 RNNs, and 31 RNNs ($m_{rnn} = \{1, 5, 11, 31\}$). The input value and output value are normalized from 0 to 1. We set the learning rate η as 0.5, a_{max} and a_{min} as $4\,\mathrm{rad}/s^2$ and $4\,\mathrm{rad}/s^2$, respectively, and θ_{max} and θ_{min} as 0.75 rad and -0.75 rad, respectively. The result is depicted in Fig. 7. By using single RNN and MRNN with m_{rnn} as 5, the stability point is not acquired until 10000 time sampling. By using 11 RNNs, the stability is acquired but the oscillation resulted is still high with oscillation amplitude is 0.4 radian. If we increase the number of RNN in MRNN system become 31, the amplitude oscillation decreased but the learning time took longer about 2000 time sampling, where MRNN with 11 RNNs took 500 time sampling. The number of RNNs effects the stability level of the robot application. In this experiment, big number of RNN indicates better stability level which has small oscillation amplitude.

Furthermore we tested the proposed stability model by putting desire point in the RNN barrier. We set the number of RNNs (m_{rnn}) as 2 and 4, and 40. The result is depicted in Fig. 8 showed better result than the previous experiment depicted in Fig. 7. In number of RNNs evaluation, there is no big different between small and big number of RNNs. Both of them are able to stabilize the proposed case in about 1000 time sampling. However, in order to learn multi desire angle, a large number of RNNs is required.

In order to prove the strength of stability system we tested the proposed system with certain number of RNNs by given the disturbance with certain power of push. In this experiment, we set the desire stability point in RNN barrier with 20 RNNs. After the robot reach the stable position without disturbance, the

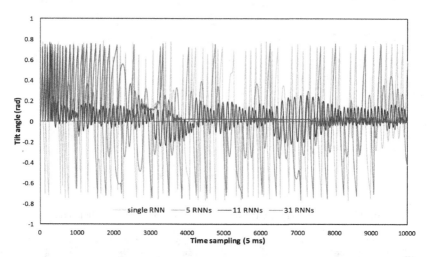

Fig. 7. The result of the learning process with different number of RNNs (desire point is inside RNN range)

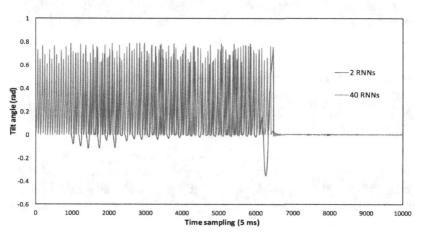

Fig. 8. The result of the learning process with different number of RNNs (desire point is in RNN barrier)

further teaching process was conducted by giving some disturbance until robot could stabilize from the disturbance. The result is depicted in Fig. 9 showed the oscillation of tilt body of robot in radian. After 15 disturbances were given or approximately 700 time sampling, the robot successfully stabilize toward the disturbance. When the robot got same disturbance like in $1600th$ and $2500th$ time sampling, the robot did not require to learn. It could stabilize because of the experience condition. If the robot frequently experience certain condition, the stabilization response is getting better. The sample of simulation of the proposed method are depicted in Fig. 10. Figure 10a shows when the robot was

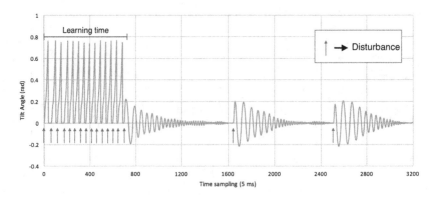

Fig. 9. The result of the learning process with given some disturbances

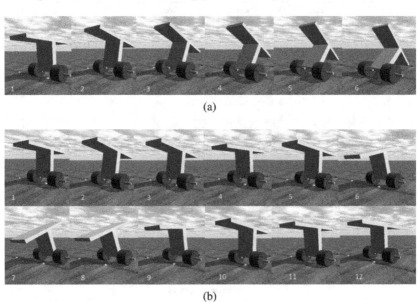

Fig. 10. Sample of simulations represent the experimental result in Fig. 7 with 11 RNNs (a) when robot was falling down ($100 < time\ sampling < 150$) (b) when robot was trying stabilize the disturbance ($300 < time\ sampling < 400$)

falling down because got some disturbance and Fig. 10b shows the robot was trying stabilize the condition from disturbance.

Although this proposed system can stabilize in certain condition, this robot could not control its movement and the direction. However, in order to control the movement and control its direction, a lot of conditions should be trained. Therefore, after training process is done, the robot can control the movement by changing the desire point of robot's tilt angle. This proposed research is the basic of the balancing system of the biped robot locomotion. In the next development, this proposed research will applied into the humanoid robot with some modifying component, which are the speed of wheel will be changed by the speed of walking step. Therefore, by applied this proposed system into humanoid robot, it is expected can develop bio-inspired robust locomotion of humanoid robot.

6   Conclusion

This paper presents the bio-inspired stability system model. MRNN consist several RNNs where the number of RNNs is dependent on the complexity cases. There is selector system that decides certain RNN system depending the current condition of the robot. If the robot often get the certain condition, then the RNN represent that certain condition has good respond for the robot. There are 2 methods for placing the range of each RNN, first, placing the desire angle point inside the RNN range and second, placing the desire angle point in the RNN barrier. There are influence of number of RNN in the first placing method. The large number of RNN indicates better stability level which has small oscillation amplitude. Single RNN was not enough to stabilize the proposed case. MRNN give better result than single RNN. In the second placing method there is no big different between small and big number of RNNs. However, big number of RNNs is required in order to learn multi-desire angle. The proposed model also successfully stabilized the robot from the disturbance in both placing methods. If the robot frequently experiences a certain condition, the stabilization response is getting better. Since the proposed stability model successfully applied in 2-wheeled robot, therefore this proposed model is able to be applied in biped robot stability.

References

1. Saputra, A.A., Sulistijono, I.A., Khalilullah, A.S., Takeda, T., Kubota, N.: Combining pose control and angular velocity control for motion balance of humanoid robot soccer EROS. In: IEEE Symposium on Robotic Intelligence in Informatically Structured Space, USA, pp. 16–21, December 2015
2. Saputra, A.A., Khalilullah, A.S., Sulistijono, I.A., Kubota, N.: Adaptive motion pattern generation on balancing of humanoid robot movement. In: Canadian Conference on Electrical and Computer Engineering, pp. 1479–1484, May 2015
3. Lin, F.-J., Shieh, P.-H., Shen, P.-H.: Robust recurrent-neural-network sliding-mode control for the x-y table of a CNC machine. IEEE Proc. Control Theor. Appl. **153**(1), 111–123 (2006)
4. Zhang, L., Quan, S., Xiang, K.: Recurrent neural network optimization for model predictive control. In: IEEE World Congress on Computational Intelligence, IEEE International Joint Conference on Neural Networks (IJCNN), pp. 751–757, June 2008
5. Li, Y., Wang, Y.: Trajectory tracking control of a redundantly actuated parallel robot using diagonal recurrent neural network. In: 5th International Conference on Natural Computation, vol. 2, pp. 292–296, August 2009

6. Tran, D.T., Koo, I.M., Lee, Y.H., Moon, H., Park, S., Koo, J.C., Choi, H.R.: Central pattern generator based reflexive control of quadruped walking robots using a recurrent neural network. Robot. Auton. Syst. **62**(10), 1497–1516 (2014)
7. Sungmoon, J., Yunjung, P., Rammohan, M., Jun, T., Minho, L.: Goal-oriented behavior sequence generation based on semantic commands using multiple timescales recurrent neural network with initial state correction. Neurocomputing **129**, 67–77 (2014)
8. Lin, C., Boldbaatar, E.: Fault accommodation control for a biped robot using a recurrent wavelet Elman neural network. Syst. J. IEEE **99**, 1–12 (2015)
9. Fukuda, T., Komata, Y., Arakawa, T.: Recurrent neural network with self-adaptive GAs for biped locomotion robot. In: International Conference on Neural Networks, vol. 3, pp. 1710–1715, June 1997
10. Saputra, A.A., Botzheim, J., Sulistijono, I.A., Kubota, N.: Biologically inspired control system for 3-D locomotion of a humanoid biped robot. IEEE Trans. Syst. Man Cybern. Syst. **46**, 898–911 (2015)
11. Cruse, H.: Neural Networks as Cybernetic Systems - 2nd and Revised Edition. Brains, Minds & Media, Bielefeld (2006)
12. Yamaguchi, S., Itakura, H.: A modular neural network for control of mobile robots. In: 6th International Conference on Neural Information Processing, vol. 2, pp. 661–666 (1999)
13. Azam, F.: Biologically inspired modular neural networks, Ph.D. dissertation, Virginia Polytechnic Institute and State University, Blacksburg, Virginia, May 2000
14. Kiros, R., Salakhutdinov, R., Zemel, R.: Multimodal neural language models. In: 31st International Conference on Machine Learning, pp. 595–603 (2014)
15. Wermter, S., Weber, C., Elshaw, M., Panchev, C., Erwin, H.R., Pulvermller, F.: Towards multimodal neural robot learning. Robot. Auton. Syst. **47**, 171–175 (2004)
16. Karras, D.A.: An improved modular neural network model for adaptive trajectory tracking control of robot manipulators. In: Köppen, M., Kasabov, N., Coghill, G. (eds.) ICONIP 2008, Part I. LNCS, vol. 5506, pp. 1063–1070. Springer, Heidelberg (2009)
17. Saputra, A.A., Sulistijono, I.A., Botzheim, J., Kubota, N.: Interconnection structure optimization for neural oscillator based biped robot locomotion. In: 2015 IEEE Symposium Series on Computational Intelligence, pp. 288–294, December 2015

A Prototype of a Laparoscope Holder Operated by a Surgeon Through Head and Jaw Movements

Shunji Moromugi[1(✉)], Tamotsu Kuroki[2], Tomohiko Adachi[3], Kotaro Oshima[4], Daiki Ito[1], Hyan Gi Kim[4], Amane Kitasato[2], and Shinichiro Ohno[3]

[1] Department of Electrical, Electric, and Communication Engineering,
Chuo University, Tokyo, Japan
moromugi@elect.chuo-u.ac.jp
[2] Department of Surgery, National Hospital Organization,
Nagasaki Medical Center, Nagasaki, Japan
[3] Graduate School of Biomedical Sciences, Nagasaki University, Nagasaki, Japan
[4] Graduate School of Science and Engineering, Chuo University, Tokyo, Japan

Abstract. A robotic device that allows a surgeon to operate a laparoscope by itself through commands based on the head and the jaw movements has been prototyped. This robotic device has a unique head-mounted interface to measure surgeon's head inclination angles in horizontal and vertical directions and detects the degree of contraction of temporal muscles that always acts under jaw movements by using sensors. Surgeons can easily operate the robotic device and control the area of scope view inside of the abdominal cavity through the head inclination angle and the voluntary effort of bite with back teeth during the surgical operations with both hands. The excellent operability of a prototype has been presented through an evaluation test under a simulated surgery task inside a training box. In addition, the high potential of the proposed system to make the surgeons possible to perform an extirpative surgery of cancer on gallbladder or liver under its own laparoscope operation through a test of an animal model.

Keywords: Robotic tool · Surgery · Laparoscope · Jaw and head movements

1   Introduction

The number of laparoscopic surgery is rapidly increasing because of its many benefits for patients such as minimal invasiveness and short-term recovery after the surgery. In Japan, at least two medical doctors are needed to conduct a laparoscopic surgery. One is a surgeon who perform surgical operations by using surgical instruments such as forceps and electric knives and the other is the so-called scopist who holds a scope inserted into patients abdominal cavity and adjusts its position so that the surgeon can perform the surgery operations under a good range of view to see forceps and target organs during the surgery. And the operations of the laparoscope is not allowed to nurses and other medical staffs other than medical doctors by law. Under this situation the laparoscopic surgery is not available at the area to where one medical doctor are allocated such as small islands and secluded areas in Japan.

© Springer International Publishing Switzerland 2016
N. Kubota et al. (Eds.): ICIRA 2016, Part I, LNAI 9834, pp. 609–618, 2016.
DOI: 10.1007/978-3-319-43506-0_53

Technological supports based on robotic technologies could be promising solutions of this issue. A surgeon could perform surgery operation under its own laparoscope operation by using a robotic tools, for example. Many studies have been reported to try to replace a scopist with a robotic device. J.M. Sackier and Y. Wang (1994) proposed a concept of robotically assisted laparoscopic surgery [1]. An endoscope is attached to the end of an arm-type robot. This robot is operated by a surgeon by a hand switch or a foot switch. E. Kobayashi et al. (1999) developed a laparoscopic manipulator system using a robot with a closed-link mechanism [2]. A gyro sensor fixed on the surgeon's head and a knee switch were used for its operation. J.A. Long et al. (2007) developed a miniaturized endoscope holder to be placed on patient's skin [3]. A voice recognition system was used for its operation. This system obtained FDA approval in 2008 and commercialized under the name ViKY. A. Mirbagheri et al. (2011) also developed a robotic laparoscope holder operated by voice recognition system [4]. One major advantage of the use of the voice recognition system is that surgeon can operate the robotic system without any additional contact to sensing devices or switches during surgery. However the voice commands are often misunderstood at a noisy environment or under situations with frequent voice communication of/with other staffs like the surgical operation. K. Tadano et al. (2015) developed a pneumatically actuated laparoscope holder controlled through combinations of head movements and operations of a foot switch [5]. This scope holder was commercialized in 2015.

Use of head movements as a control channel could be one of the best ways for scope operation. However, most of the scope holders operated by head movements use a foot switch to activate/deactivate the scope operations. However, several foot switches are often used in the operation room to operate surgical equipment such as the one for energization of an electrical knife therefore most of surgeons don't want to have additional foot switches for scope operation. And one disadvantage of the foot switch is that surgeon's foot position is constrained through the surgery.

Therefore, authors propose a robotic laparoscope holder with an innovative interface that doesn't need foot switches for its control so that the more comfortable laparoscope operation can be available for surgeon during the surgery.

2   Proposed System

Figure 1 shows a sketch of laparoscopic surgery under the surgeon's own scope operations by using the proposed system. This system is composed of a robotic device and a head-mounted interface. The robotic device is placed near the surgeon and holds a laparoscope inserted into patient's abdominal cavity instead of a scopist. An interface is installed to the surgeon's head to detect user's commands for operation of the robotic device during the surgery. This interface detects surgeon's head movements and jaw exercises by using sensors. The proposed system allows the surgeon to easily handle a laparoscope and adjusts a scope view inside the abdominal cavity through commands based on the head and the jaw movements at the time of surgery requiring the surgical tools operations with both hands and foot pedals operations.

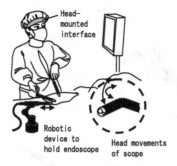

Fig. 1. A sketch of laparoscopic surgery under surgeon's own scope operation by using the proposed system

3 Prototyped System

A prototype has been built. This prototype is designed to have simple and compact structure and its function is limited to what the system really needs to support the scope operations at a target surgery operation. The target operation of this prototype is extirpative operation of gallbladder. This prototyped system is composed of a robotic device with three degree-of-freedom that can hold a flexible scope at required positions for the target surgery and a head-mounted interface to measure user's head inclination angles in two directions, the vertical and the horizontal directions, and detects user's voluntary efforts of bite with back teeth. Each components of the prototype are described in detail in this section.

3.1 Robotic Device

Figure 2 shows a photo of the robotic device prototyped. This robotic device is composed of an upper unit, a lower unit, and a supporting stand. The upper unit is in charge of holding a scope and operating two levers on it. A flexible scope, OLYMPUS LTF TYPE V3, is used in this system. This scope has two levers on its grip for directing the end of scope to desired directions. One is located in right side of the grip and works for lateral movements of the scope end and the other is in the left side of the grip and works for

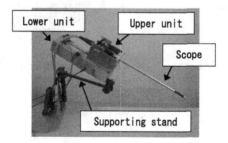

Fig. 2. Robotic device prototyped

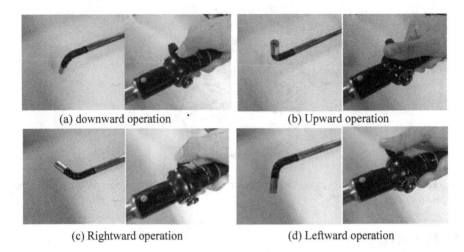

(a) downward operation   (b) Upward operation

(c) Rightward operation   (d) Leftward operation

Fig. 3. The movements of the scope end under each lever operations

horizontal movements of scope end. The movements of the scope end under each lever operations are presented in Fig. 3.

An 3D image of the upper unit made by CAD software and its photo are shown in Figs. 4 and 5, Fig. 4(a) shows a status of the upper unit without the scope. Figure 4(b) shows a status of the upper unit after installation of the scope. A pressor bar which is connected with a hinge at the top of the front wall of the upper unit let users to easily fix a scope in the right position. Two hooks with a pair of claws are used to hold the levers. Each of hooks are inserted into a straight gaps of a wheel to be rotated by a servo motor after placing the scope into the upper unit. Therefore the rotation angles of the levers can be finely adjusted by controlling the servo motors.

(a) Upper unit without the scope   (b) Upper unit after installation of the scope

Fig. 4. 3D image of the upper unit

Fig. 5. A photo of the upper unit

Figure 6 shows a 3D image of the lower unit. One of the side panels are taken off in this figure to show the actuation mechanism inside. There are a pair of liner rails on the top face of the body. The upper unit can moves forward and backward without wobbling because of these rails. The lower unit has a liner reciprocating actuation mechanism based on the combination of a servo motor and a pulley and belt mechanism inside. A bracket to be connected to the upper unit is fixed to the belt. The stroke of the reciprocating actuation is designed to be 10 [cm].

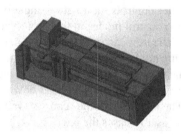

Fig. 6. 3D image of the lower unit

Based on above this mechanism, the scope can be easily replaceable from the robotic device and the pan tilt operations and the zoom in/out operations of the scope are available.

The supporting stand has one translational joint and 4 rotational joins to be fixed/released by rotary knobs. The robotic device is fixed at a desired position and posture near the surgeon by using this supporting stand.

3.2 Head-Mounted Interface

An innovative head-mounted interface has also been developed and used to control the prototype. This interface is easily set to user's head like a head phone. Figure 7 shows photos of the head-mounted interface. The surgeon's head inclination angles in horizontal and vertical directions are measured by using the combination of gyro sensors and accelerometers. In addition, the degree of contraction of temporal muscles which are located at both sides of the head and work for jaw closure is detected by using a sensor developed by authors [6]. This sensor has a sheet-like structure and is composed of a soft materials. The sensor is attached to the skin around the temporal area of user's head by wearing the interface. The temporal muscle starts pop up based on its contraction level after it exceeds a certain degree. Therefore, the sensor is squashed only when the user intentionally bite with back teeth. The degree of squash is evaluated through capacitance measurement between two conductive sheets sandwiching the sensor structure. The details of this sensor is described in the referenced paper.

Fig. 7. Head-mounted interface

Based on these information of user's movements at the head and the jaw detected by the interface, the controller controls the rotation angle of each servo motor of the robotic device. There are two control modes in this prototype. One is swing mode and the other is zooming mode. The user can switch between the two modes by biting with both sides of jaw. In the swing mode, the user can change the direction of the end of the scope through turning the head to the desired direction. In the zooming mode, the user can zooming up the scope view by moving its head downward and zooming out by moving the head upward. The operation of the scope is available only when the user is biting. It stops moving if the user releases the jaw. The verbal communications doesn't interfere the scope operations.

4   Evaluation Tests

Authors conducted two kinds of evaluation tests using the prototype to see the efficacy of the proposed system. The operability of the prototype was evaluated through a dry lab test first. Then, the applicability of the prototype was evaluated through a wet lab test.

4.1   Operability Evaluation Test Using a Training Box

By using the proposed robotic system, it is expected that surgeons will be able to conduct laparoscopic surgery without a scopist. However, it is also expected that surgeons will take a longer time to complete the surgery operations under its own laparoscope operations than the case of usual laparoscopic surgery under scopist's scope operations because the surgeon has to do the jobs for two during the surgery. In this study, the target surgery operations is limited to the extirpative operation of gallbladder. Considering this condition, authors have set a baseline, 50 [%] of increase rate of time for the surgery operation as an acceptable value, to judge if the system has adequate functions and operability to support surgeons to surely and safely complete the target surgery operation based on the opinions from some of authors who are richly experienced on the laparoscopic surgeries.

A test was conducted to see the operability of the prototyped system using a training box. A cork board was set inside the training box and a work space for the task was defined by 6 pins of 2-by-3 arrangement as shown in Figs. 8 and 9. The size of the work space depicted by the 6 pins was decided by reference to the common size of adult people's gallbladder. Five subjects who is not medical doctors were recruited for this test. The subjects were requested to perform a forceps operation task in the work space inside the training box. The task was to repeat transferring a rubber ring from one pin to the next pin and let the rubber ring go around a circle. The time for getting back to the start point after going round was recorded.

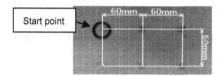

Fig. 8. Pin configuration for the task

Fig. 9. Task performed under the own scope operation by using the prototype in the test

The subjects perform the task twice under the different conditions of scope operation as 1 trial. The task was firstly performed under the scopist's scope operation (Case A) and secondly performed under the own scope operation by using the robotic device (Case B). In this test the scopist is not a medical doctor but just a person who holds a scope for the subjects. Each subject conducts 5 trials in series. The data recorded in this test are shown in Fig. 10. The average of 5 trials of each case and increase rate of Case B to Case A are shown in Table 1. It is observed that Case B always takes a little longer than Case A. The increase rate of the time for completing the task ranges from 10 to 51 [%]. The increase rate of Sub 3 is 51 [%] which is higher than the baseline, 50 [%], but that of other 4 subjects are below the baseline. In addition, the average of all 5 subjects is 31 which is much lower than the baseline.

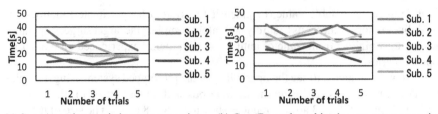

(a) Case A: under scopist's scope operation   (b) Case B: under subject's own scope operation

Fig. 10. Obtained data in the operability evaluation tests

Table 1. Results of operability evaluation test

	Sub 1	Sub 2	Sub 3	Sub 4	Sub 5	Ave
Case A: Time needed for the task under scopist's scope operation [s]	16.1	29.1	21.1	13.9	23.3	20.7
Case B: Time needed for the task under own scope operation [s]	20.7	35.8	31.9	20.2	25.5	26.8
Increase rate of Case B compared to Case A [%]	29	23	51	45	10	31

4.2   Applicability Evaluation with Animal Model

The prototype has applied to a surgery of an animal model to see its applicability as shown in Fig. 11. In this test, 3 surgeons performed extirpative operation of gallbladder and a part of liver simulating a gallbladder. Two of three surgeons successfully removed the target tissue under the own scope operation by using the prototype. However, one of the surgeon couldn't complete the task under scope operation of the prototype because the surgeon could not separately bite the back teeth on right and left sides. The two surgeons who successfully performed the surgery under own scope operation reported that its operation was very easy and the prototype is enough functional for practical use.

Fig. 11. Evaluation test with an animal model

5 Discussions

In the operability evaluation test, the increase rate of the time for completing the task ranges from 10 to 51. The increase rate of one of five subjects is 51 [%] and this is higher than the baseline, 50 [%], that authors selected but that of other 4 subjects are below the baseline. In addition, the average of all 5 subjects is 31 [%] which is much lower than the baseline. With the consideration for the advantage of capability to deduce the number of medical doctors at the surgery operation, it can be said that the prototype has desired functions and adequate operability.

In the applicability test by using an animal model, 3 surgeons performs extirpative operation of gallbladder and a part of liver simulating a gallbladder. Two of three surgeons were able to successfully remove the target tissue under the own scope operation by using the prototype and they reported that its operation was very easy and the prototype is enough functional for practical use. Therefore, it was shown that the proposed robotic scope holder has a high potential to reduce the number of surgeons at the laparoscopic surgery at least for extirpative surgeries of cancer on gallbladder or liver. However one of the three surgeons couldn't have a good command of the system because it had a trouble on separately biting back teeth which is required for controlling the prototype. Basically separate biting of back teeth is available for anybody after some training. But the result of this applicability evaluation test shows that the proposed system has more space to improve the way of command from a view point of useability for anybody.

6 Conclusions

Authors have proposed a robotic laparoscope holder with a unique interface. A prototype has been built to achieve surgeon's own scope operations at a target surgery operation, the extirpative operation of gallbladder. Its excellent operability and high potential ability are demonstrated through a dry lab test and a wet lab test, respectively.

Acknowledgement. Authors would like to acknowledge that this work was supported by JSPS KAKENHI Grant Number 15H03058.

References

1. Sackier, J.M., Wang, Y.: Robotically assisted laparoscopic surgery. Surg. Endosc. **8**, 63–66 (1994)
2. Kobayashi, E., Masamune, K., Sakuma, I., Dohi, T., Hashimoto, D.: A new safe laparoscopic manipurator system with a five-bar linkage mechanism and an optical zoom. Comput. Aided Surg. **4**(4), 182–192 (1999)
3. Long, J.A., Cinquin, P., Troccaz, J., Voros, S., Berkelman, P., Descotes, J.L., Letoublon, C., Rambeaud, J.J.: Development of the miniaturised endoscope holder LER (Light Endoscope Robot) for Laparoscopic surgery. J. Endourol. **21**(8), 911–914 (2007)

4. Mirbagheri, A., Farahmanda, F., Meghdaria, A., Karimianc, F.: Design and development of an effective low-cost robotic cameraman for laparoscopic surgery: RoboLens. Sci. Iranica B **18**(1), 105–114 (2011)
5. Tadano, K., Kawashima, K.: A pneumatic laparoscope holder controlled by head movement. Int. J. Med. Robot. Comput. Assist. Surg. **11**, 331–340 (2015)
6. Kudo, S., Oshima, K., Arizono, M., Hayashi, Y., Moromugi, S., Higashi, T., Takeoka, A., Ishihara, M., Ishimatsu, T.: Electric-powered glove for CCI patients to extend their upper-extremity function. In: Proceedings of the 2014 IEEE/SICE International Symposium on System Integration, Chuo University, Tokyo, Japan, 13–15 December 2014

The Design and the Gait Planning Analysis of Hexapod Wall-Climbing Robot

Yongjie Li, Jiaxin Zhai, Weixin Yan, and Yanzheng Zhao[(✉)]

Robotics Institute of Shanghai Jiao Tong University, Minhang, Shanghai, China
{lyjl9680517,zhaijiaxin,xiaogu4524,
yzh-zhao}@sjtu.edu.cn

Abstract. In this paper, the authors manage to design and control a kind of hexapod wall-climbing robot, which is a radial symmetric, with 18 actuators - 3 steering motors allocated at each leg - supposed to be effectively mobile. Suckers, solenoid valves and vacuum pump are equipped to enable the function of wall climbing. In order to reach the target, mathematic model is optimized in advance. And it is designed to be light but meanwhile has a certain capacity of load-carrying. By alternating the gait between the tripod and the non-tripod one, or adjusting the open-loop control strategy, the load-carrying capacity can be extended a lot at the price of reducing the mobility.

Keywords: Hexapod robot · Wall-climbing · Radial symmetry · Model optimization Mobility · Load carrying capacity · Tripod gait · Non-tripod gait

1 Introduction

Different from the continuous path of wheeled and tracked robots, multi-legged robots works by leaving discontinuous path or rather the spot exactly, minimizing the damage to the environment and much more flexible in dealing with complicated circumstance. It has one or several individual rotational joints or translational joints at each leg. Several legs like that working together make the flexibility and diversity of the robot greatly enhanced. Since the robot can modify the pose and the orthocenter position by adjusting the legs' layout and forms, it is not easy for the robot to be tipped over. Even if it happened in unanticipated and undesirable events, for example, like earthquake or intentional attacks, the robot can easily fix it in the way which insects do.

The hexapod wall-climbing robot is designed based on those assets, but required to operate not mainly on the ground. Fortunately, in recent years, climbing robots have become lighter, more adaptable to a wide variety of surfaces, and much more sophisticated in their functional capabilities. These advances have been driven by improved manufacturing techniques, increased microcontroller computational power, and novel strategies for wall-climbing [1].

In this paper, we report on the procedure and principle we design. And a prototype is produced to test the gaits.

N. Kubota et al. (Eds.): ICIRA 2016, Part I, LNAI 9834, pp. 619–631, 2016.
DOI: 10.1007/978-3-319-43506-0_54

2   Design and Modeling

2.1   Concept and Morphology

Tripod gait is quite effective and stable, which has been evolved for billion years in nature by insects. So we prefer to imitate insects in both gait and structure, but with a little difference as depicted in Fig. 1, which is actually the difference between the symmetric configuration and the radial symmetric configuration.

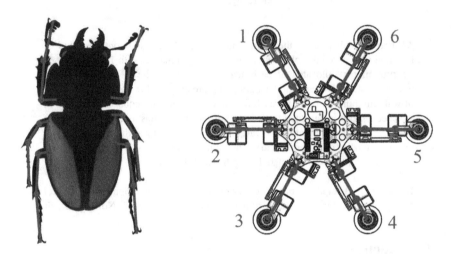

Fig. 1. Symmetric configuration and radial symmetric configuration

The benefits of radial symmetry if compared to symmetry are obvious:

1. Each leg is equivalent to one another, which stands for the universality that they could share exactly the same structure and could be fabricated in one assembly line. It is cost effective in mass production.
2. And moreover, each gait we designed is appropriate for 6 direction. It is meaningless for a robot to tell the head from the tail. We can order it to walk even in a way which is impossible for a real insect, for example, backward. Just a single gait enables the robot to reach almost any position on a plane.

2.2   Mathematic Modeling

Now let's consider the robot's mathematic model regardless of the concrete realization and mechanism. Here are four dimensions need to be decided, which are the distance of each adjacent two hip joints L0, the length of the coxa L1, the length of the femur L2 and the length of the tibia L3.

Normal Solution of Kinematics. First we mark each leg from No. 1 to No. 6, as showed in Fig. 1.

For the convenience to describe, a coordinate established at the center of the robot as the body frame, and the same to each part of legs. Homogeneous transfer matrix is the main tool we used. Here we consider the Leg No. 6 only, and define three variables which are theta, phi and psi. Specifically, let theta stand for the angular displacement between the body frame and the coxa, while phi stands for the one between coxa and femur and psi stands for the one between femur and tibia (Fig. 2).

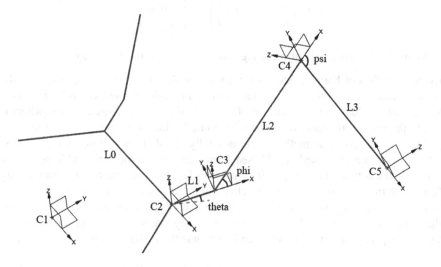

Fig. 2. Coordinates established on Leg No. 6

According to these coordinates, the Homogeneous transfer matrix of C5 with reference to C1 could be easily calculated.

$$
{}_5^1T_6 = {}_2^1T_6 \bullet {}_3^2T_6 \bullet {}_4^3T_6 \bullet {}_5^4T_6 =
\begin{bmatrix}
X & X & X & \#14 \\
X & X & X & \#24 \\
X & X & X & \#34 \\
0 & 0 & 0 & 1
\end{bmatrix}
\tag{1}
$$

Where alphabet X is used to stand for the part that we don't care.

$$
\#14 = \frac{L0 + [L1 + L3\cos(\varphi + \psi) + \cos\varphi](\cos\theta - \sqrt{3}\sin\theta)}{2},
$$

$$
\#24 = \frac{\sqrt{3}L0 + [L1 + L3\cos(\varphi + \psi) + L2\cos\theta](\sin\theta + \sqrt{3}\cos\theta)}{2},
$$

$$
\#34 = -L3\sin(\varphi + \psi) - L2\sin\varphi
$$

Single leg can be considered as a mechanical arm with 3 DOF (Degree Of Freedom)s. On this occasion, all the three DOFs are used to control the position of the tip of the leg. Another three DOFs about orientation are constrained by the mechanism to fit the surface well. Now we only concentrate on the position vector, and let the rotation matrix alone.

Equation (2) is a function about theta, phi and psi. If we input the angular displacements, the position information will be given by this equation. That's the job normal solution does. It will be helpful in further research.

$$P_6 = \begin{bmatrix} p_x \\ p_y \\ p_z \end{bmatrix} = \begin{bmatrix} \#14 \\ \#24 \\ \#34 \end{bmatrix} = \begin{bmatrix} f(\theta, \varphi, \psi) \\ g(\theta, \varphi, \psi) \\ h(\varphi, \psi) \end{bmatrix} \tag{2}$$

Similarly, equations about other legs are available in the same way.

Inverse Solution of Kinematics. What if we want the tip of the leg to reach a specific position? Here comes the inverse solution. It is the basis of robot control, expressing the mapped relation from the position of the tip to the angular displacements, which is exactly the reverse process compared with normal solution, as the name implies.

The inverse solution methods are generally divided into closed-form solution methods and numerical methods. Closed-form solution method is desirable because it is faster and easily identify all possible solutions [2]. But only the robot, which satisfy Pieper Criterion, can be described by closed solution. Pieper outlined two conditions for finding a closed-form joint solution to a robot manipulator in which either three adjacent joint axes are parallel to one another or they intersect at a single point [3]. No doubt the leg has closed solution, since it only has three DOFs. Now we are going to find it out.

First, multiply a matrix on both sides of Eq. (1), as follows:

$$({}_2^1T_6 \bullet {}_3^2T_6)^{-1} \bullet {}_5^1T_6 = ({}_2^1T_6 \bullet {}_3^2T_6)^{-1} \bullet {}_2^1T_6 \bullet {}_3^2T_6 \bullet {}_4^3T_6 \bullet {}_5^4T_6$$

$$\begin{bmatrix} X & X & X & \#14 \\ X & X & X & \#24 \\ X & X & X & \#34 \\ 0 & 0 & 0 & 1 \end{bmatrix} = \begin{bmatrix} X & X & X & L1 + L3\cos(\varphi + \psi) + L2\cos\varphi \\ X & X & X & 0 \\ X & X & X & -L3\sin(\varphi + \psi) - L2\sin\varphi \\ 0 & 0 & 0 & 1 \end{bmatrix} \tag{3}$$

In Eq. (3), we use alphabet X to represent the part that we don't care, too. And here:

$$\#14 = \frac{p_x(\cos\theta - \sqrt{3}\sin\theta)}{2} + \frac{p_y(\sin\theta + \sqrt{3}\cos\theta)}{2} - L0\cos\theta$$

$$\#24 = \frac{-p_x(\sin\theta + \sqrt{3}\cos\theta)}{2} + \frac{p_y(\cos\theta - \sqrt{3}\sin\theta)}{2} + L0\sin\theta$$

$$\#34 = p_z$$

A monadic equation #24 = 0 could be found. So we can solve theta:

$$\theta = \arctan\left(\frac{p_y - \sqrt{3}p_x}{p_x + \sqrt{3}p_y - 2L0}\right) \tag{4}$$

Another extraneous root has been removed, for the angular displacement range is narrower than (−90, 90) degree. Since Eq. (4) has been solved, we have:

$$\begin{cases} L3\cos(\varphi + \psi) + L2\cos\varphi = A \\ -L3\sin(\varphi + \psi) - L2\sin\varphi = p_z \end{cases} \tag{5}$$

Where

$$A = \#14 - L1 = \frac{p_x(\cos\theta - \sqrt{3}\sin\theta)}{2} + \frac{p_y(\sin\theta + \sqrt{3}\cos\theta)}{2} - L0\cos\theta - L1$$

We can square both sides of the equations, and plus them together. And we can get:

$$\psi = \pm\arccos\left(\frac{A^2 + p_z^2 - L3^2 - L2^2}{2L3L3}\right) \tag{6}$$

With Eqs. (5) and (6), the last variable phi can be also deduced. But on account of the dependency on psi, phi also has two situations:

$$\varphi = \arcsin\left(\frac{-p_z}{\sqrt{(L3\cos\psi + L2)^2 + (L3\sin\psi)^2}}\right) - \arctan\left(\frac{L3\sin\psi}{L3\cos\psi + L2}\right)$$

Now the inverse solution for Leg No. 6 has been finished. Other legs can be dealt with in the same way. We will not repeat the process.

Model Optimization. Up to now, variables L0, L1, L2 and L3 are considered as known quantities all along. But the truth is that they are not. Now we are going to discuss about these variables, to make the robot be faster and lighter.

Variable L0 controls the size of body. The body is the container of devices such as the vacuum pump or the circuit. It has nothing to do with the velocity directly but has relation to the weight of the robot. Theoretically, the smaller L0 is the lighter the body is, provided that it is large enough to contain all the devices required and will not cause the interference of legs. Based on this principle it is determined to be 64 mm.

Variables L1, L2 and L3 directly impact the velocity it moves. The speed of hexapod can be increased by increasing the stride frequency and also increasing the stride length within workspace limits of the joints [4]. So, there are two ways:

1. Reduce the cycle time or enhance the frequency of walking.
2. Increase the interval of each step.

It is not easy for a robot to enhance the frequency if the mechanism and actuator are kept unchanged, for that actuators are required to reciprocate much faster and the process is accompanied by acceleration and deceleration. It is not possible and profitable for a specific motor. So we try in the second way.

The interval of steps is constrained by the physical features of the sucker, which enables the robot to walk on the wall. It is set at the tip of each leg, and generates vacuum with the vacuum pump together. Pneumatic adhesion technique may be regarded as the best choice for climbing a vertical wall with higher payload [5]. Spherical hinge on this kind of sucker allow the leg to rotate about the tip in three DOFs. These passive DOFs and other three DOFs which are talked in normal solution and inverse solution above, constitute the complete six DOFs.

However, the spherical hinge does not have unlimited working space, as Fig. 3 shows, usually 35° or 30°, which depends on specific types. Attempt to attach the wall in a too small intersection angle with the wall may result in the plunge of the whole robot.

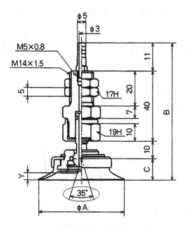

Fig. 3. Overall dimension of the sucker

So L1, L2 and L3 need to be designed to find a largest interval of step in the constraint of the suckers, showed by Fig. 4.

Set h = 25 mm, s = 60 mm. So that the main body is close enough to the wall to reduce the torque generated by gravity, and the robot looks not too big or too small. Benefit from the development of computer science, we utilize MATLAB to calculate the combination which is almost close to the best one. Here is the pseudo-code:

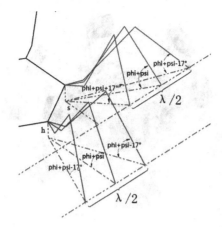

Fig. 4. Sketch of half cycle

```
function SIZE = Traverse(s, h)
L0 = 64;//determined by the mechanism
best = 0;// to record global max step
//to record the according dimension
bestL1 = 0; bestL2 = 0; bestL3 = 0;
for L1 = 24:1:35//set range of L1
    for L2 = 50:1:100//set range of L2
        for L3 = 80:1:110//set range of L3
            temp = Max step with temporary dimensions
            if temp>best best = temp;
            bestL1 = L1; bestL2 = L2; bestL3 = L3;
end end end end
```

The way to calculate max step with temporary dimensions is based on the normal and inverse solutions we discussed before.

The fact is that L1 tends to be small while L3 tends to be large, if we want to maximize the interval. Within the constraint of mechanism, the best combination seems to be L0 = 64 mm, L1 = 24 mm, L2 = 94 mm, and L3 = 110 mm. And the longest interval by calculation equals 109.2 mm, which means it can walk 218.4 mm within one cycle by the tripod gait.

2.3 Virtual and Real Prototypes

Virtual Modeling. Suckers, and steering motors are selected in a repetitive iteration process by virtual modeling, as Fig. 5 depicts. Leg parts are made of sheet metal, exactly the aluminum, to strengthen rigidity. And in order to lighten the leg, holes are dug.

Fig. 5. Virtual prototype built in solidworks

The designed weight is about 2.4 kg. For a robot to climb on a wall with vacuum suckers, the suckers must generate sufficiently big force to support. The suction force is determined by several main factors such as gauge pressure, the number, the diameter and the arrangement of the suckers [6]. According to the checking formula, let's check whether the suckers are qualified:

Here static friction coefficient $\mu \approx 0.5$, numbers of legs in support phase $n = 3$, vacuum degree $\Delta P = 60$ KPa, diameter of suckers $D = 50$ mm, safety factor $S = 4$, total mass $M = 2.4$ kg, distance from orthocenter to wall $H \leq 50$ mm.

$$\begin{cases} \mu \cdot n \cdot \Delta P \cdot \pi \cdot \frac{D^2}{4} \geq S \cdot M \cdot g \\ \Delta P \cdot \pi \cdot \frac{D^2}{4} \cdot \left(3s + \frac{3}{2}L0\right) \geq M \cdot g \cdot H \end{cases} \qquad (7)$$

The first inequation of In Eq. (7) aims to check the whether the maximum friction force is able to hold the gravitational force when in tripod gait. While the second one attempts to check whether the torque generated by 3 suckers and vacuum pump is able to hold the gravitational torque with one leg up and two legs down, which is the most dangerous situation in tripod gait.

The next step is to check the steering motors. Suppose the robot attaches on a vertical wall and tries to move one step upward. Simulation time is 5 s, so the average velocity is about 22 cm/s. The process is showed by Fig. 6.

The maximum torque showed in Fig. 7 is about 1254 Nmm which is less than rated torque of the steering motor, 1300 Nmm. By the way, the Steering Motor No. 9 is installed on the joint which articulates femur and tibia of Leg No. 3.

Prototype. The prototype is made as Fig. 8 depicts. The real weight is 2.307 kg. The photo is taken when the robot is attaching on wardrobe door with six suckers working and all steering motors kept motionless.

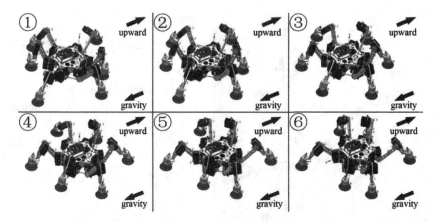

Fig. 6. The simulation of one step by tripod gait

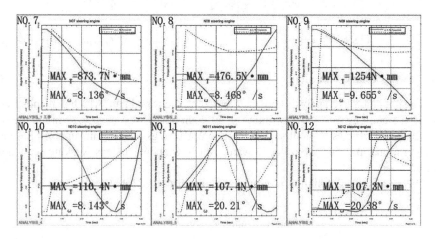

Fig. 7. The torque curve and angular velocity curve of part of steering motors

An Arduino system is put to use as a slave computer, while a laptop running a series of LabVIEW program as a master computer. Commands are transferred by serial line.

Gas route chart is showed in Fig. 9.

3 Gait Planning Analysis

A gait is a cyclic motion pattern that produces locomotion through a sequence of foot contacts with the ground. The legs provide support for the body of the robot meanwhile propel the robot forward. Gaits can differ in a variety of ways, and different gaits produces different styles of locomotion [7].

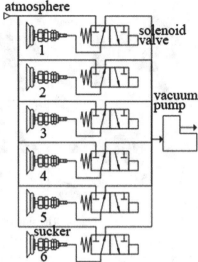

atmosphere

Fig. 8. Prototype

Fig. 9. Gas route chart

3.1 Swing Phase Planning

All the simulation and analysis above are based on the tripod gait, which demands Leg No. 1, 3, 5 and Leg No. 2, 4, 6 working in different phase alternatively, swing phase or support phase. It goes without saying that the tips of legs working in support phase is drawing a line, pushing the robot forward. But how about the legs working in swing phase?

There are several common motion curves are listed in Table 1:

Table 1. Common motion curves

Curve type	Width	Height	Starting angle	Landing angle	Arc length
Parabola	a	0.25a	45°	45°	1.321a
Cycloid	a	0.318a	90°	90°	1.273a
Cardioid	a	0.65a	90°	0°	2a
Line	a	h	90°	90°	a + 2 h

The height-width ratio reflect the locomotivity. Normally, the larger ratio means a better locomotivity but a slower speed. While the arc length is related to the time cost in swing phase. Considering the contact quality between the sucker and surface and other factors, cycloid seems to be the most suitable for the motion curves in swing phase. The outline of Leg No. 6 in swing phase is showed as Fig. 10.

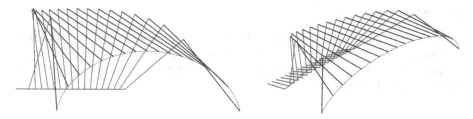

Fig. 10. Cycloid in swing phase

3.2 Tetrapod Gait

The load carrying capacity with tripod gait is very small, but moves quite fast. Empirically, the load carrying capacity is direct proportion to the quantity of legs working in support phase, while the speed is inversely proportion to it.

If we assume legs in swing phase cost the same time, then speed will decrease to the 1/3 of the one when tripod gait is applied. Legs in support phase shares a better force condition if uniform distributed as far as possible. That's why we choose that order showed in Fig. 11.

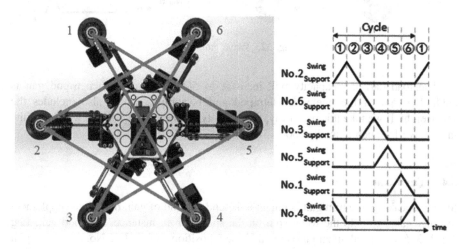

Fig. 11. Tetrapod gait order

But because there are always 4 legs in support phase in any time, the load carrying capacity will increase to 4/3 of the one when tripod gait is applied, which includes the self-weight. If we take this part away, the capacity is about 800 g.

3.3   Wave Gait

We further on this way. Now if there are always 5 legs in support phase, the speed will decrease to the 1/6 of the one when tripod gait is applied or 1/2 of the one when tetrapod gait is applied. The order does not matter anymore in this case, here order showed in Fig. 12 is chosen.

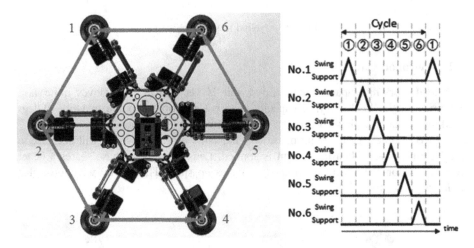

Fig. 12. Wave gait order

The load carrying capacity will increase to 5/3 of the one when tripod gait is applied or 5/4 of the one when tetrapod gait is applied, which also includes the self-weight. If we take this part away, the capacity reaches about 1600 g, which is quite impressive for such a small robot.

3.4   Turn Gait

Strictly speaking, the turn gait is not an independent kind of gait, more like supplement to other gaits, to help the robot spin on the spot. For an instance, to tripod gait, Leg No. 1, 3, 5 are raised up and move to their new positions, while Leg No. 2, 4, 6 keep in support phase. Secondly, the Leg No. 1, 3, 5 turn into support phase then Leg No. 2, 4, 6 raise up to move to the new positions [8]. It is almost the same for the robot in other gait, but only the order changed accordingly to remain the number of legs in support phase, preventing the robot from falling down.

4 Conclusions

In this paper, a hexapod wall-climbing robot has been proposed and made. It has 18 steering motors, 6 suckers and 1 vacuum pump. Mathematic model has been studied and optimized. Tripod, tetrapod and wave gait are planned. Tripod gait has been tested on the prototype, and the experiment expresses the validity of the normal and inverse solution, as well as the stability of the tripod gait. More attention could be paid to the non-tripod gait and adaptive gait in the future.

Acknowledgment. This work was supported by the National Natural Science Foundation of China under Grant No. U1401240, 6110510, 61273342, 51475305 and 61473192.

References

1. Provancher, W.R., Jensen-Segal, S.I., Fehlberg, M.A.: ROCR: an energy-efficient dynamic wall-climbing robot. IEEE/ASME Trans. Mechatron. **16**(5), 897–906 (2011)
2. Feng, Y., Yao-Nan, W., Shu-Ning, W.: Inverse kinematic solution for robot manipulator based on electromagnetism-like and modified DFP algorithms. Acta Automatica Sinica **37**(1), 74–82 (2011)
3. Ali, M.A., Park, H.A., Lee, C.S.G.: Closed-form inverse kinematic joint solution for humanoid robots. In: 2010 IEEE/RSJ International Conference on Intelligent Robots and Systems (IROS), pp. 704–709. IEEE (2010)
4. Kottege, N., Parkinson, C., Moghadam, P., et al.: Energetics-informed hexapod gait transitions across terrains. In: 2015 IEEE International Conference on Robotics and Automation (ICRA), pp. 5140–5147. IEEE (2015)
5. Das, A., Patkar, U.S., Jain, S., et al.: Design principles of the locomotion mechanism of a wall climbing robot. In: Proceedings of the 2015 Conference on Advances In Robotics, p. 13. ACM (2015)
6. Zhu, H., Guan, Y., Cai, C., et al.: W-Climbot: a modular biped wall-climbing robot. In: 2010 International Conference on Mechatronics and Automation (ICMA), pp. 1399–1404. IEEE (2010)
7. Haynes, G.C., Rizzi, A.A.: Gaits and gait transitions for legged robots. In: Proceedings 2006 IEEE International Conference on Robotics and Automation, ICRA 2006, pp. 1117–1122. IEEE (2006)
8. Duan, X., Chen, W., Yu, S., et al.: Tripod gaits planning and kinematics analysis of a hexapod robot. In: IEEE International Conference on Control and Automation, ICCA 2009, pp. 1850–1855. IEEE (2009)

Smart Material Based Systems

Adaptive Control of Magnetostrictive-Actuated Positioning Systems with Input Saturation

Zhi Li[1(✉)] and Chun-Yi Su[2]

[1] Department of Electrical Engineering, Eindhoven University of Technology, 5612AZ Eindhoven, Netherlands
zhi.li@tue.nl
[2] College of Automation Science and Engineering, South China University of Technology, Guangzhou 510641, China
cysu@scut.edu.cn

Abstract. Magnetostrictive actuators are high-force low-displacement actuators, which are profitably utilized in many engineering applications such as, high dynamic servo valves, micro/nano-positioning systems and optical systems. Nevertheless, magnetostrictive actuators are subject to hysteresis effects and input saturation, which lead poor system performances, e.g. inaccuracy and strong oscillations. To mitigate these effects, in this paper an adaptive controller with an anti-windup technique is developed. The anti-windup technique is particularly used for dealing with the input saturation effect. The simulation results demonstrate the effectiveness of the proposed controller.

Keywords: Input saturation · Hysteresis · Adaptive control · Anti-windup technique · Magnetostrictive actuators

1 Introduction

Magnetostrictive actuators exhibit dominant hysteresis behaviors between the input (current) and the output (displacement) [1]. Such nonlinearities limit the actuating precision and performance, and may cause undesirable inaccuracies or oscillations in the output, specifically when used in closed-loop control systems [2]. The common approach to compensate for the hysteretic behaviors is to construct its feedforward inverse compensator. However, this inverse compensation is open-loop based, which is vulnerable to the model uncertainty and system disturbance, etc. To overcome this disadvantage, the feedback control approaches are adopted, such as sliding mode control [3], adaptive backstepping control [4], neural network control [5], model predictive control [6], etc.

Apart from the hysteresis behaviors, the actuator saturation effect is also a common phenomenon appearing in the smart material based actuators. In practice, actuators are always subject to the magnitude limit. Thus, the applied control signal to the actuator should always be maintained within a certain range during the operation of the actuator. However, once the actuator reaches to its

© Springer International Publishing Switzerland 2016
N. Kubota et al. (Eds.): ICIRA 2016, Part I, LNAI 9834, pp. 635–645, 2016.
DOI: 10.1007/978-3-319-43506-0_55

saturation limit without being properly addressed, the performance of the actuator will be degraded. In the literature, approaches for dealing with the input saturation can be classified into two categories: the direct design method, where the saturation is considered in the construction of the Lyapunov function when synthesizing the designed controllers. The other category is the anti-windup design method [7,8], where a separate anti-windup block is implemented to deal with the limitation of the saturation. The advantage of the anti-windup technique is that it is independent of the controller design of the unconstrained system (without considering the input saturation) and therefore it is more feasible and easier to implement in practice. In this paper, an anti-windup control strategy combined with an adaptive controller is developed to address the input saturation and hysteresis nonlinearity in the magnetostrictive actuator. The simulations are studied to validate the effectiveness of the proposed control strategy.

2   Dynamic Modeling of the Magnetostrictive Actuated Systems

In [9], a dynamic model based on the principle of operation of the magnetostrictive actuator has been proposed, which comprehensively considers the electric, magnetic and mechanical domain as well as the interactions among them. The complete set of electric-magnetic-mechanical equations is as follows:

$$i(t) = i_a(t) + i_R(t) + i_H(t) \tag{1}$$

$$i_a(t) = \Phi_L(t)/L_A \tag{2}$$

$$i_R(t) = N\frac{\dot{\Phi}(t)}{R_0} \tag{3}$$

$$i_H(t) = \Pi[x] \tag{4}$$

$$\Phi(t) = \Phi_L(t) + \Phi_T(t). \tag{5}$$

$$\Phi_T(t) = T_{Mm}x(t) \tag{6}$$

$$m\ddot{x}(t) + b_s\dot{x}(t) + k_sx(t) = F_a(t) \tag{7}$$

$$F_a(t) = T_{em}i_a(t) \tag{8}$$

where $i(t)$ is the supplied current to the actuator; $i_H(t)$ denotes the hysteresis current loss, $\Pi[x]$ is a hysteresis operator which will be explained in the following development; i_R is the eddy current loss, with R_0 being the equivalent resistor of the eddy current effect, N denotes the number of turns of the solenoid; $i_a(t)$ is the actual applied current considering the hysteresis current loss and eddy current loss; $\Phi(t)$ is the magnetic flux flowing through the actuator; L_A denotes the equivalent inductor of the winding coils and $\Phi_L(t)$ is the magnetic flux flowing through the inductor L_A; $\Phi_T(t) = T_{Mm}x(t)$ is transformed from the mechanical side which is similar to the back-emf in piezoelectric actuator [10,11], T_{Mm} is magnetomechanical transduction coefficient; m is the equivalent mass of the moving part, b_s is the equivalent damping coefficient and k_s is the equivalent stiffness of the preloaded spring; F_a denotes the applied force; $T_{em} = AE^H d_{33} N_a$ denotes the electromechanical transduction coefficient, where A is the cross section area of the magnetostrictive rod, E^H is the Young's modulus at constant value of magnetic field H, d_{33} is the slope of the strain versus the magnetic field, N_a denotes the number of turns of the solenoid per unit length.

By summarizing above equations, the general dynamic model of the magneto-strictive-actuated system can be written as:

$$m\ddot{x}(t) + b_s \dot{x}(t) + (k_s + \frac{T_{em}T_{Mm}}{L_a})x(t) = \frac{T_{em}}{L_a}\Phi(t) \tag{9}$$

$$NL_a\frac{\dot{\Phi}(t)}{R_0} + \Phi(t) - T_{Mm}x(t) = L_a(i(t) - \Pi[x](t)) \tag{10}$$

3 Adaptive Control Design for the Magnetostrictive System with Input Saturation

As mentioned in the introduction, the hysteresis nonlinearity and the input saturation degrade the tracking performance of the actuator and cause oscillations and even instabilities in the actuated systems. The existing controllers are designed either by keeping the control input not to reach the saturation limit or directly ignoring the saturation nonlinearity. In this section, an adaptive control strategy combined with an anti-windup technique is investigated for the purpose of improving the tracking performance of the system.

3.1 The Dynamic Model with Input Saturation

For the control purpose, the dynamic system in (9) and (10) is rewritten in the canonical form as

$$\dddot{x}(t) + \rho_2\ddot{x}(t) + \rho_1\dot{x}(t) + \rho_0 x(t) = b\Gamma[i](t) \tag{11}$$

where $\rho_2 = \frac{NL_a b_s + R_0 m}{NL_a m}$, $\rho_1 = \frac{NL_a k_s + NT_{em}T_{Mm} + R_0 b_s}{NL_a m}$, $\rho_0 = \frac{k_s R_0}{NL_a m}$, $b = \frac{R_0 T_{em}}{NL_a m}$.
Because the displacement $x(t)$ can be represented as a function of supplied current $i(t)$, the term $i(t) - \Pi[x](t)$ in (10) can be defined as a new hysteresis nonlinearity $\Gamma[i](t)$

$$\Gamma[i](t) = u(t) = i(t) - \Pi[x](t) \tag{12}$$

Due to presence of the input saturation block, the input current $i(t)$ to the actuator becomes $i_{sat}(t) = sat_{\alpha,\beta}(i(t))$, where $sat_{\alpha,\beta}(i(t))$ is defined as

$$sat_{\alpha,\beta}(i(t)) = \begin{cases} \alpha, & \text{if } i(t) < \alpha \\ i(t), & \text{if } \alpha \leq i(t) \leq \beta \\ \beta, & \text{if } i(t) > \beta \end{cases}$$

Hence, the state space expression of the magnetostrictive-actuated dynamic system (11) can be expressed as

$$\dot{x}_1 = x_2$$
$$\dot{x}_2 = x_3$$
$$\dot{x}_3 = -\rho_2 x_3 - \rho_1 x_2 - \rho_0 x_1 + bu(t) \tag{13}$$

with

$$u(t) = \Gamma[sat_{\alpha,\beta}(i)](t) \tag{14}$$

3.2 Parameters Identification of the Magnetostrictive-Actuated Dynamic System in Absence of Input Saturation

From the model expression in (13) and (14), the dynamic model shows a cascading structure in which a nonlinear component, i.e. a hysteresis formulation $\Gamma[\cdot]$, is followed by a linear system. To identify this cascading structure, normalization should be conducted first. Without loss of generality, in this section, the dynamic part in (13) is normalized as follows

$$\dot{x}_1 = x_2$$
$$\dot{x}_2 = x_3$$
$$\dot{x}_3 = -\rho_2 x_3 - \rho_1 x_2 - \rho_0 x_1 + \rho_0 u(t) \tag{15}$$

with

$$u(t) = \Gamma_b[i](t) = \frac{b}{\rho_0}\Gamma[i](t) \tag{16}$$

Thus, the identification procedure is taken two steps as follows.
Step 1: Identification of the hysteresis component $\Gamma_b[i](t)$.
From experimental tests, hysteresis effects exhibited in the magnetostrictive actuator show asymmetric characteristics. To describe the asymmetric hysteresis behavior, an ASPI model [12] is employed in this paper. The numerical ASPI model is expressed as

$$\begin{aligned}
\Gamma_b[i](t) &= P[i](t) + H[i](t) \\
&= P[i](t) + \Psi[i](t) + g(i)(t) \\
&= p_0 i(t) + \sum_{j=1}^{n} p_j F_{r_j}[i](t) + \sum_{j=1}^{M} q_j \Psi_{c_j}[i](t) + g(i)(t)
\end{aligned} \tag{17}$$

where p_j denotes the weight of the play operator; $F_{r_j}[i](t)$ is the play operator, which is defined as

$$F_r[i](0) = f_r(i(0), 0) \tag{18}$$
$$F_r[i](t) = f_r(i(t), F_r[i](t_j)) \tag{19}$$

for $t_j < t \leq t_{j+1}, 0 \leq j \leq N - 1$, with

$$f_r(i, w) = \max(i - r, \min(i + r, w)) \tag{20}$$

r_j in (17) denotes the threshold of the play operator, and n is the number of the play operators, q_j denotes the weight of the elementary shift operator, $\Psi_{c_j}[i](t)$ is the elementary shift operator, defined as

$$\Psi_c[i](0) = \psi_c(i(0), 0) \tag{21}$$
$$\Psi_c[i](t) = \psi_c(i(t), \psi_c[i](t_j)) \tag{22}$$

for $t_j < t \leq t_{j+1}, 0 \leq j \leq N - 1$, with

$$\psi_c(i, w) = \max(ci, \min(i, w)) \tag{23}$$

c_j in (17) denotes the slope of the shift operator, and M is the number of the elementary shift operators.

$g(i)(t)$ is selected as

$$g(i)(t) = -a_3 i^3(t) - a_2 i^2(t) - a_1 i(t) - a_0 \tag{24}$$

The thresholds r_j are selected as $r_j = 0.3j$ $(j = 1, 2, ...n)$. The weights p_j, q_j, and a_0, ..., a_3 in (17) and (24) can be found using the nonlinear least-square optimization toolbox in MATLAB. The identified results [9] are shown in Table 1.

Table 1. Coefficients of the ASPI model

Numbers	r_j	p_j	c_j	q_j	a_j
0	0	0.9002			0
1	0.3	0.8445	1.1	1.3809	0
2	0.6	0.4276	1.2	0	0.3106
3	0.9	1.4821	1.3	0	0.0417
4	1.2	0.6097	1.4	0	
5	1.5	1.3596	1.5	0	
6	1.8	1.2051	1.6	0	
7	2.1	1.0574	1.7	0	
8	2.4	0.2835	1.8	1.0056	
9	2.7	0.1636			

Step 2: Identification of the dynamic part to find ρ_0, ρ_1 and ρ_2.
The s domain expression between $U(s)$ and $X(s)$ in (15) is expressed as

$$G(s) = \frac{X(s)}{U(s)} = \frac{\rho_0}{s^3 + \rho_2 s^2 + \rho_1 s + \rho_0} \tag{25}$$

To facilitate the identification, $G(s)$ is further decomposed as

$$G(s) = \frac{\tau}{s + \tau} \cdot \frac{\omega_n^2}{s^2 + 2\xi\omega_n s + \omega_n^2} \tag{26}$$

The objective is to identify the parameters of τ, ξ, ω_n in (26). To this end, a frequency response (1 to 500 Hz) of the magnetostrictive-actuated dynamic system is obtained in Fig. 1 (after normalization) with a 16 Kg mechanical load.

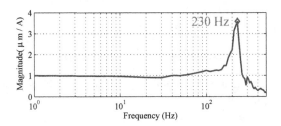

Fig. 1. Magnitude characteristics of the system

From the magnitude response in Fig. 1, we can find that $\omega_n = 230 \times 2\pi rad/s$. The other two parameters can also be determined as $\xi = 0.13$ and $\tau = 800 \times 2\pi$. Substituting these parameters in (26) yields

$$G(s) = \frac{1.05 \times 10^{10}}{s^3 + 5402s^2 + 3.98 \times 10^6 s + 1.05 \times 10^{10}} \tag{27}$$

Hence, ρ_0, ρ_1, and ρ_2 can be easily determined. The interested readers are referred to [9] for detailed identification procedure.

3.3 Controller Design with the Input Saturation

The control objective is to eliminate the hysteresis effect in the magnetostrictive actuator subject to the input saturation in order to improve the tracking performance of the positioning system. Towards this target, a backstepping control strategy combined with an anti-windup technique is developed. Figure 2 illustrates the control scheme.

Due to the existence of the hysteresis formulation in (16), the controller can not be directly designed. To achieve the controller design, the expression of the ASPI model considering the input saturation is reformulated as

$$u(t) = p_s sat_{\alpha,\beta}(i) - d(t) \tag{28}$$

where $p_s = p_0 + \sum_{j=1}^n p_j$, $d(t) \le D_s$, and D_s is a constant. The detailed proof of the boundedness of $d(t)$ may refer to [9]. Define $b_p = \rho_0 p_s$ and $d_p(t) = \rho_0 d(t)$. The controller design is summarized in Table 2.

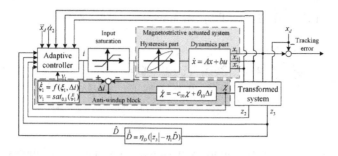

Fig. 2. The control diagram

Table 2. Adaptive backstepping control with anti-windup technique

Change of Coordinates:	
$z_1 = x_1 - x_d$	(T.1)
$z_2 = x_2 - \dot{x}_d - \alpha_1$	(T.2)
$z_3 = x_3 - \ddot{x}_d - \alpha_2 + \chi$	(T.3)
$\alpha_1 = -c_1 z_1$	(T.4)
$\alpha_2 = -c_2 z_2 + \dot{\alpha}_1 - z_1$	(T.5)
where c_1, c_2 are positive designed constants.	
Control Laws:	
$i(t) = \frac{1}{b_p}(-(c_3 - c_{30}v_1)z_3 - z_2 + \rho_2 x_3 + \rho_1 x_2 + \rho_0 x_1$	
$\qquad -\hat{D}sgn(z_3) + \ddot{x}_d + \dot{\alpha}_2 + c_{10}\chi)$	(T.6)
$v_1 = sat_{0,1}(\xi_1)$	(T.7)
$\dot{\xi}_1 = sat_{-\rho^-,\rho^+}(c_L(sat_{0,\kappa}(c_L|\Delta i|) - \xi_1))$	(T.8)
$\dot{\chi} = -c_{10}\chi + \Delta i$	(T.9)
$\Delta i = i - sat_{\alpha,\beta}(i)$	
where $b_p = \rho_0 p_s$. c_{10}, c_L and θ_{10} are positive parameters.	
Parameter Update Law:	
$\dot{\hat{D}} = \eta_D(|z_3| - \eta_1\hat{D})$	(T.10)
where $\hat{D}$ is the estimation of D, with $D = \rho_0 D_s$.	
η_D, η_1 are positive designed constants.	

The stability of the closed-loop system is established in the following theorem.

Theorem 1. For the system (15) preceded by the ASPI model (17), the adaptive controller presented by (T.6)–(T.9) guarantees that the tracking error remains bounded.

Proof. From (15), and (T.1)–(T.5), we have

$$z_1\dot{z}_1 = z_1 z_2 - c_1 z_1^2 \tag{29}$$

$$z_2\dot{z}_2 = z_2 z_3 - c_2 z_2^2 - z_1 z_2 - \chi z_2 \tag{30}$$

$$z_3\dot{z}_3 = z_3(-\rho_2 x_3 - \rho_1 x_2 - \rho_0 x_1 + b_p sat_{\alpha,\beta}(i(t)))$$
$$-d_b(t) - \dot{\alpha}_2 - \ddot{x}_d + \dot{\chi}) \tag{31}$$

where $d_p(t) = \rho_0 d(t)$.

Let $\tilde{D} = D - \hat{D}$. The Lyapunov function is constructed as follows

$$V(t) = \frac{1}{2}z_1^2 + \frac{1}{2}z_2^2 + \frac{1}{2}z_3^2 + \frac{1}{2\eta_D}\tilde{D}^2 + \frac{1}{2}\chi^2 \tag{32}$$

The time derivative of $V(t)$ along with (29)–(31) is given by

$$\dot{V}(t) = -c_1 z_1^2 - c_2 z_2^2 + z_2 z_3 - \chi z_2 - \rho_2 x_3 z_3 - \rho_1 x_2 z_3$$
$$-\rho_0 x_1 z_3 + b_p i(t) z_3 - b_p \Delta i z_3 - d_b(t) z_3 - \dot{\alpha}_2 z_3$$
$$-\ddot{x}_d z_3 + \dot{\chi} z_3 + \frac{1}{\eta_D}\tilde{D}\dot{\tilde{D}} + \chi\dot{\chi} \tag{33}$$

Substituting the control law $i(t)$ (T.6) into (33), one has

$$\dot{V}(t) = -c_1 z_1^2 - c_2 z_2^2 - (c_3 - c_{30}v_1)z_3^2 - \chi z_2 - b_p \Delta i z_3$$
$$-\hat{D}sgn(z_3)z_3 + c_{10}\chi z_3 - d_b(t)z_3 + \dot{\chi} z_3 + \frac{1}{\eta_D}\tilde{D}\dot{\tilde{D}} + \chi\dot{\chi} \tag{34}$$

Considering the definition of v_1 in (T.7), it is obvious that $v_1 \leq 1$. Besides, according to $-d_b(t)z_3 \leq D|z_3|$ and (T.10), one has

$$-d_b(t)z_3 - \hat{D}sgn(z_3)z_3 + \frac{1}{\eta_D}\tilde{D}\dot{\tilde{D}} \leq \tilde{D}|z_3| + \frac{1}{\eta_D}\tilde{D}\dot{\tilde{D}} = \eta_1\tilde{D}\hat{D} \tag{35}$$

Hence, the inequality of $\dot{V}(t)$ considering (35) and (T.9) can be written as

$$\dot{V}(t) \leq -c_1 z_1^2 - c_2 z_2^2 - (c_3 - c_{30})z_3^2 - b_p \Delta i z_3$$
$$-c_{10}\chi^2 + \theta_{10}\Delta i\chi + \theta_{10}\Delta i z_3 - \chi z_2 + \eta_1\tilde{D}\hat{D} \tag{36}$$

According to the Young's inequality, we have

$$(\theta_{10} - b_p)\Delta i z_3 \leq (\theta_{10} - b_p)z_3^2 + \sigma_0 \tag{37}$$

$$\theta_{10}\chi\Delta i \leq \theta_{10}\chi^2 + \sigma_1 \tag{38}$$

$$-\chi z_2 \leq \chi^2 + \frac{1}{4}z_2^2 \tag{39}$$

$$\eta_1\tilde{D}\hat{D} \leq -\frac{\eta_1}{2}\tilde{D}^2 + \frac{\eta_1}{2}D^2 \tag{40}$$

where $\sigma_0 = \frac{1}{4}(\theta_{10} - b_p)\Delta i^2$, $\sigma_1 = \frac{1}{4}\theta_{10}\Delta i^2$.

Considering the preceding inequalities in (37)–(40), one has

$$\dot{V}(t) \leq -c_1 z_1^2 - (c_2 - \frac{1}{4})z_2^2 - (c_3 - c_{30} - \theta_{10} + b_p)z_3^2$$
$$- \frac{\eta_1}{2}\tilde{D}^2 - (c_{10} - \theta_{10} - 1)\chi^2 + \sigma_2 \qquad (41)$$

where $\sigma_2 = \sigma_0 + \sigma_1 + \frac{\eta_1}{2}D^2$. Then we have

$$\dot{V} \leq -2\lambda V + \sigma_2 \qquad (42)$$

where $\lambda = \min\{c_1, c_2 - \frac{1}{4}, c_3 - c_{30} - \theta_{10} + b_p, \frac{\eta_1 \eta_D}{2}, c_{10} - \theta_{10} - 1\}$. Integrating it over $[0, t]$, one has

$$V(t) \leq \frac{\sigma_2}{2\lambda} + (V(0) - \frac{\sigma_2}{2\lambda})e^{-2\lambda t} \qquad (43)$$

It should be noted that for any $\gamma > 0$, the set $B_r = \{z, \chi, \tilde{D} : V(z, \chi, \tilde{D}) \leq \gamma\}$ is a compact set, and assuming $\|\Delta i\|$ has a maximum on the set B_r. Therefore, the inequality in (43) shows that $z_1, z_2, z_3, \tilde{D}$ and χ are bounded as $t \to \infty$. The proof has been finished.

4   Simulation Results

To validate the effectiveness of the developed control approach, the simulation is conducted via MATLAB/SIMULINK. The control objective is to force the system output to follow a desired signal $x_d = 10\sin(2\pi t)$. The parameters in the controller and adaptive laws are selected as $c_1 = 1500$, $c_2 = 2000$, $c_3 = 90000$, $c_{10} = 1000$, $c_{30} = 60000$, $-\rho^- = -10$, $\rho^+ = 10$, $c_L = 1000$, $\kappa = 10$, $\theta_{10} = 1000$, $\eta_D = 100$, $\eta_1 = 1000$. The initial values are selected as $x_1(0) = 1$, $x_2(0) = 0$, $x_3(0) = 0$, $\xi_1(0) = 0$, $\chi(0) = 0$, $\hat{D}(0) = 0$. The saturation threshold value in (28) are set as $\alpha = -1.7A$ and $\beta = 1.7A$.

To illustrate the effectiveness of the anti-windup block, the comparisons are made among the following three conditions

- unconstrained systems
- constrained systems without anti-windup block
- constrained systems with anti-windup block

Figure 3(a)–(d) represent the comparison results. In Fig. 3(a), in order to follow the desired signal, for the unconstraint case, the controller generates a large current spike (around $-9.4A$) at the beginning, see the green dashed line in Fig. 3(a), which may burn out the magnetostrictive actuator in the real application. If a saturation block is directly applied (the case of constrained systems without anti-windup block), severe oscillations between the upper bound ($1.7A$) and lower bound ($-1.7A$) occur in the controller output, also leading large tracking errors, see the blue dotted line in Fig. 3(a) and (b). In the case of constrained systems with anti-windup block, although at the beginning the control strategy shows a large tracking error, the tracking error then maintains within 4% and

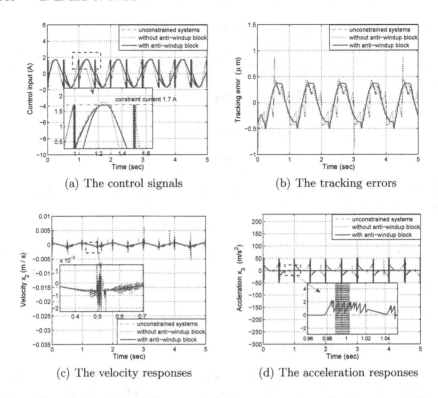

(a) The control signals   (b) The tracking errors

(c) The velocity responses   (d) The acceleration responses

Fig. 3. Simulation results with unconstrained systems, constrained systems without an anti-windup block and constrained systems with an anti-windup block (Color figure online)

no oscillations generated in the control signals, see the red solid line in Fig. 3(a) and (b). Figure 3(c) and (d) illustrate the comparisons of the velocity and acceleration at the end point of the magnetostrictive-actuated dynamic system. For the constrained system without the anti-windup block, the output of the actuator shows continuous oscillations, which might damage the actuator and reduce its lifespan. From above comparison results, it clearly demonstrates the effectiveness of the developed controller with the anti-windup block. For the future research, the output feedback control strategy combining the anti-windup block will be studied since in practical applications the velocity and acceleration of the actuator are not accessible.

5   Conclusion

Input saturation and hysteresis nonlinearity are two main problems that limit the performance of the magnetostrictive actuators. Towards these two problems, in this paper, an adaptive control approach combined with an anti-windup block is developed. From the simulation validation, the designed controller can effectively suppress the oscillation caused by the input saturation meanwhile reducing the tracking error and hysteresis error to an acceptable range. In addition, the anti-windup block is independent of the adaptive controller, which is more feasible in practical implementation.

References

1. Li, Z., Su, C.-Y., Chai, T.: Compensation of hysteresis nonlinearity in magnetostrictive actuators with inverse multiplicative structure for preisach model. IEEE Trans. Autom. Sci. Eng. **11**(2), 613–619 (2014)
2. Gu, G., Zhu, L., Su, C.-Y.: Modeling and compensation of asymmetric hysteresis nonlinearity for piezoceramic actuators with a modified prandtl-ishlinskii model. IEEE Trans. Ind. Electron. **61**(3), 1583–1595 (2014)
3. Abidi, K., Sabanovic, A.: Sliding-mode control for high-precision motion of a piezostage. IEEE Trans. Ind. Electron. **54**(1), 629–637 (2007)
4. Shieh, H.-J., Hsu, C.-H.: An adaptive approximator-based backstepping control approach for piezoactuator-driven stages. IEEE Trans. Ind. Electron. **55**(4), 1729–1738 (2008)
5. Chen, M., Ge, S.: Adaptive neural output feedback control of uncertain nonlinear systems with unknown hysteresis using disturbance observer. IEEE Trans. Ind. Electron. **62**(12), 7706–7716 (2015)
6. Nikdel, N., Nikdel, P., Badamchizadeh, M.A., Hassanzadeh, I.: Using neural network model predictive control for controlling shape memory alloy-based manipulator. IEEE Trans. Ind. Electron. **61**(3), 1394–1401 (2014)
7. Teel, A.R., Zaccarian, L., Marcinkowski, J.J.: An anti-windup strategy for active vibration isolation systems. Control Eng. Pract. **14**(1), 17–27 (2006)
8. Sun, W., Zhao, Z., Gao, H.: Saturated adaptive robust control for active suspension systems. IEEE Trans. Ind. Electron. **60**(9), 3889–3896 (2013)
9. Li, Z.: Modeling and Control of Magnetostrictive-actuated Dynamic Systems. Ph.D. thesis, Concordia University, Canada, February 2015
10. Goldfarb, M., Celanovic, N.: Modeling piezoelectric stack actuators for control of micromanipulation. IEEE Control Syst. **17**(3), 69–79 (1997)
11. Adriaens, H., Koning, W., Banning, R.: Modeling piezoelectric actuators. IEEE/ASME Trans. Mechatron. **5**(4), 331–341 (2000)
12. Li, Z., Su, C.-Y., Chen, X.: Modeling and inverse adaptive control of asymmetric hysteresis systems with applications to magnetostrictive actuator. Control Eng. Pract. **33**, 148–160 (2014)

Adaptive Dynamic Surface Inverse Output Feedback Control for a Class of Hysteretic Systems

Xiuyu Zhang[1], Dan Liu[1], Zhi Li[2], and Chun-Yi Su[2(✉)]

[1] School of Automation Engineering, Northeast Dianli University, Jilin 132012, China
zhangxiuyu80@163.com, 13804428706@139.com
[2] Department of Mechanical and Industrial Engineering, Concordia University,
Montreal, QC H3G 1M8, Canada
gavinlizhi@gmail.com, chun-yi.su@concordia.ca

Abstract. In this paper, an robust neural adaptive output-feedback inverse control scheme for a class of hysteretic nonlinear systems is proposed. Firstly, by designing a high-gain observer to estimate the states of the system and cope with the uncertainties of the system, only the output of the control system is required to be measurable. Secondly, the nonlinear function in the systems can totally unknown due to the utilization of the neural networks approximator. Finally, the arbitrarily small $\mathcal{L}_\infty$ norm of the tracking error is achieve by adjusting the initial conditions of the unknown parameters.

Keywords: Inverse control · PI hysteresis model · $\mathcal{L}_\infty$ performance

1 Introduction

Recently, the smart-material based actuators are widely used in the tuning metal cutting system and other ultrahigh-precision positioning devices [1–3]. However, the existence of the hysteresis highly prohibit the control precision [4,5].

The construction of the inverse model of the hysteresis is the commonly method dealing with hysteresis [4–11]. The robust adaptive control method without constructing the hysteresis inverse [5,12–19,30] is the other method. Though there are some existing results [1,12,16–25,28] of modeling and control for the practicable hysteretic nonlinear systems, an output-feedback inverse control scheme is still missing.

In this paper, an adaptive neural output-feedback inverse control is proposed. Firstly, by designing a high-gain observer to estimate the states of the system and cope with the uncertainties of the system, only the output of the control system is required to be measurable. Secondly, the nonlinear function in the systems

This work was supported by the National Natural Science Foundation of China under Grants 61304015.

N. Kubota et al. (Eds.): ICIRA 2016, Part I, LNAI 9834, pp. 646–662, 2016.
DOI: 10.1007/978-3-319-43506-0_56

can totally unknown due to the utilization of the neural networks approximator. Finally, the arbitrarily small $\mathcal{L}_\infty$ norm of the tracking error is achieve by
adjusting the initial conditions of the unknown parameters.

The rest of this paper is organized as follows. In Sect. 2, the problem statement, the assumptions and the control objective. Section 3 presents the design
procedure. The stability analysis are given in Sect. 4 and the simulation results
is shown to illustrate the effectiveness of the proposed method.

2   Problem Statement

We consider a class of nonlinear system preceded by hysteresis as follows:

$$\dot{x}_i = x_{i+1} + f_i(\bar{x}_i) + d_i(t),$$
$$\dot{x}_n = b_0 w(u) + f_n(\bar{x}_n) + d_n(t),$$
$$y = x_1, \, i = 0, 1, \cdots, n-1, \tag{1}$$

where $\bar{x}_i := [x_1, x_2, \cdots, x_i]^T \in \mathbb{R}^i$ is the state vector; $f_i(\bar{x}_i)$, $i = 0, 1, \cdots, n$ are
the unknown smooth nonlinear functions. τ_i are unknown time delays. $d_i(t)$ are
external disturbances. b_0 is an unknown constant parameter. $w \in R$ represents
the unknown hysteresis which can be expressed as

$$w(u) = P(u(t)) \tag{2}$$

with u being the input signal of the actuator and Π being the hysteresis operator
which will be discussed in details below.

For the system (1), the following assumptions are required:

A1: The disturbances $d_i(t)$, $i = 1, \cdots, n$, satisfy

$$|d_i(t)| \leq \bar{d}_i, \tag{3}$$

where $\bar{d}_i$ are some unknown positive constants.

A2: The desired trajectory y_r is smooth and available with $y_r(0)$ at designer's
disposal; $[y_r, \dot{y}_r, \ddot{y}_r]^T$ belongs to a known compact set for all $t \geq 0$.

A3: The sign of b_0 is known, without loss of generality, we assume that $b_0 > 0$
for simplicity.

2.1   The Prandtl-Ishlinskii (PI) Model and Its Inverse

Though a large number of hysteresis models have been reported, in this paper,
the PI model which is suitable for describing the hysteresis phenomena in piezoelectric actuators is employed and its corresponding inverse is adopted to mitigate the effects of the hysteresis phenomenon [26].

$$w(t) = P[u](t) \tag{4}$$

with $P[u](t)$ being defined as [26]

$$P[u](t) = p_0 u(t) + \int_0^\Lambda p(r) F_r[u](t) dr \tag{5}$$

where r represents the threshold, $p(r)$ is a given density function satisfying $p(r) > 0$ with $\int_0^\infty p(r) dr < \infty$, for convenience, $p_0 = \int_0^D p(r) dr$ is a constant decided by density function $p(r)$. Λ denotes the upper limit of the integration. Let f_r: $\mathbb{R} \to \mathbb{R}$ be defined by

$$f_r(u, w) = \max(u - r, \min(u + r, w)). \tag{6}$$

Then, the play operator $F_r[u](t)$ satisfies

$$\begin{aligned}
F_r[u](0) &= f_r(u(0), 0), \\
F_r[u](t) &= f_r(u(t), F_r[u](t_i)), \\
&\text{for } t_i < t \le t_{i+1} \text{ and } 0 \le i \le N - 1,
\end{aligned} \tag{7}$$

where $0 = t_0 < t_1 < \cdots < t_N = t_E$ is a partition of $[0, t_E]$ such that the function u is monotone (nondecreasing or non-increasing) on each of the sub-intervals $(t_i, t_{i+1}]$.

To compensate the hysteresis nonlinearities $w(u)$ in (1), the inverse of the PI model is constructed as follows [9]:

$$u(t) = P^{-1} \circ P[u(t)] = P^{-1}[w](t), \tag{8}$$

where $\circ$ denotes the composition operator; $P^{-1}[\cdot]$ is the inverse operator of the PI model with

$$P^{-1}[u](t) = \bar{p}_0 u(t) + \int_0^{\bar{\Lambda}} \bar{p}(r) F_r[u](t) dr, \tag{9}$$

where $\bar{\Lambda}$ is a constant denoting the upper-limit of the integration in (9) and

$$\begin{aligned}
\bar{p}_0 &= \frac{1}{p_0}, \\
\bar{p}(r) &= (\varphi^{-1})''(r), \\
\varphi(r) &= \bar{p}_0 r + \int_0^r \bar{p}(\xi)(r - \xi) d\xi.
\end{aligned} \tag{10}$$

Since in practice, the hysteresis is unknown which implies the density function $p(r)$ needs to be estimated based on the measured data. Here, we use $\hat{p}(r)$ and $\hat{P}[u](t)$, which can be got from experiments data, denotes the estimation of $p(r)$ and $P[u](t)$, respectively. Thus, as that in [26], by applying the composition theorem to the $P[\cdot](t)$ and $\hat{P}^{-1}[\cdot](t)$ as in [9], yields

$$P \circ \hat{P}^{-1}[u_d](t) = \phi'(0)u_d(t)$$
$$+ \int_0^\Lambda \phi''(r)F_r[u_d](t)dr, \tag{11}$$

with u_d being the control signal to be designed. $\phi(r) = p \circ \hat{p}^{-1}(r)$, $p(r)$ and $\hat{p}^{-1}(r)$ being the initial loading curves of the $P[\cdot](t)$ and $\hat{P}^{-1}[\cdot](t)$.

Considering (11) and the equality $F_r[u_d](t) + E_r[u_d](t) = u_d(t)$ given in [26], it follows that

$$w(t) = \phi'(\Lambda)u_d + d_b(t), \tag{12}$$

where $\phi'(\Lambda)$ is a positive constant, $E_r(\cdot)$ is the stop operator of PI model. Due to $|E_r(\cdot)| < \Lambda$ (see [9]), the term $d_b(t) = -\int_0^\Lambda \phi''(r)E_r[u_d](t)dr$ is bounded and satisfies

$$|d_b(t)| \leq D \tag{13}$$

with D being a positive constant. Therefore, from (11) and (12), the analytical error $e(t)$ expression can be obtained as follows

$$e(t) = w(t) - u_d(t)$$
$$= [\phi'(\Lambda) - 1]u_d + d_b(t). \tag{14}$$

Now, substituting (12) into (1), we have

$$\dot{x}_i = x_{i+1} + f_i(\bar{x}_i) + d_i(t), \ i = 0, 1, \cdots, n-1$$
$$\dot{x}_n = b_\Lambda u_d + f_n(\bar{x}_n) + b_0 d_b(t) + d_n(t),$$
$$y = x_1, \tag{15}$$

where b_Λ is a positive constant satisfying

$$b_\Lambda = b_0 \phi'(\Lambda). \tag{16}$$

2.2 Radial Basis Function Neural Networks

In this paper, the radial basis function neural network (RBFNNs) with a linear in the weights property will be employed to approximate a continuous function on the compact sets under the following Lemma 1.

Lemma 1 [27]: RBFNNs are universal approximators in the sense that given any real continuous function $f : \Omega_\xi \to \mathbb{R}$ being a compact set with $\Omega_\xi \subset \mathbb{R}^q$, ξ being the NNs input and q denoting the input dimension. For any $\varepsilon_m > 0$, by appropriately choosing σ and $\zeta_k \in \mathbb{R}^q$, $k = 1, \ldots, N$, then, there exists an RBFNN such that

$$f(\xi) = \psi^T(\xi)\vartheta^* + \varepsilon,$$
$$\forall \xi \in \Omega_\xi \subset \mathbb{R}^n, |\varepsilon| \leq \varepsilon_m, \tag{17}$$

where ϑ^* is an optimal weight vector of $\vartheta = [\vartheta_1, \ldots, \vartheta_N] \in \mathbb{R}^N$ and defined as

$$\vartheta^* = \arg \min_{\vartheta \in \mathbb{R}^n} \left\{ \sup_{\xi \in \Omega_\xi} |Y(\xi) - f(\xi)| \right\}, \tag{18}$$

$\psi(\xi) = [\psi_1(\xi), \ldots, \psi_N(\xi)] \in \mathbb{R}^N$ is an basis function vector. Generally, the so-called Gaussian function is used as basis function in the following form:

$$\psi_k(\xi) = \exp\left(-\frac{\|\xi - \zeta_k\|}{2\sigma^2}\right),$$
$$\text{with } \sigma > 0, k = 1, \ldots, N, \tag{19}$$

where $\zeta_k \in \mathbb{R}^n$ is a constant vector called the center of the basis function, and σ is a real number called the width of the basis function and ε being approximation error, satisfying

$$\varepsilon = f(\xi) - \vartheta^{*\mathrm{T}} \psi(\xi). \tag{20}$$

Then, by using Lemma 1 and (17), the RBFNNs are used as the approximators to approximate the unknown continuous functions in (17) as follows:

$$f_i(\bar{x}_i) = \psi_i^T(\xi_i)\vartheta_i^* + \varepsilon_i,$$
$$\text{for } i = 1, \cdots, N \tag{21}$$

with ε_i being any positive constants denoting the neural networks approximated errors and

$$\xi_i := (\hat{\bar{x}}_1, \cdots, \hat{\bar{x}}_i,), \ i = 1, \cdots, n, \tag{22}$$

where $\hat{\bar{x}}_1, \cdots, \hat{\bar{x}}_i$ are the estimations of the state variables $x_1, \cdots, x_i$, and will be introduced in the following section.

Now, substituting (21) into (15), we have

$$\dot{x}_i = x_{i+1} + \psi_i^T(\xi_i)\vartheta_i^* + \delta_{i0}$$
$$+\varepsilon_i + d_i(t),$$
$$\dot{x}_n = b_\Lambda u_d + \psi_n^T(\xi_n)\vartheta_n^* + \delta_{n0} + \varepsilon_n$$
$$b_0 d_b(t) + d_n(t),$$
$$y = x_1, \ i = 0, 1, \cdots, n-1 \tag{23}$$

from which system (1) eventually can be expressed as the following state-space form

$$\dot{x} = Ax + \Psi^T(\xi)\vartheta^* + bu_d + D_b$$
$$+\delta_0 + \varepsilon + d,$$
$$y = e_1^T x, \tag{24}$$

where

$$
A = \begin{bmatrix} 0 & 1 & & 0 \\ 0 & & \ddots & \\ \vdots & & & 1 \\ 0 & \cdots & 0 & 0 \end{bmatrix}, \ b = \begin{bmatrix} 0 \\ \vdots \\ 0 \\ b_\Lambda \end{bmatrix},
$$

$$
d = \begin{bmatrix} d_1(t) \\ \vdots \\ d_{n-1}(t) \\ d_n(t) \end{bmatrix}, \ \varepsilon = \begin{bmatrix} \varepsilon_1 \\ \vdots \\ \varepsilon_{n-1} \\ \varepsilon_n \end{bmatrix},
$$

$$
\delta_0 = \begin{bmatrix} \delta_{10} \\ \vdots \\ \delta_{n-10} \\ \delta_{n0} \end{bmatrix}, \ D_b = \begin{bmatrix} 0 \\ \vdots \\ 0 \\ d_b(t) \end{bmatrix},
$$

$$
\vartheta^* = \begin{bmatrix} \vartheta_1^* \\ \vdots \\ \vartheta_{n-1}^* \\ \vartheta_n^* \end{bmatrix} \subset \mathbb{R}^{\Sigma_{i=1}^n N_n},
$$

$$
\Psi^T(\xi) = \begin{bmatrix} \psi_1 & & \\ & \ddots & \\ & & \psi_n \end{bmatrix} \tag{25}
$$

with $\psi_1 = [\psi_{1,1}(\xi_1), \cdots, \psi_{1,N_1}(\xi_1)]$, $\psi_n = [\psi_{n,1}(\xi_n), \cdots, \psi_{n,N_n}(\xi_n)]$, and N_i, $i = 1, \cdots, n$ being defined in (19).

The control objective is to develop an adaptive neural output-feedback dynamic surface inverse control scheme for a class of nonlinear hysterestic system such that the output y well tracks the reference signal y_r with the $\mathcal{L}_\infty$ norm of the tracking error and all the signals of the closed loop system are uniformly bounded.

3 Observer Based Adaptive DSIC Design

3.1 High-Gain K-Filter Observer

Now, (24) can be transformed as the following

$$
\dot{x} = A_0 x + qy + \Psi^T(\xi)\vartheta^* + bu_d + B,
$$
$$
y = e_1^T x, \tag{26}
$$

by letting

$$
B = D_b + \delta_0 + \varepsilon + d, \tag{27}
$$

and

$$A_0 = A - qe_1^T$$

$$= \begin{bmatrix} -q_1 & 1 & & \\ -q_2 & & \ddots & \\ \vdots & & & 1 \\ -q_n & 0 & \cdots & 0 \end{bmatrix}$$

$$\text{with } q = \begin{bmatrix} q_1 \\ q_2 \\ \vdots \\ q_n \end{bmatrix}, \tag{28}$$

where A_0 is a Hurwitz matrix by properly choosing the vector q.

Inspired by the previous work [13,14], the following high-gain K-Filter is construct to estimate the states x in systems (26).

$$\dot{v}_0 = kA_0v_0 + \Phi^{-1}e_nu_d, \tag{29}$$

$$\dot{\xi}_0 = kA_0\xi_0 + kqy, \tag{30}$$

$$\dot{\Xi} = kA_0\Xi + \Phi^{-1}\Psi^T, \tag{31}$$

where $k \geq 1$ is a positive design parameter, e_n denotes the n-th coordinate vector in $\mathbb{R}^n$, and

$$\Phi = diag\{1, k, \cdots, k^{n-1}\}. \tag{32}$$

From (29)–(32), the estimated states vector is as the following:

$$\hat{x} = \Phi\xi_0 + \Phi b_\Lambda v_0 + \Phi\Xi\vartheta^*. \tag{33}$$

To proceed, we define the estimation error

$$\epsilon = x - \hat{x}. \tag{34}$$

It is easy to verify that

$$A = k\Phi A\Phi^{-1},$$
$$k\Phi qe_1^T\Phi^{-1} = k\Phi qe_1^T \tag{35}$$

with $e_1^T = [1, 0, \cdots, 0]^T$. Then, we have

$$\dot{\epsilon} = A\epsilon - k\Phi qe_1 + B, \tag{36}$$

where ϵ_1 is the first entry of ϵ and B is defined in (27).

Lemma 2: Let the high-gain K-filters be defined by (29)–(31) and the quadratic function

$$V_\epsilon := \epsilon^T P\epsilon, \tag{37}$$

where $P = (\Psi^{-1})^T \bar{P} \Psi^{-1}$ with $\bar{P} = \bar{P}^T$ is positive define matrix $(\bar{P} > 0)$ satisfying

$$A_0^T \bar{P} + \bar{P} A_0 = -2I, \tag{38}$$

where A_0 is defined by (28). Let

$$\zeta_\epsilon := \frac{k}{\lambda_{\max}(\bar{P})},$$

$$\delta_\epsilon := k \left(\frac{\|\bar{P}\| \|B\|_{\max}}{k^n} \right)^2, \tag{39}$$

where $\|B\|_{\max}$ is the maximum value of $\|B\|$. Then, for any $k \geq 1$, we have

$$\dot{V}_\epsilon \leq -\zeta_\epsilon V_\epsilon + \delta_\epsilon. \tag{40}$$

Proof: See [13] for more details.

It should be noted that because b_Λ and ϑ^* in (33) are unknown, $\hat{x}(t)$ is unavailable. Therefore, the actual state estimation is

$$\hat{\hat{x}} = \Phi \xi_0 + \Phi \hat{b}_\Lambda v_0 + \Phi \Xi \hat{\vartheta}, \tag{41}$$

where $\hat{b}_\Lambda$ and $\hat{\vartheta}$ are the estimations of b_Λ and ϑ^*, and will be given in details in the next section.

3.2   Dynamic Surface Inverse Controller Design

By using the states observer in (29)–(31), a robust adaptive dynamic surface inverse control scheme will be given with the following structure of the controller (Fig. 1).

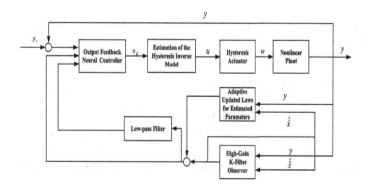

Fig. 1. The structure of the proposed control scheme

Based on the above controller structure, the procedures of the controller design are as follows:

Step 1: Define the first surface error as

$$S_1 = y - y_r, \tag{42}$$

whose time derivative by considering (24) is

$$
\begin{aligned}
\dot{S}_1 &= \dot{y} - \dot{y}_r \\
&= x_2 + \psi_1^T(\xi_1)\vartheta_1^* + \delta_{10} \\
&\quad + \varepsilon_1 + d_1 - \dot{y}_r.
\end{aligned} \tag{43}
$$

From (34), we have

$$
\begin{aligned}
x_2 &= \hat{x}_2 + \epsilon_2 \\
&= k\xi_{0,2} + b_\Lambda k v_{0,2} + k\Xi_{(2)}\vartheta^* + \epsilon_2,
\end{aligned} \tag{44}
$$

where $\Xi_{(2)}$ denotes the second row of Ξ. Then, it follows that

$$
\begin{aligned}
\dot{S}_1 &= k\xi_{0,2} + b_\Lambda k v_{0,2} + k\Xi_{(2)}\vartheta^* \\
&\quad + \vartheta_1^{*T}\psi_1(\xi_1) - \dot{y}_r + \epsilon_2 + \delta_{10} \\
&\quad + \varepsilon_1 + d_1
\end{aligned} \tag{45}
$$

with ξ_1 being defined in (22). Note that $\vartheta_1^{*T}\psi_1(\xi_1) = \vartheta^{*T}\Psi_{(1)}^T$ with $\Psi_{(1)}^T$ denoting the first row of Ψ^T, then, (45) can be rewritten as

$$
\begin{aligned}
\dot{S}_1 &= k\xi_{0,2} + b_\Lambda k(v_{0,2} - \bar{v}_{0,2}) + b_\Lambda k\bar{v}_{0,2} \\
&\quad + \vartheta^{*T}(k\Xi_{(2)} + \Psi_{(1)})^T - \dot{y}_r + \epsilon_2 \\
&\quad + \delta_{10} + \varepsilon_1 + d_1,
\end{aligned} \tag{46}
$$

where $\bar{v}_{0,2}$ is the first virtual control signal to be designed. Let $\bar{v}_{0,2}$ be of the following form

$$\bar{v}_{0,2} = \hat{\zeta}\bar{v}_{0,2}' \tag{47}$$

with $\hat{\zeta}$ being the estimation of $\zeta = 1/b_\Lambda$ and

$$
\begin{aligned}
\bar{v}_{0,2}' &= [-l_1 S_1 - k\xi_{0,2} - \hat{\vartheta}^T(k\Xi_{(2)} \\
&\quad + \Psi_{(1)})^T + \dot{y}_r]/k,
\end{aligned} \tag{48}
$$

where $\hat{\vartheta}$, is the estimate of ϑ^*. The updated laws of $\hat{\zeta}$ and $\hat{\vartheta}$ are as follows

$$\dot{\hat{\zeta}} = -\gamma_\zeta(k\bar{v}_{0,2}' S_1 + \sigma_\zeta\hat{\zeta}) \tag{49}$$

$$\dot{\hat{\vartheta}} = \gamma_\vartheta[(k\Xi_{(2)} + \Psi_{(1)})^T S_1 - \sigma_\vartheta\hat{\vartheta}] \tag{50}$$

Let $\bar{v}_{0,2}$ pass through the a first-order filter to obtain a new state variable z_2

$$\tau_2 \dot{z}_2 + z_2 = \bar{v}_{0,2}, \ z_2(0) = \bar{v}_{0,2}(0), \tag{51}$$

where τ_2 is the time constant of the first-order filter.

Step 2: Define the second surface error

$$S_2 = v_{0,2} - z_2, \tag{52}$$

whose time derivative by considering $\dot{v}_{0,2}$ in (29) is

$$\begin{aligned}
\dot{S}_2 &= -kq_2 v_{0,1} + k v_{0,3} - \dot{z}_2 \\
&= -kq_2 v_{0,1} + k(v_{0,3} - \bar{v}_{0,3}) \\
&\quad + k\bar{v}_{0,3} - \dot{z}_2.
\end{aligned} \tag{53}$$

Then the virtual control $\bar{v}_{n-\rho,i+1}$ is chosen as

$$\bar{v}_{0,3} = (-l_2 S_2 + kq_2 v_{0,1} + \dot{z}_2 - \hat{b}_\Lambda k S_1)/k, \tag{54}$$

where l_2 is a positive design parameters and $\hat{b}_\Lambda$ is the estimation of b_Λ defined in (16). The updated law of $\hat{b}_\Lambda$ is designed as

$$\dot{\hat{b}}_\Lambda = \gamma_b (k S_1 S_2 - \sigma_b \hat{b}_\Lambda) \tag{55}$$

Let $\bar{v}_{0,3}$ pass through the a first-order filter to obtain a new state variable z_3:

$$\tau_3 \dot{z}_3 + z_3 = \bar{v}_{0,3}, \ z_3(0) = \bar{v}_{0,3}(0). \tag{56}$$

where τ_3 is the time constant of the first-order filter.

Step i $(3 \le i \le n-1)$: Define the i-th surface error

$$S_i = v_{0,i} - z_i, \tag{57}$$

whose time derivative by considering $\dot{v}_{0,i}$ in (29) is

$$\begin{aligned}
\dot{S}_i &= -kq_i v_{0,1} + k v_{0,i+1} - \dot{z}_i \\
&= -kq_i v_{0,1} + k(v_{0,i+1} - \bar{v}_{0,i+1}) \\
&\quad + k\bar{v}_{0,i+1} - \dot{z}_i.
\end{aligned} \tag{58}$$

Then the virtual control $\bar{v}_{0,i+1}$ is chosen as

$$\bar{v}_{0,i+1} = (kq_i v_{0,1} + \dot{z}_i - l_i S_i)/k, \tag{59}$$

where l_i, $i = 3, \cdots, n-1$, are positive design parameters. Let $\bar{v}_{0,i+1}$ pass through the a first-order filter to obtain a new state variable z_{i+1}:

$$\begin{aligned}
\tau_{i+1} \dot{z}_{i+1} + z_{i+1} &= \bar{v}_{0,i+1}, \\
z_{i+1}(0) &= \bar{v}_{0,i+1}(0).
\end{aligned} \tag{60}$$

where τ_{i+1} is the time constant of the first-order filter.

Step n: Define the n-th surface error

$$S_n = v_{0,n} - z_n, \tag{61}$$

whose derivative by considering $\dot{v}_{0,n}$ in (29) is

$$\dot{S}_n = -kq_n v_{0,1} + k^{1-n} u_d - \dot{z}_n. \tag{62}$$

The actual control u_d appears in this step and is chosen as

$$u_d = k^{n-1}(kq_n v_{0,1} + \dot{z}_n - l_n S_n), \tag{63}$$

where l_n is a positive design parameter.

4   Stability and $\mathcal{L}_\infty$ Tracking Performance Analysis

In this section, the stability and performance analysis for the proposed adaptive output feedback DSIC scheme will be discussed. Now, we are ready to present the main theorem of this paper to analyize the stability and achieve the $\mathcal{L}_\infty$ performance of the tracking error.

Theorem 1: Consider the closed loop system including the time-delay system (1) with hysteresis nonlinearity described by (4), the updated laws of the unknown parameters (36), (37), (55), and the control law (63) with respect to Assumptions A1–A3. Then, for any given positive number p, if $V(0)$ in (68) satisfies $V(0) \le p$,

(a) all the signals of the closed loop system are uniformly bounded and can be made arbitrarily small by properly choosing the design parameters $k, l_1, \cdots, l_n$, the time constant $\tau_2, \cdots, \tau_n$, and the update law parameters γ_ϑ, σ_ϑ, γ_ζ, σ_ζ, γ_b, σ_b.

(b) the $\mathcal{L}_\infty$ performance the of the tracking error S_1 can be obtained and arbitrary small and satisfy

$$\|S_1\|_\infty \le \sqrt{\frac{C_2}{C_1} + \frac{2}{k^2}\lambda_{\max}(\bar{P})\|\epsilon(0)\|^2} \tag{64}$$

where C_1 is a design parameter and C_2 is a positive constant that will be given in the proof of Theorem 1.

Proof: First of all, define

$$y_i = z_i - \bar{v}_{n-\rho,i}, \ i = 2, \cdots, \rho. \tag{65}$$

Then, we have

$$\left| \dot{y}_2 + \frac{y_2}{\tau_2} \right| \le B_2(S_1, \ldots, S_n, y_2, \cdots, y_n,$$

$$\tilde{b}_\Lambda, \tilde{\zeta}, \tilde{\vartheta}, y_r, \dot{y}_r, \ddot{y}_r, \epsilon), \tag{66}$$

where B_2 is a continuous function. Similarly, it can be verified that for $i = 2, \cdots, n-1$,

$$\left| \dot{y}_{i+1} + \frac{y_{i+1}}{\tau_{i+1}} \right| \le B_{i+1}(S_1, \ldots, S_\rho, y_2, \cdots, y_n,$$
$$\tilde{b}_\Lambda, \tilde{\zeta}, \tilde{\vartheta}, y_r, \dot{y}_r, \ddot{y}_r, \epsilon), \qquad (67)$$

where B_{i+1} are some continuous functions. For the analysis of stability, define the Lyapunov function as:

$$V = \frac{1}{2} \sum_{i=1}^{n} S_i^2 + \frac{1}{2} \sum_{i=1}^{n-1} y_{i+1}^2 + \frac{1}{2\gamma_\vartheta} \tilde{\vartheta}^T \tilde{\vartheta}$$
$$+ \frac{b_\Lambda}{2\gamma_\varsigma} \tilde{\zeta}^2 + \frac{1}{2\gamma_b} \tilde{b}_\Lambda^2 + V_\epsilon \qquad (68)$$

where $\tilde{b}_\Lambda = \hat{b}_\Lambda - b_\Lambda$, $\tilde{\vartheta}$ and $\tilde{\zeta}$ have been introduced as in [13], V_ϵ is a quadratic function concerning the high-gain K-Filter observer error ϵ which has been given in Lemma 2.

Define the following compact sets

$$\Omega_1 = \left\{ (y_r, \dot{y}_r, \ddot{y}_r) : y_r^2 + \dot{y}_r^2 + \ddot{y}_r^2 \le G_0 \right\},$$
$$\Omega_2 = \left\{ \begin{array}{l} \sum_{i=1}^{n} S_i^2 + \sum_{i=1}^{n-1} y_{i+1}^2 + \frac{1}{\gamma_\vartheta} \tilde{\vartheta}^2 \\ \frac{1}{\gamma_b} \tilde{b}_\Lambda + \frac{b_\Lambda}{\gamma_\varsigma} \tilde{\zeta}^2 + 2\epsilon^T P\epsilon \le 2p \end{array} \right., \qquad (69)$$

where G_0 and p are positive constant. Note that $\Omega_1 \times \Omega_2$ is also compact. Therefore, as that in [29], the continuous functions B_{i+1}, in (66) and (67) have maximum values on $\Omega_1 \times \Omega_2$, say, $M_{i+1}, i = 1, \cdots, n-1$. Then, from (??)–(67), it follows that

$$y_{i+1}\dot{y}_{i+1} \le -\frac{y_{i+1}^2}{\tau_{i+1}} + \frac{y_{i+1}^2 M_{i+1}^2}{2\varsigma} + \frac{\varsigma}{2},$$
$$i = 1, \ldots, \rho - 1, \qquad (70)$$

where ς is any positive constant.

From (68), the time derivative of the Lyapunov function V is

$$\dot{V} = \sum_{i=1}^{n} S_i \dot{S}_i + \sum_{i=1}^{n-1} y_{i+1}\dot{y}_{i+1} + \frac{1}{\gamma_\vartheta} \tilde{\vartheta}^T \dot{\tilde{\vartheta}}$$
$$+ \frac{b_\Lambda}{\gamma_\varsigma} \tilde{\zeta}\dot{\tilde{\zeta}} + \frac{1}{\gamma_b} \tilde{b}_\Lambda \dot{\hat{b}}_\Lambda + \dot{V}_\epsilon \qquad (71)$$

Then, by choosing the design parameters as

$$k \geq \lambda_{\max}(\bar{P})C_1 + \frac{\lambda_{\max}(\bar{P})}{2\lambda_{\min}(\bar{P})},$$

$$l_1 \geq (\frac{b_\Lambda k}{2} + \frac{k^2}{2} + C_1),$$

$$l_2 \geq k + C_1, l_n \geq \frac{k}{2} + C_1,$$

$$l_i \geq \frac{3k}{2} + C_1, \ i = 3, \cdots, n-1,$$

$$\frac{1}{\tau_2} \geq \frac{b_\Lambda k}{2} + \frac{M_2^2}{2\varsigma} + C_1,$$

$$\frac{1}{\tau_{i+1}} \geq \frac{k}{2} + \frac{M_{i+1}^2}{2\varsigma} + C_1,$$

$$i = 2, \cdots, n-1,$$

$$\sigma_\vartheta \gamma_\vartheta \geq 2C_1, \sigma_b \gamma_b \geq 2C_1,$$

$$\sigma_\varsigma \gamma_\varsigma \geq 2C_1, \tag{72}$$

where C_1 is a positive constant. Then, it follows that

$$\dot{V} \leq -2C_1 V + C_2 \tag{73}$$

with

$$C_2 = \frac{(n-1)\varsigma}{2} + \frac{\sigma_\vartheta}{2}\vartheta^{*T}\vartheta^*$$

$$+\frac{b_\Lambda \sigma_\varsigma}{2}\zeta^2 + \frac{\sigma_b}{2}b_\Lambda^2$$

$$+\frac{1}{2}(\delta_{10}^2 + \varepsilon_1^2 + d_1^2) + \delta_\epsilon, \tag{74}$$

and C_1 satisfying

$$C_1 \geq \frac{C}{2p}. \tag{75}$$

Then, $\dot{V} \leq 0$ when $V = p$, which implies that $V(t) \leq p$ is an invariant set or in other words, if $V(0) \leq p$, then $V(t) \leq p$, for all $t \geq 0$. Therefore, Thus, all the signals of the closed loop system are uniformly bounded.

Furthermore, let $y_r(0) = y(0)$. Then, $S_1(0) = 0$. Now, we set the initial condition of the K-Filter as

$$v_0(0) = 0,$$

$$\xi_{0,1}(0) = y(0),$$

$$\Xi(0) = 0, \tag{76}$$

and $\hat{\vartheta}(0) = 0, \hat{\zeta}(0) = 0, \hat{b}_\Lambda(0) = 0$ in the updated laws. Then, it follows that

$$V(t) \leq \frac{C_2}{2C_1} + \frac{1}{k^2}\lambda_{\max}(\bar{P}) \|\epsilon(0)\|^2. \tag{77}$$

Therefore, the $\mathcal{L}_\infty$ norm of the tracking error satisfies

$$\|S_1\|_\infty = \sup_{t \geq 0} |S_1| = \|x_1 - y_r\|_\infty$$

$$\leq \sqrt{\frac{C_2}{C_1} + \frac{2}{k^2}\lambda_{\max}(\bar{P})\,\|\epsilon(0)\|^2}. \tag{78}$$

(78) implies the $\mathcal{L}_\infty$ norm of the tracking error $\|S_1\|_\infty$ can be arbitrarily small by choosing sufficient large design parameters in (72). This completes the proof. ∎

5 Simulation Results

We consider the following general second-order system:

$$\dot{x}_1 = x_2 + 0.8x_1^2 + 0.2x_1x_1(t - 0.4) + 0.1\cos(t),$$
$$\dot{x}_2 = w + 2x_1x_2 + 0.5x_1(t - 0.5)x_2(t - 1),$$
$$y = x_1, \tag{79}$$

where w is the ASPI hysteresis described by (4) with density functions being selected as $p(r) = 0.4e^{-0.015r^2}$, $r \in [0, 10]$; the initial value of state-variables are chosen as $x_1(0) = x_2(0) = 0$. The high-gain K-filters are as follows

$$\dot{v}_0 = kA_0v_0 + \Phi^{-1}e_nu_d, \quad v(0) = 0,$$
$$\dot{\xi}_0 = kA_0\xi_0 + kqy, \quad \xi_0(0) = 0,$$
$$\dot{\Xi} = kA_0\Xi + \Phi^{-1}\Psi^T, \quad \Xi(0) = 0, \tag{80}$$

where

$$k = 2, q = \begin{bmatrix} q_1 \\ q_2 \end{bmatrix} = \begin{bmatrix} 3 \\ 2 \end{bmatrix},$$

$$A_0 = \begin{bmatrix} -q_1 & 1 \\ -q_2 & 0 \end{bmatrix}, \Phi^{-1} = \begin{bmatrix} 1 & 0 \\ 0 & 1/k \end{bmatrix}. \tag{81}$$

where $\psi(\xi) = [\psi^1(\xi), \psi^2(\xi), \ldots, \psi^N(\xi)]^T \in R^N$ is RBFNNs function vector. For NNs $\psi_1(\xi_1)$, we choose 5 nodes with the centers of the basis functions ζ_j, $j = 1, \cdots, 5$, being evenly spaced in $[-1, +1]$, and the width $\eta_j = 1, j = 1, \cdots, 5$; and $\xi_1 = \hat{x}_1$. For NNs $\psi_2(\xi_2)$, we choose 11 nodes with the centers of the basis functions ζ_j, $j = 1, \cdots, 11$, being evenly spaced in $[-2, +2]$, and the width $\eta_j = 1, j = 1, \cdots, 11$; and $\xi_1 = (\hat{x}_1, \hat{x}_2)$. Then, $\Psi^T(\xi) = diag\{\psi_1, \psi_2\}$.

In this simulation, the design parameter are chosen as $l_1 = 30$, $l_2 = 40$, $k = 1.5$, $\gamma_\zeta = 3$, $\sigma_\zeta = 0.004$, $\gamma_\vartheta = 10$, $\sigma_\vartheta = 0.05$, $\gamma_b = 9$, $\sigma_b = 0.0002$. The initial value of the system states are selected as $x_1(0) = x_2(0) = 0$. The initial value of the updated parameters are chosen as $\hat{\zeta}(0) = \hat{\vartheta}(0) = \hat{b}_\Lambda(0) = 0$. The control objective of this simulation is to make the output of the control system follows the desired trajectory $y_r = \sin t$.

According to the above design procedures and the selections of the parameters, the simulation results are shown in Figs. 2, 3 4 and 5. From Fig. 2, the output of the control system $y = x_1$ well tracks the desired trajectory $y_r = \sin t$ when the ASPI hysteresis inverse compensator described in (8)–(10) is applied. Figure 3 shows the trajectories of the tracking errors under two circumstances: with (solid

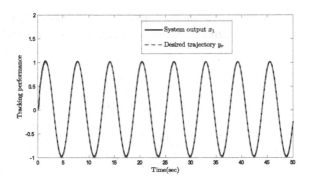

Fig. 2. Tracking performance

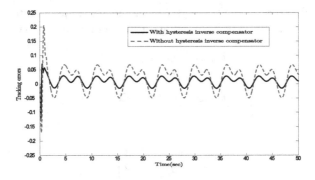

Fig. 3. Tracking errors

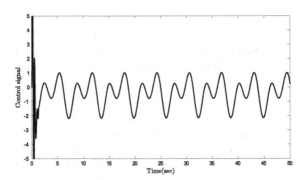

Fig. 4. Control signal

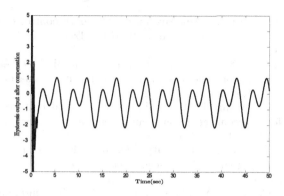

Fig. 5. Hysteresis output after compensation

line) and without (dashed line) considering the ASPI hysteresis inverse compensator. Figure 4 illustrates the trajectory of control signal u_d. Figure 5 illustrates the hysteresis output w after ASPI inverse compensation.

References

1. Gu, G.Y., Zhu, L.M., Su, C.-Y.: Modeling and compensation of asymmetric hysteresis nonlinearity for piezoceramic actuators with a modified Prandtl-Ishlinskii model. IEEE Trans. Ind. Electron. **61**(3), 1583–1595 (2014)
2. Xu, Q., Li, Y.: Micro-/nanopositioning using model predictive output integral discrete sliding mode control. IEEE Trans. Ind. Electron. **59**(2), 1161–1170 (2012)
3. Xu, Q.: Robust impedance control of a compliant microgripper for high-speed position/force regulation. IEEE Trans. Ind. Electron. **62**(2), 1201–1209 (2015)
4. Tao, G., Kokotovic, P.V.: Adaptive control of plants with unknown hysteresis. IEEE Trans. Autom. Control **40**(12), 200–212 (1995)
5. Su, C.Y., Wang, Q.Q., Chen, X.K., Rakheja, S.: Adaptive variable structure control of a class of nonlinear systems with unknown Prandtl-Ishlinskii hysteresis. IEEE Trans. Autom. Control **50**(12), 2069–2074 (2005)
6. Liu, S., Su, C.-Y., Li, Z.: Robust adaptive inverse control of a class of nonlinear systems with Prandtl-Ishlinskii hysteresis model. IEEE Trans. Autom. Control **59**(8), 2170–2175 (2014)
7. Xie, W., Fu, J., Yao, H., Su, C.-Y.: Neural network-based adaptive control of piezoelectric actuators with unknown hysteresis. Int. J. Adapt. Control Signal Process. **23**, 30–54 (2009)
8. Chen, X., Hisayama, T., Su, C.-Y.: Adaptive control for uncertain continuous-time systems using implicit inversion of Prandtl-Ishlinskii hysteresis representation. IEEE Trans. Autom. Control **55**(10), 2357–2363 (2010)
9. Krejci, P., Kuhnen, K.: Inverse control of systems with hysteresis and creep. Proc. Inst. Elect. Eng. **148**(3), 185–192 (2001)
10. Li, Z., Su, C.-Y., Chen, X.: Modeling and inverse adaptive control of asymmetric hysteresis systmes with applications to magnetostrictive actuator. Control Eng. Pract. **33**(12), 148–160 (2014)

11. Zhou, J., Wen, C., Li, T.: Adaptive output feedback control of uncertain nonlinear systems with hysteresis nonlinearity. IEEE Trans. Autom. Control **57**(10), 2627–2633 (2012)
12. Gu, G.Y., Zhu, L.M., Su, C.Y., Ding, H.: Motion control of piezoelectric positioning stages: modeling, controller design, and experimental evaluation. IEEE/ASME Trans. Mechatron. **18**(5), 1459–1471 (2013)
13. Zhang, X., Lin, Y., Wang, J.: High-gain observer based decentralised output feedback control for interconnected nonlinear systems with unknown hysteresis input. Int. J. Control **86**(6), 1046–1059 (2013)
14. Zhang, X., Lin, Y.: An adaptive output feedback dynamic surface control for a class of nonlinear systems with unknown backlash-like hysteresis. Asian J. Control **15**(2), 489–500 (2013)
15. Zhang, X., Su, C.-Y., Lin, Y., Ma, L., Wang, J.: Adaptive neural network dynamic surface control for a class of time-delay nonlinear systems with hysteresis inputs and dynamic uncertainties. IEEE Trans. Neural Netw. Learn. Syst. **26**, 2844–2860 (2015). doi:10.1109/TNNLS.2015.2397935
16. Huang, S., Tan, K.K., Lee, T.H.: Adaptive sliding-mode control of piezoelectric actuators. IEEE Trans. Ind. Electron. **56**(9), 3514–3522 (2009)
17. Shieh, H.-J., Hsu, C.-H.: An adaptive approximator-based backstepping control approach for piezoactuator-driven stages. IEEE Trans. Ind. Electron. **55**(5), 1729–1738 (2008)
18. Xu, Q.: Identification and compensation of piezoelectric hysteresis without modeling hysteresis inverse. IEEE Trans. Ind. Electron. **60**(9), 3927–3937 (2013)
19. Chen, X., Hisayama, T.: Adaptive sliding-mode position control for piezo-actuated stage. IEEE Trans. Ind. Electron. **55**(11), 3927–3934 (2008)
20. Wong, P.K., Xu, Q., Vong, C.M., Wong, H.C.: Rate-dependent hysteresis modeling and control of a piezostage using online support vector machine and relevance vector machine. IEEE Trans. Ind. Electron. **59**(4), 1988–2001 (2012)
21. Tang, H., Li, Y.: Development and active disturbance rejection control of a compliant micro-/nanopositioning piezostage with dual mode. IEEE Trans. Ind. Electron. **61**(3), 1475–1492 (2014)
22. Abidi, K., Sabanovic, A.: Sliding-mode control for high-precision motion of a piezostage. IEEE Trans. Ind. Electron. **54**(1), 629–637 (2007)
23. Huang, D., Xu, J.-X., Venkataramanan, V., Huynh, T.C.T.: High-performance tracking of piezoelectric positioning stage using current-cycle iterative learning control with gain scheduling. IEEE Trans. Ind. Electron. **61**(2), 1085–1098 (2014)
24. Goldfarb, M., Celanovic, N.: Modeling piezoelectric stack actuators for control of micromanipulation. IEEE Control Syst. **17**(3), 69–79 (1997)
25. Adriaens, H.J.M.T.S., De Koning, W.L., Banning, R.: Modeling piezoelectric actuators. IEEE/ASME Trans. Mechatron. **5**(4), 331–341 (2000)
26. Brokate, M., Sprekels, J.: Hysteresis and Phase Transitions. Springer, New York (1996)
27. Sanner, R.M., Slotine, J.-J.E.: Gaussian networks for direct adaptive control. IEEE Trans. Neural Netw. **3**(6), 837–863 (1992)
28. Tong, S.C., Li, Y.: Adaptive fuzzy output feedback tracking backstepping control of strict-feedback nonlinear systems with unknown dead zones. IEEE Trans. Fuzzy Syst. **20**(1), 168–180 (2012)
29. Swaroop, D., Hedrick, J.K., Yip, P.P., Gerdes, J.C.: Dynamic surface control for a class of nonlinear systems. IEEE Trans. Autom. Control **45**(10), 1893–1899 (2000)
30. Zhong, J., Yao, B.: Adaptive robust precision motion control of apiezoelectric positioning stage. IEEE Trans. Control Syst. Technol. **16**(5), 1039–1046 (2008)

A Neural Hysteresis Model
for Smart-Materials-Based Actuators

Yu Shen[1], Lianwei Ma[2($\boxtimes$)], Jinrong Li[2], Xiuyu Zhang[3],
Xinlong Zhao[4], and Hui Zheng[2]

[1] Department of Applied Physics, Zhejiang University of Science
and Technology, Hangzhou 310023, China
shenyu@zust.edu.cn
[2] Department of Automation, Zhejiang University of Science
and Technology, Hangzhou 310023, China
chris5257@163.com
[3] School of Automation Engineering, Northeast Dianli University,
Jilin 132012, China
[4] College of Mechanical Engineering Automation,
Zhejiang Sci-Tech University, Hangzhou 310018, China

Abstract. In this paper, a constraint factor (CF) is presented. The CF and an odd m-order polynomial form a new hysteretic operator (HO) together. And then, an expanded input space is constructed based on the proposed HO. In the expanded input and output spaces, the one-to-multiple mapping of hysteresis is transformed into a one-to-one mapping so that a neural network can be used to develop a neural hysteresis model. The model parameters are computed by using the least square method. Finally, the neural hysteresis model is employed to approximate a real data from a magnetostrictive actuator in an experiment. The experimental results demonstrate the proposed approach is effective.

Keywords: Hysteresis · Hysteretic operator (HO) · Constraint factor (CF) · Smart materials · Neural networks

1 Introduction

In the past decades, smart materials, such as piezoelectric materials, magnetostrictive materials, shape memory alloys, etc., have been widely used in many fields. The piezo-electric and magnetostrictive actuators have been specially used for micro-displacement systems [1, 2]. Since smart-materials-based actuators have advantage in the output force, position resolution, and response speed [3], they have been attached importance in ultra-precision positioning systems [4]. However, the inherent hysteresis nonlinearity in

This work is supported in part by National Natural Science Foundation of China (Grant nos. 11304282, 61540034, 61304015, and 61273184); Zhejiang Provincial Natural Science Foundation (Grant nos. LQ14F050002, LQ16F030002, and LY15F030022); Science Technology Department of Zhejiang Province (Grant no. 2014C31020); Pre-research Special Foundation for Interdisciplinary Subject at Zhejiang University of Science and Technology (Grant no. 2014JC03Y).

© Springer International Publishing Switzerland 2016
N. Kubota et al. (Eds.): ICIRA 2016, Part I, LNAI 9834, pp. 663–671, 2016.
DOI: 10.1007/978-3-319-43506-0_57

smart-materials-based actuators, which is non-differentiable and multi-valued mapping, frequently leads to undesired tracking errors, oscillations, and even instability [5]. The model-based scheme for hysteresis compensation is currently popular in control systems [6]. A large amount of hysteresis models have been proposed in the past decades, such as PI model [7], KP model [8], Preisach model [9], Maxwell slip model [10], Jiles-Atherton model [11], Duhem model [12], Bouc-Wen model [13, 14], and so on. However, the ultra-precision positioning systems need more accurate hysteresis models so as to meet the requirement of science and technology.

Three-layer feed-forward neural networks (NNs) have been regarded as one of the best ways to model nonlinear systems because they can implement all kinds of non-linear mapping. However, the mapping of hysteresis consists of one-to-multiple and multiple-to-one mappings, while the NNs are incapable of identifying one-to-multiple mapping [15], so the one-to-multiple mapping has to be eliminated so that the neural approaches can be used to model hysteresis. Ma [16] proposed a hysteretic operator (HO), expanded the input space of NN from 1-dimension to 2-dimension based on the HO so that an NN-based hysteresis model was established, and named the method as expanded-space method. Zhao [17], Dong [18], Zhang [19] and Ma [20, 21] proposed respectively new HOs and constructed neural hysteresis models, thereby improving the expanded-space method.

In this paper, a new HO, which is made up of a constraint factor (CF) and an odd m-order polynomial, is proposed to expand the input space of NN. And then, based on the proposed HO, the one-to-multiple mapping of hysteresis is transformed into one-to-one mapping so that the neural approach can be used to identify the expanded mapping. Finally, a NN-based hysteresis model is developed and used to approximate a set of real data from a magnetostrictive actuator. The experimental results demonstrate that the proposed model is effective.

2   HO Construction

2.1   HO Definition

In this paper, the HO is consisted of a CF and an odd m-order polynomial with the constant term. The function, $c(x) = 1 - e^{-x}$, is used as the CF of HO. The role of CF is to constrain the amplitude of HO curve and ensure the curve passes through the origin in every minor coordinate system. Therefore, in the ith minor coordinate system, the HO is defined as follows:

$$f(x_i) = (1 - e^{-x_i})(a_0 + \sum_{j=1}^{m} a_j x_i^{2j-1}) \tag{1}$$

where x_i and f are respectively any input and the corresponding output of HO in the ith minor coordinate system.

In the main coordinate system, the HO is described as

$$h(x) = \begin{cases} h(x_{ei}) + f(x - x_{ei}) & x > x_{ei} \\ h(x_{ei}) - f(x_{ei} - x) & x < x_{ei} \end{cases} \tag{2}$$

where $[x_{ei}, h(x_{ei})]$ are the coordinates of the origin of the ith minor coordinate system in the main coordinate system, x is any input and h is the corresponding output of HO.

2.2 Parameter Computation

As known to all, the best method of determining polynomial coefficients is the least square method. Thus, the least square method is adopted to compute the HO parameters based on the samples used for training neural network.

In terms of the least square method, the residual δ_i is written as

$$\delta_i = y_i - f(x_i) \tag{3}$$

Consequently, the sum of square residuals is shown as follows:

$$S = \sum_{i=1}^{n} \delta_i^2 = \sum_{i=1}^{n} [y_i - f(x_i)]^2 = \sum_{i=1}^{n} [y_i - (1 - e^{-x_i})(a_0 + \sum_{j=1}^{m} a_j x_i^{2j-1})]^2 \tag{4}$$

To minimize S, the partial derivatives of S with regard to $a_0, a_1, \ldots, a_m$ should be set to zeros. Therefore, the $(m + 1)$ partial derivative equations are given as follows:

$$\begin{cases} \frac{\partial S}{\partial a_0} = -2 \sum_{i=1}^{n} [(1 - e^{-x_i}) \cdot y_i - (1 - e^{-x_i})^2 (a_0 + \sum_{j=1}^{m} a_j x_i^{2j-1})] = 0 \\ \frac{\partial S}{\partial a_k} = -2 \sum_{i=1}^{n} [(1 - e^{-x_i}) \cdot x_i^{2k-1} \cdot y_i - (1 - e^{-x_i})^2 (a_0 x_i^{2k-1} + \sum_{j=1}^{m} a_j x_i^{2(j+k-1)})] = 0, \ k = 1, 2, \cdots, m \end{cases} \tag{5}$$

Rearranging the Eq. (5) gives

$$\begin{cases} a_0 \cdot \sum_{i=1}^{n} (1 - e^{-x_i})^2 + \sum_{j=1}^{m} a_j \cdot \sum_{i=1}^{n} (1 - e^{-x_i})^2 x_i^{2j-1} = \sum_{i=1}^{n} (1 - e^{-x_i}) \cdot y_i \\ a_0 \cdot \sum_{i=1}^{n} (1 - e^{-x_i})^2 x_i^{2k-1} + \sum_{j=1}^{m} a_j \cdot \sum_{i=1}^{n} (1 - e^{-x_i})^2 x_i^{2(j+k-1)} = \sum_{i=1}^{n} (1 - e^{-x_i}) \cdot x_i^{2k-1} \cdot y_i, k = 1, 2, \cdots, m \end{cases} \tag{6}$$

The expansion form of the Eq. (6) is given as follows:

$$\begin{cases} a_0 \sum_{i=1}^{n}(1-e^{-x_i})^2 + a_1 \sum_{i=1}^{n}(1-e^{-x_i})^2 x_i + \cdots + a_m \sum_{i=1}^{n}(1-e^{-x_i})^2 x_i^{2m-1} = \sum_{i=1}^{n}(1-e^{-x_i})y_i \\ a_0 \sum_{i=1}^{n}(1-e^{-x_i})^2 x_i + a_1 \sum_{i=1}^{n}(1-e^{-x_i})^2 x_i^2 + \cdots + a_m \sum_{i=1}^{n}(1-e^{-x_i})^2 x_i^{2m} = \sum_{i=1}^{n}(1-e^{-x_i})x_i y_i \\ \cdots\cdots\cdots\cdots\cdots\cdots\cdots\cdots\cdots\cdots\cdots\cdots\cdots\cdots\cdots\cdots\cdots\cdots \\ a_0 \sum_{i=1}^{n}(1-e^{-x_i})^2 x_i^{2m-1} + a_1 \sum_{i=1}^{n}(1-e^{-x_i})^2 x_i^{2m} + \cdots + a_m \sum_{i=1}^{n}(1-e^{-x_i})^2 x_i^{2(2m-1)} = \sum_{i=1}^{n}(1-e^{-x_i})x_i^{2m-1}y_i \end{cases}$$

$$(7)$$

The Eq. (7) is written as the following matrix Eq.

$$\begin{bmatrix} \sum_{i=1}^{n}(1-e^{-x_i})^2 & \cdots & \sum_{i=1}^{n}(1-e^{-x_i})^2 x_i^{2m-1} \\ \vdots & \vdots & \vdots \\ \sum_{i=1}^{n}(1-e^{-x_i})^2 x_i^{2m-1} & \cdots & \sum_{i=1}^{n}(1-e^{-x_i})^2 x_i^{2(2m-1)} \end{bmatrix} \begin{bmatrix} a_0 \\ \vdots \\ a_m \end{bmatrix}$$

$$= \begin{bmatrix} \sum_{i=1}^{n}(1-e^{-x_i})y_i \\ \vdots \\ \sum_{i=1}^{n}(1-e^{-x_i})x_i^{2m-1}y_i \end{bmatrix}$$

$$(8)$$

i.e.,

$$XA = Y \tag{9}$$

The HO parameters are obtained by solving Eq. (9),

$$A = X^{-1}Y \tag{10}$$

3   Experimental Verification

In the following, two verification experiments are implemented. In the experiments, the presented neural hysteresis model is compared with the PI model by approximating a set of real data from a smart-material-based actuator so as to validate the effectiveness of the proposed approach.

The experimental setup is comprised of a magnetostrictive actuator (MFR OTY77), a current source, a dSPACE control board with 16-bit analog-to-digital and digital-to-analog converters, and a PC. A set of data containing 1916 input-output pairs is obtained.

The data is equally divided into two groups. One group is used for training neural networks, and another group is used for model verification.

3.1   The Proposed Model

In this section, the neural hysteresis model is employed to approximate the real data. The activation function of the hidden layer is the sigmoid function and that of the output layer is the linear function. The comparison of the different orders shows that the 9-order polynomial is most suitable to the HO in this experiment. The HO parameters are listed in the Table 1.

Table 1. HO Parameters

Parameter	Value
a_0	−13.1053
a_1	3.5024e02
a_2	−4.7518e03
a_3	3.8646e04
a_4	−1.9549e05
a_5	6.3845e05
a_6	−1.3467e06
a_7	1.7721e06
a_8	−1.3230e06
a_9	4.2793e05

To determine the optimal number of hidden nodes, the number from 1 to 100 is tried in this experiment. The best 3 performances are listed in the Table 2. It can be seen from the Table 2, that the neural hysteresis model has the best performance when the number of hidden nodes becomes 5. Therefore, a three-layer feed-forward neural network with two input nodes, 5 hidden nodes and one output node was employed to approximate the real data in this experiment. After 229 iterations, the training procedure finishes. The mean square error (MSE) of model prediction is 0.0611. The Figs. 1 and 2 illustrate the model prediction and absolute error respectively.

Table 2. The top 3 performances of NN with different number of hidden neurons

No. of hidden nodes	MSE
5	0.0611
6	0.0677
2	0.0690

3.2   PI Model

In addition, to compare with the proposed model, the PI model was also applied to approximate the measured data. The model thresholds were calculated via the following formula:

$$r_i = \frac{i-1}{N}[\max(x(k)) - \min(x(k))] \tag{11}$$

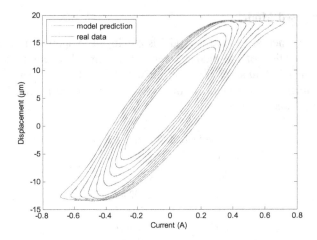

Fig. 1. Comparison of the proposed model prediction and the real data

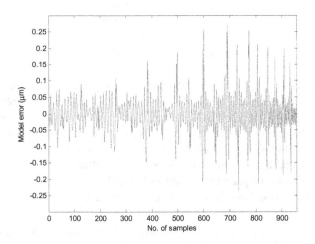

Fig. 2. The absolute error of the proposed model

where N is the number of backlash operators and $i = 1, 2, ..., N$.

The Matlab nonlinear optimization tool was used to determine the weights of backlash operators. However, the calculation time of the PI model increases along with the increase of N, so only the PI models containing 1–5000 backlash operators were tried. The top three performances are listed in the Table 3. Therefore, N = 4994 is selected in this experiment, which leads to 20-hour calculation time. The MSE of model prediction is 0.4327. Figures 3 and 4 display the model prediction and absolute errors respectively.

3.3   Comparison

In the above experiments, the MSE of the proposed neural hysteresis model is 85.88 % smaller than that of the PI model, which demonstrates that the proposed neural model can better approximate the real data measured from the magnetostrictive actuator than the PI model.

Table 3. Performances of different no. of backlash operators

No. of backlash operators	Performance
4994	0.4327
4997	0.4330
5000	0.4348

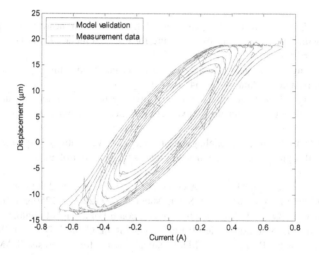

Fig. 3. Comparison of the PI model prediction and the real data

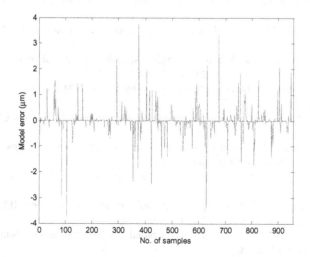

Fig. 4. The absolute error of the PI model

4 Conclusions

In this paper, a new HO is proposed. The HO consists of two components: a CF and an odd m-order polynomial. And then, based on the constructed HO, the one-to-multiple mapping of hysteresis is transformed into a continuous one-to-one mapping by expanding the input space of NN. In this way, the expanded mapping only contains one-to-one and multiple-to-one mappings, which can be identified using the neural approaches. Finally, an experiment is implemented to verify the proposed hysteresis model. The verification performance approves the proposed approach.

References

1. Li, Z., Su, C.Y., Chai, T.: Compensation of hysteresis nonlinearity in magnetostrictive actuators with inverse multiplicative structure for Preisach model. IEEE Trans. Autom. Sci. Eng. **11**(2), 613–619 (2014)
2. Chen, X., Feng, Y., Su, C.Y.: Adaptive control for continuous-time systems with actuator and sensor hysteresis. Automatica **64**, 196–207 (2016)
3. Zhu, Y., Ji, L.: Theoretical and experimental investigations of the temperature and thermal deformation of a giant magnetostrictive actuator. Sens. Actuators A Phys. **218**, 167–178 (2014)
4. Li, Z., Su, C.Y., Chen, X.: Modeling and inverse adaptive control of asymmetric hysteresis systems with applications to magnetostrictive actuator. Control Eng. Pract. **33**, 148–160 (2014)
5. Gu, G.Y., Li, Z., Zhu, L.M., et al.: A comprehensive dynamic modeling approach for giant magnetostrictive material actuators. Smart Mater. Struct. **22**(12), 125005 (2013)
6. Liu, S., Su, C.Y., Li, Z.: Robust adaptive inverse control of a class of nonlinear systems with Prandtl-Ishlinskii hysteresis model. IEEE Trans. Autom. Control **59**(8), 2170–2175 (2014)
7. Macki, J.W., Nistri, P., Zecca, P.: Mathematical models for hysteresis. SIAM Rev. **35**(1), 94–123 (1993)
8. Krasnosel'skii, M.A., Pokrovskii, A.V.: Systems with Hysteresis. Springer, New York (1989)
9. Mayergoyz, I.D.: Mathematical models of hysteresis. IEEE Trans. Magn. **22**(5), 603–608 (1986)
10. Ferretti, G., Magnani, G., Rocco, P.: Single and multistate integral friction models. IEEE Trans. Autom. Control **49**(12), 2292–2297 (2004)
11. Jiles, D., Atherton, D.: Theory of ferromagnetic hysteresis. J. Magn. Magn. Mater. **61**(1–2), 48–60 (1986)
12. Duhem, P.: Die dauernden Aenderungen und die Thermodynamik. Z. Phys. Chem. **22**, 543–589 (1879)
13. Bouc, R.: Forced vibration of mechanical systems with hysteresis. In: Proceedings of 4th Conference on Nonlinear Oscillations (1967)
14. Wen, Y.K.: Method for random vibration of hysteretic systems. ASCE J. Eng. Mech. Div. **102**(2), 249–263 (1976)
15. Wei, J.D., Sun, C.T.: Constructing hysteretic memory in neural networks. IEEE Trans. Syst. Man Cybern. Part B Cybern. **30**(4), 601–609 (2000)
16. Ma, L., Tan, Y., Chu, Y.: Improved EHM-based NN hysteresis model. Sens. Actuators A Phys. **141**(1), 6–12 (2008)

17. Zhao, X., Tan, Y.: Modeling hysteresis and its inverse model using neural networks based on expanded input space method. IEEE Trans. Control Syst. Technol. **16**(3), 484–490 (2008)

18. Dong, R., Tan, Y., Chen, H., et al.: A neural networks based model for rate-dependent hysteresis for piezoceramic actuators. Sens. Actuators A Phys. **143**(2), 370–376 (2008)

19. Zhang, X., Tan, Y., Su, M., et al.: Neural networks based identification and compensation of rate-dependent hysteresis in piezoelectric actuators. Phys. B **405**(12), 2687–2693 (2010)

20. Ma, L., Shen, Y., Li, J., et al.: A modified HO-based model of hysteresis in piezoelectric actuators. Sens. Actuators A Phys. **220**, 316–322 (2014)

21. Ma, L., Shen, Y.: A neural model of hysteresis in amorphous materials and piezoelectric materials. Appl. Phys. A **116**(2), 715–722 (2014)

Modeling of Rate-Dependent Hysteresis in Piezoelectric Actuators Using a Hammerstein-Like Structure with a Modified Bouc-Wen Model

Chun-Xia Li, Lin-Lin Li, Guo-Ying Gu$^{(\boxtimes)}$, and Li-Min Zhu

State Key Laboratory of Mechanical System and Vibration,
School of Mechanical Engineering, Shanghai Jiao Tong University,
Shanghai 200240, China
{lichunxia,lilinlin321,guguoying,zhulm}@sjtu.edu.cn

Abstract. This paper presents a modified Bouc-Wen (MBW) model based Hammerstein-like structure to describe the hysteresis in piezoelectric actuators (PEAs) with asymmetric and rate-dependent characteristics. Firstly, a MBW model with a third-order input function is proposed to characterize the hysteresis with asymmetric feature. Then, to describe the rate-dependent behavior of the hysteresis in PEAs, the MBW model is cascaded with a linear dynamics model as a Hammerstein-like model structure. To derive the parameters of this model structure, three identification steps are performed with different input signals: (i) the delay time of the PEA-actuated nanopositioning stage is tested with a step signal input; (ii) the linear dynamics model is identified with a low-amplitude white noise input; (iii) with the identified delay time and linear dynamics model, nonlinear least squares optimization method is adopted to derive the parameters of the MBW model using a multiple-amplitude triangular signal input with low frequency. Finally, to evaluate the Hammerstein-like model structure, experiments are carried out on a PEA-actuated nanopositioning stage. The experimental results verify that the predicted responses of the MBW model based Hammerstein-like structure well match the system responses.

Keywords: Piezoelectric actuator · Rate-dependent hysteresis · Bouc-Wen model · Hammerstein-like structure

1 Introduction

With the merits of large actuation force, high positioning resolution and quick response time, piezoelectric actuators (PEAs) have been extensively applied in various nanopositioning equipments, such as scanning probe microscopes [1], micro-/nano-manipulators [2], and nano-manufacturing devices [3]. However, the intrinsic hysteresis nonlinearity of the PEAs can seriously damage the positioning/tracking accuracy of these nanopositioning equipments [4]. In the

© Springer International Publishing Switzerland 2016
N. Kubota et al. (Eds.): ICIRA 2016, Part I, LNAI 9834, pp. 672–684, 2016.
DOI: 10.1007/978-3-319-43506-0_58

nanopositioning stages driven by PEAs, the hysteresis effect can induce positioning/tracking errors up to 15 % of the travel range [4].

To deal with the hysteresis of the PEAs, a number of control methods have been developed [5–7]. Among the existing methods, feedforward control is most commonly used, which is on the basis of a mathematical model that can accurately characterize the hysteresis nonlinearity. A lot of such mathematical models have been developed in the literature, such as the Dahl model [2], the Preisach model [8], the Prandtl-Ishlinskii model [7], and the Bouc-Wen (BW) model [9]. In these models, the BW model owns the benefit of simplicity for computing, as it only requires one differential equation with a few parameters [9]. In addition, it is capable of describing many categories of hysteresis nonlinearity [9]. However, the classical Bouc-Wen (CBW) model is limited to the symmetric hysteresis description [10]. Moreover, the hysteresis curve generated by the CBW model does not change with the input frequencies [11], which makes it impossible for the CBW model to characterize the rate-dependent behavior of the hysteresis. Hence, when the CBW model is utilized to represent the hysteresis effect of the PEAs which exhibits asymmetric and rate-dependent characteristics, large modeling errors would occur [11]. To describe the asymmetric hysteresis, investigations have been performed to modify the CBW model by introducing a non-odd input function [12] or an asymmetric term into the differential equation [10]. To represent the rate-dependent characteristic, various efforts have also been made in the literature. In [13], experimental data in a certain range are utilized to identify the CBW model. In [11], a factor in the frequency domain is introduced to characterize the rate-dependent behavior. In [14], a Hammerstein structure with CBW model is developed to describe the hysteresis effect with rate-dependent behavior. It can be found that, nowadays, design of BW models to describe the hysteresis effect of PEAs still attracts attention. However, the existing hysteresis descriptions with BW models are not completely satisfactory. For instance, the modeling errors in [13] are still large and the model in [14] cannot describe the asymmetric hysteresis behavior. Hence, there is a necessity to further research on how to model the hysteresis with asymmetric and rate-dependent characteristics using the BW model.

In this work, a modified Bouc-Wen (MBW) model with a third-order input function is introduced to characterize the hysteresis nonlinearity in PEAs with asymmetric behavior. Then, considering that the rate-dependent characteristic of the hysteresis can be treated as the coupled effect of the rate-independent hysteresis and the linear dynamics [15], the proposed MBW model is cascaded with the linear dynamics model of the PEA-actuated stage as a Hammerstein-like structure to characterize the rate-dependent hysteresis effect. The parameters of the hysteresis model, linear dynamics model and delay time are derived with three identification steps using the specific inputs and the corresponding system outputs. Experiments are conducted on a PEA-actuated nanopositioning stage to validate the effectiveness and feasibility of the proposed model structure. The experimental results validate that the predicted responses with the identified model well match the system responses.

The rest of this paper is organized as follows. Section 2 presents the model description of the PEA-actuated nanopositioning stage. Section 3 presents the experimental platform and the model identification. Experimental results are presented in Sect. 4, and Sect. 5 draws the conclusion.

2   Model Description

The BW model was firstly proposed by Bouc [16] and modified by Wen [17]. It is capable of describing many categories of hysteresis and it also owns the benefit of simplicity in computing. The CBW model can be expressed in the form of

$$w(t) = pv(t) + h(t) \tag{1}$$

$$\dot{h} = A\dot{v} - \beta \left| \dot{v} \right| \left| h \right|^{n-1} h - \gamma \dot{v} \left| h \right|^n \tag{2}$$

where $w(t)$ is the hysteresis output, $v(t)$ is the input voltage to the PEAs, $h(t)$ is a variable about the hysteresis, $\dot{h}$ and $\dot{v}$ are the first derivative of h and v with respect to time, respectively, A, β, γ and n are the shape parameters of hysteresis curves. Due to the properties and characteristics of the component material in the PEAs, $n = 1$ is generally utilized to characterize the hysteresis of the PEA-actuated nanopositioning stages [9].

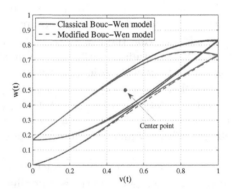

Fig. 1. Comparison of hysteresis curves generated by the CBW model and MBW model.

The hysteresis generated by the CBW model exhibits symmetrical characteristic about the center point [10], as can be observed from Fig. 1. Thus, it cannot accurately characterize the asymmetric hysteresis in PEAs. In this work, a MBW model will be developed to describe the hysteresis effect in PEAs with asymmetric characteristic.

2.1 MBW Model

Inspired by our previous work [7], to describe the hysteresis in PEAs which exhibits asymmetric characteristic, a third-order input function is utilized to develop the MBW model by modifying (1) as follows:

$$w(t) = p_1 v(t)^3 + p_2 v(t) + h(t) \tag{3}$$

It can be found that, when $p_1 = 0$, the proposed MBW model is simplified as the CBW model. Hence, the CBW model can be regarded as a special case of the MBW model. To test the ability of the proposed MBW model to characterize the hysteresis effect with asymmetric feature, the output of MBW model under sinusoidal input signal is shown in Fig. 1. The parameters of the MBW model are $A = -0.7$, $\beta = 3$, $\gamma = -1$, $n = 1$, $p_1 = -0.1$, and $p_2 = 1$. As a comparison, the output of the CBW model with $p = 1$ is also shown in Fig. 1. The other parameters of the CBW model are the same with those of the MBW model. It can be found that the hysteresis curve of the CBW model is symmetrical about the center point, while the hysteresis curve generated with the MBW model exhibits asymmetric characteristic.

It should be mentioned that, in this work the input voltage is always positive value. Hence, the introduction of the third-order polynomial term of input function can result in the asymmetric characteristic of the MBW model. When the input voltage is symmetrical about the origin, non-odd polynomial term can be added to the model to produce the asymmetric characteristic [12].

2.2 Hammerstein-Like Model Structure

The rate-dependent hysteresis can be treated as the coupled effect of the rate-independent hysteresis and the linear dynamics of the PEA-actuated nanopositioning stage [15]. Hence, for the purpose of characterizing the rate-dependent behavior of the hysteresis effect, the dynamics model of the PEA-actuated nanopositioning stage is described as a cascaded model of the hysteresis nonlinearity and the linear dynamics model, as shown in Fig. 2, which is actually a Hammerstein-like model structure. In this work, the hysteresis effect $H(\cdot)$ is represented by the proposed MBW model in Sect. 2.1. The linear dynamics model $G(z)$ can be written as

$$G(z) = \frac{B(z)}{A(z)} z^{-d} \tag{4}$$

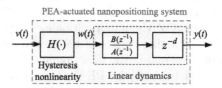

Fig. 2. The cascaded model (Hammerstein-like structure) of the PEA-actuated nanopositioning systems.

with

$$A(z) = 1 + a_1 z^{-1} + a_2 z^{-2} + \cdots + a_{na} z^{-na}$$
$$B(z) = b_1 + b_2 z^{-1} + \cdots + b_{nb} z^{-nb+1} \tag{5}$$

where z^{-1} represents the one step time delay, z^{-d} is the delay steps of the nanopositioning stage with d being a positive integer, and $A(z)$ and $B(z)$ are denominator and numerator of $G(z)$, respectively. The degrees of $A(z)$ and $B(z)$ are na and $nb - 1$, respectively.

It should be mentioned that in [14] a BW model based Hammerstein model is also used for the description of the hysteresis nonlinearity in PEAs with rate-dependent behavior. The differences between this work and [14] lie in that: (i) the MBW model in this work is modified to be capable of describing the asymmetric characteristic of the hysteresis; (ii) the delay time is considered in this work, which is not considered in [14]; and (iii) the identification procedure in this work is quite different from that in [14] where the hysteresis nonlinearity is identified first by regarding $w(t) = y(t)$. Our identification steps will be presented in the next section.

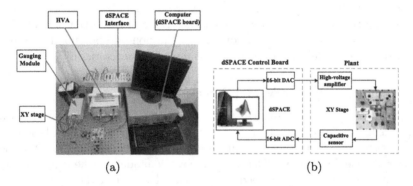

(a)   (b)

Fig. 3. The experimental setup. (a) The experimental platform. (b) Block diagram of the experimental platform.

3   Model Identification

There are three types of parameters in the system model to be identified, i.e. the parameters of the MBW model, the coefficients a_i and b_i of the linear dynamics model, and the delay steps d of the nanopositioning stage. The delay time of the PEA-actuated nanopositioning stage can be derived directly by comparing the input signal and output signal in the time domain. Besides, when the input with low amplitude is used to drive the PEA-actuated nanopositioning stage, the influence of the hysteresis can be neglected [15], which makes it possible to identify the linear dynamics model. After identifying the delay steps and the coefficients of $A(z)$ and $B(z)$, the hysteresis output $w(t)$ can be determined when the system output $y(t)$ under low-frequency input is measured. Then, the parameters of the MBW model can be optimized with the low-frequency input data and the hysteresis output data. This three-step identification procedure is similar to that in our previous work [15].

3.1   Experimental Setup

A PEA-actuated XY nanopositioning stage developed in our previous paper [18] is adopted to evaluate the proposed model structure in this work. Figure 3(a) shows the experimental platform. A dSPACE-DS1103 board is used to transmit the control voltage signal and the displacement signal between the software and the experimental devices. The 16-bit DACs of the dSPACE interface provide the control voltage which is calculated in the software of the computer to the high-voltage amplifier (HVA). Then, the HVA amplifies the control voltage with a gain of 20 and provides it to the PEAs. Capacitive sensors (Probe 2823 and Gauging Module 8810 from MicroSense) are utilized to measure the displacements of the end-effector of the nanopositioning stage. The 16-bit ADCs of the dSPACE interface acquire the sensor output signal and then transmit it to the computer for the control signal calculation. In this work, the sample time is set as 0.00002 s. The block diagram of the experimental setup is shown in Fig. 3(b).

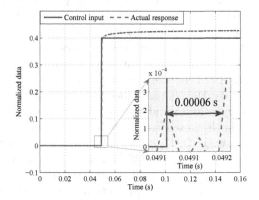

Fig. 4. Experimental results of the step response.

3.2   Identification of the Delay Time

The step signal input and the corresponding system output is used to identify the delay time of the nanopositioning system. The experimental results of the step response of the nanopositioning stage are shown in Fig. 4. By comparing the system response with the input signal, it is found that the delay time between the output signal and input signal is 0.00006 s. It should be noted that several experiments under step signal inputs are conducted with different sample times. All the experimental results show the same delay time, i.e. 0.00006 s. As the sample time in this work is 0.00002 s, the time delay term $d = 3$ is derived.

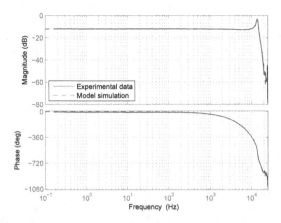

Fig. 5. The frequency responses of the experimental data and the identified model.

3.3   Identification of the Linear Dynamics

To identify the linear dynamics model of the nanopositioning stage, a band-limited white noise signal with low amplitude is utilized to excite the system. The input voltage and output signal are simultaneously acquired by the dSPACE. Then, they are used to identify the coefficients of a_i and b_i of the linear dynamics model with the System Identification Toolbox of Matlab. ARX algorithm is employed for the identification and the Least squares method is adopted for the parameter optimization. It should be mentioned that, the higher order of na and nb, the more accuracy of the identified linear dynamics model, but the more computational complexity. Hence, the tradeoff of the identification accuracy and the computational complexity should be made during the identification. In this work, na and nb are both chosen as 8. The coefficients of $A(z)$ and $B(z)$ are finally identified as $a_1 = -0.05847$, $a_2 = 0.2423$, $a_3 = -0.05833$, $a_4 = -0.4917$, $a_5 = 0.4170$, $a_6 = -0.1945$, $a_7 = 0.07748$, $a_8 = -0.01198$, $b_1 = 0.004481$, $b_2 = 0.04629$, $b_3 = 0.1320$, $b_4 = 0.1154$, $b_5 = -0.02198$, $b_6 = -0.06236$, $b_7 = -0.0007691$, and $b_8 = 0.01408$. The prediction error of the identified model is 2.534e-09. Besides, the frequency responses of the identified model and the experimental results are shown in Fig. 5, which also verifies the accuracy of the identified linear dynamics model.

3.4   Identification of the MBW Model

To obtain the parameters of the MBW model, a triangular signal with multiple amplitudes is utilized as the input to the nanopositioning stage. The fundamental frequency of the triangular signal is 1 Hz. With this frequency, the linear dynamics of the nanopositioning stage can be treated as a dc gain with a fixed time delay. As $G(j2\pi) = 0.2464$, the hysteresis output $w(t)$ as shown in Fig. 2 can be derived as $w(i) = y(i + 3)/0.2464$. Then, the parameters of the MBW model can be identified using the input data $v(t)$ and hysteresis output data $w(t)$.

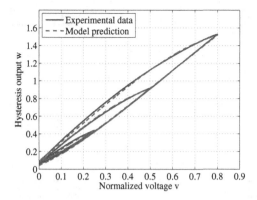

Fig. 6. Comparison of the hysteresis nonlinearity of the experimental results and the predicted response with the identified MBW model.

Table 1. The RMS prediction errors of the dynamic model (Hammerstein-like structure) and the MBW model under sinusoidal reference input with various frequencies.

Frequency (Hz)	1	50	100	200	300	400
Dynamic model (%)	0.90	1.80	2.39	3.30	3.54	4.02
B-W model (%)	0.90	2.29	3.60	5.95	7.98	10.79

For convenience, both of the input voltage and the system output are normalized. A discrete form of the MBW model is used for the parameters identification, which can be expressed as

$$w_i = p_1 v_i^3 + p_2 v_i + h_i \tag{6}$$

$$h_i = h_{i-1} + A\Delta v_i - \beta |\Delta v_i| h_{i-1} - \gamma \Delta v_i |h_{i-1}| \tag{7}$$

with $\Delta v_i = v_i - v_{i-1}$. In this work, the nonlinear least squares method is adopted to optimize the parameters of the MBW model. The identified parameters of the MBW model is $A = -0.7691$, $\beta = 4.6075$, $\gamma = -0.8725$, $p_1 = -0.1085$, and $p_2 = 2.1925$. Figure 6 shows the simulated results with the identified MBW model in comparison with the experimental results. It can be seen that the hysteresis curves generated by the identified MBW model well match the experimental response of the PEA-actuated nanopositioning stage. The RMS prediction error is 0.73 %, which also validates the accuracy of the identified MBW model.

4   Experimental Evaluation

To evaluate the effectiveness of the MBW model based Hammerstein-like structure to represent the hysteresis in PEAs with rate-dependent characteristic,

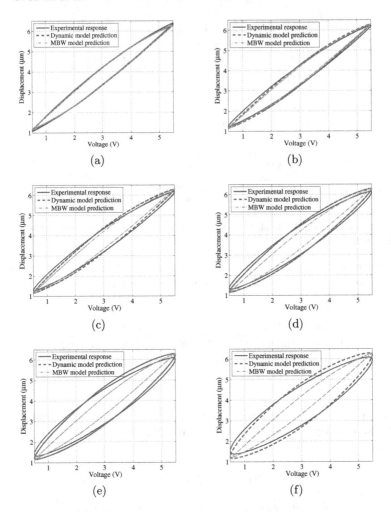

Fig. 7. The experimental results of the PEA-actuated nanopositioning stage and the prediction results of the dynamic model (Hammerstein-like structure) and MBW model under sinusoidal reference input with frequencies of (a) 1 Hz; (b) 50 Hz; (c) 100 Hz; (d) 200 Hz; (e) 300 Hz and (f) 400 Hz.

experiments are performed with sinusoidal signals and triangular signals with different frequencies.

Firstly, the input signals are chosen as $v(t) = 3.5 + 2.5sin(2\pi ft)$ (V) with frequencies of $f = 1$, 50, 100, 200, 300, and 400 Hz. Figure 7 shows the comparison of the experimental results and the predicted results with the dynamic model (Hammerstein-like model structure) and the MBW model. Figure 8 shows the comparison of the predicted errors of these two models. It can be observed from Fig. 7 that the predicted results with MBW model exhibit the rate-independent characteristic and it can only characterize the hysteresis effect of

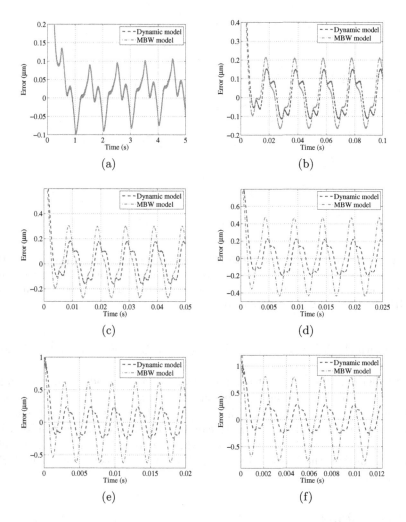

Fig. 8. The prediction errors of the dynamic model (Hammerstein-like structure) and the MBW model under sinusoidal reference with frequencies of (a) 1 Hz; (b) 50 Hz; (c) 100 Hz; (d) 200 Hz; (e) 300 Hz; and (f) 400 Hz.

the PEA-actuated nanopositioning stage with the input frequency below 100 Hz. This is why we propose the dynamic model which cascades the MBW model with the linear dynamics of the nanopositioning stage. From Fig. 7, it can be observed that the dynamic model accurately describes the rate-dependent hysteresis non-linearity under different input frequencies. To quantize the comparison, Table 1 lists the RMS prediction errors of the Hammerstein-like model structure and the MBW model, which also demonstrates the advantage of the proposed dynamic model structure. Hence, the effectiveness of the proposed Hammerstein-like

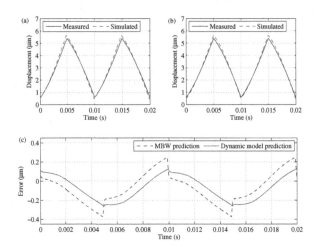

Fig. 9. The experimental responses of the PEA-actuated nanopositioning stage and the simulated results of (a) MBW model and (b) dynamic model under triangular signal input with fundamental frequency of 100 Hz. (c) is the comparison of prediction errors with MBW model and the dynamic model.

structure with the MBW model to represent the rate-dependent hysteresis nonlinearity in PEAs is validated.

To further demonstrate the feasibility of the proposed model structure under multiple harmonics inputs, the comparisons of experimental responses and model prediction results under triangular signal input with the fundamental frequency of 100 Hz are illustrated in Fig. 9. It can be observed that the results with dynamic model structure match the experimental responses more accurately, which also validates the effectiveness of the Hammerstein-like structure with the MBW model.

5 Conclusions

This paper proposes a MBW model with a third-order input function to describe the asymmetric hysteresis effect in the PEAs. Then, the MBW model is cascaded with a linear dynamics model as a Hammerstein-like structure to represent the hysteresis nonlinearity in PEAs with rate-dependent characteristic. The proposed Hammerstein-like model is identified with three steps using the specific input signals and the corresponding output responses of the nanopositioning stage. Experimental results on a PEA-actuated nanopositioning stage demonstrate that the proposed model can well predict the rate-dependent hysteresis nonlinearity. This work lays a foundation for the design of controllers that require system models.

Acknowledgment. This work was supported by the National Natural Science Foundation of China under Grant No. 51405293 and the Specialized Research Fund for the Doctoral Programme of Higher Education under Grant No. 20130073110037.

References

1. Yong, Y.K., Moheimani, S.O.R., Kenton, B.J., Leang, K.K.: Invited review article: high-speed flexure-guided nanopositioning: mechanical design and control issues. Rev. Sci. Instrum. **83**(12), 121101 (2012)
2. Xu, Q.S., Li, Y.M.: Dahl model-based hysteresis compensation and precise positioning control of an XY parallel micromanipulator with piezoelectric actuation. ASME J. Dyn. Syst. Meas. Control **132**(4), 041011 (2010)
3. Tian, Y.L., Zhang, D.W., Shirinzadeh, B.: Dynamic modelling of a flexure-based mechanism for ultra-precision grinding operation. Precis. Eng. **35**(4), 554–565 (2011)
4. Gu, G.-Y., Zhu, L.-M., Su, C.-Y., Ding, H., Fatikow, S.: Modeling and control of piezo-actuated nanopositioning stages: a survey. IEEE Trans. Autom. Sci. Eng. **13**(1), 313–332 (2016)
5. Fleming, A.J., Leang, K.K.: Charge drives for scanning probe microscope positioning stages. Ultramicroscopy **108**(12), 1551–1557 (2008)
6. Li, C.-X., Gu, G.-Y., Yang, M.-J., Zhu, L.-M.: High-speed tracking of a nanopositioning stage using modified repetitive control. IEEE Trans. Autom. Sci. Eng. **PP**(99), 1–11 (2015)
7. Gu, G.-Y., Zhu, L.-M., Su, C.-Y.: Modeling and compensation of asymmetric hysteresis nonlinearity for piezoceramic actuators with a modified Prandtl-Ishlinskii model. IEEE Trans. Ind. Electron. **61**(3), 1583–1595 (2014)
8. Li, Z., Su, C.-Y., Chai, T.: Compensation of hysteresis nonlinearity in magnetostrictive actuators with inverse multiplicative structure for preisach model. IEEE Trans. Autom. Sci. Eng. **11**(2), 613–619 (2014)
9. Rakotondrabe, M.: Bouc-Wen modeling and inverse multiplicative structure to compensate hysteresis nonlinearity in piezoelectric actuators. IEEE Trans. Autom. Sci. Eng. **8**(2), 428–431 (2011)
10. Zhu, W., Wang, D.-H.: Non-symmetrical Bouc-Wen model for piezoelectric ceramic actuators. Sens. Actuat. A Phys. **181**, 51–60 (2012)
11. Zhu, W., Rui, X.-T.: Hysteresis modeling and displacement control of piezoelectric actuators with the frequency-dependent behavior using a generalized bouc-wen model. Precis. Eng. **43**, 299–307 (2016)
12. Wang, G., Chen, G., Bai, F.: Modeling and identification of asymmetric Bouc-Wen hysteresis for piezoelectric actuator via a novel differential evolution algorithm. Sens. Actuat. A Phys. **235**, 105–118 (2015)
13. Gomis-Bellmunt, O., Ikhouane, F., Montesinos-Miracle, D.: Control of a piezoelectric actuator considering hysteresis. J. Sound Vibr. **326**(3), 383–399 (2009)
14. Wang, Z., Zhang, Z., Mao, J., Zhou, K.: A hammerstein-based model for rate-dependent hysteresis in piezoelectric actuator. In: 2012 24th Chinese Control and Decision Conference (CCDC), pp. 1391–1396. IEEE (2012)
15. Gu, G.-Y., Li, C.-X., Zhu, L.-M., Su, C.-Y.: Modeling and identification of piezoelectric-actuated stages cascading hysteresis nonlinearity with linear dynamics. IEEE/ASME Trans. Mechatron. **21**(3), 1792–1797 (2016)
16. Bouc, R.: Forced vibration of mechanical systems with hysteresis. In: Proceedings of the Fourth Conference on Non-linear Oscillation, Prague, Czechoslovakia (1967)

17. Wen, Y.-K.: Method for random vibration of hysteretic systems. J. Eng. Mech. Div. **102**(2), 249–263 (1976)
18. Li, C.-X., Gu, G.-Y., Yang, M.-J., Zhu, L.-M.: Design, analysis and testing of a parallel-kinematic high-bandwidth XY nanopositioning stage. Rev. Sci. Instrum. **84**(12), 125111 (2013)

A High Efficiency Force Predicting Method of Multi-axis Machining Propeller Blades

Zerun Zhu[1], Rong Yan[1], Fangyu Peng[2(✉)], Kang Song[1], Zepeng Li[1], Chaoyong Guo[1], and Chen Chen[1]

[1] National NC System Engineering Research Center, School of Mechanical Science and Engineering, Huazhong University of Science and Technology, Wuhan 430074, China
[2] State Key Laboratory of Digital Manufacturing Equipment and Technology, School of Mechanical Science and Engineering, Huazhong University of Science and Technology, 1037 Luoyu Road, Hongshan District, Wuhan 430074, Hubei, People's Republic of China
zwm8917@263.net

Abstract. Accurate cutting force prediction is a vital factor to analyze vibration, deformation, and residual stress in the machining. Meanwhile, it provides important reference for feed speed and tool orientation optimization. The problems of predicting accuracy and efficiency of cutting forces are serious especially for the 5-axis machining. Thus a rapid cutting force prediction method based on the classic mechanical model is proposed to calculate the 5-axis machining cutting forces. In the method, the cutting forces are directly calculated by elementary function integral, which can improve the calculating efficiency, compared with discretization and accumulation method. The experiments were conducted to verify the validity of the proposed method. The results show that the proposed method increases efficiency nearly 86 times compared with the z-map method and 20 times compared with the cutter/workpiece engagement boundary method and keeps a good accuracy.

Keywords: Multi-axis machining · Cutting forces · Highly efficiency prediction

1 Introduction

Accurate cutting force prediction is a vital factor to analyze vibration, deformation, and residual stress in the machining. Meanwhile, it provides important reference for feed rate and tool orientation optimization. The accuracy and efficiency of cutting forces prediction have great influences on the reliability and practicality of those works mentioned above. As a result, it is necessary to provide a rapid cutting force prediction method in the 5-axis machining.

In order to model instantaneous undeformed chip thickness(IUCT) which reflected the changes of the attitude angle in five-axis machining, as analyzing in reference [1] two methods are used: one is the vector projection method [2–4] from the total feed vector including linear and angular feed mode to the unit vector on the tool envelope

© Springer International Publishing Switzerland 2016
N. Kubota et al. (Eds.): ICIRA 2016, Part I, LNAI 9834, pp. 685–696, 2016.
DOI: 10.1007/978-3-319-43506-0_59

surface normal direction, the other is geometrical calculation method [5–8] of calculating the segment length on the IUCT defined line between the cutting edge element (CEE) and the cutter edge trajectory sweeping surface of the front cutting edge. In reference [1], the parametric model of IUCT is deduced and expressed as three sub models about tool orientation, tool orientation change and cutter runout, and this model is used in this paper for higher computational efficiency.

Determining the engaged cutting edge is also a main limitation in 5-axis milling. *Larue* and *Altintas* [9] used ACIS solid modeling environment to determine the engagement region for force simulations of flank milling, *Kim* et al. [10] and Zhu [5] determined the cutter contact area from the Z-map of the surface geometry and current cutter location. *Ozturk* and *Budak* [4] presented a complete geometry and force model for 5-axis milling operations using ball-end mills, and analyzed the effect of lead and tilt angles on the process geometry in detail. In the work of *Ozturk* [11] three boundary of CWE area were modeled and used to simulated cutting forces efficiently. Here, the boundary model was used and developed to improve the efficiency further more.

Compared with the previous work, the main contribution of this paper are: (1) the upper and lower boundary of the in-cutting cutter flute were calculated based on the boundary model, (2) the elementary integral model was developed and utilized to simulate propeller machining cutting forces. The remaining part of this paper is organized as follows. In Sect. 2, the elementary integral model was deduced and established. In Sect. 3, experiments were conducted to calibrate cutting force coefficients and verify the effectiveness of the proposed model. In addition, the propeller baldes machining forces were simulated with two other cutting force models and the simulation efficiency and accuracy were compared.

2   The Elementary Function Integral Model of 5-Axis Machining Forces

2.1   The Deducing Process Based on Mechanical Model

In the light of the assumption that milling force is proportional to the instantaneous undeformed chip thickness, to each differential element of cutting edge, the mechanical milling force model is expressed in the form of Eq. (1).

$$dF_{q,j}(z) = K_q h \, db(z) \tag{1}$$

Where $q = r, t, a$, $K_q(h)$ is radial, tangential and axial cutting force coefficient, $h_j(\psi, z, t)$ is undeformed chip thickness and $db(z)$ is undeformed chip width.

The IUCT model only considering fixed orientation translational feed mode [1]:

$$h(\theta_L, \theta_T) = \frac{F}{nN} \cos \theta_L \sin\kappa \sin\varphi - \frac{F}{nN} \sin \theta_L \cos\kappa \tag{2}$$

where, θ_L is the lead angle of cutter axis. κ, φ are axial and radial contact angle locating the cutting edge element in CCS. N is the number of cutter flute. n is the RPM. F is the cutter feed rate.

The instantaneous undeformed chip width:

$$db(z) = \frac{dz}{sin\kappa} \tag{3}$$

Subsequently, the resulting force components are obtained by summing up the elemental cutting forces for all in-cutting tool elements by performing a numerical integration along the engaged cutting edge in machining process,

$$\left[F_{x,y,z}(\varphi) \right] = \int_{j=1}^{N} \left\{ \int_{z_{1,j}}^{z_{2,j}} A \, dF_{q,j}(z) \right\} \tag{4}$$

where, A is a matrix defined as

$$A = \begin{bmatrix} -\sin\kappa\sin\varphi & -\cos\varphi & -\cos\kappa\sin\varphi \\ -\sin\kappa\cos\varphi & \sin\varphi & -\cos\kappa\cos\varphi \\ \cos\kappa & 0 & -\sin\kappa \end{bmatrix}$$

substituting Eqs. (1), (2) and (3) into Eq. (4), comprehensive considering these constitutive relations and Eq. (4) was simplified into

$$\begin{aligned} \left[F_{x,y,z}(j,\varphi) \right] &= \sum_{j=1}^{N} \int_{z_{1,j}}^{z_{2,j}} A \, dF_{q,j}(z) \\ &= \sum_{j=1}^{N} \int_{z_{1,j}}^{z_{2,j}} A \, K_q \left(\frac{F}{nN} \cos\theta_L \sin\kappa \sin\varphi - \frac{F}{nN} \sin\theta_L \cos\kappa \right) \frac{dz}{sin\kappa} \\ &= \sum_{j=1}^{N} \frac{F}{nN} \left\{ \cos\theta_L P_\varphi - \sin\theta_L P_\kappa \right\} \end{aligned} \tag{5}$$

where: P_φ and P_κ are middle parameters related to the position of cutter flute and the corresponding forces coefficients, and described as

$$P_\varphi = \int_{z_{1,j}}^{z_{2,j}} A \, K_q \sin\varphi \, dz$$

$$P_\kappa = \int_{z_{1,j}}^{z_{2,j}} A \, K_q \cot\kappa \, dz.$$

2.2   The Cutter Flute Model

Figure 1 shows the bull-nose milling cutter model. The observation of a variety of tools shows that the radial contact angle of the cutter edges in the arc segment of the bull and ball nose milling cutter change very little. So it could be assumed that the radial contact angle φ of the cutter edges in the arc segment involved in cutting is constant. Besides, the side edges' axial contact angle κ of the cutter are constants. The calculation of κ and φ of the cutter edges in the arc segment and the side edges are as follows:

The axial contact angle κ of the side edges of the cutter is constant. And in the arc segment, the relationship of axial contact angle κ and axial high z of the cutter edges is described in Eqs. (6) and (7).

$$cos\kappa = \frac{r-z}{r}, \ sin\kappa = \frac{\sqrt{r^2 - (r-z)^2}}{r} \tag{6}$$

$$z = r - r\, cos\kappa, \ dz = r\, sin\kappa d\kappa \tag{7}$$

Where, the r represents bull-nose fillet radius.

For the cutters with constant pitch, the radial contact angle φ of the cutter edges in the arc segment and the side edges of the cutter is described in Eq. (8).

$$d\varphi = mdz, \ \varphi = \varphi_0 + mz, \ m = -\frac{2\, tan\beta_0}{D} \tag{8}$$

Where, the φ_0 represents the sum of spindle rotation angle and the teeth angle, m describes the pitch.

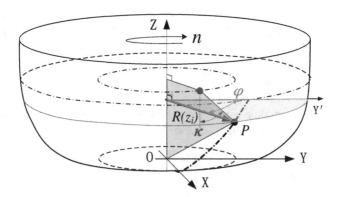

Fig. 1. The bull-nose milling cutter model

2.3   The Calculation of P_φ and P_κ

The calculation of P_φ is deduced as expressed in Eq. 9 according to the cutting force model in Sect. 2.1. The integral results are listed in Table 1 for every sub-part of P_φ.

$$
\begin{aligned}
P_\varphi &= \int_{z_{1,j}}^{z_{2,j}} A\,K_q\,sin\varphi dz = \int_{z_{1,j}}^{z_{2,j}}
\begin{bmatrix}
-\sin\kappa\sin\varphi & -\cos\varphi & -\cos\kappa\sin\varphi \\
-\sin\kappa\cos\varphi & \sin\varphi & -\cos\kappa\cos\varphi \\
\cos\kappa & 0 & -\sin\kappa
\end{bmatrix}
\begin{bmatrix}
K_r \\ K_t \\ K_a
\end{bmatrix}
sin\varphi dz \\
&=
\begin{bmatrix}
-\int_L K_r \sin\kappa\sin^2\varphi\,dz - \int_L K_t \cos\varphi\,sin\varphi dz - \int_L K_a \cos\kappa\sin^2\varphi\,dz \\
-\int_L K_r \sin\kappa\cos\varphi\,sin\varphi dz + \int_L K_t \sin^2\varphi\,dz - \int_L K_a \cos\kappa\cos\varphi\,sin\varphi dz \\
\int_L K_r \cos\kappa\,sin\varphi dz - \int_L K_a \sin\kappa\,sin\varphi dz
\end{bmatrix} \quad (9)\\
&=
\begin{bmatrix}
p_{\varphi x1} + p_{\varphi x2} + p_{\varphi x3} \\
p_{\varphi y1} + p_{\varphi y2} + p_{\varphi y3} \\
p_{\varphi z1} + p_{\varphi z2} + p_{\varphi z3}
\end{bmatrix}\Biggr|_{z_{1,j}}^{z_{2,j}}
\end{aligned}
$$

Where, K_r, K_t and K_a are radial, tangential and axial cutting force coefficient, φ and κ are the radial and axial cutter contact angle of the edge element, dz is the axial discrete element length. The relationship of φ, κ z and dz of the elements on the cutter flute are expressed in Eqs. 6, 7 and 8. The sub-parts $p_{\varphi n1}$, $p_{\varphi n2}$ and $p_{\varphi n3}$ (n = x, y, z) represent the n-direction parts of P_φ.

Table 1. Part integrals of P_φ

	The arc zone of cutter flute	The side zone of cutter flute
$p_{\varphi x1}$	$\frac{K_r \sin^2\bar\varphi}{r}\left(\frac{(r-z)}{2}\sqrt{r^2-(r-z)^2}+\frac{r^2}{2}arcsin\frac{r-z}{r}\right)$	$-\frac{K_r \sin\kappa}{2m}\left((\varphi_0+mz)-\frac{1}{2}sin(2\varphi_0+2mz)\right)$
$p_{\varphi x2}$	$\frac{K_t}{4m}cos(2\varphi_0+2mz)$	$\frac{K_t}{4m}cos(2\varphi_0+2mz)$
$p_{\varphi x3}$	$-\frac{K_a}{2r}\left(rz-\frac{1}{2}z^2\right)+\frac{K_a}{4mr}(r-z)sin(2\varphi_0+2mz)$ $-\frac{K_a}{8m^2r}cos(2\varphi_0+2mz)$	$-\frac{K_a \cos\kappa}{2m}\left((\varphi_0+mz)-\frac{1}{2}sin(2\varphi_0+2mz)\right)$
$p_{\varphi y1}$	$\frac{K_r}{2r}sin(2\bar\varphi)\left(\frac{(r-z)}{2}\sqrt{r^2-(r-z)^2}+\frac{r^2}{2}arcsin\frac{r-z}{r}\right)$	$\frac{K_r}{4m}\sin\kappa\,cos(2\varphi_0+2mz)$
$p_{\varphi y2}$	$\frac{K_t}{2}\left(z-\frac{1}{2m}sin(2\varphi_0+2mz)\right)$	$\frac{K_t}{2}\left(z-\frac{1}{2m}sin(2\varphi_0+2mz)\right)$
$p_{\varphi y3}$	$\frac{K_a}{4mr}(r-z)cos(2\varphi_0+2mz)+\frac{K_a}{8m^2r}sin(2\varphi_0+2mz)$	$\frac{K_a}{4m}\cos\kappa\,cos(2\varphi_0+2mz)$
$p_{\varphi z1}$	$-\frac{K_r}{mr}(r-z)cos(\varphi_0+mz)-\frac{K_r}{m^2r}sin(\varphi_0+mz)$	$-\frac{K_r}{m}\cos\kappa\,cos(\varphi_0+mz)$
$p_{\varphi z2}$	0	0
$p_{\varphi z3}$	$\frac{K_a}{r}sin\bar\varphi\left(\frac{(r-z)}{2}\sqrt{r^2-(r-z)^2}+\frac{r^2}{2}arcsin\frac{r-z}{r}\right)$	$\frac{K_a}{m}\sin\kappa\,cos(\varphi_0+mz)$

Note: $\bar\varphi$ is the mean value of φ_{z_1} and φ_{z_2}.

The calculation of P_κ is deduced as expressed in Eq. 9 according to the cutting force model in Sect. 2.1. The integral results are listed in Table 2 for every sub-part of P_κ.

$$P_\kappa = \int_{z_{1,j}}^{z_{2,j}} A\, K_q\, cot\kappa\, dz = \int_{z_{1,j}}^{z_{2,j}} \begin{bmatrix} -\sin\kappa\sin\varphi & -\cos\varphi & -\cos\kappa\sin\varphi \\ -\sin\kappa\cos\varphi & \sin\varphi & -\cos\kappa\cos\varphi \\ \cos\kappa & 0 & -\sin\kappa \end{bmatrix} \begin{bmatrix} K_r \\ K_t \\ K_a \end{bmatrix} cot\kappa\, dz$$

$$= \begin{bmatrix} -\int_L K_r \cos\kappa\sin\varphi\, dz - \int_L K_t cot\kappa \cos\varphi\, dz - \int_L K_a \cos\kappa\, cot\kappa\sin\varphi\, dz \\ -\int_L K_r \cos\kappa\cos\varphi\, dz + \int_L K_t cot\kappa \sin\varphi\, dz - \int_L K_a \cos\kappa\, cot\kappa\cos\varphi\, dz \\ \int_L K_r \cos\kappa\, cot\kappa\, dz - \int_L K_a \cos\kappa\, dz \end{bmatrix}$$

$$= \begin{bmatrix} p_{\kappa x1} + p_{\kappa x2} + p_{\kappa x3} \\ p_{\kappa y1} + p_{\kappa y2} + p_{\kappa y3} \\ p_{\kappa z1} + p_{\kappa z2} + p_{\kappa z3} \end{bmatrix} \Bigg|_{z_{1,j}}^{z_{2,j}}$$

$$(10)$$

Where, K_r, K_t and K_a are radial, tangential and axial cutting force coefficient, φ and κ are the radial and axial cutter contact angle of the edge element, dz is the axial discrete element length. The relationship of φ, κ z and dz of the elements on the cutter flute are expressed in Eqs. 6, 7 and 8. The sub-parts $p_{\kappa n1}$, $p_{\kappa n2}$ and $p_{\kappa n3}$ (n = x, y, z) represent the n-direction parts of P_κ.

Table 2. Part integrals of P_κ

	The arc zone of cutter flute	The side zone of cutter flute
$p_{\kappa x1}$	$\frac{K_r}{mr}(r-z)cos(\varphi_0+mz) + \frac{K_r}{m^2 r}sin(\varphi_0+mz)$	$\frac{K_r}{m}cos\kappa\, cos(\varphi_0+mz)$
$p_{\kappa x2}$	$-K_t cos\bar\varphi\sqrt{r^2-(r-z)^2}$	$-\frac{K_t}{m}cot\kappa\, sin(\varphi_0+mz)$
$p_{\kappa x3}$	$-\frac{K_a \sin\bar\varphi}{2r}\left((r-z)\sqrt{r^2-(r-z)^2} + r^2 arccos\left(\frac{r-z}{r}\right)\right)$	$\frac{K_a}{m}cos\kappa\, cot\kappa\, cos(\varphi_0+mz)$
$p_{\kappa y1}$	$-\frac{K_r}{mr}(r-z)sin(\varphi_0+mz) + \frac{K_r}{m^2 r}cos(\varphi_0+mz)$	$-\frac{K_r}{m}cos\kappa\, sin(\varphi_0+mz)$
$p_{\kappa y2}$	$K_t sin\bar\varphi\sqrt{r^2-(r-z)^2}$	$-\frac{K_t}{m}cot\kappa\, cos(\varphi_0+mz)$
$p_{\kappa y3}$	$-\frac{K_a \cos\bar\varphi}{2r}\left((r-z)\sqrt{r^2-(r-z)^2} + r^2 arccos\left(\frac{r-z}{r}\right)\right)$	$-\frac{K_a}{m}cos\kappa\, cot\kappa\, sin(\varphi_0+mz)$
$p_{\kappa z1}$	$\frac{K_r}{2r}\left((r-z)\sqrt{r^2-(r-z)^2} + r^2 arccos\left(\frac{r-z}{r}\right)\right)$	$K_r cos\kappa\, cot\kappa\, z$
$p_{\kappa z2}$	0	0
$p_{\kappa z3}$	$-K_a\left(z - \frac{1}{2r}z^2\right)$	$-K_a\cos\kappa\, z$

Note: $\bar\varphi$ is the mean value of φ_{z_1} and φ_{z_2}.

2.4 The Calculation of Upper and Lower Limits of Integration

In order to verify the validity of the method of cutting force calculation, a modeling method of ball-end mill instantaneous engagement region given by B. Ozturk and I. Lazoglu [11] was referred to get three engagement region boundaries. Figure 2 shows the representation of top, front and detail views of boundaries. In the literature, boundary-1 is obtained from the intersection of the upper face of the workpiece and the semi spherical cutter, boundary-2 is obtained from the intersection of the currently machined surface and the semi spherical cutter, and boundary-3 is the intersection of previously machined surface and the semi spherical cutter. In this paper, three engagement region boundaries were expressed as the functions of radial contact angle φ of cutting edge element based on the expression method of cutting edge element. As with the assumption that Sect. 2.2 of the cutting edge line, the change of radial contact angle of element of the same cutting edge was neglected. However, for the engagement region boundary model in literature [11] was constrained in ball-end part

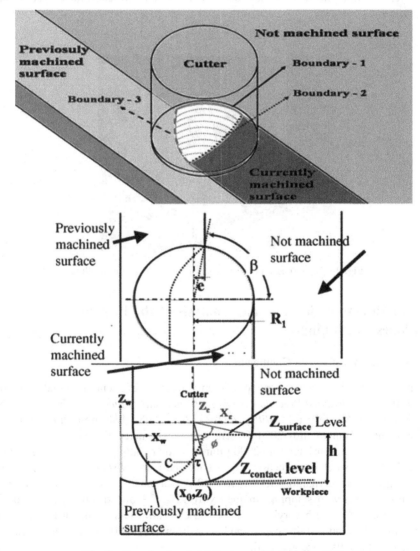

Fig. 2. Top, front and detail views of boundaries [11].

of the cutter and not applied to the side part, the method of calculating the upper and lower limits of integration for ball-end part was given as follows:

For this case, for the specific spindle angle φ_i, the corresponding point of $b_1(\varphi_i)$ of boundary-1, $b_2(\varphi_i)$ of boundary-2 and $b_3(\varphi_i)$ of boundary-3 could be obtained as shown in Fig. 3. When the lead angle $\theta_L \geq 0$, the minimum value between $b_1(\varphi_i)$ and $b_3(\varphi_i)$ was the upper Z values limit of the in-cutting edge in the tool coordinate system and described as z_2, and the $b_2(\varphi_i)$ was the lower Z values limit of the in-cutting edge in the tool coordinate system and described as z_1. When the lead angle $\theta_L < 0$, the maximum value between $b_1(\varphi_i)$ and $b_3(\varphi_i)$ was the lower Z values limit of the in-cutting edge in the tool coordinate system and described as z_1, and the $b_2(\varphi_i)$ was the upper Z values limit of the in-cutting edge in the tool coordinate system and described as z_2.

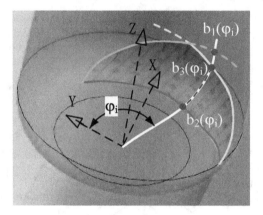

Fig. 3. The upper and lower limits of integration on the flute

3   Experiment Validation and Analysis of the Propeller Blades Machining

3.1   Calibration of Force Coefficients

A single calibration experiment of cutting force coefficient is conducted on *GMC1600H/2 High-speed Precision 5-axis Linkage Gantry Machining Center* with the combination of *Sandvik CoroMill® R216F-12A16C-085* ball nose milling cutter (Table 3) and aluminum alloy 2024-T6. The cutting parameters are set as cutting depth 3 mm, feed rate 500 mm/min, rpm 2000 r/min and down milling with immersion width 5 mm. The force singles are measured and collected using *Kistler 9257A* three-phase piezoelectric crystal dynamometer, *Kistler 5070 type* charge amplifier, and *NI* data acquisition system with sampling frequency 10 kHz. Calibrated milling force coefficients corresponding with varying undeformed chip thickness is shown in Fig. 4, in order to meet the model of this paper, the IUCTs from 0.005 to 0.075 are calculated and the mean coefficients were acquired:

$$\begin{cases} K_t = 2088 \\ K_r = 1653.8 \\ K_a = -161.6 \end{cases} \tag{12}$$

Table 3. cutter parameters

diameter (mm)	Fillet radius (mm)	Flute number	body half taper angle (deg.)	orthogonal rake angle (deg.)
12	6	2	1.33	0

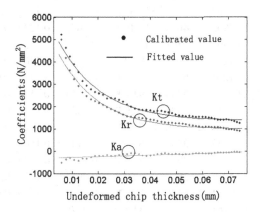

Fig. 4. The cutting force coefficients curve

3.2   Cutting Forces Experiment Validation of Propeller Blades Machining

In order to validate the effectiveness and accuracy of the proposed cutting force model, propeller blades surface was machined with lead angel and tilt angel using ball end mill in this experiment. The model of blades and experiment site are shown in Fig. 5. Cutting parameters were set as follows: cutting width 1.602 mm, feed rate 500 mm/min, spindle speed 2000 r/min. one of the tool paths was pictured in Fig. 5 and recorded in Table 4, which listed the cutter location and orientation.

Three models were utilized to predict the cutting forces: the model established by R. Zhu in literature [5], which using z-map method to determine CWE state, the model establish by B. Ozturk in literature [11], which using CWE boundary method to determine edge element cutting state, and the model in this paper. These models were computed with an identical discrete size but for no discrete required in axial direction of cutters in the method of this paper. The simulation results were drawn in Fig. 6 and the simulation time were recorded in Table 5. As shown in Fig. 6, good prediction accuracy was acquired for all of these three models, especially for the model using z-map method. This is mainly due to the fact that the CWE boundary method in

Ozturk's model and this paper simplify the workpiece surface to a normal plane of local coordinate system. From Table 5 which drawn the simulation time, we could find Ozturk's model is superior to R. Zhu's model in predictive efficiency, and the model in this paper had the best performance, which is faster nearly 20 times and 86 times compared with these two models. Mainly because the model in this paper is based on Ozturk's boundary model and superior to that for the cutting forces are calculated in a cutting flute but not infinitesimal element.

Table 4. The cutter location (CL) file of a tool path

CL No.	X	Y	Z	I	J	K
1	9.4555	35.1119	−13.3606	−0.1683	−0.1135	0.9792
2	12.2993	32.4690	−12.4651	−0.1658	−0.1232	0.9784
3	15.1896	30.0161	−11.5884	−0.1616	−0.1322	0.9780
4	18.1281	27.7572	−10.7281	−0.1559	−0.1403	0.9778
5	21.1201	25.6936	−9.8804	−0.1491	−0.1474	0.9778
6	24.1721	23.8255	−9.0400	−0.1414	−0.1534	0.9780
7	27.2914	22.1526	−8.2020	−0.1330	−0.1585	0.9784
8	30.4834	20.6736	−7.3626	−0.1240	−0.1628	0.9788
9	33.7532	19.3867	−6.5185	−0.1146	−0.1665	0.9794
10	37.1058	18.2908	−5.6661	−0.1052	−0.1696	0.9799
11	40.5464	17.3848	−4.8021	−0.0957	−0.1724	0.9804
12	44.0811	16.6686	−3.9226	−0.0864	−0.1750	0.9810
13	47.7161	16.1421	−3.0235	−0.0775	−0.1776	0.9810
14	49.5723	15.9493	−2.5657	−0.0732	−0.1791	0.9811
15	50.1100	15.9026	−2.4331	−0.0720	−0.1796	0.9811

Fig. 5. The model of blades and experiment site

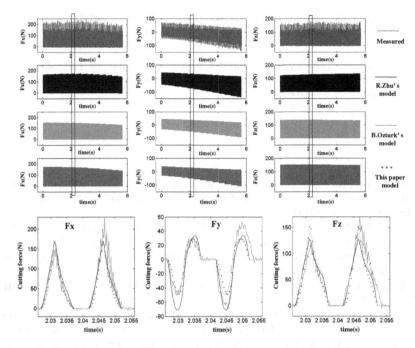

Fig. 6. Comparison between simulated and measured cutting forces

Table 5. Simulation time

Simulation model	Discrete size		Simulation time (s)
	Axial (mm)	Radial (Deg.)	
R. Zhu [5]	0.01	5	534.37
B. Ozturk [11]	0.01	5	123.54
This paper	No discrete	5	6.27

4   Conclusions

This paper proposed an efficient prediction method of multi-axis machining cutting forces, which creatively accomplished the one-off calculation of the cutting force along the whole cutting edge by using integrals of elementary functions in place of classical discretization and accumulation method and avoided both the judgment of cutting state of CEE and repeated calculation of the cutting force for each CEE. Furthermore, the parametric IUCT model considered the cutter orientation angle was used to improve the accuracy in predicting five-axis machining cutting force when transforming classical mechanical force model. Through experiments and simulations on machining of propeller blades, the practicability and efficiency of the method proposed here was verified. Nevertheless, experiments and simulations were confined to the ball-end cutter in this work, please keep on tracking our research for the modeling of the engagement boundary model for bull-nose mills including the side zone, and the verification of more complex cutting tools like bull nose milling cutters.

Acknowledgements. This research is supported by Major Scientific and Technological Innovation Project of Hubei under Grant no. 2015AAA002, National Natural Science Foundation of China under Grant no. 51421062.

References

1. Zhu, Z., Yan, R., Peng, F., et al.: Parametric chip thickness model based cutting forces estimation considering cutter runout of five-axis general end milling. Int. J. Mach. Tools Manuf **101**, 35–51 (2015)
2. Ferry, W.B., Altintas, Y.: Virtual five-axis flank milling of jet engine impellers—Part I: mechanics of five-axis flank milling. J. Manuf. Sci. Eng. **130**(1), 011005 (2008)
3. Budak, E., Ozturk, E., Tunc, L.T.: Modeling and simulation of 5-axis milling processes. CIRP Ann.-Manufact. Technol. **58**(1), 347–350 (2009)
4. Ozturk, E., Budak, E.: Modeling of 5-axis milling processes. Mach. Sci. Technol. **11**(3), 287–311 (2007)
5. Zhu, R., Kapoor, S.G., DeVor, R.E.: Mechanistic modeling of the ball end milling process for five-axis machining of free-form surfaces. J. Manufact. Sci. Eng. **123**(3), 369–379 (2001)
6. Dongming, G., Fei, R., Yuwen, S.: An approach to modeling cutting forces in five-axis ball-end milling of curved geometries based on tool motion analysis. J. Manufact. Sci. Eng. **132**(4), 041004 (2010)
7. Sun, Y., Guo, Q.: Numerical simulation and prediction of cutting forces in five-axis milling processes with cutter run-out. Int. J. Mach. Tools Manuf **51**(10), 806–815 (2011)
8. Li, Z.-L., Niu, J.-B., Wang, X.-Z., Zhu, L.-M.: Mechanistic modeling of five-axis machining with a general end mill considering cutter runout. Int. J. Mach. Tools Manuf **96**, 67–79 (2015)
9. Larue, A., Altintas, Y.: Simulation of flank milling processes. Int. J. Mach. Tools Manuf **45**(4), 549–559 (2005)
10. Kim, G.M., Kim, B.H., Chu, C.N.: Estimation of cutter deflection and form error in ball-end milling processes. Int. J. Mach. Tools Manuf **43**(9), 917–924 (2003)
11. Ozturk, B., Lazoglu, I.: Machining of free-form surfaces. Part I: analytical chip load. Int. J. Mach. Tools Manuf **46**(7), 728–735 (2006)

A High-Flexible ACC/DEC Look-Ahead Strategy Based on Quintic Bézier Feed Rate Curve

Hui Wang, Chao Liu, Jianhua Wu, Xinjun Sheng, and Zhenhua Xiong[✉]

State Key Laboratory of Mechanical System and Vibration,
School of Mechanical Engineering, Shanghai Jiao Tong University,
Shanghai 200240, China
{351582221,aalon,wujh,xjsheng,mexiong}@sjtu.edu.cn

Abstract. To realize real-time generation of a feed rate profile during NC machining process, the look-ahead strategy utilizing bidirectional scanning algorithm is widely adopted. However, the strategy is time-consuming since the acceleration/deceleration (ACC/DEC) scheduling is called frequently. To overcome the deficiency, a high-flexible ACC/DEC look-ahead strategy is proposed in this paper, which is composed of a ACC/DEC scheduling and a backward scanning and forward revision (BSFR) algorithm. Firstly, the ACC/DEC scheduling based on quintic Bézier curve is presented, and a total of 20 types of feed rate profiles are deduced. The scheduling has the characteristics of jerk continuity, simple calculation, and ease of implementation. After that, to deduce the number of times that the ACC/DEC scheduling is called, the BSFR algorithm is put forward. Meanwhile, constraints such as machine's kinematics, chord error, and command feed rate are also taken into account. Experiments are performed at last. Not only is the optimum feed rate profile generated in real time, but also it has nearly 40 % decrement on the number of calls of scheduling, which demonstrates the proposed strategy is valid and feasible in NC machining.

Keywords: Look-ahead strategy · ACC/DEC scheduling · Bézier feed rate curve · Backward scanning · Forward revision · Feed rate profile

1 Introduction

In NC machining, linear segments are widely applied to describe complicated surface and tool paths for their ease of construction and flexibility. While the tangent vectors and the curvatures are not the same at the junctions, which leads to sudden velocity changes when tools move across the junctions. To solve the problem, a path-smoothing scheme is put forward, which employs a series of parametric curves to smooth linear segments [1]. Thus, a smooth NC trajectory is generated. After that, the trajectory is divided into several ACC/DEC scheduling units (ASUs), and every ASU implements the ACC/DEC scheduling

© Springer International Publishing Switzerland 2016
N. Kubota et al. (Eds.): ICIRA 2016, Part I, LNAI 9834, pp. 697–708, 2016.
DOI: 10.1007/978-3-319-43506-0_60

to calculate the feed rate profile [2]. When the machine tool moves along the trajectory under the guidance of the feed rate profile, a smooth movement can be realized. Ordinarily, the smoothness of movement is evaluated according to the derivative of the trajectory with respect to the time. Accordingly, the feed rate continuity, acceleration continuity, and jerk continuity are defined. Generally speaking, the higher order continuity, the smoother movement. Therefore, how to achieve higher order continuity with simple calculation has become a focus in current research [3]. On the other hand, considering the computation constraints, the ACC/DEC scheduling cannot generate the entire feed rate profile at a time. So the look-ahead strategy is broadly utilized, which generates the feed rate profile gradually in each interpolation cycle. However, every ASU needs to carry out the ACC/DEC scheduling respectively, which increases the calculation burden. Meanwhile, the ASU is too short to implement the high-speed machining due to frequent acceleration and deceleration. Therefore, how to improve the look-ahead strategy to realize high-speed machining is another issue in current research [4].

By adopting ACC/DEC scheduling, the machine tool can realize the smooth movement. Normally, a S-shape ACC/DEC scheduling, based on piecewise function, is widely used. Tsai et al. proposed the 7-segment ACC/DEC profile with acceleration continuity [5]. However, the feed rate profile was not including all the seven regions. Thus, Wang et al. deduced the 17 types of feed rate profiles in all [6], while the calculation was complicated. And then, a simplified S-shape ACC/DEC profile was presented [7]. Although acceleration continuity has already been 'flexible' enough in general, jerk discontinuity can generate undesired effects on the kinematic chains and the inertial loads in some applications. Although the scheduling with jerk continuity could be deduced by piecewise function, many more judgment conditions are introduced. So a scheduling based on trigonometric function was proposed [8]. While the result is not accurate due to the truncation error. Biagiotti et al. utilized the quintic polynomial to construct a ACC/DEC scheduling [9]. However, the calculated maximal acceleration is not consistent with the predetermined maximum. To deal with the problem, Shi et al. presented the equivalent acceleration and gave the conversion formula [10]. While the method couldn't obtain the corresponding feed rate profile when the absolute value of acceleration is greater than that of jerk. Therefore, a ACC/DEC scheduling based on quintic Bézier feed rate curve is proposed in this paper, and the 20 types of feed rate profiles are presented in detail.

Through look-ahead strategy, the feed rate profile can be generated in real time. To satisfy the constraints such as machine's kinematics, chord error, and command feed rate, a bidirectional scanning algorithm is broadly applied into look-ahead strategy. Dong et al. firstly proposed this algorithm, which is intended to realize the profile with the shortest processing time [11]. And then, the constrained feed rate optimization algorithm was put forward [12]. While it was only applied for parametric curves. Mattmüller et al. modified the algorithm and made it suitable for the linear segments [13]. The flowchart of those algorithms is described as follows: firstly, backward scanning is utilized and the reverse feed rate profile is obtained, which can meet the requirements of machine's kinematics characteristics. And then forward scanning is implemented to generate the forward feed rate profile, which can satisfy the machine's dynamics constraints. At last, by the intersection of the generated feed rate profiles, the optimum feed rate profile is gained eventually. While the algorithm requires calling the ACC/DEC scheduling twice at each ASU. In this paper, the BSFR algorithm is put forward, which utilizes the forward revision instead of the forward scanning to reduce the number of times that the ACC/DEC scheduling is called. Meanwhile, the above-mentioned constraints are also well considered.

This article is organized as follows. In Sect. 2, the ACC/DEC scheduling based on Bézier feed rate curve is given. Section 3 gives the selection criteria for ASU. The BSFR look-ahead strategy is proposed in Sect. 4. In Sect. 5, experiments are presented to evaluate the performance of the proposed ACC/DEC scheduling and the BSFR algorithm. Finally, a conclusion is given in Sect. 6.

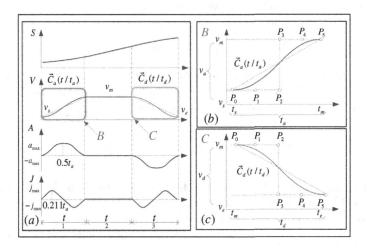

Fig. 1. The diagram of (a) the ACC/DEC scheduling with jerk continuity, (b) the acceleration region, and (c) the deceleration region

2 An ACC/DEC Scheduling Based on Bézier Feed Rate Curve

In this section, a standard S-shape feed rate profile is divided into three regions, which are the acceleration, the constant velocity, and the deceleration. The constant velocity region is denoted a straight line. The acceleration region and deceleration region are expressed as two quintic Bézier curves receptively. The quintic Bézier curve $\vec{C}(u)$ will be defined as:

$$C(u) = \sum_{i=0}^{5} B_{i,5}(u)P_i \tag{1}$$

Where $B_{i,5}$ are 5-th degree Bernstein polynomials [14], P_i are the control points, and u is the parameter with a range of [0,1]. As can be seen in Eq. 1, when control points P_i are given, the Bézier curve can be determined. And the first and second derivative of $\overrightarrow{C}(u)$ are as follows:

$$C'(u) = 5 \sum_{i=0}^{4} B_{i,4}(P_{i+1} - P_i)$$
$$C''(u) = 20 \sum_{i=0}^{3} B_{i,3}(P_{i+2} - 2P_{i+1} + P_i) \tag{2}$$

Figure 1(a) shows a ACC/DEC scheduling with jerk continuity, where $S(t)$, $V(t)$, $A(t)$, and $J(t)$ denote the displacement, feed rate, acceleration, and jerk profile. Figure 1(b) and (c) illustrate the enlarged drawings of acceleration and deceleration region. As seen in Fig. 1(b), v_s indicates the initial velocity at the instant time t_s, v_m denotes the maximal velocity at t_m. Thus the velocity change value is $v_a = v_m - v_s$ and the elapsed time is $t_a = t_m - t_s$. Here, the expression of the acceleration region is:

$$\overrightarrow{C}(t/t_a) = \{x(t/t_a), V(t/t_a)\}, t \in [0, t_a] \tag{3}$$

Where $x(t/t_a)$ denotes the relationship between the coordinate x and the time t, $V(t/t_a)$ is the velocity value v with respect to t. By eliminating the component of factor $x(t/t_a)$, a simplified Bézier curve $\overrightarrow{C}_a(t/t_a)$ is achieved.

$$\overrightarrow{C}_a(t/t_a) = V(t/t_a) = \sum_{i=0}^{5} B_{i,5}(t/t_a)P_i \tag{4}$$

In order to realize jerk continuity, the feed rate profile needs to achieve C^2 continuity [9]. That is to say, at junctions, the Bézier curves' first and second derivatives should be equal to the lines' ones. As seen in Fig. 1(a), when the derivatives at junctions are set to be zeros, the C^2 continuity can be guaranteed. Thus, according to Eqs. 1 and 2, the control points can be calculated.

$$P_0 = P_1 = P_2 = v_s$$
$$P_3 = P_4 = P_5 = v_m \tag{5}$$

According Eqs. 1 and 5, the expression of acceleration region is gained. Meanwhile, the deceleration region is similar with the acceleration region. By integrating $\overrightarrow{C}_a(t/t_a)$, $S(t)$ can be achieved. Similarly, $A(t)$ and $J(t)$ can also be obtained by taking the derivatives of the $\overrightarrow{C}_a(t/t_a)$.

$$S(t) = \int_0^t \overrightarrow{C}_a(t/t_a)dt$$
$$= (v_s t + 5t^4/2t_a^3 - 3t^5/t_a^4 + t^6/t_a^5)(v_m - v_s) \tag{6}$$

$$A(t) = \overrightarrow{C}_a'(t/t_a) \times (t/t_a)'$$
$$= \sum_{i=0}^{4} B_{i,4}(t/t_a) \times (P_{i+1} - P_i) \times (t/t_a)'$$
$$= 30(t^2/t_a^3)(1 - t/t_a)^2(v_m - v_s) \tag{7}$$

$$J(t) = \overrightarrow{C_a'}(t/t_a) \times (t/t_a)'' + \overrightarrow{C_a''}(t/t_a)(1/t_a)^2$$
$$= \sum_{i=0}^{3} B_{i,3}(t/t_a) \times (P_{i+2} + 2P_{i+1} - P_i) \times (1/t_a)^2 \qquad (8)$$
$$= 60t(t - t_a)(2t - t_a)(v_m - v_s)/t_a^5$$

When t is equal to $0.5t_a$, the maximum of $A(t)$ can be achieved and it is $a_{max} = 1.875v_a/t_a$. Accordingly, the maximum of $J(t)$ is $10v_a/\sqrt{3}t_a$ when t is $0.211t_a$. So t_a can be determined by the following equation:

$$t_a = \max(\frac{1.875v_a}{a_{max}}, \sqrt{\frac{5.733v_a}{j_{max}}}) \qquad (9)$$

Therefore, the ACC/DEC scheduling based on the Bézier feed rate curve has been realized. If the initial velocity v_s is greater than or equal to the ending velocity v_e, there will be 10 types of S-shape feed rate profiles. Otherwise, another 10 types will be obtained. The type-determinant process is shown in Fig. 2, and the expressions of all types of S-shape feed rate profiles are listed in Appendix.

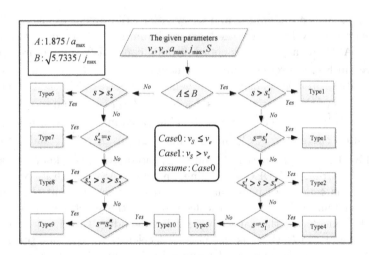

Fig. 2. The flow chart of the type-determinant process

3   The Selection Criteria for ACC/DEC Scheduling Unit

In NC machining, path-smoothing scheme is generally adopted to generate a smooth NC trajectory, which employs a series of transition curves to smooth a series of linear segments. Thus, chord error δ is introduced during the parametric interpolation process. In order to guarantee the machining surface quality, δ should be bounded within a given tolerance. According to Zhao's method [2], each midpoint of the curve should have a curvature extremum and the corresponding feedrate should be a local minimum. Thus, the trajectory is divided into

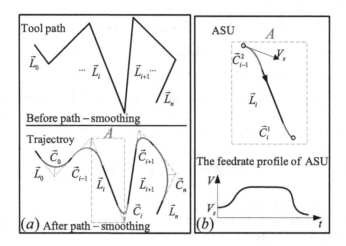

Fig. 3. The diagram of (a) a smooth trajectory and (b) one of ASU

several ASUs at the midpoints, and every ASU will implement the ACC/DEC scheduling respectively. Hence, the local minimum feedrate is restricted at each end of the ASU. As a result, the chord errors δ can satisfy the requirements.

As seen in Fig. 3, a series of linear segments $\overrightarrow{L}_0 \cdots \overrightarrow{L}_i, \overrightarrow{L}_{i+1} \cdots \overrightarrow{L}_n$ denotes a NC tool path. After path-smoothing scheme, a smooth trajectory is generated, which consists of the linear segments $\overrightarrow{L}_0 \cdots \overrightarrow{L}_i, \overrightarrow{L}_{i+1} \cdots \overrightarrow{L}_n$ and the transition curves $\overrightarrow{C}_0 \cdots \overrightarrow{C}_i, \overrightarrow{C}_{i+1} \cdots \overrightarrow{C}_n$. One ASU is indicated in Frame A, which contains two transition curves $\overrightarrow{C}_{i-1}^2, \overrightarrow{C}_i^1$ and one line segment $\overrightarrow{L}_i$. $\overrightarrow{C}_{i-1}^2$ and $\overrightarrow{C}_i^1$ can be achieved by dividing the curves $\overrightarrow{C}_i$ and $\overrightarrow{C}_{i-1}$ into halves using knot intersection method [14]. After ACC/DEC scheduling, one of the ASU's feedrate profile, seen in Fig. 3(b), is generated. Note that the sharp corners occur at the junctions between different ASUs, so only the feedrates at the junctions V_s need to be restricted. Considering the various constrains including the chord error δ, the given feed command F, the maximal normal acceleration A_{max}, and the maximal normal jerk J_{max}, V_s is therefore determined using the following formula:

$$V_s = \min(\frac{2}{T}\sqrt{\rho_{max}^2 - (\rho_{max} - \sigma_{max})^2}, \sqrt{\rho_{max}A_{max}}, \sqrt[3]{\rho_{max}^2 J_{max}}, F) \qquad (10)$$

Where ρ_{max} denotes the curvature extremum of ASU, and T is sampling period. Here, the maximal normal acceleration and jerk are supposed be equal to the tangential ones.

4   Real-Time BSFR Look-Ahead Strategy

To generate the feed rate profile in real time, the look-ahead strategy is broadly utilized. In this paper, the core of the strategy is the BSFR algorithm, which aims to increase the computation efficiency and achieve jerk continuity. Here, a floating window is employed. The window covers a set of ASUs, the number of which is dependent on the computational performance of the NC system. At the beginning of look-ahead strategy, the path-smoothing scheme is performed within the window. A series of ASUs' ideal velocity at the starting point V_s are calculated in advance by Eq. 10. Then, both V_s and the length of the ASUs S are stored into the buffer. After that, the BSFR algorithm is implemented, and two steps are engaged.

Step I. In the first step, when the first interpolation cycle is arrived, the date of $ASU_0, \cdots ASU_i$, as seen in Fig. 4(a), are imported into window 0. Then, backward scanning is performed from the end of the window. From Frame A in Fig. 4(b), it can be seen that the ending velocity of ASU_n is set to be zero. Within every ASU, the ACC/DEC scheduling is applied from the endpoint to the starting point. The velocity at the endpoint is expressed as v_{end}. Thus, by substituting v_{end} into Eq. 11, the velocity at the starting point v_{start} can be calculated. However, the ideal velocity at the starting point is V_s, it may not equal to the v_{start}. In this case, if v_{end} is less than v_s and v_{start} is larger than V_s, the actual velocity at the starting point will be set as V_s. If both v_{end} and v_{start} are less than v_s, the actual velocity will be equal to v_{start}. Similarity, if v_{end} is larger than V_s and v_{start} is less than V_s, the actual velocity will be set as V_s. If both v_{end} and v_{start} are larger than v_s, the actual velocity at the starting point will be equal to V_s, and v_{end} will be reset as $v_{end} - v_{start} + V_s$. After that, the acutual velocity is set to be the start one of the next ASU. The process is performed recursively until the start of the window is reached. Therefore, a reverse feed rate profile is acquired, which is depicted in red line in Frame A.

$$
\begin{cases}
v_{start} = \sqrt{\frac{16a_{\max}S}{15} + v_{end}^2}, & if\ v_{end} < V_s\ and\ \frac{1.875}{a_{\max}} \leq \sqrt{\frac{5.7335}{j_{\max}}} \\
\frac{v_{start}+V_s}{2}\sqrt{\frac{10(V_s-v_{start})}{\sqrt{3}j_{\max}}} - s = 0 & if\ v_{end} < V_s\ and\ \frac{1.875}{a_{\max}} > \sqrt{\frac{5.7335}{j_{\max}}} \\
v_{start} = \sqrt{v_{end}^2 - \frac{16a_{\max}S}{15}} & if\ v_{end} > V_s\ and\ \frac{1.875}{a_{\max}} \leq \sqrt{\frac{5.7335}{j_{\max}}} \\
\frac{v_{start}+V_s}{2}\sqrt{\frac{10(v_{start}-V_s)}{\sqrt{3}j_{\max}}} - s = 0 & if\ v_{end} < V_s\ and\ \frac{1.875}{a_{\max}} > \sqrt{\frac{5.7335}{j_{\max}}}
\end{cases}
\tag{11}
$$

where $a_{\max}$ and $j_{\max}$ denote the maximal tangential acceleration and the maximal tangential jerk respectively.

Step II. Once Step I is completed, forward revision is implemented from the start of the window, which can be seen in Frame B. The initial velocity is set to be the ending velocity of the previous ASU. The ACC/DEC scheduling is applied once again from the starting point to the endpoint. Through Eq. 11, the velocity at the endpoint is re-calculated and the result may not equal to the ideal one that has been achieved form step I. If the result is less than the one that achieved form step I, the velocity at the endpoint will be set as the result. This process is performed recursively until one of ASU is reached that the calculated velocity at the endpoint is equal to the one that acquired form step I. Thus a final feed rate profile is obtained, which is depicted in blue line in Frame B.

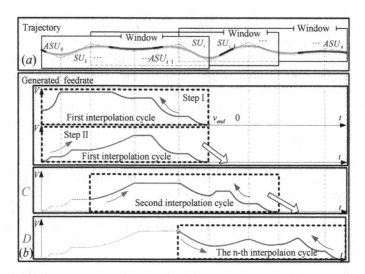

Fig. 4. Example of generating (a) the trajectory's (b) feed rate profile using the BSFR look-ahead strategy

From Frame C in Fig. 4(b), it can be seen that the window is moved to the position of the third ASU when the next cycle is arrived. Repeat the above steps until all ASUs are processed. As seen in Fig. 4(a), since the windows overlap each other and the initial velocity is set to be the ending velocity of the previous ASU, the machine tool don't need to slow down to zero at junction. Thus the continuous tool movement with jerk continuity can be realized, which can be seen in Frame D. Meanwhile, because not all the segments need to carry out the ACC/DEC scheduling twice, the computation efficiency is increased.

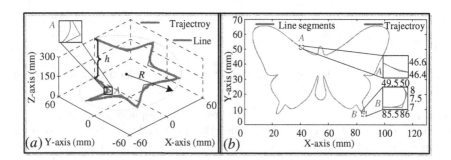

Fig. 5. Linear segments and the trajectories of (a) the 3D pentagram (b) the butterfly graphic

5   Experiments

5.1   ACC/DEC Scheduling Comparison

An instance of a 3D pentagram is shown in Fig. 5(a), which consists of 10 linear segments. The radius of the circumscribed circle is $R = 60\,\text{mm}$, and the height is $h = 300\,\text{mm}$. To evaluate the performance of the ACC/DEC scheduling, the BSFR look-ahead strategy is utilized. Firstly, four linear segments are read in the floating window. After path-smoothing scheme [2], a smooth trajectory is obtained. Then, both the proposed ACC/DEC scheduling and the transitional 7-segment ACC/DEC scheduling are carried out respectively. When the next interpolation cycle arrives, the window is slid to the fourth ASU and the process repeats until all ASUs are processed. The generated feed rate, acceleration, and the jerk profiles are shown in Fig. 6(a) and (b). From Fig. 6(a), it can be seen that the feed rate profile and acceleration profile are continuous, which is obtained by the traditional scheduling. While the jerk profile is discontinuous and the step change occurs in every segment. On the contrary, the continuous jerk profile is obtained by adopting the proposed scheduling, which can be seen in Fig. 6(b). Meanwhile, the fluctuation of jerk profile is less than that in Fig. 6(a). That is to say. The proposed ACC/DEC scheduling can efficiently decrease the demand of the drive capability. Although the ACC/DEC scheduling with quintic polynomial also can achieve jerk continuity, it is only applied for the condition that a_{max} is less than j_{max}. So it is not suitable for the instance of a 3D pentagram. Compared with the quintic polynomial, the proposed scheduling gives 20 types of S-shape feed rate profiles and all conditions are considered. Thus the ACC/DEC scheduling based on quintic Bézier curve is more flexible in application.

5.2   The Performance of the BSFR Algorithm

The other instance of a butterfly graphic is shown in Fig. 5(b), which is composed of 250 linear segments. Accordingly, 250 ASUs is generated using path-smoothing scheme [2]. Here, two look-ahead strategies, the BSFR algorithm and the traditional bidirectional scanning algorithm, are utilized respectively. In each interpolation cycle, five linear segments are read in the floating window, and the window is slid to the fourth linear segments when the next period arrives. Therefore, the look-ahead process is required to be called 63 times, and a total of 312 ASUs need to be applied the ACC/DEC scheduling. Using the proposed ACC/DEC scheduling, the jerk continuity, as is shown in Fig. 7, can be achieved. When the traditional algorithm is adopted, the ACC/DEC scheduling is needed to be called 624 times in all, which can be seen in Fig. 8(a), On the contrary, the BSFR algorithm uses the forward revision process to revised the reverse feed rate profile. Therefore, not all the ASUs need to implement the scheduling twice. From Fig. 8(a), utilizing the BSFR algorithm only requires calling the scheduling 384 times, which can nearly save up to 40 % execution time while the generated feed rate profile is the same as that is generated by the traditional algorithm.

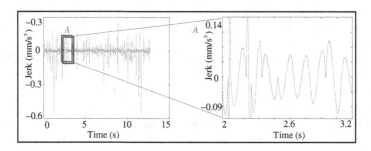

Fig. 6. The feed rate, acceleration, and jerk profiles achieved by (a) the 7-segment ACC/DEC scheduling and (b) the proposed ACC/DEC scheduling

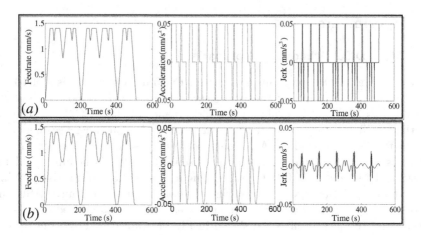

Fig. 7. The jerk profile achieved by the proposed ACC/DEC scheduling

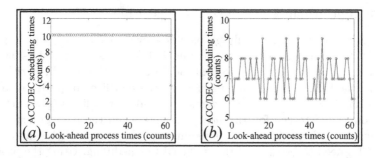

Fig. 8. The number of the ACC/DEC scheduling calls by (a) the traditional bidirectional scanning algorithm (b) and the BSFR algorithm

6   Conclusions

A high-flexible ACC/DEC look-ahead strategy is proposed in this paper, which realizes real-time generation of a feed rate profile during NC machining process.

The strategy includes the ACC/DEC scheduling and the BSFR algorithm. Based on quintic Bézier curve, the proposed ACC/DEC scheduling can achieve jerk continuity with simple calculation. To decrease the number of times that the ACC/DEC scheduling is called, the BSFR algorithm is deduced, which improves the computation efficiently. At last, experiments are carried out. The results reveal that the proposed strategy can nearly save up to 40 % execution time. Besides, the generated feed rate profile still satisfies the constraints of machine's kinematics, chord error, and command feed rate. Thus, the proposed strategy is valid and feasible for performing high-speed NC machining.

Acknowledgements. This research was supported in part by National Key Basic Research Program of China under Grant 2013CB035804, National Natural Science Foundation of China under Grant U1201244, and China Postdoctoral Science Foundation under Grant 2015M80325.

Appendix

See Table 1

Table 1. The data of the time series and the formula for the maximal actual feed rate for each type

Type	Calculation formula and time series
Type 1	$t_1 = 1.875\frac{v_m - v_s}{a}$, $t_2 = \frac{16as + 15v_s^2 + 15v_e^2 - 30v_m^2}{16av_m}$, $t_3 = 1.875\frac{v_m - v_e}{a}$
Type 2	$t_1 = 1.875\frac{v_m - v_s}{a}$, $t_2 = 0$, $t_3 = 1.875\frac{v_m - v_e}{a}$
Type 3	$t_1 = 1.875\frac{f_{\max} - v_s}{a}$, $t_2 = 0$, $t_3 = 1.875\frac{f_{\max} - v_e}{a}$ $f_{\max} = \sqrt{\frac{8as}{15} + \frac{v_s^2 + v_e^2}{2}}$
Type 4	$t_1 = 1.875\frac{v_e - v_s}{a}$, $t_2 = 0$, $t_3 = 0$
Type 5	$t_1 = 1.875\frac{f_{\max} - v_s}{a}$, $t_2 = 0$, $t_3 = 0$
Type 6	$t_1 = \sqrt{\frac{10(v_m - v_s)}{\sqrt{3}j}}$, $t_2 = \frac{s - 0.5(v_s + v_m)t_1 - 0.5(v_m + v_e)t_3}{v_m}$, $t_3 = \sqrt{\frac{10(v_m - v_e)}{\sqrt{3}j}}$,
Type 7	$t_1 = \sqrt{\frac{10(v_m - v_s)}{\sqrt{3}j}}$, $t_2 = 0$, $t_3 = \sqrt{\frac{10(v_m - v_e)}{\sqrt{3}j}}$
Type 8	$t_1 = \sqrt{\frac{10(f_{\max} - v_s)}{\sqrt{3}j}}$, $t_2 = 0$, $t_3 = \sqrt{\frac{10(f_{\max} - v_e)}{\sqrt{3}j}}$ $\frac{v_s + f_{\max}}{2}\sqrt{\frac{10(f_{\max} - v_s)}{\sqrt{3}j}} + \frac{v_e + f_{\max}}{2}\sqrt{\frac{10(f_{\max} - v_e)}{\sqrt{3}j}} - s = 0, f_{\max} \in [v_e, v_m]$
Type 9	$t_1 = \sqrt{\frac{10(v_m - v_s)}{\sqrt{3}j}}$, $t_2 = 0$, $t_3 = 0$
Type 10	$t_1 = \sqrt{\frac{10(f_{\max} - v_s)}{\sqrt{3}j}}$, $t_2 = 0$, $t_3 = 0$ $\frac{v_s + f_{\max}}{2}\sqrt{\frac{10(f_{\max} - v_s)}{\sqrt{3}j}} - s = 0$, $v_e = f_{\max}$, $f_{\max} \in [v_s, v_e]$

References

1. Jin, Y., Bi, Q., Wang, Y.: Dual-bezier path smoothing and interpolation for five-axis linear tool path in workpiece coordinate system. Adv. Mech. Eng. **7**(7) (2015)
2. Zhao, H., Zhu, L., Ding, H.: A real-time look-ahead interpolation methodology with curvature-continuous B-spline transition scheme for CNC machining of short line segments. Int. J. Mach. Tools Manuf. **65**(2), 88–98 (2013)
3. Qiao, Z., Wang, H., Liu, Z., Wang, T.: Nanoscale trajectory planning with flexible ACC/DEC and look-ahead method. Int. J. Adv. Manuf. Technol. **79**(5–8), 1–11 (2015)
4. Wang, Y., Yang, D., Liu, Y.: A real-time look-ahead interpolation algorithm based on Akima curve fitting. Int. J. Mach. Tools Manuf. **85**(5), 122–130 (2014)
5. Tsai, M., Cheng, M., Lin, K.: On acceleration, deceleration before interpolation for CNC motion control. In: International Conference on Mechatronics (2005)
6. Wang, H., Liu, C., Wu, J., Sheng, X., Xiong, Z.: Design of a NURBS interpolator with predicted tangent constraints. In: Liu, H., Kubota, N., Zhu, X., Dillmann, R., Zhou, D. (eds.) ICIRA 2015, Part II. LNCS, vol. 9245, pp. 597–608. Springer, Switzerland (2015)
7. Dai, Z., Sheng, X., Hu, J., Wang, H., Zhang, D.: Design and implementation of Bézier curve trajectory planning in DELTA parallel robots. In: Liu, H., Kubota, N., Zhu, X., Dillmann, R., Zhou, D. (eds.) ICIRA 2015, Part II. LNCS, vol. 9245, pp. 420–430. Springer, Switzerland (2015)
8. Luo, F., Zhou, Y., Yin, J.: A universal velocity profile generation approach for high-speed machining of small line segments with look-ahead. Int. J. Adv. Manuf. Technol. **35**(5), 505–518 (2007)
9. Biagiotti, L., Melchiorri, C.: Trajectory Planning for Automatic Machines and Robots. Springer, Heidelberg (2008)
10. Shi, J., Bi, Q., Wang, Y., Liu, G.: Development of real-time look-ahead methodology based on quintic PH curve with G2 continuity for high-speed machining. Appl. Mech. Mater. **464**, 258–264 (2013)
11. Dong, J., Stori, J.: A generalized time-optimal bidirectional scan algorithm for constrained feed-rate optimization. J. Dyn. Syst. Measur. Control **128**(2), 725–739 (2006)
12. Dong, J., Stori, J., Stori, J.: Optimal feed-rate scheduling for high-speed contouring. J. Manuf. Sci. Eng. **129**(1), 497–513 (2007)
13. Mattmller, J., Gisler, D.: Calculating a near time-optimal jerk-constrained trajectory along a specified smooth path. Int. J. Adv. Manuf. Technol. **45**(9), 1007–1016 (2009)
14. Piegl, L., Tiller, W.: The NURBS Book. Springer, Heidelberg (1995)

Data-Driven Feedforward Decoupling Filter Design for Parallel Nanopositioning Stages

Zhao Feng, Jie Ling, Min Ming, and Xiaohui Xiao[(⊠)]

School of Power and Mechanical Engineering,
Wuhan University, Wuhan 430072, China
{fengzhaozhao7,jamesling,
mingmin_whu,xhxiao}@whu.edu.cn

Abstract. Cross-coupling effect severely hinder fast and accurate tracking for parallel piezo nanopositioning stages. In this paper, a data-driven feedforward decoupling filter (DDFDF) is proposed to reduce the cross-coupling caused errors. Traditional control methods for coupled system could achieve good performance on the premise that the dynamic model is accurate and no non-minimum phase zeros exist. The proposed method is totally data-driven with the advantage of no need for accurate identified model and model structure by Gauss-Newton gradient-based algorithm. The DDFDF for eliminating cross-coupling errors was verified on a 2-DOF coupled nanopositioning stage through simulations. Results show the effectiveness of the proposed controller by comparing with open-loop simulations and the well-designed feedback controller.

Keywords: Parallel nanopositioning stage · Cross-coupling effect · Data-driven · Gradient-based algorithm

1 Introduction

The rapid development in nanoscience and nanotechnology has increased the urgent demand for high-speed and high-performance nanopositioning systems [1]. The emergence of flexure-guided, piezoelectric stack-actuated, compact and light nanopositioner that provides repeatable, reliable, and smooth motions meets the requirements for these related applications, such as scanning probe microscopy (SPM) [2], atomic force microscope (AFM) [3], micromanipulation system [4] and so on. There are two kinds of configurations of piezo nanopositioning stage: serial and parallel [5]. For serial-kinematic configuration, only one axis can achieve the high mechanical bandwidth [6]. Parallel structures offer higher resonance frequencies and stiffness on all axes. Therefore, parallel nanopositioning stages have been widely used in commercial design [5].

For parallel nanopositioning stages, cross-coupling among axes inevitably appears and becomes one of the main obstacles for achieving excellent servo performance, especially at high scanning speed, which can be observed from the experiment data presented in [7]. Many special mechanical structures have been designed to reduced cross-coupling effect [8, 9], and the interactive effect can be suppressed significantly at

© Springer International Publishing Switzerland 2016
N. Kubota et al. (Eds.): ICIRA 2016, Part I, LNAI 9834, pp. 709–720, 2016.
DOI: 10.1007/978-3-319-43506-0_61

low frequencies. However, during high-speed motion, the cross-coupling effect still cannot be ignored as inertial effects, especially near the first resonant mode.

Various approaches were proposed to handle the problem. Multiple-input multiple-output (MIMO) damping controllers using reference model matching approach [10, 11] and mixed negative-imaginary and small-gain approach [12] have been designed to damp the first resonant mode and minimize cross-coupling effect simultaneously. Yuen Kuan Yong [13] proposed a H_∞ controller design strategy for each axis regarding the interactive effect as deterministic output disturbance based on the accurate system modeling. To reduce dependence on model, some data-driven controllers using errors obtained from measurement data have been implemented. In [14, 15], iterative feedback tuning (IFT) has been proposed for industrial process control and virtual reference feedback tuning (VRFT) [16, 17] has been implemented to MIMO system with the advantage of one-trial convergence. However, we should note that the controllers above are MIMO feedback design. A more natural configuration would utilize feedforward decoupling part and leave the single-input single-output (SISO) controller part intact.

To avoid feedback loop redundancy and simplify the control system, a feedforward decoupling controller has been designed to compensate for vibration due to interaction in hard disk drives at the cost of accurate modeling in [18]. However, if the plant has non-minimum phase zeros, the controller may be unstable. Besides, the frequency-domain characteristics encompass significant variation from machine-to-machine which cannot be properly modelled. Therefore, we introduce the data-driven feedforward decoupling filter with the advantage of no need for accurate modeling and plant knowledge to suppress the cross-coupling effect. In this paper, the Gauss-Newton algorithm [19] is used to obtain the coefficients of the finite impulse response (FIR) filters by utilize the measurement data like IFT and VRFT. The DDFDF can alleviate problems due to non-minimum phase zeros and the selection of model structure [20] with the aim to provide the conditions that validate a SISO control approach in coupled parallel piezo nanopositioning stage.

The rest of the paper is organized as follows. The cross-coupling effect problems and control scheme are formulated in Sect. 2. The controllers design procedure, including feedback controller, DDFDF are presented in Sect. 3. Simulation results and comparison are presented in Sect. 4 and conclusions are given in Sect. 5.

2 Problem Formulation

In the section, we present the description of the cross-effect as well as the control scheme for corresponding issue.

2.1 Dynamics Model

To simplify the presentation of the coupled parallel piezo nanopositioning stage, a 2×2 diagonal domain plant is considered in this paper, and described as

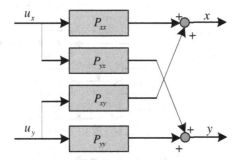

Fig. 1. Dynamics model of a 2 DOFs coupled parallel system

$$P(j\omega) = \begin{bmatrix} P_{xx}(j\omega) & P_{xy}(j\omega) \\ P_{yx}(j\omega) & P_{yy}(j\omega) \end{bmatrix} \tag{1}$$

Figure 1 shows the internal relationship from input to output. Therefore, the coupled parallel dynamics can be modeled as

$$x(\omega) = P_{xx}(j\omega)u_x(\omega) + P_{xy}(j\omega)u_y(\omega) \tag{2}$$

$$y(\omega) = P_{yy}(j\omega)u_y(\omega) + P_{yx}(j\omega)u_x(\omega) \tag{3}$$

Where $x(w)$, $u_x(\omega)$, $y(\omega)$, $u_y(\omega)$ denote the Fourier transforms of $x(t)$, $u_x(t)$, $y(t)$ and $u_y(t)$, respectively. $P_{xx}(j\omega)$ presents the open-loop dynamics of system output x due to the x axis input, and $P_{xy}(j\omega)$ presents the open-loop, cross-coupling effect dynamics under control input $u_y(t)$. Similar definitions are for y axis. To simplify, ω and $|\omega$ are tacitly omitted for conciseness. As we can see from (2) and (3), the output of the individual axis depends on both the diagonal domain dynamics and the cross-coupling dynamic, i.e. non-diagonal domain dynamics, especially at high frequency.

2.2 Control Scheme

For cross-coupling systems, various decoupling feedback methods are proposed to address interactive effect problems [21, 22]. Nevertheless, these existing decoupling feedback control schemes are usually too complex to realize in practical applications, and multi-intersected feedback paths also may render internal uncertainly. Therefore, a common control configuration is a combination SISO feedback control with decoupling feedforward control. The SISO feedback part is to guarantee system robust stability, while the decoupling feedforward part is to attenuate the cross-coupling effect for excellent performance. Figure 2 depicts the configuration of the control system. D_x and D_y are the decoupling feedforward controllers for x axis and y axis. The x axis control input u_x is obtained by subtracting feedback input u_{fbx} from the decoupling feedforward control input u_{dfx} as described in Fig. 2

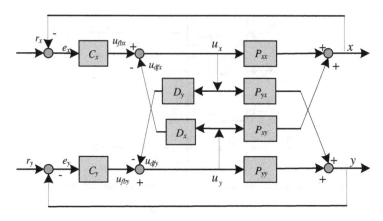

Fig. 2. Configuration of SISO feedback control combined with decoupling feedforward control

$$u_x = u_{fbx} - u_{dfx} \tag{4}$$

The x axis output of the close-loop system can be deduced as

$$x = T_x r_x + S_x P_{xy} u_y - S_x P_{xx} u_{dfx} \tag{5}$$

where T_x, S_x are the complementary sensitivity function and sensitivity function, i.e.

$$T_x = \frac{P_{xx} C_x}{1 + P_{xx} C_x}, \quad S_x = \frac{1}{1 + P_{xx} C_x} \tag{6}$$

Therefore, the error of x axis can be described as

$$e_x = S_x r_x - S_x P_{xy} u_y + S_x P_{xx} u_{dfx} \tag{7}$$

Hence, the cross-coupling effect due to u_y can be canceled if

$$-S_x P_{xy} u_y + S_x P_{xx} u_{dfx} = 0 \tag{8}$$

from Eq. (8) the equation can be represented as

$$u_{dfx} = \frac{P_{xy}}{P_{xx}} u_y \tag{9}$$

Therefore, the decoupler of x and y axis is derived as

$$D_x = \frac{P_{xy}}{P_{xx}}, \quad D_y = \frac{P_{yx}}{P_{yy}} \tag{10}$$

Through the two decouplers, the parallel system is decoupled into two SISO systems. However, we should note that if P_{xx} or P_{yy} has non-minimum phase zeros,

D_x or D_y is unstable. Therefore, DDFDF is introduced to design the decouplers and a H_∞ controller is implemented to retain robust stability.

3 Controllers Design

In this section, we review the H_∞ controller design briefly. Then a DDFDF is proposed to design the decouplers using the collected data breaking through the limit of accurate modeling and non-minimum phase zeros.

3.1 Feedback Controller

In this paper, the H_∞ controller is designed with the advantage of performance, resolution, and robust to model uncertainty directly considered in the frequency domain via appropriate weighting functions [23]. Now, we consider the H_∞ controller design for x axis. Same is the design for y axis.

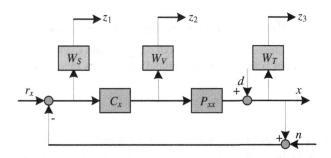

Fig. 3. H_∞ design weighting functions

In H_∞ design the goal is to optimize competing objectives of reference-to-error sensitivity S_x, reference-to-output sensitivity T_x, i.e. complementary sensitivity, and reference-to-control sensitivity $C_x S_x$ simultaneously. The controller C_x is obtained through an iterative design of weighting function to minimize

$$\gamma_{feedback} = \left\| \begin{array}{c} W_S S_x \\ W_T T_x \\ W_V C_x S_x \end{array} \right\|_\infty \tag{11}$$

where $\|\cdot\|_\infty$ is H_∞ norm. The weighting function W_S, W_T, W_V, which can be seen in Fig. 3, penalize the error, output, and the controller output, respectively; these functions are chosen properly for shaping and obtaining a required close loop transfer function of the controlled system. The details on weighting functions design can be found in [23].

3.2   Data-Driven Feedforward Decoupling Filter Design

In this paper, we chose the FIR filter as the feedforward controllers because of its linear characteristic, which is the essential condition for making the optimal problem convex. The FIR filter structure is defined as

$$D_{ij}(z) = p_0^{ij} + p_1^{ij}z^{-1} + p_2^{ij}z^{-2} + \cdots + p_n^{ij}z^{-n}, \; i,j \in \{x,y\} \text{ and } i \neq j \qquad (12)$$

With n is the filter order and $p_0^{ij}, p_1^{ij}, p_2^{ij}, \cdots p_n^{ij}$ are its coefficients. Because Eq. (12) represents a numerator polynomial, it has the ability to create zeros inside the unit circle that can approximating the inverse plant dynamics. All the poles located at the origin, which makes the filter stable.

The data-driven feedforward decoupling filter is optimized by a Gauss-Newton gradient-based algorithm. To compensate the cross-coupling errors, the coefficients of FIR filter are obtained by minimizing the objection function. Herein, the objection criterion is chosen as

$$J(k) = e(p^k)^T \lambda e(p^k) \qquad (13)$$

where $e(p^k) = \left[e(p^k)_x^T e(p^k)_y^T \right]^T$ donates the cross-coupling errors with respect to the coefficients to be optimized. λ is a diagonal weight matrix. k refers to the iteration number. The algorithm is to obtain the optimal coefficients that satisfies

$$p_n := \arg \min_p J(k) \qquad (14)$$

The value of coefficients for each of control directions separately update by a gradient-based algorithm, and the update law can be express as [14]

$$p^{k+1} = p^k - \gamma^k R^{-1} \nabla J|_{p^k} \qquad (15)$$

Where R is a certain position definite matrix, i.e. Hessian matrix, and γ donates the step size. The gradient can be derived from object functions as

$$\nabla J|_{p^k} = 2\nabla e(p^k)^T \lambda e(p^k) \qquad (16)$$

Herein, we chose the Gauss-Newton method is due to its high convergence rate an accuracy. Therefore, the Hessian matrix can be described as

$$R = \nabla(\nabla J|_{p^k}) \approx 2\nabla e(p^k)^T \lambda \nabla e(p^k) \qquad (17)$$

Hence, by substituting (16) and (17), the update law (15) becomes

$$p^{k+1} = p^k - \gamma^k (\nabla e(p^k)^T \lambda \nabla e(p^k))^{-1} \nabla e(p^k)^T \lambda e(p^k) \qquad (18)$$

The update law can be seen as strictly data-driven. It requires only the error signal $e(k)$ and the gradient error matrix ∇e which can be obtained from experiment data. Because of the FIR filter coefficients appear affine in the error signals than only one step is need to obtain the optimal set of coefficients from arbitrary initial conditions [24]. To make the optimal method data-driven, the gradient

$$\nabla e(p^k) = [\nabla e(p_0^k) \nabla e(p_1^k) \cdots \nabla e(p_n^k)] \qquad (19)$$

is obtained from perturbed-parameter experiment. The first-order Taylor series expansion of the error $e(p^k)$ about the user-defined parameter perturbation difference Δp_i i.e.,

$$e(p^{k+1}) = e(p^k) + \nabla e_i(p^k)\Delta p_i + O(\Delta p_i)^2 \qquad (20)$$

Here $e(p^{k+1})$ and $e(p^k)$ are the cross-coupling errors with the perturbation $p_i + \Delta p_i$ and p_i, respectively. Hence, the resulting gradient approximation is given by

$$e_i(p^k) \approx \frac{e(p^{k+1}) - e_i(p^k)}{\Delta p_i} \qquad (21)$$

However, the numbers of experiment are large as the coefficient numbers increase because the one experiment is need for each coefficient. To make the optimization technique more suited for practical application, the choice of FIR filter structure is critical. The structure Eq. (12) has a benefit of obtaining the gradient. Once the term $\nabla e_0(p^k)$ is obtained, the whole gradient $\nabla e(p^k)$ can be calculated immediately since the term $\nabla e_n(p^k)$ is equal the term $\nabla e_0(p^k)$ by a delay of i sampling times, i.e.

$$\nabla e_n(p^k) = \nabla e_0(p^k)z^{-i} \qquad (22)$$

Hence, only two experiments are made for each direction. This feature simplifies the practical implementation of the algorithm. In summary, the follow procedure is adopted to obtain the FIR filter coefficients.

1. Set the initial FIR filter coefficients p to $p_0^k = 0$.
2. Execute a task and obtain the error signals e_0 from the time-intervals.
3. Perturb the coefficients p^{xy} with Δp_0, excuse the task, and store the signals $e_{xy}(p_0^{k+1})$
4. Use Eq. (21) compute the error gradient $\nabla e_{xy}(p_0^k)$.
5. Apply the time delay Eq. (22) and get $\nabla e_{xy}(p^k)$.
6. Use Eq. (18) obtain the optimal coefficients of D_{xy}.
7. Repeat the 3 $\sim$ 6 to calculate coefficients of D_{yx}.

4 Evaluation

The following section evaluates the DDFDF through simulation on a 2-DOF parallel piezo nanopositioning stage. The proposed MFDF was verified via the elimination of cross-coupling effect on y axis when triangle signals with difference frequencies were excited to x axis from time domain and frequency domain.

4.1 System Description

A 2-DOF parallel piezo-actuated nanopositioning stage with a stroke of 100 μm × 100 μm was used as the controlled objective. Each of the x and y direction is actuated by a PZT and The displacement of each axis is detected by a capacitive sensor with the close loop resolution of 10 nm. Because the structure of the nanopositioning stage is symmetric, the frequency response of the system is also symmetric as described in Fig. 4. It can be seen the cross coupling in non-diagonal plots are achieved from −65 dB to −20 dB at low frequency (from 1 Hz to 70 Hz). However, the magnitude tends to be positive with the increase of scan speed, which results in strong cross coupling effect on tracking, especially at the resonant frequency of 123 Hz. This limits the positioning accuracy of the stage. The normalized transfer function of the MIMO system from the identification process is displayed in Eq. (23). When implementation in simulation, the normalized model and controllers were discretized by Tustin method with sampling interval of 0.0004 s.

$$\begin{cases} G_{xx} = \dfrac{146.6s^5 + 7.9\times10^5 s^4 + 9.8\times10^8 s^3 + 2.1\times10^{12}s^2 + 7.3\times10^{14}s + 9.4\times10^{17}}{s^6 + 1009s^5 + 3.8\times10^6 s^4 + 1.8\times10^9 s^3 + 3.5\times10^{12}s^2 + 7.1\times10^{14}s + 9.4\times10^{17}} \\[2mm] G_{xy} = \dfrac{104.1s^5 - 3.6\times10^4 s^4 + 8.9\times10^7 s^3 - 1.7\times10^{11}s^2 + 8.2\times10^{13}s - 1.6\times10^4}{s^6 + 1009s^5 + 3.8\times10^6 s^4 + 1.8\times10^9 s^3 + 3.5\times10^{12}s^2 + 7.1\times10^{14}s + 9.4\times10^{17}} \\[2mm] G_{yx} = \dfrac{104.1s^5 - 3.6\times10^4 s^4 + 8.9\times10^7 s^3 - 1.7\times10^{11}s^2 + 8.2\times10^{13}s - 1.6\times10^4}{s^6 + 1009s^5 + 3.8\times10^6 s^4 + 1.8\times10^9 s^3 + 3.5\times10^{12}s^2 + 7.1\times10^{14}s + 9.4\times10^{17}} \\[2mm] G_{yy} = \dfrac{146.6s^5 + 7.9\times10^5 s^4 + 9.8\times10^8 s^3 + 2.1\times10^{12}s^2 + 7.3\times10^{14}s + 9.4\times10^{17}}{s^6 + 1009s^5 + 3.8\times10^6 s^4 + 1.8\times10^9 s^3 + 3.5\times10^{12}s^2 + 7.1\times10^{14}s + 9.4\times10^{17}} \end{cases} \tag{23}$$

4.2 Suppression of Cross-Coupling Effect

To evaluate the effect of DDFDF for eliminating cross-coupling errors, triangle signals with difference frequencies are input into x axis and the output data from y axis were measured to obtain cross-coupling errors. Figure 5 shows the cross-coupling errors with the 10 Hz, 20 Hz, 50 Hz and 80 Hz triangle input for y axis with the peak-to-peak amplitude of 2 μm. It can be observed that the cross-coupling errors increase with the input frequencies increasing. At low frequency of 10 Hz, the RMS error is below 13 nm and the MAX error is below 35 nm. However, the errors with FB & DDFDF are 7 nm for RMS and 16 nm for MAX from 10 Hz to 50 Hz which verifies the effect of MFDF to suppress cross-coupling errors. The RMS errors and MAX errors for y axis were recorded in Table 1. We should note that at 40 Hz, the cross-coupling errors of open-loop and FB are larger than 50 Hz because the resonant frequency of non-diagonal term is 123 Hz, which is about twofold of fundamental frequency for the

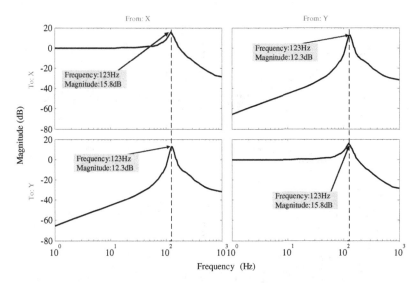

Fig. 4. Frequency responses of the stage. The resonant peak is 15.8 dB at 123 Hz for diagonal frequency responses and 12.3 dB at 123 Hz for non-diagonal frequency responses.

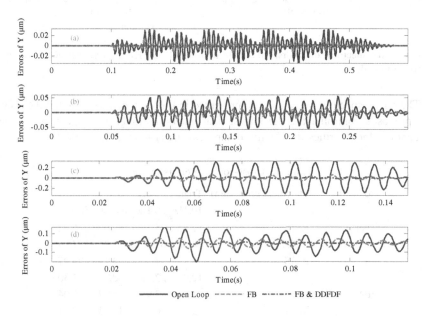

Fig. 5. Cross-coupling errors of *y* axis. (a) 10 Hz triangle. (b) 20 Hz triangle. (c) 40 Hz triangle. (d) 50 Hz triangle.

40 Hz input. At higher frequency of 50 Hz, the proposed control strategy reduces the RMS errors by 92.09 % (from 76.32 nm to 6.04 nm) and 76.74 % (from 25.97 nm to 6.04 nm) with respect to open-loop and FB, respectively.

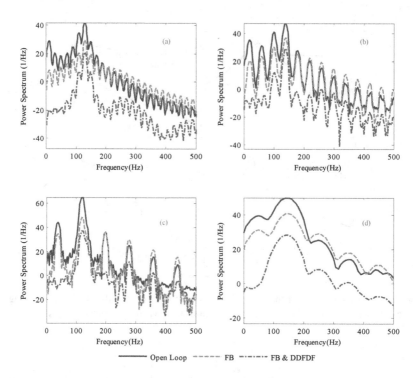

Fig. 6. Power spectrum of errors. (a) 10 Hz triangle. (b) 20 Hz triangle. (c) 40 Hz triangle. (d) 50 Hz triangle

The power spectrums of errors are presented in Fig. 6. It can be seen that the power with proposed controller are the lowest than others, although near the first resonant frequency, which verifies the ability to suppress the cross-coupling effect.

Table 1. Cross-coupling errors of y axis

Controller	Statistical errors(nm)	Triangle wave			
		10 Hz	20 Hz	40 Hz	50 Hz
Open Loop	RMS	12.85	25.19	167.00	76.32
	MAX	34.80	62.59	338.31	197.84
FB	RMS	3.13	7.56	23.14	25.97
	MAX	12.60	22.21	62.47	57.30
FB&DDFDF	RMS	1.01	2.95	5.92	6.04
	MAX	2.27	6.37	15.53	15.41

5 Conclusions

In this paper, the data-driven feedforward decoupling filter was introduced to compensate errors resulting from cross-coupling effects for coupled parallel piezo nanopositioning stages. The coefficients of DDFDF were obtained by Gauss-Newton gradient-based algorithm with the superiority of no need for accurate identified model and model structure. The simulations based on a parallel piezo nanopositioning stage show that cross-coupling errors were suppressed significantly especially at high frequencies by implementing DDFDF when input signals of one axis were triangle wave with different frequencies.

Acknowledgments. This research was sponsored by National Natural Science Foundation of China (NSFC, Grant No.51375349).

References

1. Devasia, S., Eleftheriou, E., Moheimani, S.O.R.: A survey of control issues in nanopositioning. IEEE Trans. Control Syst. Technol. **15**(5), 802–823 (2007)
2. Salapaka, S.M., Salapaka, M.V.: Scanning probe microscopy. IEEE Control Syst. **28**(2), 65–83 (2008)
3. Binnig, G., Quate, C.F., Gerber, C.: Atomic force microscope. Phys. Rev. Lett. **56**(9), 930 (1986)
4. Rakotondrabe, M., Haddab, Y., Lutz, P.: Development, modeling, and control of a micro-/nanopositioning 2-DOF stick–slip device. IEEE/ASME Trans. Mechatron. **14**(6), 733–745 (2009)
5. Yong, Y.K., Moheimani, S.O.R., Kenton, B.J., et al.: Invited review article: High-speed flexure-guided nanopositioning: Mechanical design and control issues. Rev. Sci. Instrum. **83**(12), 121101 (2012)
6. Kenton, B.J., Leang, K.K.: Design and control of a three-axis serial-kinematic high-bandwidth nanopositioner. IEEE/ASME Trans. Mechatron. **17**(2), 356–369 (2012)
7. Bhikkaji, B., Ratnam, M., Moheimani, S.O.R.: PVPF control of piezoelectric tube scanners. Sens. Actuators A **135**(2), 700–712 (2007)
8. Li, Y., Xu, Q.: Development and assessment of a novel decoupled XY parallel micropositioning platform. IEEE/ASME Trans. Mechatron. **15**(1), 125–135 (2010)
9. Yao, Q., Dong, J., Ferreira, P.M.: Design, analysis, fabrication and testing of a parallel-kinematic micropositioning XY stage. Int. J. Mach. Tools Manuf. **47**(6), 946–961 (2007)
10. Das, S.K., Pota, H.R., Petersen, I.R.: Multivariable negative-imaginary controller design for damping and cross coupling reduction of nanopositioners: a reference model matching approach. IEEE/ASME Trans. Mechatron. **20**(6), 3123–3134 (2015)
11. Das, S.K., Pota, H.R., Petersen, I.R.: A MIMO double resonant controller design for nanopositioners. IEEE Trans. Nanotechnol. **14**(2), 224–237 (2015)
12. Das, S.K., Pota, H.R., Petersen, I.R.: Resonant controller design for a piezoelectric tube scanner: a mixed negative-imaginary and small-gain approach. IEEE Trans. Control Syst. Technol. **22**(5), 1899–1906 (2014)
13. Yong, Y.K., Liu, K., Moheimani, S.O.R.: Reducing cross-coupling in a compliant XY nanopositioner for fast and accurate raster scanning. IEEE Trans. Control Syst. Technol. **18**(5), 1172–1179 (2010)
14. Hjalmarsson, H., Gevers, M., Gunnarsson, S., et al.: Iterative feedback tuning: theory and applications. IEEE Control Syst. **18**(4), 26–41 (1998)
15. Hjalmarsson, H.: Efficient tuning of linear multivariable controllers using iterative feedback tuning. Int. J. Adapt. Control Signal Process. **13**(7), 553–572 (1999)

16. Campi, M.C., Lecchini, A., Savaresi, S.M.: Virtual reference feedback tuning: a direct method for the design of feedback controllers. Automatica **38**(8), 1337–1346 (2002)
17. Campi, M.C., Lecchini, A., Savaresi, S.M.: An application of the virtual reference feedback tuning method to a benchmark problem. Eur. J. Control **9**(1), 66–76 (2003)
18. Zheng, J., Guo, G., Wang, Y.: Feedforward decoupling control design for dual-actuator system in hard disk drives. IEEE Trans. Magn. **40**(4), 2080–2082 (2004)
19. Boyd, S., Vandenberghe, L.: Convex Optimization. Cambridge University Press, Cambridge (2004)
20. Teo, Y.R., Eielsen, A.A., Gravdahl, J.T., et al.: Discrete-time repetitive control with model-less FIR filter inversion for high performance nanopositioning. In: IEEE/ASME International Conference on Advanced Intelligent Mechatronics (AIM), pp. 1664–1669 (2014)
21. Hu, C., Yao, B., Wang, Q.: Coordinated adaptive robust contouring control of an industrial biaxial precision gantry with cogging force compensations. IEEE Trans. Industr. Electron. **57**(5), 1746–1754 (2010)
22. Chen, C.S., Chen, L.Y.: Robust cross-coupling synchronous control by shaping position commands in multiaxes system. IEEE Trans. Industr. Electron. **59**(12), 4761–4773 (2012)
23. Skogestad, S., Postlethwaite, I.: Multivariable Feedback Control: Analysis and Design. Wiley, New York (2007)
24. Huusom, J.K., Poulsen, N.K., Jørgensen, S.B.: Improving convergence of iterative feedback tuning. J. Process Control **19**(4), 570–578 (2009)

Study on the Relationship Between the Stiffness of RV Reducer and the Profile Modification Method of Cycloid-pin Wheel

Jianing Wang[1(✉)], Jingjun Gu[2], and Yonghua Yan[1]

[1] School of Mechanical Engineering, Institute of Robotics,
Shanghai Jiao Tong University, Shanghai 200240, China
jennylining@126.com
[2] Nan Tong Zhen Kang Welding Mechanical and Electrical Co., Ltd.,
Jiangsu 226153, China
zgjshmggjj@163.com

Abstract. This paper briefly introduces the status of RV reducer in industrial robots. It also shows the necessity of profile modification of cycloid-pin wheel and its influence on the stiffness of RV reducer. At the same time, the method for determining the number of meshing tooth is given. Based on the initial gap distribution, the size of the initial gaps produced by four kinds of combined profile modification methods is compared. As a result, the best profile modification method with the highest meshing stiffness can be determined, which has a great significance on mass production of RV reducer.

Keywords: Cycloid-pin wheel · Stiffness · Initial gap · Profile modification method · RV reducer

1 Introduction

In order to ensure product quality, improve production efficiency and realize industrial automation, industrial robots play a more and more important role in production. As is known to all, robot joint is the most important part of industrial robot, and it is also the core of transmission. At present, RV reducer is one of the most widely used two kinds of reducers in industrial robot joint, and it has a trend to replace harmonic reducer as well.

The wide application of industrial robots makes the production of RV reducer into business model, which brings the problem that theoretical research and actual production of RV reducer is relatively out of touch. The profile modification of cycloid-pin wheel is the core technology of RV reducer mass production, and it is also the most difficult technology. What's more, the profile modification curve of cycloid-pin wheel directly determines whether the performance index of RV reducer can reach the

This paper is based on research project supported by the National High Technology Research and Development Program of China (863 Program) No.2015AA043003.

N. Kubota et al. (Eds.): ICIRA 2016, Part I, LNAI 9834, pp. 721–735, 2016.
DOI: 10.1007/978-3-319-43506-0_62

international advanced level. Therefore, mastering the key to the manufacture of RV reducer—the best profile modification method of cycloid-pin wheel, is of great strategic importance for the mass production of RV reducer.

RV transmission is a new type of multi-stage planetary transmission developed on the basis of cycloid-pin wheel planetary transmission. It has a series of advantages, such as small size, light weight, high precision, wide range of transmission ratio and so on. As one of the most important performance indicators of RV reducer, stiffness directly affects the working performance of the robot. The stiffness of RV reducer includes bearing stiffness and meshing stiffness. Meshing stiffness is composed of the meshing stiffness of involute gear and cycloid-pin wheel. Accordingly, the number of cycloid-pin wheel meshing tooth is an important factor that affects the meshing stiffness.

Cycloid-pin wheel planetary transmission is non-backlash meshing. In theory, at the same time, the number of meshing tooth can be half of the number of pin gear. In fact, in order to compensate manufacturing errors and ensure good lubrication, the profile modification of cycloid-pin wheel is necessary. After profile modification, there will be initial gaps between cycloid gear and pin wheel along the common normal line direction of theoretical meshing point (the meshing point before modification). It is found that initial gap directly affects the number of meshing tooth. Hence, it also determines the meshing stiffness of cycloid-pin wheel. So by choosing a proper profile modification method to reduce initial gap has a great sense of improving the stiffness of RV reducer.

In China, researches on RV transmission started later, so the corresponding research achievements are still rather few. Yu Lei [3] analyzes reasons for profile modification of cycloid-pin wheel, and introduces three basic profile modification methods and their effects. Zhang Dawei et al. [4, 5] establish the meshing stiffness model of cycloid-pin wheel of RV reducer. Li Lihang [6] illustrates the basic principle of judging the number of meshing tooth at the same time. Yu Ying and Yu Bo [7] present a calculation formula for determining the number of meshing tooth of cycloid-pin wheel. He Weidong et al. [8] study the initial gap distribution of different profile modification methods. Yu Ying [9] compares the size of initial gaps between modification of equidistance and modification of moved distance. Chen Pengfei et al. [10, 11] discuss the influence of on the stiffness of cycloid-pin wheel. Wei Bo et al. [12] calculate the initial gap and meshing force of cycloid-pin wheel with the same profile modification method but different modification amount.

In literatures, there is no further research on the relationship among profile modification method of cycloid-pin wheel, meshing stiffness as well as initial gap distribution. Meanwhile, the best profile modification method which can provide theoretical guidance for the actual production of RV reducer is not proposed either. Therefore, by exploring the influence of initial gap on meshing stiffness and comparing initial gaps of four kinds of combined profile modification methods, this paper presents an optimal profile modification method with the highest meshing stiffness. As a result, the performance of RV reducer is improved, which greatly promotes the industrialization of RV reducer.

2   Meshing Stiffness Model of Cycloid-pin Wheel of RV Reducer

In theory, the transmission of cycloid-pin wheel is through linear contact. In fact, because of the existence of elastic deformation, the meshing transmission of cycloid-pin wheel passes surface contact. It is known that the material of cycloid gear is GCr15, which is the same as pin wheel. So Young's modulus and Poisson's ratio are:

$$E_1 = E_2 = 210Gpa$$

$$\mu_1 = \mu_2 = 0.3$$

The contact deformation of cycloid-pin wheel is shown in Fig. 1. Assuming that the elastic deformation zone is rectangular, its length is $2L$, and width is b.

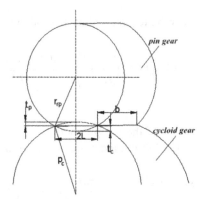

Fig. 1. The contact deformation of cycloid-pin wheel

Based on Hertz formula [1], the expression of L_i is as follows:

$$L_i = \sqrt{\frac{8F_i\rho_i(1-\mu^2)}{\pi bE}} \tag{1}$$

where
  F_i is the force of the i-th gear tooth.
  ρ_{ci} is the radius of curvature of cycloid gear tooth profile at φ_i:

$$\rho_{ci} = \frac{r_p S^{\frac{3}{2}}}{T_i} + r_{rp} \tag{2}$$

$$T_i = k(1+z_p)cos\varphi_i - (1-z_pk^2) \tag{3}$$

$$\rho_i = \frac{r_{rp}^2 T_i}{r_p S^{\frac{3}{2}}} + r_{rp} \tag{4}$$

r_p is the radius of pin wheel.
r_{rp} is the radius of pin tooth.
a is the distance between the center of cycloid gear and center of pin wheel.
z_p is the number of pin teeth.
k is the short width coefficients of cycloid gear, $k = az_p/r_p$ and $k \in (0, 1)$.
φ_i is the phase angle of meshing, $\varphi_i \in [0, \pi]$.
S is a function of k and φ_i, $S = 1 + k^2 - 2kcos\varphi_i$.
For a single pin tooth, according to the Pythagorean theorem, the following equation is obtained as:

$$\left(r_p - t_{pi}\right)^2 + L_i^2 = r_p^2$$

After simplification, t_{pi} can be represented by the following expression:

$$t_{pi} = r_p \left(1 - \sqrt{1 - \left(\frac{L_i}{r_{rp}}\right)^2}\right) = \frac{4F_i\rho_i(1 - \mu^2)}{\pi bE} \tag{5}$$

where
t_{pi} is the radial deformations of the i-th pin tooth.
Thereby, the stiffness of the i-th pin tooth K_{pi} is defined as:

$$K_{pi} = \frac{F_i}{t_{pi}} = \frac{\pi bEr_p S^{\frac{3}{2}}}{4(1 - \mu^2)\left(r_p S^{\frac{3}{2}} + r_{rp}T_i\right)} \tag{6}$$

Similarly, the radial deformations of the i-th cycloid gear tooth t_{ci} can be expressed in the form of:

$$t_{ci} = \frac{4F_i\rho_i(1 - \mu^2)}{\pi bE\rho_{ci}} \tag{7}$$

So the stiffness of the i-th cycloid gear tooth K_{ci} is determined according to the following relation:

$$K_{ci} = \frac{F_i}{t_{ci}} = \frac{\pi bEr_p S^{\frac{3}{2}}}{4(1 - \mu^2)r_{rp}T_i} \tag{8}$$

In summary, meshing stiffness of the i-th pairs of teeth can be written as:

$$K_i = \frac{K_{pi}K_{ci}}{K_{pi}+K_{ci}} = \frac{\pi b E r_p S^{\frac{3}{2}}}{4(1-\mu^2)\left(r_p S^{\frac{3}{2}}+2r_{rp}T_i\right)} \tag{9}$$

Meshing stiffness of cycloid-pin wheel is a function of angle and it is also related to the number of meshing teeth. Consequently, meshing stiffness of cycloid-pin wheel is not a simple addition of single tooth meshing stiffness. Suppose under the function of load torque, pin teeth from $i=m$ to $i=n$ are meshing simultaneously. The meshing stiffness formula of cycloid-pin wheel is:

$$K = \sum_{i=m}^{n} K_i L_i^2 \tag{10}$$

3 Determination of the Number of Meshing Tooth of Modified Cycloid-pin Wheel

As Fig. 2 shows, when cycloid gear and pin wheel are meshing under the function of load torque, there will be some deformation. Cycloid gear will also turn a certain angle. If the deformations of cycloid gear body, pin wheel housing and rotary arm all can be ignored, the deformations of the common normal line direction of theoretical meshing point or the displacement along the normal direction of meshing point is:

$$\delta_i = l_i\beta = sin\varphi_i S^{-\frac{1}{2}}\delta_{max}\left(i=1,2,\ldots,\frac{z_p}{2}\right) \tag{11}$$

where
l_i is the distance between the center of cycloid gear and the common normal line or the normal line of meshing points [6].

$$l_i = r_c \frac{sin\varphi_i}{\sqrt{1+k^2-2kcos\varphi_i}} = r_c sin\varphi_i S^{-\frac{1}{2}} \tag{12}$$

r_c is the radius of pitch circle of cycloid gear, $r_c = a z_c$.
z_c is the number of cycloid gear tooth, $z_c = z_p - 1$.
β is the angle caused by deformations of the parts, $\beta = \frac{\delta_{max}}{l_{max}} = \frac{\delta_{max}}{r_c}$.
δ_{max} is the deformations of a pair of cycloid gear tooth and pin tooth with the maximum force F_{max}.
In theory, assume that the force of each tooth is linear proportional to the corresponding deformations when cycloid-pin wheel is running.

$$F_i = \frac{\delta_i}{\delta_{max}}F_{max} = \frac{l_i}{r_c}F_{max} \tag{13}$$

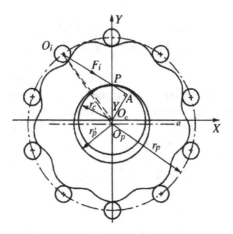

Fig. 2. The force analysis of standard cycloid-pin wheel

Actually, there is a initial gap between modified cycloid gear and pin wheel. Considering the influence of initial gaps, the stress of the i-th pairs of teeth which participate in meshing can be expressed by the following formula [6]:

$$F_i = \frac{\delta_i - \Delta(\varphi_i)}{\delta_{max}} F_{max} \tag{14}$$

What's more, the relationship between the torque on the output shaft M_v and the force of cycloid-pin wheel F_i is known as:

$$\sum_{i=m}^{n} F_i l_i = 0.55 M_v \tag{15}$$

Substituting (13) and (14) in (15), the final expression of F_{max} is:

$$F_{max} = \frac{0.55 M_v}{\sum_{i=m}^{n} \left[\left(\frac{l_i}{r_c} - \frac{\Delta(\varphi_i)}{\delta_{max}} \right) l_i \right]} \tag{16}$$

When phase angle of meshing $\varphi_i = arccos k$, the maximum deformation δ_{max} is the sum of contact deformation W_{max} and bending deformation f_{max} in common normal direction of the meshing point of cycloid-pin wheel:

$$\delta_{max} = W_{max} + f_{max} \tag{17}$$

Substituting $\varphi_i = arccos k$ in the formula of contact deformation, W_{max} is written as [2]:

$$W_{max} = \frac{2(1 - \mu^2)}{E} \cdot \frac{F_{max}}{\pi b} \left(\frac{2}{3} + ln \frac{16 r_{rp} |\rho|}{C^2} \right) \tag{18}$$

where

$$C = 4.99 \times 10^{-3} \sqrt{\frac{2(1-\mu^2)}{E} \cdot \frac{F_{max}}{b} \cdot \frac{2r_{rp}|\rho|}{r_{rp}+|\rho|}} \tag{19}$$

$$\rho = \frac{r_p(1+k^2-2kcos\varphi_i)^{\frac{3}{2}}}{k(1+z_p)cos\varphi_i - (1+z_pk^2)} + r_{rp} \tag{20}$$

Meanwhile, bending deformation f_{max} in common normal direction of the meshing point of cycloid-pin wheel:

$$f_{max} = \frac{F_{max}L^3}{48EJ}B \tag{21}$$

where

L is the span of wheel pin.

J is the axis moment of inertia of wheel pin, $J = \frac{\pi d_{sp}^4}{64}$.

d_{sp} is the diameter of wheel pin.

Two pivot wheel pin, $B = 31/64$.

Three pivot wheel pin, $B = 7/126$.

In order to calculate δ_{max} by (17)–(21), we should know F_{max}. While to calculate F_{max} by (16), we should know δ_{max}. For this reason, we may assign F_{max} an initial value F_{max0}:

$$F_{max0} = \frac{0.55M_v}{\sum_{i=m}^{n} l_i sin\varphi_i S^{-\frac{1}{2}}} \tag{22}$$

Plugging F_{max0} into (17)–(21), δ_{max0} is available. Then plugging δ_{max0} into (16), F_{max1} is available either. If $|F_{max1} - F_{max0}| \geq 1\% F_{max1}$, iteration will continue. Otherwise, calculation will not stop until the result of k-th iteration meets the condition that $|F_{maxk} - F_{maxk-1}| < 1\% F_{maxk}$.

The maximum force F_{max} is formed as:

$$F_{max} = \frac{1}{2}(F_{maxk} + F_{maxk-1}) \tag{23}$$

Substituting (23) and (17) in (11), then the expression of δ_i is available.

When modified cycloid-pin wheel rotates under the load torque, if the deformation δ_i is larger than initial gap $\Delta(\varphi_i)$ somewhere, the pin gear will mesh. If not, the pin gear will not mesh. Thus, the two angles (represented by φ_m and φ_n respectively) corresponding to intersection points of deformation distribution curve and initial gap distribution curve of cycloid-pin wheel are the angles in which cycloid gear and pin wheel mesh at the same time (Fig. 3).

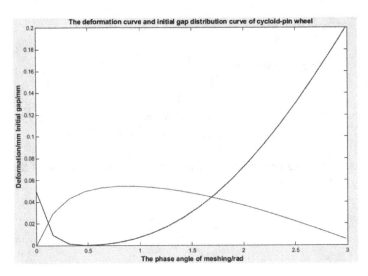

Fig. 3. The deformation curve and initial gap distribution curve of cycloid-pin wheel

So the number of meshing tooth Z_T can be determined in the form of:

$$Z_T = Int\left(\frac{\Delta\varphi/2\pi}{z_p}\right) \tag{24}$$

where

$$\Delta\varphi = \varphi_n - \varphi_m$$

Finally, Z_T is in the interval $\left[Int\left(\frac{\Delta\varphi/2\pi}{z_p}\right), Int\left(\frac{\Delta\varphi/2\pi}{z_p}\right) + 1\right]$ [7].

4   The Initial Gap Distribution of Profile Modification of Cycloid-pin Wheel

At present, there are three common profile modification methods for cycloid-pin wheel: modification of equidistance, modification of moved distance and modification of rotated angle. Because of the complexity of modification of rotated angle, it is generally not used in industrial production. So factories often use combined profile modification of equidistance and moved distance in production.

The initial gap distribution of profile modification of cycloid-pin wheel is [8]:

$$\Delta(\varphi_i) = \Delta r_{rp}\left(1 - sin\varphi_i S^{-\frac{1}{2}}\right) + \Delta r_p\left(1 - kcos\varphi_i - \sqrt{1 - k^2}sin\varphi_i\right)S^{-\frac{1}{2}} \tag{25}$$

where

Δr_{rp} is the amount of modification of equidistance.

Δr_p is the amount of modification of moved distance.

After calculation, the first-order derivative of $\Delta(\varphi_i)$ is shown in the following expression:

$$\frac{d\Delta(\varphi_i)}{d\varphi_i} = \Delta r_{rp}k\sin^2\varphi_i S^{-\frac{3}{2}} - \Delta r_{rp}\cos\varphi_i S^{-\frac{1}{2}} - \Delta r_p k\sin\varphi_i S^{-\frac{3}{2}} + \Delta r_p k\sin\varphi_i S^{-\frac{1}{2}}$$

$$+ \Delta r_p k^2 \sin\varphi_i \cos\varphi_i S^{-\frac{3}{2}} - \Delta r_p\sqrt{1-k^2}\cos\varphi_i S^{-\frac{1}{2}} + \Delta r_p k\sqrt{1-k^2}\sin^2\varphi_i S^{-\frac{3}{2}} \qquad (26)$$

Set first-order derivative of $\Delta(\varphi_i)$ to be zero, $\varphi_i = arccosk$.

Plugging $\varphi_i = arccosk$ into (25), $\Delta(\varphi_i)|_{\varphi_i=arccosk} = 0$.

It means that if cycloid-pin wheel rotates in non-load condition, the pin gear in the place or near the place where phase of meshing $\varphi_i = arccosk$ will mesh with cycloid gear first.

Taking the derivative of (26), the second-order derivative of $\Delta(\varphi_i)$ is as follows:

$$\frac{d^2\Delta(\varphi_i)_3}{d\varphi_i^2} = \Delta r_{rp}\left(3k\sin\varphi_i\cos\varphi_i S^{-\frac{3}{2}} - 3k\sin^3\varphi_i S^{-\frac{5}{2}} + \sin\varphi_i S^{-\frac{1}{2}}\right)$$

$$+ \Delta r_p\left(k\cos\varphi_i + \sqrt{1-k^2}\sin\varphi_i\right)S^{-\frac{1}{2}}$$

$$- \Delta r_p\left(k\cos\varphi_i + 2k^2\sin^2\varphi_i - k^2\cos^2\varphi_i - 3k\sqrt{1-k^2}\sin\varphi_i\cos\varphi_i\right)S^{-\frac{3}{2}}$$

$$+ \Delta r_p\left(3k^2\sin^2\varphi_i - 3k^3\sin^2\varphi_i\cos\varphi_i - 3k^2\sqrt{1-k^2}\sin^3\varphi_i\right)S^{-\frac{5}{2}} \qquad (27)$$

Substituting $\varphi_i = arccosk$ in (27), the second-order derivative of $\Delta(\varphi_i)$ is represented by the relation:

$$\frac{d^2\Delta(\varphi_i)_3}{d\varphi_i^2} = \Delta r_{rp} + \Delta r_p\frac{1}{\sqrt{1-k^2}} \qquad (28)$$

Therefore, the initial gap of combined profile modification has two kinds of situations:

- $\frac{d^2\Delta(\varphi_i)_3}{d\varphi_i^2} = \Delta r_{rp} + \Delta r_p\frac{1}{\sqrt{1-k^2}} > 0$. $\Delta(\varphi_i)$ is the minimum value, so the initial gap is positive.

- $\frac{d^2\Delta(\varphi_i)_3}{d\varphi_i^2} = \Delta r_{rp} + \Delta r_p\frac{1}{\sqrt{1-k^2}} < 0$. $\Delta(\varphi_i)$ is the maximum value, so the initial gap is negative.

It is widely known that the radial gap between cycloid gear and pin wheel should be positive after profile modification:

$$\Delta r = \Delta r_{rp} + \Delta r_p > 0$$

While using negative equidistance & negative moved distance modification, $\Delta r_{rp} + \Delta r_p < 0$. Therefore, there are only three combined profile modifications meeting the requirement:

- Positive equidistance & positive moved distance modification ($\Delta r_{rp1} > 0$, $\Delta r_{p1} > 0$)

$$\Delta(\varphi_i)_1 = \Delta r_{rp1}\left(1 - sin\varphi_i S^{-\frac{1}{2}}\right) + \Delta r_{p1}\left(1 - kcos\varphi_i - \sqrt{1 - k^2}sin\varphi_i\right)S^{-\frac{1}{2}} \quad (29)$$

- Positive equidistance & negative moved distance modification ($\Delta r_{rp2} > 0$, $\Delta r_{p2} < 0$)

$$\Delta(\varphi_i)_2 = \Delta r_{rp2}\left(1 - sin\varphi_i S^{-\frac{1}{2}}\right) + \Delta r_{p2}\left(1 - kcos\varphi_i - \sqrt{1 - k^2}sin\varphi_i\right)S^{-\frac{1}{2}} \quad (30)$$

- Negative equidistance & positive moved distance modification ($\Delta r_{rp3} < 0$, $\Delta r_{p3} > 0$)

$$\Delta(\varphi_i)_3 = \Delta r_{rp3}\left(1 - sin\varphi_i S^{-\frac{1}{2}}\right) + \Delta r_{p3}\left(1 - kcos\varphi_i - \sqrt{1 - k^2}sin\varphi_i\right)S^{-\frac{1}{2}} \quad (31)$$

5 The Cycloid-pin Wheel Profile Modification Method of the Best Meshing Stiffness

Under the condition that the deformation of cycloid-pin wheel is a fixed value, the initial gap is smaller, the number of meshing tooth is greater. In this way, the meshing stiffness of cycloid-pin wheel is higher, the stiffness of RV reducer is larger. So the corresponding profile modification method is more reasonable.

5.1   The Best Profile Modification Method for Positive Initial Gap

When the initial gap of combined profile modification is positive, these three kinds of combined profile modification are satisfied with the following conditions:

$$\begin{cases} \Delta r_{rp} > -\Delta r_p \frac{1}{\sqrt{1-k^2}} \\ \Delta r_{rp} > -\Delta r_p \end{cases}$$

Comparison between the initial gap of positive equidistance & positive moved distance modification and positive equidistance & negative moved distance modification.

The difference between (29) and (30) is represented according to the following relation:

$$\Delta_1 = \Delta(\varphi_i)_1 - \Delta(\varphi_i)_2$$

$$= (\Delta r_{rp1} - \Delta r_{rp2}) \left[\left(1 - sin\varphi_i S^{-\frac{1}{2}} \right) - \left(1 - kcos\varphi_i - \sqrt{1 - k^2} sin\varphi_i \right) S^{-\frac{1}{2}} \right] \quad (32)$$

Suppose

$$f(\varphi_i) = \left(1 - sin\varphi_i S^{-\frac{1}{2}} \right) - \left(1 - kcos\varphi_i - \sqrt{1 - k^2} sin\varphi_i \right) S^{-\frac{1}{2}} \quad (33)$$

Then the first-order derivative of $f(\varphi_i)$ is as follows [9]:

$$\frac{df}{d\varphi_i} = ksin\varphi_i \left(S^{-\frac{1}{2}} - 1 \right) + cos\varphi_i \left(1 - \sqrt{1 - k^2} \right) \quad (34)$$

Set the first-order derivative of $f(\varphi_i)$ to be zero, $\varphi_i = arccosk$.
Substituting $\varphi_i = arccosk$ in (33), $f(\varphi_i)$ is equal to:

$$f(\varphi_i)|_{\varphi_i = arccosk} = k^2 - \left(1 - \sqrt{1 - k^2} \right) \sqrt{1 - k^2} + \sqrt{1 - k^2} - 1 = 0$$

It has been calculated that the second-order derivative of $f(\varphi_i)$ is written as [9]:

$$\frac{d^2 f}{d\varphi_i^2} = -k^2 S^{-\frac{3}{2}} sin^2\varphi_i + k \left(S^{-\frac{1}{2}} - 1 \right) cos\varphi_i + \left(1 - \sqrt{1 - k^2} \right) sin\varphi_i \quad (35)$$

Plug $\varphi_i = arccosk$ into (35), the second-order derivative of $f(\varphi_i)$ is equal to:

$$\frac{d^2 f}{d\varphi_i^2}|_{\varphi_i = arccosk} = \sqrt{1 - k^2} - 1 < 0$$

Thereby, when $\varphi_i = arccosk$, $f(\varphi_i)$ gets its maximum value:

$$f(\varphi_i)_{max} = f(arccosk) = 0$$

$$f(\varphi_i) = \left(1 - sin\varphi_i S^{-\frac{1}{2}} \right) - \left(1 - kcos\varphi_i - \sqrt{1 - k^2} sin\varphi_i \right) S^{-\frac{1}{2}} \leq 0 \quad (36)$$

Notice that $\Delta r_{rp1} + \Delta r_{p1} = \Delta r_{rp2} + \Delta r_{p2}$, $\Delta r_{p1} > 0$ and $\Delta r_{p2} < 0$. Therefore, the first part of (32) is negative:

$$\Delta r_{rp1} - \Delta r_{rp2} = \Delta r_{p2} - \Delta r_{p1} < 0 \tag{37}$$

For this reason, Δ_1 which equals (37) multiplied by (36) is:

$$\Delta_1 = \Delta(\varphi_i)_1 - \Delta(\varphi_i)_2 = (\Delta r_{rp1} - \Delta r_{rp2}) f(\varphi_i) \geq 0$$

In conclusion, when the amount of profile modification of two methods is equal, except in the place where $\varphi_i = arccosk$, the initial gaps caused by two profile modifications are the same. In other place, the initial gaps of positive equidistance & positive moved distance modification are always larger than the initial gaps of positive equidistance & negative moved distance modification (Fig. 4).

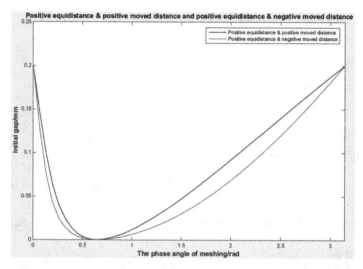

Fig. 4. Initial gaps of positive equidistance & positive moved distance modification and positive equidistance & negative moved distance modification

Comparison between the initial gap of positive equidistance & positive moved distance modification and negative equidistance & positive moved distance modification.

The difference between (29) and (31) can be expressed in the form of:

$$\Delta_2 = \Delta(\varphi_i)_1 - \Delta(\varphi_i)_3$$

$$= (\Delta r_{rp1} - \Delta r_{rp3}) \left[\left(1 - sin\varphi_i S^{-\frac{1}{2}} \right) - \left(1 - kcos\varphi_i - \sqrt{1 - k^2 sin\varphi_i} \right) S^{-\frac{1}{2}} \right] \tag{38}$$

Notice that $\Delta r_{rp1} + \Delta r_{p1} = \Delta r_{rp3} + \Delta r_{p3}$, $\Delta r_{rp1} > 0$ and $\Delta r_{rp3} < 0$. Then the first part of (38) is positive:

$$\Delta r_{rp1} - \Delta r_{rp3} > 0 \tag{39}$$

Therefore, the result of the difference between (29) and (31) is up to (39) multiplied by (36):

$$\Delta_2 = \Delta(\varphi_i)_1 - \Delta(\varphi_i)_3 = \left(\Delta r_{rp1} - \Delta r_{rp3}\right) f(\varphi_i) \leq 0$$

To sum up, when the amount of profile modification of two methods is equal, except in the place where $\varphi_i = arccosk$, the initial gaps caused by two profile modifications are the same. In other place, the initial gaps of negative equidistance & positive moved distance modification are always larger than the initial gaps of positive equidistance & positive moved distance modification (Fig. 5).

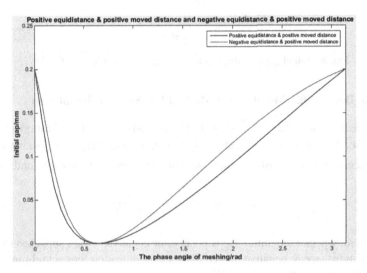

Fig. 5. Initial gaps of positive equidistance & positive moved distance modification and negative equidistance & positive moved distance modification

In summary, the relationship between initial gaps of three kinds of combined profile modifications can be described as the following inequality (Fig. 6):

$$\Delta(\varphi_i)_2 \leq \Delta(\varphi_i)_1 \leq \Delta(\varphi_i)_3 \tag{40}$$

Figure 6 indicates that the initial gap of positive equidistance & negative moved distance modification is the smallest. Thus, the best profile modification method for positive initial gap is the positive equidistance & negative moved distance modification.

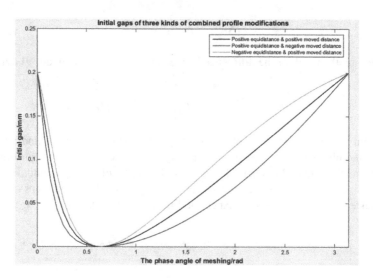

Fig. 6. Initial gaps of three kinds of combined profile modifications

5.2 The Best Profile Modification Method for Negative Initial Gap

It has already known that $\sqrt{1-k^2} \in (0,1)$. The amount of modification of moved distance $\Delta r_p < 0$ for the reason that $\frac{1}{\sqrt{1-k^2}} > 1$. Consequently, only positive equidistance & negative moved distance modification can meet the following condition:

$$-\Delta r_p < \Delta r_{rp} < -\Delta r_p \frac{1}{\sqrt{1-k^2}}$$

In a word, whether the initial gap is positive or negative, we are supposed to choose positive equidistance & negative moved distance modification for cycloid-pin wheel in order to get the highest stiffness of RV reducer.

6 Conclusion

RV reducer has the advantages of high rigidity, high efficiency, accurate positioning and strong controllability. Therefore, it has been widely applied in the joint of industrial robots to achieve accurate transmission of motions and instructions as well as to complete the precise and complicated work. As the core technology of RV reducer, profile modification of cycloid-pin wheel not only determines the working performance of RV reducer, but also affects its practical application and commercial production.

According to the formula of single tooth meshing stiffness of cycloid-pin wheel, this paper establishes the meshing stiffness model of modified cycloid-pin wheel. Then, the curves of tooth deformation and initial gap are drawn via MATLAB to determine the number of meshing tooth. After that, the accurate meshing stiffness of modified cycloid-pin wheel is obtained. Finally, by comparing the initial gap distribution of four kinds of combined profile modification methods, it is concluded that RV reducer will have the highest meshing stiffness when cycloid-pin wheel is modified by positive equidistance & negative moved distance modification.

References

1. Ximeng, Y.: Elastic Plastic Mechanics. Mechanical Industry Press, Beijing (1987)
2. Hao, X.: Machine Design Handbook. Mechanical Industry Press, Beijing (1992)
3. Lei, Y.: Research on the modification of RV with high precision. In the form of Tianjin University (2011)
4. Dawei, Z., Gang, W., Tian, H., et al.: Dynamic modeling and structural parameters analysis of RV gear reducer. J. Mech. Eng. **37**(1), 69–74 (2001)
5. Yinghui, Z., Junjun, X., Weidong, H.: The meshing stiffness calculation of the RV gear reducer for robot. J. Dalian Jiaotong Univ. **31**(2), 20–23 (2010)
6. Lihang, L.: Modification of the tooth shape and the force analysis of the planetary gear transmission with the pin pin gear. J. Dalian Inst. Railway Technol. **4**, 40–49 (1984)
7. Ying, Y., Bo, Y.: Study on the mechanical characteristics of the modified tooth profile of the gear wheel and pin wheel. J. Jiamusi Univ.: Nat. Sci. Ed. **20**(3), 257–259 (2002)
8. Weidong, H., Lihang, L., Xin, L.: The robot speed reducer optimization new cycloid gear shaped. J. Mech. Eng. High Precis. RV **36**(3), 51–55 (2000)
9. Ying, Y., Zhongxi, S., Yuan, L.: Study on the initial clearance and optimal modification of the modified. J. Jiamusi Univ.: Nat. Sci. Ed. **25**(6), 797–800 (2007)
10. Pengfei, C., Wei, Q., Bo, X.: Equivalent contact torsion stiffness of the meshing transmission of the pin gear meshing drive. Mech. Sci. Technol. **33** (4) (2014)
11. Pengfei,C.: Study on meshing stiffness of cycloid-pin gear planetary transmission. Chongqing University (2013)
12. Bo, W., Guangwu, Z., Rongsong, Y., et al.: Comparative study on the method of gear tooth profile modification of RV reducer. Mech. Des. Res. **1**, 41–44 (2016)
13. Zhenyu, C., Zhaoguang, S., Yuhu, Y.: Analysis on meshing characteristics of the planetary transmission mechanism of the pin wheel planetary transmission. Mech. Sci. Technol. **10**, 9–14 (2015)
14. Wentao, W., Honghai, X., Xueao, L., et al.: Theoretical study on the characteristics of the short range profile and its profile. Mech. Des. Manuf. **1**, 94–97 (2016)

Mechatronics Systems
for Nondestructive Testing

Proposal of Carburization Depth Inspection of Both Surface and Opposite Side on Steel Tube Using 3D Nonlinear FEM in Consideration of the Carbon Concentration

Saijiro Yoshioka[(✉)] and Yuji Gotoh

Department of Mechanical and Energy Systems Engineering,
Faculty of Engineering, Oita University, 700 Dannoharu, Oita 870-1192, Japan
v15f1003@oita-u.jp

Abstract. In the steel tube of a heating furnace inside an oil-refining plant, its both surface and opposite side is carburized. If these carburization depths are increased, the steel tube will be exploded suddenly and a big accident may occur. Therefore, the inspection of these carburization depths is important. The conductivity of the layer with carburization is larger than the layer without carburization in the steel, and its permeability is smaller than the layer without carburization. Therefore, the estimation of both carburization depths is possible by using the differences of these electromagnetic properties. In this paper, the new technique of measuring the both depths by using two kinds of alternating magnetic field is proposed. The both depths are obtained by evaluating the flux density in layers with and without carburization steel tube using the 3-D non-linear FEM. It is shown that the inspection of both depths is possible by using the differential electromagnetic characteristics.

Keywords: Electromagnetic non-destructive inspection · Carburization steel tube · Both surface and opposite side depth · 3-D finite element method

1 Introduction

In recent years, the necessity for inspection of the deteriorated oil-refining plant is increased. In particular, the non-destructive inspection of a heating furnace steel tube in the plant is important. In the steel tube, its both surface and opposite side are carburized. If these carburization depths are increased, the steel tube will be exploded suddenly and a big accident may occur. Therefore, the inspection for these depths is important. The conductivity of the layer with carburization is larger than the layer without carburization in the steel tube, and its permeability is smaller than the layer without carburization [1]. Then, the evaluation of these depths is possible by detecting the difference of these electromagnetic characteristics [2–5].

In this paper, the electromagnetic inspection method for these carburization depths is proposed. In this method, the alternating magnetic field of the two kinds of exciting frequency using one sensor is applied to the examined steel tube. These both depths are

© Springer International Publishing Switzerland 2016
N. Kubota et al. (Eds.): ICIRA 2016, Part I, LNAI 9834, pp. 739–748, 2016.
DOI: 10.1007/978-3-319-43506-0_63

obtained by evaluating the flux density in layers with and without carburization inside steel tube using the 3-D nonlinear FEM. In addition, the experimental verification is also carried out.

2 Inspection Model and Method of Analysis

2.1 Measurement of Electromagnetic Properties With and Without Carburization Steel

Figure 2 shows the example of measurement result of carbon concentration inside steel tube using electron probe-micro analysis (EPMA) when the surface carburization depth d_s and the opposite side depth d_o are equal to 1 mm and 2.5 mm, respectively. The domain that the carbon concentration is more than 0.27 % is defined as the carburization layer. The figure denotes that the carbon concentration in the carburization layer is nonlinearly distributed from about 3.8 % to 0.27 % (layer without carburization).

The macroscopic magnetization curves of the rectangular specimens with and without carburization steel (STFA26) are measured using a magnetization equipment as shown in Fig. 2. The average carbon concentration in the rectangular specimen without carburization is 0.27 %. The rectangular specimen with carburization is a full carburized material, and the average carbon concentration is 3.8 %. A rectangular specimen of the steel is placed between magnetic yokes, and magnetized by magnetic field of 0.1 Hz (The minimum exciting frequency in the magnetizing equipment). The flux density in the specimen is measured using a search coil (B coil) wound around the specimen. The magnetic field strength is measured using a thin search coil (H coil) arranged on the specimen [4]. The output voltages of these search coils are amplified with a small-signal amplifier with noise filter, and calculated by the integration of a computer. Figure 3 shows B-H curves [6] of the steel tube with and without carburization layer. The carbon concentration in the carburization layer is equal to 3.8 %.

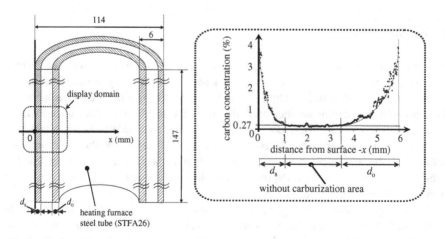

Fig. 1. Distribution of carbon concentration using electron probe-micro analysis (EPMA) when d_s and d_o are equal to 1 mm and 2.5 mm.

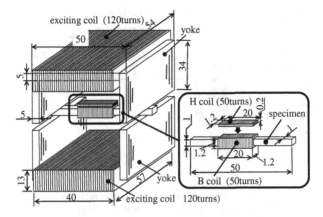

Fig. 2. Magnetization equipment for measuring magnetic properties of the rectangular speci mens with and without carburization steel (STFA26 steel, exciting frequency: 0.1 Hz)

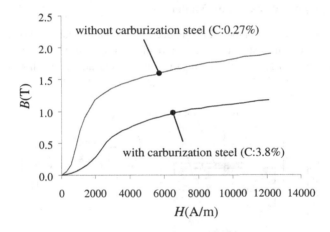

Fig. 3. *B-H* curves of with and without carburization steel (STFA26 steel).

The figure denotes that the permeability in the steel is decreased by the carburization. This is, because the magnetic property of the carburized domain is changed to the structure of strain crystal. Since many lattice defects are generated in the strain crystal, the magnetic wall motion is interfered by the lattice defects [1]. The conductivity of the steel with and without carburization is measured using the Kelvin bridge circuit. Figure 4 shows the conductivities of the steel with and without carburization. The figure denotes that the conductivity of the steel is decreased with the carburization. This is, because it is thought that the conductivity is increased since the carbon permeated into the steel. The conductivity of the steel without carburization is about 1.28 times of that of the steel with carburization.

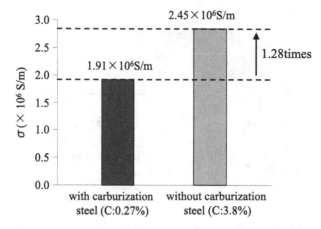

Fig. 4. Conductivities of with and without carburization steel (STFA26 steel).

2.2 Electromagnetic Inspection Model

Figure 5 shows the proposed model for inspecting the both surface and opposite side of the carburization depth in steel tube. This model is composed of the yokes (lamination of silicon steel plates) for ac (alternating) magnetic field and a search coil. As for the

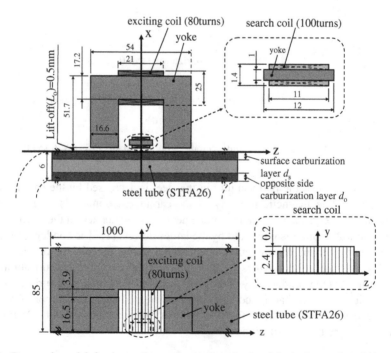

Fig. 5. Proposed model for inspecting carburization depth of both the surface (d_s) and the opposite side (d_o) in steel tube.

search coil in this sensor, the z-direction flux density (B_z) of the magnetic field on the surface of the steel tube is detected. The distance (lift-off: L_o) between the sensor and the surface of steel tube is equal to 0.5 mm. The exciting ampere-turns is 16AT. The exciting frequency of 500 Hz and 15 Hz by one electromagnetic sensor is used to inspect both surface and opposite side of carburized layer in the steel tube.

2.3   3-D Electromagnetic FEM Analysis Method

3-D FEM using the 1st order hexahedral edge element is applied. The flux and eddy current are analyzed by the step-by-step method taking account of the non-linearity of the steel tube. Moreover, an initial magnetization curve of the magnetic yoke is also taken into consideration, but the eddy current is neglected. In order to get the steady state result, the calculation is carried out during 3 periods (=96 steps). The basic equation of eddy current analysis using the A-ϕ method is given by

$$rot(vrotA) = J_0 - \sigma \left(\frac{\partial A}{\partial t} + grad\phi \right) \tag{1}$$

$$div\left\{ -\sigma \left(\frac{\partial A}{\partial t} + grad\phi \right) \right\} = 0 \tag{2}$$

where A is the magnetic vector potential, ϕ is the scalar potential, v is the reluctivity, J_o is the current density and σ is the conductivity.

The carbon concentration is changed non-linearly as it approaches to without carburization layer. Therefore, the numerical value of B-H curves and conductivity are different by with carburization layer location. Analysis is obtained by interpolating the carbon concentration using Fig. 1, and performs the nonlinear calculation. For non-linear calculations is calculated by changing the B-H curves to v-B^2 curves of the magnetic resistivity and flux density. Figure 6 shows v-B^2 curves of carbon concentration 0.27 %

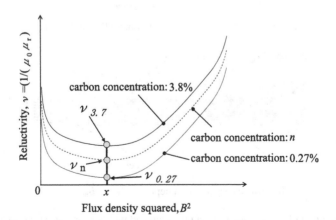

Fig. 6. v-B^2 curves of with and without carburization steel (STFA26 steel).

and 3.8 % used non-linear electromagnetic analysis. It performs the calculation of the *B-H* curves and conductivity using this interpolation method for all elements in the with carburization layer.

3   Inspection of Carburization Depth of Both the Surface and the Opposite Side in Steel Tube

In this research, the carburizations depths of both the surface d_s and the opposite side d_o are inspected by the difference of these electromagnetic characteristics using two kinds of alternating magnetic fields.

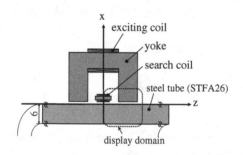

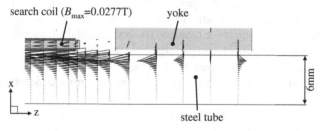

(a) surface carburization depth d_s=0mm (B_{max}=0.096T)

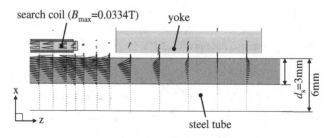

(b) surface carburization depth d_s=3mm (B_{max}=0.036T)

Fig. 7. Distribution of flux density in steel tube with and without surface carburization (500 Hz, 16AT).

3.1 Inspection of Surface Carburization Depth

At first, an alternating magnetic field of 500 Hz is impressed to the steel tube, and the surface carburization depth is inspected. Figure 7 shows the distribution of flux density in the steel tube when the surface carburization depth d_s is equal to 0 mm and 3 mm, respectively. These figures illustrate that the flux density is distributed on the surface of the steel tube by a skin effect, since exciting frequency is high as 500 Hz. Then, the maximum flux density in steel tube is decreased when the surface carburization depth d_s is increased since the permeability in the carburization layer is lower than that of the non-carburization layer.

Figure 8 shows the effect of the change of flux density B_z in a search coil by the 3-D nonlinear FEM when only the surface depth d_s is changed. The figure denotes the rate of change from the measured value B_z of the steel tube without the carburization. The figure illustrates that B_z is increased when the surface depth d_s is increased. This is because, the leakage flux on surface of the steel tube is increased when the surface depth d_s is increased, since the permeability of carburization layer is lower than the non-carburization layer [2]. Therefore, only the surface depth d_s is inspected using the high exciting frequency of 500 Hz as shown in Fig. 8.

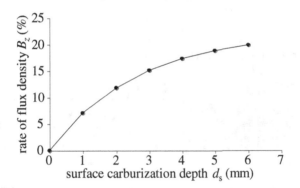

Fig. 8. Effect of the change of flux density B_z in a search coil by the 3-D nonlinear FEM when only the surface depth d_s is changed. (500 Hz, 16AT, calculated).

3.2 Inspection of Opposite Side Carburization Depth

Next, an alternating magnetic field of 15 Hz is impressed to the steel tube, and the opposite side carburization depth d_o is inspected. Figure 9 shows the distribution of flux density in the steel tube when the opposite side carburization depth d_o is equal to 0 mm and 3 mm, respectively. These figures illustrate that the flux density is distributed to the deep domain near the opposite side of steel tube when the opposite side depth d_o is equal to 0 mm, since exciting frequency is low as 15 Hz. However, the flux density is distributed in surface steel tube when the opposite side depth d_o is increased since the permeability in the carburization layer is lower than that the non-carburization layer.

Figure 10 shows the effect of change of flux density B_z in a search coil by the 3-D nonlinear magnetic FEM when only the opposite side depth d_o is changed. The figure denotes the rate of change from the measured value B_z of the steel tube without the carburization. Moreover, the rate of flux density B_z for each opposite side depth d_o is also shown when each surface depth d_s is in constant depth. The figure illustrates that B_z is increased when the opposite side depth d_o is increased. This is because, the leakage flux on surface of the steel tube is increased, since the flux density in surface steel tube is increased when the opposite side depth d_o is increased as shown in Fig. 9. The opposite side depth d_o is inspected using linear interpolation of the curves corresponding to each surface depth d_s as shown in Fig. 10, since the surface depth d_s was inspected using the high exciting frequency of 500 Hz as shown in Fig. 8.

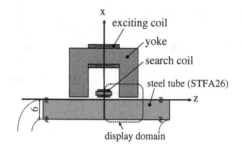

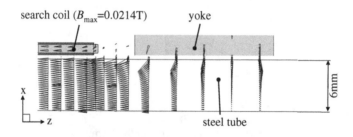

(a) opposite side carburization depth d_o=0mm (B_{max}=0.047T)

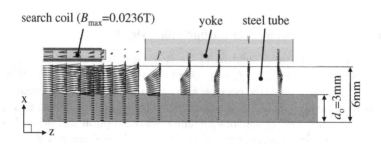

(b) opposite side carburization depth d_o=3mm (B_{max}=0.056T)

Fig. 9. Distribution of flux density in steel tube with and without opposite side carburization (15 Hz, 16AT).

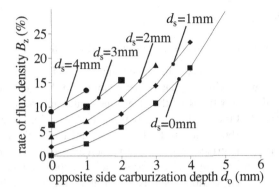

Fig. 10. Effect of change of flux density B_z in a search coil by the 3-D nonlinear FEM when only the opposite side depth do is changed(15 Hz, 16AT, calculated).

4   Verification Evaluation by Actual Carburized Steel Tube in the Oil-Refining Plant

The surface and opposite sides of carburization depth in actual steel tube inside the oil-refining plant are inspected by the proposed method. The heating furnace steel tube which continued being used for 30 years inside the oil-refining plant in Japan was inspected. Table 1 shows the inspection results of the surface and opposite sides of carburization depth. The domain that the carbon concentration is more than 0.27 % by using EPMA is defined as the actual carburization depth. In the proposed method, the carburization depth of the surface and opposite side is searched by the linear interpolation of the calculated values as shown in Figs. 8 and 10. This table denotes that the actual carburization depth of the surface and opposite side is in agreement with the obtained ones. Therefore, it will be possible to detect the carburization depth of both surface and opposite side depth by using the proposed electromagnetic inspection method.

Table 1. Obtained results of the carburization depth of both the surface and the opposite in actual heating furnace steel tube

depth side	actual depth (mm)	inspection depth (mm)	error (mm)
surface (d_s)	0.91	1.00	0.09
opposite (d_o)	0.64	0.63	0.01
surface (d_s)	1.00	0.99	0.01
opposite (d_o)	1.87	1.86	0.01
surface (d_s)	0.96	0.95	0.01
opposite (d_o)	1.50	1.55	0.05
surface (d_s)	1.16	1.06	0.10
opposite (d_o)	1.72	1.60	0.12
surface (d_s)	1.45	1.50	0.05
opposite (d_o)	1.91	2.07	0.16

5 Conclusions

The results obtained by this research are summarized as follows:

(1) The permeability in the steel tube of the material STFA26 is decreased by the carburization. The maximum relative permeability in the steel tube without carburization is about 3.3 times of the steel tube with carburization. And the conductivity in the steel tube is decreased with the carburization. The conductivity of the steel without carburization is 1.28 times of the steel with carburization.

(2) The leakage flux on surface of steel tube is increased using alternating magnetic field of 500 Hz when the surface carburization depth is increased, since the permeability of carburization layer is lower than the non-carburization layer. Then, the leakage flux is increased using alternating magnetic field of 15 Hz when the opposite depth is increased, since the flux density is distributed in surface steel tube when the opposite side carburization depth with the low permeability domain is increased. Therefore, the non-contacting inspection of the carburization depth of both surface and opposite side steel tube is possible by detecting the rate of the change of leakage flux in a search coil.

(3) It is possible to estimate the carburization depth of both the surface and the opposite side in steel plate by the two kinds of exciting frequency using one proposed inspection sensor. Moreover, the calculated results using the 3-D nonlinear FEM are in agreement with the measured values using actual steel tube in an oil-refining plant.

References

1. Kamada, Y., Takahashi, S., Kikuchi, H., Kobayashi, S., Ara, K., Echigoya, J., Tozawa, Y., Watanabe, K.: Effect of predeformation on the precipitation process and magnetic properties of Fe-Cu model alloys. J. Mater. Sci. **44**, 949–953 (2009)
2. McMaster, R.C., McIntrire, P., Mester, M.L.: "Electromagnetic Testing", Nondestructive Testing Handbook, vol. 5, 3rd edn. American Society for Nondestructive Testing, Columbus (2004)
3. Gotoh, Y., Sasaguri, N., Takahashi, N.: Evaluation of electromagnetic inspection of hardened depth of spheroidal graphite cast iron using 3-D nonlinear FEM. IEEE Trans. Magn. **46**(8), 3137–3144 (2010)
4. Gotoh, Y., Matsuoka, A., Takahashi, N.: Measurement of thickness of nickel-layer on steel using electromagnetic method. IEEE Trans. Magn. **43**(6), 2752–2754 (2007)
5. Gotoh, Y., Matsuoka, A., Takahashi, N.: Electromagnetic inspection technique of thickness of nickel-layer on steel plate without influence of lift-off between steel and inspection probe. IEEE Trans. Magn. **47**(5), 950–953 (2011)
6. Nakata, T., Kawase, Y., Nakano, N.: Improvement of measuring accuracy of magnetic field strength in single sheet testers by using two H coils. IEEE Trans. Magn. **23**(5), 2596–2598 (1987)

Evaluation for Electromagnetic Non-destructive Inspection of Hardened Depth Using Alternating Leakage Flux on Steel Plate

Kazutaka Nishimura[✉] and Yuji Gotoh

Department of Mechanical and Energy Systems Engineering,
Oita University, 700 Dannoharu, Oita 870-1192, Japan
v15e1034@oita-u.ac.jp

Abstract. Many surface hardening steels are used for the crankshaft or machine parts, etc. in the automobile engine. The inspection of surface hardening depth is important in the quality assurance of the strength or their parts. Particularly, in order to raise the productivity of these parts, the non-destructive inspection method is needed for the evaluation of the hardened depth. The permeability and the conductivity of the hardened domain are smaller than the non-hardened domain in surface hardening-steel. Evaluation of the surface hardening depth is possible by detecting of difference of these electromagnetic characteristics. In this paper, the electromagnetic inspection method using the detecting of the leakage flux on the surface of the steel material is investigated. The leakage flux and flux density in steel material are estimated by 3D nonlinear finite element method (FEM). The usefulness of this proposal inspection method is shown also from comparison with a experimental verification.

Keywords: Leakage flux · 3-D non-linear finite element method · Hardened depth · Surface hardening steel

1 Introduction

High frequency hardening is the heat treating method for hardening the surface domain of the steel materials for a short time. Therefore, this heat treating method is widely used in the manufacture process of the many parts of engine or crankshaft in automobile parts, etc. In order to guarantee the quality of these parts, it is necessary to inspection the hardened depth. Generally, as for inspection of these hardening depth, the destructive inspection method, such as Vickers or Rockwell hardness tester, is used. However, total inspection such as automobile parts isn't able to be carried out with these inspection method. Therefore, the non-destructive and the non-contacting inspection method is needed for the evaluation of the hardened depth.

The conductivity and permeability of the hardened steel are decreased compared with those of the without hardening steel. For this reason, the hardened depth is possible to estimate by measurement in the difference of the electromagnetic characteristics [1]. The magnetic yoke with an exciting coil is approached to the surface hardened steel, and the inspection method of evaluating a hardening depth by detecting

© Springer International Publishing Switzerland 2016
N. Kubota et al. (Eds.): ICIRA 2016, Part I, LNAI 9834, pp. 749–758, 2016.
DOI: 10.1007/978-3-319-43506-0_64

the change in magnetic flux density in the yoke is proposed [2]. However, the inspection sensitivity of the hardened depth using this technique is not so high.

There is a flaw inspection of steel by detecting the leakage flux [3]. In this paper, the high sensitivity inspection method of the hardened depth in surface hardening steel plate by detecting the leakage flux using the low frequency magnetic field is investigated. The initial magnetization curves and conductivities of the layer with and without hardening of the steel are measured. Then, the evaluation of the flux density and eddy current between the layer with and without hardening inside the steel is calculated by the 3-D electromagnetic finite element method (FEM). In this magnetic characteristic nonlinear FEM with and without hardening steel, the initial B-H curves and conductivities inside the middle layer between the layer with and without hardening is calculated taken into consideration of linear interpolation using these electromagnetic properties. Moreover, the usefulness of this proposal inspection method is shown from comparison with the inspection method examined before [2]. Moreover, it was carried out verification experiment of this inspection method.

2 Measurement of Magnetization Curve and Conductivity of Steel with and Without Hardening

Table 1 shows the chemical component of the steel plate (SCM440 steel) without hardening. Figure 1 shows the measured results of the hardness using the Vickers hardness tester when the effective hardened depths D are 1 mm, 3 mm and 5 mm, respectively. In Japanese Industrial Standards (JIS), the larger hardness domain than 400 HV inside SCM440 steel is defined as the effective hardness depth D. In this paper, the effective hardness depth is 2–4 mm. Therefore, the target of the hardened depth inspection is to discriminate 2–4 mm.

The magnetization curves of the SCM440 steel with and without hardening are measured using an electromagnet shown in Fig. 2 [4]. The average hardness in the steel without hardening is 275 HV. The SCM440 steel with hardening is a full hardened material, and the average hardness is 650 HV. The rectangular steel is placed between pole pieces, and magnetized by magnetic field of 0.1 Hz. The flux density B(T) in the steel is measured using a search coil. The magnetic field strength H(A/m) is measured using a transverse type Hall element.

Figure 3 shows initial magnetization curves of the steel with and without hardening. The figure denotes that the maximum permeability is decreased by the hardening. This is, because the magnetic property of the hardened domain (martensite domain) is changed to the structure of strain crystal. Since many lattice defects are generated in the strain crystal, the magnetic wall motion is interfered by the lattice defects [5].

The conductivity of the steel with and without hardening is measured using the Kelvin bridge circuits. The conductivities of the layers with and without hardening of the steel are 3.61×10^6 S/m and 3.98×10^6 S/m, respectively.

As a result, the maximum relative permeability and conductivity are both decreased with the increase of the hardness.

Table 1. Chemical component of SCM440 steel without hardening

C	Si	Mn	P	S	Ni	Cr	Mo	Cu (wt%)
0.4	0.16	0.61	0.017	0.003	0.08	1.03	0.16	0.2

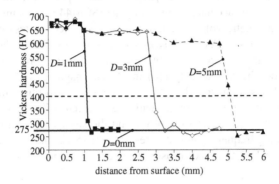

Fig. 1. Distribution of the hardness using Vickers hardness tester (SCM440 steel, Vickers load = 0.3 kgf).

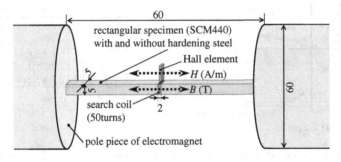

Fig. 2. Electromagnet for measuring magnetic properties of the rectangular specimens with and without hardening steel (SCM440 steel, exciting frequency: 0.1 Hz).

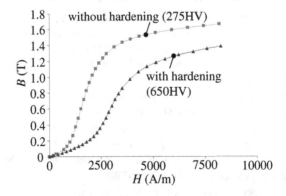

Fig. 3. Initial magnetization curves of the steel with and without hardening (SCM440 steel).

3   Inspection Model and Calculation Method

3.1   Electromagnetic Inspection Model

Figure 4 shows the 1/2 domain of the proposed the inspection model for evaluating hardened depth in the surface hardening steel plate (SCM440). This inspection probe is composed of the magnetic yoke (lamination of silicon steel plates) with an exciting coil, and two search coils. As for a search coil-A in this probe, the x-direction of leakage flux (B_x) on the surface of the steel plate is detected. Moreover, as for a search coil-B in the probe, the flux density (B_z) inside the magnetic yoke is detected. The arrangement of this search coil-B is the conventional inspection method [2]. The comparison of the inspection sensitivity with these search coils A and B is evaluated. The distance (lift-off: L_o) between the probe and the surface of steel plate is equal to 0.5 mm. The exciting frequency and ampere-turns are 15 Hz and 126 AT, respectively. The conditions of calculation and measurement are shown in Table 2.

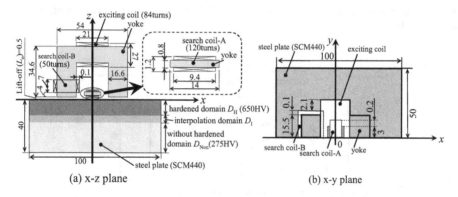

(a) x-z plane   (b) x-y plane

Fig. 4. Inspection model of the surface hardening steel plate (1/2 domain).

Table 2. Conditions of calculation and measurement

Exciting coil	exciting frequency: 15 Hz, ampere-turns of exciting coil: 126 AT
Search coil	Search coil-A: 120 turns Search coil-B: 50 turns
Lift-off (Lo)	0.5 mm
Dimension of specimen	SCM440 steel plate 100 × 100 × 40 mm
Hardening depth	0 mm, 1 mm, 3 mm, 5 mm
Conductivity	hardening domain: 3.61×10^6 S/m non-hardening domain: 3.98×10^6 S/m
Nodes and elements	109980, 102300
Convergence criterion	N-R method: 1.0×10^{-4} T, ICCG method: 1.0×10^{-4}

3.2   Method of Analysis Considering Interpolation of Magnetization Curves and Conductivities

3-D FEM using the 1st order hexahedral edge element is applied. The flux and eddy current are analyzed by the step-by-step method taking account of the initial magnetization curves of the layers with and without hardening as shown in Fig. 3. Moreover, an initial magnetization curve of the magnetic yoke is also taken into consideration, but the eddy current is neglected because laminated yoke. Get the result of the steady state by during 2 periods (= 32 steps) calculation. The time interval Δt of the step-by-step method is the 4.167×10^{-3} s obtained from the excitation frequency 15 Hz.

The basic equation of eddy current analysis using the $A\text{-}\phi$ method is given by

$$rot(v \ rot \ A) = J_o - \sigma\left(\frac{\partial A}{\partial t} + grad \ \phi\right) \tag{1}$$

$$div\left\{-\sigma\left(\frac{\partial A}{\partial t} + grad \ \phi\right)\right\} = 0 \tag{2}$$

where A is the magnetic vector potential, ϕ is the scalar potential, v is the reluctivity, J_o is the current density and ϕ is the conductivity. The Newton-Raphson ($N\text{-}R$) method is used for the nonlinear iteration. The changes δA and $\delta\phi$ of A and ϕ are obtained by solving the following equation:

$$\begin{bmatrix} \frac{\partial G_i}{\partial A_k} & \frac{\partial G_i}{\partial \phi_j} \\ \frac{\partial \eta_m}{\partial A_k} & \frac{\partial \eta_m}{\partial \phi_j} \end{bmatrix} \begin{Bmatrix} \delta A_k \\ \delta \phi_j \end{Bmatrix} = \begin{Bmatrix} -G_i \\ -\eta_m \end{Bmatrix} \tag{3}$$

where G_i and η_m are the residues of (1) and (2), respectively. $\partial G_i/\partial A_k$ includes the derivative term of reluctivity $\partial v/\partial B^2$. The $N\text{-}R$ iterations are carried out using the initial magnetization curves shown in Fig. 3. The magnetization curves and conductivity in the hardened domain, non-hardened domain and region between them are obtained as follows:

Figure 5 shows an example of the measurement result of the hardness using the Vickers hardness tester when the effective hardening depth D is 3 mm. The figure denotes that the hardness in 2.75 mm depth from the surface is about 650 HV, the hardness in the domain from 2.75 mm to 3.25 mm (interpolation domain) depth is decreased rapidly, and that of more than 3.25 mm depth becomes about 275 HV. The initial magnetization curve of the interpolation domain depth is obtained by the linear interpolation using the initial magnetization curves of the layer with (0 mm–2.75 mm depth) and without hardening (more than 3.25 mm depth). Figure 6 illustrates the $v\text{-}B$ curves corresponding to the initial magnetization curves in Fig. 3. The figure denotes the method of linear interpolation of the reluctivitiy v_I when the flux density is equal to B_n using the $v\text{-}B$ curve (v_H) with hardening and that (v_{Non}) without hardening. The reluctivity v_I of the interpolated curve at the point of which the depth is equal to D_I

shown in Fig. 5 is obtained by the v-B curve (v_H) with hardening and that (v_{Non}) without hardening. The reluctivity v_I of the interpolated curve at the point of which the depth is equal to D_I shown in Fig. 5 is obtained by

$$\frac{v_I - v_{Non}}{v_H - v_I} = \frac{D_{Non} - D_I}{D_I - D_H} \tag{4}$$

where, D_H is the depth of the edge of hardened domain and D_{Non} is that without hardened domain.

The conductivity σ in the interpolation domain in Fig. 7 is also interpolated similarly. Figure 7 shows the distribution of conductivity when the effective hardened depth D is equal to 3 mm. The conductivity (σ_I) of the interpolation domain is obtained by the linear interpolation using the conductivity (σ_H) of the layer with hardening and that (σ_{Non}) without hardening. The conductivity σ_I of the interpolated value at the point of which the depth is equal to D_I shown in Fig. 7 is obtained by

$$\frac{\sigma_{Non} - \sigma_I}{\sigma_I - \sigma_H} = \frac{D_{Non} - D_I}{D_I - D_H} \tag{5}$$

where, D_H is the depth of the edge of hardened domain and D_{Non} is that without hardened domain.

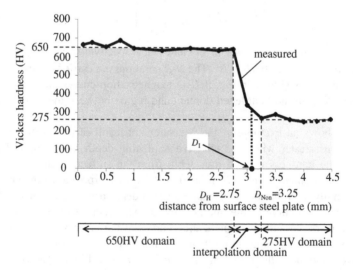

Fig. 5. Hardness measured using Vickers hardness tester when the effective hardened depth is equal to 3 mm.

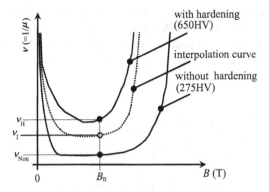

Fig. 6. Interpolation of v-B curves.

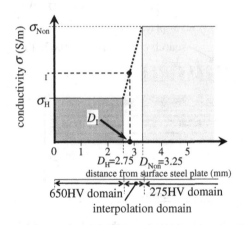

Fig. 7. Distribution of conductivity when hardened depth is equal to 3 mm.

4   Inspection of Hardened Depth in Steel Plate

4.1   Distribution of Leakage Flux, Flux Density and Eddy Current Density in Steel Plate

Figure 8 shows the distribution of leakage flux B_x near a search coil-A and flux density inside surface hardened steel plate when the hardened depth is equal to 0 mm and 3 mm, respectively. Figure 8(a) shows the display domain and Fig. 8(b), (c) is an enlarged view of the Fig. 8(a). This figure illustrates the difference between the way of the magnetic flux penetration due to with and without hardening. Then, the leakage flux B_x is increased when the hardened depth is increased. This is because, permeability and conductivity of the surface domain of the steel plate is decreased when the hardening depth is increased.

The eddy current density in the case of hardened domain having a low conductivity is increased in the vicinity of the non-hardened domain, it shows a larger value than the eddy current density of the steel plate has non-hardened of the same depth as shown in Fig. 9. Therefore, the flux can more easily penetrate deep position in the steel plate.

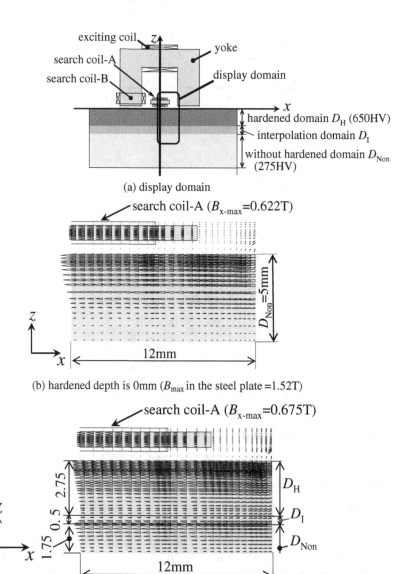

(a) display domain

search coil-A ($B_{x\text{-}max}$=0.622T)

(b) hardened depth is 0mm (B_{max} in the steel plate =1.52T)

search coil-A ($B_{x\text{-}max}$=0.675T)

(c) hardened depth is 3mm (B_{max} in the steel plate =1.24T)

Fig. 8. Distribution of leakage flux and flux density near surface hardened steel plate (15 Hz, 126 AT).

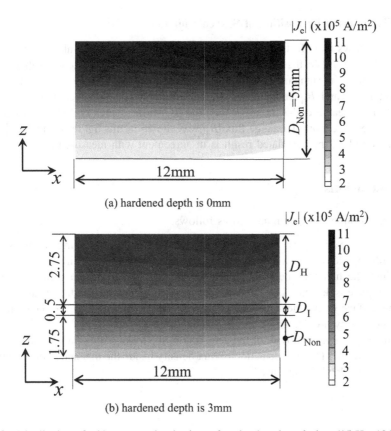

(a) hardened depth is 0mm

(b) hardened depth is 3mm

Fig. 9. Distribution of eddy current density in surface hardened steel plate (15 Hz, 126 AT).

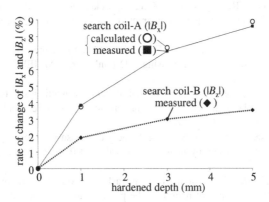

Fig. 10. Effect of hardened depth on flux density $|B_x|$ and $|B_z|$ of the search coil-A and -B (15 Hz, 126 AT).

4.2 Comparison by Position of Search Coil

Figure 10 shows the effect of the hardened depth on the absolute value of the change rate of $|B_x|$ and $|B_z|$ in these search coils A and B. The measurement result took the average value. The standard deviation of measurement result is about 0.1. The figure denotes that $|B_x|$ and $|B_z|$ are increased when the hardened depth is increased. Moreover, the $|B_x|$ in the search coil-A is larger than the $|B_z|$ in the search coil-B. Therefore, the usefulness of detection of leakage flux $|B_x|$ was shown in this figure. As for the $|B_x|$ in the search coil-A, the calculated result is in agreement with measurement.

5 Conclusions

The results obtained are summarized as follows:

(1) The permeability and conductivity of the layer with hardening in the surface hardened steel are lower than that of the layer without hardening.
(2) When the hardened depth is increased, the flux density inside surface hardened steel plate is decreased by the proposed inspection method, but the leakage flux on the surface of the steel plate is increased. This is, because the permeability and conductivity are decreased in the hardened layer.
(3) The inspection sensitivity of the detecting of the leakage flux on surface hardened steel is higher than the detection of flux density inside the magnetic yoke when the hardened depth is increased.

References

1. Chan, S.C., Grimberg, R., Hejase, J.A., Zeng, Z., Lekeakatakunju, P., Udpa, L., Udpa, S.S.: Nonlinear eddy current technique for characterizing case hardening profiles. IEEE Trans. Magn. **46**(6), 1821–1824 (2010)
2. Gotoh, Y., Sasaguri, N., Takahashi, N.: Evaluation of electromagnetic inspection of hardened depth of spheroidal graphite cast iron using 3-D nonlinear FEM. IEEE Trans. Magn. **46**(8), 3137–3144 (2010)
3. Gotoh, Y., Fujioka, H., Takahashi, N.: Proposal of electromagnetic inspection method of outer side defect on steel tube with steel support plate using optimal differential search coils. IEEE Trans. Magn. **47**(5), 1006–1009 (2011)
4. Nakata, T., Kawase, Y., Nakano, N.: Improvement of measuring accuracy of magnetic field strength in single sheet testers by using two H coils. IEEE Trans. Magn. **23**(5), 2596–2598 (1987)
5. Kamada, Y., Takahashi, S., Kikuchi, H., Kobayashi, S., Ara, K., Echigoya, J., Tozawa, Y., Watanabe, K.: Effect of predeformation on the precipitation process and magnetic properties of Fe-Cu model alloys. J. Mater. Sci. **44**, 949–953 (2009)

Estimation of Mutual Induction Eddy Current Testing Method Using Horizontal Exciting Coil and Spiral Search Coil

Tatsuya Marumoto[1](✉), Yuta Motoyasu[1], Yuji Gotoh[1], and Tatsuo Hiroshima[2]

[1] Oita University, 700 Dannoharu, Oita 870-1192, Japan
Marupa.com@gmail.com
[2] Hokuto Electronics, INC., 2-36, Najiotoukubo, Nishinomiya, Hyogo 669-1148, Japan

Abstract. The mutual induction type eddy current testing using a spiral type search coil is one of the eddy current testing (ECT). In this type ECT probe, since the thickness of a spiral search coil is very thin, the distance (lift-off: L_o) between the ECT probe and the specimen is able to set small. Therefore, the high detection sensitivity is obtained. However, since the wire is coiled in the shape of spiral, the evaluation of the flux density inside the spiral search coil is made difficult by experiment. Moreover, the phenomenon elucidation of inspection using this ECT probe is not carried out. In this paper, the inspection characteristic of a surface defect on an aluminum plate using this probe was investigated by 3-D alternating electromagnetic finite element method (FEM). Moreover, the evaluation in consideration of the characteristic of the spiral search coil in this ECT probe is also investigated by (1 + 1) evolution strategy. It is shown that high detection sensitivity is obtained using the spiral search coil.

Keywords: Mutual induction type eddy current testing · Spiral search coil · (1 + 1) evolution strategy · 3-D FEM

1 Introduction

The eddy current testing (ECT) is one of the electromagnetic inspection methods [1, 2] of surface defect in a non-magnetic material, etc. There are a self-induction type and a mutual induction type in the ECT. In the self-induction type, the exciting current is passed to one coil and the impedance of the coil is measured. As for this type, both excitation and detection are carried out with one coil. Since it is necessary to make the coil small when detecting small defect using this type, it is made difficult to enlarge the exciting current. As for the mutual induction type, the exciting coil and the search coil have been independent, respectively. Therefore, a large magnetic field is able to impress to the specimen using this type.

In order to increase the detection sensitivity of the defect in the specimen using the mutual induction type, it is necessary to decrease the distance (lift-off: Lo) between the ECT probe and the specimen [3]. In recent years, the manufacture technology of a semiconductor was applied, and the spiral search coil film in the ECT probe was

© Springer International Publishing Switzerland 2016
N. Kubota et al. (Eds.): ICIRA 2016, Part I, LNAI 9834, pp. 759–769, 2016.
DOI: 10.1007/978-3-319-43506-0_65

created. Since the thickness of the spiral coil is very thin, the Lo is able to set small. However, since the wire is coiled in the shape of spiral, the evaluation of the flux density inside the spiral search coil is made difficult by experiment. Moreover, the phenomenon elucidation of the ECT probe using the spiral search coil is not carried out.

In this paper, the inspection principle of the mutual induction type ECT probe using the spiral search coil was investigated by 3D electromagnetic finite element method (FEM). Moreover, the measured flux density inside the spiral search coil is calculated by the inversion method using the (1 + 1) evolution strategy [4–6], and it is compared with the calculated result of 3D FEM.

2 Procedure for Paper Submission

2.1 Inspection Model

Figure 1 shows the inspection model of 1/2 domain of the ECT probe using a spiral search coil. The structure is a rectangle exciting coil and a spiral search coil. The impression magnetic field by the exciting coil is distributed in the x-directions. The diameter, the thickness and number of turns of the spiral search coil are 2 mm, 0.1 mm and 44 turns, respectively. As for the search coil, the z-direction flux density (Bz) on the surface of specimen is detected. An aluminum plate of specimen is magnetized by exciting coil of 25 kHz and 4.6 A-turns. The distance (lift-off: Lo) from the bottom of the spiral search coil to the surface of the aluminum plate is equal to 1 mm. In this research, since the elucidation of the inspection principle of this ECT is the main purposes, the size of a defect is set as rectangular large dimensions as shown in Fig. 1.

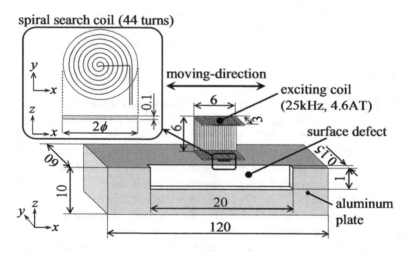

Fig. 1. Inspection model for surface defect on aluminum plate (1/2domain).

This ECT probe is moved in the x-direction on the aluminum plate, and the surface defect is inspected. The conductivity of the aluminum specimen is 3.5×10^7 S/m.

2.2   3-D Electromagnetic FEM Analysis

3-D FEM using the 1st order hexahedral edge element is applied. The flux density is analyzed by the step-by-step method taking account of eddy current in the aluminum plate. In order to get the steady state result, the calculation is carried out during 3 periods (=96steps). The time interval Δt of the step-by-step method is chosen as 1.25×10^{-6} s.

The basic equation of eddy current analysis using the A-ϕ method is given by

$$rot(v\, rot\, A) = J_o - \sigma\left(\frac{\partial A}{\partial t} + grad\varphi\right) \tag{1}$$

$$div\left\{-\sigma\left(\frac{\partial A}{\partial t} + grad\,\phi\right)\right\} = 0 \tag{2}$$

where A is the magnetic vector potential, ϕ is the scalar potential, v is the reluctivity, J_o is the current density and σ is the conductivity.

3   Elucidation of Inspection Principle

The elucidation of the inspection principle of this ECT probe is carried out using this 3-D FEM. Figure 2 shows the distribution of calculated Bz in the spiral search coil by moving the ECT probe in the x-direction on the aluminum plate with surface defect. The Bz is displayed on plus, when the waveform of both exciting current and Bz in the search coil is same phase. And it is displayed on minus when the phase is reverse. The figure denotes that each peak value is obtained near two edges of the defect. Therefore, a surface defect is detectable from these two peak values.

Figure 3 shows the distribution of magnetic flux near a surface defect when the ECT probe is located at the centre (x = 0 mm) of the defect as shown in Fig. 2. This figure illustrates that the impression magnetic field from an exciting coil is distributed in the about x-directions. Figure 4 shows only z-component of magnetic flux near a surface defect. These figures denotes that since the plus and minus values of z-component of the flux inside a search coil are almost equal, the output voltage is hardly generated by a search coil.

Figure 5 shows the distribution of magnetic flux near a surface defect when the ECT probe is located in the edge part (x = 9.5 mm) of the defect as shown in Fig. 2. Then, Fig. 6 shows the distribution of eddy current density near a surface defect inside the aluminium plate. Figure 6(a) shows the distribution of eddy current inside the aluminium plate in x-z plane. And, Fig. 6(b) shows the vector distribution of eddy current surface inside the aluminium plate in x−y plane. Figure 5 illustrates that the impressed magnetic field is hardly distributed inside the aluminium plate near the edge part of a

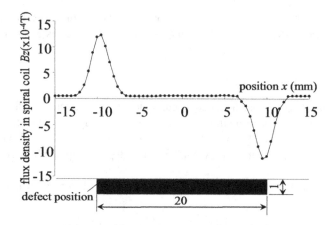

Fig. 2. Inspection waveform of flux density Bz (calculated, 25 kHz, 4.6 AT.)

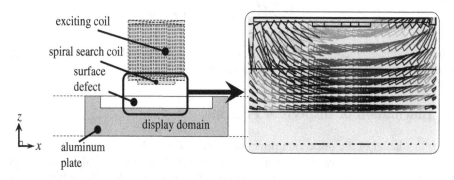

Fig. 3. Distribution of magnetic flux near a surface defect when the ECT probe is located in the centre of the defect ($B_{max} = 5.98 \times 10^{-4}$T).

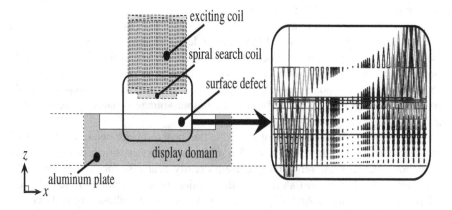

Fig. 4. Distribution of z-component of magnetic flux near a surface defect when the ECT probe is located in the centre of the defect.

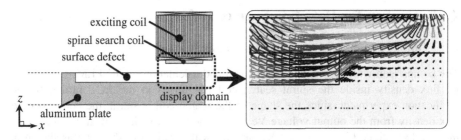

Fig. 5. Distribution of magnetic flux near a surface defect when the ECT probe is located in the edge part of the defect ($B_{max} = 6.16 \times 10^{-4}$T).

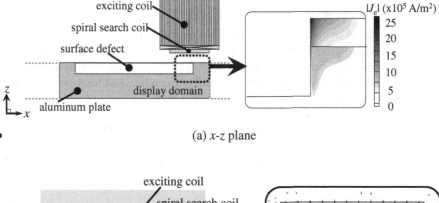

(a) x-z plane

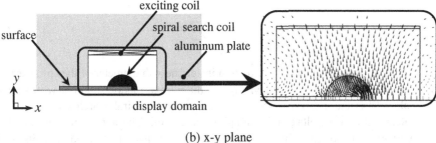

(b) x-y plane

Fig. 6. Distribution of eddy current near a surface defect when the ECT probe is located in the edge part of the defect.

defect, and is bypassed to z-direction. This is, because the eddy current is generated inside the edge portion as shown in Fig. 6. Figure 7 shows only z-component of magnetic flux near a surface defect. This figure denotes that the z-component of flux near the edge part of the defect is strongly distributed, since the eddy current is distributed inside the aluminum plate near the edge of defect. Therefore, the output voltage is generated by a search coil near this position.

4 Evaluation for Detecting Surface Defect

4.1 Output Signal Analysis Using Evolution Strategy

The comparison with verification experiment and calculation by 3D FEM is evaluated by flux density inside the spiral search coil (44 turns) in the ECT probe. In the verification experiment, since the shape of the search coil is spiral, the conversion to flux density from the output voltage Ve of the search coil is difficult. In this research, the spiral search coil is assumed to be the coils with 44 kinds of diameter as shown in Fig. 8. Then, the measured flux density inside the spiral search coil is calculated by the (1 + 1) evolution strategy. Various techniques are proposed for solving the inverse problem. Although large number of calculations is necessary in the (1 + 1) evolution strategy, it is adopted because a global optimal solution can be obtained.

The iteration process of calculation is as follows:

(1) Calculation of output voltage in the spiral search coil
The flux density Bz inside 44 kinds of diameter of the search coils is chosen as design variable. Then, total output voltage Vc is calculated. The formula of flux density Bz inside each diameter using random number, and total output voltage Vc are given by

$$B_z^n = B_z^n + N(0, \sigma^2) \quad (n = 1, 2, 3, \ldots, 44) \tag{3}$$

$$V_c^n = -S^n \frac{dB_z^n}{dt} \quad (n = 1, 2, 3, \ldots, 44) \tag{4}$$

$$V_c = \sum_{n=1}^{44} V_c^n \tag{5}$$

where, B_z^n is flux density of z-direction inside the n-th diameter of the spiral search coil, $N(0, \sigma^2)$ (σ: standard deviation) is the normal random number with weight value, S^n is the n-th cross-section area (m^2) in the spiral search coil, and V_c is calculated output voltage in total spiral search coil. The normal random number $N(0, \sigma^2)$ with weight value is added to each flux density in 44 kinds of coils. And the total output voltage of the spiral search coil is calculated from the sum total of the output voltage inside each diameter.

(2) Calculation of objective function
The following objective function W is calculated by

$$W = |V_c - V_e| \tag{6}$$

where, Vc is calculated output voltage and Ve is measured output voltage in the spiral search coil. The residual of the measured output voltage Ve and the calculated output voltage Vc is an objective function W. If the calculated objective function W is less than the previous one, W is updated. The initial value of the standard deviation σ and the convergence criterion of the objective function W are 2.25 and 1×10^{-3} mV, respectively.

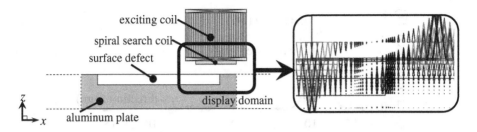

Fig. 7. Distribution of z-component of magnetic flux near a surface defect when the ECT probe is located in the edge part of the defect.

Above process is iterated until the final result is obtained. Then, the final flux density Bz in the spiral search coil is calculated by

$$B_z = \sum_{n=1}^{44} B_z^n \tag{7}$$

Where, Bzn is flux density of z-direction inside the n-th diameter of the spiral search coil and Bz is final flux density of z-direction in the spiral search coil.

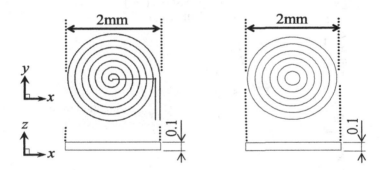

Fig. 8. Spiral search coil (44 turns) and assumed coils with 44 kinds of diameter.

4.2   Inspection of Surface Defect

Figure 9 shows the distribution of calculated and measured values of Bz in the spiral search coil. The calculated result Bz by 3D FEM is shown in the "calculated ($\triangle$)" inside this figure. And, the "measured ($\blacklozenge$)" in this figure is the value which changed measured output voltage Ve into flux density Bz using the $(1 + 1)$ evolution strategy. Bz is obtained by moving the ECT probe in the x-direction on the aluminum plate. The figure denotes that each peak value is obtained near two edges of the defect. Figures 10 and 11 show the distribution of calculated and measured values of Bz in the spiral search coil when the lengths of x-direction of the surface defect are set to 10 mm and 5 mm, respectively. The width of y-direction and the depth of z-direction of these

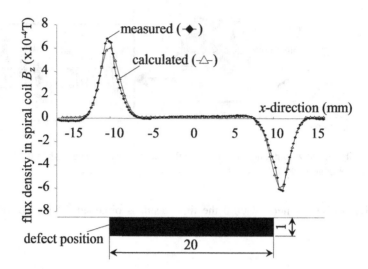

Fig. 9. Distribution of calculated and measured values of B_z in the spiral search coil (25 kHz, 4.6 AT).

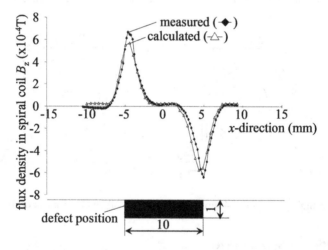

Fig. 10. Distribution of calculated and measured values of B_z in the spiral search coil when the length of the surface defect is 10 mm (25 kHz, 4.6 AT).

defects are same dimensions, and are 0.3 mm and 1 mm, respectively. These figures denote that the length of x-direction of the surface defect is presumed from the distance between each peak value of Bz. However, even if the length of x-direction of the surface defect is changed, the amplitude of each peak value of Bz is not almost changed. The calculated results are in agreement with measurement.

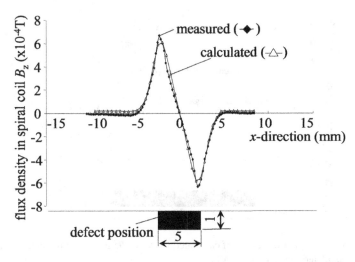

Fig. 11. Distribution of calculated and measured values of B_z in the spiral search coil when the length of the surface defect is 5 mm (25 kHz, 4.6 AT).

4.3   Effect of Spiral Search Coil

In this research, the comparison of the detecting sensitivity of a usual search coil and a spiral search coil in this ECT probe is investigated by 3D FEM. The diameter, the thickness and number of turns of the usual search coil are the same as the spiral coil, and are 2 mm, 0.1 mm and 44 turns, respectively. The structure of the usual coil is that the wire is rolled 44 times with the same diameter of 2 mm as shown in Fig. 12. Figure 13 shows the calculated Bz in the spiral search coil and usual search coil. The figure denotes that each peak value near two edges of the defect by the spiral coil is higher than that by the usual search coil. In the usual search coil, the average value of the magnetic flux distributed inside of the search coil is detected. On the other hand, in the spiral search coil, total value of the intensity distribution of the magnetic flux inside the coil is detected. Therefore, the flux detection sensitivity of the spiral coil is higher than that of the usual search coil.

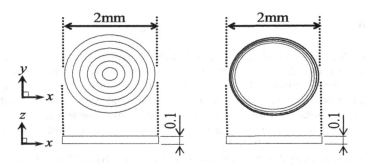

Fig. 12. Model of spiral search coil and usual search coil (diameter: 2 mm, thickness: 0.1 mm, 44 turns).

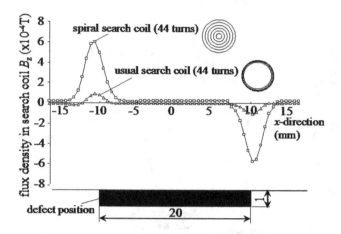

Fig. 13. Inspection signal waveform of calculated values of Bz in the spiral search coil and usual search coil (calculated, 25 kHz, 6.51 AT).

5 Conclusions

The results obtained are summarized as follows:

(1) In this mutual induction ECT probe, the impressed magnetic field is distributed to bypass the edge portion of the surface defect, since eddy current is generated inside the aluminum plate near the edge portion of the defect. Therefore, output signal in the search coil is obtained near two edges of the defect.

(2) It is possible to estimate the magnetic flux density from measured output voltage in the spiral search coil of this ECT probe using evolution strategy. Moreover, it is shown that the detection sensitivity of the spiral search coil is higher than that of the usual search coil.

The investigation of the detectable size and form of the defect by this inspection method is a future subject.

References

1. Goldfine, N.J.: Magnetometers for improved materials characterization in aerospace applications. Mater. Eval. **51**, 396–405 (1993). The American Society for Nondestructive Testing
2. McMaster, R.C., McIntrire, P., Mester, M.L.: Electromagnetic Testing. Nondestructive Testing Handbook, vol. 5, 3edn. American Society for Nondestructive Testing (2004)
3. Gotoh, Y., Matsuoka, A., Takahashi, N.: Electromagnetic inspection technique of thickness of nickel-layer on steel plate without influence of lift-off between steel and inspection probe. IEEE Trans. Magn. **47**(5), 950–953 (2011)
4. Bäck, T.: Evolutionary Algorithms in Theory and Practice. Oxford University Press, Oxford (1996)

5. Horii, M., Takahashi, N., Narita, T.: Investigation of evolution strategy and optimization of induction heating model. IEEE Trans. Magn. **36**(4), 1085–1088 (2000)
6. Gotoh, Y., Fujioka, H., Takahashi, N.: Proposal of electromagnetic inspection method of outer side defect on steel tube with steel support plate using optimal differential search coils. IEEE Trans. Magn. **47**(5), 1006–1009 (2011)

Examination of Evaluation Method of Power Generation Current Using Static Magnetic Field Around Polymer Electrolyte Fuel Cell

Daiki Nagata[1(✉)], Yuji Gotoh[1], Ryota Naganoma[1],
and Masaaki Izumi[2]

[1] Department of Mechanical and Energy Systems Engineering,
Oita University, 700 Dannoharu, Oita 870-1192, Japan
v16e1028@oita-u.ac.jp
[2] The University of Kitakyushu, 1-1 Hibikino, Wakamatsu-ku,
Kitakyushu, Fukuoka 808-0135, Japan

Abstract. A polymer electrolyte fuel cell (PEFC) is the clean energy converters, it is developed as a power supply for electric vehicle. In order to raise the power generation efficiency of PEFC, it is important to know the power generation state inside MEA (membrane electrode assembly). In this paper, an measurement method for monitoring the distribution of the power generation current in the MEA of PEFC from static magnetic field around the fuel cell is investigated. The inverse problem analysis of the power generation current distribution in the MEA using 3D finite element method is examined, and the effectiveness of this method is investigated.

Keywords: Polymer Electrolyte Fuel Cell (PEFC) · MEA (membrane electrode assembly) · Power generation current · Heuristic search · Inverse problem analysis · 3-D finite element method

1 Introduction

The fuel cell is the clean energy converters by electrochemical reaction between hydrogen and oxygen. In the fuel cell, there is no emission of air pollutants such as carbon dioxide. Therefore, contributing to energy problems and environmental problems are expected by the development of the improvement in the power generation efficiency of the fuel cell. Since the polymer Electrolyte Fuel Cell (PEFC) is one of the fuel cell with a quick power generation, it is used for the power supply of an electric vehicle or home, etc. The power generation efficiency of the fuel cell is influenced by movement and distribution of hydrogen, oxygen and steam in it. Therefore, it is important to clarify the distribution of the power generation current that has close relation to there mass transfers. Especially, the measurement of the power generation current inside MEA (membrane electrode assembly) is necessary [1]. The generation current distribution was measured by taking out currents externally through divided elements of MEA or a separator [2].

© Springer International Publishing Switzerland 2016
N. Kubota et al. (Eds.): ICIRA 2016, Part I, LNAI 9834, pp. 770–779, 2016.
DOI: 10.1007/978-3-319-43506-0_66

The static magnetic field around the fuel cell is generated by the power generation current in the fuel cell. Therefore, the distribution of power generation current inside MEA may be measured using the static magnetic field around the fuel cell.

In this paper, the measurement method of the distribution of power generation current inside MEA using of the magnetic field around the PEFC is proposed. In this research, the static magnetic field calculated [3–6] by the forward electromagnetic finite element method is used as the measured value in the proposed inverse problem analysis using 3D heuristic search method. Then the power generation current distribution inside MEA is determined. In addition, the equivalent experimental verification is also carried out.

2   Model and Method of Analysis

Model of PEFC and Measurement Domain of Flux Density

Figure 1 shows the structure of a single cell type fuel cell of the PEFC. This is composed of a pair of copper end plates, a pair of separators made from carbon, which is the passage of hydrogen and oxygen, and a sheet of MEA (membrane electrode assembly).

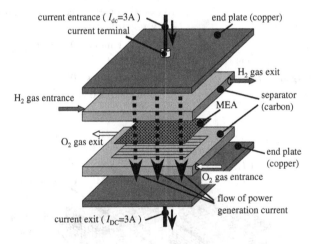

Fig. 1. Single type of polymer electrolyte fuel cell

Figure 2 shows the analyzed model of the fuel cell using the 3D electro magnetic FEM. The MEA is a sheet and its dimension is 50 mm × 50 mm × 1 mm.

The total amount of power generation in the PEFC is 3A. The static flux density around the fuel cell is generated by power generation current in the fuel cell. Therefore, the distribution of power generation current inside the MEA may be determined using the static magnetic field around the fuel cell. The x-, y-, and z-components of the magnetic field, B_x, B_y and B_z along the line a-b-c-d-a shown in Fig. 3, obtained by the forward analysis, are used in 3D inverse problem analysis method. MEA is divided into one-hundred elements and its currents are treated as unknown variables by the inverse problem analysis. Table 1 shows the conditions of the 3D FEM analysis.

Inverse Problem Analysis Using Heuristic Search Method

The measurement of the distribution of power generation current inside MEA in PEFC is an inverse problem analysis using 3D FEM. There are various techniques in the inverse problem method. When the Evolution Strategy [7, 8] using Tikhonov's method is used, the optimal solution was not obtained if the number of design variables is large.

Table 1. Conditions of 3D FEM Analysis

Total power generation current	$I_{dc} = 3A$
End plate (copper)	Width in the x-direction = 100 mm,
	Length in the y-direction = 100 mm,
	Thickness in the z-direction = 3 mm
	$\sigma = 5.9 \times 10^7$ S/m
Separator (carbon)	Width in the x-direction = 100 mm,
	Length in the y-direction = 100 mm,
	Thickness in the z-direction = 15 mm
	$\sigma = 8.1 \times 10^4$ S/m
MEA	$50 \times 50 \times 1$ mm (100 elements)
Nodes and elements	95220, 101614
Convergence criterion	N-R method: 1.0×10^{-6} T
	ICCG method: 1.0×10^{-9}

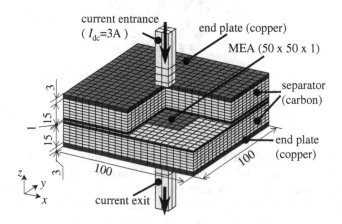

Fig. 2. Model of the PEFC using 3D FEM.

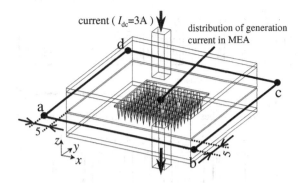

current (I_{dc}=3A)

distribution of generation current in MEA

Fig. 3. Position where static magnetic field is measured.

In this research, the inverse problem analysis using the heuristic search method which determines an "ON" or "OFF" domain of the power generation current in MEA which satisfies the specified flux distribution is introduced. The iteration calculation process of the inverse problem analysis is shown in Fig. 4 and the algorithm is as follows:

(I) Process 1
    One "OFF" element of no power generation current is generated as shown in Fig. 4. Then put it in one hundred kinds of MEA from the left bottom to the right top as shown in Fig. 4(a). The following objective function W at the kth iteration is calculated:

$$W^k = \sum_{i=1}^{n} \left\{ (B_{ix} - B_{0x})^2 + (B_{iy} - B_{0y})^2 + (B_{iz} - B_{0z})^2 \right\} (k = 1, 2, 3......)$$

(1)

where n is the number of elements along the line a-b-c-d-a around the fuel cell shown in Fig. 3. B_{ix}, B_{iy} and B_{iz} are the x-, y-, and z-components of the flux density at a point i calculated along the line a-b-c-d-a around the fuel cell using 3D electromagnetic FEM. "k" of the superscript of W shows the calculation process k. B_{0x}, B_{0y} and B_{0z} are the x-, y-, and z-components of flux density calculated along the line a-b-c-d-a around the fuel cell by the forward analysis. The power generation current distribution that minimizes the objective function W in (1) is the desired value.

(I) Calculation process 1
    The W in the position of the power generation current "OFF" domain by one hundred places inside MEA is calculated. Then, the smallest five which are defined as $W_{No.1}^1$ to $W_{No.5}^1$, are chosen from all W^1. The element of black shows an "OFF" element of the power generation current in MEA as shown in Fig. 4.

(II) Calculation Process 2
    The W in the position of the power generation current of the additional one "OFF" domain by one hundred places inside MEA is calculated as shown in

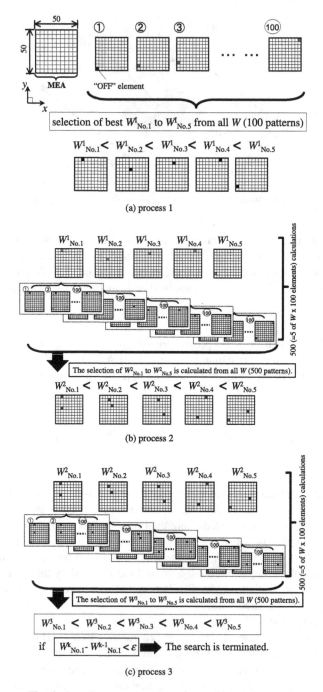

Fig. 4. Iteration process using proposed heuristic search.

Fig. 4(b). Then, new power generation current distributions of $W^2_{No.1}$ to $W^2_{No.5}$ are selected to five hundred patterns.

(III) Calculation Process 3

The calculation of Calculation process 2 is iterated as shown in Fig. 4(c). "K" in Fig. 4(c) shows the number of the iteration times. If $W^k_{No.1} - W^{k-1}_{No.1}$ is less than ε ($= 1.0 \times 10^{-20}$), the search is terminated.

3   Forward and Inverse Problem Analysis

In order to confirm the usefulness of the proposed inverse problem analysis of heuristic search method, the distribution of power generation current inside MEA using forward problem analysis is estimated by the inverse problem analysis. The distribution of the "OFF" power generation currents of five domain in MEA as shown in Fig. 5 is examined by the proposed inverse problem analysis. In forward problem analysis, the power generation current of five domain inside MEA is calculated as "OFF". Figure 6 shows the whole distribution of power generation current in MEA when the total value of the power generation current is 3A.

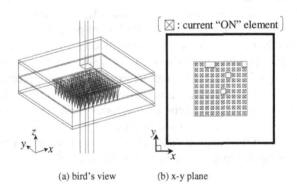

($\boxtimes$: current "ON" element)

(a) bird's view   (b) x-y plane

Fig. 5. Distribution of current density set up in forward analysis (total I_{dc} = 3A).

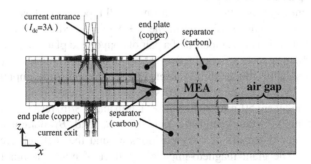

Fig. 6. Whole distribution of power generation current in the fuel cell.

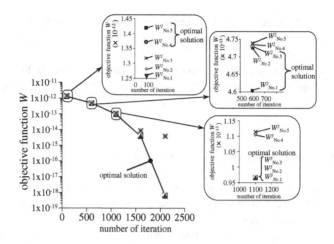

Fig. 7. Calculation process of an objective function W.

The figure illustrate that the current generated within MEA flows through a separator, an end plate, and a current terminal wire. As the conductivity of the end plate of copper is larger than that of the separator, the current density in the end plate is the largest within the fuel cell. The values of $|B|_{max}$ and $|B|_{min}$ of magnetic field around the fuel cell are 0.1151×10^{-4} T and 0.1023×10^{-4} T, respectively.

Figure 7 shows the calculation process of objective function W at each iteration by the proposed inverse problem analysis of the heuristic search method. The figure illustrates that $W^1_{No.4}$ and $W^1_{No.5}$ at the calculation process 1 are reached to the final optimal solution. The inverse problem analysis is converged by about 2100 iterations.

Figure 8 shows the comparison between the power generation current distribution inside MEA by the forward analysis and that by the inverse problem analysis. The figure denotes that the "OFF" domain inside MEA is presumed by the proposed inverse problem analysis using the magnetic field around the fuel cell.

4 Equivalent Experimental Verification

Equivalent Experimental Model

An equivalent experimental model of a single cell type fuel cell as shown in Fig. 9 was made and the current distribution in the MEA composed of a conductive rubber board was estimated. This is composed of a pair of copper end plates, a pair of separators made from carbon, a sheet of conductive rubber board that imitates MEA and sixty magnetic sensors. In this model, direct current of 8A is compulsorily impressed from the current entrance by a direct-current power supply. The dimension and conductivity of imitated MEA model are 50 mm × 50 mm × 1 mm and 4 S/m, respectively. The x-, y-, and z-components of static magnetic field, B_x, B_y and B_z obtained by the MI (Magneto-impedance) sensors of sixty positions around the fuel cell are used in 3D heuristic search. The giant magneto-impedance effect of magnetic amorphous metal wire is used in the MI sensor and can detect up to 1×10^{-10} T of the minute-magnetic fields. The case when the currents of two elements in MEA are "OFF" domain as shown in Fig. 9(a) is examined. The total current in MEA is 8A (dc). $|B|_{max}$ and $|B|_{min}$ at the measured points of flux density of sixty MI sensors are 0.3222×10^{-4} T and 0.2783×10^{-4} T, respectively.

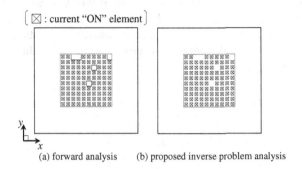

(a) forward analysis   (b) proposed inverse problem analysis

Fig. 8. Distribution of current density in MEA.

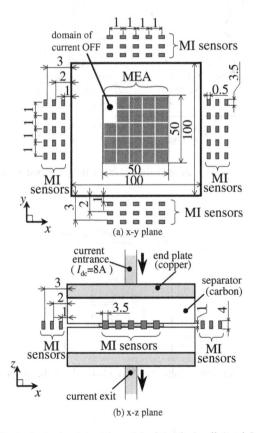

Fig. 9. Equivalent experimental model of the fuel cell (total I_{dc} = 8A).

Results and Discussion

Figure 10 shows the change of objective function W at each iteration. The history of reduction of five kinds of objective functions $W^1_{No.1}$ to $W^1_{No.5}$, which were selected at the process 1, are monitored. The figure illustrates that $W^1_{No.1}$ and $W^1_{No.2}$ at the process 1 are reached to the final optimal solution. The heuristic search is converged by 275 iterations. Figure 11 shows the current distributions in MEA obtained by the heuristic search. Two "OFF" elements in MEA in Fig. 10 obtained by the heuristic search are the same with Fig. 9(a). The figure shows that the current distribution in MEA of the fuel cell can be obtained by the proposed heuristic search using the flux distribution around the fuel cell.

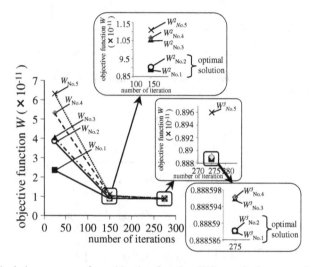

Fig. 10. Calculation process of an objective function W by equivalent experimental model.

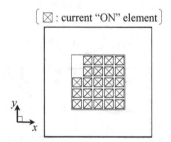

Fig. 11. Distribution of current density in MEA obtained by the proposed inverse problem analysis.

5 Conclusions

The inverse problem analysis of the heuristic search method for measurement of the distribution of the power generation current inside MEA using the static magnetic field around the fuel cell is proposed. The power generation situation inside MEA may be able to presume by this proposed technique, without stopping power generation of the PEFC. The increase in the division of the current distribution in MEA and the evaluation by the fuel cell in a stack state are a future research subject.

References

1. Bender, G., Wison, M.S., Zawodzinski, T.: Further refinements in the segmented cell approach to diagnosing performance in polymer electrolyte fuel cells. J. Power Sources **123**, 163–171 (2003)
2. Hamalaine, M., Hari, R., Ilmoniemi, R.J., Knuutila, J., Lounasmaa, O.V.: Magneto encephalography-theory, instrumentation, and applications to noninvasive studies of the working human brain. Rev. Mod. Phys. **65**(2), 413–497 (1993)
3. Gotoh, Y., Takahashi, N.: Evaluation of detecting method with AC and DC excitations of opposite side defect in steel using 3D non-linear FEM taking account of minor loop. IEEE Trans. Magn. **44**(6), 1622–1625 (2008)
4. Gotoh, Y., Sasaguri, N., Takahashi, N.: Evaluation of electromagnetic inspection of hardened depth of spheroidal graphite cast iron using 3-D nonlinear FEM. IEEE Trans. Magn. **46**(8), 3137–3144 (2010)
5. Gotoh, Y., Kiya, A., Takahashi, N.: Electromagnetic Inspection of outer side defect on steel tube with steel support using 3D nonlinear FEM considering non-uniform permeability and conductivity. IEEE Trans. Magn. **46**(8), 3145–3148 (2010)
6. Gotoh, Y., Takahashi, N.: Proposal of detecting method of outer side crack by alternating flux leakage testing using 3-D non-linear FEM. IEEE Trans. Magn. **42**(4), 1415–1418 (2006)
7. Takahashi, N., Kitamura, T., Horii, M., Takehara, J.: Optimal design of tank shield model of transformer. IEEE Trans. Magn. **36**(4), 1089–1093 (2000)
8. Horii, M., Takahashi, N., Narita, T.: Investigation of evolution strategy and optimization of induction heating model. IEEE Trans. Magn. **36**(4), 1085–1088 (2000)

Evaluation of Nondestructive Inspecting Method of Surface and Opposite Side Defect in Steel Plate Using Large Static and Minute Alternating Magnetic Field

Akira Fujii[✉] and Yuji Gotoh

Department of Mechanical and Energy Systems Engineering,
Faculty of Engineering, Oita University, 700 Dannoharu, Oita 870-1192, Japan
Fujii.Akira@jeed.or.jp

Abstract. In the inspection of the steel wall or bottom steel plate inside the tank in the petrochemical plant etc., it is necessary to detect the opposite side defect from the surface of the steel plate. In this paper, the electromagnetic nondestructive inspection method using large static and minute alternating magnetic field is proposed. And, this proposed inspection method is investigated by verification experiment and 3-D nonlinear finite-element method (FEM) taking account of hysteresis (minor loop) and eddy current in steel. It is shown that the detection of both surface and opposite side defect is possible by the differential permeability of minor loop of which the position on the hysteresis loop is affected by the existence of defect.

Keywords: 3-D nonlinear finite element method · Electromagnetic non-destructive inspection · Surface and opposite side defect in steel plate · Large static and minute alternating magnetic field · Minor loop magnetism property

1 Introduction

The inspection of the defect in the steel wall or bottom steel plate inside the oil tank in the petrochemical plant, etc. is important for safe maintenance. Especially, it is necessary to detect opposite side defect from inside the oil tank. Generally, an ultrasonography method is used for inspecting the opposite side defect in steel plate. However, since this method is the contacting inspection using the media such as water, a long inspection time is needed. On the other hand, the electromagnetic method is useful for the inspection. This is, because the high-speed inspection is enabled, since non-contacting inspection is carried out by this method.

In this paper, in order to detect the opposite side defect of the steel wall, the electromagnetic inspection method using large static (dc) and minute alternating (ac) magnetic field [1] is proposed. Y. Gotoh [2] et al. proposed the electromagnetic sensor which consisted of both a dc magnetic yoke and an ac magnetic yoke. However, the detection sensitivity of the defect by this sensor is not so good. In this research, the electromagnetic sensor by which the dc exciting coil and the ac exciting coil were

© Springer International Publishing Switzerland 2016
N. Kubota et al. (Eds.): ICIRA 2016, Part I, LNAI 9834, pp. 780–789, 2016.
DOI: 10.1007/978-3-319-43506-0_67

wound around one magnetic yoke is proposed. Moreover, in order to increase the detection sensitivity, the search coil with which the leakage flux on the surface of the steel plate is detected is used in this sensor. This sensor is investigated by verification experiment and 3-D non-linear FEM taking account of minor loop magnetism property in steel.

2   Inspection Model and Conditions

Figure 1 shows proposed electromagnetic inspection model. This probe sensor is composed of the magnetic yoke for dc and ac magnetic field and a search coil. As for the search coil in this probe sensor, the x-direction leakage flux (B_x) of surface on the steel plate is detected. The distance (Lift-off) between the search coil and the surface of steel plate is equal to 0.5 mm. The opposite side defect is assumed to be the shape of a slit. The depth (D_d) of z-direction, width (D_w) of x-direction and length (D_l) of y-direction of the defect are 1 mm, 0.5 mm and 100 mm, respectively. The conditions of 3-D nonlinear FEM and verification experiment are shown in Table 1. The dc exciting current is 142 A-turns (AT). The ac exciting current is 14.2AT(rms) and the exciting frequency is 500 Hz.

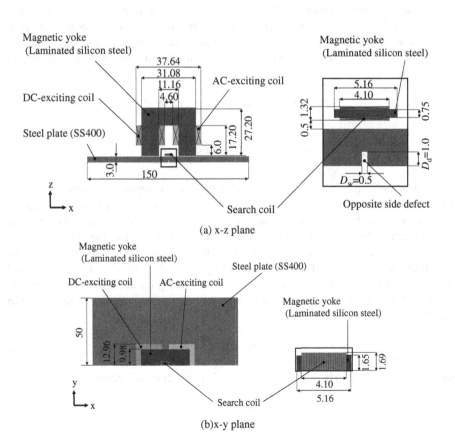

(a) x-z plane

(b)x-y plane

Fig. 1. Inspection model of steel plate with opposite side defect (1/2domain).

Table 1. Conditions of 3-D nonlinear FEM and verification experiment

DC exciting coil	71turns, 2A
AC exciting coil	71turns, 0.2A(rms), 500 Hz
Search coil	Width of x-direction: 4.1 mm, Thickness of z-direction: 1.32 mm, Length of y-direction: 3.38 mm, 60turns, Lift-off: 0.5 mm
Steel plate	SS400, $\sigma = 7.505 \times 10^6$ S/m, Width of x-direction: 150 mm, Thickness of z-direction: 3 mm, Length of y-direction: 100 mm
Opposite side defect	Width(D_w): 0.5 mm, Depth(D_d): 1.0 mm, Length(D_l): 100 mm
Nodes and elements	29835, 26752
Convergence criterion	N-R method: 1.0×10^{-4} T, ICCG method: 1.0×10^{-3}

3 3-D Nonlinear FEM Taking Account of Minor Loop

3.1 Method of Analysis

The magnetization phenomenon caused by dc and ac magnetic field needs taking account of hysteresis (minor loop) [3]. Figure 2 shows the hysteresis loops of SS400 steel obtained by experimental measurement. The flux density B in the steel is analyzed using these hysteresis loops and 3-D edge-based hexahedral nonlinear FEM and the step-by-step method. The basic equation of eddy current analysis using the A-ϕ method is given by

$$\mathrm{rot}(v\,\mathrm{rot}\,\boldsymbol{A}) = \boldsymbol{J}_0 - \sigma\left(\frac{\partial \boldsymbol{A}}{\partial t} + \mathrm{grad}\varphi\right) \tag{1}$$

$$\mathrm{div}\left\{-\sigma\left(\frac{\partial \boldsymbol{A}}{\partial t} + \mathrm{grad}\varphi\right)\right\} = 0 \tag{2}$$

where A is the magnetic vector potential, ϕ is the scalar potential, v is reluctivity, J_0 is the current density, and σ is the conductivity. Figure 3 shows explanation of calculation of minor loop. At first, the upper minor curve is modeled by linear interpolation using these hysteresis loops shown in Fig. 2 [4]. Then, it is assumed that the obtained H and B are at the point b (H_{min}, B_{min}) on the upper minor curve as shown in Fig. 3. If the calculated flux density B_c at N-R iteration is larger than B_{min}, then B_c should be located at the point d (H_d, B_c) on the lower minor curve. The H_d on the lower minor curve is given by the following equation if the upper minor curve is symmetrical with respect to the middle point e:

$$H_d = H_f + 2\left(H_e - H_f\right) \tag{3}$$

the v_d and $\partial v_d / \partial B_c^2$, which are necessary in N-R iteration, are given by

$$v_d = \frac{H_d}{B_c} \qquad (4)$$

$$\frac{\partial v_d}{\partial B_c^2} = \frac{1}{2B_c^2}\left(\frac{\partial H_g}{\partial B_g} - v_d\right) \qquad (5)$$

where $\partial v_d / \partial B_c^2$ at the point d is obtained using the property that $\partial v_d / \partial B_c^2$ at the point d is equal to that at the point f.

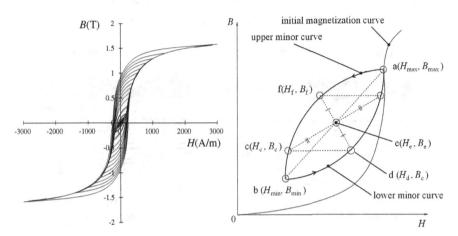

Fig. 2. Hysteresis curves of steel plate (SS400 steel).

Fig. 3. Explanation of minor loop.

3.2   Results of Analysis

Figure 4 shows the vector distribution of synthetic magnetic field of dc and ac flux density in steel plate with and without the opposite side defect. This figure denotes that the flux density inside surface portion in the steel plate near the defect is increased when there is the defect. Moreover, the flux density B_x inside the search coil of leakage flux on the steel plate is increased when there is defect.

Figure 5 shows the vector distribution of only dc flux density in steel plate with and without defect. The dc flux is uniformly distributed in the steel plate when there is not the defect as shown in Fig. 5(a). The dc flux density bypasses the defect when there is the defect as shown in Fig. 5(b). Therefore, the dc flux density inside surface portion in the steel plate near the defect is increased from $B_{x\text{-dc}} = 0.51$T to $B_{x\text{-dc}} = 0.64$T.

Figure 6 shows the vector distribution of only ac flux density in steel plate with and without the defect. This figure denotes that the ac flux density is distributed in the surface of steel plate, and the ac flux density near the defect in steel plate surface is decreased from $B_{x\text{-ac}} = 0.59$T to $B_{x\text{-ac}} = 0.52$T. This is, because the ac flux density is difficult to penetrate to the steel plate and the ac leakage flux is increased, since the dc flux density bypasses the defect and approximates magnetic saturation.

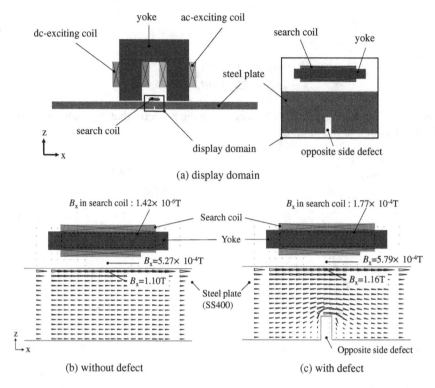

(a) display domain

B_x in search coil : 1.42×10^{-6}T B_x in search coil : 1.77×10^{-4}T

(b) without defect   (c) with defect

Fig. 4. Distribution of dc and ac flux density with and without an opposite side defect in steel plate (dc = 142AT, ac = 500 Hz, 14.2AT).

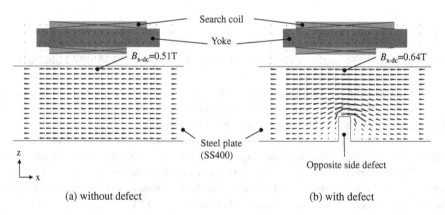

(a) without defect   (b) with defect

Fig. 5. Distribution of only dc flux density with and without an opposite side defect in steel plate (dc = 142AT).

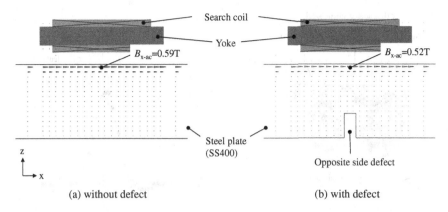

Fig. 6. Distribution of only ac flux density with and without an opposite side defect in steel plate when the dc and ac magnetic field are impressed (dc = 142AT, ac = 500 Hz, 14.2AT).

4 Verification Experiment and Discussion

4.1 Inspection of Opposite Side Defect

The verification experiment is carried out as shown in Fig. 7. In this experiment, the center of the opposite side defect is set as x = 0 mm, and the inspection sensor is moved in the x-direction from −10 mm to 10 mm by 0.5 mm steps, keeping the Lift-off at 0.5 mm. The dc exciting current is 2A and the ac exciting current is 0.2A (rms). The exciting frequency of the coil is 500 Hz.

Figure 8 shows the measured inspection result of the leakage flux B_x detected by search coil. Figure 8 also shows the result of analysis. This figure denotes that this inspection sensor is possible to detect the opposite side defect, since B_x is increased at the position of the defect.

Figure 9 shows the measured inspection result of the leakage flux B_x detected by search coil when D_d is 0.5 mm and 1.0 mm. This figure denotes that the peak value of B_x is increased when D_d is increased from 0.5 mm to 1.0 mm.

Figure 10 shows the measured inspection result of the leakage flux B_x detected by search coil when only ac magnetic field and dc and ac magnetic field are impressed. This figure denotes that it is impossible to detect the opposite side defect using only ac magnetic field. This is, because ac magnetic field is distributed in the surface of steel plate and not affected by the opposite side defect.

4.2 Inspection of Surfaced Defect

Figure 11 shows the calculated results when there is a surface defect on the steel plate. The surface defect is assumed to be the same shape of a slit as the opposite side defect. The depth (D_d) of z-direction, width (D_w) of x-direction and length (D_l) of y-direction of the defect are 1 mm, 0.5 mm and 100 mm, respectively. The conditions of exciting current are also same condition as inspection of opposite side defect. The dc exciting

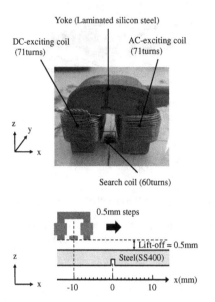

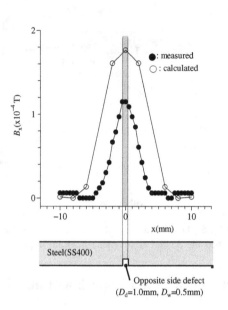

Fig. 7. Verification experiment model using proposed inspection sensor.

Fig. 8. Inspection result of B_x in a search coil (measured, dc = 142AT, ac = 500 Hz, 14.2AT).

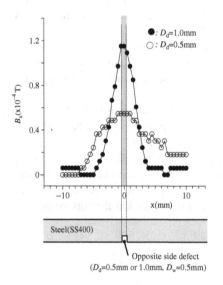

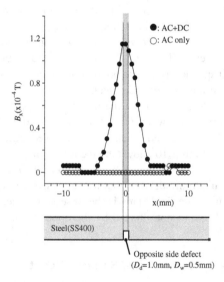

Fig. 9. Inspection result when D_d is changed (measured, dc = 142AT, ac = 500 Hz, 14.2AT).

Fig. 10. Inspection result when only ac magnetic field and dc and ac magnetic field are impressed (measured, dc = 142AT, ac = 500 Hz, 14.2AT).

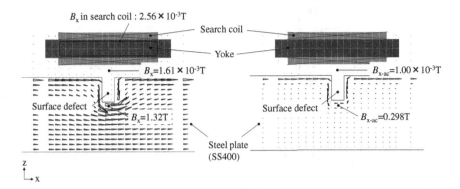

(a) Distribution of synthetic magnetic field
of dc and ac flux density

(b) Distribution of only ac flux density
when the dc and ac magnetic field are
impressed.

Fig. 11. Results of analysis when a defect exists at surface steel plate. (dc = 142AT, ac = 500 Hz, 14.2AT).

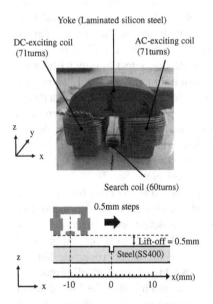

Fig. 12. Verification experiment model using inspection sensor when a defect exists at surface steel plate (measured, dc = 142AT, ac = 500 Hz, 14.2AT).

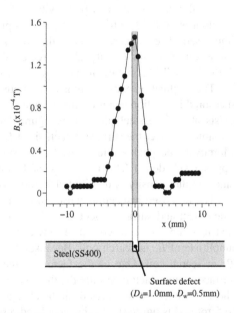

Fig. 13. Inspection result when a defect exists surface steel plate (measured, dc = 142AT, ac = 500 Hz, 14.2AT).

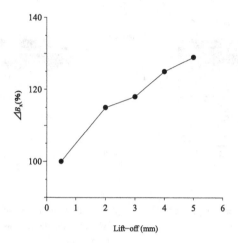

Fig. 14. Effect of the Lift-off on B_x in a search coil.

current is 2A and the ac exciting current is 0.2A (rms). The exciting frequency of the coil is 500 Hz. Figure 11(a) shows the vector distribution of synthetic magnetic field of dc and ac flux density in steel plate with surface defect. Figure 11 (b) shows the vector distribution of only ac flux density in steel plate when the dc and ac magnetic field are impressed. These figures denote that the flux density bypasses the surface defect. Moreover, the flux density B_x in search coil is increased from 1.77×10^{-4} T to 2.56×10^{-3} T in comparison with inspection of opposite side defect as shown Fig. 4(c).

The verification experiment is carried out as shown in Fig. 12. Figure 13 shows the measured inspection result of the leakage flux B_x detected by search coil when a defect exists at surface steel plate. This figure denotes that proposed inspection sensor is possible to detect the surface defect, since B_x is increased at the position of the defect. Moreover, the peak value of B_x by the surface defect is larger than the B_x by the opposite side defect. This is, because ac flux density is more affected by surface defect, since ac flux density is distributed in the surface of steel plate.

The proposed inspection sensor in this research is possible to detect both opposite side defect and surface defect as shown Figs. 8 and 13. However, there is a problem that it is difficult to discriminate between opposite side defect and surface defect by amplitude of B_x in search coil. Because, there are plural factors to affect amplitude of B_x in search coil. One of the factors is the position of the defect (opposite side or surface). Other factors are the depth of defect as shown Fig. 9 and Lift-off. Figure 14 shows the effect of the Lift-off on the change of B_x in a search coil at the Lift-off = 0.5 to 5 mm. This figure shows the calculated results of the change of B_x in the search coil. The rate ΔB_x is defined by

$$\Delta B_x = \frac{B_x(\text{each Lift} - \text{off}) - B_x(\text{Lift} - \text{off} = 0.5\,\text{mm})}{B_x(\text{Lift} - \text{off} = 0.5\,\text{mm})} \times 100 \qquad (6)$$

where B_x (Lift-off = 0.5 mm) is absolute value of B_x in the search coil when the Lift-off is equal to 0.5 mm. This figure denotes that ΔB_x is increased almost linearly by the increased in Lift-off. Therefore, the method of discriminate between opposite side defect and surface defect is the future subject.

5 Conclusions

The results obtained by this research are summarized as follows:

(1) The inspection of the opposite side defect in steel plate using dc and ac magnetic field is possible. Because, the ac flux in the surface steel plate is decreased, and the ac leakage flux is increased near the opposite side defect due to the saturation caused by the dc flux.

(2) The inspection of the surface defect in steel plate using dc and ac magnetic field is also possible. Moreover, the signal of surface defect is larger than the signal of the opposite side defect. This is, because ac flux density is more affected by surface defect, since ac flux density is distributed in the surface of steel plate.

References

1. McMaster, R.C., McIntrire, P., Mester, M.L.: Electromagnetic testing. In: Nondestructive Testing Handbook, American Society for Nondestructive Testing, vol. 5, 3rd edn (2004)
2. Gotoh, Y., Takahashi, N.: Evaluation of detecting method with AC and DC excitations of opposite side defect in steel using 3D non-linear FEM taking account of minor loop. IEEE Trans. Magn. **44**(6), 1622–1625 (2008)
3. Vertesy, G., Meszaros, I., Tomas, I.: Nondestructive indication of plastic deformation of cold-rolled stainless steel by magnetic minor hysteresis loops measurement. J. Magn. Magn. Mater. **285**(3), 335–342 (2005)
4. Miyata, K., Ohashi, K., Muraoka, A., Takahashi, N.: 3-D magnetic field analysis of permanent-magnet type MRI taking account of minor loop. IEEE Trans. Magn. **42**(4), 1452–1454 (2006)

Evaluation of Positions of Thinning Spots in a Dual Structure Using ECT with DC Magnetization

Daigo Kosaka[✉]

Polytechnic University, 2-32-1, Ogawanishimachi, Kodaira-shi, Tokyo 187-0035, Japan
kosaka@uitec.ac.jp

Abstract. Ultrasonic thickness meters are generally used in corrosion evaluations of pipes. On the other hand, there are pipes with gusset plates. Then the meters using a simple reflection cannot be used, because the ultrasonic is attenuated by an air gap between pipes and gusset plates. For the pipes with the dual structure, new method with multiple reflections is being studied. When the method is used, it is important to know number and positions of thinning spots.

This work is motivated by the desire to evaluate positions of thinning spots in a Dual Structure of steel nondestructively. Our approach is based on the Eddy Current Testing with DC magnetization. The ECT with DC magnetization has been studied. Recent studies have revealed that the Hall coefficient is sensitive to the presence of elastic strain. However, there are not still studies which applied the ECT to the structures. In this paper, the ECT is applied to an evaluation of thinning spots in the dual structure.

This paper attempts a first step toward visualizing magnetized states of specimens with an electromagnet using the 2D nonlinear electromagnetic field analysis. Here, needed magnetic force is obtained to detect the thinning spots on the back surface. Then, the measurement principle is explained. In the next section, the ECT with DC magnetization is applied to a specimen with a thinning spot. From the results, the usability of this method is evaluated.

Keywords: ECT · SLOFEC · Dual structure

1 Introduction

Ultrasonic thickness meters are generally used in corrosion evaluations of pipes. On the other hand, there are pipes with gusset plates. Then the meters using a simple reflection cannot be used, because the ultrasonic is attenuated by an air gap between pipes and gusset plates. For the pipes with the dual structure, new method with multiple reflections is being studied [1]. When the method is used, it is important to know number and positions of thinning spots.

This work is motivated by the desire to evaluate positions of thinning spots in a dual Structure of steel nondestructively. Our approach is based on the Eddy Current Testing with DC magnetization. The ECT with DC magnetization has been studied [2]. Recent studies have revealed that the usability of ECT is evaluated. However, there are not still

© Springer International Publishing Switzerland 2016
N. Kubota et al. (Eds.): ICIRA 2016, Part I, LNAI 9834, pp. 790–796, 2016.
DOI: 10.1007/978-3-319-43506-0_68

studies which applied the ECT to the structures as far as I know. In this paper, the ECT is applied to an evaluation of thinning spots in the dual structure.

This paper attempts a first step toward visualizing magnetized states of specimens with an electromagnet using the 2D nonlinear electromagnetic field analysis. Here, needed magnetic force is obtained to detect the thinning spots on the back surface. Then, the measurement principle is explained. In the next section, the ECT with DC magnetization is applied to specimens with 25 or 75 % thinning spot. From the results, the usability of this method is evaluated.

2   Simulations of Magnetized States of Specimens with DC Magnetization

In this work, the 2D FEM with vector potential A was used in order to calculate magnetic statuses of specimens. Distributions of magnetic flux density and permeability in the specimens were obtained from this. Figure 1 illustrates a model and conditions of the simulation. There is an electromagnet on the dual structure. The width and height of the electromagnet are 300 and 190 mm. There is 1 mm air gap between plates of the dual structure. The bottom plate has a thinning spot on the back surface.

Figure 2 shows distributions of the magnetic flux density and permeability which are obtained from the simulations. When the specimen was magnetized by the electromagnet, distributions of the statuses were changed. The magnetic permeability on the surface of the top plate responds to changes in the thinning shape. This result suggests that measurements of the permeability on the surface are useful in order to detect positions of the thinning on the back surface from the surface of the top plate. Figure 3(a) shows distributions of the permeability on the surface of the top plate. When the specimen had no thinning, the line that showed the permeability was flat. However, the permeability of the thinning area was lower than the permeability of other areas. Figure 3(b) shows a correlation between the permeability changing and the thinning depth. The simulations were also performed in order to evaluate an effect of the thinning depth. The simulation results show that the changing became larger as the depth was larger.

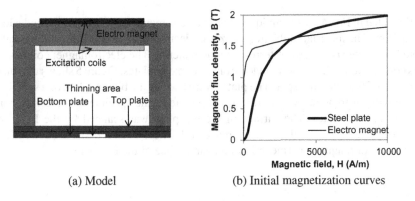

(a) Model   (b) Initial magnetization curves

Fig. 1. Simulation conditions

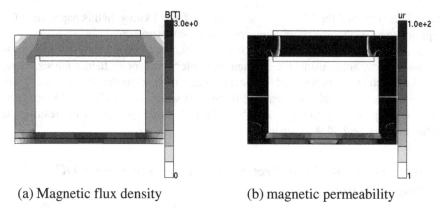

(a) Magnetic flux density   (b) magnetic permeability

Fig. 2. Simulation results of Distribution of the 20 % thinning specimen at 3200A magnet force.

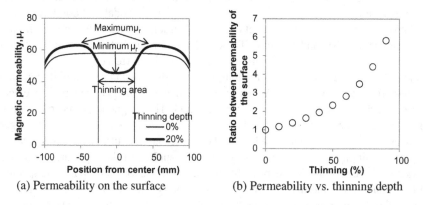

(a) Permeability on the surface   (b) Permeability vs. thinning depth

Fig. 3. Distributions of magnet permeability on near surfaces of the 20 % thinning specimen at 3200A.

3   Experimental Results

As illustrated in Fig. 4, this technique uses an electromagnet, two ECT coils, and a lock-in amplifier. The material of the magnetizer is electromagnetic soft iron which is set on a couple of plate specimens of ferromagnetic material. The electromagnet coil has 640 turns coil. Used specimens were made from a couple of plates. Figure 5 shows two plates with a 25 or 75 % thinning spot. The plates were used as the bottom plate of a specimen. A plate without thinning was used as the top plate of the specimen. There was 1 mm gap between the plates. The surface of the top plate was scanned by the ECT coil. Another ECT coil was used as the reference coil. Then, 100 kHz was used in order to detect status of magnetic permeability on near surface of the top plate.

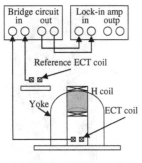

(a) Schematic illustration.

(b) Photograph of the electromagnet on a specimen.

Fig. 4. Experimental setup.

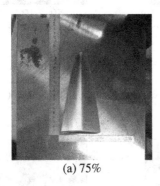

(a) 75%

(b) 25%

Fig. 5. Photographs of specimens with 25 or 75 % thinning spot.

First of all, I performed experiments in order to evaluate an effect of the magnet force. Figure 6 shows measurement results of the specimens with 25 % thinning spot at 100 kHz. The color bar at each figure shows detection signals of the ECT. In Fig. 6(a), you can seem like that the signal captured the status of surface of the top plate when the magnet force was low. A part at a center of Fig. 6(c) has high value. There was a 25 % thinning spot under the part. From the simulation results, the permeability of the part on the thinning spot becomes lower than the permeability of other parts as the magnet force is higher. It is known that magnetic permeability has an impact on the signal of the ECT. Therefore, the part with high value means that there is a thinning spot under the part. As shown Fig. 6(a), (b) and (c), the contrast between the part with high value and other parts became higher as the magnet force was higher. On the other hand, Fig. 7 shows measurement results of the specimens with 75 % thinning spot. These results also show that the contrast became higher as the magnet force was higher. However, you can detect the 75 % thinning spot using lower magnet force than magnet force of the 25 % thinning spot. In other words, these results show that the contrast between the thinning spot part and other parts became higher as the thinning depth was deeper.

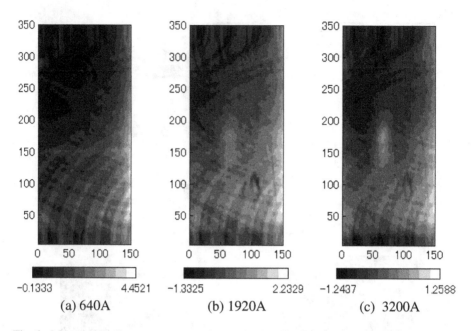

Fig. 6. Measured in-phase components of a specimen with a 25 % thinning spot at 100 kHz as each magnet force.

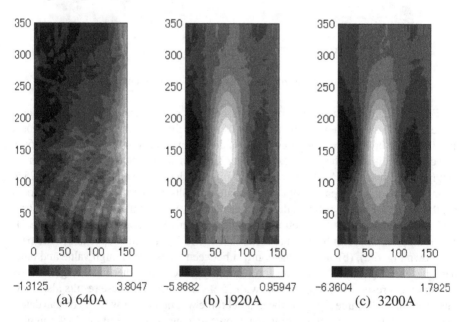

Fig. 7. Measured in-phase components of a specimen with a 75 % thinning spot at 100 kHz.

In the next place, I performed experiments in order to evaluate an effect of frequency of the ECT. Figure 8 shows measurement results of the specimens with 25 % thinning spot at 5120A magnet force. In Fig. 8(a), you can seem like that the contrast of the thinning spot became higher as the frequency was higher. It is possible that one of the origins of a difference between the results at each frequency is the skin effect. However, the impedance of coils changes depending on the frequency. It is also possible that one of the origins is the impedance matching between the coils and the input impedance of the LIA. The effect of the frequency invites closer analysis.

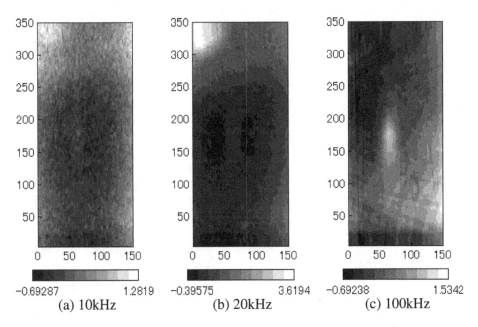

Fig. 8. Measured in-phase components of a specimen with a 25 % thinning spot at 5120A magnetizing force.

4   Conclusion

I tried to detect a spot thinning on the back surface of the dual structure by using the ECT with DC magnetization. The contrast of magnetic permeability on the surface of the top plate has been studied in the simulations. The depth of the thinning had an impact on the contrast. This trend of the simulation results was applicable to those of measurement results. Therefore, this method can detect thinning spots on the back surface of the dual structure with an air gap.

It is expected that the width of the thinning spot, magnetic characteristic of specimens and the gap length also would have an impact on this method. My future research will be focused on experiments and evaluation of the effects.

References

1. Shiroshita, S., et al.: Study of a thickness measurement under gusset plates. In: Proceedings of 9th Annual Conference, Japan Society of Maintenology (2012). (In Japanese)
2. Bönisch, A.: Magnetic flux and SLOFEC inspection of thick walled components. The e-Journal of Nondestructive Testing (2000). http://www.ndt.net/article/wcndt00/papers/idn352/idn352.htm

Author Index

Printed in the United States
By Bookmasters